HONDA ... DEL SOL
... AIR MANUAL

CHILTON'S

Covers all U.S. and Canadian models of Honda Civic, CRX and del Sol

by Jaffer A. Ahmad

CHILTON *Automotive Books*
PUBLISHED BY HAYNES NORTH AMERICA, Inc.

Manufactured in USA
© 1995 Haynes North America, Inc.
ISBN 0-8019-8683-4
Library of Congress Catalog Card No. 94-069432
11 12 13 14 15 9 8 7 6 5 4 3 2 1 0

Haynes Publishing Group
Sparkford Nr Yeovil
Somerset BA22 7JJ England

Haynes North America, Inc
861 Lawrence Drive
Newbury Park
California 91320 USA

ABCDE
FGHIJ
KLMNO
F   2

6E1

Chilton is a registered trademark of W.G. Nichols, Inc., and has been licensed to Haynes North America, Inc.

Contents

1 GENERAL INFORMATION AND MAINTENANCE

- 1-2 HOW TO USE THIS BOOK
- 1-2 TOOLS AND EQUIPMENT
- 1-4 SERVICING YOUR VEHICLE SAFELY
- 1-6 FASTENERS, MEASUREMENTS AND CONVERSIONS
- 1-9 HISTORY
- 1-9 SERIAL NUMBER IDENTIFICATION
- 1-13 ROUTINE MAINTENANCE
- 1-27 FLUIDS AND LUBRICANTS
- 1-34 TRAILER TOWING
- 1-34 TOWING THE VEHICLE
- 1-35 JACKING
- 1-36 MAINTENANCE INTERVAL CHARTS

2 ENGINE PERFORMANCE AND TUNE-UP

- 2-2 TUNE-UP PROCEDURES
- 2-5 FIRING ORDERS
- 2-5 ELECTRONIC IGNITION (1984-87)
- 2-12 PGM-IG IGNITION SYSTEM (1988-95)
- 2-15 IGNITION TIMING
- 2-18 VALVE LASH
- 2-20 IDLE SPEED AND MIXTURE

3 ENGINE AND ENGINE OVERHAUL

- 3-2 ENGINE ELECTRICAL
- 3-5 ENGINE MECHANICAL
- 3-52 EXHAUST SYSTEM

4 EMISSION CONTROLS

- 4-2 EMISSION CONTROLS
- 4-9 ELECTRONIC ENGINE CONTROLS
- 4-15 DIAGNOSTIC CHARTS
- 4-21 VACUUM DIAGRAMS

5 FUEL SYSTEM

- 5-2 BASIC FUEL SYSTEM DIAGNOSIS
- 5-2 CARBURETED FUEL SYSTEM
- 5-10 PROGRAMMED FUEL INJECTION (PGM-FI) SYSTEM

6 CHASSIS ELECTRICAL

- 6-2 UNDERSTANDING AND TROUBLESHOOTING ELECTRICAL SYSTEMS
- 6-8 COMPONENT LOCATIONS
- 6-9 SUPPLEMENTAL RESTRAINT SYSTEM (SRS)
- 6-11 HEATING AND AIR CONDITIONING
- 6-17 CRUISE CONTROL
- 6-18 ENTERTAINMENT SYSTEM
- 6-21 WINDSHIELD WIPERS AND WASHERS
- 6-23 INSTRUMENTS AND SWITCHES
- 6-28 LIGHTING
- 6-30 CIRCUIT PROTECTION
- 6-32 WIRING DIAGRAMS

Contents

7-2 MANUAL TRANSAXLE	**7-22** DRIVELINE	**DRIVE TRAIN 7**
7-12 CLUTCH	**7-25** REAR AXLE	
7-16 AUTOMATIC TRANSAXLE		

8-2 WHEELS	**8-13** REAR SUSPENSION	**SUSPENSION AND STEERING 8**
8-2 FRONT SUSPENSION	**8-20** STEERING	

9-2 BRAKE OPERATING SYSTEM	**9-17** REAR DISC BRAKES	**BRAKES 9**
9-9 FRONT DISC BRAKES	**9-19** PARKING BRAKE	
9-14 REAR DRUM BRAKES	**9-21** ANTI-LOCK BRAKE (ALB/ABS) SYSTEM	

10-2 EXTERIOR	**10-12** INTERIOR	**BODY AND TRIM 10**

10-21 GLOSSARY

GLOSSARY

10-25 MASTER INDEX

MASTER INDEX

SAFETY NOTICE

Proper service and repair procedures are vital to the safe, reliable operation of all motor vehicles, as well as the personal safety of those performing repairs. This manual outlines procedures for servicing and repairing vehicles using safe, effective methods. The procedures contain many NOTES, CAUTIONS and WARNINGS which should be followed, along with standard procedures to eliminate the possibility of personal injury or improper service which could damage the vehicle or compromise its safety.

It is important to note that repair procedures and techniques, tools and parts for servicing motor vehicles, as well as the skill and experience of the individual performing the work vary widely. It is not possible to anticipate all of the conceivable ways or conditions under which vehicles may be serviced, or to provide cautions as to all possible hazards that may result. Standard and accepted safety precautions and equipment should be used when handling toxic or flammable fluids, and safety goggles or other protection should be used during cutting, grinding, chiseling, prying, or any other process that can cause material removal or projectiles.

Some procedures require the use of tools specially designed for a specific purpose. Before substituting another tool or procedure, you must be completely satisfied that neither your personal safety, nor the performance of the vehicle will be endangered.

Although information in this manual is based on industry sources and is complete as possible at the time of publication, the possibility exists that some car manufacturers made later changes which could not be included here. While striving for total accuracy, the authors or publishers cannot assume responsibility for any errors, changes or omissions that may occur in the compilation of this data.

PART NUMBERS

Part numbers listed in this reference are not recommendations by Haynes North America, Inc. for any product brand name. They are references that can be used with interchange manuals and aftermarket supplier catalogs to locate each brand supplier's discrete part number.

SPECIAL TOOLS

Special tools are recommended by the vehicle manufacturer to perform their specific job. Use has been kept to a minimum, but where absolutely necessary, they are referred to in the text by the part number of the tool manufacturer. These tools can be purchased, under the appropriate part number, from your local dealer or regional distributor, or an equivalent tool can be purchased locally from a tool supplier or parts outlet. Before substituting any tool for the one recommended, read the SAFETY NOTICE at the top of this page.

ACKNOWLEDGMENTS

The publisher expresses appreciation to Honda Motor Co. for their generous assistance.

All rights reserved. No part of this book may be reproduced or transmitted in any form or by any means, electronic or mechanical, including photocopying, recording or by any information storage or retrieval system, without permission in writing from the copyright holder.

While every attempt is made to ensure that the information in this manual is correct, no liability can be accepted by the authors or publishers for loss, damage or injury caused by any errors in, or omissions from, the information given.

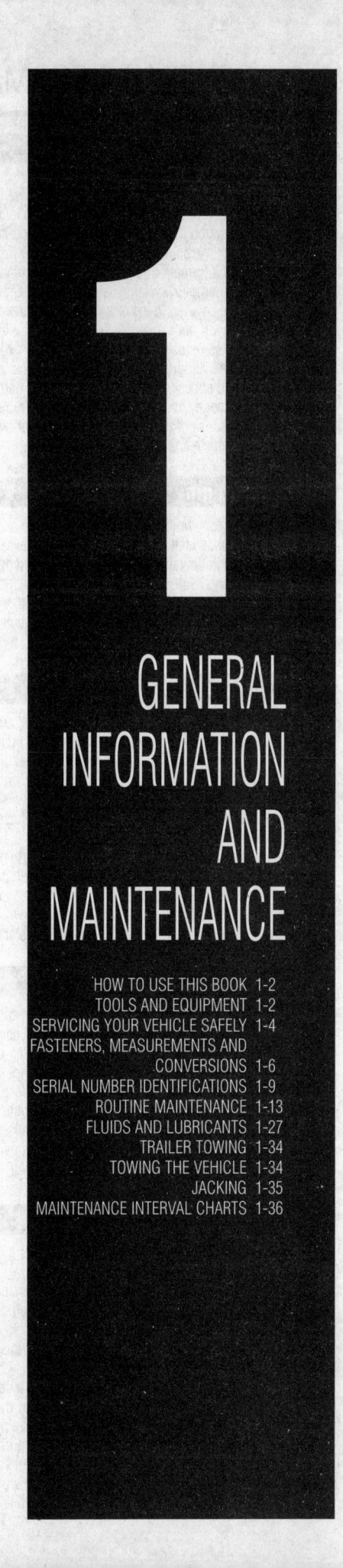

1 GENERAL INFORMATION AND MAINTENANCE

HOW TO USE THIS BOOK 1-2
WHERE TO BEGIN 1-2
AVOIDING TROUBLE 1-2
MAINTENANCE OR REPAIR? 1-2
AVOIDING THE MOST COMMON
 MISTAKES 1-2
TOOLS AND EQUIPMENT 1-2
SPECIAL TOOLS 1-4
**SERVICING YOUR VEHICLE
 SAFELY 1-4**
DO'S 1-4
DON'TS 1-5
**FASTENERS, MEASUREMENTS AND
 CONVERSIONS 1-6**
BOLTS, NUTS AND OTHER THREADED
 RETAINERS 1-6
TORQUE 1-6
 TORQUE WRENCHES 1-7
 TORQUE ANGLE METERS 1-7
STANDARD AND METRIC
 MEASUREMENTS 1-7
HISTORY 1-9
COVERAGE 1-9
**SERIAL NUMBER
 IDENTIFICATION 1-9**
VEHICLE IDENTIFICATION (CHASSIS)
 NUMBER 1-9
ENGINE SERIAL NUMBER 1-9
TRANSAXLE SERIAL NUMBER 1-9
ROUTINE MAINTENANCE 1-13
AIR CLEANER 1-13
 REMOVAL & INSTALLATION 1-13
FUEL FILTER 1-14
 REMOVAL & INSTALLATION 1-15
POSITIVE CRANKCASE VENTILATION
 (PCV) 1-16
 INSPECTION 1-16
 REMOVAL & INSTALLATION 1-16
EVAPORATIVE CHARCOAL
 CANISTER 1-16
 SERVICING 1-16
BATTERY 1-16
 PRECAUTIONS 1-16
 GENERAL MAINTENANCE 1-16
 BATTERY FLUID 1-17
 CABLES 1-17
 CHARGING 1-18
 REPLACEMENT 1-18
BELTS 1-18
 INSPECTION 1-18
 ADJUSTING TENSION 1-19
 REMOVAL & INSTALLATION 1-19
TIMING BELTS 1-20
 INSPECTION 1-20
HOSES 1-20
 INSPECTION 1-20
 REMOVAL & INSTALLATION 1-20
CV-BOOTS 1-23
 INSPECTION 1-23
AIR CONDITIONING SYSTEM 1-23
 SYSTEM SERVICE & REPAIR 1-23
 PREVENTIVE MAINTENANCE 1-23
 SYSTEM INSPECTION 1-24
WINDSHIELD WIPERS 1-24
 ELEMENT (REFILL) CARE &
  REPLACEMENT 1-24
TIRES AND WHEELS 1-24
 TIRE ROTATION 1-25
 TIRE DESIGN 1-25
 TIRE STORAGE 1-26
 INFLATION & INSPECTION 1-26
FLUIDS AND LUBRICANTS 1-27
FLUID DISPOSAL 1-27
FUEL AND ENGINE OIL
 RECOMMENDATIONS 1-27
 FUEL 1-27
 OIL 1-27
ENGINE 1-28
 OIL LEVEL CHECK 1-28
 OIL AND FILTER CHANGE 1-29
MANUAL TRANSAXLE 1-30
 FLUID RECOMMENDATIONS 1-30
 LEVEL CHECK 1-30
 DRAIN AND REFILL 1-30
AUTOMATIC TRANSAXLES 1-30
 FLUID RECOMMENDATIONS 1-30
 LEVEL CHECK 1-30
 DRAIN AND REFILL 1-30
TRANSFER CASE 1-31
DRIVE AXLE—4WD WAGON 1-30
 FLUID RECOMMENDATIONS 1-30
 LEVEL CHECK 1-31
 DRAIN AND REFILL 1-31
COOLING SYSTEM 1-31
 FLUID RECOMMENDATIONS 1-31
 LEVEL CHECK 1-31
 DRAIN AND REFILL 1-31
 FLUSHING AND CLEANING THE
  SYSTEM 1-32
BRAKE MASTER CYLINDER 1-32
 FLUID RECOMMENDATIONS 1-32
 LEVEL CHECK 1-32
CLUTCH MASTER CYLINDER 1-32
 FLUID RECOMMENDATIONS 1-33
 LEVEL CHECK 1-33
POWER STEERING PUMP 1-33
 FLUID RECOMMENDATIONS 1-33
 RESERVOIR LEVEL CHECK 1-33
MANUAL STEERING GEAR 1-33
 INSPECTION 1-33
 FLUID RECOMMENDATIONS 1-33
WINDSHIELD WIPER FLUID 1-33
CHASSIS GREASING 1-34
BODY LUBRICATION 1-34
WHEEL BEARINGS 1-34
TRAILER TOWING 1-34
TOWING THE VEHICLE 1-34
JACKING 1-35
**MAINTENANCE INTERVAL
 CHARTS 1-36**
SPECIFICATIONS CHARTS
 ENGINE IDENTIFICATION 1-12
 CAPACITIES 1-40

HOW TO USE THIS BOOK 1-2
TOOLS AND EQUIPMENT 1-2
SERVICING YOUR VEHICLE SAFELY 1-4
FASTENERS, MEASUREMENTS AND
 CONVERSIONS 1-6
SERIAL NUMBER IDENTIFICATIONS 1-9
ROUTINE MAINTENANCE 1-13
FLUIDS AND LUBRICANTS 1-27
TRAILER TOWING 1-34
TOWING THE VEHICLE 1-34
JACKING 1-35
MAINTENANCE INTERVAL CHARTS 1-36

1-2 GENERAL INFORMATION AND MAINTENANCE

HOW TO USE THIS BOOK

Chilton's Total Car Care Manual for Honda's Civic series (the Civic, CRX and del Sol) is intended to help you learn more about the inner workings of your vehicle while saving you money on its upkeep and operation.

The beginning of the book will likely be referred to the most, since that is where you will find information for maintenance and tune-up. The other sections deal with the more complex systems of your vehicle. Systems (from engine through brakes) are covered to the extent that the average do-it-yourselfer can attempt. This book will not explain such things as rebuilding a differential because the expertise required and the special tools necessary make this uneconomical. It will, however, give you detailed instructions to help you change your own brake pads and shoes, replace spark plugs, and perform many more jobs that can save you money and help avoid expensive problems.

A secondary purpose of this book is a reference for owners who want to understand their vehicle and/or their mechanics better.

Where to Begin

Before removing any bolts, read through the entire procedure. This will give you the overall view of what tools and supplies will be required. So read ahead and plan ahead. Each operation should be approached logically and all procedures thoroughly understood before attempting any work.

If repair of a component is not considered practical, we tell you how to remove the part and then how to install the new or rebuilt replacement. In this way, you at least save labor costs.

Avoiding Trouble

Many procedures in this book require you to "label and disconnect . . ." a group of lines, hoses or wires. Don't be think you can remember where everything goes—you won't. If you hook up vacuum or fuel lines incorrectly, the vehicle may run poorly, if at all. If you hook up electrical wiring incorrectly, you may instantly learn a very expensive lesson.

You don't need to know the proper name for each hose or line. A piece of masking tape on the hose and a piece on its fitting will allow you to assign your own label. As long as you remember your own code, the lines can be reconnected by matching your tags. Remember that tape will dissolve in gasoline or solvents; if a part is to be washed or cleaned, use another method of identification. A permanent felt-tipped marker or a metal scribe can be very handy for marking metal parts. Remove any tape or paper labels after assembly.

Maintenance or Repair?

Maintenance includes routine inspections, adjustments, and replacement of parts which show signs of normal wear. Maintenance compensates for wear or deterioration. Repair implies that something has broken or is not working. A need for a repair is often caused by lack of maintenance. for example: draining and refilling automatic transmission fluid is maintenance recommended at specific intervals. Failure to do this can shorten the life of the transmission/transaxle, requiring very expensive repairs. While no maintenance program can prevent items from eventually breaking or wearing out, a general rule is true: MAINTENANCE IS CHEAPER THAN REPAIR.

Two basic mechanic's rules should be mentioned here. First, whenever the left side of the vehicle or engine is referred to, it means the driver's side. Conversely, the right side of the vehicle means the passenger's side. Second, screws and bolts are removed by turning counterclockwise, and tightened by turning clockwise unless specifically noted.

Safety is always the most important rule. Constantly be aware of the dangers involved in working on an automobile and take the proper precautions. Please refer to the information in this section regarding SERVICING YOUR VEHICLE SAFELY and the SAFETY NOTICE on the acknowledgment page.

Avoiding the Most Common Mistakes

Pay attention to the instructions provided. There are 3 common mistakes in mechanical work:

1. Incorrect order of assembly, disassembly or adjustment. When taking something apart or putting it together, performing steps in the wrong order usually just costs you extra time; however, it CAN break something. Read the entire procedure before beginning. Perform everything in the order in which the instructions say you should, even if you can't see a reason for it. When you're taking apart something that is very intricate, you might want to draw a picture of how it looks when assembled in order to make sure you get everything back in its proper position. When making adjustments, perform them in the proper order. One adjustment possibly will affect another.

2. Overtorquing (or undertorquing). While it is more common for overtorquing to cause damage, undertorquing may allow a fastener to vibrate loose causing serious damage. Especially when dealing with aluminum parts, pay attention to torque specifications and utilize a torque wrench in assembly. If a torque figure is not available, remember that if you are using the right tool to perform the job, you will probably not have to strain yourself to get a fastener tight enough. The pitch of most threads is so slight that the tension you put on the wrench will be multiplied many times in actual force on what you are tightening.

There are many commercial products available for ensuring that fasteners won't come loose, even if they are not torqued just right (a very common brand is Loctite®). If you're worried about getting something together tight enough to hold, but loose enough to avoid mechanical damage during assembly, one of these products might offer substantial insurance. Before choosing a threadlocking compound, read the label on the package and make sure the product is compatible with the materials, fluids, etc. involved.

3. Crossthreading. This occurs when a part such as a bolt is screwed into a nut or casting at the wrong angle and forced. Crossthreading is more likely to occur if access is difficult. It helps to clean and lubricate fasteners, then to start threading the bolt, spark plug, etc. with your fingers. If you encounter resistance, unscrew the part and start over again at a different angle until it can be inserted and turned several times without much effort. Keep in mind that many parts have tapered threads, so that gentle turning will automatically bring the part you're threading to the proper angle. Don't put a wrench on the part until it's been tightened a couple of turns by hand. If you suddenly encounter resistance, and the part has not seated fully, don't force it. Pull it back out to make sure it's clean and threading properly.

Be sure to take your time and be patient, and always plan ahead. Allow yourself ample time to perform repairs and maintenance.

TOOLS AND EQUIPMENT

▶ See Figures 1 thru 15

Without the proper tools and equipment it is impossible to properly service your vehicle. It would be virtually impossible to catalog every tool that you would need to perform all of the operations in this book. It would be unwise for the amateur to rush out and buy an expensive set of tools on the theory that he/she may need one or more of them at some time.

The best approach is to proceed slowly, gathering a good quality set of those tools that are used most frequently. Don't be misled by the low cost of bargain tools. It is far better to spend a little more for better quality. Forged wrenches, 6 or 12-point sockets and fine tooth ratchets are by far preferable to their less expensive counterparts. As any good mechanic can tell you, there are few worse experiences than trying to work on a vehicle with bad tools. Your monetary savings will be far outweighed by frustration and mangled knuckles.

Begin accumulating those tools that are used most frequently: those associated with routine maintenance and tune-up. In addition to the normal assortment of screwdrivers and pliers, you should have the following tools:
• Wrenches/sockets and combination open end/box end wrenches in sizes 1/8–3/4 in. and/or 3mm–19mm 13/16 in. or 5/8 in. spark plug socket (depending on plug type).

➡ If possible, buy various length socket drive extensions. Universal-joint and wobble extensions can be extremely useful, but be careful when using them, as they can change the amount of torque applied to the socket.

• Jackstands for support.
• Oil filter wrench.

GENERAL INFORMATION AND MAINTENANCE 1-3

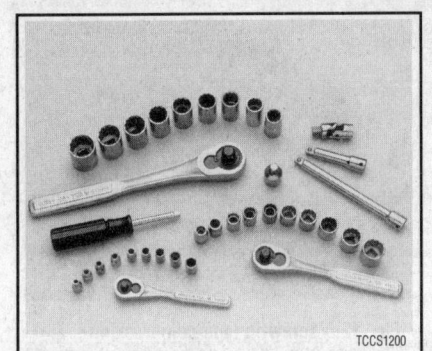

Fig. 1 All but the most basic procedures will require an assortment of ratchets and sockets

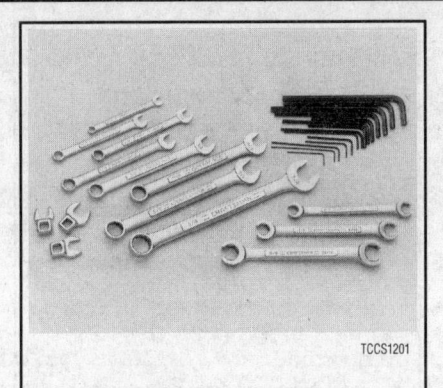

Fig. 2 In addition to ratchets, a good set of wrenches and hex keys will be necessary

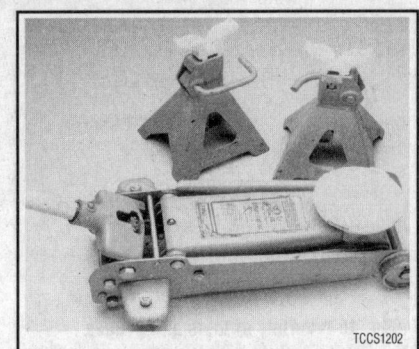

Fig. 3 A hydraulic floor jack and a set of jackstands are essential for lifting and supporting the vehicle

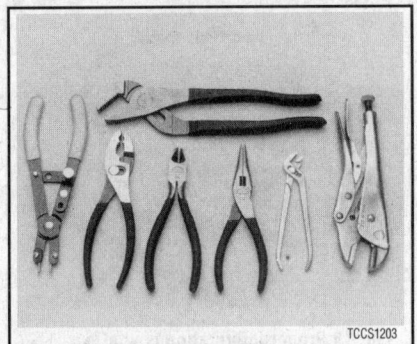

Fig. 4 An assortment of pliers, grippers and cutters will be handy for old rusted parts and stripped bolt heads

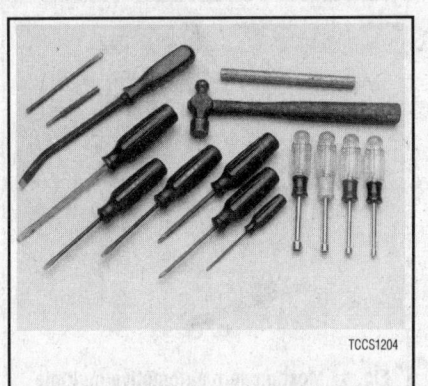

Fig. 5 Various drivers, chisels and prybars are great tools to have in your toolbox

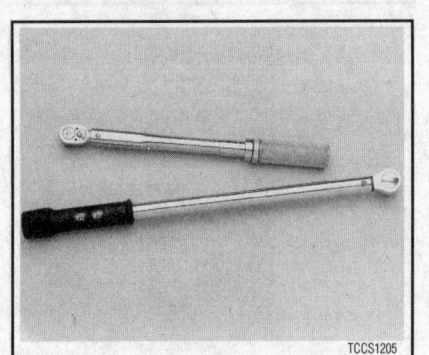

Fig. 6 Many repairs will require the use of a torque wrench to assure the components are properly fastened

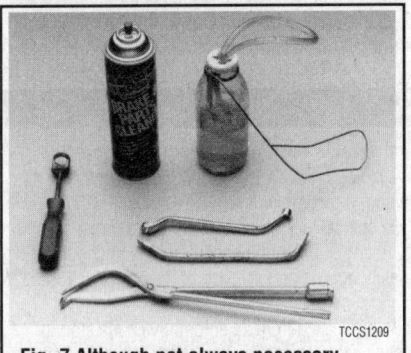

Fig. 7 Although not always necessary, using specialized brake tools will save time

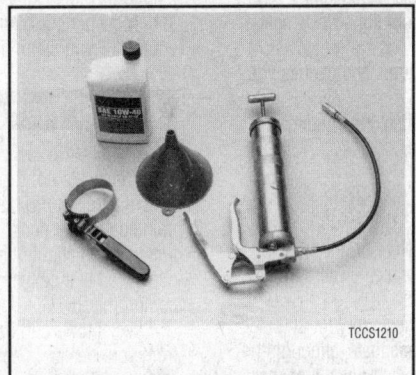

Fig. 8 A few inexpensive lubrication tools will make maintenance easier

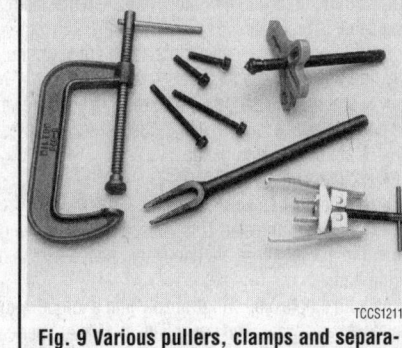

Fig. 9 Various pullers, clamps and separator tools are needed for many larger, more complicated repairs

- Spout or funnel for pouring fluids.
- Grease gun for chassis lubrication (unless your vehicle is not equipped with any grease fittings)
- Hydrometer for checking the battery (unless equipped with a sealed, maintenance-free battery).
- A container for draining oil and other fluids.
- Rags for wiping up the inevitable mess.

In addition to the above items there are several others that are not absolutely necessary, but handy to have around. These include an equivalent oil absorbent gravel, like cat litter, and the usual supply of lubricants, antifreeze and fluids. This is a basic list for routine maintenance, but only your personal needs and desire can accurately determine your list of tools.

After performing a few projects on the vehicle, you'll be amazed at the other tools and non-tools on your workbench. Some useful household items are: a large turkey baster or siphon, empty coffee cans and ice trays (to store parts), a ball of twine, electrical tape for wiring, small rolls of colored tape for tagging lines or hoses, markers and pens, a note pad, golf tees (for plugging vacuum lines), metal coat hangers or a roll of mechanic's wire (to hold things out of the way), dental pick or similar long, pointed probe, a strong magnet, and a small mirror (to see into recesses and under manifolds).

A more advanced set of tools, suitable for tune-up work, can be drawn up easily. While the tools are slightly more sophisticated, they need not be outrageously expensive. There are several inexpensive tach/dwell meters on the market that are every bit as good for the average mechanic as a professional model. Just be sure that it goes to a least 1200–1500 rpm on the tach scale and that it works on 4, 6 and 8-cylinder engines. The key to these purchases is to make them with an eye towards adaptability and wide range. A basic list of tune-up tools could include:

- Tach/dwell meter.
- Spark plug wrench and gapping tool.
- Feeler gauges for valve adjustment.
- Timing light.

The choice of a timing light should be made carefully. A light which works on the DC current supplied by the vehicle's battery is the best choice; it should

1-4 GENERAL INFORMATION AND MAINTENANCE

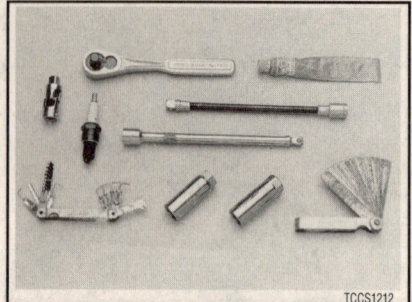

Fig. 10 A variety of tools and gauges should be used for spark plug gapping and installation

Fig. 11 Inductive type timing light

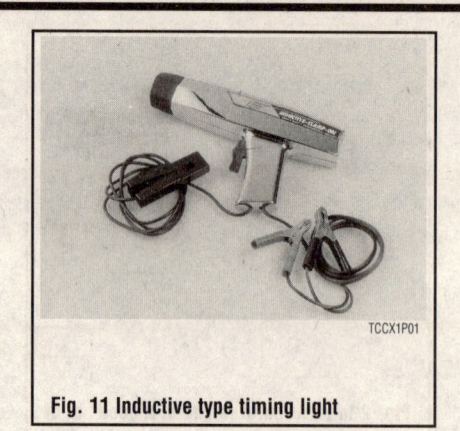

Fig. 12 A screw-in type compression gauge is recommended for compression testing

Fig. 13 A vacuum/pressure tester is necessary for many testing procedures

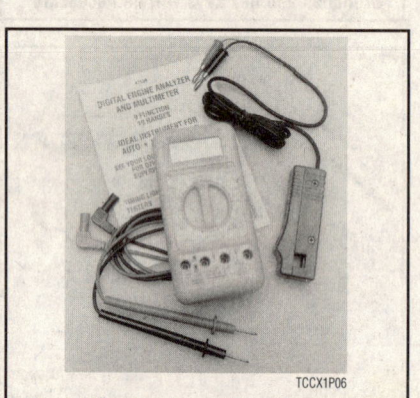

Fig. 14 Most modern automotive multimeters incorporate many helpful features

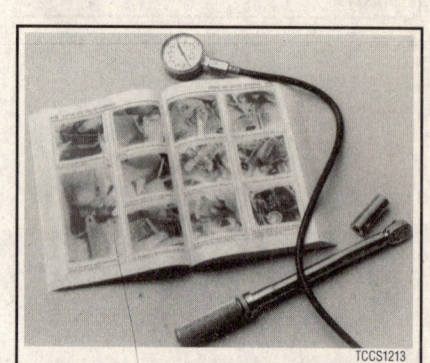

Fig. 15 Proper information is vital, so always have a Chilton Total Car Care manual handy

have a xenon tube for brightness. On any vehicle with an electronic ignition system, a timing light with an inductive pickup that clamps around the No. 1 spark plug cable is preferred.

In addition to these basic tools, there are several other tools and gauges you may find useful. These include:
- Compression gauge. The screw-in type is slower to use, but eliminates the possibility of a faulty reading due to escaping pressure.
- Manifold vacuum gauge.
- 12V test light.
- A combination volt/ohmmeter
- Induction Ammeter. This is used for determining whether or not there is current in a wire. These are handy for use if a wire is broken somewhere in a wiring harness.

As a final note, you will probably find a torque wrench necessary for all but the most basic work. The beam type models are perfectly adequate, although the newer click types (breakaway) are easier to use. The click type torque wrenches tend to be more expensive. Also keep in mind that all types of torque wrenches should be periodically checked and/or recalibrated. You will have to decide for yourself which better fits your pocketbook, and purpose.

Special Tools

Normally, the use of special factory tools is avoided for repair procedures, since these are not readily available for the do-it-yourself mechanic. When it is possible to perform the job with more commonly available tools, it will be pointed out, but occasionally, a special tool was designed to perform a specific function and should be used. Before substituting another tool, you should be convinced that neither your safety nor the performance of the vehicle will be compromised.

Special tools can usually be purchased from an automotive parts store or from your dealer. In some cases special tools may be available directly from the tool manufacturer.

SERVICING YOUR VEHICLE SAFELY

▶ See Figures 16, 17 and 18

It is virtually impossible to anticipate all of the hazards involved with automotive maintenance and service, but care and common sense will prevent most accidents.

The rules of safety for mechanics range from "don't smoke around gasoline," to "use the proper tool(s) for the job." The trick to avoiding injuries is to develop safe work habits and to take every possible precaution.

Do's

- Do keep a fire extinguisher and first aid kit handy.
- Do wear safety glasses or goggles when cutting, drilling, grinding or prying, even if you have 20–20 vision. If you wear glasses for the sake of vision, wear safety goggles over your regular glasses.

GENERAL INFORMATION AND MAINTENANCE   1-5

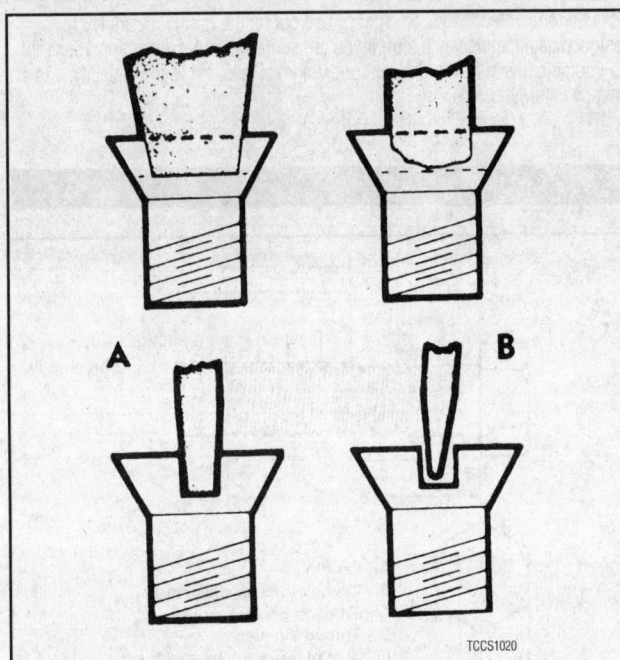

Fig. 16 Screwdrivers should be kept in good condition to prevent injury or damage which could result if the blade slips from the screw

Fig. 17 Using the correct size wrench will help prevent the possibility of rounding off a nut

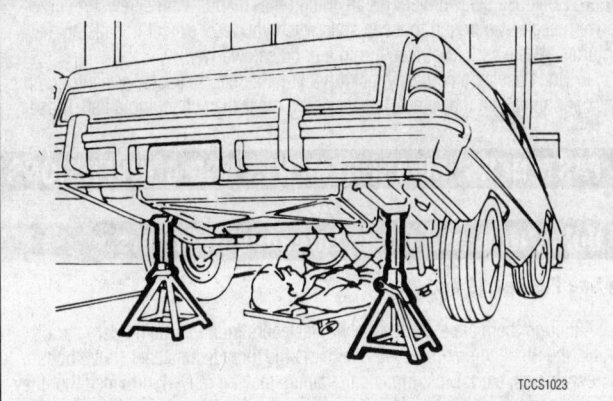

Fig. 18 NEVER work under a vehicle unless it is supported using safety stands (jackstands)

- Do properly maintain your tools. Loose hammerheads, mushroomed punches and chisels, frayed or poorly grounded electrical cords, excessively worn screwdrivers, spread wrenches (open end), cracked sockets, slipping ratchets, or faulty droplight sockets can cause accidents.
- Likewise, keep your tools clean; a greasy wrench can slip off a bolt head, ruining the bolt and often harming your knuckles in the process.
- Do use the proper size and type of tool for the job at hand. Do select a wrench or socket that fits the nut or bolt. The wrench or socket should sit straight, not cocked.
- Do, when possible, pull on a wrench handle rather than push on it, and adjust your stance to prevent a fall.
- Do be sure that adjustable wrenches are tightly closed on the nut or bolt and pulled so that the force is on the side of the fixed jaw.
- Do strike squarely with a hammer; avoid glancing blows.
- Do set the parking brake and block the drive wheels if the work requires a running engine.

Don'ts

- Don't run the engine in a garage or anywhere else without proper ventilation—EVER! Carbon monoxide is poisonous; it takes a long time to leave the human body and you can build up a deadly supply of it in your system by simply breathing in a little at a time. You may not realize you are slowly poisoning yourself. Always use power vents, windows, fans and/or open the garage door.
- Don't work around moving parts while wearing loose clothing. Short sleeves are much safer than long, loose sleeves. Hard-toed shoes with neoprene soles protect your toes and give a better grip on slippery surfaces. Watches and jewelry is not safe working around a vehicle. Long hair should be tied back under a hat or cap.
- Don't use pockets for toolboxes. A fall or bump can drive a screwdriver deep into your body. Even a rag hanging from your back pocket can wrap around a spinning shaft or fan.
- Don't smoke when working around gasoline, cleaning solvent or other flammable material.
- Don't smoke when working around the battery. When the battery is being charged, it gives off explosive hydrogen gas.
- Don't use gasoline to wash your hands; there are excellent soaps available. Gasoline contains dangerous additives which can enter the body through a cut or through your pores. Gasoline also removes all the natural oils from the skin so that bone dry hands will suck up oil and grease.
- Don't service the air conditioning system unless you are equipped with the necessary tools and training. When liquid or compressed gas refrigerant is released to atmospheric pressure it will absorb heat from whatever it contacts. This will chill or freeze anything it touches.

- Do shield your eyes whenever you work around the battery. Batteries contain sulfuric acid. In case of contact with, flush the area with water or a mixture of water and baking soda, then seek immediate medical attention.
- Do use safety stands (jackstands) for any undervehicle service. Jacks are for raising vehicles; jackstands are for making sure the vehicle stays raised until you want it to come down.
- Do use adequate ventilation when working with any chemicals or hazardous materials. Like carbon monoxide, the asbestos dust resulting from some brake lining wear can be hazardous in sufficient quantities.
- Do disconnect the negative battery cable when working on the electrical system. The secondary ignition system contains EXTREMELY HIGH VOLTAGE. In some cases it can even exceed 50,000 volts.
- Do follow manufacturer's directions whenever working with potentially hazardous materials. Most chemicals and fluids are poisonous.

1-6 GENERAL INFORMATION AND MAINTENANCE

- Don't use screwdrivers for anything other than driving screws! A screwdriver used as an prying tool can snap when you least expect it, causing injuries. At the very least, you'll ruin a good screwdriver.
- Don't use an emergency jack (that little ratchet, scissors, or pantograph jack supplied with the vehicle) for anything other than changing a flat! These jacks are only intended for emergency use out on the road; they are NOT designed as a maintenance tool. If you are serious about maintaining your vehicle yourself, invest in a hydraulic floor jack of at least a 1½ ton capacity, and at least two sturdy jackstands.

FASTENERS, MEASUREMENTS AND CONVERSIONS

Bolts, Nuts and Other Threaded Retainers

▶ See Figures 19 and 20

Although there are a great variety of fasteners found in the modern car or truck, the most commonly used retainer is the threaded fastener (nuts, bolts, screws, studs, etc.). Most threaded retainers may be reused, provided that they are not damaged in use or during the repair. Some retainers (such as stretch bolts or torque prevailing nuts) are designed to deform when tightened or in use and should not be reinstalled.

Whenever possible, we will note any special retainers which should be replaced during a procedure. But you should always inspect the condition of a retainer when it is removed and replace any that show signs of damage. Check all threads for rust or corrosion which can increase the torque necessary to achieve the desired clamp load for which that fastener was originally selected. Additionally, be sure that the driver surface of the fastener has not been compromised by rounding or other damage. In some cases a driver surface may become only partially rounded, allowing the driver to catch in only one direction. In many of these occurrences, a fastener may be installed and tightened, but the driver would not be able to grip and loosen the fastener again.

If you must replace a fastener, whether due to design or damage, you must ALWAYS be sure to use the proper replacement. In all cases, a retainer of the same design, material and strength should be used. Markings on the heads of most bolts will help determine the proper strength of the fastener. The same material, thread and pitch must be selected to assure proper installation and safe operation of the vehicle afterwards.

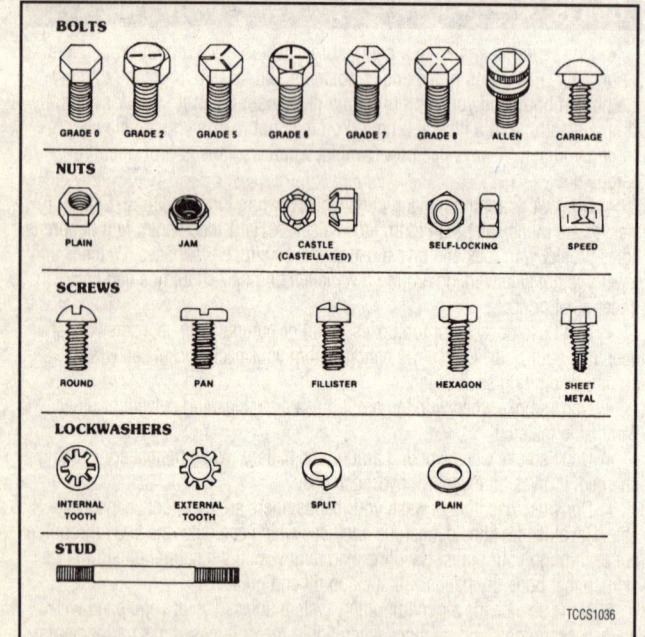

Fig. 19 There are many different types of threaded retainers found on vehicles

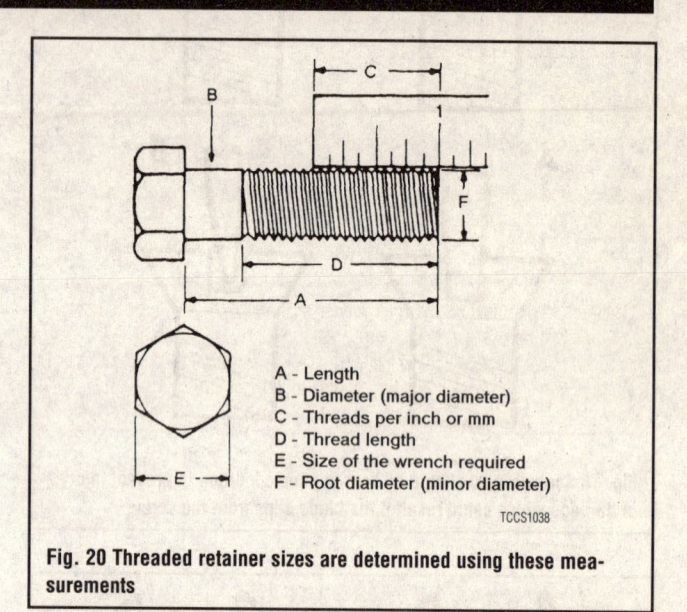

Fig. 20 Threaded retainer sizes are determined using these measurements

A - Length
B - Diameter (major diameter)
C - Threads per inch or mm
D - Thread length
E - Size of the wrench required
F - Root diameter (minor diameter)

Thread gauges are available to help measure a bolt or stud's thread. Most automotive and hardware stores keep gauges available to help you select the proper size. In a pinch, you can use another nut or bolt for a thread gauge. If the bolt you are replacing is not too badly damaged, you can select a match by finding another bolt which will thread in its place. If you find a nut which threads properly onto the damaged bolt, then use that nut to help select the replacement bolt.

✱✱ WARNING

Be aware that when you find a bolt with damaged threads, you may also find the nut or drilled hole it was threaded into has also been damaged. If this is the case, you may have to drill and tap the hole, replace the nut or otherwise repair the threads. NEVER try to force a replacement bolt to fit into the damaged threads.

Torque

Torque is defined as the measurement of resistance to turning or rotating. It tends to twist a body about an axis of rotation. A common example of this would be tightening a threaded retainer such as a nut, bolt or screw. Measuring torque is one of the most common ways to help assure that a threaded retainer has been properly fastened.

When tightening a threaded fastener, torque is applied in three distinct areas, the head, the bearing surface and the clamp load. About 50 percent of the measured torque is used in overcoming bearing friction. This is the friction between the bearing surface of the bolt head, screw head or nut face and the base material or washer (the surface on which the fastener is rotating). Approximately 40 percent of the applied torque is used in overcoming thread friction. This leaves only about 10 percent of the applied torque to develop a useful clamp load (the force which holds a joint together). This means that friction can account for as much as 90 percent of the applied torque on a fastener.

GENERAL INFORMATION AND MAINTENANCE   1-7

TORQUE WRENCHES

♦ See Figure 21

In most applications, a torque wrench can be used to assure proper installation of a fastener. Torque wrenches come in various designs and most automotive supply stores will carry a variety to suit your needs. A torque wrench should be used any time we supply a specific torque value for a fastener. Again, the general rule of "if you are using the right tool for the job, you should not have to strain to tighten a fastener" applies here.

Beam Type

The beam type torque wrench is one of the most popular types. It consists of a pointer attached to the head that runs the length of the flexible beam (shaft) to a scale located near the handle. As the wrench is pulled, the beam bends and the pointer indicates the torque using the scale.

Click (Breakaway) Type

Another popular design of torque wrench is the click type. To use the click type wrench you pre-adjust it to a torque setting. Once the torque is reached, the wrench has a reflex signaling feature that causes a momentary breakaway of the torque wrench body, sending an impulse to the operator's hand.

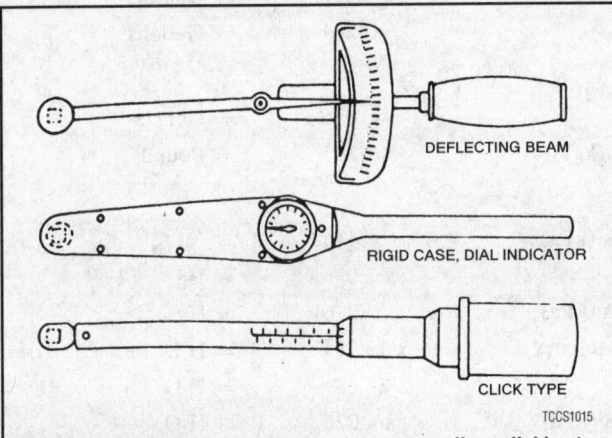

Fig. 21 Various styles of torque wrenches are usually available at your local automotive supply store

Pivot Head Type

♦ See Figure 22

Some torque wrenches (usually of the click type) may be equipped with a pivot head which can allow it to be used in areas of limited access. BUT, it must be used properly. To hold a pivot head wrench, grasp the handle lightly, and as you pull on the handle, it should be floated on the pivot point. If the handle comes in contact with the yoke extension during the process of pulling, there is a very good chance the torque readings will be inaccurate because this could alter the wrench loading point. The design of the handle is usually such as to make it inconvenient to deliberately misuse the wrench.

➡ It should be mentioned that the use of any U-joint, wobble or extension will have an effect on the torque readings, no matter what type of wrench you are using. For the most accurate readings, install the socket directly on the wrench driver. If necessary, straight extensions (which hold a socket directly under the wrench driver) will have the least effect on the torque reading. Avoid any extension that alters the length of the wrench from the handle to the head/driving point (such as a crow's foot). U-joint or wobble extensions can greatly affect the readings; avoid their use at all times.

Rigid Case (Direct Reading)

A rigid case or direct reading torque wrench is equipped with a dial indicator to show torque values. One advantage of these wrenches is that they can be

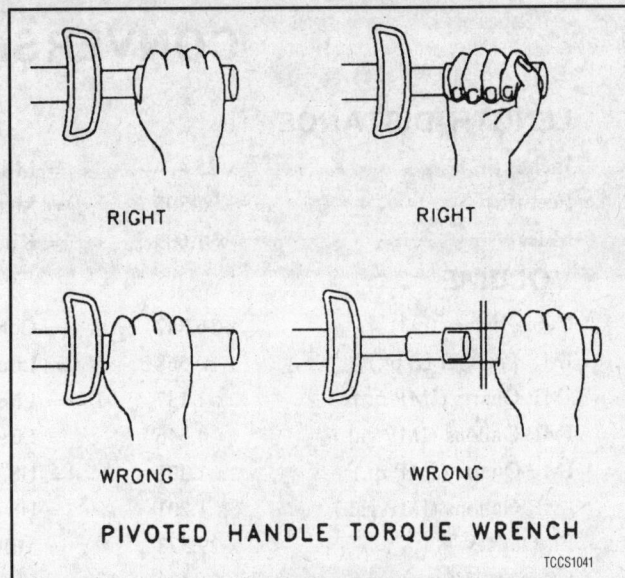

Fig. 22 Torque wrenches with pivoting heads must be grasped and used properly to prevent an incorrect reading

held at any position on the wrench without affecting accuracy. These wrenches are often preferred because they tend to be compact, easy to read and have a great degree of accuracy.

TORQUE ANGLE METERS

Because the frictional characteristics of each fastener or threaded hole will vary, clamp loads which are based strictly on torque will vary as well. In most applications, this variance is not significant enough to cause worry. But, in certain applications, a manufacturer's engineers may determine that more precise clamp loads are necessary (such is the case with many aluminum cylinder heads). In these cases, a torque angle method of installation would be specified. When installing fasteners which are torque angle tightened, a predetermined seating torque and standard torque wrench are usually used first to remove any compliance from the joint. The fastener is then tightened the specified additional portion of a turn measured in degrees. A torque angle gauge (mechanical protractor) is used for these applications.

Standard and Metric Measurements

♦ See Figure 23

Throughout this manual, specifications are given to help you determine the condition of various components on your vehicle, or to assist you in their installation. Some of the most common measurements include length (in. or cm/mm), torque (ft. lbs., inch lbs. or Nm) and pressure (psi, in. Hg, kPa or mm Hg). In most cases, we strive to provide the proper measurement as determined by the manufacturer's engineers.

Though, in some cases, that value may not be conveniently measured with what is available in your toolbox. Luckily, many of the measuring devices which are available today will have two scales so the Standard or Metric measurements may easily be taken. If any of the various measuring tools which are available to you do not contain the same scale as listed in the specifications, use the accompanying conversion factors to determine the proper value.

The conversion factor chart is used by taking the given specification and multiplying it by the necessary conversion factor. For instance, looking at the first line, if you have a measurement in inches such as "free-play should be 2 in." but your ruler reads only in millimeters, multiply 2 in. by the conversion factor of 25.4 to get the metric equivalent of 50.8mm. Likewise, if the specification was given only in a Metric measurement, for example in Newton Meters (Nm), then look at the center column first. If the measurement is 100 Nm, multiply it by the conversion factor of 0.738 to get 73.8 ft. lbs.

1-8 GENERAL INFORMATION AND MAINTENANCE

CONVERSION FACTORS

LENGTH–DISTANCE

Inches (in.)	x 25.4	= Millimeters (mm)	x .0394	= Inches
Feet (ft.)	x .305	= Meters (m)	x 3.281	= Feet
Miles	x 1.609	= Kilometers (km)	x .0621	= Miles

VOLUME

Cubic Inches (in3)	x 16.387	= Cubic Centimeters	x .061	= in3
IMP Pints (IMP pt.)	x .568	= Liters (L)	x 1.76	= IMP pt.
IMP Quarts (IMP qt.)	x 1.137	= Liters (L)	x .88	= IMP qt.
IMP Gallons (IMP gal.)	x 4.546	= Liters (L)	x .22	= IMP gal.
IMP Quarts (IMP qt.)	x 1.201	= US Quarts (US qt.)	x .833	= IMP qt.
IMP Gallons (IMP gal.)	x 1.201	= US Gallons (US gal.)	x .833	= IMP gal.
Fl. Ounces	x 29.573	= Milliliters	x .034	= Ounces
US Pints (US pt.)	x .473	= Liters (L)	x 2.113	= Pints
US Quarts (US qt.)	x .946	= Liters (L)	x 1.057	= Quarts
US Gallons (US gal.)	x 3.785	= Liters (L)	x .264	= Gallons

MASS–WEIGHT

Ounces (oz.)	x 28.35	= Grams (g)	x .035	= Ounces
Pounds (lb.)	x .454	= Kilograms (kg)	x 2.205	= Pounds

PRESSURE

Pounds Per Sq. In. (psi)	x 6.895	= Kilopascals (kPa)	x .145	= psi
Inches of Mercury (Hg)	x .4912	= psi	x 2.036	= Hg
Inches of Mercury (Hg)	x 3.377	= Kilopascals (kPa)	x .2961	= Hg
Inches of Water (H_2O)	x .07355	= Inches of Mercury	x 13.783	= H_2O
Inches of Water (H_2O)	x .03613	= psi	x 27.684	= H_2O
Inches of Water (H_2O)	x .248	= Kilopascals (kPa)	x 4.026	= H_2O

TORQUE

Pounds–Force Inches (in–lb)	x .113	= Newton Meters (N·m)	x 8.85	= in–lb
Pounds–Force Feet (ft–lb)	x 1.356	= Newton Meters (N·m)	x .738	= ft–lb

VELOCITY

Miles Per Hour (MPH)	x 1.609	= Kilometers Per Hour (KPH)	x .621	= MPH

POWER

Horsepower (Hp)	x .745	= Kilowatts	x 1.34	= Horsepower

FUEL CONSUMPTION*

Miles Per Gallon IMP (MPG)	x .354	= Kilometers Per Liter (Km/L)
Kilometers Per Liter (Km/L)	x 2.352	= IMP MPG
Miles Per Gallon US (MPG)	x .425	= Kilometers Per Liter (Km/L)
Kilometers Per Liter (Km/L)	x 2.352	= US MPG

*It is common to covert from miles per gallon (mpg) to liters/100 kilometers (l/100 km), where mpg (IMP) x l/100 km = 282 and mpg (US) x l/100 km = 235.

TEMPERATURE

Degree Fahrenheit (°F) = (°C x 1.8) + 32
Degree Celsius (°C) = (°F – 32) x .56

Fig. 23 Standard and metric conversion factors chart

GENERAL INFORMATION AND MAINTENANCE 1-9

HISTORY

In 1946, after the close of World War II, Soichiro Honda established his Technical Research Institute in Hamamatsu, Japan, to develop auxiliary gasoline engines for bicycles. The young company successfully marketed gasoline-engined bicycles and primitive motorcycles. Within five years, the company established a production record within the Japanese motorcycle industry by producing 130 units per day at its new Tokyo plant.

Through the 1950's and into the next decade, the company grew steadily, pioneering and refining many phases of motorcycle engineering and production, leading to the opening of a separate research and development center. Honda furthered its reputation by participating in many domestic and international motorcycle racing events. Additionally, the company adapted its small engine technology to non-road vehicle use by introducing engines for farm equipment, generators and marine use. Manufacturing and support facilities were established throughout Japan, Asia and Europe.

By August, 1963, Honda had produced a small utility truck powered by a 360cc engine. This was closely followed by a series of passenger coupes and sedans, all powered by air-cooled engines displacing well below one liter. As an adjunct to the passenger car program, a Honda-powered Formula One (Grand Prix) car competed in the 1964–1968 seasons.

The 15 millionth motorcycle was produced in March, 1971. The CVCC engine was perfected in the same year and hailed as the first engine to meet the 1975 U.S Clean Air Act specifications. Honda won many awards for design, engineering and manufacturing as well as industrial pollution control and community betterment. Mr. Honda retired as President of the company in 1973 but remained as Supreme Advisor.

The first Honda passenger cars began appearing in North America in the early 1970's, lead by the 600 family and followed closely by the Civic, Honda's first water-cooled vehicle. The year 1976 saw the introduction of the Accord, a larger five passenger car powered by the CVCC engine that was an immediate hit in North America. In September, 1979 the company's success was further bolstered by a U.S. motorcycle manufacturing plant in Marysville, Ohio. And that same year, the Prelude automobile was introduced as "An advanced engineering platform." By 1980, Honda had produced 4,000,000 power products (generators, lawn mowers, etc.), two million Civics, and over 30 million motorcycles.

Through the 1980's, Honda Motor Co. continued to grow in size and competitiveness. In 1983 automobile manufacture was begun at the expanded Marysville plant. By August of the same year, the plant was producing 300 cars per day. Around this same time, Honda re-entered international Formula One racing, winning many events and manufacturer's championships with both Team Williams and Team McLaren. Engine manufacture was begun at Marysville in September of 1986.

By the end of the 1980's, Honda Motor Co. was well-established as a major manufacturer in the world market. Its automotive products have been highly praised by owners, media, and competitors alike. In 1990, the Accord had established itself as the best selling car in the United States for several consecutive quarters.

Soichiro Honda died August 5th, 1991 at the age of 82. His company now has motorcycle or automobile manufacturing facilities in over 30 countries. The first Accord produced at the Ohio plant is on display at the Henry Ford Museum in Dearborn, Michigan.

Coverage

This manual covers all versions of the Civic, Civic Wagon, CRX and del Sol produced from 1984 to 1995. This includes all of the various trim designations (Std., DX, LX, EX, HF, Si, VTEC, etc.). The following is a reference list illustrating the generations of the Civic family of Honda vehicles.

- Civic 1st generation: 1973–79
- Civic 2nd generation: 1980–83
- Civic 3rd generation: 1984–87
- Civic 4th generation: 1988–92
- CRX 1st generation: 1984–87
- CRX 2nd generation: 1988–91
- Civic 5th generation: 1992–95
- del Sol 1st generation: 1993–95

SERIAL NUMBER IDENTIFICATION

Vehicle Identification (Chassis) Number

▶ See Figures 24, 25 and 26

Vehicle identification numbers are mounted on the top edge of the instrument panel and are visible through the glass from the outside. There are also identification plates located on the firewall and the rear door jamb of the driver's door.

Engine Serial Number

▶ See Figure 27

The engine serial number is stamped into the clutch casing on all vehicles. For models produced from 1984–86, the first three numbers provide engine model identification. For models produced from 1987–91, the first five numbers provide engine model identification. For models produced from 1992–95, the fourth, fifth and sixth numbers provide the engine model identification. The remaining numbers refer to production sequence. This same number may also be stamped into an identification plate mounted on the radiator support.

Transaxle Serial Number

▶ See Figures 28, 29 and 30

The transaxle serial number is stamped on the top of the transaxle/clutch case.

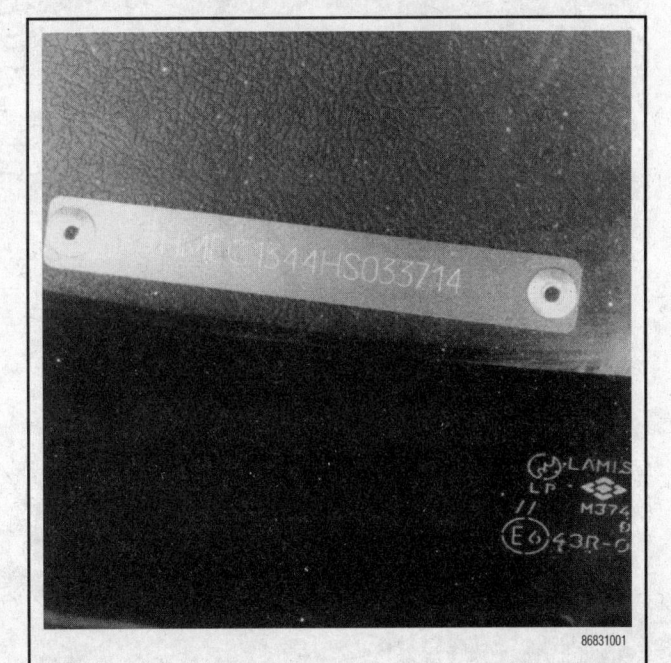

Fig. 24 The vehicle identification number can be viewed through the driver's side windshield

1-10 GENERAL INFORMATION AND MAINTENANCE

Fig. 25 The vehicle identification number is also stamped into the top of the firewall . . .

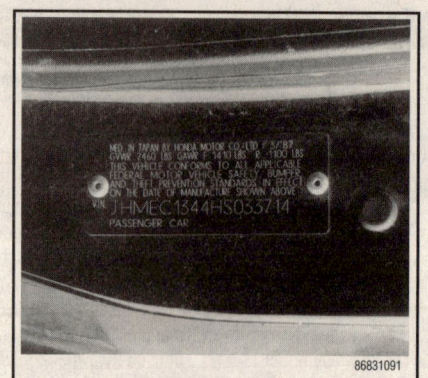

Fig. 26 . . . and into the plate located below driver's door striker plate

Fig. 27 The engine serial number may also be found stamped on the radiator support

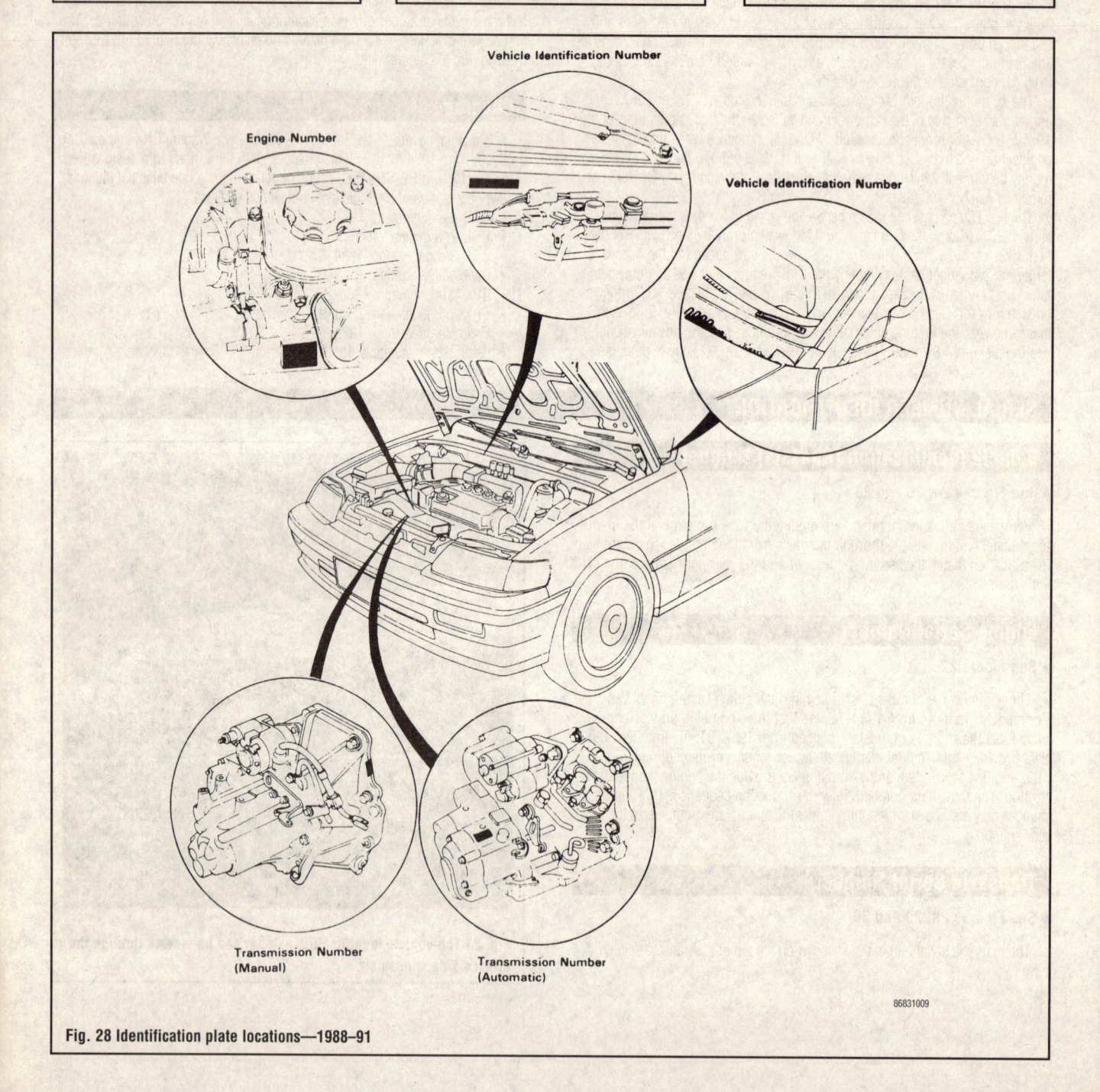

Fig. 28 Identification plate locations—1988–91

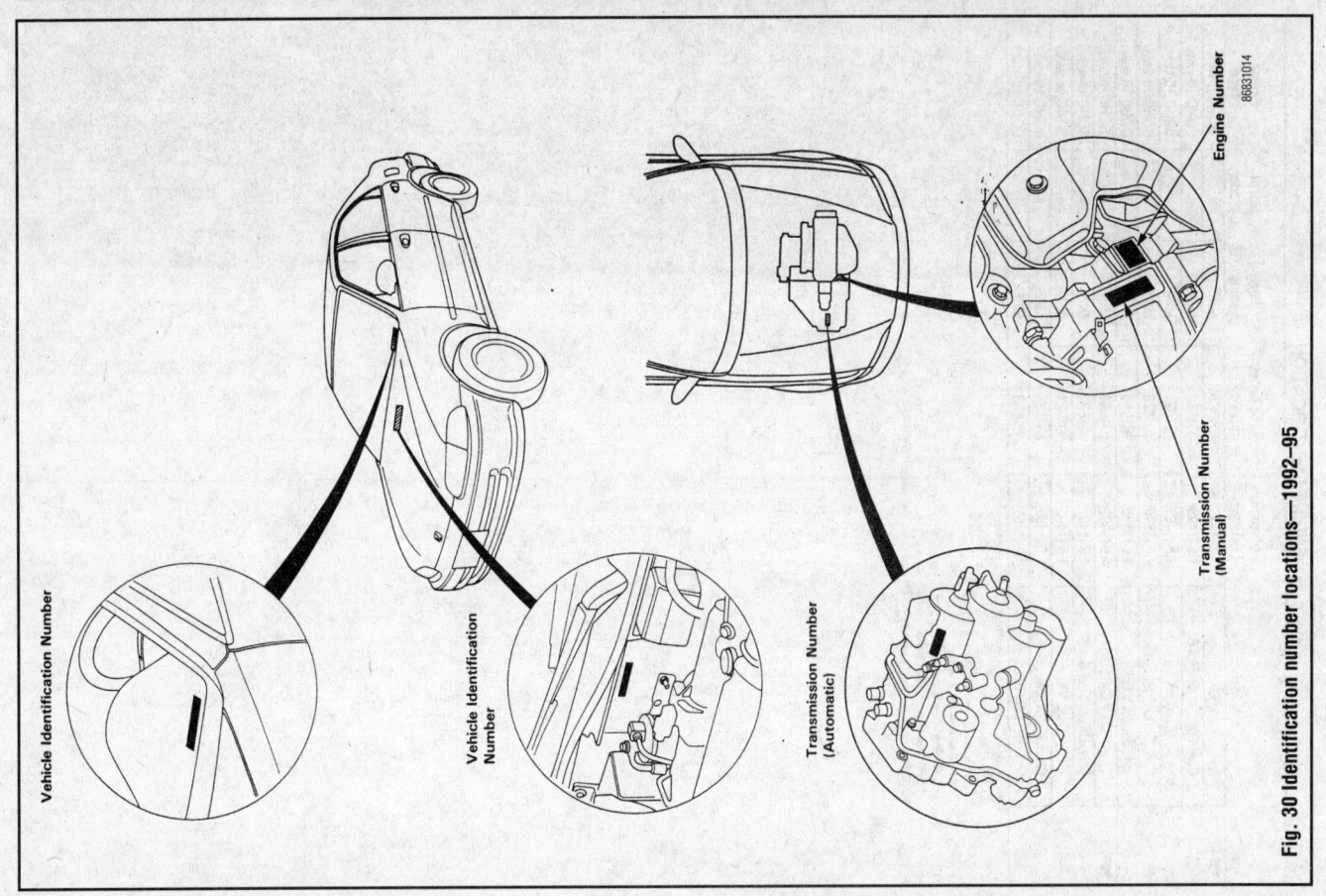

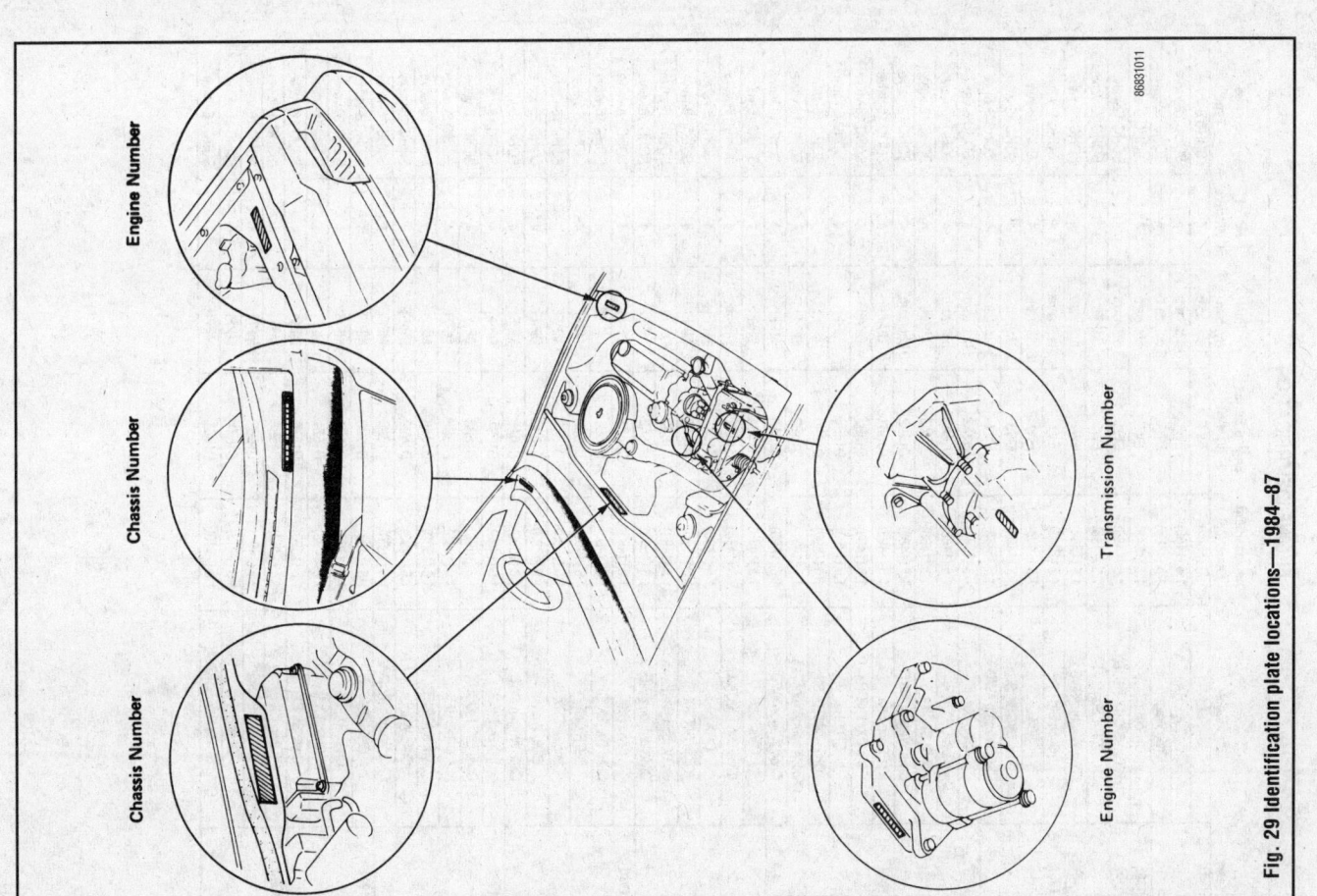

GENERAL INFORMATION AND MAINTENANCE 1-11

Fig. 30 Identification number locations—1992–95

Fig. 29 Identification plate locations—1984–87

1-12 GENERAL INFORMATION AND MAINTENANCE

ENGINE IDENTIFICATION

Year	Model	Engine Displacement Liters (cc)	Engine Series (ID/VIN)	Fuel System ②	No. of Cylinders	Engine Type
1994	Civic CX	1.5 (1493)	D15B8	MPFI	4	SOHC 8-valve
	Civic DX, LX, del Sol S ①	1.5 (1493)	D15B7	MPFI	4	SOHC 16-valve
	Civic VX	1.5 (1493)	D15Z1	MPFI	4	SOHC 16-valve
	Civic Si, EX, del Sol Si	1.6 (1590)	D16Z6	MPFI	4	SOHC 16-valve
	del Sol VTEC	1.6 (1595)	B16A3	MPFI	4	DOHC 16-valve
1995	Civic CX	1.5 (1493)	D15B8	MPFI	4	SOHC 8-valve
	Civic DX, LX, del Sol S ①	1.5 (1493)	D15B7	MPFI	4	SOHC 16-valve
	Civic VX	1.5 (1493)	D15Z1	MPFI	4	SOHC 16-valve
	Civic Si, EX, del Sol Si	1.6 (1590)	D16Z6	MPFI	4	SOHC 16-valve
	del Sol VTEC	1.6 (1595)	B16A3	MPFI	4	DOHC 16-valve

① "Civic DX" applies to both Hatchback and 4-door models.
② Fuel System Terms:
MPFI: Multipoint Fuel Injection
DP-FI: Dual Point Fuel Injection
SOHC: Single Overhead Camshaft
DOHC: Dual Overhead Camshafts
NOTE: Honda's trade name for fuel injection is "Programmed Fuel Injection" (PGM-FI)

ENGINE IDENTIFICATION

Year	Model	Engine Displacement Liters (cc)	Engine Series (ID/VIN)	Fuel System	No. of Cylinders	Engine Type
1984	Civic Std.	1.3 (1342)	EV1	3 bbl. carb.	4	SOHC 8-valve
	Civic DX/CRX ①	1.5 (1488)	EW1	3 bbl. carb.	4	SOHC 12-valve
	Civic Wagon	1.5 (1488)	EW1	3 bbl. carb.	4	SOHC 12-valve
1985	Civic Std.	1.3 (1342)	EV1	3 bbl. carb.	4	SOHC 8-valve
	Civic DX, Si/CRX HF, DX ①	1.5 (1488)	EW1	3 bbl. carb.	4	SOHC 12-valve
	CRX Si	1.5 (1488)	EW3	MPFI	4	SOHC 12-valve
1986	Civic Std.	1.3 (1342)	EV1	3 bbl. carb.	4	SOHC 8-valve
	Civic Wagon, 4WD	1.5 (1488)	EW1	3 bbl. carb.	4	SOHC 12-valve
	Si ①	1.5 (1488)	EW3	MPFI	4	SOHC 12-valve
	Civic Si/CRX Si	1.5 (1488)	EW1	3 bbl. carb.	4	SOHC 8-valve
	Civic Wagon, 4WD	1.5 (1488)	EW1	3 bbl. carb.	4	SOHC 8-valve
1987	Civic Std.	1.3 (1342)	D13A2	3 bbl. carb.	4	SOHC 12-valve
	Civic DX/CRX HF, DX ①	1.5 (1488)	D15A2	3 bbl. carb.	4	SOHC 12-valve
	Civic Si/CRX Si	1.5 (1488)	D15A3	MPFI	4	SOHC 12-valve
	Civic Wagon, 4WD	1.5 (1488)	D15A2	3 bbl. carb.	4	SOHC 12-valve
1988	Civic Std.	1.5 (1493)	D15B1	DP-FI	4	SOHC 16-valve
	Civic DX, LX/CRX ①	1.5 (1493)	D15B2	DP-FI	4	SOHC 16-valve
	CRX HF	1.5 (1493)	D15B6	MPFI	4	SOHC 8-valve
	Civic Si/CRX Si	1.6 (1590)	D16A6	MPFI	4	SOHC 16-valve
	Civic Wagon/Wagovan	1.5 (1493)	D15B2	DP-FI	4	SOHC 16-valve
	Civic Wagon 4WD	1.6 (1590)	D16A6	MPFI	4	SOHC 16-valve
1989	Civic DX, LX/CRX ①	1.5 (1493)	D15B1	DP-FI	4	SOHC 16-valve
	CRX HF	1.5 (1493)	D15B6	MPFI	4	SOHC 8-valve
	Civic Si/CRX Si	1.6 (1590)	D16A6	MPFI	4	SOHC 16-valve
	Civic Wagon	1.5 (1493)	D15B2	DP-FI	4	SOHC 16-valve
	Civic Wagon 4WD	1.6 (1590)	D16A6	MPFI	4	SOHC 16-valve
1990	Civic Std.	1.5 (1493)	D15B1	DP-FI	4	SOHC 16-valve
	Civic DX, LX ①	1.5 (1493)	D15B2	DP-FI	4	SOHC 16-valve
	CRX HF	1.5 (1493)	D15B6	MPFI	4	SOHC 8-valve
	Civic Wagon	1.6 (1590)	D16A6	MPFI	4	SOHC 16-valve
1991	Civic Std.	1.5 (1493)	D15B1	DP-FI	4	SOHC 16-valve
	Civic DX, LX/CRX ①	1.5 (1493)	D15B2	DP-FI	4	SOHC 16-valve
	CRX, HF	1.5 (1493)	D15B6	MPFI	4	SOHC 8-valve
	Civic Si/CRX Si	1.6 (1590)	D16A6	MPFI	4	SOHC 16-valve
	Civic Wagon	1.5 (1493)	D15B2	DP-FI	4	SOHC 16-valve
	Civic Wagon 4WD	1.6 (1590)	D16A6	MPFI	4	SOHC 16-valve
1992	Civic CX	1.5 (1493)	D15B8	MPFI	4	SOHC 8-valve
	Civic DX, LX ①	1.5 (1493)	D15B7	MPFI	4	SOHC 16-valve
	Civic VX	1.5 (1493)	D15Z1	MPFI	4	SOHC 16-valve
	Civic Si, EX	1.6 (1590)	D16Z6	MPFI	4	SOHC 16-valve
1993	Civic CX	1.5 (1493)	D15B8	MPFI	4	SOHC 8-valve
	Civic DX, LX, del Sol S ①	1.5 (1493)	D15B7	MPFI	4	SOHC 16-valve
	Civic VX	1.5 (1493)	D15Z1	MPFI	4	SOHC 16-valve
	Civic Si, EX, del Sol Si	1.6 (1590)	D16Z6	MPFI	4	SOHC 16-valve

GENERAL INFORMATION AND MAINTENANCE   1-13

ROUTINE MAINTENANCE

Air Cleaner

REMOVAL & INSTALLATION

▶ See Figures 31 thru 38

On carbureted models, the air cleaner is housed above the carburetor, and on fuel injected models, in a housing at the front of the vehicle. The recommended change interval for the air filter is 30,000 miles (48,300 km) under normal circumstances. Severe service requires more frequent changes.

1. Remove the fasteners (wing nut(s), bolts and/or spring clips) securing the air cleaner cover.
2. Remove the air cleaner cover and element.

To install:

3. Using a clean rag, wipe out the air cleaner housing.
4. Position the element in the air cleaner housing.

➯**Air cleaner elements are not interchangeable, although they may appear to be. Make sure you have the proper element for the year and model of your vehicle.**

5. Install the air cleaner cover and any applicable retaining fasteners.

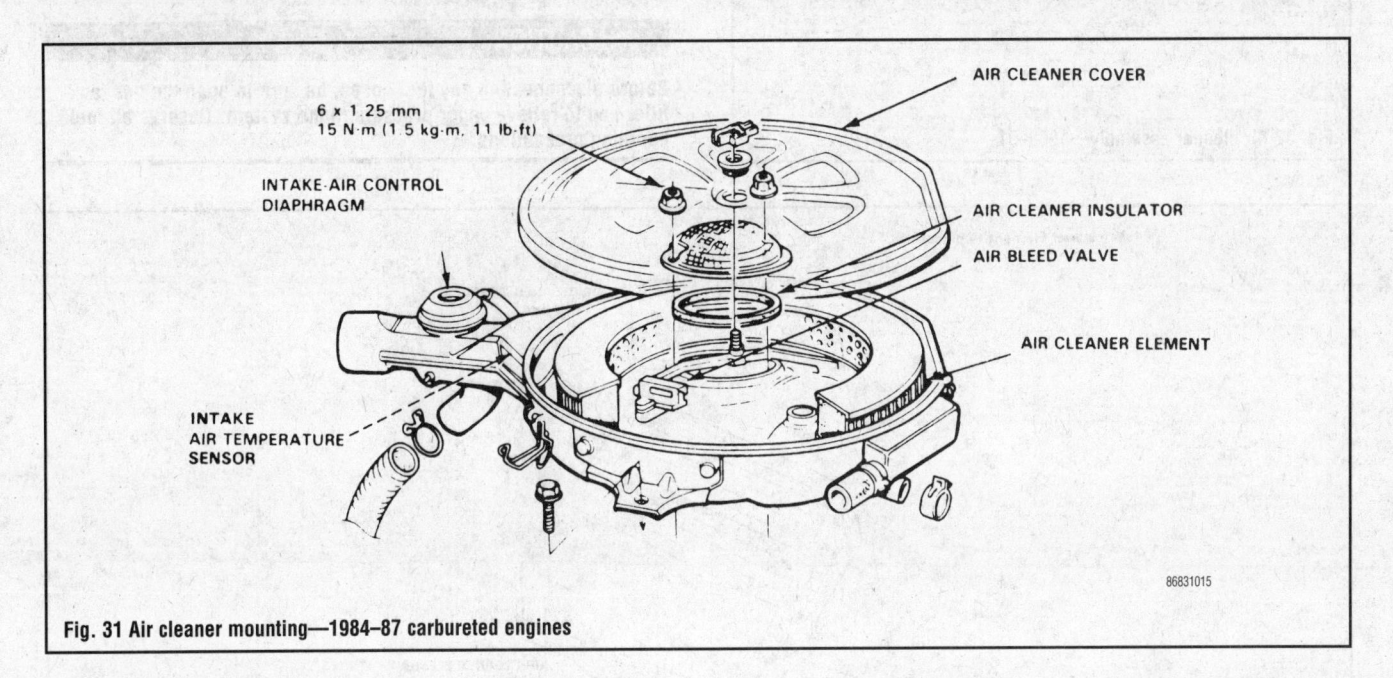

Fig. 31 Air cleaner mounting—1984–87 carbureted engines

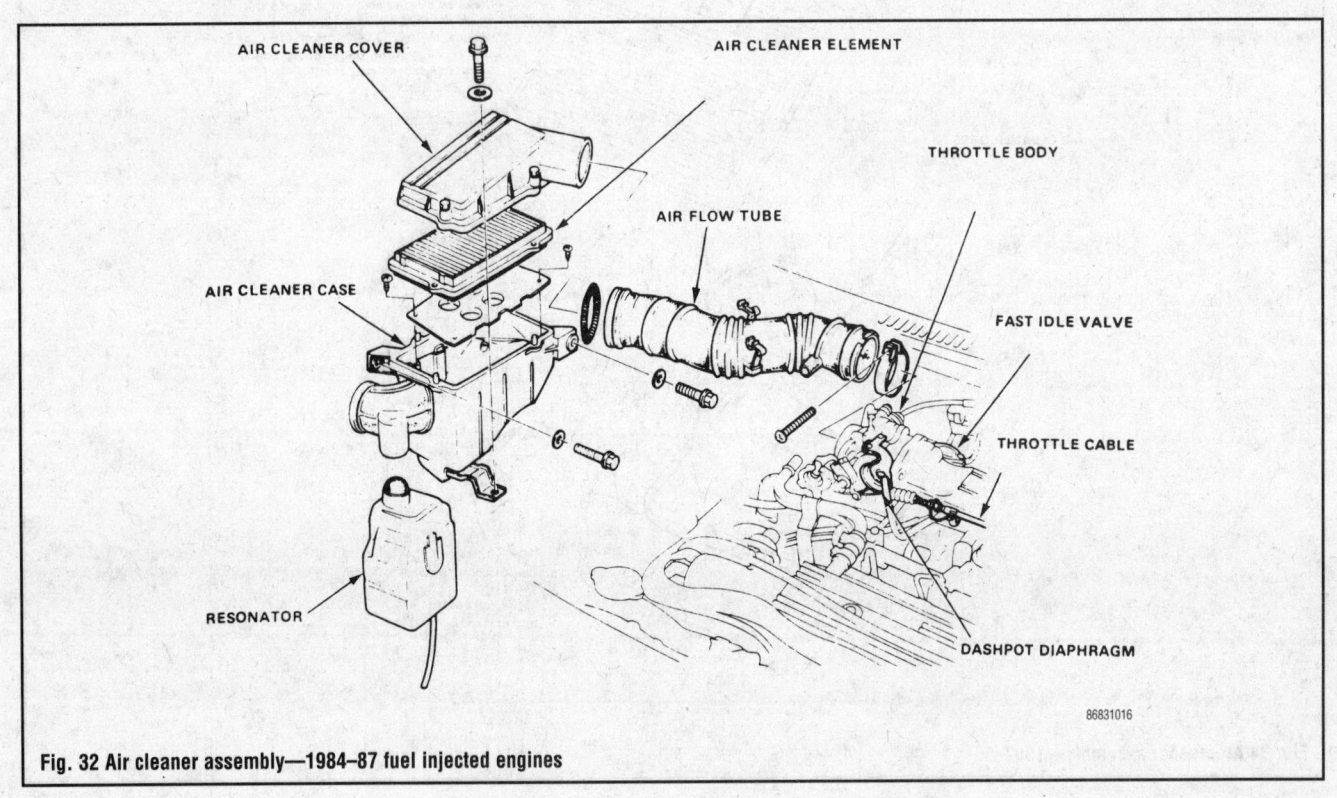

Fig. 32 Air cleaner assembly—1984–87 fuel injected engines

1-14 GENERAL INFORMATION AND MAINTENANCE

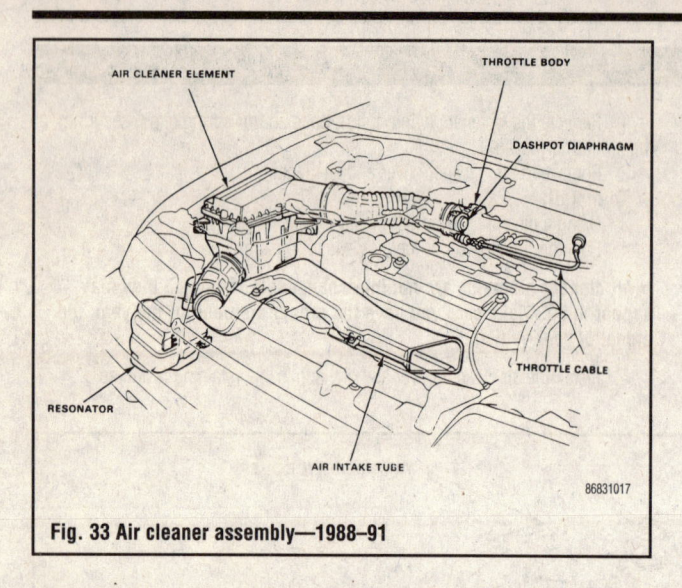

Fig. 33 Air cleaner assembly—1988–91

Fuel Filter

The recommended change interval for the fuel filter(s) on all models is every four years or 60,000 miles (96,500 km) under normal operating conditions. Severe service will require more frequent changing of the fuel filter.

Carbureted Civics have two fuel filters. All fuel injected Civics have one filter. On carbureted cars one filter is located in a holder on the underside of the car near the fuel tank. The other is located in the passenger's side of the engine compartment. Fuel injected cars also have their filter located in this area.

The fuel filter is a simple apparatus with one inlet (feed) and one outlet fitting designed to screen out contaminants from the fuel before it reaches the combustion chamber. The feed and output fittings on carbureted Civics are attached by clamps. Fuel injected models with higher line pressure, employ banjo bolts or a screw-in pipe may be used for the feed side.

❋❋ CAUTION

Before disconnecting any fuel hoses, be sure to open the gas tank filler cap to relieve vapor pressure in the system. Observe all fuel handling precautions.

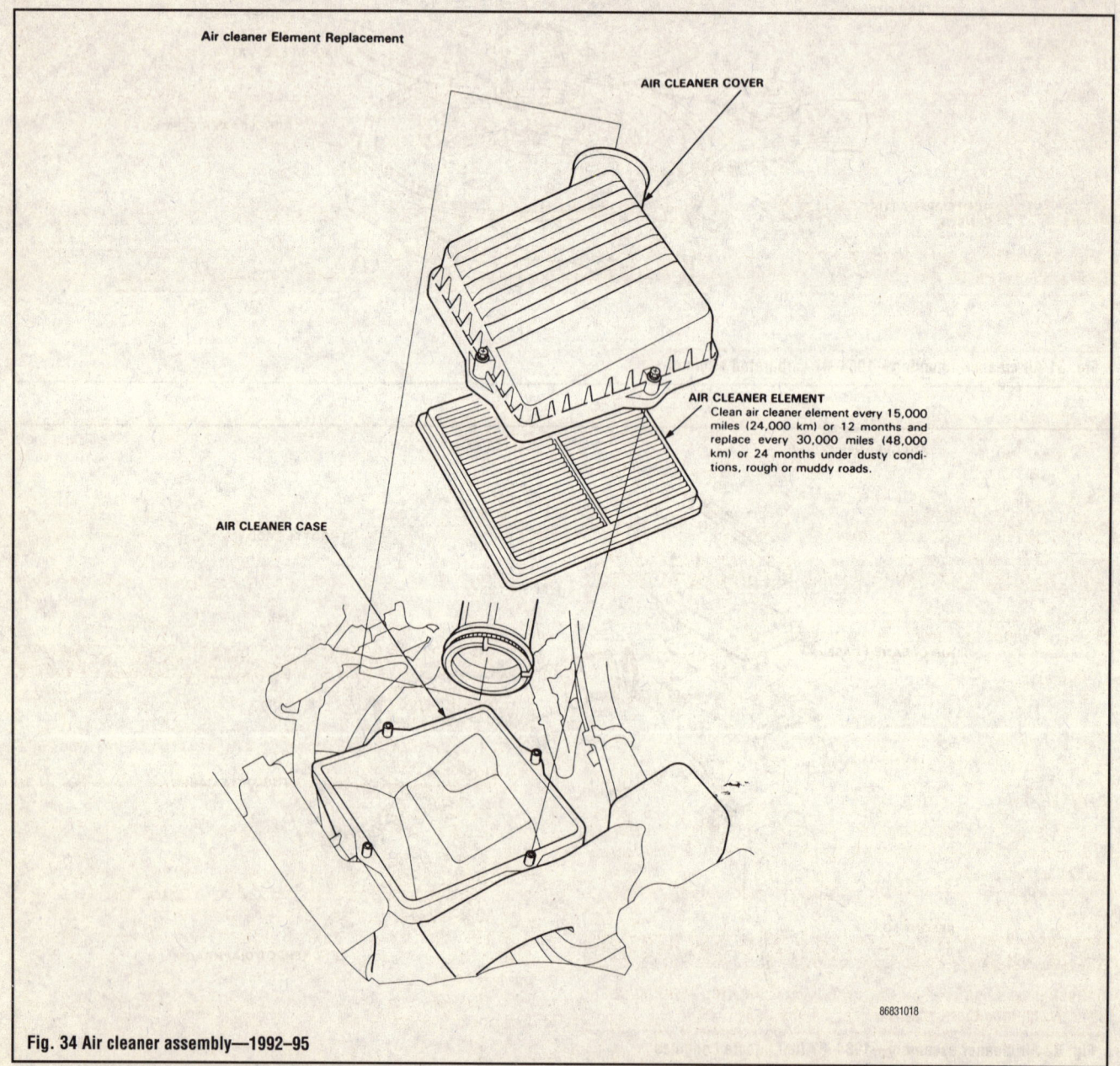

Fig. 34 Air cleaner assembly—1992–95

GENERAL INFORMATION AND MAINTENANCE 1-15

Fig. 35 On fuel injected models, the air cleaner housing is usually secured by screws

Fig. 36 Age and heat can cause the air cleaner element to warp

Fig. 37 When installing the filter, make sure the correct side is facing up

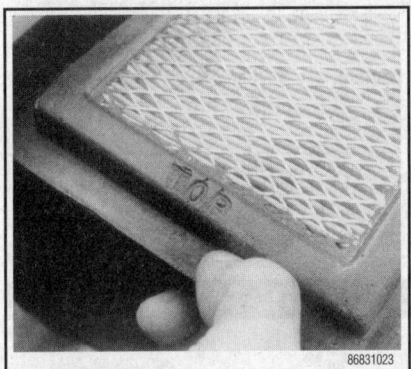

Fig. 38 Most filters are stamped to ensure correct installation

Fig. 39 Wrap a rag around the fuel filter when releasing the fuel pressure

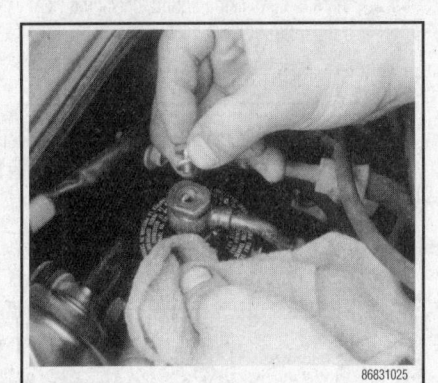

Fig. 40 Be careful to not lose the service screw from atop the fuel filter

REMOVAL & INSTALLATION

1984–1987 Carbureted Models

FRONT FILTER

1. Disconnect the negative battery cable.
2. Use fuel line clamps or equivalent locking pliers to pinch the fuel hoses closed.
3. Loosen the fuel hose clamps.
4. Disconnect the fuel lines and remove the fuel filter.

※※ WARNING

To avoid damaging the fuel lines when disconnecting them, slide back the clamps then twist the lines as you pull.

To install:
5. Remove the filter cover from the old filter and install it on the new filter.
6. Install the new fuel filter and tighten the hose clamps.
7. Remove the fuel line clamps/locking pliers.
8. Connect the negative battery cable.

REAR FILTER

1. Raise the rear of the car and place jackstands in the proper locations.
2. Push in the tab of the fuel filter to release it from the holder, then remove the filter from its bracket.
3. Use fuel line clamps or equivalent locking pliers to pinch the fuel hoses closed.

※※ WARNING

To avoid damaging the fuel lines when disconnecting them, slide back the clamps then twist the lines as you pull.

4. Install the new fuel filter.
5. Remove the fuel line clamps.

Fuel Injected Models

▶ See Figures 39 and 40

1. Disconnect the negative battery cable.
2. Use fuel line clamps or equivalent locking pliers to pinch the fuel lines closed.
3. Wrap a shop towel around the fittings. Bleed off the fuel system pressure by slowly loosening the service bolt at the top of the fuel outlet fitting, while holding the special banjo bolt with another wrench.
4. Support the fuel filter with a wrench, then remove the banjo bolts or the fuel inlet pipe (as applicable) from the filter.
5. Loosen the fuel filter clamp and remove the filter from the vehicle.

※※ CAUTION

It is very important that ALL of the fuel line banjo bolt washers be replaced every time the banjo bolts are loosened. If the washers are not replaced, the fuel lines may leak pressurized fuel causing the risk of fire or explosion.

To install:
6. Install the new filter in position and tighten the clamp mounting bolt to 7 ft. lbs. (10 Nm).
7. Attach the banjo fittings using new washers, and tighten the banjo bolts or fuel inlet pipe (as applicable) to the proper torque specification.
- **1984–1990 cars:** tighten both banjo bolts to 16 ft. lbs. (22 Nm).
- **1991 cars:** tighten the banjo bolt attaching the fuel outlet hose (the one with the service bolt on top) to 16 ft. lbs. (22 Nm); tighten the banjo bolt to the fuel feed hose to 26 ft. lbs. (34 Nm).
- **1992–95 cars:** tighten the banjo bolt to 16 ft. lbs. (22 Nm), and the fuel inlet pipe to 28 ft. lbs. (38 Nm).
8. Start the vehicle and check immediately for leaks.

1-16 GENERAL INFORMATION AND MAINTENANCE

Positive Crankcase Ventilation (PCV)

♦ See Figures 41 and 42

The Civics are equipped with a Positive Crankcase Ventilation (PCV) system. This system prevents blow-by gas from escaping to the atmosphere by recycling the gases to the combustion chamber through the intake manifold and/or the air cleaner.

The maintenance interval for replacing the PCV valve is 60,000 miles (96,500 km) under normal use. Under severe operating conditions, the PCV valve may need to be changed sooner.

INSPECTION

1. Check the PCV valve hoses. They should not leak or be clogged.
2. At idle, make sure the PCV valve clicks when the hose between the PCV valve and intake manifold is lightly pinched with your fingers or pliers.
3. If there is no clicking sound, check the PCV valve grommet for damage or cracks. If the grommet is intact, replace the PCV valve and recheck.

REMOVAL & INSTALLATION

1. The PCV valve is mounted in the intake manifold and has a hose connected to it from the crankcase breather chamber. The valve is removed by pulling it from the manifold and hose.

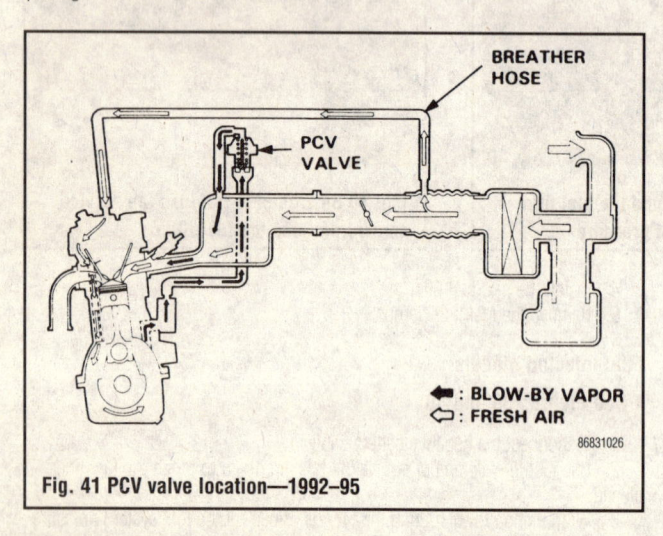

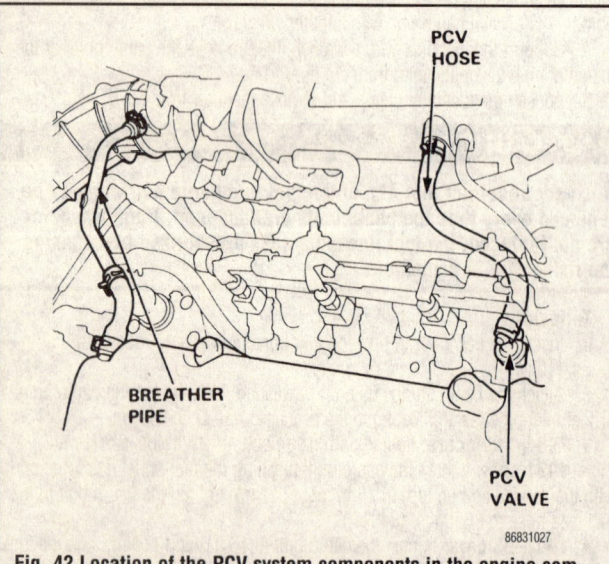

Fig. 41 PCV valve location—1992–95

Fig. 42 Location of the PCV system components in the engine compartment

2. Check for loose, disconnected, clogged or deteriorated tubes and replace as necessary. Make sure the hoses are clean inside, cleaning them with a safe solvent, if necessary.
3. If the system has a condensation chamber attached to the bottom of the air cleaner, unscrew and remove the chamber, clean and replace it; when removing the top gasket from the chamber, note the angle of installation. Reinstall it in the same position to provide proper airflow.
4. Check the valve by pulling it out of the manifold with the engine idling. Cover the open end of the valve with your finger so airflow is stopped. If the valve clicks, it is OK; if not, replace it.

➡ For PCV valve testing refer to Section 4 of this manual.

Evaporative Charcoal Canister

The charcoal canister is part of the Evaporative Emission Control System. This system prevents the escape of raw gasoline vapors from the fuel tank and carburetor (if equipped).

The charcoal canister is designed to absorb fuel vapors under certain conditions.

➡ For a more detailed description, see Section 4.

SERVICING

The canister does not require periodic replacement. The entire system requires a careful operational check with a vacuum gauge at 60,000 miles (96,500 km). See Section 4 for testing procedures.

Battery

PRECAUTIONS

Always use caution when working on or near the battery. Never allow a tool to bridge the gap between the negative and positive battery terminals. Also, be careful not to allow a tool to provide a ground between the positive cable/terminal and any metal component on the vehicle. Either of these conditions will cause a short circuit, leading to sparks and possible personal injury.

Do not smoke or all open flames/sparks near a battery; the gases contained in the battery are very explosive and, if ignited, could cause severe injury or death.

All batteries, regardless of type, should be carefully secured by a battery hold-down device. If not, the terminals or casing may crack from stress during vehicle operation. A battery which is not secured may allow acid to leak, making it discharge faster. The acid can also eat away at components under the hood.

Always inspect the battery case for cracks, leakage and corrosion. A white corrosive substance on the battery case or on nearby components would indicate a leaking or cracked battery. If the battery is cracked, it should be replaced immediately.

GENERAL MAINTENANCE

Always keep the battery cables and terminals free of corrosion. Check and clean these components about once a year.

Keep the top of the battery clean, as a film of dirt can help discharge a battery that is not used for long periods. A solution of baking soda and water may be used for cleaning, but be careful to flush this off with clear water. DO NOT let any of the solution into the filler holes. Baking soda neutralizes battery acid and will de-activate a battery cell.

Batteries in vehicles which are not operated on a regular basis can fall victim to parasitic loads (small current drains which are constantly drawing current from the battery). Normal parasitic loads may drain a battery on a vehicle that is in storage and not used for 6–8 weeks. Vehicles that have additional accessories such as a phone or an alarm system may discharge a battery sooner. If the vehicle is to be stored for longer periods in a secure area and the alarm system is not necessary, the negative battery cable should be disconnected to protect the battery.

Remember that constantly deep cycling a battery (completely discharging and recharging it) will shorten battery life.

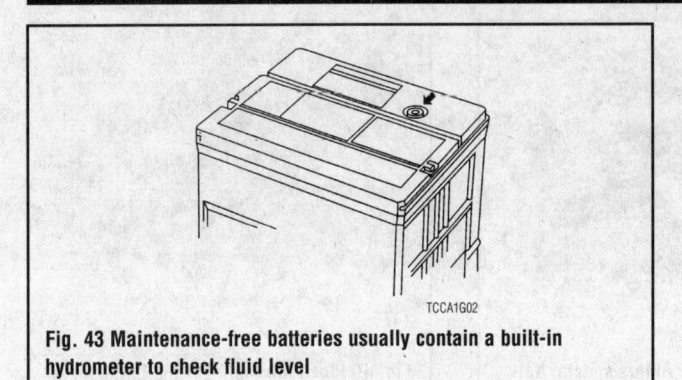

GENERAL INFORMATION AND MAINTENANCE 1-17

Fig. 43 Maintenance-free batteries usually contain a built-in hydrometer to check fluid level

BATTERY FLUID

See Figure 43

Check the battery electrolyte level at least once a month, or more often in hot weather or during periods of extended vehicle operation. On non-sealed batteries, the level can be checked either through the case (if translucent) or by removing the cell caps. The electrolyte level in each cell should be kept filled to the split ring inside each cell, or the line marked on the outside of the case.

If the level is low, add only distilled water through the opening until the level is correct. Each cell must be checked and filled individually. Distilled water should be used, because the chemicals and minerals found in most drinking water are harmful to the battery and could significantly shorten its life.

If water is added in freezing weather, the vehicle should be driven several miles to allow the water to mix with the electrolyte. Otherwise, the battery could freeze.

Although some maintenance-free batteries have removable cell caps, the electrolyte condition and level on all sealed maintenance-free batteries must be checked using the built-in hydrometer "eye." The exact type of eye will vary. But, most battery manufacturers, apply a sticker to the battery itself explaining the readings.

➥Although the readings from built-in hydrometers will vary, a green eye usually indicates a properly charged battery with sufficient fluid level. A dark eye is normally an indicator of a battery with sufficient fluid, but which is low in charge. A light or yellow eye usually indicates that electrolyte has dropped below the necessary level. In this last case, sealed batteries with an insufficient electrolyte must usually be discarded.

Checking the Specific Gravity

See Figures 44, 45 and 46

A hydrometer is required to check the specific gravity on all batteries that are not maintenance-free. On batteries that are maintenance-free, the specific gravity is checked by observing the built-in hydrometer "eye" on the top of the battery case.

✲✲ CAUTION

Battery electrolyte contains sulfuric acid. If you should splash any on your skin or in your eyes, flush the affected area with plenty of clear water. If it lands in your eyes, get medical help immediately.

The fluid (sulfuric acid solution) contained in the battery cells will tell you many things about the condition of the battery. Because the cell plates must be kept submerged below the fluid level in order to operate, the fluid level is extremely important. And, because the specific gravity of the acid is an indication of electrical charge, testing the fluid can be an aid in determining if the battery must be replaced. A battery in a vehicle with a properly operating charging system should require little maintenance, but careful, periodic inspection should reveal problems before they leave you stranded.

At least once a year, check the specific gravity of the battery. It should be between 1.20 and 1.26 on the gravity scale. Most auto stores carry a variety of inexpensive battery hydrometers. These can be used on any non-sealed battery to test the specific gravity in each cell.

The battery testing hydrometer has a squeeze bulb at one end and a nozzle at the other. Battery electrolyte is sucked into the hydrometer until the float is lifted from its seat. The specific gravity is then read by noting the position of the float. If gravity is low in one or more cells, the battery should be slowly charged and checked again to see if the gravity has come up. Generally, if after charging, the specific gravity between any two cells varies more than 50 points (0.50), the battery should be replaced, as it can no longer produce sufficient voltage to guarantee proper operation.

CABLES

See Figures 47, 48, 49 and 50

Once a year (or as necessary), the battery terminals and the cable clamps should be cleaned. Loosen the clamps and remove the cables, negative cable first. On top post batteries, the use of a puller specially made for this purpose is recommended. These are inexpensive and available in most parts stores. Side terminal battery cables are secured with a small bolt.

Clean the cable clamps and the battery terminal with a wire brush, until all corrosion, grease, etc., is removed and the metal is shiny. It is especially important to clean the inside of the clamp thoroughly (an old knife is useful here), since a small deposit of oxidation there will prevent a sound connection and inhibit starting or charging. Special tools are available for cleaning these parts, one type for conventional top post batteries and another type for side terminal batteries. It is also a good idea to apply some dielectric grease to the terminal, as this will aid in the prevention of corrosion.

After the clamps and terminals are clean, reinstall the cables, negative cable last; DO NOT hammer the clamps onto battery posts. Tighten the clamps securely, but do not distort them. Give the clamps and terminals a thin external coating of grease after installation, to retard corrosion.

Check the cables at the same time that the terminals are cleaned. If the cable insulation is cracked or broken, or if the ends are frayed, the cable should be replaced with a new cable of the same length and gauge.

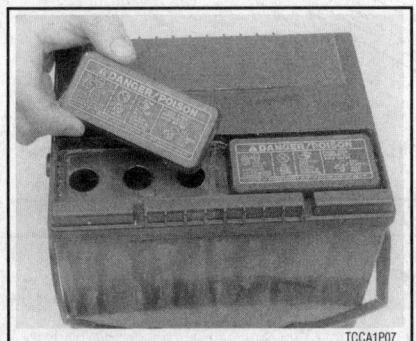

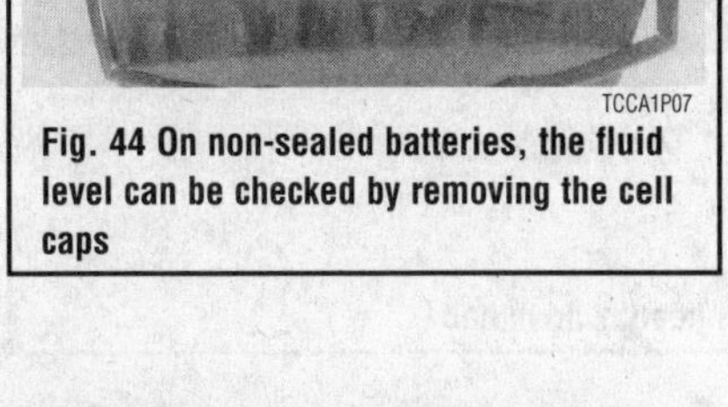

Fig. 44 On non-sealed batteries, the fluid level can be checked by removing the cell caps

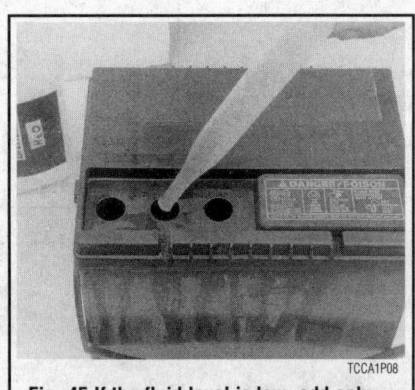

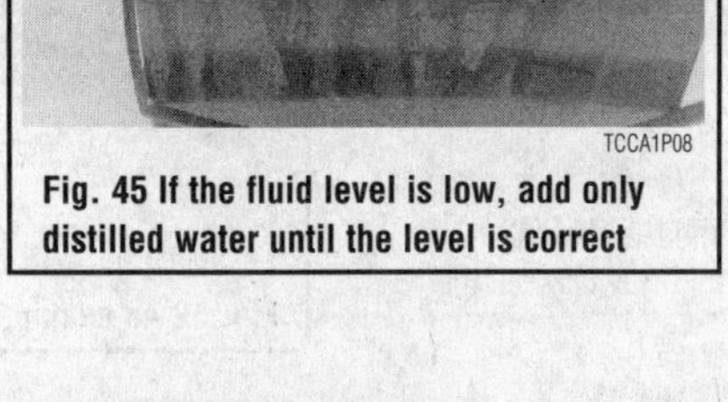

Fig. 45 If the fluid level is low, add only distilled water until the level is correct

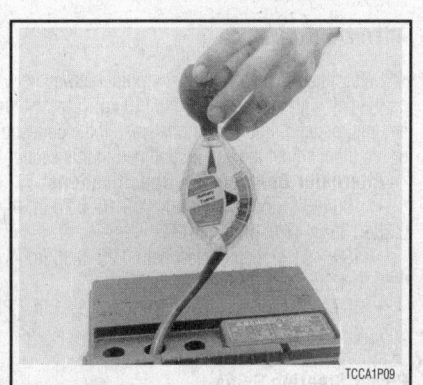

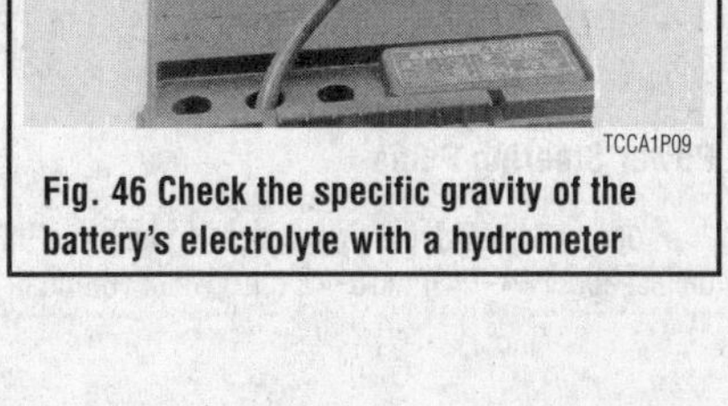

Fig. 46 Check the specific gravity of the battery's electrolyte with a hydrometer

1-18 GENERAL INFORMATION AND MAINTENANCE

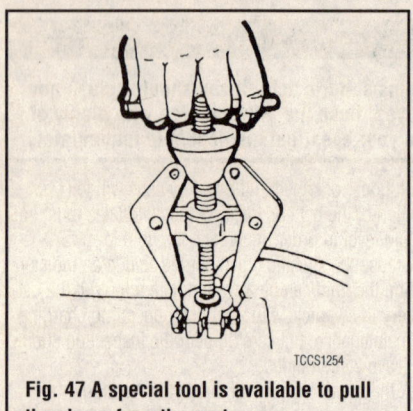

Fig. 47 A special tool is available to pull the clamp from the post

Fig. 48 The underside of this special battery tool has a wire brush to clean post terminals

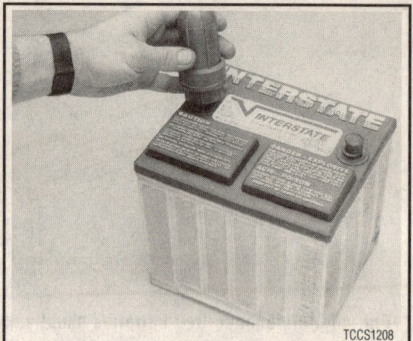

Fig. 49 Place the tool over the battery posts and twist to clean until the metal is shiny

CHARGING

> **CAUTION**
>
> The chemical reaction which takes place in all batteries generates explosive hydrogen gas. A spark can cause the battery to explode and splash acid. To avoid personal injury, be sure there is proper ventilation and take appropriate fire safety precautions when working with or near a battery.

A battery should be charged at a slow rate to keep the plates inside from getting too hot. However, if some maintenance-free batteries are allowed to discharge until they are almost "dead," they may have to be charged at a high rate to bring them back to "life." Always follow the charger manufacturer's instructions on charging the battery.

REPLACEMENT

When it becomes necessary to replace the battery, select one with an amperage rating equal to or greater than the battery originally installed. Deterioration and just plain aging of the battery cables, starter motor, and associated wires makes the battery's job harder in successive years. This makes it prudent to install a new battery with a greater capacity than the old.

Belts

INSPECTION

▶ See Figures 51 thru 56

Inspect the drive belt(s) every 30,000 miles (48,300 km) or 24 months. Check the belts to see that they are not cracked or worn. Be sure that their surfaces are free of grease and oil. Also, check the belt tension as follows.

Alternator

Determine the belt tension at a point halfway between the pulleys by pressing on the belt with a force of 22 lbs. (10 kg). The belt should deflect a certain amount, depending on the model year. If the deflection is found to be too much or too little, adjust the belt as outlined in this section.

Alternator Belt Tension Specifications:
- 1984–87 Models—New belt: 0.16–0.20 in. (4.0–6.5 mm). Used belt: 0.28–0.39 in. (7.0–10.0 mm).
- 1988–91 Models—New belt: 0.25–0.35 in. (7.0–9.0 mm). Used belt: 0.35–0.43 in. (9.0–11.0 mm).
- 1992–95 Models—New belt: 0.22–0.31 in. (5.5–8.0 mm). Used belt: 0.28–0.41 in. (7.0–10.5 mm).

Power Steering Pump

Properly adjusted, the power steering belt for 1984–87 models should deflect about ¼–⅞ in. (18–22 mm) when you push on it mid-way between the

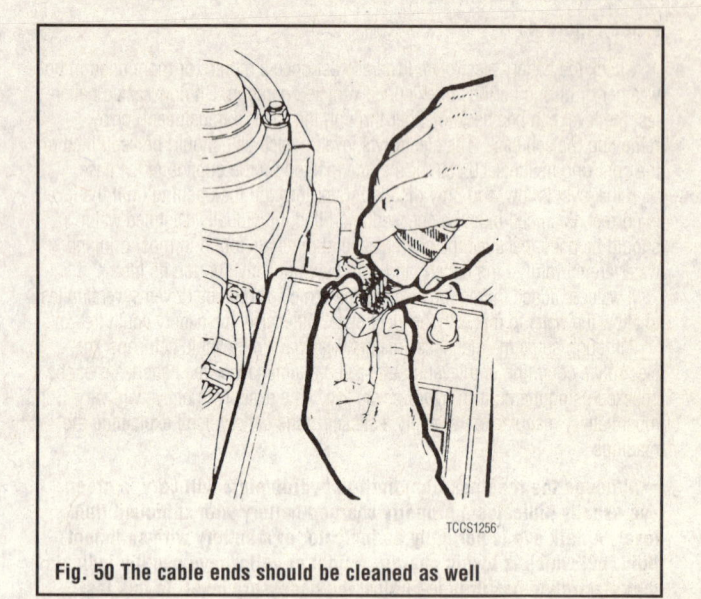

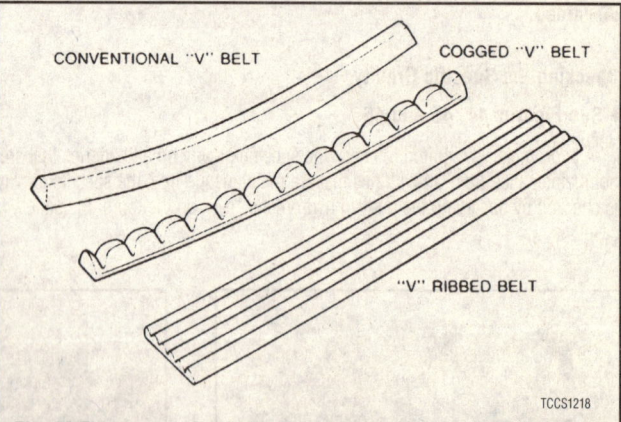

Fig. 50 The cable ends should be cleaned as well

Fig. 51 There are typically 3 types of accessory drive belts found on vehicles today

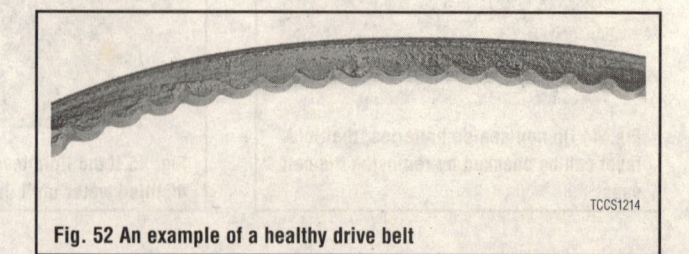

Fig. 52 An example of a healthy drive belt

GENERAL INFORMATION AND MAINTENANCE   1-19

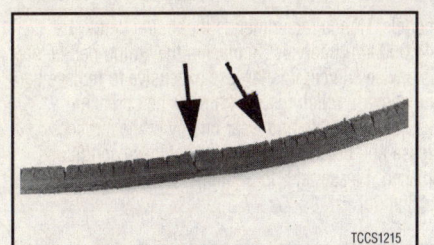

Fig. 53 Deep cracks in this belt will cause flex, building up heat that will eventually lead to belt failure

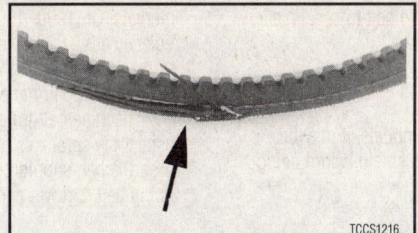

Fig. 54 The cover of this belt is worn, exposing the critical reinforcing cords to excessive wear

Fig. 55 Installing too wide a belt can result in serious belt wear and/or breakage

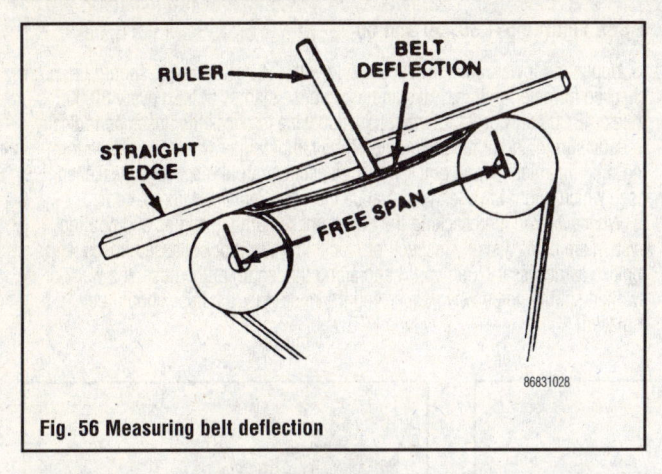

Fig. 56 Measuring belt deflection

pulleys with a force of about 22 lbs. (10 kg). Using the same force, the belt should deflect 0.35–0.47 in. (9–12 mm) for 1988–91 models and the deflection for a new belt for 1992–95 models should be 0.24–0.37 in. (6.0–9.5 mm). If reusing a belt for 1992–95 models, the deflection should be 0.31–0.47 in. (8.0–12.0 mm).

Air Conditioning Compressor

Determine the belt tension at a point halfway between the pulleys by pressing on the belt with a force of 22 lbs. (10 kg). The belt should deflect a certain amount, depending on the model year and type of compressor. If the deflection is found to be too much or too little, adjust the belt as outlined in this section.
- 1984–87 (Honda/Keihin type)—9/32–11/32 in. (7.0–9.0 mm).
- 1984–87 (Sanden type)—5/16–25/64 in. (8.0–10.0mm).
- 1988–91 (Matsushita or Sanden type)—New belt: 0.35–0.43 in. (9.0–11.0 mm). Used belt: 0.28–0.35 in. (7.0–9.0mm).
- 1992–95—New belt: 0.20–0.28 in. (5.0–7.0 mm). Used belt: 0.28–0.35 in. (7.0–9.0mm).

ADJUSTING TENSION

Alternator Belt

1. If the belt tension requires adjustment, loosen the top mounting bolt and then turn the adjusting nut outboard of the alternator. This will reposition the alternator without prying.
2. When the belt tension is correct, tighten the upper mounting bolt to about 16 ft. lbs. (24 Nm). Recheck the tension, correcting the adjustment if necessary.

➡ Do not apply pressure to any other part of the alternator.

Power Steering Belt

1984–87

1. Loosen the four bolts that hold the pump in place.
2. Slide the pump body to tension the belt.
3. When the tension is correct, tighten the bolts to 33 ft. lbs. (45 Nm).
4. Start the engine and let it idle. Turn the wheel lock-to-lock several times. Recheck the belt tension and readjust as required.

1988–91

1. Loosen the pump pivot bolt and the flange bolt.
2. Using the appropriate size square drive ratchet or breaker bar, pry the pump away from the engine until proper tension is achieved.
3. Retighten the pivot bolt to 17 ft. lbs. (24 Nm) and the flange bolt to 29 ft. lbs. (40 Nm).
4. Start the engine and turn the wheel lock-to-lock several times. Recheck the belt tension and readjust as required.

1992–95

1. Loosen the pump pivot bolt and adjusting nut.
2. Turn the adjusting bolt to get the proper belt tension.
3. Retighten the bolts when the proper tension is achieved.
4. Start the engine and turn the wheel lock-to-lock several times. Recheck the belt tension and readjust as required.

Air Conditioning Compressor Belt

1. Loosen the adjuster bolt and locknut.
2. Apply a force of 22 lbs. (10 kg) to the midway point of the belt between the pulleys and adjust the compressor so that the tension on the belt is within specification for the model year car you are working on.
3. When the correct tension is achieved, tighten the adjuster bolt and locknut.

REMOVAL & INSTALLATION

Alternator Belt

1. Loosen the component-to-mounting bracket bolts.
2. Rotate the component to relieve the tension on the drive belt. If the engine uses more than one belt, it may be necessary to remove other belts that are in front of the one being removed.
3. Slip the drive belt from the component pulley and remove it from the engine.
4. Reverse the removal procedures. Adjust the component drive belt tension to specifications.

Power Steering Belt

1984–1987

1. Loosen the four bolts that secure the pump in place.
2. Slide the pump body and remove the belt.
3. To install, reverse the removal procedures. Tension the belt to specification and tighten the mounting bolts to 33 ft. lbs. (45 Nm).

1988–1991

1. Loosen the pump pivot bolt and the flange bolt.
2. Using the appropriate size square drive ratchet or breaker bar, pry the pump away from the engine and remove the belt.
3. To install, reverse the removal procedure. Tension the belt to specification and tighten the pivot bolt to 17 ft. lbs. (24 Nm) and the flange bolt to 29 ft. lbs. (40 Nm).

1992–1995

1. Loosen the pump pivot bolt and adjusting nut.
2. Turn the adjusting bolt to loosen the belt, then remove the belt.

1-20 GENERAL INFORMATION AND MAINTENANCE

3. To install, reverse the removal procedure. Tension the belt to specification and tighten the pivot bolt and adjusting nut.

Air Conditioning Compressor Belt

1. Loosen the adjuster bolt and locknut. On 1984–87 models, it may be necessary to raise the vehicle and remove the lower splash shield before being able to get at the compressor bolts.
2. Remove the compressor belt.
3. Install the new compressor belt and adjust the tension. When the proper tension is achieved, tighten the locknut.

Timing Belts

INSPECTION

> **WARNING**
> Severe engine damage will occur if the timing belt should break. Honda uses an "interference" engine design. If the timing belt breaks, the valves will contact the pistons and become dam-aged.

On models up to 1989, Honda does not specify a recommended inspection or replacement interval. However, since these engines utilize an interference design, it is highly recommended to at least inspect the timing belt at 60,000 mile (96,000 km) intervals. It is an even better idea to replace the belt at these intervals. On 1990–1994 models, Honda recommends the timing belt to be replaced at 90,000 mile (144,000 km) intervals. Replacing the timing belt before it's recommended interval is a wise choice. It is far less expensive to replace the belt than to repair the engine damage which results from the belt breaking.

Inspect the belt for cracks, missing teeth and wear on any of the surfaces. Inspect the sprockets for grease and other deposits. If any of these conditions exist, the belt should be replaced. Please refer to Section 3 for procedures on timing belt removal and installation.

Hoses

INSPECTION

▶ See Figures 57, 58, 59 and 60

Upper and lower radiator hoses, along with the heater hoses, should be checked for deterioration, leaks and loose hose clamps at least every 30,000 miles (48,300 km) or 24 months. It is also wise to check the hoses periodically in early spring and at the beginning of the fall or winter when you are performing other maintenance. A quick visual inspection could discover a weakened hose which might have left you stranded if it had remained unrepaired.

Whenever you are checking the hoses, make sure the engine and cooling system are cold. Visually inspect for cracking, rotting or collapsed hoses, and replace as necessary. Run your hand along the length of the hose. If a weak or swollen spot is noted when squeezing the hose wall, the hose should be replaced.

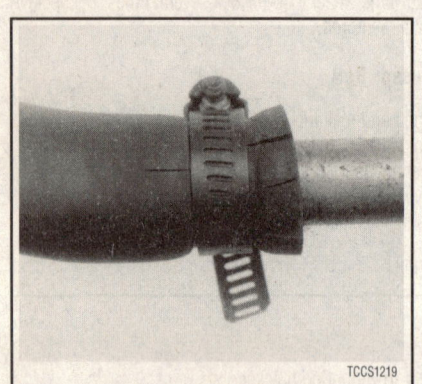

Fig. 57 The cracks developing along this hose are a result of age-related hardening

Fig. 58 A hose clamp that is too tight can cause older hoses to separate and tear on either side of the clamp

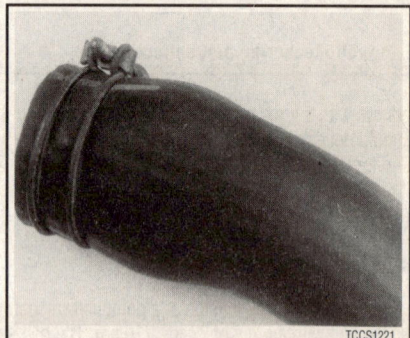

Fig. 59 A soft spongy hose (identifiable by the swollen section) will eventually burst and should be replaced

REMOVAL & INSTALLATION

▶ See Figures 61, 62, 63 and 64

1. Drain the cooling system.

> **CAUTION**
> When draining the coolant, keep in mind that cats and dogs are attracted by ethylene glycol antifreeze, and are quite likely to drink any that is left in an uncovered container or in puddles on the ground. This will prove fatal in sufficient quantities. Always drain the coolant into a sealable container. Coolant may be reused unless it is contaminated or several years old.

2. Loosen the hose clamps at each end of the hose.
3. Working the hose back and forth, slide it off its connection.

→When replacing the heater hoses, maintain a 1½ in. (38mm) clearance from any surface. Draw the hoses tight to prevent sagging or rubbing against other components; route the hoses through the clamps as installed originally. Always make sure that the hose clamps are beyond the component bead and placed in the center of the clamping surface before tightening them.

4. To install, reverse the removal procedures. Refill the cooling system.

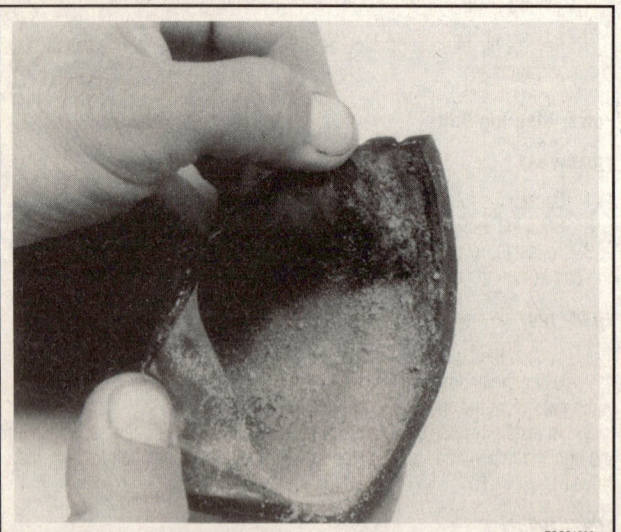

Fig. 60 Hoses are likely to deteriorate from the inside if the cooling system is not periodically flushed

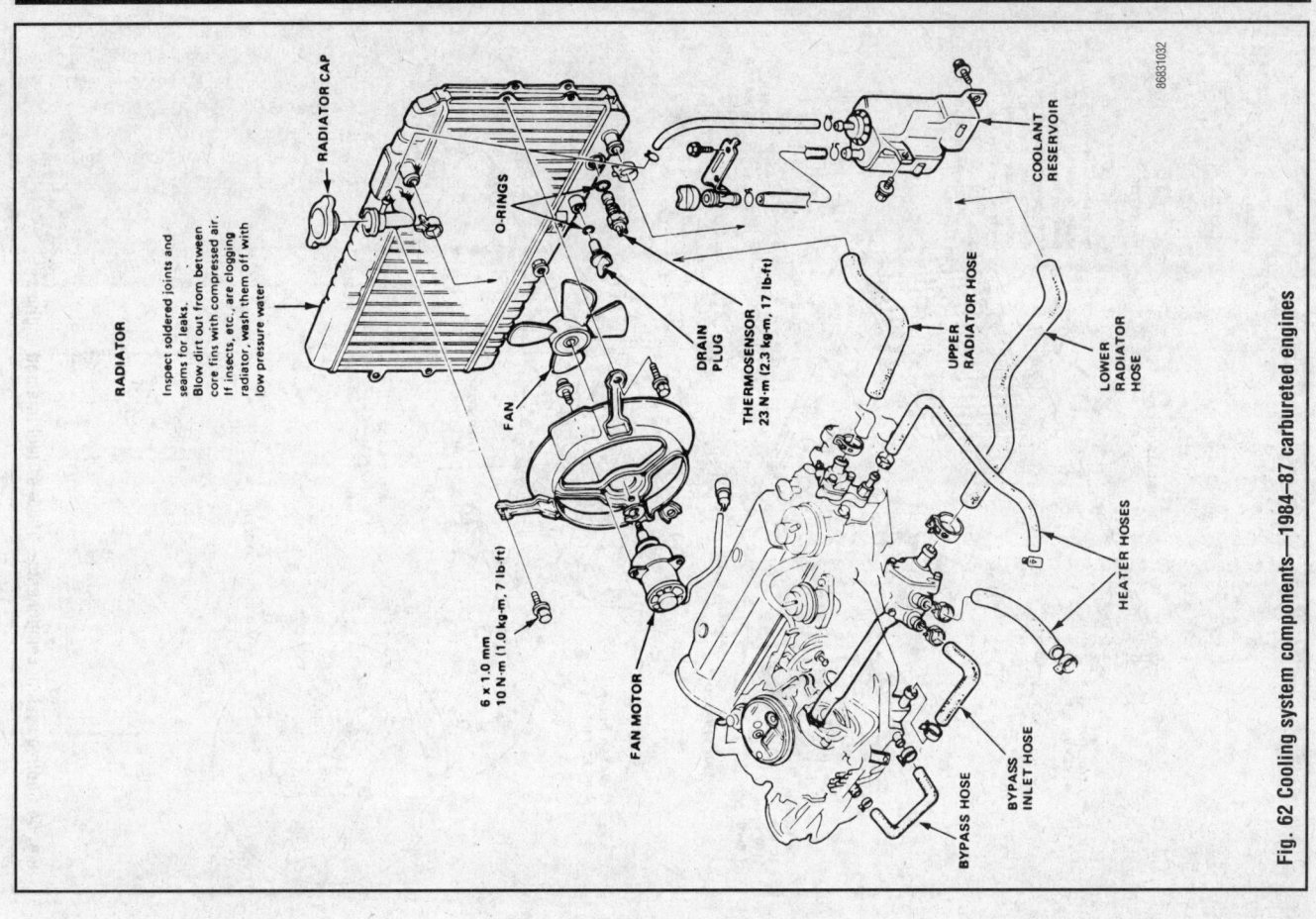

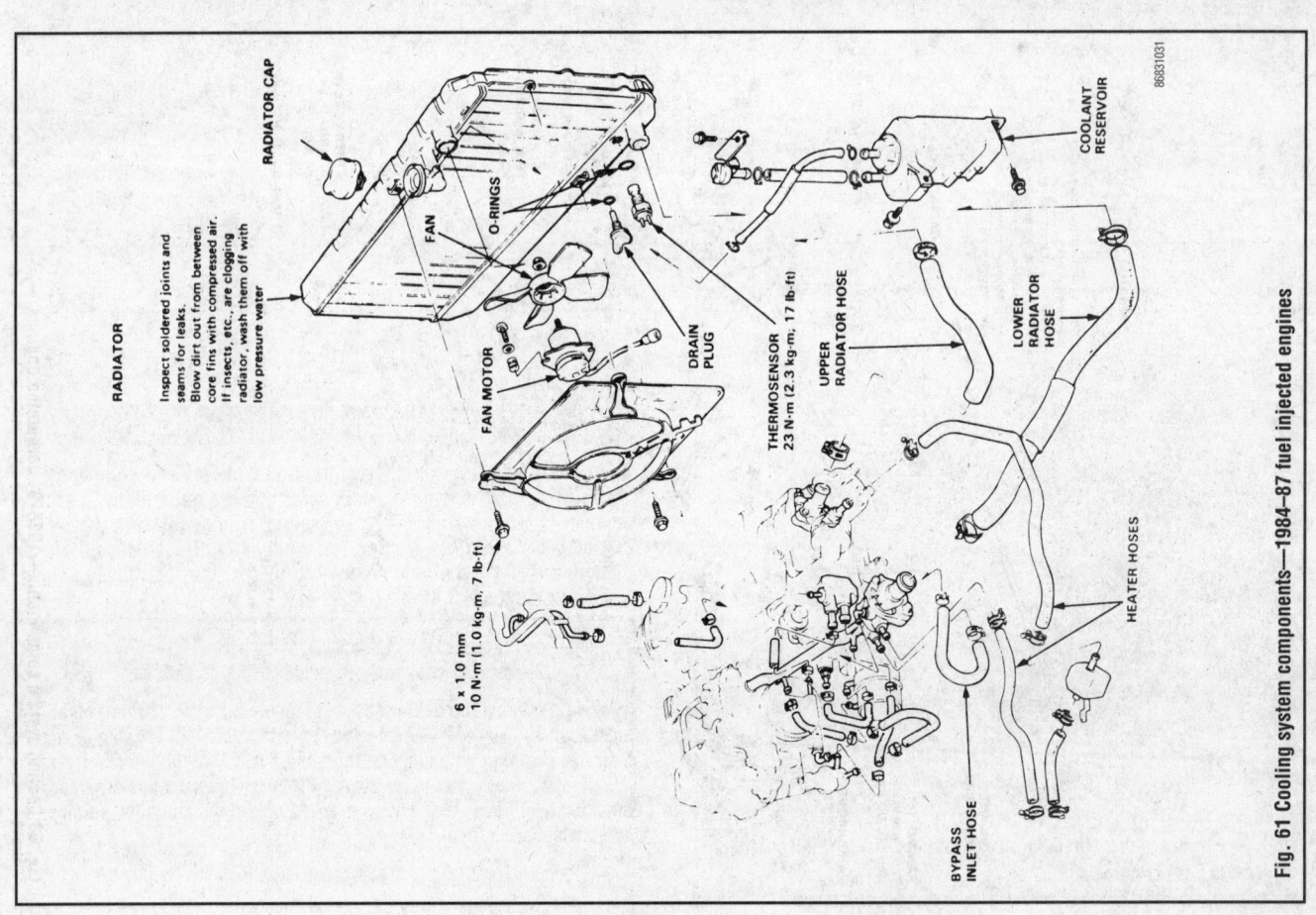

GENERAL INFORMATION AND MAINTENANCE 1-21

Fig. 62 Cooling system components—1984–87 carbureted engines

Fig. 61 Cooling system components—1984–87 fuel injected engines

1-22 GENERAL INFORMATION AND MAINTENANCE

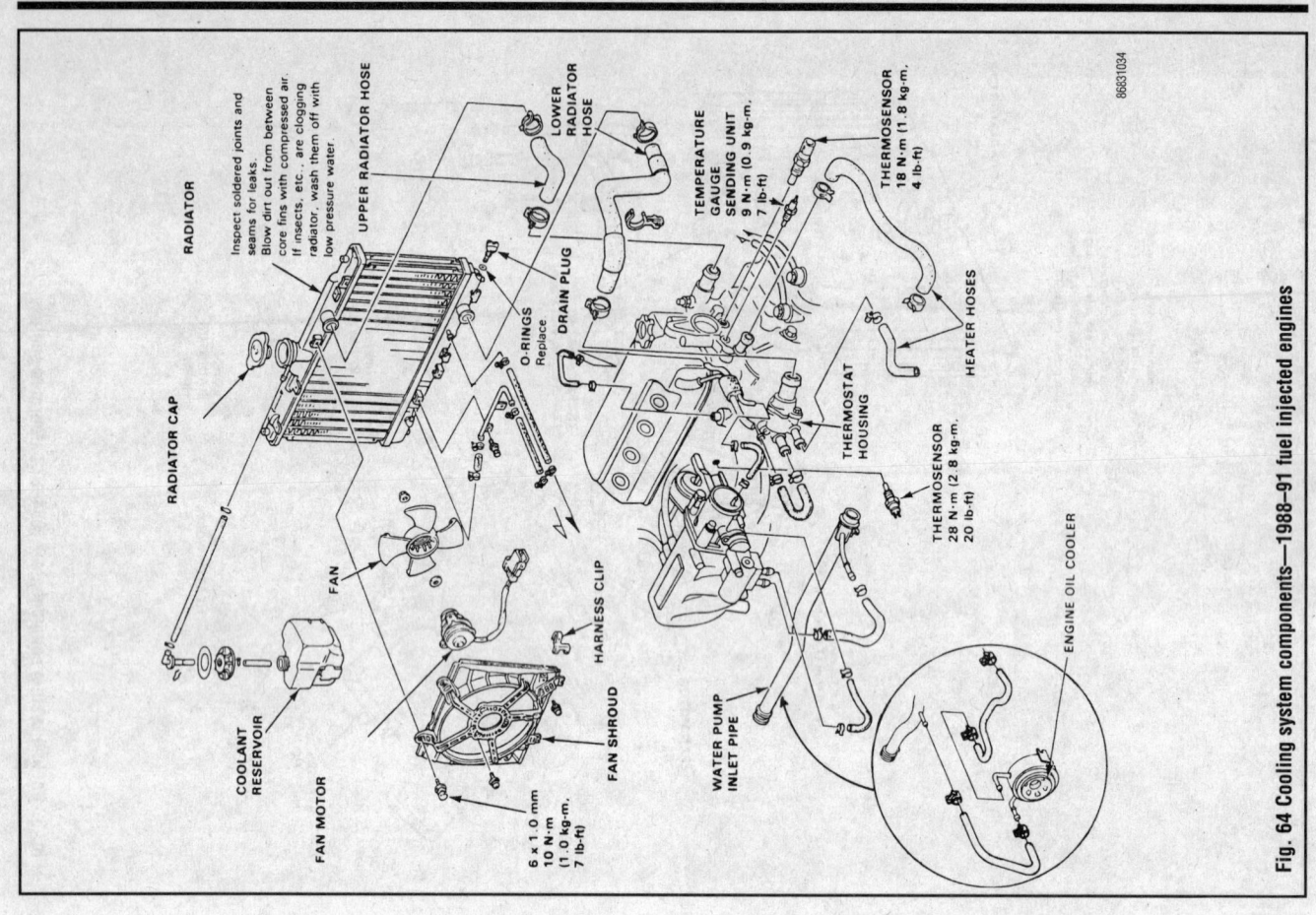

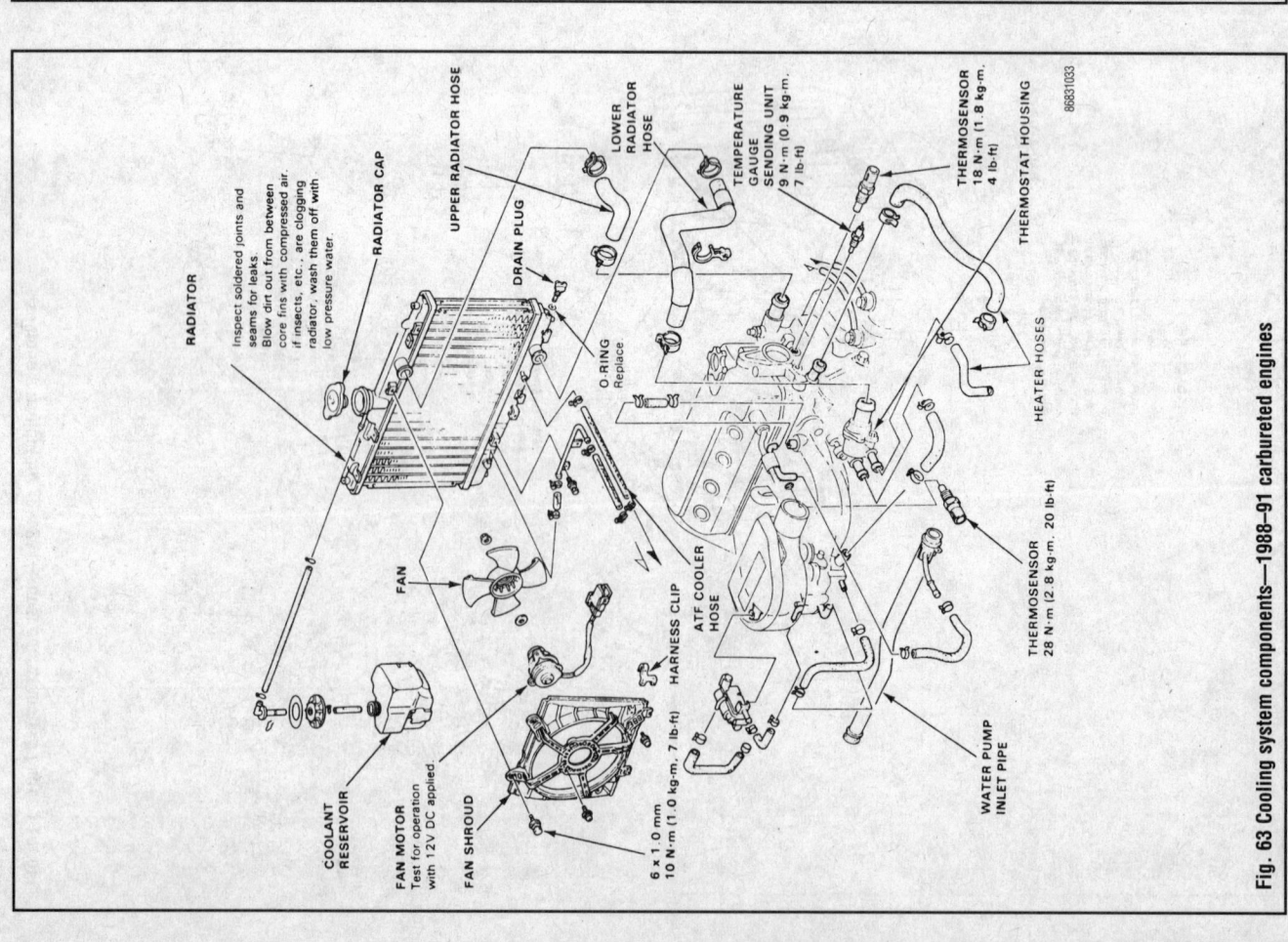

Fig. 64 Cooling system components—1988-91 fuel injected engines

Fig. 63 Cooling system components—1988-91 carbureted engines

CV-Boots

The Constant Velocity (CV) joint boot encloses the constant velocity (CV) joint, a critical component of your front-wheel-drive car. The CV-joints are flexible couplings on each end (inner and outer) of the transaxle halfshafts. They allow the halfshafts to adjust to up-and-down and back-and-forth movements of the suspension and steering while simultaneously transmitting the engine's power to the front wheels. A CV-joint is more complex than a traditional universal joint as it must allow the wheels to turn smoothly under driving and braking loads. Because CV-joints are precision machined parts, they must be packed in lubricating grease and sealed from road grime in a protective rubber boot.

If a CV-boot tears, it will quickly let dirt enter the joint at the same time its grease is spun out onto the undercarriage and road. Thus contaminated, the under-lubricated joint will quickly wear and deteriorate. For this reason, the boots should be periodically inspected.

INSPECTION

♦ See Figures 65 and 66

The proper way to inspect CV-boots with the vehicle raised and supported on jackstands. Rotate the front wheels slowly one at a time while looking at and feeling the texture of the boots for cracks and tears.

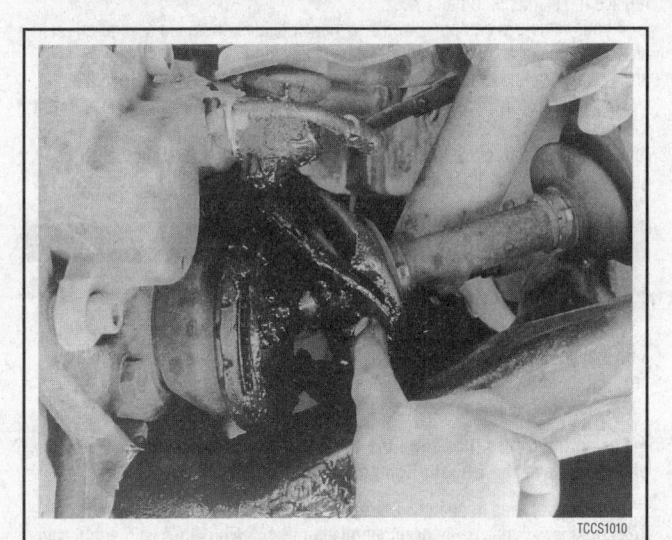

Fig. 65 This boot is torn and must be replaced immediately

Fig. 66 This boot is still in good condition

✱✱ CAUTION

Always wear appropriate eye protection while under the car.

This inspection should be performed every time the vehicle is raised for any kind of service (such as oil and filter changes). If a CV-boot has a tear or is starting to crack and look worn, replace it. In the case of an already torn boot, you will want to inspect the joint itself. For information on these procedures, see Section 7, under CV-Joint and CV-Boot Removal & Installation.

Air Conditioning System

SYSTEM SERVICE & REPAIR

➥It is recommended that the A/C system be serviced by an EPA Section 609 certified automotive technician utilizing a refrigerant recovery/recycling machine.

The do-it-yourselfer should not service his/her own vehicle's A/C system for many reasons, including legal concerns, personal injury, environmental damage and cost.

According to the U.S. Clean Air Act, it is a federal crime to service or repair (involving the refrigerant) a Motor Vehicle Air Conditioning (MVAC) system for money without being EPA certified. It is also illegal to vent R-12 and R-134a refrigerants into the atmosphere. State and/or local laws may be more strict than the federal regulations, so be sure to check with your state and/or local authorities for further information.

➥Federal law dictates that a fine of up to $25,000 may be levied on people convicted of venting refrigerant into the atmosphere.

When servicing an A/C system you run the risk of handling or coming in contact with refrigerant, which may result in skin or eye irritation or frostbite. Although low in toxicity (due to chemical stability), inhalation of concentrated refrigerant fumes is dangerous and can result in death; cases of fatal cardiac arrhythmia have been reported in people accidentally subjected to high levels of refrigerant. Some early symptoms include loss of concentration and drowsiness.

➥Generally, the limit for exposure is lower for R-134a than it is for R-12. Exceptional care must be practiced when handling R-134a.

Also, some refrigerants can decompose at high temperatures (near gas heaters or open flame), which may result in hydrofluoric acid, hydrochloric acid and phosgene (a fatal nerve gas).

It is usually more economically feasible to have a certified MVAC automotive technician perform A/C system service on your vehicle.

R-12 Refrigerant Conversion

If your vehicle still uses R-12 refrigerant, one way to save A/C system costs down the road is to investigate the possibility of having your system converted to R-134a. The older R-12 systems can be easily converted to R-134a refrigerant by a certified automotive technician by installing a few new components and changing the system oil.

The cost of R-12 is steadily rising and will continue to increase, because it is no longer imported or manufactured in the United States. Therefore, it is often possible to have an R-12 system converted to R-134a and recharged for less than it would cost to just charge the system with R-12.

If you are interested in having your system converted, contact local automotive service stations for more details and information.

PREVENTIVE MAINTENANCE

Although the A/C system should not be serviced by the do-it-yourselfer, preventive maintenance should be practiced to help maintain the efficiency of the vehicle's A/C system. Be sure to perform the following:

• The easiest and most important preventive maintenance for your A/C system is to be sure that it is used on a regular basis. Running the system for five minutes each month (no matter what the season) will help ensure that the seals and all internal components remain lubricated.

1-24 GENERAL INFORMATION AND MAINTENANCE

➡ Some vehicles automatically operate the A/C system compressor whenever the windshield defroster is activated. Therefore, the A/C system would not need to be operated each month if the defroster was used.

• In order to prevent heater core freeze-up during A/C operation, it is necessary to maintain proper antifreeze protection. Be sure to properly maintain the engine cooling system.

• Any obstruction of or damage to the condenser configuration will restrict air flow which is essential to its efficient operation. Keep this unit clean and in proper physical shape.

➡ Bug screens which are mounted in front of the condenser (unless they are original equipment) are regarded as obstructions.

• The condensation drain tube expels any water which accumulates on the bottom of the evaporator housing into the engine compartment. If this tube is obstructed, the air conditioning performance can be restricted and condensation buildup can spill over onto the vehicle's floor.

SYSTEM INSPECTION

Although the A/C system should not be serviced by the do-it-yourselfer, system inspections should be performed to help maintain the efficiency of the vehicle's A/C system. Be sure to perform the following:

The easiest and often most important check for the air conditioning system consists of a visual inspection of the system components. Visually inspect the system for refrigerant leaks, damaged compressor clutch, abnormal compressor drive belt tension and/or condition, plugged evaporator drain tube, blocked condenser fins, disconnected or broken wires, blown fuses, corroded connections and poor insulation.

A refrigerant leak will usually appear as an oily residue at the leakage point in the system. The oily residue soon picks up dust or dirt particles from the surrounding air and appears greasy. Through time, this will build up and appear to be a heavy dirt impregnated grease.

For a thorough visual and operational inspection, check the following:
• Check the surface of the radiator and condenser for dirt, leaves or other material which might block air flow.
• Check for kinks in hoses and lines. Check the system for leaks.
• Make sure the drive belt is properly tensioned. During operation, make sure the belt is free of noise or slippage.
• Make sure the blower motor operates at all appropriate positions, then check for distribution of the air from all outlets.

➡ Remember that in high humidity, air discharged from the vents may not feel as cold as expected, even if the system is working properly. This is because moisture in humid air retains heat more effectively than dry air, thereby making humid air more difficult to cool.

Windshield Wipers

ELEMENT (REFILL) CARE & REPLACEMENT

♦ See Figures 67, 68 and 69

For maximum effectiveness and longest element life, the windshield and wiper blades should be kept clean. Dirt, tree sap, road tar and so on will cause streaking, smearing and blade deterioration if left on the glass. It is advisable to wash the windshield carefully with a commercial glass cleaner at least once a month. Wipe off the rubber blades with the wet rag afterwards. Do not attempt to move wipers across the windshield by hand; damage to the motor and drive mechanism will result.

To inspect and/or replace the wiper blade elements, place the wiper switch in the **LOW** speed position and the ignition switch in the **ACC** position. When the wiper blades are approximately vertical on the windshield, turn the ignition switch to **OFF**.

Examine the wiper blade elements. If they are found to be cracked, broken or torn, they should be replaced immediately. Replacement intervals will vary with usage, although ozone deterioration usually limits element life to about one year. If the wiper pattern is smeared or streaked, or if the blade chatters across the glass, the elements should be replaced. It is easiest and most sensible to replace the elements in pairs.

If your vehicle is equipped with aftermarket blades, there are several different types of refills and your vehicle might have any kind. Aftermarket blades and arms rarely use the exact same type blade or refill as the original equipment.

Regardless of the type of refill used, be sure to follow the part manufacturer's instructions closely. Make sure that all of the frame jaws are engaged as the refill is pushed into place and locked. If the metal blade holder and frame are allowed to touch the glass during wiper operation, the glass will be scratched.

Tires and Wheels

♦ See Figure 70

Common sense and good driving habits will afford maximum tire life. Make sure that you don't overload the vehicle or run with incorrect pressure in the tires. Either of these will increase tread wear. Fast starts, sudden stops and sharp cornering are hard on tires and will shorten their useful life span.

➡ For optimum tire life, keep the tires properly inflated, rotate them often and have the wheel alignment checked periodically.

Inspect your tires frequently. Be especially careful to watch for bubbles in the tread or sidewall, deep cuts or underinflation. Replace any tires with bubbles in the sidewall. If cuts are so deep that they penetrate to the cords, discard the tire. Any cut in the sidewall of a radial tire renders it unsafe. Also look for uneven

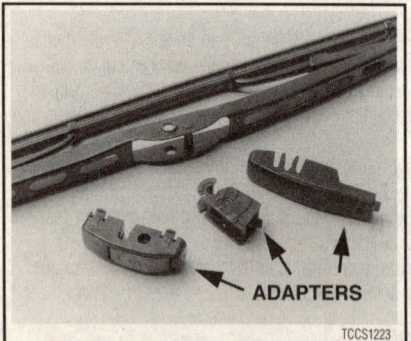

Fig. 67 Most aftermarket blades are available with multiple adapters to fit different vehicles

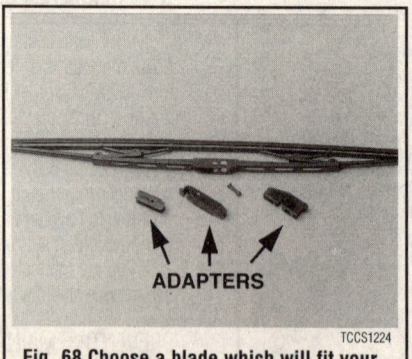

Fig. 68 Choose a blade which will fit your vehicle, and that will be readily available next time you need blades

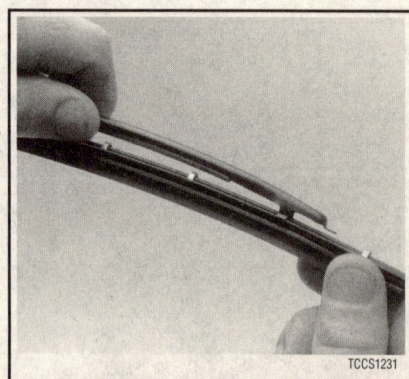

Fig. 69 When installed, be certain the blade is fully inserted into the backing

GENERAL INFORMATION AND MAINTENANCE   1-25

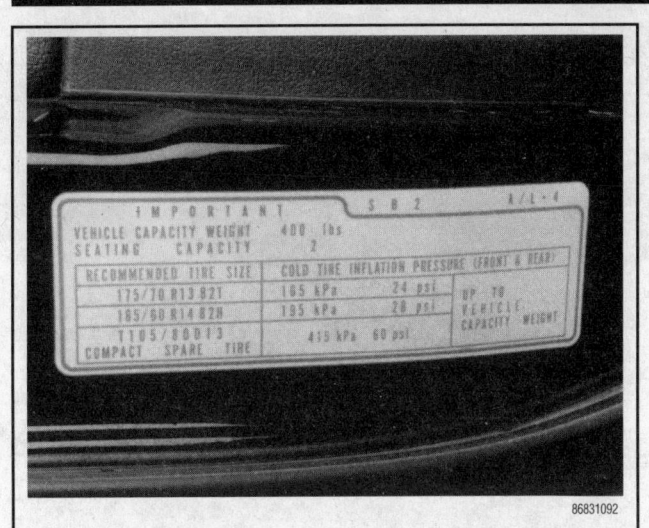

Fig. 70 Tire information can be found on a sticker located near the driver's door striker plate

tread wear patterns that may indicate the front end is out of alignment or that the tires are out of balance.

TIRE ROTATION

♦ See Figure 71

Tires must be rotated periodically to equalize wear patterns that vary with a tire's position on the vehicle. Tires will also wear in an uneven way as the front steering/suspension system wears to the point where the alignment should be reset.

Rotating the tires will ensure maximum life for the tires as a set, so you will not have to discard a tire early due to wear on only part of the tread. Regular rotation is required to equalize wear.

When rotating "unidirectional tires," make sure that they always roll in the same direction. This means that a tire used on the left side of the vehicle must not be switched to the right side and vice-versa. Such tires should only be rotated front-to-rear or rear-to-front, while always remaining on the same side of the vehicle. These tires are marked on the sidewall as to the direction of rotation; observe the marks when reinstalling the tire(s).

Some styled or "mag" wheels may have different offsets front to rear. In these cases, the rear wheels must not be used up front and vice-versa. Furthermore, if these wheels are equipped with unidirectional tires, they cannot be rotated unless the tire is remounted for the proper direction of rotation.

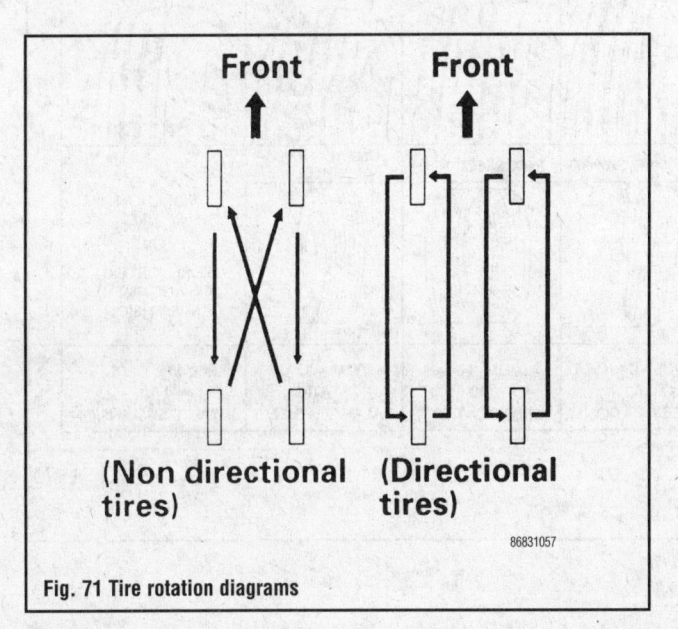

Fig. 71 Tire rotation diagrams

➥The compact or space-saver spare is strictly for emergency use. It must never be included in the tire rotation or placed on the vehicle for everyday use.

TIRE DESIGN

♦ See Figure 72

For maximum satisfaction, tires should be used in sets of four. Mixing of different brands or types (radial, bias-belted, fiberglass belted) should be avoided. In most cases, the vehicle manufacturer has designated a type of tire on which the vehicle will perform best. Your first choice when replacing tires should be to use the same type of tire that the manufacturer recommends.

When radial tires are used, tire sizes and wheel diameters should be selected to maintain ground clearance and tire load capacity equivalent to the original specified tire. Radial tires should always be used in sets of four.

✲✲ CAUTION

Radial tires should never be used on only the front axle.

When selecting tires, pay attention to the original size as marked on the tire. Most tires are described using an industry size code sometimes referred to as P-Metric. This allows the exact identification of the tire specifications, regardless of the manufacturer. If selecting a different tire size or brand, remember to check the installed tire for any sign of interference with the body or suspension while the vehicle is stopping, turning sharply or heavily loaded.

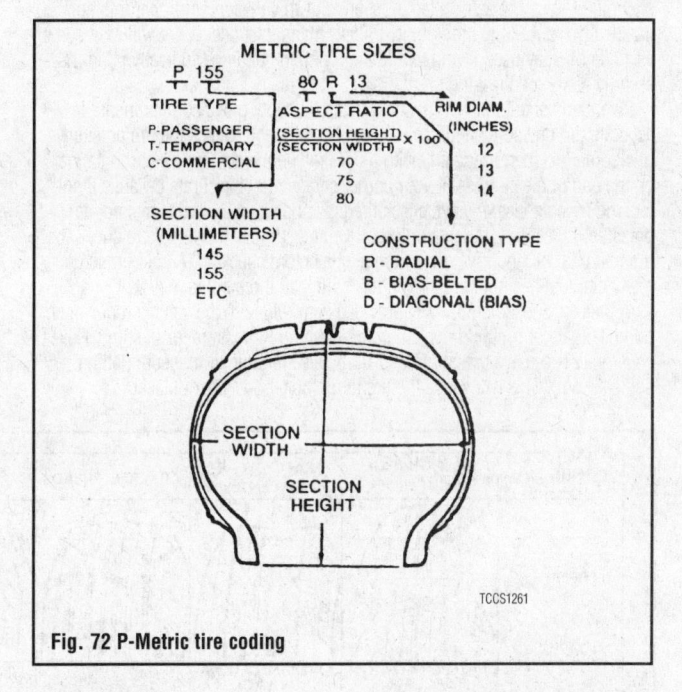

Fig. 72 P-Metric tire coding

Snow Tires

Good radial tires can produce a big advantage in slippery weather, but in snow, a street radial tire does not have sufficient tread to provide traction and control. The small grooves of a street tire quickly pack with snow and the tire behaves like a billiard ball on a marble floor. The more open, chunky tread of a snow tire will self-clean as the tire turns, providing much better grip on snowy surfaces.

To satisfy municipalities requiring snow tires during weather emergencies, most snow tires carry either an M + S designation after the tire size stamped on the sidewall, or the designation "all-season." In general, no change in tire size is necessary when buying snow tires.

Most manufacturers strongly recommend the use of 4 snow tires on their vehicles for reasons of stability. If snow tires are fitted only to the drive wheels, the opposite end of the vehicle may become very unstable when braking or turning on slippery surfaces. This instability can lead to unpleasant endings if the driver can't counteract the slide in time.

1-26 GENERAL INFORMATION AND MAINTENANCE

Note that snow tires, whether 2 or 4, will affect vehicle handling in all non-snow situations. The stiffer, heavier snow tires will noticeably change the turning and braking characteristics of the vehicle. Once the snow tires are installed, you must re-learn the behavior of the vehicle and drive accordingly.

→ Consider buying extra wheels on which to mount the snow tires. Once done, the "snow wheels" can be installed and removed as needed. This eliminates the potential damage to tires or wheels from seasonal removal and installation. Even if your vehicle has styled wheels, see if inexpensive steel wheels are available. Although the look of the vehicle will change, the expensive wheels will be protected from salt, curb hits and pothole damage.

TIRE STORAGE

If they are mounted on wheels, store the tires at proper inflation pressure. All tires should be kept in a cool, dry place. If they are stored in the garage or basement, do not let them stand on a concrete floor; set them on strips of wood, a mat or a large stack of newspaper. Keeping them away from direct moisture is of paramount importance. Tires should not be stored upright, but in a flat position.

INFLATION & INSPECTION

♦ See Figures 73 thru 78

The importance of proper tire inflation cannot be overemphasized. A tire employs air as part of its structure. It is designed around the supporting strength of the air at a specified pressure. For this reason, improper inflation drastically reduces the tire's ability to perform as intended. A tire will lose some air in day-to-day use; having to add a few pounds of air periodically is not necessarily a sign of a leaking tire.

Two items should be a permanent fixture in every glove compartment: an accurate tire pressure gauge and a tread depth gauge. Check the tire pressure (including the spare) regularly with a pocket type gauge. Too often, the gauge on the end of the air hose at your corner garage is not accurate because it suffers too much abuse. Always check tire pressure when the tires are cold, as pressure increases with temperature. If you must move the vehicle to check the tire inflation, do not drive more than a mile before checking. A cold tire is generally one that has not been driven for more than three hours.

A plate or sticker is normally provided somewhere in the vehicle (door post, hood, tailgate or trunk lid) which shows the proper pressure for the tires. Never counteract excessive pressure build-up by bleeding off air pressure (letting some air out). This will cause the tire to run hotter and wear quicker.

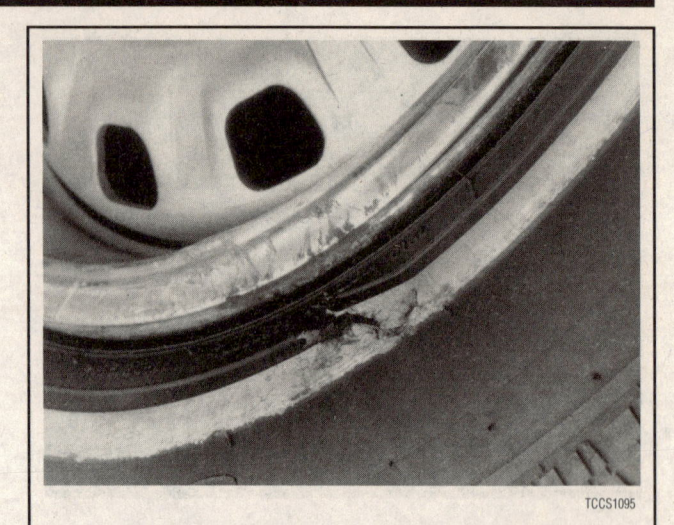

Fig. 73 Tires with deep cuts, or cuts which bulge, should be replaced immediately

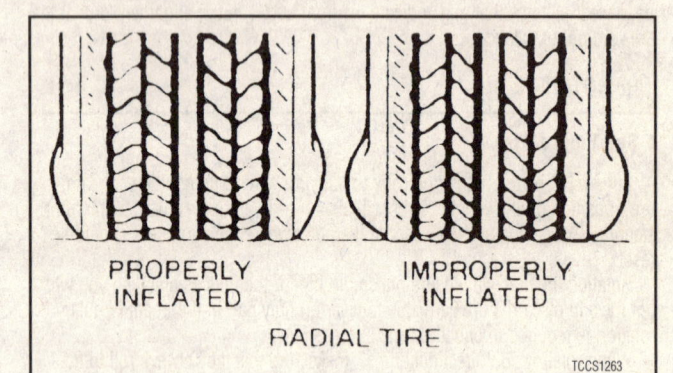

Fig. 74 Radial tires have a characteristic sidewall bulge; don't try to measure pressure by looking at the tire. Use a quality air pressure gauge

CONDITION	RAPID WEAR AT SHOULDERS	RAPID WEAR AT CENTER	CRACKED TREADS	WEAR ON ONE SIDE	FEATHERED EDGE	BALD SPOTS	SCALLOPED WEAR
EFFECT							
CAUSE	UNDER-INFLATION OR LACK OF ROTATION	OVER-INFLATION OR LACK OF ROTATION	UNDER-INFLATION OR EXCESSIVE SPEED*	EXCESSIVE CAMBER	INCORRECT TOE	UNBALANCED WHEEL OR TIRE DEFECT*	LACK OF ROTATION OF TIRES OR WORN OR OUT-OF-ALIGNMENT SUSPENSION.
CORRECTION	ADJUST PRESSURE TO SPECIFICATIONS WHEN TIRES ARE COOL ROTATE TIRES			ADJUST CAMBER TO SPECIFICATIONS	ADJUST TOE-IN TO SPECIFICATIONS	DYNAMIC OR STATIC BALANCE WHEELS	ROTATE TIRES AND INSPECT SUSPENSION

*HAVE TIRE INSPECTED FOR FURTHER USE.

Fig. 75 Common tire wear patterns and causes

GENERAL INFORMATION AND MAINTENANCE   1-27

※※ CAUTION

Never exceed the maximum tire pressure embossed on the tire! This is the pressure to be used when the tire is at maximum loading, but it is rarely the correct pressure for everyday driving. Consult the owner's manual or the tire pressure sticker for the correct tire pressure.

Once you've maintained the correct tire pressures for several weeks, you'll be familiar with the vehicle's braking and handling personality. Slight adjustments in tire pressures can fine-tune these characteristics, but never change the cold pressure specification by more than 2 psi. A slightly softer tire pressure will give a softer ride but also yield lower fuel mileage. A slightly harder tire will give crisper dry road handling but can cause skidding on wet surfaces. Unless you're fully attuned to the vehicle, stick to the recommended inflation pressures.

All automotive tires have built-in tread wear indicator bars that show up as ½ in. (13mm) wide smooth bands across the tire when ¹⁄₁₆ in. (1.5mm) of tread remains. The appearance of tread wear indicators means that the tires should be replaced. In fact, many states have laws prohibiting the use of tires with less than this amount of tread.

You can check your own tread depth with an inexpensive gauge or by using a Lincoln head penny. Slip the Lincoln penny (with Lincoln's head upside-down) into several tread grooves. If you can see the top of Lincoln's head in 2 adjacent grooves, the tire has less than ¹⁄₁₆ in. (1.5mm) tread left and should be replaced. You can measure snow tires in the same manner by using the "tails" side of the Lincoln penny. If you can see the top of the Lincoln memorial, it's time to replace the snow tire(s).

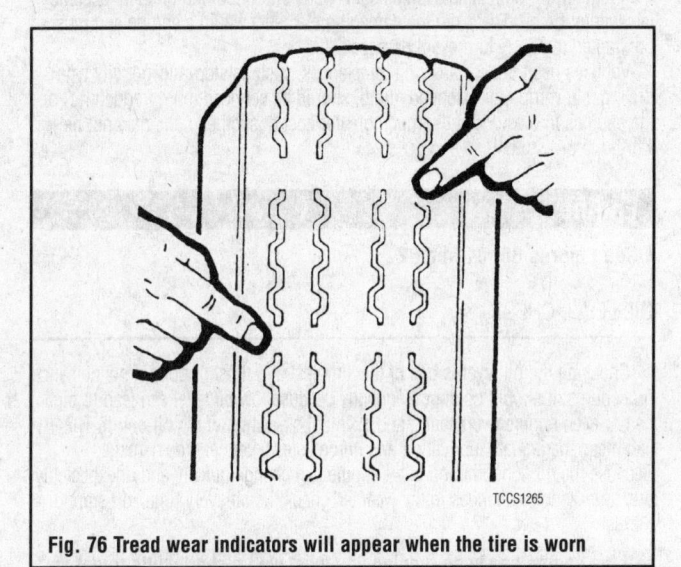

Fig. 76 Tread wear indicators will appear when the tire is worn

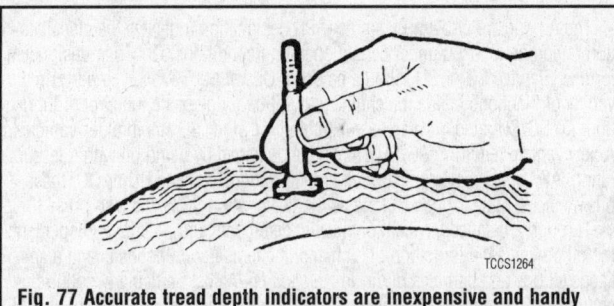

Fig. 77 Accurate tread depth indicators are inexpensive and handy

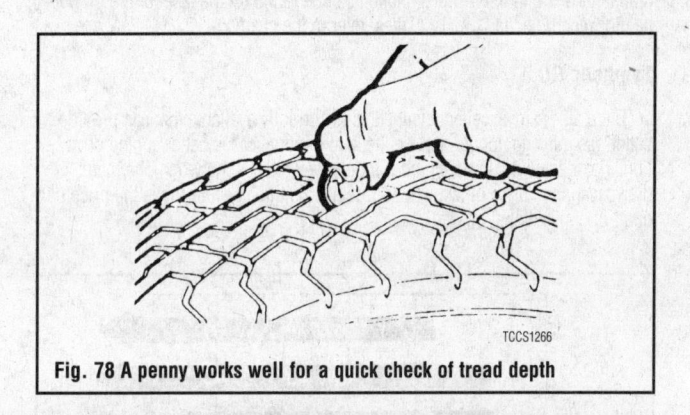

Fig. 78 A penny works well for a quick check of tread depth

FLUIDS AND LUBRICANTS

Fluid Disposal

Used fluids such as engine oil, transmission fluid, antifreeze, and brake fluid are hazardous wastes and must be disposed of properly. Before draining any fluids, consult with the local authorities. In many areas, waste oil, etc. is being accepted as part of recycling programs. A number of service stations and auto parts stores are also accepting waste fluids for recycling.

Be sure of the recycling center's policies before draining any fluids, as many will not accept different fluids that have been mixed together, such as oil and antifreeze.

Fuel and Engine Oil Recommendations

FUEL

All Honda vehicles covered in this manual are designed to run on unleaded fuel only.

You should be careful to use quality fuels having an octane rating of at least 86 when measured by the R/M method, which is an average of the "Research" and "Motor" octane ratings. Too low an octane rating may produce combustion knock, which may prove to be damaging to the engine. Always buy fuel from a reputable dealer, preferably where a regular volume is pumped so that the fuel is always fresh.

OIL

▶ See Figure 79

➡ **Motor oil is often called the "life blood" of an engine. It is primarily responsible for lubricating the dozens of fast-moving, precision metal parts that make up an internal combustion engine. A sufficient supply of clean, high-quality oil is probably the single greatest factor in the longevity of any given engine. If the level dips too low, or the oil is old and becomes "broken down" (in molecular composition), severely accelerated wear and even sudden catastrophic engine failure could be the result. It pays to check your oil often and to change it on time.**

The Society of Automotive Engineers (SAE) grade number indicates the viscosity of the engine oil (it's resistance to flow at a given temperature). The lower the SAE grade number, the lighter the oil. For example, the mono-grade oils begin with SAE 5 weight, which is a thin, light oil, and continue in viscosity up to SAE 80 or 90 weight, which are heavy gear lubricants. These oils are also known as "straight weight", meaning they are of a single viscosity, and do not vary with engine temperature.

Multi-viscosity oils offer the important advantage of being adaptable to temperature extremes. These oils have designations such as 10W-40, 20W-50, etc. For example, 10W-40 means that in winter (the "W" in the designation) the oil acts like a thin, 10 weight oil, allowing the engine to spin easily when cold and offering rapid lubrication. Once the engine has warmed up, however, the oil acts

1-28 GENERAL INFORMATION AND MAINTENANCE

like a straight 40 weight, maintaining good lubrication and protection for the engine's internal components. A 20W-50 oil would therefore be slightly heavier and not as ideal in cold weather as the 10W-40, but would offer better protection at higher rpm and temperatures because, when warm, it acts like a 50 weight oil. Whichever oil viscosity you choose when changing the oil, make sure you are anticipating the temperatures your engine will be operating in until the oil is changed again. Refer to the oil viscosity chart for oil recommendations according to temperature.

➥**Honda does not recommend the use of any oil additive or supplement in the engine. A normal engine does not need them. If the engine is worn or damaged, it's usually too late for any benefit.**

The American Petroleum Institute (API) designation indicates the classification of engine oil used under certain operating conditions. Only oils designated for use "Service SG or SH" should be used. Oils of the SG/SH type perform a variety of functions inside the engine in addition to the basic function of lubrication. An SG/SH rated oil may be substituted for SF or SE oils in older vehicles. A new vehicle requiring SG/SH oil may be damaged by using oil with a lesser rating. API labels may also carry other letter ratings such as CD or CC; these oils are acceptable for use as long as the designation SG/SH is also present.

Through a balanced system of metallic detergents and polymeric dispersants, the oil prevents the formation of high and low temperature deposits and also keeps sludge and particles of dirt in suspension. Acids, particularly sulfuric acid, as well as other by-products of combustion, are neutralized. Both the SAE grade number and the API designation can be found on the label of the oil bottle. For recommended oil viscosities, refer to the chart.

Synthetic Oil

There are many excellent synthetic oils currently available that can provide better gas mileage, longer service life, and in some cases better engine protection. These benefits do not come without a few hitches, however. The main drawback is the price of synthetic oils, which is three or four times the price per quart of conventional oil.

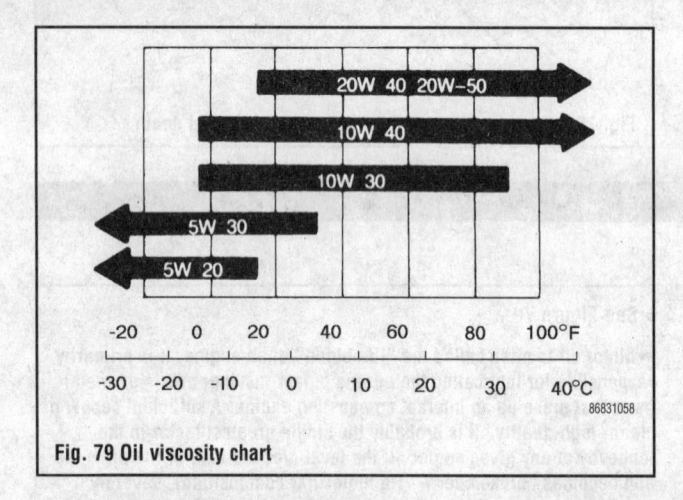

Synthetic oil is not for every car and every type of driving, so you should consider your engine's condition and your driving situation. Also, check your vehicle's warranty conditions regarding the use of synthetic oils.

Both brand new engines and older, high mileage engines are the wrong candidates for synthetic oil. The synthetic oils are so slippery that they can prevent the proper break-in of new engines; most oil manufacturers recommend that you wait until the engine is properly broken in (3,000 miles) before using synthetic oil. Older engines with wear have a different problem with synthetics. The slippery synthetic oils get past these worn parts easily.

Consider your type of driving. If most of your accumulated mileage is high speed, highway type driving, the more expensive synthetic oils may be a benefit. Extended highway driving gives the engine a chance to warm up, accumulating less acids in the oil and putting less stress on the engine over the long run. Cars with synthetic oils may show increased fuel economy in highway driving, due to less internal friction.

If synthetic oil is used, it should still be replaced at regular intervals as stated in the maintenance schedule. While the oil itself will last much longer than regular oil, pollutants such as soot, water and unburned fuel still accumulate within the oil. These are the damaging elements within a engine and must be drained regularly to prevent damage

Vehicles used under harder circumstances, such as stop-and-go, city type driving, short trips, or extended idling, should be serviced more frequently. For the engines in these cars, the much greater cost of synthetic oils may not be worth the investment.

Engine

▶ See Figures 80, 81 and 82

OIL LEVEL CHECK

Checking the oil level is one of the simplest and most important maintenance procedures. It should be done frequently because low oil level can lead to oil pan or even engine overheating and eventual starvation of the oil pump. This can mean inadequate lubrication and immediate, severe engine damage. Because the oil consumption of an engine can change quickly and unexpectedly due to leakage or changes in the weather, check the oil every time you stop for fuel.

➥**If the engine has been running you must wait a short while to get an accurate dipstick reading. Allow the oil a few minutes to drain back into the sump before checking the oil level.**

1. Park the car on a level surface.
2. Raise the hood. Pull the oil dipstick from the engine and wipe it clean.
3. Insert the dipstick into the engine until it is fully seated, then, remove it and check the reading.

➥**The oil level should register within the crosshatch design on the dipstick or between the two lines, depending on the type of stick.**

4. Oil is added through the capped opening of the valve cover. Do not add oil if the level is significantly above the lower line or the lower edge of the crosshatch. If the level is near or below the crosshatch or lower line, add oil but

Fig. 79 Oil viscosity chart

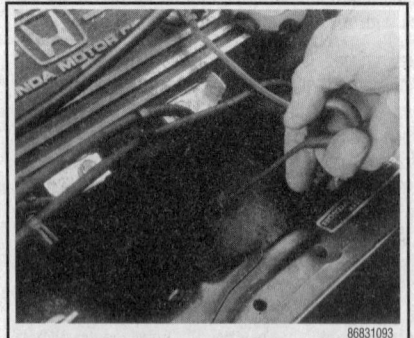

Fig. 80 The dipstick to check the motor oil level is usually brightly colored and prominent for easy identification

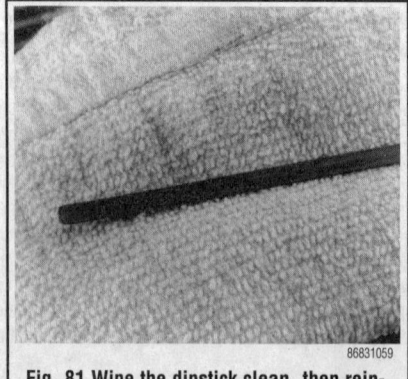

Fig. 81 Wipe the dipstick clean, then reinsert it

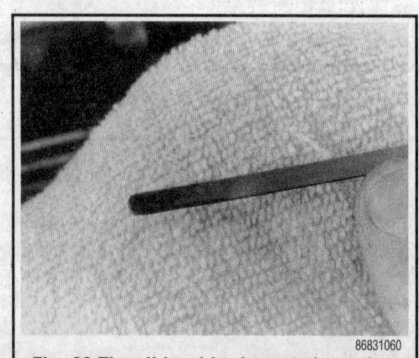

Fig. 82 The oil level is shown to be at the lower mark. This indicates the need for about one quart of oil

GENERAL INFORMATION AND MAINTENANCE 1-29

do not overfill. The length covered by the crosshatching on the dipstick is roughly equivalent to one quart of oil.

5. If oil has been added, wipe and insert the dipstick to recheck the level. It is important to avoid overfilling the crankcase. Doing so will cause the oil to be whipped and become foamy by the rotation of the crankshaft. This will affect lubrication and may also harm the engine oil seals.

OIL AND FILTER CHANGE

▶ See Figures 83 thru 88

✼✼ CAUTION

The EPA warns that prolonged contact with used engine oil may cause a number of skin disorders, including cancer! You should make every effort to minimize your exposure to used engine oil. Protective gloves should be worn when changing the oil. Wash your hands and any other exposed skin areas as soon as possible after exposure to used engine oil. Soap and water, or waterless hand cleaner should be used.

After the initial (break-in period) oil and filter change at 600 miles (966 km), the oil should be changed every 7,500 miles (12,000 km)/6 months, whichever comes first. Under severe driving conditions (stop-and-go city traffic, delivery service, or in very dusty environments, etc.) change the oil and filter more frequently. Be certain to use a high quality oil and filter.

➡While some manufacturers have at times recommended the oil filter be changed with every other oil change during non-severe vehicle use, we recommend that you replace it any time the oil is changed. The added expense of a new filter is small when compared with the price of the parts it protects. The old filter may contain a pint or more of used engine oil and the added benefit of fresh oil is quickly lost if the old filter is dirty or clogged.

1. Before changing the oil, see that the vehicle is situated on a flat surface with the engine warmed; warm oil will flow more freely from the oil pan.

2. Turn the engine **OFF**, open the hood and remove the oil filler cap from the top of the engine valve cover.

✼✼ CAUTION

Hot oil can burn you. Keep an inward pressure on the plug until the last thread is cleared, then, quickly remove it.

3. Place a container under the oil pan large enough to catch the oil. A large, flat drain pan is the most desirable.
4. Remove the oil drain plug. Allow the oil to drain completely.
5. When the oil has finished draining, install the drain plug, then tighten until just snug. Use a new washer (if applicable) on the plug.
6. Move the oil drain pan or use a second pan under the oil filter; the filter retains some oil which may drain when it is removed.
7. Using an oil filter wrench, loosen the filter, then unscrew it by hand.
8. Using a clean cloth, wipe the filter mounting surface of the cylinder block. Apply a thin coat of clean oil to the new filter gasket and install the filter. Only use filters which have an integral bypass.

➡**Hand-tighten the filter only; DO NOT use a filter wrench for tightening. If the filter has instructions printed on it as to how far it should be tightened (for example ½ or ¾ turn), mark the filter, then tighten it accordingly.**

9. Remove the drain pan from under the engine.

➡**Used engine oil is an extreme environmental hazard. Proper disposal of used engine oil is very important. Rather than dumping the oil in storm drains or throwing it in the trash, you should take it to an approved recycling center. Many shops will accept used engine oil. If you are uncertain where to take the oil, many local municipalities can direct you.**

10. Add the correct amount of recommended oil into the filler hole on top of the valve cover. Be sure that the oil level registers near the full line on the oil dipstick.
11. Install the filler cap. Start the engine without touching the accelerator pedal and allow it to idle. The oil pressure light on the instrument panel should go out after a few seconds.

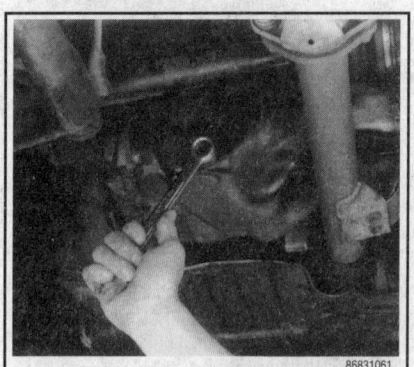

Fig. 83 Removing the oil pan drain plug

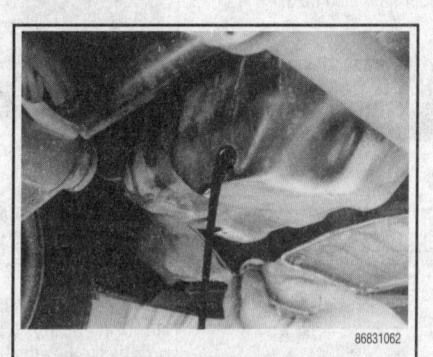

Fig. 84 Deftly remove plug to avoid draining oil on hands

Fig. 85 Use a filter wrench to loosen the oil filter. Be sure to have a drain pan ready

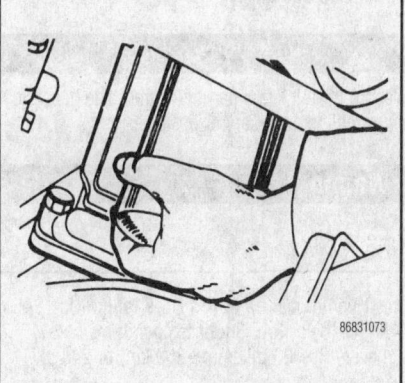

Fig. 86 Install the new oil filter by hand

Fig. 87 The oil filler cap screws in atop the valve cover

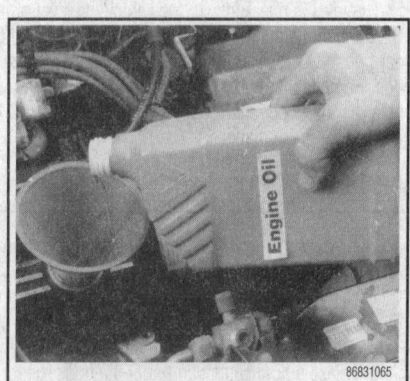

Fig. 88 A funnel to avoid spilling the oil over the engine

1-30 GENERAL INFORMATION AND MAINTENANCE

12. Run the engine for 1–2 minutes and check for leaks. Turn the engine **OFF**, then, after a few minutes, recheck the oil level; add oil. Be sure to check for oil leaks. If you find any, correct them immediately.
13. Recheck and/or refill the engine with oil.

➡ If the filter leaks, make sure the old filter gasket is not on the mounting flange. If the old gasket is not on the flange, tighten the filter a little more. If it still leaks, try another filter.

Manual Transaxle

FLUID RECOMMENDATIONS

On the Civic manual transaxles, use an engine oil labeled for SG use or better. Use 10W-30 or 10W-40 viscosity; the 10W-40 viscosity is the safer recommendation, especially if you expect to be driving at highway speeds for prolonged periods in hot weather.

LEVEL CHECK

1. Remove the oil level check bolt from the side of the transaxle. If oil runs out, the level is ok. Retighten the bolt to 33 ft. lbs. (45 Nm).
2. If the level was not ok, loosen the filler plug and pour oil in slowly until it begins to run out the level check bolt. Tighten the bolt and filler plug to 33 ft. lbs. (45 Nm).

DRAIN AND REFILL

▶ See Figures 89 and 90

The transaxle fluid should be changed every 30,000 miles (48,300 km).
1. Raise and safely support the front of the vehicle.
2. Place a fluid catch pan under the transaxle.
3. Remove the upper and lower plugs, then drain the fluid.
4. Using a new washer, install the bottom plug tightening to 33 ft. lbs. (45 Nm). Refill the transaxle, until the oil is level with the upper filler plug hole. Install the filler plug and tighten to 33 ft. lbs. (45 Nm).

➡ Be sure to properly dispose of any waste oil from the transmission.

Automatic Transaxles

FLUID RECOMMENDATIONS

All Civics with automatic transaxles, including the older Hondamatics use Dexron®II automatic transmission fluid, or Dexron®III which has superceded it. These fluids replace straight Dexron® fluid which was the actual recommendation for older models. All versions of Dexron® are compatible and the can be mixed together.

LEVEL CHECK

The level is checked with the vehicle on level ground and the engine hot, but OFF. The early models used a screw in dipstick and the later models use a standard push in dipstick. Remove the dipstick and wipe it clean, reinstall it in position. Remove the dipstick and check the oil level on the stick, it should be between the upper and lower marks on the dipstick. If needed, add fluid until it is in between the marks.

DRAIN AND REFILL

The recommended change interval for the fluid in the automatic transaxle is 30,000 miles (48,300 km).
1. Drive the vehicle to bring the transaxle fluid up to operating temperature.
2. Raise and safely support the front of the vehicle.
3. Place a fluid catch pan under the transaxle.
4. Remove the drain plug, located on the bottom of the transaxle housing, and drain the transaxle.

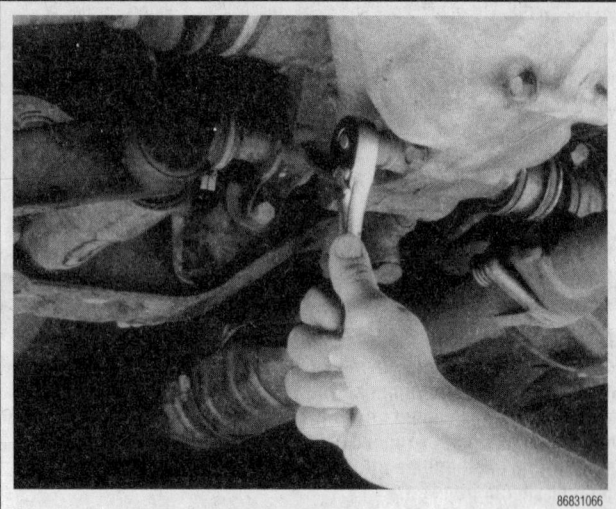

Fig. 89 Removing the transaxle drain plug

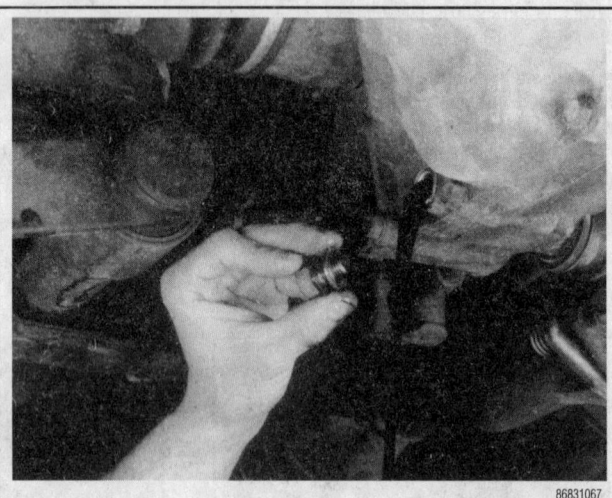

Fig. 90 Use inward pressure while unscrewing the drain bolt, then quickly pull away to avoid the hot oil

5. Using a new washer, install the drain plug and tighten it to 29 ft. lbs. (40 Nm). Using Dexron®II automatic transmission fluid, or equivalent, refill the transaxle through the fluid dipstick hole or filler cap, until fluid reaches the full mark on the dipstick; DO NOT overfill the transaxle.

➡ The amount of fluid you should add is always slightly less than the specified quantity, because there will inevitably be some fluid remaining in the recesses of the transaxle housing.

Transfer Case

The transfer case (4WD Wagon) is an integral part of the transaxle housing. No separate fluid check, drain or refill is necessary or possible.

Drive Axle—4WD Wagon

FLUID RECOMMENDATIONS

The rear drive axle requires Hypoid gear oil of API GL-5 rating. Use SAE 90 for temperatures above 41°F (5°C) or SAE 80 for temperatures below 41°F (5°C). Replace the fluid every 30,000 miles (48,300 km) or 24 months.

GENERAL INFORMATION AND MAINTENANCE 1-31

LEVEL CHECK

1. Raise and safely support the entire vehicle on a level surface.
2. Remove the oil filler plug from the drive axle housing.
3. Check that the oil is level with the bottom edge of the filler hole; if not, add oil.

➡ If the oil level is low, inspect the companion flange and drive axle housing for signs of leakage.

4. Using a new aluminum washer, install the filler plug. Torque the plug to 33 ft. lbs. (45 Nm).

DRAIN AND REFILL

1. Raise and safely support the entire vehicle on a level surface.
2. Place a drain pan under the drive axle housing.
3. Remove the oil filler and drain plugs from the drive axle housing, and allow the fluid to drain.
4. Using a new aluminum washer, install the drain plug. Torque the plug to 33 ft. lbs. (45 Nm).
5. Using fresh gear oil, pour it through the filler hole until it is level with the bottom edge of the filler hole.
6. Using a new aluminum washer, install the filler plug. Torque the plug to 33 ft. lbs. (45 Nm).

Cooling System

FLUID RECOMMENDATIONS

Use a good quality ethylene glycol-based engine coolant specifically recommended for use with vehicles utilizing aluminum engine parts. Note that some coolants, although labeled for use in such vehicles, actually may fail to provide effective corrosion protection. It is best to buy a top quality product that is known to work effectively under such conditions. Always add coolant mixed with the proper amount of clean water. Never add either water or coolant alone. Consult the chart on the antifreeze container and utilize the proportions recommended for the lowest expected temperatures in your area. Coolant is usually mixed at a 50:50 ratio.

LEVEL CHECK

To check the coolant level, simply discern whether the coolant is up to the FULL line on the expansion tank. Add coolant to the expansion tank if the level is low, being sure to mix it first with clean water. Never add cold water or coolant to a hot engine as damage to both the cooling system and the engine could result.

The radiator cap should be removed only for the purpose of cleaning or draining the system.

➡ If any coolant spills on painted portions of the body, rinse it off immediately.

DRAIN AND REFILL

▶ See Figures 91 thru 96

✱✱ CAUTION

The cooling system is under pressure when hot. Removing the radiator cap when the engine is warm or overheated will cause coolant to spill or shoot out, possibly causing serious burns. The system should be allowed to cool before attempting removal of the radiator cap or hoses.

The coolant should be changed every 45,000 miles (72,400 km); thereafter, replace every 24 months or 30,000 miles (48,300 km). When following this procedure, be sure to follow the proper precautions.
1. Remove the radiator cap and splash shield if it obstructs access.
2. Slide a drain pan under the radiator. Loosen the drain bolt at the base of the radiator and allow the coolant to drain. If equipped, loosen the drain bolt on the drain cock on the side of the block.

✱✱ CAUTION

When draining the coolant, keep in mind that cats and dogs are attracted by ethylene glycol antifreeze, and are quite likely to drink any that is left in an uncovered container or in puddles on the ground. This will prove fatal in sufficient quantity. Always drain the coolant into a sealable container.

3. Drain the coolant in the reservoir tank by unclipping and disconnecting the hose.

➡ It may be necessary to remove the thermostat to completely drain the engine.

4. Mix a solution of 50% ethylene glycol (designed for use in aluminum engines) and 50% clean water. Tighten the drain bolt(s) and refill the radiator all the way to the filler mouth. Reconnect the overflow tank connecting tube.

➡ Honda recommends that antifreeze be maintained at a 50/50 ratio year round.

5. Loosen the cooling system bleed bolt to purge air from the system. When coolant flows out of the bleed port, close the bolt and refill the radiator with coolant up to the mouth.
6. To purge any air trapped in other parts of the cooling system, set the heater control to Hot, start the engine, set it to fast idle and allow it to reach

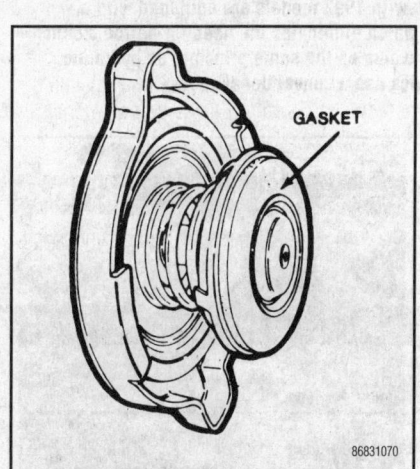

Fig. 91 Check the radiator cap gasket for cracks or wear. If necessary, replace the cap

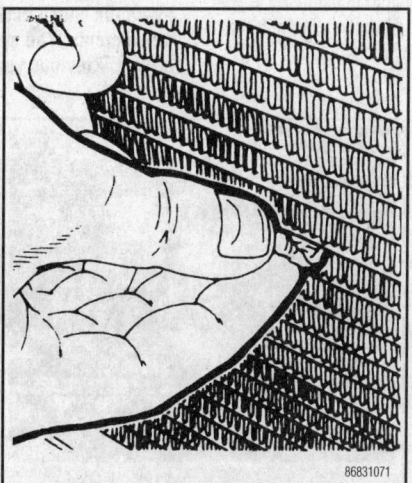

Fig. 92 Keep the radiator fins clear of debris for maximum cooling

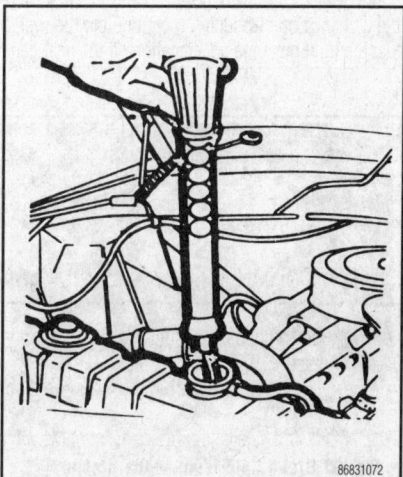

Fig. 93 A coolant tester is available at most auto parts stores

1-32 GENERAL INFORMATION AND MAINTENANCE

Fig. 94 After the cooling system is drained, fill the system through the radiator

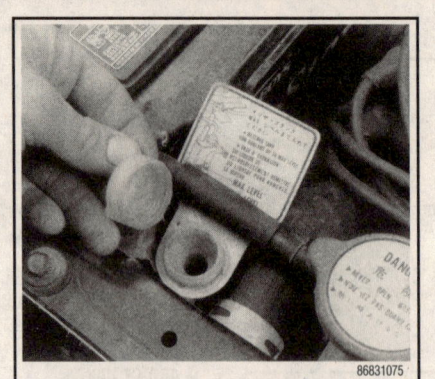

Fig. 95 The cooling system should be topped off through the expansion tank

Fig. 96 A funnel should be used to prevent spilling coolant over the engine and painted surfaces

normal operating temperatures. DO NOT tighten the radiator cap and leave the heater control in the Hot position. When the engine reaches normal operating temperatures, top off the radiator and keep checking until the level stabilizes; then, refill the coolant reservoir to the Full mark and make sure that the radiator cap is properly tightened.

FLUSHING AND CLEANING THE SYSTEM

1. Remove the thermostat from the engine (refer to section 3).
2. Using a high pressure water hose, force fresh water into the thermostat housing opening, allowing the water to back-flush into the engine, heater and radiator. Back-flush the system until the water flowing from the radiator hose is clear.
3. After cleaning, reverse the removal procedures. Torque the thermostat mounting bolts to 7 ft. lbs. (10 Nm.) Refill the cooling system with fresh coolant.

➥Honda recommends that you use new gaskets and O-rings when reassembling.

Brake Master Cylinder

FLUID RECOMMENDATIONS

➥DOT 3 and DOT 4 brake fluid is "hygroscopic." It is prone to water contamination and absorbs humidity right out of the air. This is why it is always recommended to use fresh fluid. If there is any doubt about the fluid you are using, buy new fluid.

Use only fresh DOT 3 or DOT 4 specification brake fluid from a sealed container. If you are unsure of the condition of the fluid (whether or not it has been tightly sealed or how long it has been in storage), use new fluid rather than taking a chance of introducing moisture into the system. It is critically important that the fluid meet the specification so the heat generated by modern disc brakes will not cause it to boil and reduce braking performance. Fluid must be moisture free for the same reason.

❋❋ WARNING

Brake fluid is a very effective paint remover. If any spills on a painted surface, wash it off immediately.

LEVEL CHECK

▶ See Figures 97, 98 and 99

Brake master cylinder fluid level should be checked every few weeks for indication of leaks or low fluid level due to normal wear of the brake lining.

On all Hondas, there is a fill line (or MIN/MAX lines) on the brake fluid reservoir(s) as well as an arrow on the reservoir cap(s) which should face forward when installed. When adding brake fluid, the following precautions should be observed:

- Use only recommended brake fluid: DOT 3 or DOT 4; SAE J 1703b HD type.
- Never reuse brake fluid and never use fluid that is dirty, cloudy or has air bubbles.
- Store brake fluid in a clean dry place in the original container. Cap tightly and do not puncture a breather hole in the container.
- Carefully remove any dirt from around the master cylinder reservoir cap before opening.
- Take special care not to spill the fluid. The painted surface of the vehicle will be damaged by brake fluid.

Clutch Master Cylinder

➥Honda Civics beginning with 1992 models are equipped with a hydraulic clutch actuator which eliminates the need for periodic clutch adjustments. The system works on the same principle as hydraulic brakes. Previous year Civics use a conventional cable.

Fig. 97 Brake fluid level in the master cylinder reservoir should be monitored regularly

Fig. 98 Clean the master cylinder reservoir cap before removing it

Fig. 99 Always add new brake fluid—be sure to wipe up any that might be spilled

GENERAL INFORMATION AND MAINTENANCE 1-33

FLUID RECOMMENDATIONS

Use only fresh DOT 3 or DOT 4 specification brake fluid from a sealed container. If you are unsure of the condition of the fluid (whether or not it has been tightly sealed), use new fluid rather than taking a chance of introducing moisture into the system.

※ WARNING

Brake fluid is a very effective paint remover. If any spills on a painted surface, wash it off without delay.

LEVEL CHECK

The fluid level should be between the MIN and MAX marks on the side of the reservoir. Low fluid level could mean that there is a leak in the clutch system. Unlike brake wear, clutch wear will not proportionally lower the fluid level in the reservoir.

When adding brake fluid, the following precautions should be observed:
- Use only recommended brake fluid: DOT 3 or DOT 4; SAE J 1703b HD type.
- Never reuse brake fluid and never use fluid that is dirty, cloudy or has air bubbles.
- Store brake fluid in a clean dry place in the original container. Cap tightly and do not puncture a breather hole in the container.
- Carefully remove any dirt from around the master cylinder reservoir cap before opening.
- Take special care not to spill the fluid. The painted surface of the vehicle will be damaged by brake fluid.

Power Steering Pump

FLUID RECOMMENDATIONS

Only genuine Honda power steering fluid or a known equivalent may be used when adding fluid. Honda states that ATF or fluids manufactured for use in other brands of vehicles (by their manufacturers or independents) are not compatible with the Honda power steering system. The use of any other fluid will cause the seals to swell and create leaks.

RESERVOIR LEVEL CHECK

♦ See Figure 100

The fluid in the power steering reservoir should be checked every few weeks for indications of leaks or low fluid level. Check the fluid with the engine cold and the vehicle parked on a level spot. The level should be between the upper and lower marks. Fluid need not be added right away unless it has dropped almost to the lower mark. DO NOT overfill the reservoir.

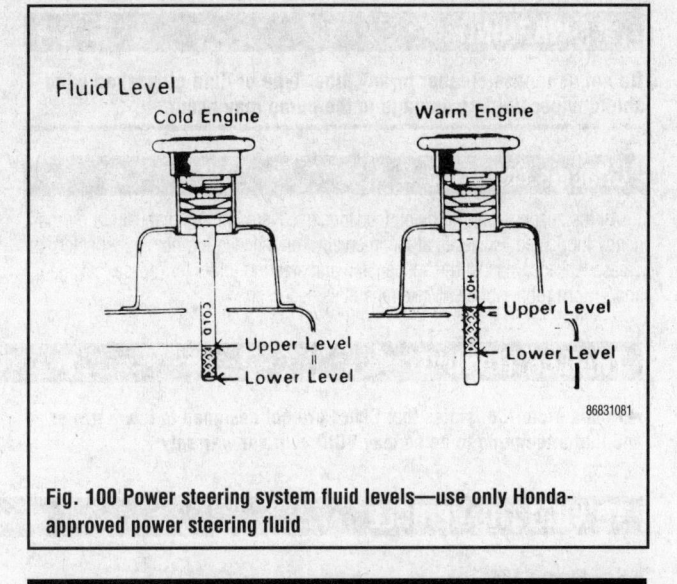

Fig. 100 Power steering system fluid levels—use only Honda-approved power steering fluid

Manual Steering Gear

INSPECTION

The manual steering gear used on Hondas is of the rack and pinion design. This unit is packed with grease and therefore does not require a periodic fluid level check. However, inspect the unit and associated rubber boot-type seals for obvious grease leaks or torn boots at 15,000 miles (24,100 km), 30,000 miles (48,300 km), then, at 30,000 mile (48,300 km) increments. Make repairs as necessary.

FLUID RECOMMENDATIONS

Repack with about 2 oz. of multipurpose grease.

Windshield Wiper Fluid

♦ See Figures 101, 102 and 103

Windshield wiper fluid is helpful year-round for clearing the windshield while driving. The wiper fluid reservoir is located next to the fender in the engine compartment and is easily identified by the windshield symbol on its cap. The bottle is translucent to enable you to easily check the level. You should always have an adequate supply. Windshield wiper fluid is inexpensive and widely available from auto parts stores and many supermarkets.

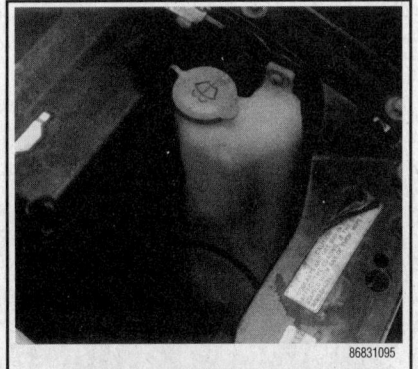

Fig. 101 The washer fluid reservoir is usually mounted on one of the inner fenders

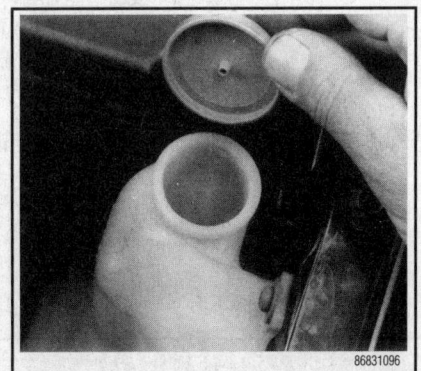

Fig. 102 Fill the reservoir through the cap to the level indicated on the side

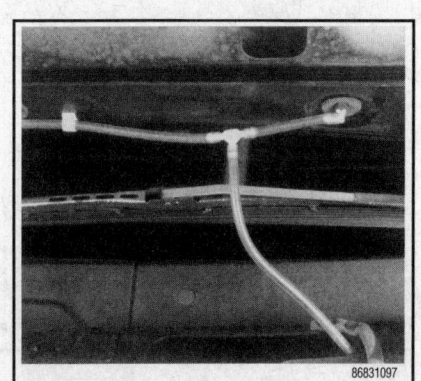

Fig. 103 Make sure that the fluid lines are not clogged

1-34 GENERAL INFORMATION AND MAINTENANCE

✳✳ WARNING

Do not use glass cleaner or any other type of fluid other than windshield wiper fluid as damage to the pump may occur.

Chassis Greasing

All the suspension fittings on the Hondas covered in this manual are permanently lubricated. However, at the interval when the steering box is inspected for grease leakage, inspect the suspension and steering joints for grease leakage and/or torn rubber boots. Make repairs as necessary.

TRAILER TOWING

➥ Honda Motor Co. states that Civics are not designed to tow a trailer and that attempting to do so may VOID your car warranty.

TOWING THE VEHICLE

▶ See Figure 104

If your Honda's rear axle is operable, you can tow your vehicle with the rear wheels on the ground. Due to its front wheel drive format, the Honda is a relatively easy vehicle to tow with the front wheels up. Before doing so, you should release the parking brake.

If the rear axle is damaged, the vehicle must then be towed with the rear wheels off the ground. Before attempting this, a dolly should be placed under the front wheels. If a dolly is not available, and you still have to tow it with the rear wheels up, then you should first shift the transaxle into Neutral and then lock the steering wheel so that the front wheels are pointing straight ahead. In such a position, the vehicle must not be towed at speeds above 35 mph or for more then 50 miles (80 km).

✳✳ WARNING

It is critical that you observe these limitations to prevent damage to your transaxle due to inadequate lubrication.

Body Lubrication

Lubricate all locks and hinges with multi-purpose grease every 6000 miles (9,700 km) or 6 months.

Wheel Bearings

The wheel bearings on these models are part of a sealed hub/bearing unit and does not require periodic maintenance. If replacement of the bearing is necessary, please refer to Section 7.

Front:

CAUTION: The car is equipped with a front spoiler, remove it before towing so it is not damaged.

TOWING HOOKS

Rear:

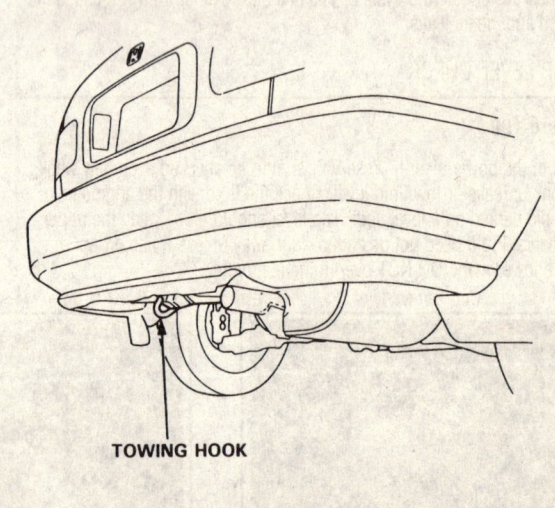

TOWING HOOK

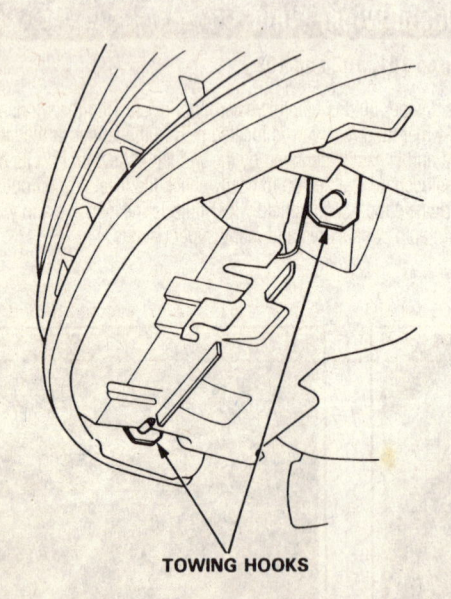

Fig. 104 Towing hooks are provided for the tie-down chains

GENERAL INFORMATION AND MAINTENANCE 1-35

JACKING

▶ See Figures 105 and 106

Your Honda comes equipped with an emergency-use scissor jack. This jack is fine for changing a flat tire or other operations where you do not have to go beneath or put your hands beneath the vehicle. There are four lifting points where this jack may be used: one behind each front wheel well and one in front of each rear wheel well in reinforced sheet metal brackets beneath the rocker panels.

A more convenient way of jacking is the use of a garage or floor jack. You may use the floor jack beneath any of the four jacking points or you can raise either the entire front or entire rear of the vehicle using the special jacking brackets beneath the front center or rear center of the vehicle. On station wagon models, the rear of the vehicle may be jacked beneath the center of the rear axle beam.

The following safety points cannot be overemphasized:
- Always block the opposite wheel or wheels to keep the vehicle from rolling off the jack.
- When raising the front of the vehicle, firmly apply the parking brake.
- When raising the rear of the vehicle, place the transaxle in First or Reverse gear.
- Always use jack stands to support the vehicle when you are working underneath (not cinder blocks which can crumble). Place the stands beneath the jacking brackets. Before climbing underneath, rock the vehicle a bit to make sure it is firmly supported.

If you are going to have your Honda serviced on a garage hoist, make sure the four hoist platform pads are placed beneath the scissors jacking brackets. These brackets are reinforced and will support the weight of the entire vehicle.

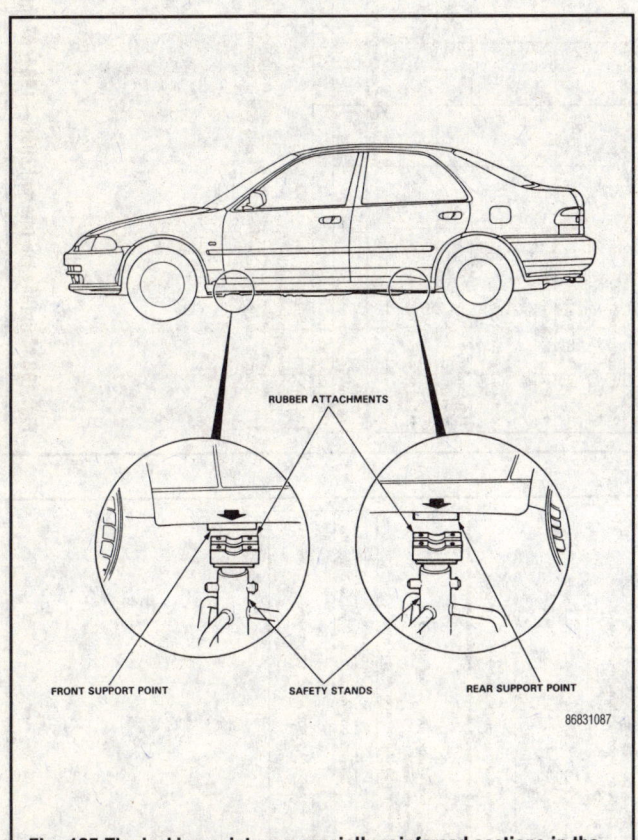

Fig. 105 The jacking points are specially reinforced sections in the body seam below the rocker panel

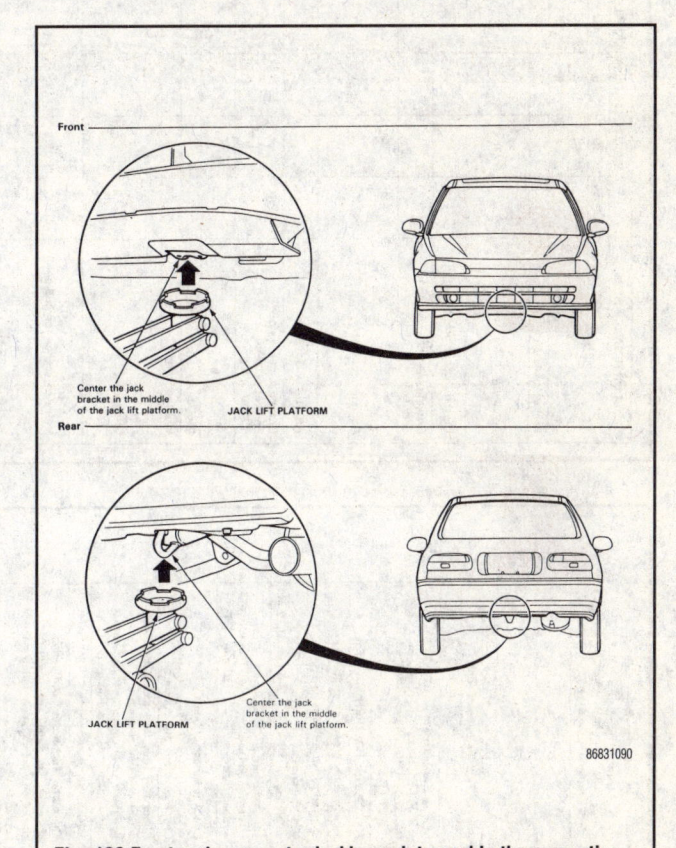

Fig. 106 Front and rear center jacking points enable the respective half of the car to be lifted

1-36 GENERAL INFORMATION AND MAINTENANCE

MAINTENANCE INTERVAL CHARTS

The maintenance listed below must be performed at mileage intervals shown to ensure safe and efficient operating conditions and proper emission levels. This checklist is arranged in hot/cold/hot sequence so you can work on a hot engine while waiting for it to cool, and then run cold and hot emissions checks.

R = Replace   C = Clean
I = Inspect
After inspection, clean, adjust, repair or replace if necessary.

MAINTENANCE ITEM	MAINTENANCE INTERVALS							NOTES
X 1,000 miles	15	30	45	60	75		
X 1,000 km	24	48	72	96	120		
months	12	24	36	48	60		
Engine oil							Replace every 7,500 miles (12,000 km) or 6 months
Engine oil filter		R		R			
Manual transmission and rear differential oil							Manual transmission: 2.3ℓ (2.4 qts.) Rear differential (Wagon 4WD only): 1.2ℓ (1.3 qts.)
Automatic transmission fluid		R		R			2.4ℓ (2.5 qts.) DEXRON® A.T.F.
Rear brakes	I	I	I	I	I		Minimum thickness: 2 mm (0.08 in.)
Parking brake	I	I	I	I	I		Fully engaged: 4-8 clicks
Front brake pads							Inspect every 7,500 miles (12,000 km) or 6 months Min. thickness: Pad (lining) 3.0 mm (0.12 in.)
Front brake discs and calipers	I	I	I	I	I		Min. thickness: Ventilated Disc 15 mm (0.6 in.) Solid Disc 10 mm (0.4 in.)
Suspension mounting bolts		R		R			Check tightness of bolts.
Exhaust pipe and muffler	I	I	I	I	I		
•● Fuel line connections							Check fuel lines for loose connections, cracks and deterioration. Retighten loose connections and replace any damaged or deformed parts.
•• Fuel filter and hoses			R		R		Same as carbureted engine
•● Fuel filters (front and rear) and engine compartment fuel hoses		R		R			The rubber fuel hoses need periodic replacement since they are subject to cracks and deterioration during a long period of use.
•● Fuel line connections	I	I	I	I	I		Same as carbureted engine
Steering operation, tie rod ends, steering gearbox and boots	••	••	••	••	••		Check rack grease and steering linkage. Check the boot for damage or leaking grease.
Front wheel alignment	I	I	I	I	I		
Power steering system	I	I	I	I	I		
Power steering pump belt	I	I	I	I	I		

•¹ Fuel-Injected Engine
•² Carbureted Engine
* Tension adjustment only.
** Recommended by manufacturer only for Fuel-Injected Engine sold in California.

CAUTION: The following items must be serviced more frequently on cars normally used under severe driving conditions. Refer to the chart below for the appropriate maintenance intervals.
"Severe driving conditions":
A: Repeated short distance driving
B: Driving in dusty conditions
C: Driving in severe cold weather
D: Driving in areas using road salt or other corrosive materials
E: Driving on rough and/or muddy roads

Condition	Maintenance item	Maintenance operation	Interval
A B • • •	Engine oil and oil filter	R	Every 3,000 miles (5,000 km) or 3 months
A B • D E	Front brake discs and calipers	I	
A B C • E	Clutch release-arm travel	I	
• • C • E			
• • • • E			Every 7,500 miles (12,000 km) or 6 months
B C • • E			

Fig. 107 Maintenance interval chart—1984–87 models

MAINTENANCE ITEM	MAINTENANCE INTERVALS							NOTES
X 1,000 miles	15	30	45	60	75		
X 1,000 km	24	48	72	96	120		
months	12	24	36	48	60		
Brake hoses and lines	I	I	I	I	I		Use only DOT 3 or 4 fluid
Brake fluid		R		R			Check that brake fluid level is between the upper and lower marks on the reservoir.
Clutch release arm travel	I	I	I	I	I		Free play at arm: 4.0–5.0 mm (5/32–13/64 in.)
Cooling system hoses and connections		I		I			Cooling system capacity 1500: 4.5ℓ (1.2 US gal) 1300: 3.4ℓ (0.9 US gal)
Radiator coolant		** R		** R			Check specific gravity for freezing point.
Alternator drive belt	I	I	I	I	I		7–10 mm (0.28–0.39 in.) @ 10 kg (22 lbs.) tension.
Crankcase emission control system —PCV valve		*** I		*** I			If clicking sound is heard as you pinch the PCV hose between the PCV valve and intake manifold. Valve is OK.
** —Blow-by filter		*** I		*** I			Replace if filter is dripping with oil or obstructed by dust.
Distributor cap and rotor	I	I	I	I	I		Maximum resistance 25,000 ohms
Ignition wiring	I	I	I	I	I		
Spark plugs		R		R			Carbureted Engine NGK: BUR5EB-11 or ND: W16EKR-S11 1.0–1.1 mm (0.039–0.043 in.) Fuel-Injected Engine
Valve clearance (engine cold)	I	I	I	I	I		In & Aux: Exhaust: 0.17–0.22 mm (0.007–0.009 in.) 0.22–0.27 mm (0.009–0.011 in.)
Air cleaner element	I	* R	I	* R	I		
*² Choke mechanism		C		C			Spray with carburetor cleaner.
*² Choke opener operation (cold)	I	I	I	I	I		Choke blade should partially open on start up.
*² Intake air control (cold)	I	I	I	I	I		Door should stay up for 3 sec. after cranking.
*² Fast idle unloader (cold)	I	I	I	I	I		Engine should hold fast idle.
Evaporative emission control system (cold)	I	I	I	I	I		Disconnect upper hose at purge control diaphragm valve and connect vacuum gauge to hose. Start engine and let idle. Vacuum should not appear.
*² Secondary air supply system (cold)	I	I	I	I	I		
*² E.G.R. System (cold)	I	I	I	I	I		Connect vacuum gauge to EGR valve hose. Vacuum should not be available at 4,500–5,000 rpm.
*² Choke coil tension (warm up) and heater	I	I	I	I	I		Warm up the engine to normal operating temperature. Choke blade should open fully as engine warms up.
*² Intake air control (hot)	I	I	I	I	I		Door should be down.

*¹ Fuel-Injected Engine *² Carbureted Engine
* Tension adjustment only.
** Thereafter, replace every 2 years or 30,000 miles (48,000 km) which comes first.
*** Recommended by manufacturer only for cars sold in California.

Fig. 108 Maintenance interval chart—1984–87 models (cont'd)

GENERAL INFORMATION AND MAINTENANCE 1-37

The maintenance listed below must be performed at mileage (km) intervals shown to ensure safe and efficient operating conditions and proper emission levels. This checklist is arranged in hot/cold/hot sequence so you can work on a hot engine while waiting for it to cool, and then run cold and hot emissions checks.

R = Replace
I = Inspect
After inspection, clean, adjust, repair or replace if necessary.

MAINTENANCE ITEM			MAINTENANCE INTERVAL					NOTE
Service at the interval listed 1,000 miles (or km) after that number of months, whichever comes first.	× 1,000 miles	15	30	45	60	75	90	
	× 1,000 km	24	48	72	96	120	144	
	months	12	24	36	48	60	72	
☆ Engine oil		Replace every 7,500 miles (12,000km) or 6 months						3.5ℓ (3.7 qts.) at oil/filter change
☆ Engine oil filter		R	R	R	R	R	R	
☆ Manual transmission oil			R		R		R	Manual transmission: 1.8ℓ (1.9 qts.)
☆ Automatic transmission fluid			R		R		R	2.4ℓ (2.5qts.) DEXRON®II A.T.F.
Rear brake drums, wheel cylinders and linings			I		I		I	Minimum thickness: 2 mm (0.08 in.)
Parking brake		I	I	I	I	I	I	Fully engaged: 6-10 clicks
Front brake pads		Inspect every 7,500 miles (12,000km) or 6 months						Min. thickness: Pad 3.0mm(0.12 in.)
☆ Front brake discs and calipers			I		R		I	Min. thickness: 19 mm (0.75 in.)
Suspension mounting bolts			I		I		I	Check tightness of bolts.
Exhaust pipe and muffler			I		I		I	Check condition and tightness.
Fuel filter and hoses					R			The rubber fuel hoses need periodic replacement since they are subject to cracks and deterioration during a long period of use.
Fuel line connections			I*³		I		I*³	
Steering operation, tie rod ends, steering gear box and boots			I		I		I	Check rack grease and steering linkage. Check the boot for damage or leaking grease.
Front wheel alignment			I		I		I	
☆ Power steering system**			I*³		I		I*³	
Power steering pump belt**			I		I		I	9-12mm (0.35-0.47in) @98N (10kg, 22 lbs.) tension.

■ : Check oil and coolant level at each fuel stop.
☆ : Service more frequently on cars normally used under severe driving conditions.
*¹ : Tension adjustment only
*² : For cars sold in California, this service is recommended only; for other areas, it is required.
** : Car equipped with power steering

CAUTION : The following items must be serviced more frequently on cars normally used under severe driving conditions. Refer to the chart below for the appropriate maintenance intervals.
Severe driving conditions include:
A : Repeated short distance driving
B : Driving in dusty conditions
C : Driving in severe cold weather
D : Driving in areas using road salt or other corrosive materials
E : Driving on rough and/or muddy roads

Condition	Maintenance item	Maintenance operation	Interval
A B · · ·	Engine oil and oil filter	R	Every 3,750 miles (6,000 km) or 3 months
A B · D E	Front brake discs and calipers	I	Every 7,500 miles (12,000 km) or 6 months
A B · · E	Clutch release-arm travel	I	Every 3,750 miles (6,000 km) or 3 months
· B · · E	Power steering system	I	Every 7,500 miles (12,000 km) or 6 months

Fig. 110 Maintenance interval chart—1988-91 models

MAINTENANCE ITEM	MAINTENANCE INTERVALS						NOTES
	× 1,000 miles	15	30	45	60	75	
	× 1,000 km	24	48	72	96	120	
	months	12	24	36	48	60	
Ignition timing			I		I		Red mark at idle. See page 25-28 for idle speeds.
Ignition timing control			I		I		Vacuum advance at idle
*² Fast idle unloader (hot)			I		I		With engine warm, idle speed should drop below 1,400 rpm
Fast Idle			I		I		*¹ 1,750 ± 500 rpm *³ 3,000 ± 500 rpm
*² Throttle controller — Dashpot			I		I		With the engine idling, disconnect hose from the throttle controller. Engine speed should rise to 1700 ± 300 rpm.
Idle speed		I	I	I	I	I	
Idle CO		I	I	I	I	I	Check with propane enrichment or CO meter.
Idle control system (A/C only)		I	I	I	I	I	Recheck idle speed with A/C on.
*² E.G.R. System (hot)			I		I		
*² Secondary air supply system (hot)			I		I		
Evaporative emission control system (hot)			I		I		
Catalytic converter heat shield			I		I		Check condition and tightness

*¹ Fuel-Injected Engine *² Carvureted Engine

Fig. 109 Maintenance interval chart—1984-87 models (cont'd)

1-38 GENERAL INFORMATION AND MAINTENANCE

Fig. 111 Maintenance interval chart—1988–91 models (cont'd)

MAINTENANCE ITEM	MAINTENANCE INTERVAL							NOTE
Service at the interval listed x 1,000 miles (or km) or after that number of months, whichever comes first.	x 1,000 miles	15	30	45	60	75	90	
	x 1,000 km	24	48	72	96	120	144	
	months	12	24	36	48	60	72	
Brake hoses and lines		I		I		I		
Brake fluid			R		R		R	Use only DOT 3 or 4 fluid. Check that brake fluid level is between the upper and lower marks on the reservoir.
☆Clutch release arm travel		Inspect every 7,500 miles (12,000 km) or 6 months						Free play at arm: 3.0–4.0 mm (0.12–0.16 in.)
Cooling system hoses and connections			I		I		I	
■Radiator coolant			R		R*²		R	Cooling system capacity Sedan and CX, DX Hatchback: M/T: 4.5ℓ (4.8 U.S.qt., 4.0 Imp.qt.) A/T: 4.4ℓ (4.7 U.S.qt., 3.9 Imp.qt.) STD Hatchback: 4.4ℓ (4.7 U.S.qt., 3.9 Imp.qt) Si and EX: M/T: 4.4ℓ (4.7 U.S.qt., 3.9 Imp.qt.) A/T: 4.9ℓ (5.2 U.S.qt., 4.3 Imp.qt.) Check specific gravity for freezing point.
Alternator drive belt			I		I		I	9–11 mm (0.35–0.43 in.) @ 98 N (10 kg, 22 lbs.) tension.
Timing belt							R	
Water pump							I	
Positive crankcase ventilation valve					I*¹			If clicking sound is heard as you pinch the PCV hose between the PCV valve and intake manifold, Valve is OK.
Distributor cap and rotor					I		I	
Ignition wiring					I		I	Maximum resistance 25,000 ohms
Spark plugs			R		R		R	NGK: BCPR6E-11 or BCPR6EY-N11 ND: Q20PR-U11 CHAMPION: RC9YCN4 (used for only 1.5ℓ engine: USA model) Gap: 1.0–1.1 mm (0.039–0.043 in.)
Valve clearance (engine cold)			I		I		I	Intake: 0.17–0.22 mm (0.007–0.009 in.) Exhaust: 0.22–0.27 mm (0.009–0.011 in.)
Air cleaner element					R		R	
Evaporative emission control system							I	Disconnect upper hose at purge control diaphragm valve and connect vacuum gauge to hose. Start engine and let idle. Vacuum should not appear.
E.G.R. system (cold)*⁵							I	See "Emission Section

Fig. 112 Maintenance interval chart—1988–91 models (cont'd)

MAINTENANCE ITEM	MAINTENANCE INTERVAL							NOTE
Service at the interval listed x 1,000 miles (or km) or after that number of months, whichever comes first.	x 1,000 miles	15	30	45	60	75	90	
	x 1,000 km	24	48	72	96	120	144	
	months	12	24	36	48	60	72	
		Warm up the engine to normal operating temperature.						
Ignition timing and control system			I		I		I	Red mark at idle. See page 23-65 for idle speeds.
Idle speed			I		I		I	See Emission Section.
Idle CO			I		I		I	Check with propane enrichment or CO meter.
E.G.R. system*⁵					I		I	See Emission Section.
Catalytic converter heat shield			I		I		I	Check condition and tightness.

■: Check oil and coolant level at each fuel stop.
☆: Service more frequently on cars normally used under severe driving conditions.
*¹ Tension adjustment only.
*² Thereafter, replace every 2 years or 30,000 miles (48,000 km), whichever comes first.
*⁵ 1.5 ℓ California models with automatic transmission.

Fig. 113 Maintenance interval chart—1992–95 models

R—Replace I—Inspect After inspection, clean, adjust, repair or replace if necessary.

Maintenance item	x 1,000 miles	7.5	15	22.5	30	37.5	45	52.5	60	67.5	75	82.5	90	97.5	105	NOTE
	x 1,000 km	12	24	36	48	60	72	84	96	108	120	132	144	156	168	
	months	6	12	18	24	30	36	42	48	54	60	66	72	78	84	
Engine and Transmission																
□ Air cleaner element									R						R	
Idle speed																U.S. Model: D15Z1: 600 ± 50 rpm (M/T: neutral) Except D15Z1: 670 ± 50 rpm (M/T: neutral) 700 ± 50 rpm (A/T: [N] or [P] position) Canada Model: Except D15Z1: 750 ± 50 rpm (M/T: neutral) 750 ± 50 rpm (A/T: [N] or [P] position)
Positive crankcase ventilation valve									I						I	If clicking sound is heard as you pinch the hose between the PCV valve and intake manifold, valve is OK
Valve clearance (cold)									I						I	Intake: 0.18–0.22 mm (0.007–0.009 in) Exhaust: 0.23–0.27 mm (0.009–0.011 in) Measured when cold.
Fuel filter									R*¹						R*¹	The rubber fuel hoses need periodic replacement since they are subject to cracks and deterioration during a long period of use.
Fuel pipes, hoses, and connections									I*¹						I*¹	Check fuel lines for loose connections, cracks and deteriorations. Retighten loose connections and replace any damaged parts.
Spark plugs									R						R	D15B6/D15Z1: NGK: ZFR4F-11 NIPPONDENSO: KJ14CR-L11 D15B7: NGK: ZFR5F-11 NIPPONDENSO: KJ16CR-L11 D16Z6: NGK: ZFR5J-11 NIPPONDENSO: KJ16CR-L11 Gap: 1.0–1.1 mm (0.039–0.043 in)
Distributor ignition cap and rotor									R						R	
Ignition wires									I						I	
□ Engine oil		R	R	R	R	R	R	R	R	R	R	R	R	R	R	Capacity for change with filter: 3.3ℓ (3.5 US qt, 2.9 Imp qt)
□ Engine oil filter			R		R		R		R		R		R		R	

*: Check oil and coolant level at each fuel stop.
□: Under severe driving conditions, service these items more often.
*¹ For cars sold in California, this service is recommended only; other areas, it is required.

GENERAL INFORMATION AND MAINTENANCE 1-39

R—Replace I—Inspect After inspection, clean, adjust, repair or replace if necessary.

Service at the interval listed x 1,000 miles (or km) or after that number of months, whichever comes first.														NOTE
	x 1,000 miles	7.5	15	22.5	30	37.5	45	52.5	60	67.5	75	82.5	90	97.5	105	
	x 1,000 km	12	24	36	48	60	72	84	96	108	120	132	144	156	168	
Maintenance item	months	6	12	18	24	30	36	42	48	54	60	66	72	78	84	
Engine and Transmission																
Alternator drive belt					I			I			I					7.0–10.5 mm (0.28–0.41 in) with 100 N (10 kg, 22 lbs) tension
Cooling system hoses and connections					I*1			I*1			I*1					
• Engine coolant							R			R			R*2			Capacity for change: M/T: D15Z1: 3.5 ℓ (0.92 US gal, 0.77 Imp gal) Except D15Z1: 3.6 ℓ (0.95 US gal, 0.79 Imp gal) A/T: D15B7: 3.5 ℓ (0.92 US gal, 0.77 Imp gal) D16Z6: 3.8 ℓ (1.00 US gal, 0.84 Imp gal) Check specific gravity for freezing point
Timing belt													R*3			
Water pump													I			
Three way catalytic converter heat shield								I								Check condition and tightness
Exhaust pipe (before catalytic converter)					I*1			I*1			I*1					Check condition and tightness
Exhaust pipe and muffler (after catalytic converter)		I	I	I	I	I	I	I			I				I	Check condition and tightness
□ Manual transmission oil					R				R				R			1.8 ℓ (1.9 US qt, 1.6 Imp qt) for change
□ Automatic transmission fluid					R				R				R			2.7 ℓ (2.9 US qt, 2.4 Imp qt) for change HONDA Premium Formula ATF or DEXRON® II ATF
Brakes																
Front brake pad		I	I	I	I	I	I	I	I	I	I	I	I	I	I	Min. thickness: 1.6 mm (0.06 in)
□ Front brake discs and calipers			I		I		I		I		I		I			Min. thickness: 19.0 mm (0.75 in)
□ Rear brake discs, calipers and pads*4			I		I		I		I		I		I			Min. thickness: Discs 8.0 mm (0.32 in) : Pads 1.6 mm (0.06 in)
Rear brake drums, wheel cylinders and linings					I				I				I			Max. diameter: Drums: M/T, Except Coupe 1.6 ℓ M/T 181 mm (7.13 in) A/T, Coupe 1.6 M/T 201 mm (7.91 in) Min. thickness: Linings 2.0 mm (0.08 in)

•: Check oil and coolant level at each fuel stop.
□: Under severe driving conditions, service these items more often.
*1: For cars sold in California, this service is recommended only; other areas, it is required.
*2: Thereafter, replace every 2 years or 30,000 miles (48,000 km), whichever comes first.
*3: This service is recommended only.
*4: For 3D Si and 4D U.S.A.: EX, Canada: Optional on EX-V.

Fig. 114 Maintenance interval chart—1992–95 models (cont'd)

R—Replace I—Inspect After inspection, clean, adjust, repair or replace if necessary.

Service at the interval listed x 1,000 miles (or km) or after that number of months, whichever comes first.														NOTE
	x 1,000 miles	7.5	15	22.5	30	37.5	45	52.5	60	67.5	75	82.5	90	97.5	105	
	x 1,000 km	12	24	36	48	60	72	84	96	108	120	132	144	156	168	
Maintenance item	months	6	12	18	24	30	36	42	48	54	60	66	72	78	84	
Brakes																
Brake hoses and lines (Including anti-lock brake system*5)			I		I		I		I		I		I		I	Check for leaks, damage, interference or twisting.
Parking brake			I		I			I			I					Fully engaged 6 to 10 clicks.
Brake fluid (Including anti-lock brake system*5)					R			R			R					Use only DOT 3 or DOT 4 fluid. Check that brake fluid level is between the upper and lower marks on the reservoir.
Anti-lock brake system operation*5					I			I			I					
Anti-lock brake system high pressure hose*5								R								
Steering, Suspension, Miscellaneous																
Front wheel alignment			I		I			I			I					
Steering operation, tie rod ends, steering gearbox and boots			I		I			I			I					Check rack grease and steering linkage. Check the boot for damage or leaking grease.
□ Power steering system			I		I		I		I		I		I			Check that the fluid level is between the upper and lower level line on the reservoir.
Power steering pump belt					I			I			I					8.0–12.0 mm (0.31–0.47 in) with 100 N (10 kg, 22 lbs) tension
Suspension mounting bolts					I			I			I				I	Check tightness of bolts.
Supplemental Restraint System	Inspect system 10 years after production															

□: Under severe driving conditions, service these items more often.
*5: For U.S.A.: EX, Canada: Optional on EX-V

Severe Driving Conditions
Items marked R or I in the chart indicate you will need some services more frequently in certain severe driving conditions.

The conditions are:
A: Driving less than 5 miles (8 km) per trip, or, in freezing temperatures, driving less than 10 miles (16 km) per trip.
B: Driving on rough or muddy roads, or de-iced roads.
C: Driving in extremely dusty conditions.
D: Extensive idling or driving long periods at slow speeds, such as a delivery vehicle.

Services for Severe Driving Conditions
- Clean the air cleaner element every 15,000 miles (24,000 km) or 12 months and replace every 30,000 miles (48,000 km) or 24 months under conditions B or C.
- Replace engine oil and oil filter every 3,750 miles (6,000 km) or 3 months under conditions A, B, C or D.
- Replace transmission oil every 15,000 miles (24,000 km) or 12 months under conditions D.
- Inspect front brake discs and calipers, and rear brake discs, calipers and pads every 7,500 miles (12,000 km) or 6 months under conditions B, C or D.
- Inspect the power steering system every 7,500 miles (12,000 km) or 6 months under conditions A, B, or C.

Fig. 115 Maintenance interval chart—1992–95 models (cont'd)

1-40 GENERAL INFORMATION AND MAINTENANCE

CAPACITIES

Year	Model	Engine ID/VIN	Engine Displacement Liters (cc)	Engine Oil with Filter	Transmission Pts. 4-Spd	Transmission Pts. 5-Spd	Transmission Pts. Auto.	Rear Drive Axle (pts.) ①	Fuel Tank (gal.)	Cooling System (qts.) Manual	Cooling System (qts.) Auto.
1984	Civic Std.	EV1	1.3 (1342)	3.7	5.0	-	-	-	11.9	4.8	-
	Civic DX	EW1	1.5 (1488)	3.7	-	5.0	5.0	-	11.9 ②	6	-
	CRX	EW1	1.5 (1488)	3.7	-	5.0	5.0	-	10.8	6	-
	Wagon	EW1	1.5 (1488)	3.7	-	5.0	5.0	-	12.1	6	-
1985	Civic Std.	EV1	1.3 (1342)	3.7	5.0	-	-	-	11.9	4.8	-
	Civic DX, S	EW1	1.5 (1488)	3.7	-	5.0	6.0	-	11.9 ②	6	-
	CRX DX, S	EW1, EW3	1.5 (1488)	3.7	-	5.0	6.0	-	10.8	6	-
	Civic Wagon	EW1	1.5 (1488)	3.7	-	5.0	6.0	-	12.1	6	-
	Civic Wagon 4WD	EW1	1.5 (1488)	3.7	-	5.0	6.0	1.4	12.1	6	-
1986	Civic Std.	EV1	1.3 (1342)	3.7	5.0	-	-	-	11.9	4.8	-
	Civic DX	EW1	1.5 (1488)	3.7	-	5.0	5.0	-	11.9 ②	6	-
	Civic Si	EW3	1.5 (1488)	3.7	-	5.0	-	-	11.9	6	-
	CRX HF	EW1	1.5 (1488)	3.7	-	5.0	-	-	10.0	4.8	-
	CRX DX	EW1	1.5 (1488)	3.7	-	5.0	5.0	-	10.8	6	-
	CRX Si	EW3	1.5 (1488)	3.7	-	5.0	-	-	10.8	6	-
	Civic Wagon	EW1	1.5 (1488)	3.7	-	5.0	6.0	-	11.9	6	-
	Civic Wagon 4WD	EW1	1.5 (1488)	3.7	-	5.0	6.0	1.4	11.9	6	-
1987	Civic Std.	D13A2	1.3 (1342)	3.7	5.0	-	-	-	11.9	4.8	-
	Civic DX	D15A2	1.5 (1488)	3.7	-	5.0	5.0	-	11.9 ②	6	-
	Civic Si	D15A2	1.5 (1488)	3.7	-	5.0	-	-	11.9	6	-
	CRX HF	D15A2	1.5 (1488)	3.7	-	4.0	-	-	10.0	4.8	-
	CRX DX	D15A2	1.5 (1488)	3.7	-	5.0	5.0	-	10.8	6	-
	CRX Si	D15A3	1.5 (1488)	3.7	-	5.0	-	-	10.8	6	-
	Civic Wagon	D15A2	1.5 (1488)	3.7	-	5.0	6.0	-	11.9	6	-
	Civic Wagon 4WD	D15A2	1.5 (1488)	3.7	-	5.0	6.0	1.4	11.9	6	-
1988	Civic Std.	D15B1	1.5 (1493)	3.7	-	5.0	-	-	11.9	4.7	-
	Civic DX, LX	D15B2	1.5 (1493)	3.7	-	5.0	5.0	-	11.9 ②	4.8	4.7
	CRX HF	D15B6	1.5 (1493)	3.7	-	4.0	-	-	10.6	5.5	-
	CRX DX	D15B2	1.5 (1493)	3.7	-	5.0	5.0	-	11.9	5.8	5.7
	CRX Si	D16A6	1.6 (1590)	3.7	-	5.0	-	-	11.9	5.7	-
	Civic Si/CRX Si	D15B2	1.5 (1493)	3.7	-	5.0	-	-	11.9	5.8	-
	Civic Wagon, Wagovan	D16A6	1.6 (1590)	3.7	-	5.0	5.0	-	11.9	5.8	6.2
	Civic Wagon 4WD	D16A6	1.6 (1590)	3.7	-	5.0	5.0	1.4	11.9	5.8	6.2
1989	Civic Std.	D15B1	1.5 (1493)	3.7	-	5.0	-	-	11.9	5.8	-
	Civic DX, LX	D15B2	1.5 (1493)	3.7	-	5.0	5.0	-	11.9	4.8	4.7
	CRX HF	D15B6	1.5 (1493)	3.7	-	4.0	-	-	10.6	5.5	-
	CRX DX	D15B2	1.5 (1493)	3.7	-	5.0	5.0	-	11.9	5.8	5.7
	Civic Si/CRX Si	D16A6	1.6 (1590)	3.7	-	5.0	-	-	11.9	5.8	-
	Civic Wagon, Wagovan	D16A6	1.6 (1590)	3.7	-	5.0	5.0	-	11.9	5.8	6.2
	Civic Wagon 4WD	D16A6	1.6 (1590)	3.7	-	5.0	5.0	1.4	11.9	5.8	6.2
1990	Civic Std.	D15B1	1.5 (1493)	3.7	-	5.0	-	-	11.9	5.8	-
	Civic DX, LX	D15B2	1.5 (1493)	3.7	-	5.0	5.0	-	11.9	4.8	4.7
	CRX HF	D15B6	1.5 (1493)	3.7	-	4.0	-	-	10.6	5.5	-
	CRX DX	D15B2	1.5 (1493)	3.7	-	5.0	5.0	-	11.9	5.8	5.7
	Civic Si/CRX Si	D16A6	1.6 (1590)	3.7	-	5.0	-	-	11.9	5.8	-
	Civic Wagon, Wagovan	D15B2	1.5 (1493)	3.7	-	5.0	5.0	-	11.9	5.8	6.2
	Civic Wagon 4WD	D16A6	1.6 (1590)	3.7	-	5.0	5.0	1.4	11.9	5.8	6.2

CAPACITIES

Year	Model	Engine ID/VIN	Engine Displacement Liters (cc)	Engine Oil with Filter	Transmission Pts. 4-Spd	Transmission Pts. 5-Spd	Transmission Pts. Auto.	Rear Drive Axle (pts.) ①	Fuel Tank (gal.)	Cooling System (qts.) Manual	Cooling System (qts.) Auto.
1991	Civic Std.	D15B1	1.5 (1493)	3.7	-	-	-	-	11.9	5.8	-
	Civic DX, LX	D15B2	1.5 (1493)	3.7	-	5.0	5.0	-	11.9	4.8	4.7
	CRX HF	D15B6	1.5 (1493)	3.7	-	4.0	5.0	-	10.6	5.5	-
	CRX DX	D15B2	1.5 (1493)	3.7	-	5.0	5.0	-	11.9	5.8	5.7
	Civic Si/CRX Si	D16A6	1.6 (1590)	3.7	-	5.0	5.0	-	11.9	5.7	-
	Civic Wagon, Wagovan	D15B2	1.5 (1493)	3.7	-	5.0	5.0	-	11.9	5.8	6.2
	Civic Wagon 4WD	D16A6	1.6 (1590)	3.7	-	5.0	5.0	1.4	11.9	5.8	6.2
1992	Civic CX	D15B8	1.5 (1493)	3.5	-	4.0	5.6	-	11.9	4.8	4.6
	Civic DX, LX	D15B7	1.5 (1493)	3.5	-	4.0	5.6	-	11.9	4.8	4.6
	Civic VX	D15Z1	1.5 (1493)	3.5	-	4.0	-	-	11.9	4.8	-
	Civic Si, EX	D16Z6	1.6 (1590)	3.5	-	4.0	5.6	-	11.9	4.8	5
1993	Civic CX	D15B8	1.5 (1493)	3.5	-	4.0	5.6	-	11.9	4.8	4.6
	Civic DX, LX	D15B7	1.5 (1493)	3.5	-	4.0	5.6	-	11.9	4.8	4.6
	Civic VX	D15Z1	1.5 (1493)	3.5	-	4.0	-	-	11.9	4.8	-
	Civic Si, EX	D16Z6	1.6 (1590)	3.5	-	4.0	5.6	-	11.9	4.8	4.6
	del Sol S	D15B7	1.5 (1493)	3.5	-	4.0	5.6	-	11.9	4.8	4.6
	del Sol Si	D16Z6	1.6 (1590)	3.5	-	4.0	5.6	-	11.9	3.8	5
1994	Civic CX	D15B8	1.5 (1493)	3.5	-	3.8	5.6	-	11.9	3.8	3.7
	Civic DX, LX	D15B7	1.5 (1493)	3.5	-	4.0	5.6	-	11.9	4.8	4.6
	Civic VX	D15Z1	1.5 (1493)	3.5	-	4.0	-	-	11.9	4.8	-
	Civic Si, EX	D16Z6	1.6 (1590)	3.5	-	4.0	5.6	-	11.9	4.8	4.6
	del Sol S	D15B7	1.5 (1493)	3.5	-	4.0	5.6	-	11.9	4.8	4.6
	del Sol Si	D16Z6	1.6 (1590)	3.5	-	3.8	5.6	-	11.9	4.8	5
	del Sol VTEC	B16A3	1.6 (1595)	4.2	-	4.6	-	-	11.9	4.1	-
1995	Civic CX	D15B8	1.5 (1493)	3.5	-	3.8	5.8	-	11.9	4.8	4.6
	Civic DX, LX	D15B7	1.5 (1493)	3.5	-	3.8	-	-	11.9	4.8	4.6
	Civic VX	D15Z1	1.5 (1493)	3.5	-	3.8	-	-	11.9	4.8	-
	Civic Si, EX	D16Z6	1.6 (1590)	3.5	-	3.8	5.8	-	11.9	4.8	5
	del Sol S	D15B7	1.5 (1493)	3.5	-	4.0	5.6	-	11.9	-	-
	del Sol Si	D16Z6	1.6 (1590)	3.5	-	4.0	5.6	-	11.9	-	-
	del Sol VTEC	B16A3	1.6 (1595)	4.2	-	4.8	-	-	11.9	4.1	-

Note: all capacities are approximate. Fill gradually and check often.
① Does not include torque converter.
② Sedan: 12.1 gallons.
③ "Civic DX" applies to both hatchback and sedan models.

TUNE-UP PROCEDURES 2-2
SPARK PLUGS 2-2
  REMOVAL & INSTALLATION 2-3
  INSPECTION & CLEANING 2-4
SPARK PLUG WIRES 2-5
  TESTING & REPLACEMENT 2-5
FIRING ORDERS 2-5
**ELECTRONIC IGNITION
  (1984–87) 2-5**
DESCRIPTION AND OPERATION 2-5
  PRECAUTIONS 2-6
ADJUSTMENT 2-6
  RELUCTOR GAP 2-6
COMPONENT REPLACEMENT 2-6
  DISTRIBUTOR CAP 2-6
  ROTOR 2-6
  RELUCTOR 2-8
  PICK-UP COIL 2-8
  IGNITER UNIT 2-9
  VACUUM ADVANCER UNIT 2-9
  DISTRIBUTOR SHAFT 2-10
TROUBLESHOOTING 2-10
  SYSTEM TEST 2-10
  IGNITION COIL TEST 2-10
  IGNITER UNIT TEST 2-11
  PICK-UP COIL TEST 2-11
**PGM-IG IGNITION SYSTEM
  (1988–95) 2-12**
DESCRIPTION AND OPERATION 2-12
  PRECAUTIONS 2-13
COMPONENT REPLACEMENT 2-13
  IGNITION COIL 2-13
  IGNITER UNIT 2-13
  ROTOR 2-14
DIAGNOSIS AND TESTING 2-14
  IGNITER UNIT 2-14
  IGNITION COIL 2-14
IGNITION TIMING 2-15
GENERAL INFORMATION 2-15
CHECKING AND ADJUSTING
  TIMING 2-15
    1984–87 MODELS 2-16
    1988–91 MODELS 2-17
    1992–95 MODELS 2-17
VALVE LASH 2-18
ADJUSTMENT 2-18
IDLE SPEED AND MIXTURE 2-20
IDLE SPEED 2-20
  ADJUSTMENT 2-20
IDLE MIXTURE 2-22
  ADJUSTMENT 2-22
SPECIFICATIONS CHARTS
  ENGINE TUNE-UP
    SPECIFICATIONS 2-23

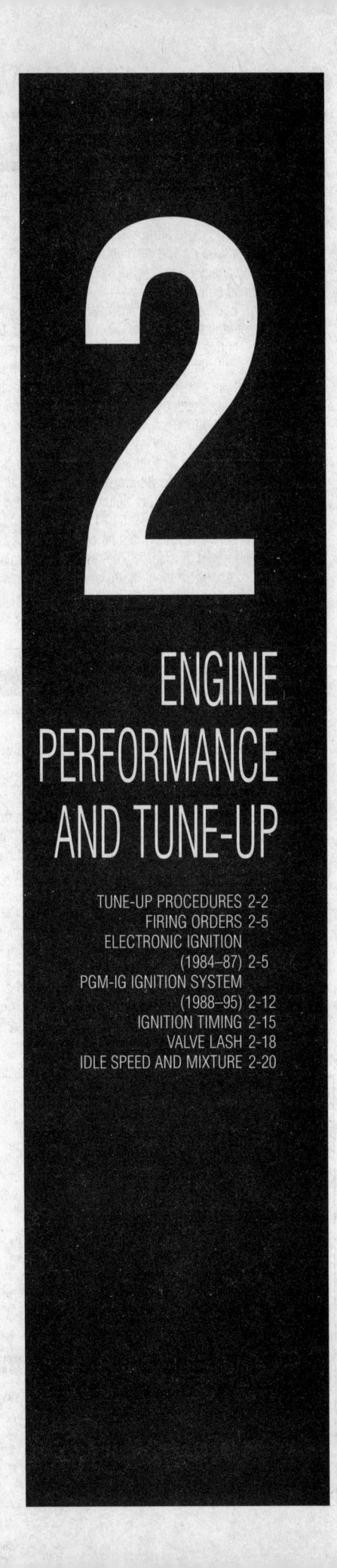

2

ENGINE PERFORMANCE AND TUNE-UP

TUNE-UP PROCEDURES 2-2
FIRING ORDERS 2-5
ELECTRONIC IGNITION
(1984–87) 2-5
PGM-IG IGNITION SYSTEM
(1988–95) 2-12
IGNITION TIMING 2-15
VALVE LASH 2-18
IDLE SPEED AND MIXTURE 2-20

2-2 ENGINE PERFORMANCE AND TUNE-UP

TUNE-UP PROCEDURES

In order to extract the best performance and economy from your engine it is essential that it be properly tuned at regular intervals. Although computerized engine controls and more durable components have reduced ignition maintenance, a regular tune-up will keep your Honda's engine running smoothly and will prevent the annoying minor breakdowns and poor performance associated with an un-tuned engine.

Federal law now requires that manufacturers certify that their vehicle's spark plugs will meet emission rules for 30,000 miles (48,300 km). The maintenance schedules reflect this, showing a mandatory replacement at this mileage. This is a minimum specification; for many vehicles, 30,000 miles (48,300 km) is too long. Plan on at least checking spark plugs and other ignition components at least once per year. Check them anytime a deterioration of engine performance is noted. Replaceable tune-up components are relatively inexpensive compared to the inconveniences of a poorly running vehicle or one that will not start.

This inspection interval should be halved if the car is operated under severe conditions, such as trailer towing, prolonged idling, continual stop and start driving, or if starting or running problems are noticed. It is assumed that the routine maintenance described in Section 1 has been kept up, as this will have a profound effect on the results of a tune-up. All of the applicable steps of a tune-up should be followed in order, as the result is a cumulative one.

If the specifications on the tune-up or emission label in the engine compartment of your Honda disagree with the tune-up specifications chart in this section, the figures on the sticker must be used. The sticker often reflects changes made during the production run.

Spark Plugs

◆ See Figures 1, 2, 3 and 4

➡ The spark plugs should be replaced every 30,000 miles (48,300 km).

Spark plugs ignite the air and fuel mixture in the cylinder as the piston reaches the top of the compression stroke. The controlled explosion that results forces the piston down, turning the crankshaft and the rest of the drive train.

The average life of a spark plug is dependent on a number of factors: the mechanical condition of the engine, the type of fuel, the driving conditions and the driver. Although the standard factory plugs will last a considerable period of time, extended life may be gained in some engines by using platinum-tipped plugs. You must decide if the benefits outweigh the extra cost.

When you remove the spark plugs, check their condition. They are a good indicator of the condition of the engine. A small deposit of light tan or gray material (or rusty red with some fuels) on a spark plug that has been used for any period of time is to be considered normal. Any other color, or abnormal amounts of deposit, indicates that there is something amiss in the engine.

The gap between the center electrode and the side or ground electrode can be expected to increase very slightly under normal conditions.

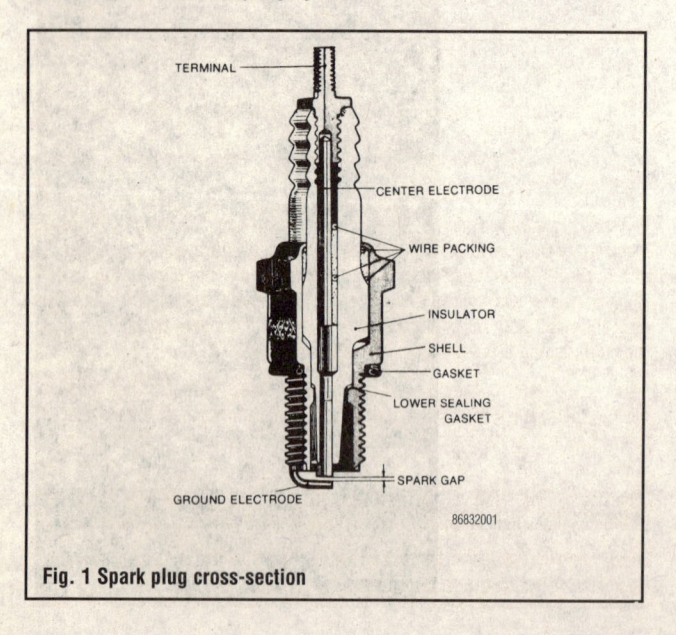

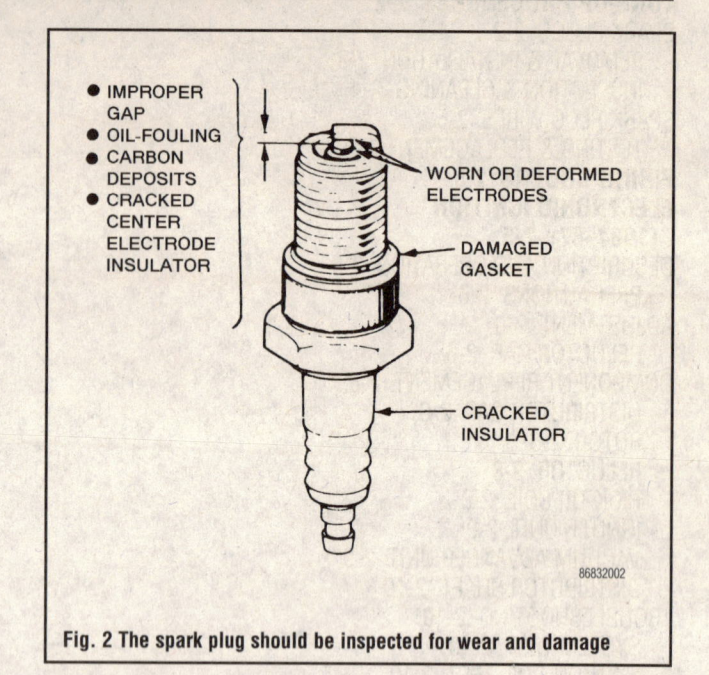

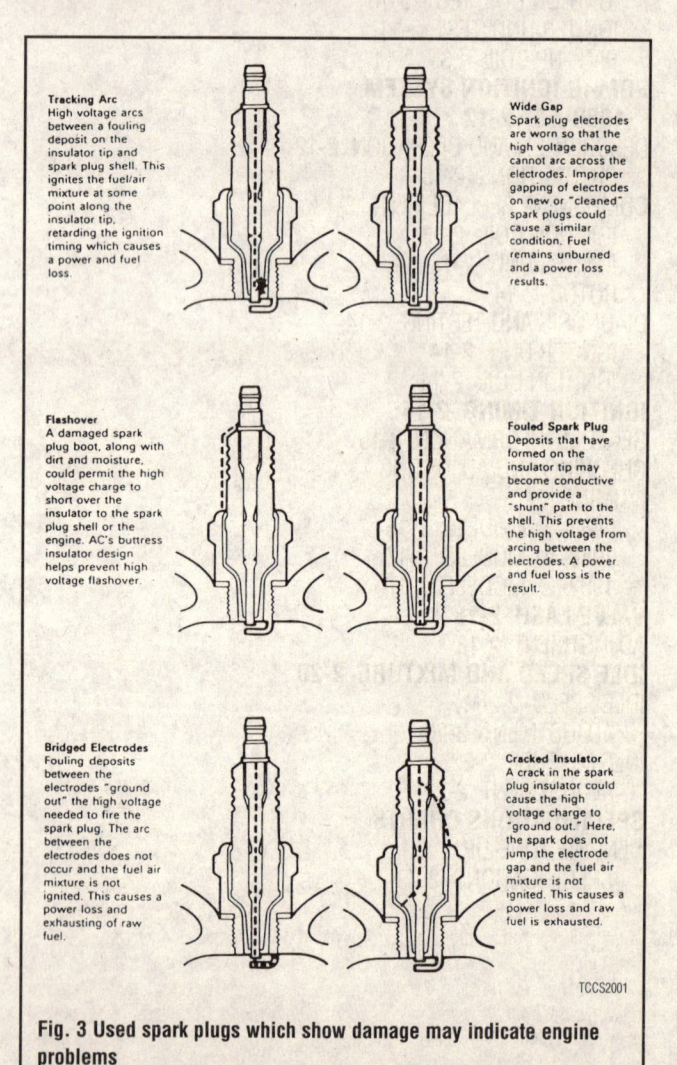

Fig. 1 Spark plug cross-section

Fig. 2 The spark plug should be inspected for wear and damage

Fig. 3 Used spark plugs which show damage may indicate engine problems

ENGINE PERFORMANCE AND TUNE-UP 2-3

A normally worn spark plug should have light tan or gray deposits on the firing tip.

A carbon fouled plug, identified by soft, sooty, black deposits, may indicate an improperly tuned vehicle. Check the air cleaner, ignition components and engine control system.

This spark plug has been **left in the engine too long**, as evidenced by the extreme gap. Plugs with such an extreme gap can cause misfiring and stumbling accompanied by a noticeable lack of power.

An oil fouled spark plug indicates an engine with worn poston rings and/or bad valve seals allowing excessive oil to enter the chamber.

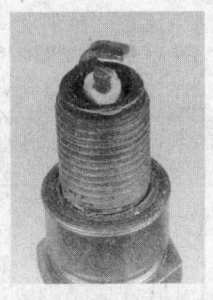

A physically damaged spark plug may be evidence of severe detonation in that cylinder. Watch that cylinder carefully between services, as a continued detonation will not only damage the plug, but could also damage the engine.

A bridged or almost bridged spark plug identified by a build-up between the electrodes caused by excessive carbon or oil build-up on the plug

Fig. 4 Inspect the spark plug to determine engine running conditions

When a spark plug is functioning normally or, more accurately, when the plug is installed in an engine that is functioning properly, the plugs can be taken out, cleaned, re-gapped, and reinstalled in the engine without doing the engine any harm. This is acceptable as an improvement or emergency measure, but new plugs are always recommended.

When, and if, a plug fouls and begins to misfire, you will have to investigate, correct the cause of the fouling, and either clean or replace the plug. Replacement is always recommended if possible.

Spark plugs suitable for use in your engine are offered in different heat ranges. The amount of heat which the plug absorbs is determined by the length of the lower insulator. The longer the insulator the hotter the plug will operate; the shorter the insulator, the cooler it will operate. A spark plug that absorbs little heat and remains too cool will accumulate deposits of lead, oil, and carbon, because it is not hot enough to burn them off. This leads to fouling and consequent misfiring.

A spark plug that absorbs too much heat will have no deposits, but the electrodes will burn away quickly and, in some cases, preignition may result. Preignition occurs when the spark plug tips get so hot that they ignite the air/fuel mixture before the actual spark fires. This premature ignition will usually cause a pinging sound under conditions of low speed and heavy load. In severe cases, the heat may become high enough to start the air/fuel mixture burning throughout the combustion chamber rather than just to the front of the plug. In this case, the resultant explosion will be strong enough to damage pistons, rings, and valves.

In most cases the factory recommended heat range is correct; it is chosen to perform well under a wide range of operating conditions. However, if most of your driving is long distance, high speed travel, you may want to install a spark plug one step colder than standard. If most of your driving is of the short trip variety, when the engine may not always reach operating temperature, a hotter plug may help burn off the deposits normally accumulated under those conditions.

REMOVAL & INSTALLATION

▶ See Figures 5 and 6

1. Place a piece of masking tape around each spark plug wire and number it according to its corresponding cylinder.
2. Pull the wires from the spark plugs, grasping the wire by the end of the rubber boot and twisting off.

➡**Avoid spark plug removal while the engine is hot. Since the cylinder head spark plug threads are aluminum, the spark plug becomes tight due to the different coefficients of heat expansion. If a plug is too tight to be removed even while the engine is cold, apply a solvent around the plug followed with an application of oil once the solvent has penetrated the threads. Do this only when the engine is cold.**

3. Loosen each spark plug with a spark plug socket. When the plug has been loosened a few turns, stop to clean any material from around the spark plug holes; compressed air is preferred. If air is not available, simply use a rag to clean the area.

✳✳ WARNING

In no case should foreign matter be allowed to enter the cylinders. Severe engine damage could result.

2-4 ENGINE PERFORMANCE AND TUNE-UP

4. Remove and inspect the spark plugs; if necessary, clean them.

To install:

5. Check the gap of the plugs with a feeler gauge. Apply an anti-seize compound to the spark plug threads and hand tighten them into the cylinder head. Torque the spark plugs into the cylinder head to 13 ft. lbs. (18 Nm).

✲✲ WARNING

Do not over-tighten the spark plugs, as this could damage the aluminum cylinder heads.

6. Connect the wires to the plugs, in the correct order, making sure that each is securely fitted.

INSPECTION & CLEANING

▶ See Figures 7, 8, 9, 10 and 11

Check the plugs for deposits and wear. If they are not going to be replaced, clean the plugs thoroughly. This may be done with very fine sandpaper or a small flat file. Do not use a wire brush. Remember that any kind of deposit will decrease the efficiency of the plug. Plugs can be cleaned on a spark plug cleaning machine, which can sometimes be found in service stations. If the plugs are cleaned, the electrodes must be filed flat. Use an ignition points file, not an emery board or the like, which will leave deposits. The electrodes must be filed perfectly flat with sharp edges; rounded edges reduce the spark plug voltage by as much as 50 percent.

Check the spark plug gap before installation. The ground electrode (the L-shaped one connected to the body of the plug) must be parallel to the center electrode and the specified-size wire gauge must fit in the gap with slight resistance.

➡ **NEVER adjust the gap on a platinum-tipped spark plug.**

Always check the gap on new plugs too. NEVER rely on so-called "pre-gapped" plugs. Do not use a flat feeler gauge when measuring the gap; the reading will be inaccurate. Wire-type plug gapping tools usually have a bending

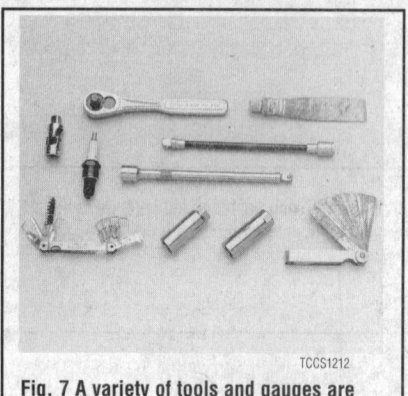

Fig. 5 A rubber insert in the socket holds the plug securely

Fig. 6 This plug can be cleaned and re-used

Fig. 7 A variety of tools and gauges are needed for spark plug service

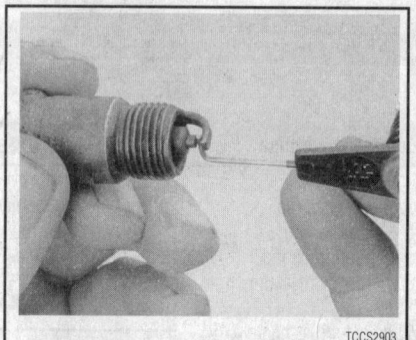

Fig. 8 Checking the spark plug gap with a feeler gauge—the gauge should drag lightly between the electrodes

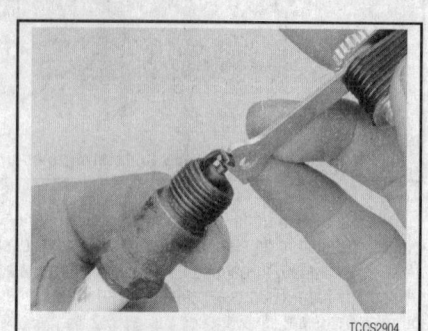

Fig. 9 Adjust the spark plug gap by bending the side electrode slightly—the adjustments you will make are only in the thousandths of an inch

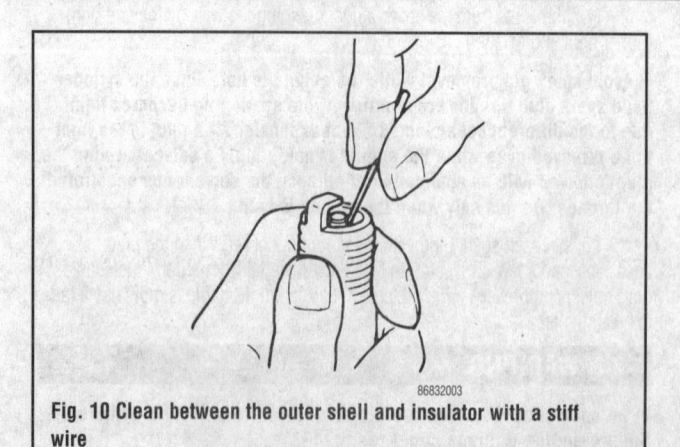

Fig. 10 Clean between the outer shell and insulator with a stiff wire

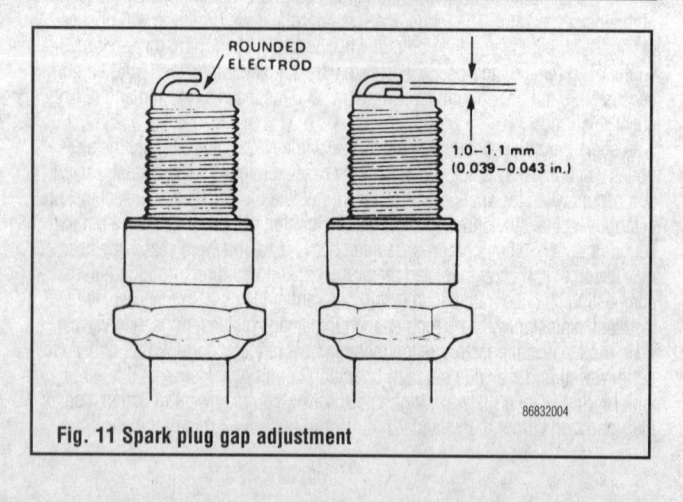

Fig. 11 Spark plug gap adjustment

ENGINE PERFORMANCE AND TUNE-UP 2-5

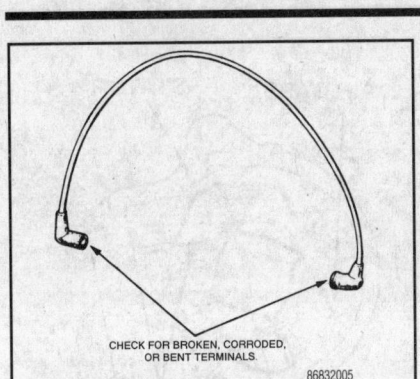

Fig. 12 Check the spark plug wires for breaks, cracked shielding or corroded terminals

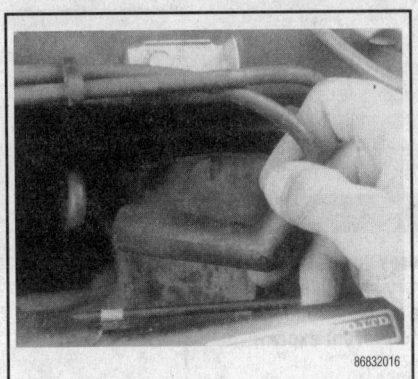

Fig. 13 Always grasp the boot of the wire, not the wire itself

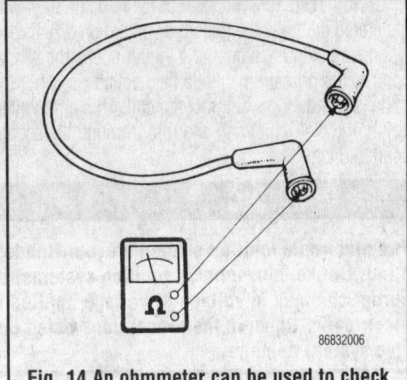

Fig. 14 An ohmmeter can be used to check the resistance of the spark plug wire

tool attached. Use that to adjust the side electrode until the proper distance is obtained. Absolutely never bend the center electrode. Also, be careful not to bend the side electrode too far or too often. It may weaken and break off within the engine, requiring removal of the cylinder head to retrieve it.

Spark Plug Wires

TESTING & REPLACEMENT

▶ See Figures 12, 13 and 14

At every tune-up, visually inspect the spark plug cables for burns, cuts, or breaks in the insulation. Check the boots and the nipples on the distributor cap and coil. Replace any damaged wiring. Always replace spark plug wiring in sets, with a coil wire as well. Length is important; get the correct set for your vehicle.

Every 36,000 miles (58,000 km) or so, the resistance of the wires should be checked with an ohmmeter. Wires with excessive resistance will cause misfiring, and may make the engine difficult to start in damp weather. Generally, the useful life of the cables is 36,000–50,000 miles (58,000–80,000 km).

To check resistance, remove the wire from the plug and the distributor cap. Look at each contact inside the wire for any sign of cracking or burning. A small amount of discoloration is normal but there should be no heavy burn marks. Connect one lead of an ohmmeter to each end of the cable. Replace any wire which shows a resistance over 25,000 ohms.

Test the high tension lead from the coil in the same fashion. If resistance is more than 25,000 ohms, replace the cable. It should be remembered that resistance is also a function of length; the longer the cable, the greater the resistance. Thus, if the cables on your car are longer than the factory originals, resistance will be higher, quite possibly outside these limits. Honda recommends the 25,000 ohm limit be observed in all cases.

When installing new cables, replace them one at a time to avoid mix-ups. Start by replacing the longest one first. Install the boot firmly over the spark plug. Route the wire over the same path as the original. Insert the nipple firmly into the tower on the cap or the coil. Make certain each cable is replaced in any holding or retaining clips along the route.

FIRING ORDERS

▶ See Figures 15 and 16

➡ To avoid confusion, remove and tag the wires one at a time.

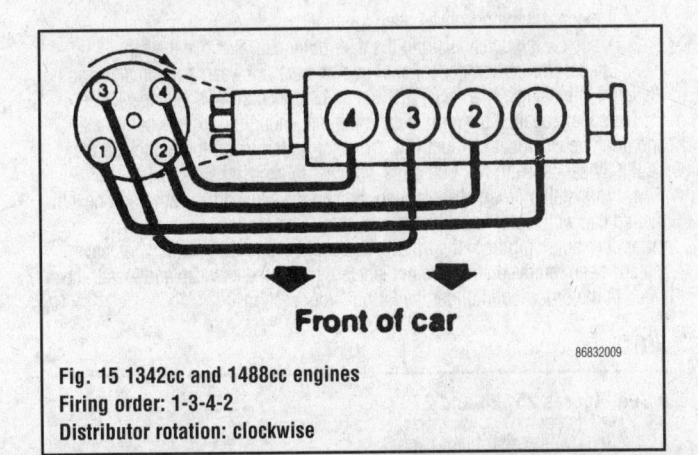

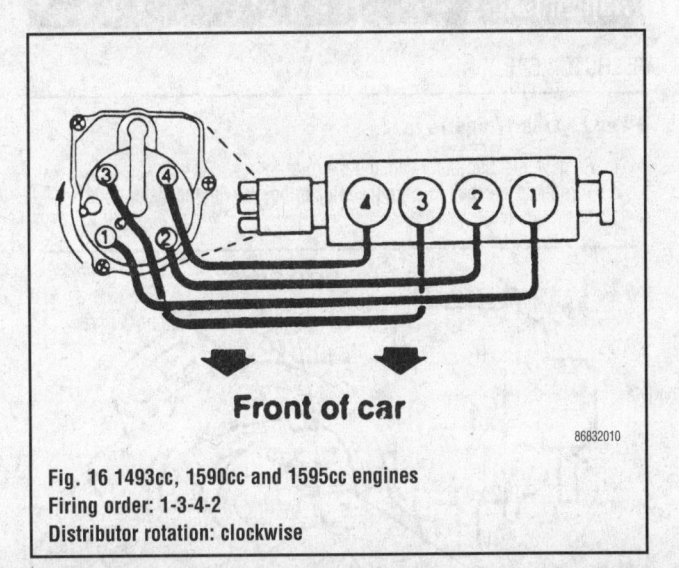

Fig. 15 1342cc and 1488cc engines
Firing order: 1-3-4-2
Distributor rotation: clockwise

Fig. 16 1493cc, 1590cc and 1595cc engines
Firing order: 1-3-4-2
Distributor rotation: clockwise

ELECTRONIC IGNITION (1984–87)

Description and Operation

All 1984–87 vehicles are equipped with a magnetic pulse-type electronic ignition system. This system eliminates the points and condenser; it requires no periodic maintenance.

The electronic ignition system uses a magnetic pulse/igniter distributor and a conventional ignition coil. The distributor cap, rotor, advance mechanism (vacuum and centrifugal) and secondary ignition wires are also of conventional design. The distributor contains the stator, reluctor and pulse generator (pick-up coil) and igniter assembly.

2-6 ENGINE PERFORMANCE AND TUNE-UP

During operation, the teeth of the reluctor align with the stator, a signal is generated by the pulse generator (pick-up coil) and sent to the igniter unit. Upon receiving the signal, the igniter opens the primary circuit of the ignition coil. As the primary magnetic field collapses, a high voltage surge is developed in the secondary windings of the coil. This high voltage surge travels from the coil to the distributor cap and rotor through the secondary ignition wires to the spark plugs.

** WARNING

The electronic ignition system on your Honda requires special handling. Unlike conventional ignition systems, it is very sensitive to abrupt changes in voltage or voltage applied in the wrong direction electrically. Observe the precautions listed below to prevent expensive system damage.

PRECAUTIONS

- Always disconnect the battery cables before doing repair work on the electronic ignition system.
- Always double check the markings on the battery and the routing of the cables before making connections, especially if the battery has been removed and might have been reinstalled in the opposite position. Hooking up the battery connections backwards will cause current to flow through the electronic ignition system in an improper way and may immediately damage it. Be careful when jumping the vehicle's battery for the same reasons.
- Do not allow the wires connected to the pulse generator to touch other ignition wiring connections.
- Abnormal voltage pulses may damage the system. Therefore, be sure to disconnect the battery before doing any work on the vehicle that involves the electrical system. This includes charging the battery and replacing small bulbs.
- Connect any electrical tachometer to the negative (−) terminal of the ignition coil—not to any other connection.
- Always double check any connection (you are making) involving the ignition system before reconnecting the battery and putting the system into operation.
- When cranking the engine for compression testing or similar purposes, disconnect the coil wire at the distributor.

Adjustment

RELUCTOR GAP

▶ See Figures 17 and 18

1. Remove the distributor cap and the rotor.
2. Turn the crankshaft to align the reluctor points with the stator ends.

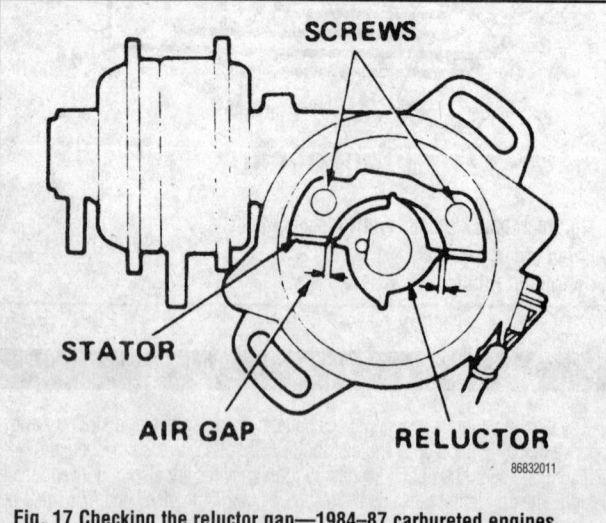

Fig. 17 Checking the reluctor gap—1984–87 carbureted engines

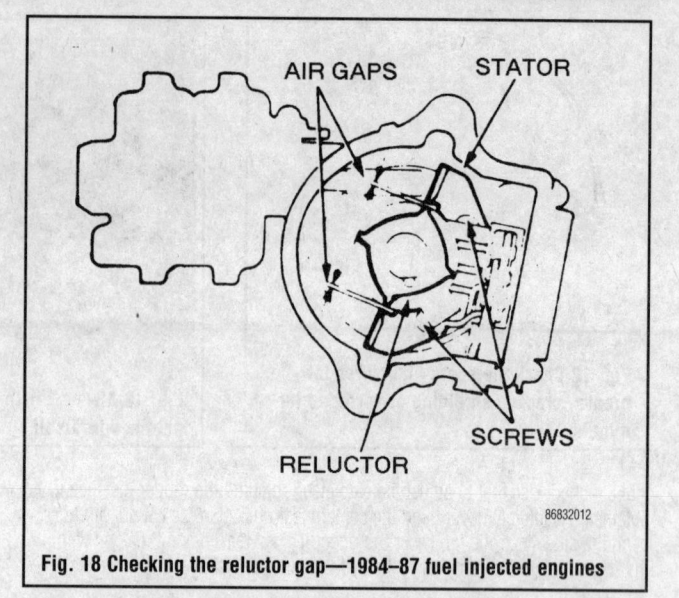

Fig. 18 Checking the reluctor gap—1984–87 fuel injected engines

3. Using a non-metallic feeler gauge, check the air reluctor-to-stator air gaps; they must be equal.
4. To adjust, loosen the stator-to-distributor screws, adjust the stator-to-reluctor air gaps and tighten the screws.
5. Recheck the air gaps.

Component Replacement

▶ See Figures 19 and 20

It is customary to replace the distributor cap and rotor at the tune-up interval or when replacing spark plug wires as part of a conservative maintenance policy. However, if these parts are in good shape, you may choose to clean and reuse. If in any doubt, replace them. The cap and rotor are vital ignition system components and inexpensive enough to warrant replacement rather than trying to make worn parts last longer.

DISTRIBUTOR CAP

▶ See Figures 21, 22, 23 and 24

1. Disconnect the negative battery terminal.
2. Unscrew the bolts securing the distributor cap. Remove the cap.
3. Inspect for carbon deposits, pitting or cracks. If wear is minimal, clean up the terminals using a fine metal file (not an emery board or sandpaper). Keep the file square against the terminal and file just enough to expose bare metal and improve electrical conductivity. The rotor button in the top of the cap may be lightly sanded with No. 600 grit sandpaper to expose bare metal.
4. To install a new distributor cap, line up the new part in your hand next to the old cap which still has the wires attached. One-by-one, swap the wires in order corresponding from their place on the old cap to the new. If wires have already been removed and were not labeled, install the new cap and install the wires in the order found under the Firing Order section.

ROTOR

▶ See Figures 25, 26 and 27

1. Remove the distributor cap.
2. Pull the rotor from the distributor shaft.
3. Inspect the rotor for burns and damage; if necessary, replace it. It may be necessary to lightly file the tip with a very fine metal file (not an emery board or sandpaper) so as to keep the edge square and to expose bare metal.
4. To install, apply a light coat of silicone grease to the rotor tip, align the rotor with the distributor shaft and push it into place.
5. Secure the distributor cap back in place.

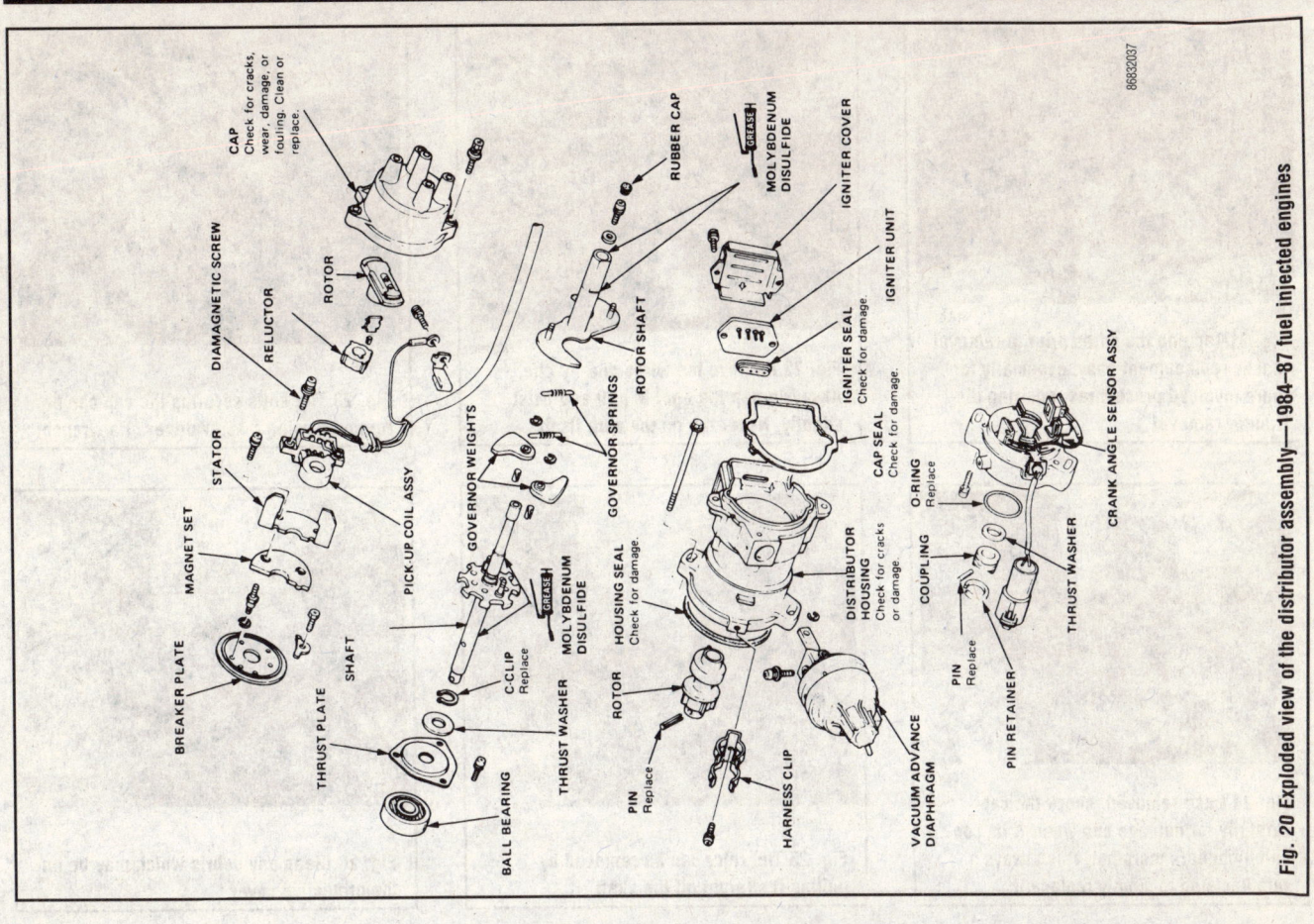

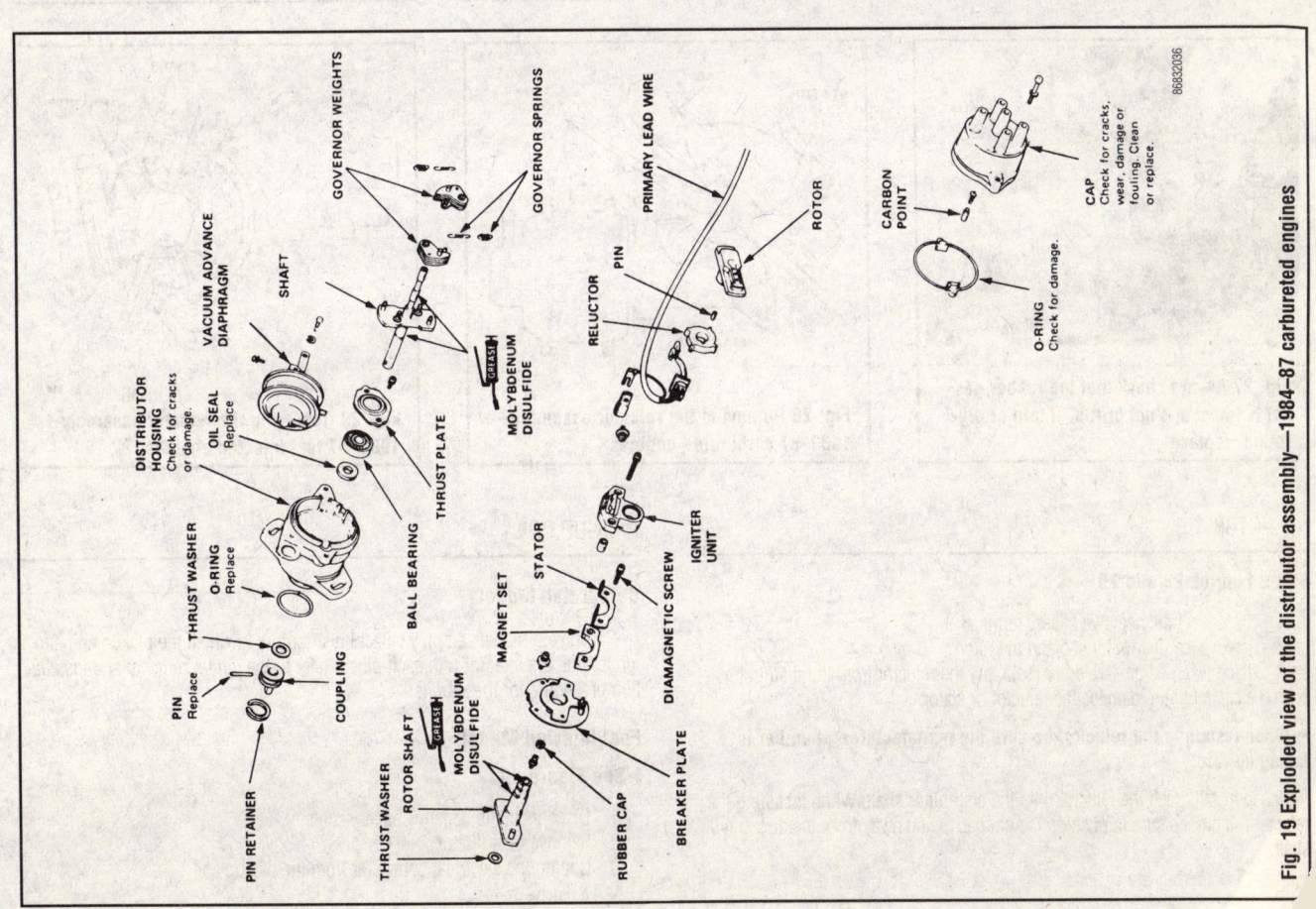

ENGINE PERFORMANCE AND TUNE-UP   2-7

Fig. 20 Exploded view of the distributor assembly—1984-87 fuel injected engines

Fig. 19 Exploded view of the distributor assembly—1984-87 carbureted engines

2-8 ENGINE PERFORMANCE AND TUNE-UP

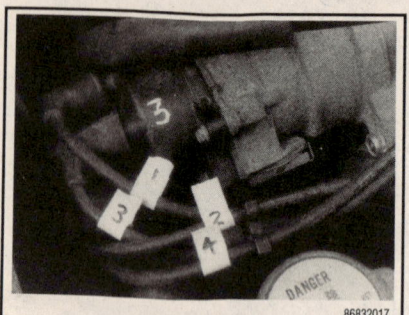

Fig. 21 Tagging the wires prior to removal makes replacement easy, especially for more involved procedures requiring distributor removal

Fig. 22 Remove the wires one-by-one. Always grasp the boot to pull and twist slightly. Never tug on the wire itself

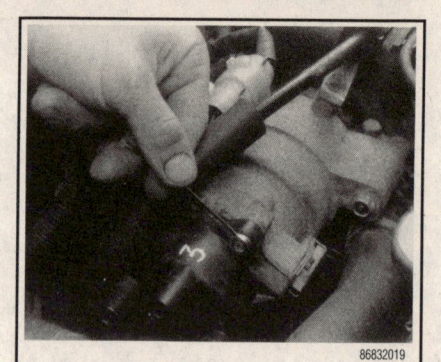

Fig. 23 The bolts securing the cap can be removed using a screwdriver or a wrench

Fig. 24 Once removed, check the cap carefully for damage and wear. If its condition appears marginal, it is always a safe decision to simply replace it

Fig. 25 The rotor can be removed by pulling it straight off the shaft

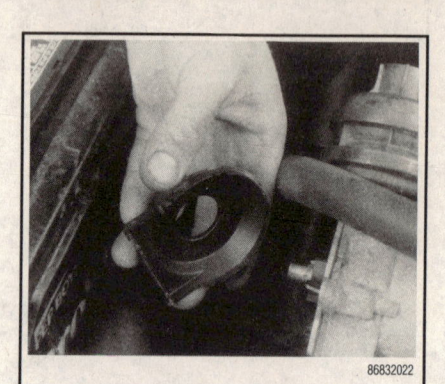

Fig. 26 Clean any debris which may be on the protective cover

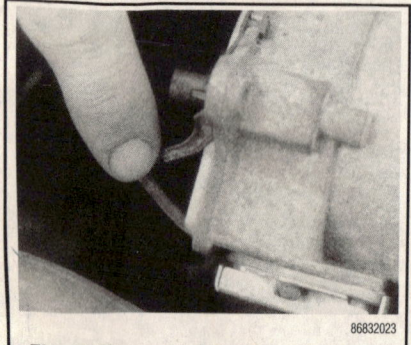

Fig. 27 Always check that the rubber gasket is intact and not brittle. If torn or dry-rotted, replace

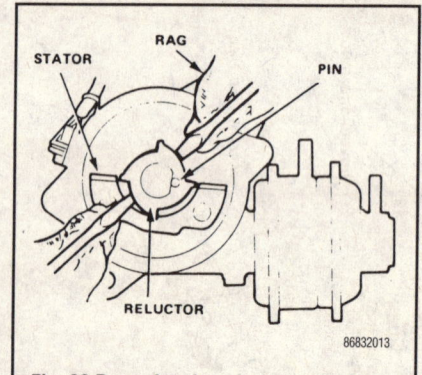

Fig. 28 Removing the reluctor assembly—1984–87 carbureted engines

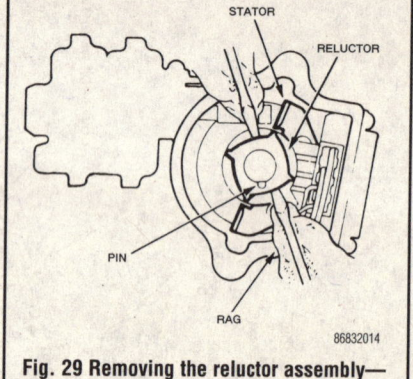

Fig. 29 Removing the reluctor assembly—1985–87 fuel injected engines

RELUCTOR

▶ See Figures 28 and 29

1. Disconnect the negative battery terminal.
2. Remove the distributor cap and the rotor.
3. Using two medium-sized pry tools, pry the reluctor from the distributor shaft; be careful not to damage the reluctor or stator.

➥When installing the reluctor, be sure the manufacturer's number is facing upward.

4. To install, push the reluctor onto the distributor shaft. When installing the rotor roll pin, be sure to position the pin gap facing away from the distributor

PICK-UP COIL

Carbureted Models

Carbureted models employ Hitachi distributors in which the pickup coil and igniter unit are integral with each other. Refer to the igniter removal and installation procedure for these units.

Fuel Injected Models

▶ See Figure 30

1. Disconnect the negative battery cable.
2. Remove the reluctor.
3. Unplug the electrical connector from the pickup coil.

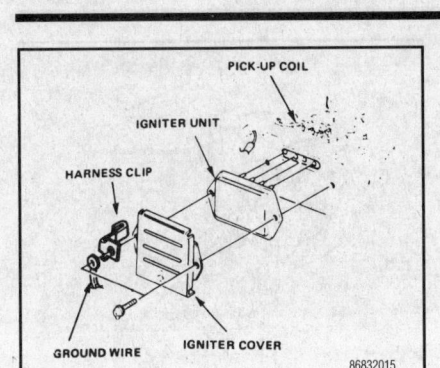

ENGINE PERFORMANCE AND TUNE-UP 2-9

Fig. 30 Removing the igniter unit from the distributor housing—1985–87 fuel injected engines

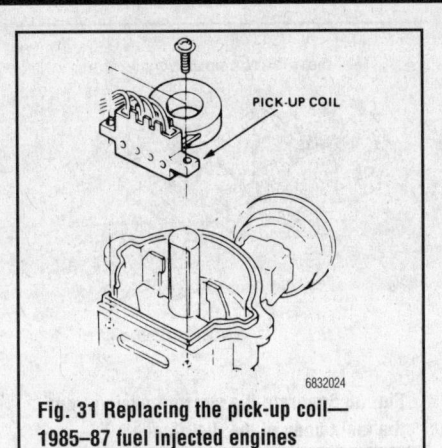

Fig. 31 Replacing the pick-up coil—1985–87 fuel injected engines

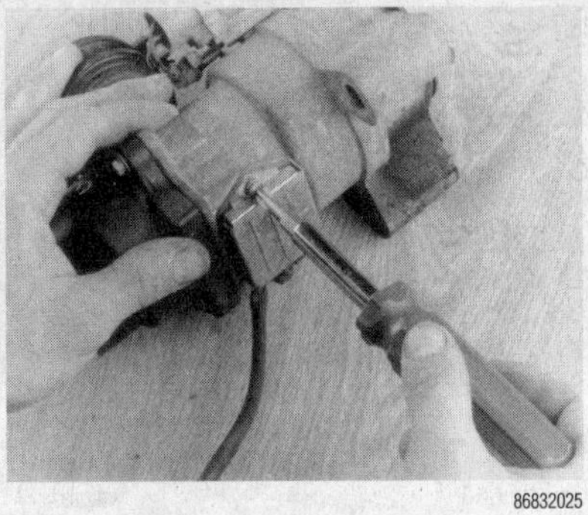

Fig. 32 Remove the screws securing the igniter cover

Fig. 33 With the screws removed, the igniter unit can be unscrewed

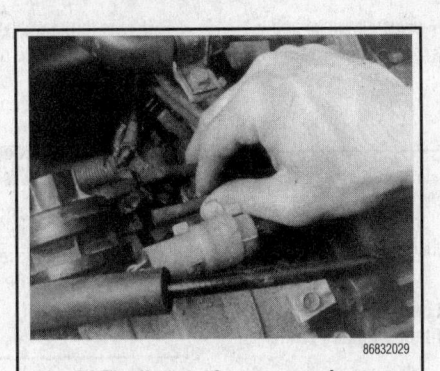

Fig. 34 The lines to the vacuum advance unit should be labeled or marked to prevent improper routing when re-installing

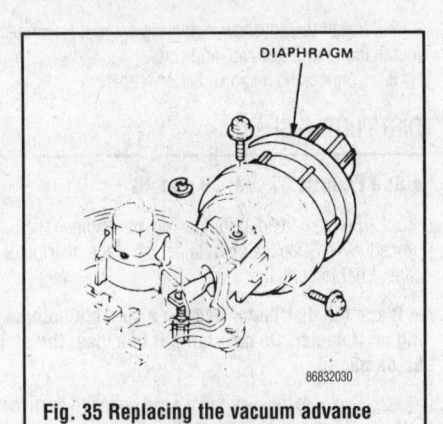

Fig. 35 Replacing the vacuum advance assembly—1985–87 fuel injected engines

4. Remove the pickup coil-to-distributor screws and pull the pickup coil from the distributor.
5. To install, position the pickup coil on the distributor and secure it with the hold-down screws. Install the remaining components in the reverse order of removal.

IGNITER UNIT

Carbureted Models

➡ The igniter is internally mounted in the distributor.

1. Remove the distributor cap, rotor and reluctor.
2. Unplug the electrical connections.
3. The igniter may be removed by removing the retaining screws. Note the position of the retaining screws; one is diamagnetic and must be reinstalled in the original location.

To install:

4. Position the igniter in the distributor, then tighten the retaining screws until snug. Make sure the diamagnetic screw is installed in its original location.
5. Engage the electrical connections.
6. Install the reluctor, then check the air gap.
7. Install the rotor and distributor cap.

Fuel Injected Models

▶ See Figures 31, 32 and 33

On the 1985–87 fuel injected engines, Honda introduced an igniter unit which is installed on the side of the distributor housing

1. Disconnect the negative battery terminal.
2. Remove the igniter cover-to-distributor screws and the cover.
3. Pull the igniter unit from the distributor.
4. If necessary, perform the igniter unit test procedures in this section.
5. Using silicone grease, apply it to the connector housing.

6. To install, place the igniter back into the distributor and screw on the cover.

VACUUM ADVANCER UNIT

▶ See Figures 34, 35 and 36

1. Disconnect the negative battery cable.
2. Remove the distributor cap and the rotor.
3. Label and disconnect the two vacuum lines from the advancer unit.
4. Remove the three advancer mounting screws.
5. Pull the advancer unit from the distributor housing, disengage the diaphragm arm carefully from inside the distributor housing.

To install:

6. Install the advancer unit on the housing and engage the diaphragm arm into the housing.

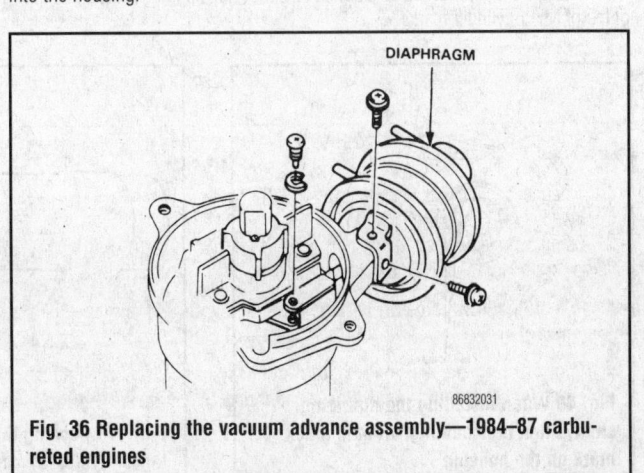

Fig. 36 Replacing the vacuum advance assembly—1984–87 carbureted engines

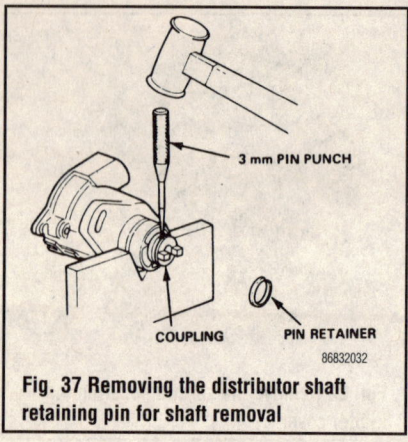

2-10 ENGINE PERFORMANCE AND TUNE-UP

Fig. 37 Removing the distributor shaft retaining pin for shaft removal

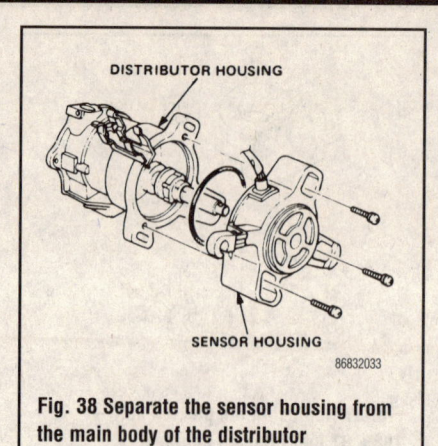

Fig. 38 Separate the sensor housing from the main body of the distributor

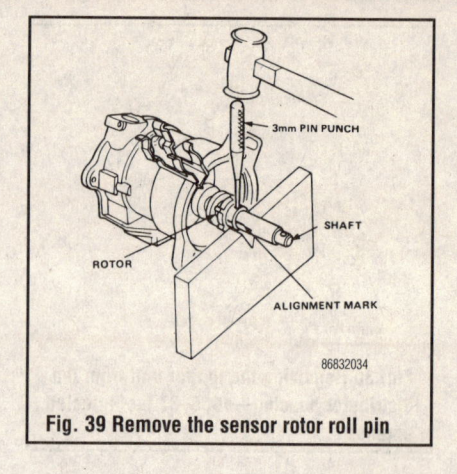

Fig. 39 Remove the sensor rotor roll pin

7. Install the advancer mounting screws. Reconnect the vacuum lines and install the distributor cap, and rotor.
8. Connect the negative battery cable.

DISTRIBUTOR SHAFT

◆ See Figures 37, 38, 39 and 40

1. Remove the distributor from the engine (see Section 3 for the removal procedures). Slide the pin retainer off of the distributor shaft coupling, being careful not to stretch it.

➡Place the distributor body in a suitable holding fixture to make working on it easier. Do not clamp it in a vise, the aluminum housing could be damaged.

2. Separate the distributor shaft coupling from the shaft by removing the roll pin with a suitable punch. Save the thrust washer (if equipped) for later use.
3. If necessary, separate the sensor housing from the distributor housing by removing the four screws. Do not remove the sensors from housing. Be careful not to damage the sensor housing seal.
4. Drive out the rotor's roll pin after scribing an alignment mark on the rotor and shaft.
5. Carefully pry up on the rotor by using suitable pry tools. Be careful to not damage the rotor.
6. Remove the distributor shaft from the housing and replace the parts as necessary. When removing the internal distributor components, be careful not to loose any of the small springs or clips.

To install:

7. To install shaft, reverse the removal procedure. When installing the sensor housing, be sure that the housing seal is correctly positioned. If the seal is damaged it must be replaced.

➡Always use new roll pins when reassembling the distributor assembly.

8. Install the rotor on the shaft, then turn the shaft so it faces the direction of the of No. 1 cylinder mark.

9. Set thrust washer and coupling on the shaft.
10. Check that the rotor is still pointing toward No. 1 cylinder, then align the index mark made on the housing with index mark on the coupling.
11. Drive in a new pin and secure it with a pin retainer.

Troubleshooting

◆ See Figures 41 and 42

➡Most test procedures for the electronic ignition system require the use of a digital volt/ohmmeter (DVOM).

SYSTEM TEST

1. First carefully pull the coil wire out of the coil. Then, hold the wire so that the metal parts of the coil tower and wire are ¼ in. (6mm) apart.
2. Have an assistant crank the engine. If there is a bright blue spark, the electronic ignition system is working, proceed with routine checks for spark at the plugs and inspection of the ignition wires, cap and rotor to be certain the rest of the ignition system is operable. If there is a bright blue spark, or if there is no spark or if the spark is weak yellow or orange spark, proceed with the component tests.

IGNITION COIL TEST

1. With the ignition switch still **ON**, switch the multimeter into the 20V range. Connect the ground (black) lead of the meter to a good, clean ground (bare metal, not covered with paint). Connect the positive lead to the positive (+) terminal of the coil. The voltage should be approximately 12V. If not, look for problems in the wiring connector at the coil or somewhere in the wiring between the ignition switch and the coil.
2. Turn the ignition switch **OFF**. Set your meter to read resistance. Choose a scale that will measure the resistance specified.
3. On 1986–87 models, disconnect the coil high tension wire and the two primary connectors.

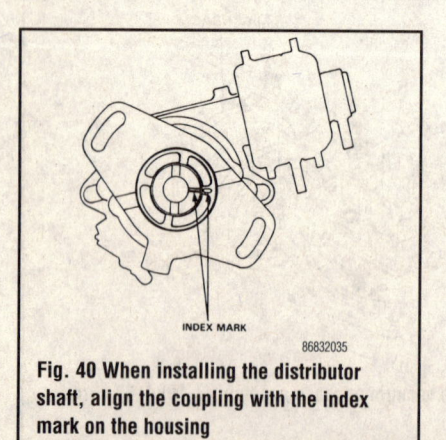

Fig. 40 When installing the distributor shaft, align the coupling with the index mark on the housing

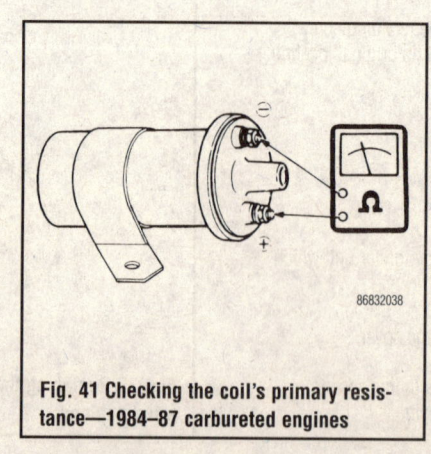

Fig. 41 Checking the coil's primary resistance—1984–87 carbureted engines

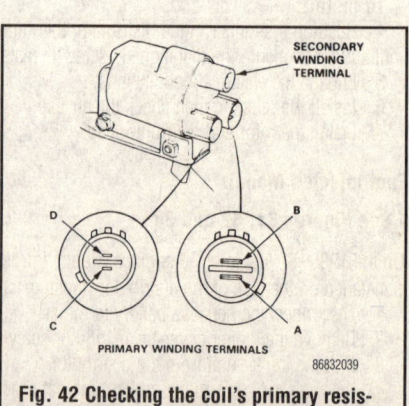

Fig. 42 Checking the coil's primary resistance—1984–87 carbureted engines

ENGINE PERFORMANCE AND TUNE-UP   2-11

4. Connect the two meter probes to the two primary (small) terminals of the coil. On 1984–85 models, these are the 2 small connectors that use nuts to retain the coil wires. On 1986–87 models, these are marked **A** and **D** in the illustration.

5. Measure the resistance; it must be 0.84–1.02 ohms (1984–85) or 1,215–1,485 ohms (1986–87).

➡ **The resistance specifications given are for an ambient temperature of 70°F (21°C). It is advisable to conduct this test in the same approximate temperature. If you are unable, slight variances in resistance will be normal due to the temperature difference.**

6. Now measure the secondary resistance. Connect the meter probes between the (**+**) primary terminal and the coil tower on 1984–85 models. On 1986–87 models, connect the ohmmeter between the A terminal and the coil tower.

7. This resistance must be 8,000–12,000 ohms on 1984–85 models. On 1986–87 models, it must be 11,074–11,526 ohms.

8. On 1986–87 coils, also measure the resistance between the **B** and **D** terminals. It should be approximately 2,200 ohms.

9. If any test is failed, replace the coil.

IGNITER UNIT TEST

Carbureted Engines

♦ See Figures 43, 44, 45 and 46

1. Disconnect the wires from the igniter unit.
2. Using a voltmeter, check the voltage between blue wire and body ground, then, black/yellow wire to body ground. The ignition switch must be **ON**. There should be battery voltage.
3. If battery voltage is not present, check wiring to the igniter unit.
4. Using an ohmmeter, set the scale on the R x 100 position.
5. Disconnect the lead wires.

6. Place the positive (red) probe on the black/yellow wire terminal and negative (black) probe on the blue wire terminal; no continuity should be read on the ohmmeter.

7. Place the positive (red) probe on the blue wire terminal and negative (black) probe on the black/yellow wire terminal; continuity should be read on the ohmmeter.

8. If the continuity readings are not as specified, replace the igniter unit.

Fuel Injected Engines

1986 MODELS

1. Disconnect lead wires from the igniter unit.
2. Using a voltmeter, check the voltage between the blue wire and ground, then, the black/yellow wire to ground, with the ignition switch **ON**; there should be battery voltage.
3. If no voltage, check wiring to igniter unit.

1987 MODELS

1. Remove igniter cover and pull out the igniter unit.
2. Using a voltmeter, check the voltage between the BLU terminal and ground, then, the BL/Y wire terminal and ground, turn the ignition switch **ON**; battery voltage should be present.

PICK-UP COIL TEST

♦ See Figures 47, 48 and 49

Fuel Injected Engines

1985 MODELS

1. Using an ohmmeter, measure the resistance between the blue and green wire terminals and measure resistance. The resistance should be approximately 750 ohms at 70°F (21°C).
2. If not within specifications, replace the primary coil.

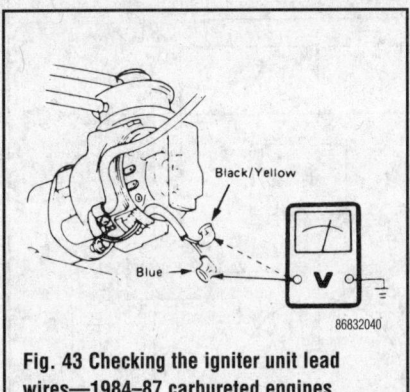

Fig. 43 Checking the igniter unit lead wires—1984–87 carbureted engines (Hitachi)

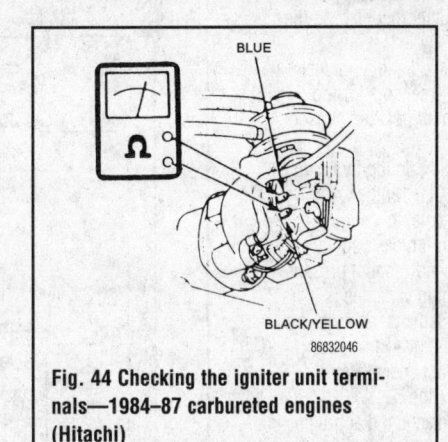

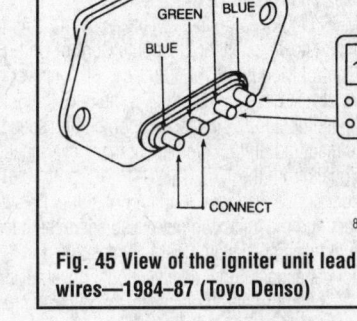

Fig. 44 Checking the igniter unit terminals—1984–87 carbureted engines (Hitachi)

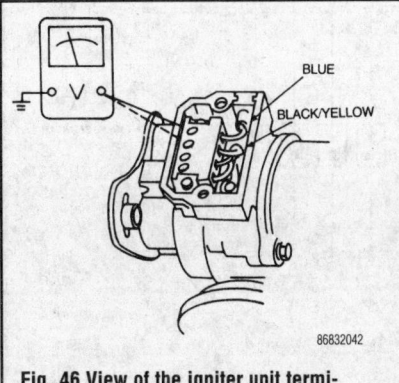

Fig. 45 View of the igniter unit lead wires—1984–87 (Toyo Denso)

Fig. 46 View of the igniter unit terminals—1984–87

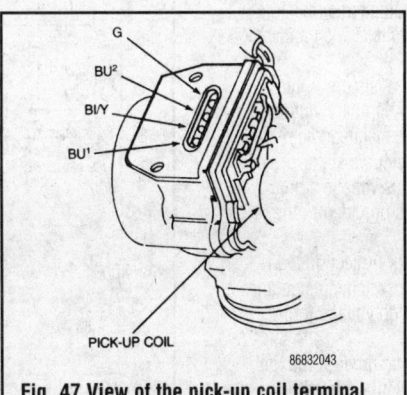

Fig. 47 View of the pick-up coil terminal locations

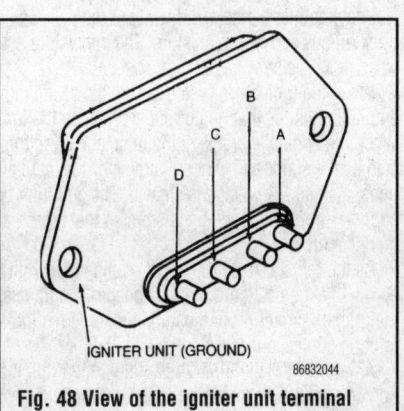

Fig. 48 View of the igniter unit terminal locations

2-12 ENGINE PERFORMANCE AND TUNE-UP

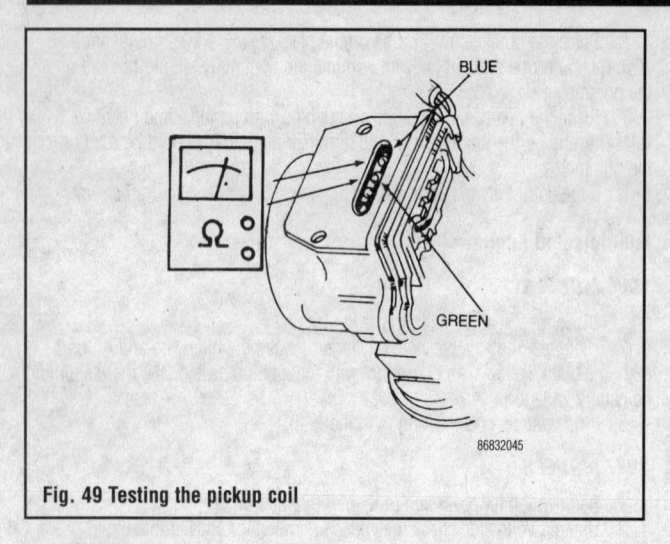

Fig. 49 Testing the pickup coil

1986 MODELS

1. Set the scale of the ohmmeter to the R x 100 position, disconnect the lead wires and check the continuity between the igniter unit terminals.

2. Place the probes of meter to black/yellow wire terminal and blue wire terminal. No continuity should be read on the ohmmeter.
3. Reverse the meter probes. Continuity should be read with the probes in one direction and no continuity in the other.
4. If the ohmmeter readings are not as specified, replace the igniter unit.

1987 MODELS

1. Using an ohmmeter, measure the resistance between the G and BLU terminals on pick-up coil. Pick-up coil resistance should be approximately 750 ohms at 70°F (21°C). Replace the pick-up coil if resistance is not within specifications.

➡ Resistance may vary slightly with the coil temperature.

2. Set the scale of the ohmmeter on the R x 100 setting.
3. Check the continuity between the A and B terminals; there should be continuity in only one direction.
4. Adjust the ohmmeter to the R x 100,000 scale.
5. Connect the positive (red) probe to the D terminal and the negative (black) probe to ground the reading should be 50,000 ohms or more at 70°F (21°C).

➡ The resistance may vary slightly with changes in the temperature.

6. If the readings are not as specified, replace the igniter unit.

PGM-IG IGNITION SYSTEM (1988–95)

Description and Operation

▶ See Figures 50 and 51

Beginning with the redesigned 1988 Civic, Honda began using a computer-controlled Programmed Ignition (PGM-IG). Only evolutionary changes to component design were made for 1992 and later models. The PGM-IG consists of a distributor driven directly from the front camshaft, an amplifier mounted on the side of the distributor and a high energy coil.

The distributor contains a reluctor mounted on a rotor shaft and a magnet mounted on a base plate. Within the distributor are also the sensors for top dead center (TDC), crank position sensor (CKP), and cylinder position sensor (CYP) on 1992 and later models.

The PGM-IG employed on these vehicles controls ignition timing by using a microcomputer known as either Electronic Control Unit (ECU) 1988–92, or Electronic Control Module (ECM) 1993-on. By whatever name, this engine management device has stored within its memory the optimum basic ignition timing for operating conditions based upon engine speed and vacuum pressure in the intake manifold. Information about these ever-changing variables is transmitted by signals from the TDC/CKP/CYP (as applicable) sensors, as well as a coolant temperature sensor and Manifold Absolute Pressure (MAP) sensor. These sensors and the microcomputer also serve the interdependent Programmed Fuel Injection system (PGM-FI) described in Section 5. The PGM-IG system is not dependent upon a conventional governor or vacuum diaphragm and is capable of controlling timing with complicated characteristics which cannot be provided by these non-computerized systems.

The amplifier connected directly to the pick-up coil contains three resistors, three diodes and two transistors. The transistors act as switches which are activated at a precise voltage.

When the ignition is switched **ON**, the switching of the transistors in the amplifier ensures that no current can flow in the ignition coil primary windings. When the engine is cranked, the reluctor moves through the magnetic field created by the stator and, when the reluctor teeth are aligned with the stator projections, a small AC voltage is created. The amplifier boosts this voltage signal and uses it to switch the transistors so that a ground path is provided to the primary circuit.

As the reluctor teeth move out of alignment with the stator projections, an abrupt change occurs in the AC voltage. The transistors are switched again and the primary circuit ground path is broken. This induces a high voltage in the ignition coil secondary winding.

A time control circuit in the amplifier controls the charging time for the ignition coil according to engine speed, this reduces consumption at low engine speeds and prevents secondary voltage drop at high engine speeds.

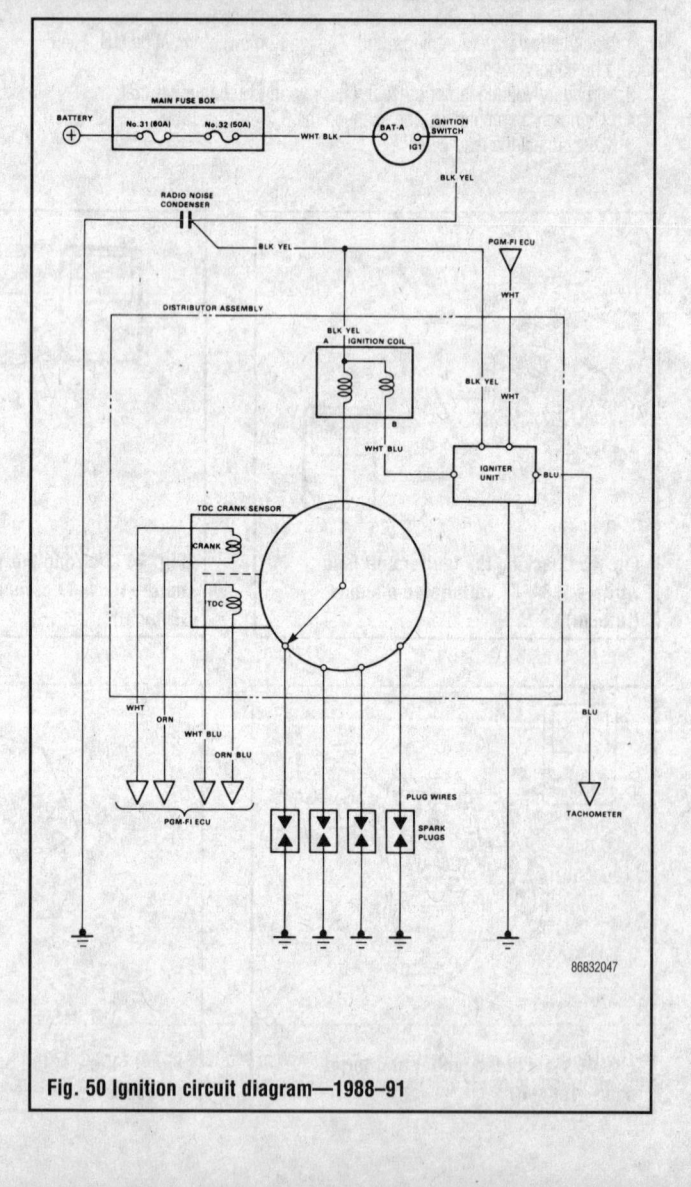

Fig. 50 Ignition circuit diagram—1988–91

ENGINE PERFORMANCE AND TUNE-UP   2-13

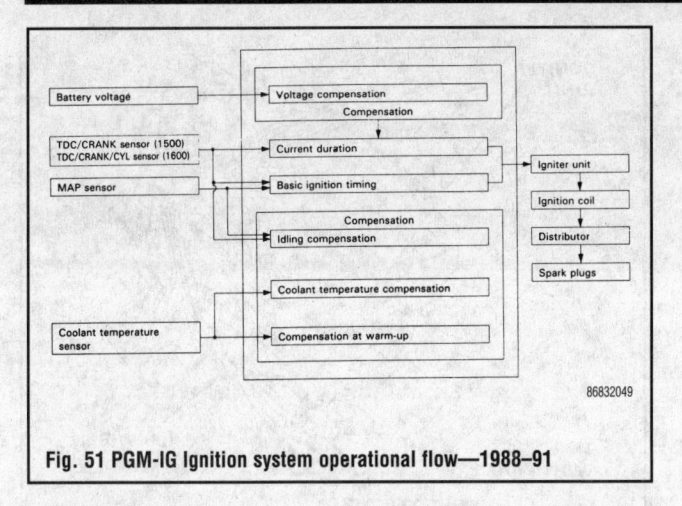

Fig. 51 PGM-IG Ignition system operational flow—1988-91

PRECAUTIONS

- Always disconnect the battery cables before doing repair work on the electronic ignition system.
- Always double check the markings on the battery and the routing of the cables before making connections, especially if the battery has been removed and might have been reinstalled in the opposite position. Hooking up the battery connections backwards will cause current to flow through the electronic ignition system and computer-controlled engine management system in an improper way and immediate damage may occur. Be careful when jumping the vehicle's battery for the same reasons.
- Do not allow the wires connected to the pulse generator to touch other ignition wiring connections.
- Abnormal voltage pulses may damage the system. Therefore, be sure to disconnect the battery before doing any work on the vehicle that involves the electrical system. This includes charging the battery and replacing small bulbs.
- Connect any electrical tachometer to the negative (−) terminal of the ignition coil—not to any other connection.
- Always double check any connection (you are making) involving the ignition system before reconnecting the battery and putting the system into operation.
- When cranking the engine for compression testing or similar purposes, disconnect the coil wire at the distributor.

Component Replacement

▶ See Figure 52

➡There are no service parts available from Honda other than the distributor cap, rotor, igniter and igniter coil. If any other parts within the distributor require replacements, the entire assembly must be replaced.

IGNITION COIL

▶ See Figures 53 and 54

1. Disconnect the negative battery cable.
2. Remove the distributor cap retaining screws and remove the distributor cap and seal. Slide off the leak cover.
3. Remove the two screws that retain the BLK/YEL and WHT/BLU wires from the terminals in the distributor.
4. Remove the screws that retain the coil and remove the coil from the distributor body.

To install:

5. Install the coil in position and tighten the screws securely.
6. Reconnect the two leads and install the leak cover.
7. Install the distributor cap and seal.

IGNITER UNIT

➡The igniter unit is called the Ignition Control Module (ICM) for 1993-95 models but is the same part. The radio noise condenser is built into it for 1992-95 models.

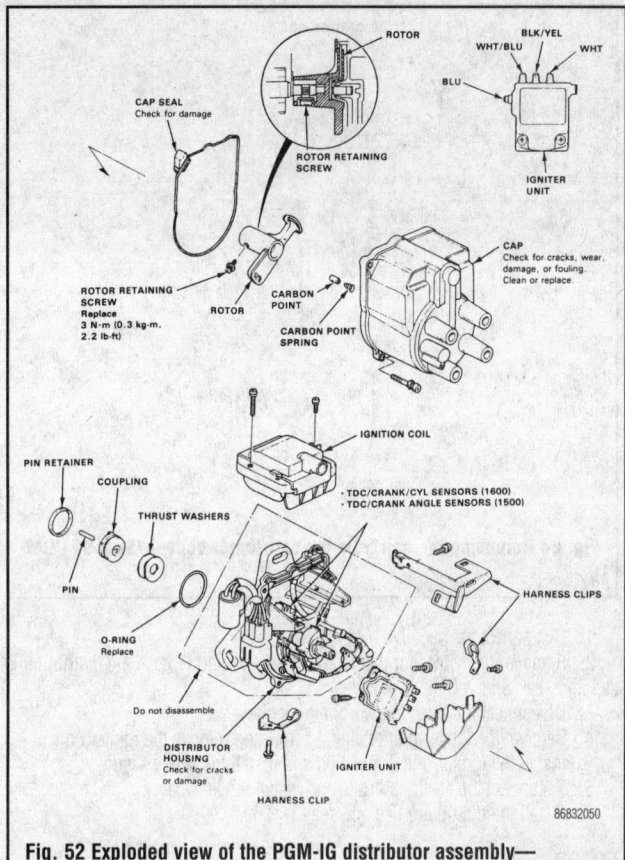

Fig. 52 Exploded view of the PGM-IG distributor assembly—1988-91

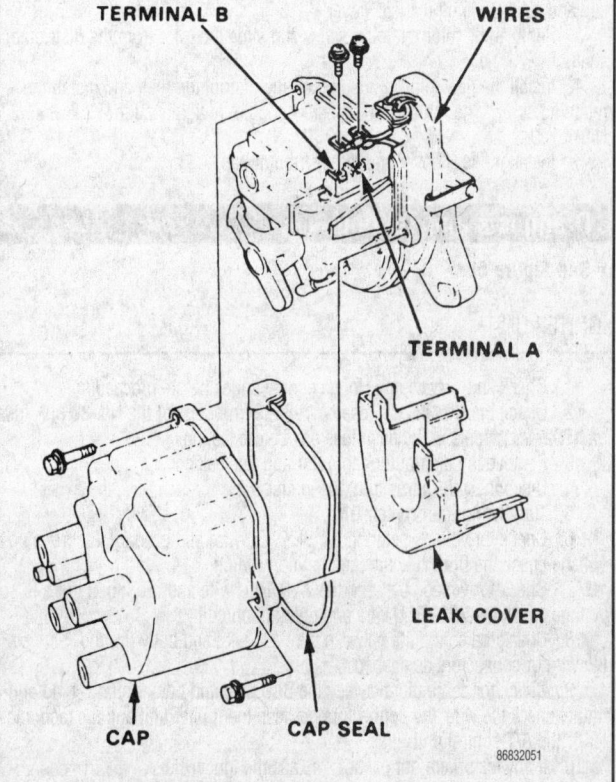

Fig. 53 Remove the screws securing the terminals to the ignition coil—1988-91 PGM-IG

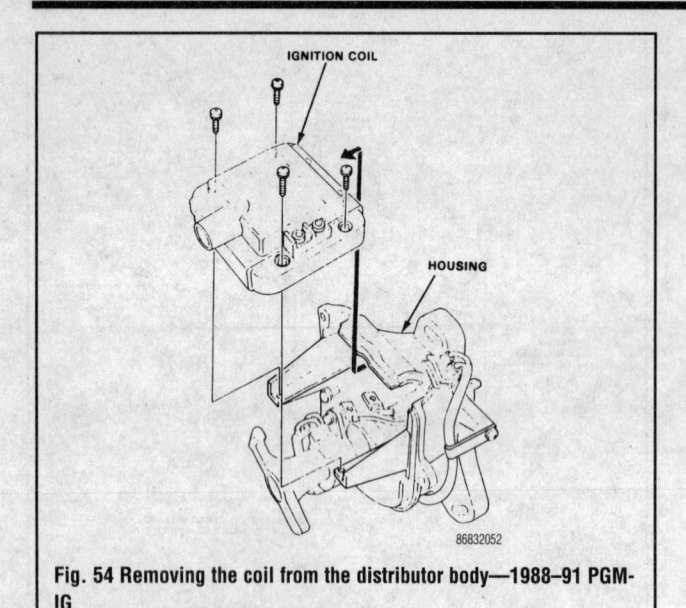

2-14 ENGINE PERFORMANCE AND TUNE-UP

Fig. 54 Removing the coil from the distributor body—1988–91 PGM-IG

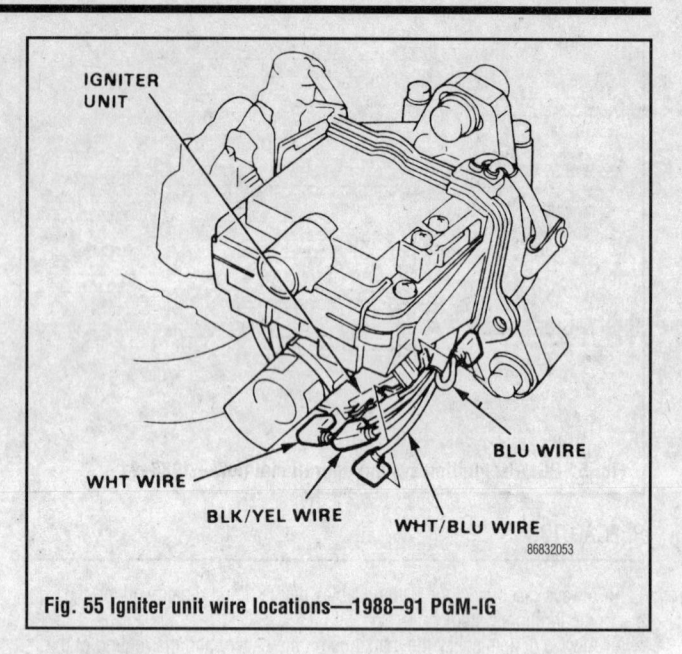

Fig. 55 Igniter unit wire locations—1988–91 PGM-IG

1. Disconnect the negative battery cable.
2. Remove the distributor cap retaining screws and remove the distributor cap and seal. Slide off the leak cover.
3. Disconnect the three leads from the igniter unit.
4. Remove the igniter unit retaining screw and remove the igniter unit.
5. Install the igniter unit in position and install the screw securely.
6. Reconnect the electrical leads and install the leak cover.
7. Install the distributor cap and seal.

ROTOR

1. Disconnect the negative battery cable.
2. Remove the distributor cap retaining screws and remove the distributor cap and seal. Slide off the leak cover.
3. Remove the rotor retaining screw and slide the rotor from the distributor shaft.
4. Install the new rotor in position on the distributor shaft and install the retaining screw. The rotor retaining screw should be tightened to 2.2 ft. lbs. (3 Nm).
5. Install the leak cover and the distributor cap.

Diagnosis and Testing

♦ See Figure 55

IGNITER UNIT

1. Check and service any Honda engine codes before proceeding.
2. Check the No. 32 50A fuse on 1985–91 models, or the No. 39 50A fuse on 1992–95 models in the dash fuse box before testing.
3. Remove the distributor cap, rotor and the leak cover.
4. Disconnect the wires from the igniter.
5. Turn the ignition switch **ON**.
6. Check for voltage between the BLK/YEL wire and ground. If there is no voltage check the BLK/YEL wire and ignition switch.
7. Check for voltage between the WHT/BLU wire and ground. If there is no voltage check the WHT/BLU wire and the ignition coil.
8. Check the continuity between the WHT wire and body ground. Service the wire or connectors as needed.
9. Check for continuity between the BLU wire and body ground. If no continuity check the wire and connectors between the igniter unit and the tachometer or the A/T control unit.
10. If any tests were not passed, replace the igniter unit.
11. Turn the ignition **OFF**.

IGNITION COIL

1988–89 Models

♦ See Figure 56

1. With ignition switch **OFF**, remove the distributor cap.
2. Remove the two screws to disconnect the BLK/YEL front terminal A and WHT/BLU from terminal B of the coil.
3. Using an ohmmeter, check resistance between terminals. Replace coil if resistance is not within specification.

➡ **Resistance will vary slightly with coil temperature. Specifications are at 70°F (21°C)**

4. Check that the primary winding resistance between terminals A and B is 0.3–0.5 ohms.
5. Check that the secondary winding resistance between terminals A and the coil tower is 9,440–14,160 ohms.
6. If not within specifications, replace the coil.

1990–91 Models

1. With ignition switch OFF, remove the distributor cap.
2. Remove the two screws to disconnect the BLK/YEL front terminal A and WHT/BLU from terminal B of the coil.

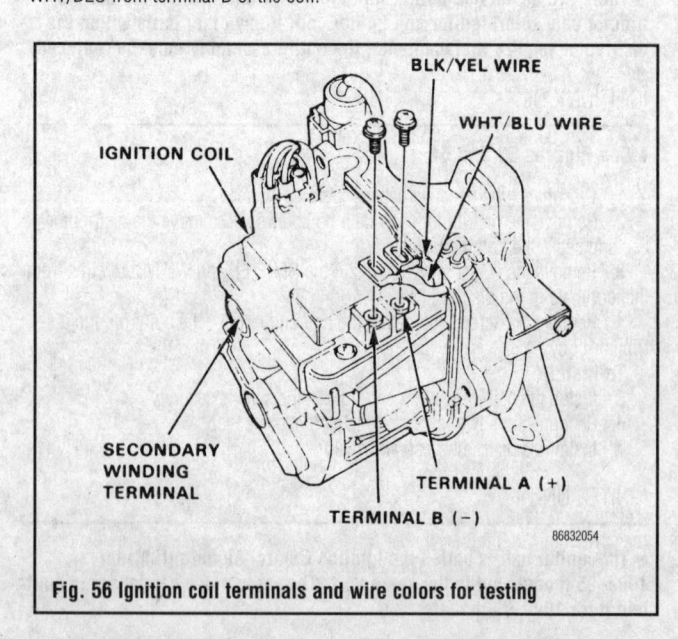

Fig. 56 Ignition coil terminals and wire colors for testing

ENGINE PERFORMANCE AND TUNE-UP

3. Using an ohmmeter, check resistance between terminals. Replace coil if resistance is not within specification.

➡ **Resistance will vary slightly with coil temperature. Specifications are at 70°F (21°C)**

4. Check that the primary winding resistance between terminals A and B is 0.6–0.8 ohms.
5. Check that the secondary winding resistance between terminals A and the coil tower is 9760–14640 ohms.
6. If not within specifications, replace the coil.

1992 Models

1. With ignition switch **OFF**, remove the distributor cap, rotor and cap seal, then remove the leak cover.
2. Remove the two screws to disconnect the BLK/YEL front terminal A and WHT/BLU from terminal B of the coil.
3. Using an ohmmeter, check resistance between terminals. Replace coil if resistance is not within specification.

➡ **Resistance will vary slightly with coil temperature. Specifications are at 70°F (21°C)**

4. Check that the primary winding resistance between terminals A and B is 0.6–0.8 ohms.
5. Check that the secondary winding resistance between terminals A and the coil tower is 13,200–19,800 ohms.
6. If not within specifications, replace the coil.

1993–95 Models

1. With ignition switch **OFF**, remove the distributor cap, rotor and cap seal, then remove the leak cover.
2. Remove the 2 screws to disconnect the BLK/YEL front terminal A and WHT/BLU from terminal B of the coil.
3. Using an ohmmeter, check resistance between terminals. Replace coil if resistance is not within specification.

➡ **Resistance will vary slightly with coil temperature. Specifications are at 70°F (21°C)**

4. Check that the primary winding resistance between terminals A and B is 0.6–0.8 ohms.
5. Check that the secondary winding resistance between terminals A and the coil tower is 12,800–19,200 ohms.
6. If not within specifications, replace the coil.

IGNITION TIMING

General Information

Ignition timing is the measurement, in degrees of crankshaft rotation, at the instant the spark plugs in the cylinders fire, in relation to the location of the piston.

Ideally, the air/fuel mixture in the cylinder will be ignited and just beginning its rapid expansion as the piston passes top dead center (TDC). If this happens, the piston will be beginning the power stroke just as the compressed and ignited air/fuel mixture starts to expand. The expansion of the air/fuel mixture will force the piston down on the power stroke and turn the crankshaft.

It takes a fraction of a second for the spark from the plug to completely ignite the mixture in the cylinder. Because of this, the spark plug must fire before the piston reaches TDC so the mixture is completely ignited as the piston passes TDC. This measurement is given in degrees of crankshaft rotation before the piston reaches top dead center (BTDC). If the ignition timing setting is 6 degrees BTDC, this means that the spark plug must fire at a time when the piston for that cylinder is 6 degrees BTDC of its compression stroke. However, this only holds true while the engine is at idle.

As you accelerate from idle, the speed of your engine increases. This increase in engine speed means that the pistons are now traveling up and down much faster. Because of this, the spark plugs will have to fire even sooner if the mixture is to be completely ignited as the piston passes TDC. To accomplish this, the system incorporates means to advance the timing of the spark as engine speed increases.

The mechanical distributor has two means of advancing the ignition timing. One is centrifugal advance and is actuated by weights in the distributor. The other is vacuum advance and is controlled in that large circular "can" on the side of the distributor. Fully electronic systems use the Engine Control Unit (ECU) to control the trigger signal to the coil. The computer can provide for much more accurate spark control than mechanical units.

In addition, some Honda distributors have a vacuum retard mechanism which is contained in the same housing on the side of the distributor as the vacuum advance. Models having two hoses going to the distributor vacuum housing have both vacuum advance and retard. The function of this mechanism is to regulate the timing of the ignition spark under certain engine conditions. This causes more complete burning of the air/fuel mixture in the cylinder and consequently lowers exhaust emissions.

The later model (1988–95) Honda PGM-IG ignition system uses various inputs from engine sensors to change the ignition timing in accordance with the needs of the engine. This is all done completely by the engine control computer, in place of the vacuum and centrifugal means of timing adjustment.

If ignition timing is set too far advanced (BTDC), the ignition and burning of the air/fuel mixture in the cylinder will try to oppose the motion of the piston in the cylinder while it is still traveling upward. This causes engine ping, a sound similar to marbles in a coffee can. If the ignition timing is too far retarded (after, or ATDC), the piston will have already started down on the power stroke when the air/fuel mixture ignites. This will cause the piston to be forced down with much less power. This will result in rough engine performance, lack of power and poor gas mileage.

Checking and Adjusting Timing

▶ **See Figures 57 and 58**

Honda recommends that the ignition timing be checked every 60,000 miles (96,500 km) and adjusted (if necessary).

On all Civic series engines the timing marks are located on the crankshaft pulley, with a pointer on the timing belt cover. All are visible while looking at the driver's side of the engine compartment. The timing is checked with the engine warmed to operating temperature—176°F (80°C)—idling in Neutral (manual transaxle) or Drive (automatic transaxle), and with all vacuum hoses connected.

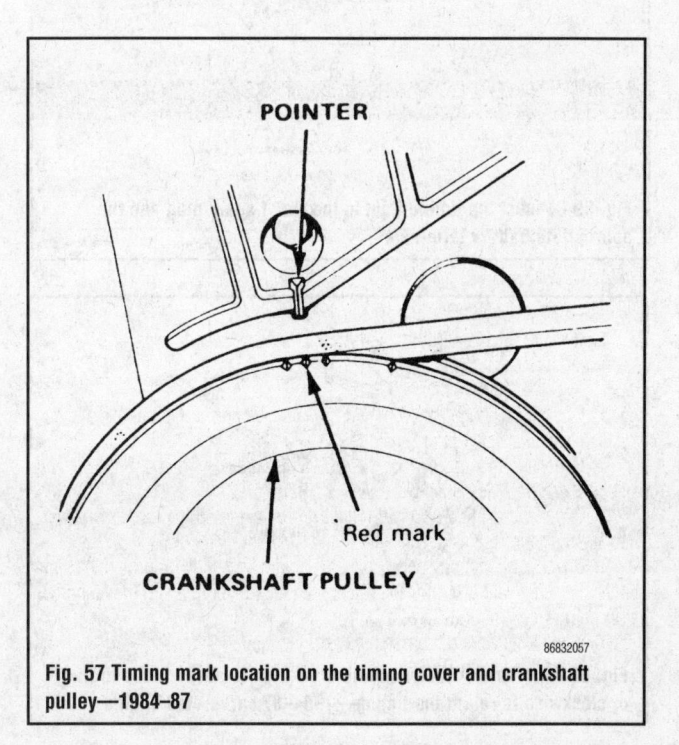

Fig. 57 Timing mark location on the timing cover and crankshaft pulley—1984–87

2-16　ENGINE PERFORMANCE AND TUNE-UP

Fig. 58 Clean the notches prior to setting the timing. This will make identifying them much easier

1984–87 MODELS

◆ See Figures 59, 60, 61 and 62

1. Stop the engine and connect a tachometer to the engine. The positive lead connects to the distributor side terminal of the ignition coil and the negative lead to a good ground, such as an engine bolt.

➡ On some models you will have to pull back the rubber ignition coil cover to reveal the terminals.

2. Following the manufacturer's instructions, install a timing light to the

Fig. 59 Connect the timing light to the No. 1 spark plug and the appropriate battery terminals

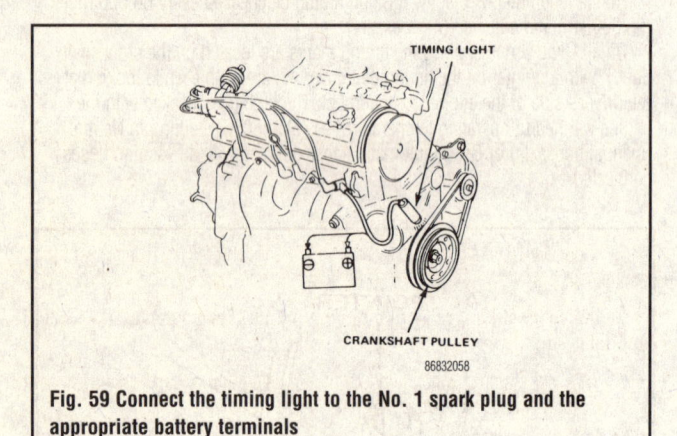

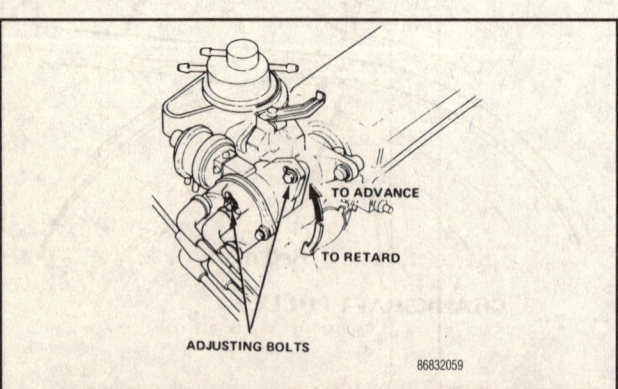

Fig. 60 Turn the distributor counterclockwise to advance the timing or clockwise to retard the timing—1984–87 carbureted engine

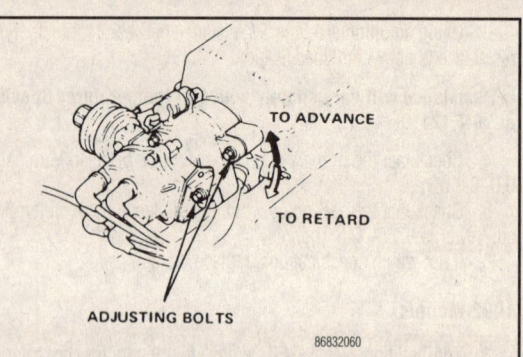

Fig. 61 Turn the distributor counterclockwise to advance the timing or clockwise to retard the timing—1984–87 fuel injected engine shown

engine. The positive and negative leads connect to their corresponding battery terminals and the spark plug lead to the No. 1 spark plug. The No. 1 spark plug is the one at the driver's side of the engine compartment.

3. Make sure that all wires are clear of the cooling fan and hot exhaust manifold.
4. Set the parking brake and block the front wheels. Start the engine. Check that the idle speed is set to specifications with the transaxle in Neutral (manual transaxle) or Drive (automatic transaxle).
5. If the distributor is equipped with a vacuum advance mechanism, disconnect the hose(s), plug it (them).

➡ At any engine speed other than specified, the distributor advance or retard mechanisms will actuate, leading to an erroneous timing adjustment.

Fig. 62 To check the timing, rotate the distributor until it is within specification

ENGINE PERFORMANCE AND TUNE-UP 2-17

6. Point the timing light at the timing marks.
7. If necessary to adjust the timing, loosen the distributor hold-down bolt(s) and/or nut. Slowly rotate the distributor counterclockwise to retard the timing or clockwise to retard the timing while observing the timing marks.

➡ Some models are equipped with two bolts, others are equipped with a bolt and a nut, which may be loosened to adjust ignition timing. If there is a smaller bolt on the underside of the distributor swivel mounting plate, it should not be loosened, unless you cannot obtain a satisfactory adjustment using the upper bolt. Its purpose is to provide an extra range of adjustment, such as in cases where the distributor was removed and then installed one tooth off.

✱✱ CAUTION

Do not grasp the top of the distributor cap while the engine is running, as you might get a shock. Instead, grasp the distributor housing to rotate.

8. To complete the adjustment operation, tighten the hold-down bolt, taking care not to disturb the adjustment. If equipped with a vacuum advance mechanism, unplug and reinstall the hose(s).

1988–91 MODELS

Except CRX

♦ See Figures 63, 64, 65 and 66

1. Start engine and allow to reach normal operating temperature (cooling fan comes on).
2. Remove the YEL rubber cap from the ignition timing adjusting connector. This is located at the rear of the engine compartment on 1988–89 models or under the right side of the dash on 1990–91 models.
3. Connect a jumper wire between the BRN and GRN/WHT terminals of the connector.
4. Connect a timing light and check timing using the red timing mark.
5. Timing should be:
- All Models Except Wagon—16–20° BTDC 675–750 rpm
- Wagon Models Except 1988 4WD—16–20° BTDC 700–800 rpm
- 1988 4WD Wagon—16–20° BTDC 675–750 rpm
6. Adjust timing as needed, by turning the distributor counterclockwise to retard the timing, clockwise to retard the timing.
7. Turn engine off, remove jumper wire and install the rubber caps.

CRX

1. Start engine and allow to reach normal operating temperature (cooling fan comes on).
2. Remove the YEL rubber cap from the from the ignition timing adjusting connector, located at the rear engine compartment on 1988–89 models or under the right side of the dash on 1990–91 models.
3. Jumper the BRN and GRN/WHT terminals of the connector.
4. Connect a timing light and check timing using the red timing mark.
5. Timing should be:
- STD Models—16–20° BTDC at 675–750 rpm
- SI Models—16–20° BTDC at 700–800 rpm
- HF Models (Except California)—12–16° BTDC at 550–650 rpm
- HF Models (California)—12–16° BTDC at 600–700 rpm
6. Adjust timing as needed, by turning the distributor counterclockwise to retard the timing, clockwise to retard the timing.
7. Turn engine off, remove jumper wire and replace rubber caps.

1992–95 MODELS

1. Start engine and allow to reach normal operating temperature (cooling fan comes on).
2. Pull out the service check connector located behind the right kick panel.
3. Connect a jumper wire between the BRN and GRN/WHT terminals of the connector.

➡ Do not connect jumper wire to three pin Data Link connector on 1993 and later cars.

4. Connect a timing light and check timing using the red timing mark.
5. Timing should be:

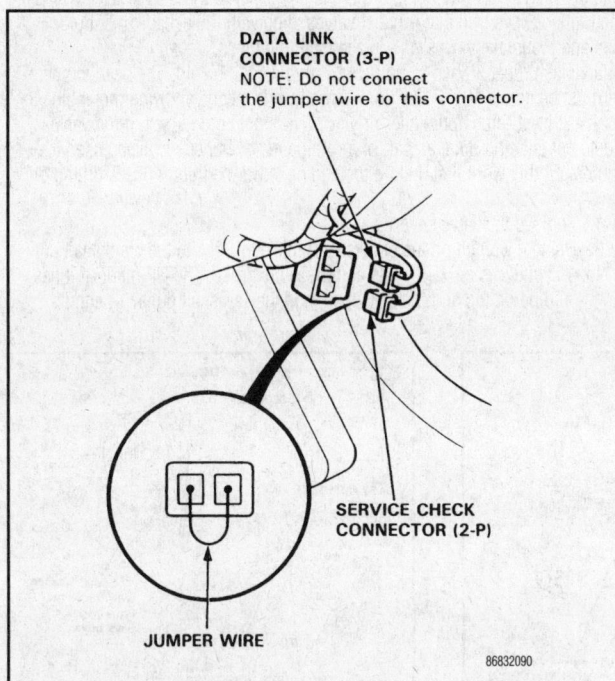

Fig. 63 Location of the service check connector for 1992–95 models—shown is post-1993 model which also includes a 3-pin data link connector

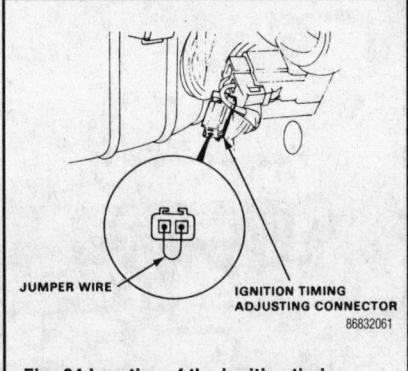

Fig. 64 Location of the ignition timing adjusting connector—1988–91

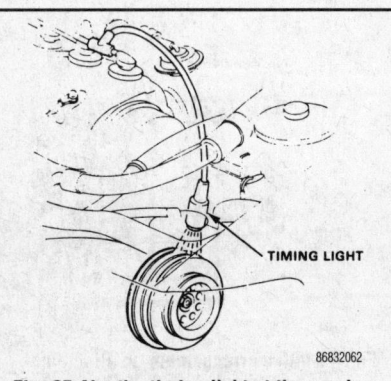

Fig. 65 Aim the timing light at the crankshaft pulley—1988–91

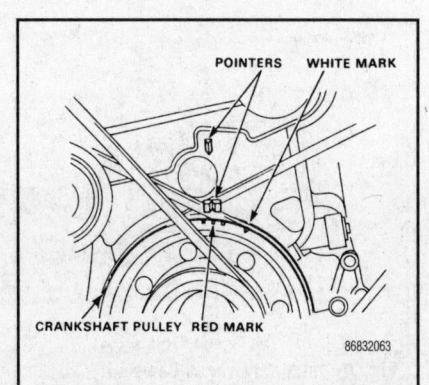

Fig. 66 Timing mark locations on the timing cover and crankshaft pulley—1988–91

2-18 ENGINE PERFORMANCE AND TUNE-UP

➡ All engine speed settings may vary 50 rpm above or below the specification.
- D15Z1 engine: M/T—16° BTDC (RED) 600 rpm (USA) or 700 rpm (Canada)
- D15B8 engine: M/T—12° BTDC (RED) 670 rpm (USA) or 750 rpm (Canada)
- D15B7/D16Z6 engine: M/T—16° BTDC (RED) at 670 rpm (USA) or 750 rpm (Canada)
- D15B7/D16Z6 engine: A/T—16° BTDC (RED) at 700 rpm (USA) or 750 rpm (Canada)

6. Adjust timing as needed, by turning the distributor housing counterclockwise to retard the timing, clockwise to retard the timing.
7. Turn engine off and remove jumper wire.

VALVE LASH

♦ See Figures 67 thru 82

➡ The adjustment procedure for intake and exhaust valves is essentially the same for all Civic engines as they share a common design, whether 2, 3 or 4-valve-per-cylinder. All engines employ overhead cam(s), overhead valve designs that adjust by a screw and locknut method. Therefore, the procedures in this section will cover all Civic series models. Just be certain to follow torque specifications for your particular model.

As part of every major tune-up or service interval, the valve clearance should be checked and adjusted if necessary.

If the valve clearance is too large, part of the lift of the camshaft will be used up in removing the excessive clearance, thus the valves will not be opened far enough. This condition makes the valve train noisy as they take up the excessive clearance, and the engine will perform poorly, since a smaller amount of air/fuel mixture will be admitted to the cylinders. The exhaust valves will not open far enough to vent the cylinder completely; retained pressure (back pressure) will restrict the entry of the next air/fuel charge.

If the valve clearance is too small, the intake and exhaust valves will not fully seat on the cylinder head when they close. This causes internal cylinder leakage and prevents the hot valve from transferring some heat to the head and cooling off. Therefore, the engine will run poorly (due to gases escaping from the combustion chamber), and the valves will overheat and warp (since they cannot transfer heat unless they are firmly touching the seat in the cylinder head).

➡ While all valve adjustments must be as accurate as possible, it is better to have the valve adjustment slightly loose than slightly tight, as burnt valves may result from overly tight adjustments.

Adjustment

Honda recommends that the valve clearance be checked at 15,000 mile (24,000 km) intervals.

Be advised that the Civic series engines covered by this manual do vary in the location of the intake valve(s), exhaust valve(s) and auxiliary (intake) valves (auxiliary for 12-valve models). If there is any uncertainty as to which valve is which, it is always true that exhaust valves align with the exhaust manifold tubes and the intake valves with intake manifold tubes.

Valve lash must always be adjusted with the engine cold. The head temperature must be below 100°F (38°C). Generally, this means allowing the engine to cool for at least three hours after driving. Overnight cold is best. If the valve adjustment is being done as part of a routine maintenance or mileage service regimen, do this work first, before the engine is warmed up to check timing or idle

1. Remove the valve cover.
2. Using a wrench on the crankshaft pulley bolt, rotate the crankshaft until the No. 1 cylinder is on TDC of its compression stroke. The UP mark (if it has one) should be on top and the TDC marks on the camshaft pulley(s) should

Fig. 67 Camshaft pulley marks on 1984–87 models

Fig. 68 Camshaft pulley marks on 1988–91 models

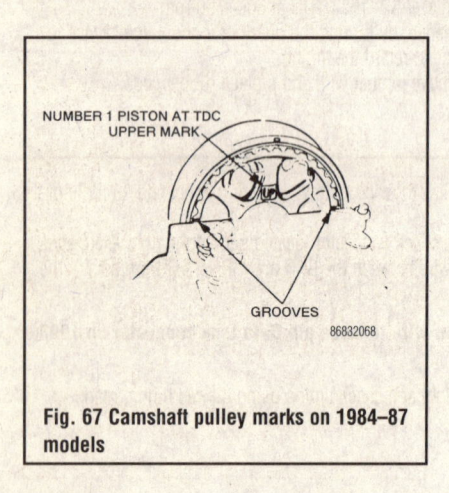

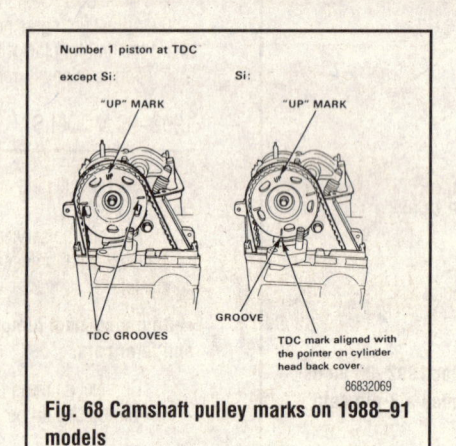

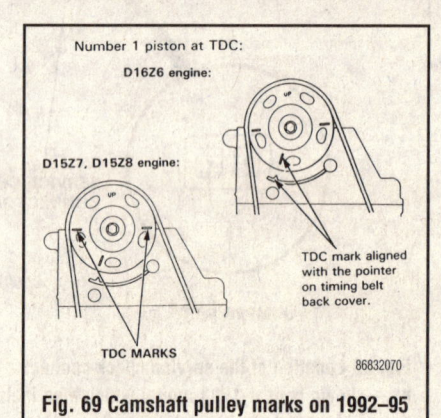

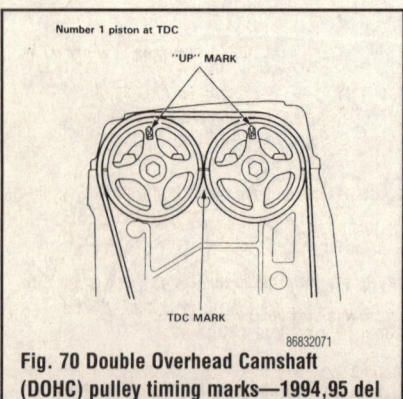

Fig. 69 Camshaft pulley marks on 1992–95 SOHC engines

Fig. 70 Double Overhead Camshaft (DOHC) pulley timing marks—1994,95 del Sol VTEC (shown)

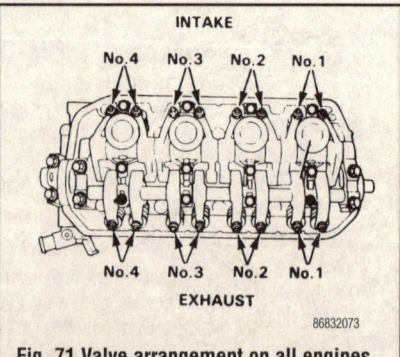

Fig. 71 Valve arrangement on all engines except 1992–95 Civic CX and 1994–95 del Sol VTEC

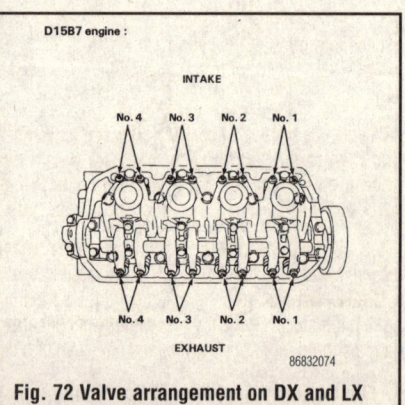

Fig. 72 Valve arrangement on DX and LX Civics—1992–95

ENGINE PERFORMANCE AND TUNE-UP 2-19

Fig. 73 The valve adjustment procedure will be a little shorter for Civic CX with just two valves per cylinder—1992–95

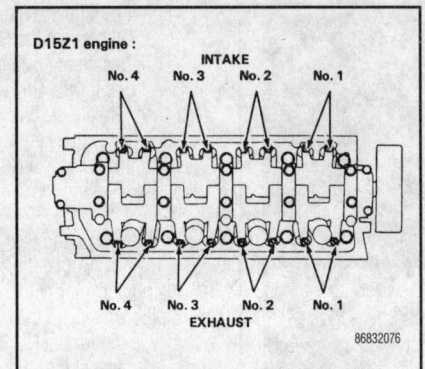

Fig. 74 Valve arrangement on high-MPG Civic VX—1992–95

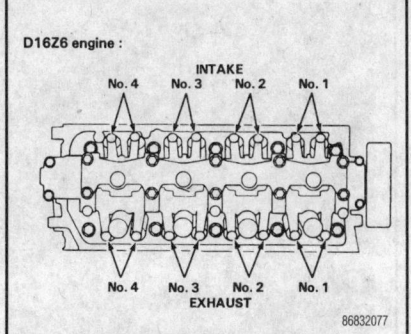

Fig. 75 This is the valve arrangement on the milder VTEC engine used on Civic Si, EX, del Sol Si—1992–95

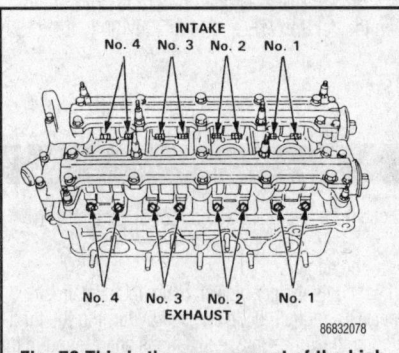

Fig. 76 This is the arrangement of the high output DOHC VTEC engine in the del Sol VTEC—1994, 95

Fig. 77 The first step to valve adjustment is access via the valve cover

Fig. 78 The breather hose shown is held on with a quick-to-remove pinch clamp

Fig. 79 Remove the nuts securing the cover

Fig. 80 Remove the valve cover. If the gasket is damaged or deteriorated, it will have to be replaced. Store the valve cover in a safe and clean place out of the way

Fig. 81 Loosen the locknut, then adjust the clearance using a feeler gauge and screwdriver

align with the top of the cylinder head surface. The crankshaft pulley should be at TDC. Also, the rotor in the distributor should point to the No. 1 spark plug wire.

➡ **If you are unfamiliar with using feeler gauges, you can practice getting the proper "feel" by running the gauge through a (metal) precision micrometer set to the same clearance as the gauge. By doing this, you will get a better idea how "too tight" and "too loose" tolerances feel.**

3. Loosen the locknut. Using a flat feeler gauge, place it between the top of the valve and the adjusting stud. If the clearance is under, turn the stud outward with a screwdriver until the blade can be inserted between the two items, then, tighten the stud very gently just until it touches the gauge. A slight drag on the gauge should be felt by moving the gauge in and out—it must not be pinched between the two parts.

4. Hold the position of the stud with a screwdriver, then, tighten the locknut until just snug—torque intake/exhaust valve locknuts for 1984–87 models to 14 ft. lbs. (20 Nm) and torque 1984–87 auxiliary valve locknut to 10 ft. lbs. (14 Nm). For all 1988–91 models, re-torque intake/exhaust valve locknuts to 10 ft. lbs. (14 Nm); for 1992–95 models, re-torque locknuts on cars with engine type D15B7/D15B8 to 10 ft. lbs. (14 Nm), and torque D16Z6/D15Z1/D16A6 locknuts to 14 ft. lbs (20 Nm).

5. Slide the gauge in and out to make sure the required clearance has been maintained. If not, readjust the valve the procedure for the remaining No. 1 cylinder valves.

➡ **As the work progresses, keep double checking that you are using the proper gauge for the type of valve being adjusted. Also use locknut torque specs (in step "d" above) for all four cylinders.**

6. Rotate the crankshaft counterclockwise 180° (the camshaft turns 90°). Now, the TDC groove on the outer edge of the camshaft pulley, if equipped,

2-20 ENGINE PERFORMANCE AND TUNE-UP

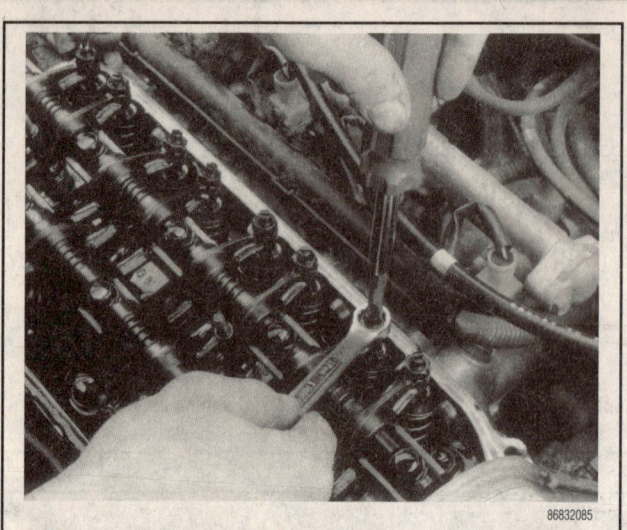

Fig. 82 When the correct clearance is achieved, hold the stud with the screwdriver and tighten the locknut

should be aligned with the indentation on the timing belt cover. The distributor rotor will point to the No. 3 cylinder plug wire; the No. 3 cylinder should be on TDC of its compression stroke. Perform the same adjustment procedures as described for the No. 1 cylinder.

7. Rotate the crankshaft counterclockwise 180°. Now the TDC grooves, if equipped, will again be visible and the distributor rotor will point to No. 4 cylinder's plug wire. The No. 4 cylinder should be at the TDC of its compression stroke. Perform the same adjustment procedures as you did for the No. 1 cylinder.

8. Rotate the crankshaft counterclockwise 180°, until the mark on the rear of the pulley, if equipped, aligns with the indentation on the belt cover and the distributor rotor points to No. 2 plug wire. The UP mark should also be visible, on the left side of the camshaft pulley. Perform the same adjustment procedures as you did for the No. 1 cylinder.

9. Install the valve cover. (Refer to Section 3 for the procedure.)

10. After the valve adjustment is complete, re-torque the crankshaft pulley nut which may have backed out during the procedure. For 1984–87 models re-torque to 83 ft. lbs. (115 Nm); for 1988–91 models re-torque to 119 ft. lbs. (165 Nm); for 1992–95 models (except DOHC engine) re-torque to 134 ft. lbs. (185 Nm); for DOHC engine on the del Sol VTEC, re-torque to 130 ft. lbs. (180 Nm).

IDLE SPEED AND MIXTURE

This section contains only adjustments which apply to an engine tune-up—namely, idle speed and mixture adjustments (on pre-1988 carbureted models). Descriptions of the carburetors/throttle bodies used and complete adjustment procedures can be found in Section 5.

Idle speed and mixture adjustment is the last step in any tune-up. Prior to making the final carburetor adjustments, make sure that the spark plugs, ignition timing and valve clearance have all been checked, serviced and adjusted (if necessary). If any of these tune-up items have been overlooked, it may be difficult to obtain a proper adjustment.

➡ All adjustments must be made with the engine fully warmed to operating temperatures.

Idle Speed

ADJUSTMENT

Carbureted Engines

▶ See Figures 83, 84, 85, 86 and 87

The idle mixture adjustment on these vehicles is controlled by the PGM-FI Engine Control Unit (ECU) and is therefore not adjustable. For further fuel system diagnosis, refer to Section 5 of this manual. The following applies to carbureted engines only.

1. Operate the engine until normal operating temperature is reached.

2. Remove the vacuum hose from the intake air control diaphragm and clamp the hose end.
3. Connect a tachometer to the engine.
4. With all accessories (headlights, heater blower, rear window defroster, cooling fan and air conditioner) OFF, adjust the idle speed by turning the throttle stop screw to the rpm listed in the Tune-Up Specifications chart found in this section or the underhood sticker.

➡ If the underhood sticker disagrees in any way with the Tune-Up specifications chart, always defer to the sticker.

5. After the idle is adjusted, remove the tachometer and reconnect the vacuum hose.

Fuel Injected Engines

1985–87 MODELS

1. Start the engine and warm it up to normal operating temperature. Allow the cooling fan to come on at least once.
2. Connect a tachometer.
3. Disconnect the No. 10 vacuum hose from the intake manifold. (Refer to the vacuum diagram in Section 4.)
4. Cap the end of the hose and intake manifold
5. Check the idle speed with no electrical system load, headlights, heater blower, rear window defogger, cooing fan and A/C should be off. Adjust the idle speed if necessary, by turning the adjusting screw on the top of the throttle body.

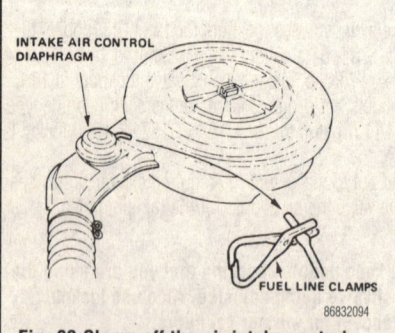

Fig. 83 Clamp off the air intake control vacuum hose—1985–87 carbureted models

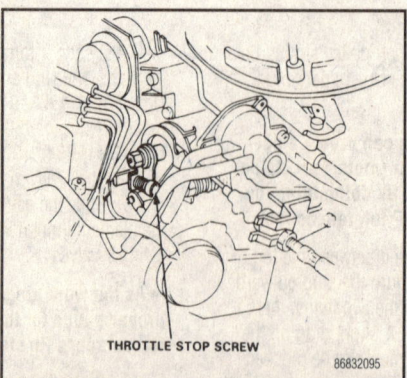

Fig. 84 Adjust the curb idle at the throttle stop screw—1984–87 carbureted models

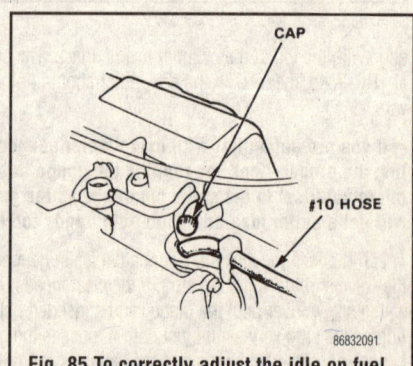

Fig. 85 To correctly adjust the idle on fuel injected models, disconnect the No. 10 hose, then cap it and the intake manifold

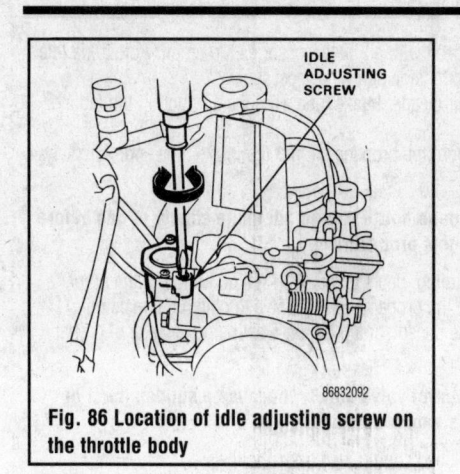

ENGINE PERFORMANCE AND TUNE-UP 2-21

Fig. 86 Location of idle adjusting screw on the throttle body

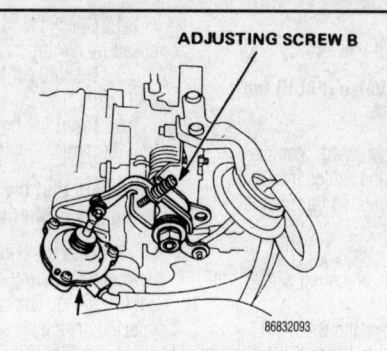

Fig. 87 Use adjusting screw B (shown) to set idle speed while A/C is on after the no-load idle adjustment has been performed

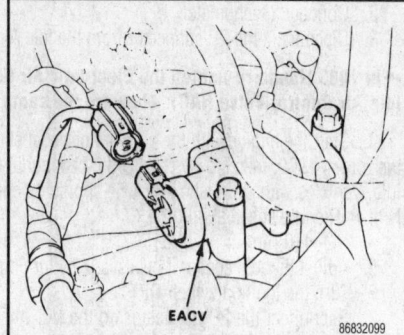

Fig. 88 Disconnect the 2P wire from the EACV to set the idle speed—1988–91 1.5L engine

Idle speed should be: 700–800 rpm.
6. Next check the idle speed with the A/C on. Idle speed should be 700–800 rpm.
7. If necessary, adjust by turning adjusting screw B.

1988–91 MODELS

◆ See Figures 88, 89, 90 and 91

1. Start the engine and warm it up to normal operating temperature. Allow the cooling fan to come on at least once.
2. Connect a tachometer.
3. Disconnect the 2P connector from the Electronic Air Control Valve (EACV).
4. Set the front wheels in the straight ahead position and check the idle speed with the headlights, blower fan, rear defroster, cooling fan and air conditioner not operating. Vehicles with automatic transaxle should be checked with the transaxle in N or P.
5. The idle speed should be as follows:

- All Civics and CRX with 1.5L engine—575–675 rpm
- CRX HF—450–550 rpm
- Civic/CRX Si, and 4WD wagon—500–600 rpm

6. Adjust the idle speed, if necessary, by turning the adjusting screw.

➡ If the idle speed is excessively high on Civic and CRX, check the throttle control system.

7. Turn the ignition switch **OFF**.
8. Reconnect the 2P connector on the EACV.
9. On Civic and CRX, remove the HAZARD fuse in the main fuse box for 10 seconds to reset the ECU.

1992–95 MODELS

◆ See Figures 92 and 93

1. Start the engine and warm it up to normal operating temperature. Allow the cooling fan to come on at least once.

Fig. 89 Idle adjustment screw—1988–91 1.5L engine

Fig. 90 Disconnect the 2P wire from the EACV to set the idle speed—1988–91 1.6L engine

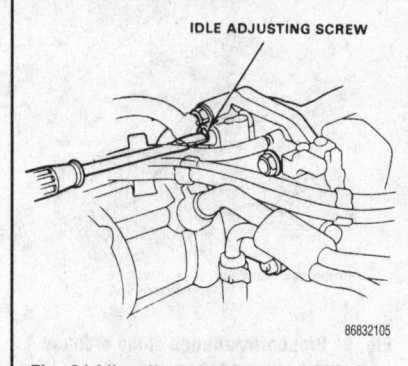

Fig. 91 Idle adjustment screw—1988–91 1.6L engine

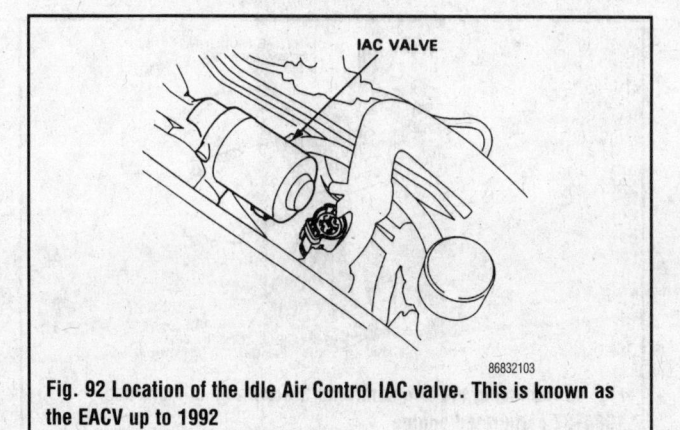

Fig. 92 Location of the Idle Air Control IAC valve. This is known as the EACV up to 1992

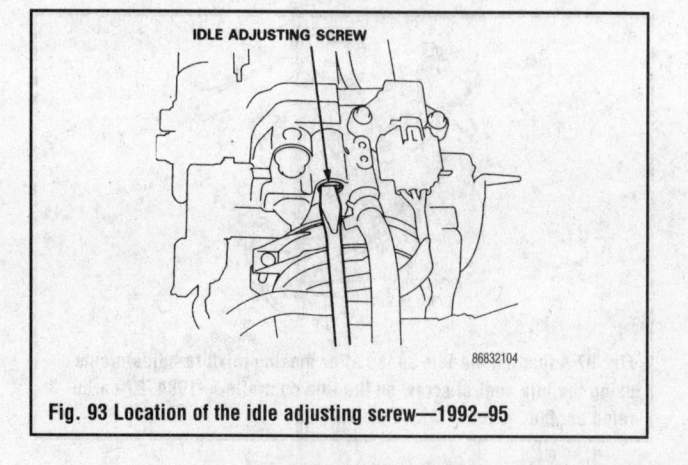

Fig. 93 Location of the idle adjusting screw—1992–95

ENGINE PERFORMANCE AND TUNE-UP

2. Connect a tachometer.
3. Disconnect the 2P connector from the Idle Air Control Valve (IAC).

➡ In 1993 Honda re-named the Electronic Air Control Valve (EACV) the Idle Air Control Valve (IAC). They are the same part.

4. Start the engine with the accelerator pedal slightly depressed. Stabilize the rpm at 1000, then slowly release the pedal until the engine settles into its idle. Vehicles with automatic transaxle should be checked with the transaxle in N or P. The idle speed should be:
   • 370–470 rpm
5. Adjust the idle speed, if necessary, by turning the idle adjusting screw.
6. Turn the ignition switch **OFF**.
7. Reconnect the 2P connector on the IAC, then remove the BACK UP (7.5A) fuse in the under hood fuse/relay box for 10 seconds to reset the ECM.

➡ In 1993 Honda re-named the Electronic Control Unit (ECU) the Engine Control Module (ECM). They are the same part.

Idle Mixture

ADJUSTMENT

◆ See Figures 94, 95, 96, 97 and 98

➡ This procedure requires a propane enrichment kit.

1. Place the vehicle in the P or N position. Apply the emergency brake and block the drive wheels. Start the engine and warm it up to normal operating temperature. The cooling fan should come on at least once.
2. Remove the No. 8 vacuum hose (refer to Section 4 for vacuum diagrams from the intake air control diaphragm and clamp the hose end.
3. Connect a suitable tachometer to the engine following the manufacturers instructions.
4. Check the idle speed with all the accessories turned off. Adjust the idle speed by turning the throttle stop screw, if necessary.
5. Disconnect the air cleaner intake tube from the air duct on the radiator bulkhead.
6. Insert the hose from the propane kit into the intake tube approximately 4 in. (102mm).

➡ Be sure that the propane bottle has an adequate supply of gas before going any further with this procedure.

7. With the engine idling, depress the push button on top of the propane device, then slowly open the propane control valve to obtain the maximum engine speed. The engine speed should increase as the percentage of propane injected goes up.

➡ Open the propane control valve slowly, because a sudden burst of propane may cause the engine to stall.

8. The engine idle speed should increase as follows:
   • 1984–88 Civic 1342cc and 1488cc HF engine with manual transmission—100–150 rpm.
   • 1984–88 Civic 1488cc (except the HF) with manual transmission—75–125 rpm.
   • 1984–88 Civic 1488cc (except the HF) with automatic transmission—30–70 rpm.
9. If the engine speed increases according to specifications, remove the propane kit, all test equipment and reconnect all disconnected vacuum hoses. If the engine speed fails to increase as specified, go on with the following steps.
10. Disconnect the vacuum hose to the fast idle unloader. Pull the throttle cable out of the bracket.
11. Remove the carburetor (refer to Section 5).
12. Remove the throttle controller bracket. Remove the mixture adjusting screw hole cap from the throttle controller bracket, then reinstall the bracket.
13. Reinstall the carburetor, reconnect the vacuum hose to the fast idle unloader. Reinstall the air cleaner.
14. Start the engine and let it warm up to normal operating temperature. Allow the cooling fan to come on at least once.

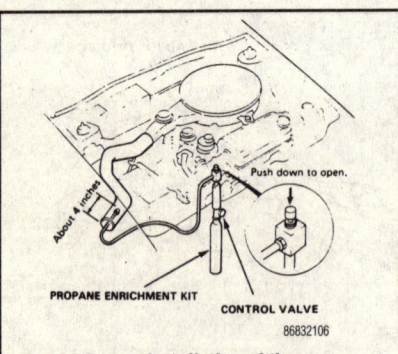

Fig. 94 Proper installation of the propane enrichment kit—1984–87 carbureted engine

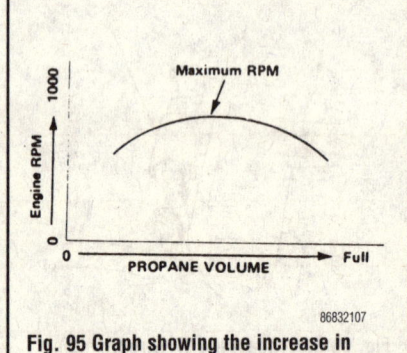

Fig. 95 Graph showing the increase in engine rpm as compared to the increase in propane—1984–87 carbureted engine

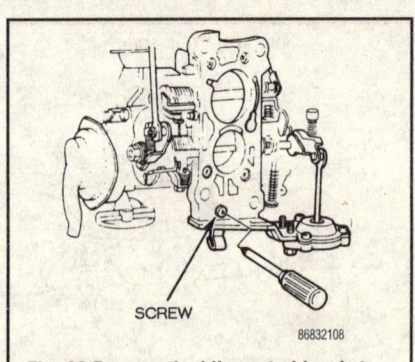

Fig. 96 Remove the idle control bracket mounting screw—1984–87 carbureted engine

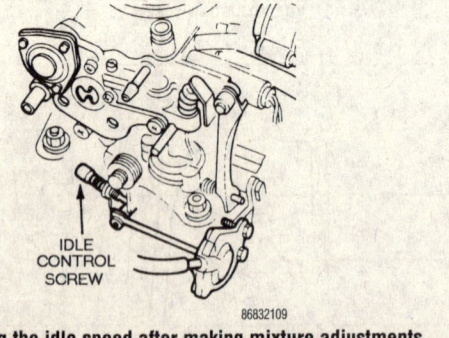

Fig. 97 Adjusting the idle speed after making mixture adjustments using the idle control screw on the idle controller—1984–87 carbureted engine

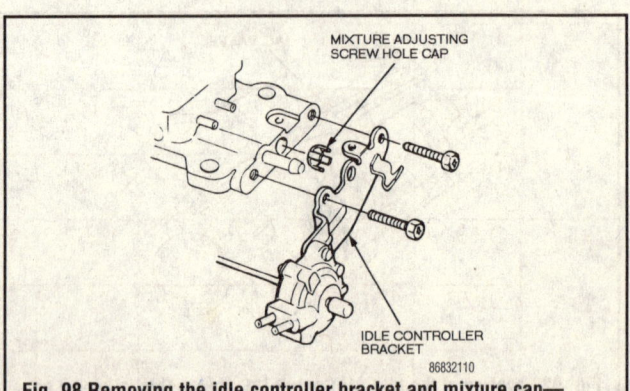

Fig. 98 Removing the idle controller bracket and mixture cap—1984–87 carbureted engine

ENGINE PERFORMANCE AND TUNE-UP 2-23

15. Remove the vacuum hose from the intake air control diaphragm and clamp the hose end. Reinstall the propane enrichment kit and recheck the maximum propane enrichment rpm.

16. If the propane enriched speed is to low, the mixture is rich. Turn the mixture screw a ¼ turn clockwise and recheck.

17. If the propane enriched speed is to high, the mixture is lean. Turn the mixture screw a ¼ turn counterclockwise and recheck.

18. Close the propane control valve. Run the engine at 2500 rpm for 10 seconds to stabilize the idle condition. Recheck the idle speed.

19. If the engine speed is set to specifications, remove the propane enrichment kit, all test equipment and reconnect all vacuum hoses and the air cleaner intake tube.

20. If the engine speed is not set to specifications, reset the idle speed, then repeat the enrichment steps.

21. Reinstall the mixture adjusting screw hole cap.

ENGINE TUNE-UP SPECIFICATIONS

Year	Engine ID/VIN	Engine Displacement Liters (cc)	Spark Plugs Gap (in.)	Ignition Timing (deg.) MT	Ignition Timing (deg.) AT	Fuel Pump (psi)	Idle Speed (rpm) MT	Idle Speed (rpm) AT	Valve Clearance In.	Valve Clearance Ex.
1984	EV1	1.3 (1342)	0.042	21B [1]	-	3.0	650-750	-	0.007-0.009	0.009-0.011
	EW1	1.5 (1488)	0.042	20B	15B [3]	3.0	650-750	650-750	0.007-0.009	0.009-0.011
	EW1	1.5 (1488)	0.042	21B [2]	-	3.0	650-750	650-750	0.007-0.009	0.009-0.011
1985	EV1	1.3 (1342)	0.042	21B [1]	-	3.0	650-750	-	0.007-0.009	0.009-0.011
	EW1	1.5 (1488)	0.042	20B	15B [3]	3.0	650-750	650-750	0.007-0.009	0.009-0.011
	EW1	1.5 (1488)	0.042	21B [2]	-	3.0	650-750	650-750	0.007-0.009	0.009-0.011
	EW3	1.5 (1488)	0.042	16B [1]	-	35	550-650	-	0.007-0.009	0.009-0.011
1986	EV1	1.3 (1342)	0.042	21B [1]	-	3.0	650-750	-	0.007-0.009	0.009-0.011
	EW1	1.5 (1488)	0.042	20B [3]	15B [3]	3.0	650-750	650-750	0.007-0.009	0.009-0.011
	EW1	1.5 (1488)	0.042	21B [3]	-	3.0	650-750	650-750	0.007-0.009	0.009-0.011
	EW3	1.5 (1488)	0.042	16B [1]	-	35	700-800	-	0.007-0.009	0.009-0.011
1987	D15A2	1.3 (1342)	0.042	21B [1]	-	3.0	650-750	-	0.007-0.009	0.009-0.011
	D13A2	1.5 (1488)	0.042	20B [1]	15B [2]	3.0	650-750	650-750	0.007-0.009	0.009-0.011
	D13A2	1.5 (1488)	0.042	26B [1]	-	3.0	650-750	650-750	0.007-0.009	0.009-0.011
	D15A3	1.5 (1488)	0.042	16B [2]	-	35	700-800	-	0.007-0.009	0.011-0.011
1988	D15B2	1.5 (1493)	0.042	14B	14B	35	600-700	700-800	0.005-0.007	0.007-0.009
	D15B2	1.5 (1493)	0.042	18B	18B	35	600-700	700-800	0.007-0.009	0.009-0.011
	D16A6	1.6 (1590)	0.042	18B	18B	35	700-800	700-800	0.007-0.009	0.009-0.011
1989	D15B2	1.5 (1493)	0.042	18B	18B	36	700-800	700-800	0.007-0.009	0.009-0.011
	D15B6	1.5 (1493)	0.042	18B	18B	36	600-650	600-650	0.005-0.007	0.007-0.009
	D16A6	1.6 (1590)	0.0042	18B	18B	36	700-800	700-800	0.007-0.009	0.009-0.011

2-24 ENGINE PERFORMANCE AND TUNE-UP

ENGINE TUNE-UP SPECIFICATIONS

Year	Engine ID/VIN	Engine Displacement Liters (cc)	Spark Plugs Gap (in.)	Ignition Timing (deg.) MT	Ignition Timing (deg.) AT	Fuel Pump (psi)	Idle Speed (rpm) MT	Idle Speed (rpm) AT	Valve Clearance In.	Valve Clearance Ex.
1990	D15B1	1.5 (1493)	0.042	18B	18B	36	700-800	700-800	0.007-0.009	0.009-0.011
	D15B2	1.5 (1493)	0.042	18B	18B	36	700-800	700-800	0.007-0.009	0.009-0.011
	D15B6	1.5 (1493)	0.042	14B	14B	36	600-700	700-800	0.005-0.007	0.007-0.009
	D16A6	1.6 (1590)	0.042	18B	18B	36	700-800	700-800	0.007-0.009	0.009-0.011
1991	D15B1	1.5 (1493)	0.042	18B	18B	36	700-800	700-800	0.007-0.009	0.009-0.011
	D15B6	1.5 (1493)	0.042	14B	14B	36	550-650	550-650	0.005-0.007	0.007-0.009
	D16A6	1.6 (1590)	0.042	18B	18B	36	700-800	700-800	0.007-0.009	0.009-0.011
1992	D15B7	1.5 (1493)	0.039-0.043	16B	16B	40	370-470	370-470	0.007-0.009	0.009-0.011
	D15B8	1.5 (1493)	0.039-0.043	12B	-	40	370-470	-	0.007-0.009	0.009-0.011
	D15Z1	1.5 (1493)	0.039-0.043	16B	16B	40	370-470	370-470	0.007-0.009	0.009-0.011
	D16Z6	1.6 (1590)	0.039-0.043	16B	16B	40	370-470	370-470	0.007-0.009	0.001-0.011
1993	D15B7	1.5 (1493)	0.039-0.043	16B	16B	31-38	620-720	650-750	0.007-0.009	0.009-0.011
	D15B8	1.5 (1493)	0.039-0.043	12B	-	31-38	620-720	-	0.007-0.009	0.009-0.011
	D15Z1	1.5 (1493)	0.039-0.043	16B	-	31-38	550-650	-	0.007-0.009	0.009-0.011
	D16Z6	1.6 (1590)	0.039-0.043	16B	16B	31-38	620-720	650-750	0.007-0.009	0.009-0.011
1994	D15B7	1.5 (1493)	0.039-0.043	16B	16B	31-38	620-720	650-750	0.007-0.009	0.009-0.011
	D15B8	1.5 (1493)	0.039-0.043	12B	-	31-38	620-720	-	0.007-0.009	0.009-0.011
	D15Z1	1.5 (1493)	0.039-0.043	16B	-	31-38	550-650	-	0.007-0.009	0.009-0.011
	D16Z6	1.6 (1590)	0.039-0.043	16B	16B	31-38	620-720	650-750	0.007-0.009	0.009-0.011
	B16A3	1.6 (1595)	0.047-0.051	16B	-	31-38	650-750	-	0.006-0.007	0.007-0.008

ENGINE TUNE-UP SPECIFICATIONS

Year	Engine ID/VIN	Engine Displacement Liters (cc)	Spark Plugs Gap (in.)	Ignition Timing (deg.) MT	Ignition Timing (deg.) AT	Fuel Pump (psi)	Idle Speed (rpm) MT	Idle Speed (rpm) AT	Valve Clearance In.	Valve Clearance Ex.
1995	D15B7	1.5 (1493)	0.039-0.043	16B	16B	31-38	620-720 ④	650-750 ④	0.007-0.009	0.009-0.011
	D15B8	1.5 (1493)	0.039-0.043	12B	-	31-38	620-720	750	0.007-0.009	0.009-0.011
	D15Z1	1.5 (1493)	0.039-0.043	16B	-	31-38	550-650 ⑥	-	0.007-0.009	0.009-0.011
	D16Z6	1.6 (1590)	0.039-0.043	16B	16B	31-38	620-720 ⑤	650-750 ④	0.007-0.009	0.009-0.011
	B16A3	1.6 (1595)	0.047-0.051	16B	-	31-38	650-750 ④	-	0.006-0.007	0.007-0.008

NOTE: The Vehicle Emission Control Information label often reflects specification changes made during production. The label figures must be used if they differ from those in this chart.

AT: Automatic Transaxle
MT: Manual Transaxle

1. Calif.: 12B
2. Std.: Calif. 16B
   MT: 20B, AT: 16B
3. Aim timing light at red mark on crankshaft pulley.
4. Canada: 700-800 rpm.
5. Canadian Civic: 650-750 rpm.
   Canadian del Sol: 700-800 rpm.
6. Canada: 650-750 rpm.

ENGINE ELECTRICAL 3-2
IGNITION COIL 3-2
DISTRIBUTOR 3-2
   REMOVAL & INSTALLATION 3-2
ALTERNATOR 3-3
   PRECAUTIONS 3-3
   REMOVAL & INSTALLATION 3-4
REGULATOR 3-4
BATTERY 3-4
   REMOVAL & INSTALLATION 3-4
STARTER 3-4
   REMOVAL & INSTALLATION 3-4
ENGINE MECHANICAL 3-5
ENGINE DESIGN 3-5
ENGINE OVERHAUL TIPS 3-5
   TOOLS 3-5
   INSPECTION TECHNIQUES 3-5
   OVERHAUL TIPS 3-5
   REPAIRING DAMAGED THREADS 3-5
CHECKING ENGINE COMPRESSION 3-6
ENGINE 3-11
   REMOVAL & INSTALLATION 3-11
CYLINDER HEAD COVER 3-19
   REMOVAL & INSTALLATION 3-19
ROCKER ARMS/SHAFTS 3-19
   REMOVAL & INSTALLATION 3-19
THERMOSTAT 3-21
   REMOVAL & INSTALLATION 3-21
THERMOSTAT HOUSING 3-22
   REMOVAL & INSTALLATION 3-22
INTAKE MANIFOLD 3-22
   REMOVAL & INSTALLATION 3-22
EXHAUST MANIFOLD 3-24
   REMOVAL & INSTALLATION 3-24
RADIATOR 3-26
   REMOVAL & INSTALLATION 3-26
CONDENSER FAN 3-27
   TESTING 3-27
   REMOVAL & INSTALLATION 3-28
COOLING FAN 3-28
   TESTING 3-28
   REMOVAL & INSTALLATION 3-28
WATER PUMP 3-28
   REMOVAL & INSTALLATION 3-28
CYLINDER HEAD 3-28
   PRECAUTIONS 3-28
   REMOVAL & INSTALLATION 3-29
   CLEANING AND INSPECTION 3-35
   RESURFACING 3-35
VALVES 3-35
   REMOVAL & INSTALLATION 3-35
   INSPECTION 3-36
   REFACING 3-36
VALVE SPRINGS 3-36
   REMOVAL & INSTALLATION 3-36
   INSPECTION 3-37
VALVE SEATS 3-37
VALVE GUIDES 3-37
   REMOVAL & INSTALLATION 3-38
   KNURLING 3-39

OIL PAN 3-39
   REMOVAL & INSTALLATION 3-39
OIL PUMP 3-41
   REMOVAL & INSTALLATION 3-41
   OIL PUMP OVERHAUL 3-42
TIMING BELT COVER AND SEAL 3-42
   REMOVAL & INSTALLATION 3-42
   OIL SEAL REPLACEMENT 3-42
TIMING BELT AND TENSIONER 3-43
   REMOVAL & INSTALLATION 3-43
CAMSHAFT SPROCKETS 3-45
   REMOVAL & INSTALLATION 3-45
CAMSHAFT 3-45
   REMOVAL & INSTALLATION 3-45
   INSPECTION 3-46
PISTONS AND CONNECTING RODS 3-46
   REMOVAL & INSTALLATION 3-46
   CLEANING AND INSPECTION 3-48
   CYLINDER BORE INSPECTION 3-48
   FITTING AND POSITIONING PISTON RINGS 3-48
REAR MAIN SEAL 3-49
   REMOVAL & INSTALLATION 3-49
CRANKSHAFT AND MAIN BEARINGS 3-49
   REMOVAL & INSTALLATION 3-49
   CLEANING AND INSPECTION BEARING OIL CLEARANCE 3-50
   CRANKSHAFT END-PLAY/ CONNECTING ROD SIDE PLAY 3-50
   CRANKSHAFT REPAIRS 3-51
FLYWHEEL/FLEX PLATE 3-51
   REMOVAL & INSTALLATION 3-51
EXHAUST SYSTEM 3-52
MUFFLER ASSEMBLY 3-52
   REMOVAL & INSTALLATION 3-52
CATALYTIC CONVERTER AND/OR PIPE ASSEMBLY 3-52
   REMOVAL & INSTALLATION 3-54
SPECIFICATIONS CHARTS
GENERAL ENGINE SPECIFICATIONS 3-7
VALVE SPECIFICATIONS 3-7
CRANKSHAFT AND CONNECTING ROD SPECIFICATIONS 3-8
PISTON AND RING SPECIFICATIONS 3-9
TORQUE SPECIFICATIONS 3-10

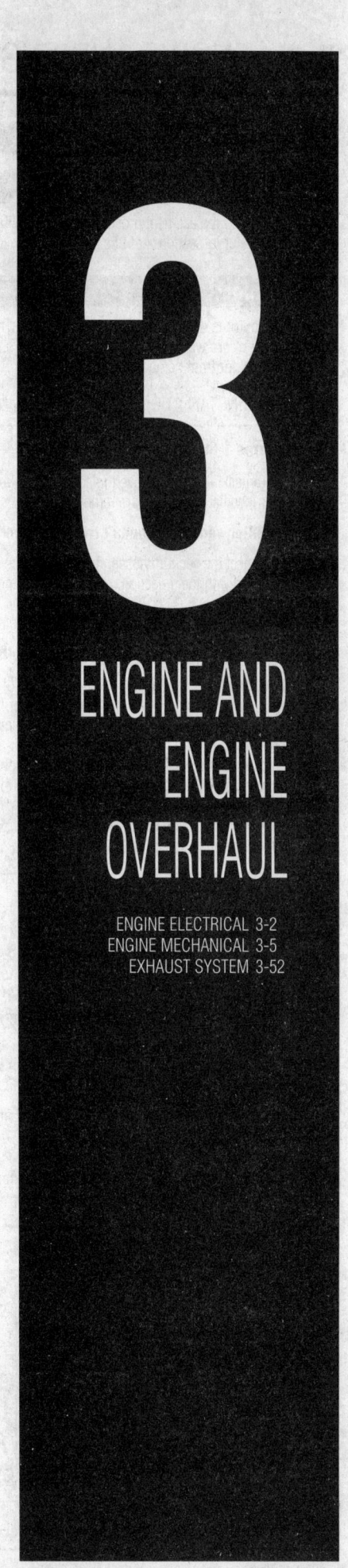

3

ENGINE AND ENGINE OVERHAUL

ENGINE ELECTRICAL 3-2
ENGINE MECHANICAL 3-5
EXHAUST SYSTEM 3-52

3-2 ENGINE AND ENGINE REBUILDING

ENGINE ELECTRICAL

Ignition Coil

For testing and replacement procedures of the ignition coil and other ignition components, refer to Section 2 of this manual.

Distributor

The distributor is driven directly by the camshaft. The end of the camshaft contains a set of grooves that engage the distributor shaft. These grooves prevent the distributor from being incorrectly installed.

REMOVAL & INSTALLATION

▶ See Figures 1 thru 7

1. Using masking tape, remove the spark plug wires from the cap and number them for installation as they are removed.

➡ From 1988-onwards, all Honda Civic series models were fuel injected.

2. Disconnect the vacuum hoses, the primary wire and the high tension wire. If equipped with fuel injection, remove the four prong connector from the crank angle sensor.
3. Remove the hold-down bolt(s) and pull the distributor from the head.

➡ On CRX with fuel injection, there are three bolts.

4. Crank the engine until No. 1 piston is at TDC of its compression stroke.
5. Using a new O-ring, install it on the distributor housing.
6. On the all models, the distributor drive gear remains engaged when the distributor is removed. To install, align the distributor and turn the shaft until the drive locks and the distributor seats in the cylinder head.
7. Loosely install the hold-down bolt(s). Tighten the bolt(s) temporarily and replace the cap.
8. Connect all wires and hoses.
9. Start the engine and adjust the ignition timing.

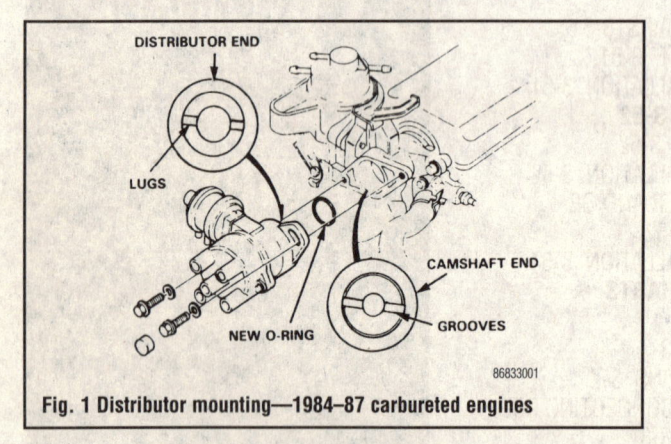

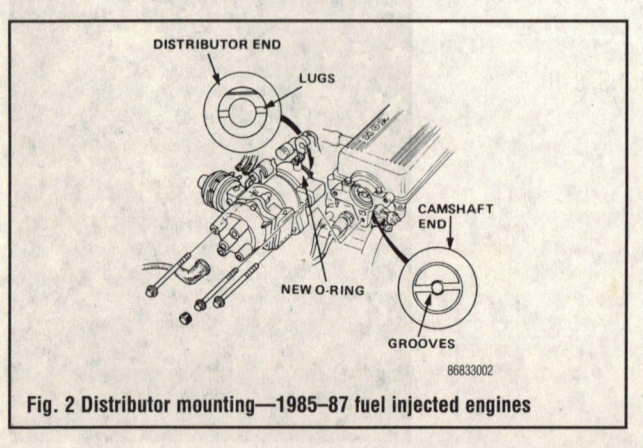

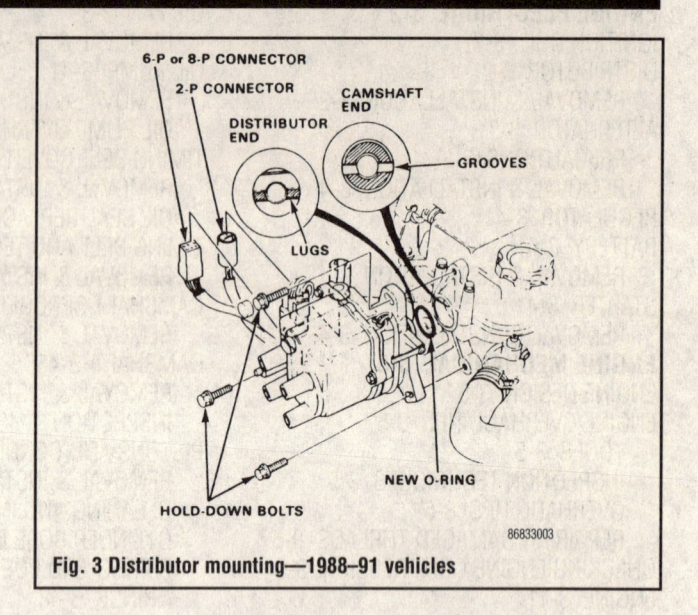

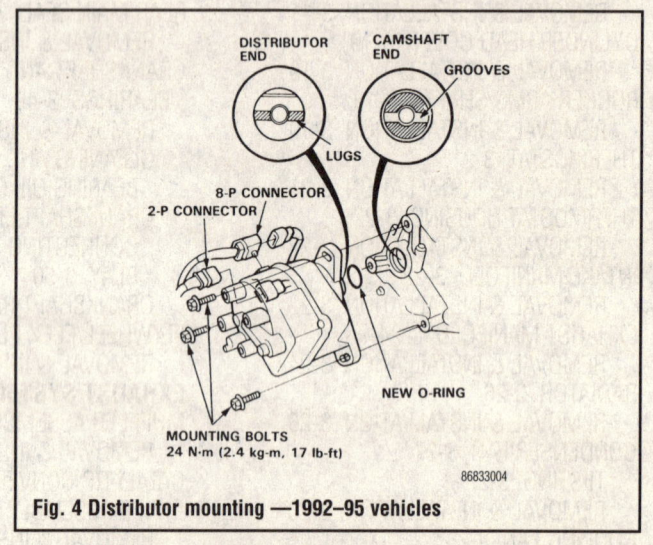

Fig. 1 Distributor mounting—1984–87 carbureted engines

Fig. 2 Distributor mounting—1985–87 fuel injected engines

Fig. 3 Distributor mounting—1988–91 vehicles

Fig. 4 Distributor mounting—1992–95 vehicles

Fig. 5 Pulling the distributor from where it mates to the camshaft at the right side of the valve cover

ENGINE AND ENGINE REBUILDING 3-3

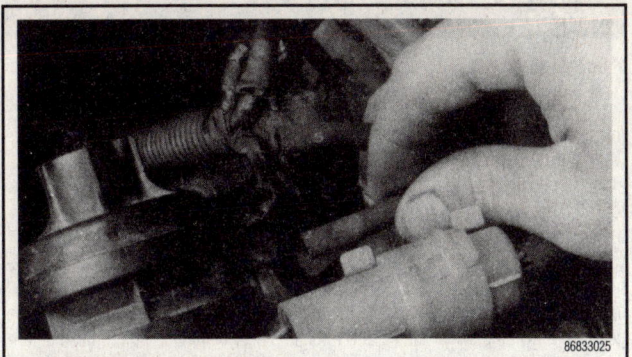

Fig. 6 Remove the distributor vacuum hoses, being careful not to get them confused. If necessary, label the hoses as performed in spark plug wire removal

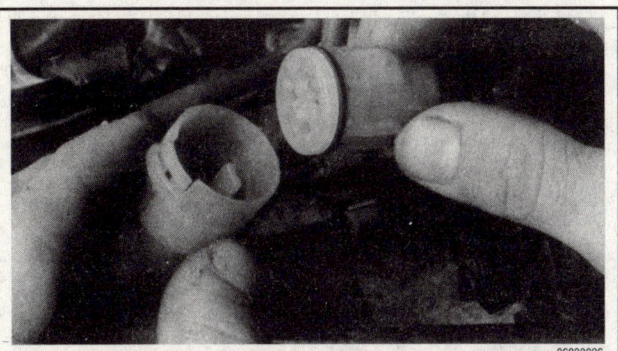

Fig. 7 Unplugging the four-prong connector from the crank angle sensor. Notice the secure snap-connection and water-resistant O-ring seal around the inner circumference

Alternator

▶ See Figures 8, 9 and 10

The alternator converts the mechanical energy which is supplied by the drive belt into electrical energy by electromagnetic induction. When the ignition switch is turned **ON**, current flows from the battery, through the charging system light or ammeter, to the voltage regulator and finally to the alternator. When the engine is started, the drive belt turns the rotating field (rotor) in the stationary windings (stator), inducing alternating current. This alternating current is converted into usable direct current by the diode rectifier. Most of this current is used to charge the battery and power the electrical components of the vehicle. A small part is returned to the field windings of the alternator

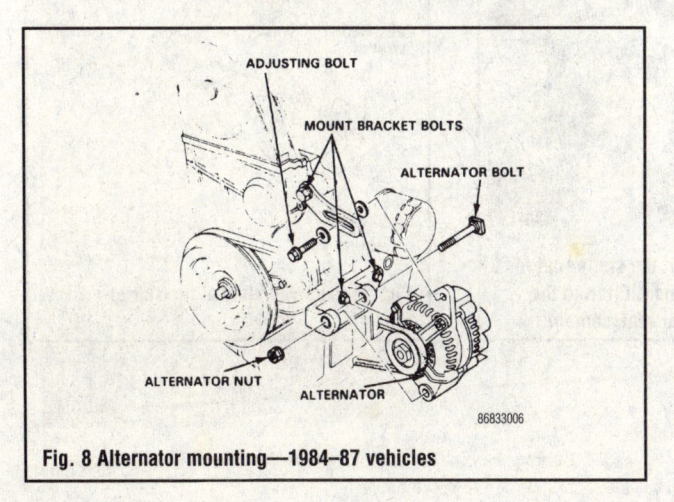

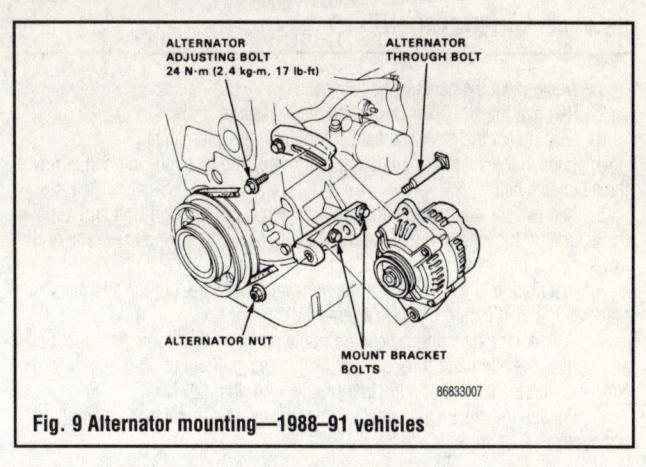

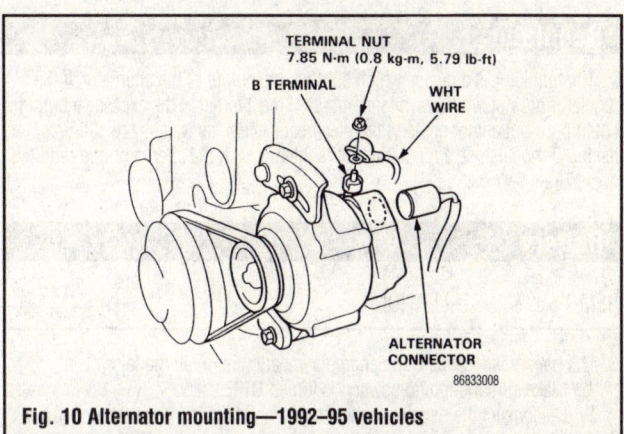

Fig. 8 Alternator mounting—1984–87 vehicles

Fig. 9 Alternator mounting—1988–91 vehicles

Fig. 10 Alternator mounting—1992–95 vehicles

enabling it to increase its output. When the current in the field windings reaches a predetermined control voltage, the voltage regulator grounds the circuit, preventing any further increase. The cycle is continued so that the voltage remains constant.

The alternator is located near the No. 1 spark plug at the front of the engine compartment. On models equipped with air conditioning, the alternator is mounted on a special vibration absorbing bracket at the driver's side of the engine compartment.

PRECAUTIONS

Several precautions must be observed with alternator equipped vehicles to avoid damage to the unit.

- If the battery is removed for any reason, make sure it is reconnected with the correct polarity. Reversing the battery connections may result in damage to the one-way rectifiers.
- When utilizing a booster battery as a starting aid, always connect the positive to positive terminals and the negative terminal from the booster battery to a good engine ground on the vehicle being started.
- Never use a fast charger as a booster to start vehicles.
- Disconnect the battery cables when charging the battery with a fast charger.
- Never attempt to polarize the alternator.
- Do not use test lamps of more than 12 volts when checking diode continuity.
- Do not short across or ground any of the alternator terminals.
- The polarity of the battery, alternator and regulator must be matched and considered before making any electrical connections within the system.
- Never separate the alternator on an open circuit. Make sure all connections within the circuit are clean and tight.
- Disconnect the battery ground terminal when performing any service on electrical components.
- Disconnect the battery if arc welding is to be done on the vehicle.

3-4 ENGINE AND ENGINE REBUILDING

REMOVAL & INSTALLATION

1. Disconnect the negative battery cable.
2. Remove the air cleaner assembly if necessary.
3. Disconnect the electrical connectors from the alternator.
4. Loosen the alternator adjusting bolt or nut and through-bolt and remove the alternator belt.
5. Remove the alternator adjusting bolt or nut and through-bolt and remove the alternator. If necessary, remove the mount bracket bolts and the upper and lower mount brackets.
6. Installation is the reverse of the removal procedure: put the alternator in place and re-bolt mount bracket (if removed).
7. Follow belt adjusting procedure in the previous procedure to adjust belt and tighten the alternator through-bolt or through-bolt nut to 33 ft. lbs. (45 Nm). Tighten the alternator adjusting bolt to 17 ft. lbs. (24 Nm).
8. Engage the electrical connectors and install the air cleaner assembly (if removed).

Regulator

The regulator is a device which controls the output of the alternator. If the regulator did not limit the voltage output of the alternator, the excessive output could burn out components of the electrical system, as well as the alternator itself. All models use a regulator that is integral with the alternator and requires no separate service.

Battery

REMOVAL & INSTALLATION

The battery is located in the engine compartment on all models.
1. Make sure the ignition switch is turned **OFF**.
2. Disconnect the negative battery cable, and then the positive cable from the battery.
3. Remove the battery hold-down clamp nuts, the clamp and the battery from the vehicle.
4. To install, reverse the removal procedures; be sure the battery is seated correctly on the battery tray.

➡ Before installing the battery, it would be a wise idea to clean the battery posts and the terminal connectors.

5. Connect the positive battery cable, and then the negative battery cable. Coat the battery terminals with a non-metallic grease; this will keep the terminals from oxidizing.

Starter

▸ See Figures 11, 12, 13 and 14

The starter is located on the firewall side of the engine block, adjacent to the flywheel or torque converter housing. Early models, use a direct drive type starter, while the later models use a gear reduction starter. Otherwise, the two units are similar in operation and service. Both starters are 4-pole, series wound, DC units to which an outboard solenoid is mounted. When the ignition switch is turned to the start position, the solenoid armature is drawn in, engaging the starter pinion with the flywheel. When the starter pinion and flywheel are fully engaged, the solenoid armature closes the main contacts for the starter, causing the starter to crank the engine. When the engine starts, the increased speed of the flywheel causes the gear to overrun the starter clutch and rotor. The gear continues in full mesh until the ignition is switched from the start to the on position, interrupting the starter current. The shift lever spring then returns the gear to its neutral position.

REMOVAL & INSTALLATION

1. Disconnect the negative battery terminal and the starter motor cable at the positive terminal.
2. Remove the engine wire harness from the harness clip on the starter motor
3. Disconnect the cable from the B terminal on the solenoid, and the BLK/WHT wire from the S terminal.
4. Remove the starter motor by loosening the two attaching bolts.
5. To install, reverse the removal procedures: re-position the starter in place, replace bolts, tighten the starter-to-engine bolts to 32 ft. lbs. (45 Nm).
6. Make sure that all wires are securely connected.

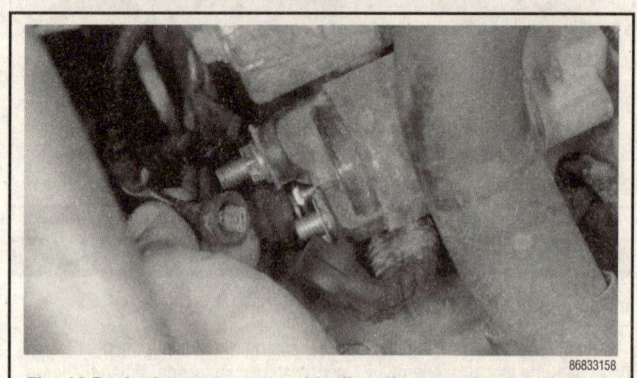

Fig. 11 Begin the starter removal by first disconnecting the wires leading to it . . .

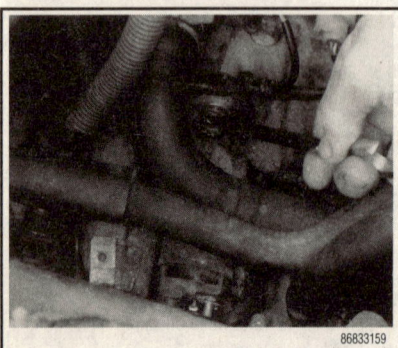

Fig. 12 . . . next unbolt the starter by loosening the through-bolts and sliding them out

Fig. 13 Wiggle and pull the starter out to disengage the gears and lift it from the engine for inspection or replacement

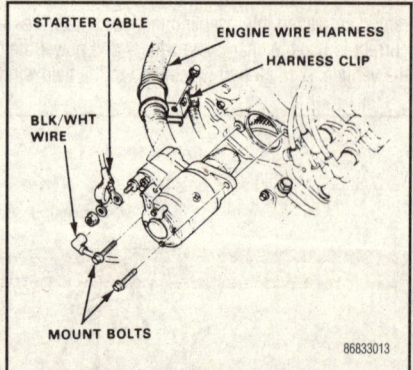

Fig. 14 Exploded view of the starter mounting

ENGINE AND ENGINE REBUILDING 3-5

ENGINE MECHANICAL

Engine Design

The engines used in the Honda Civic models are water cooled, single overhead cam (except for the DOHC del Sol VTEC), transversely mounted, inline four cylinder power plants. These engines are an evolutionary design built upon what Honda has learned from its engines over the past two decades.

In the 1970's, Honda offered two types of engine: Compound Vortex Controlled Combustion (CVCC) and Non-CVCC engines. The non-CVCC engines have been offered in two displacements; 1170cc (1973 only), and 1237cc (1974–79). These engines were unusually advanced for their time, both the engine and the cylinder head were cast in aluminum. The cylinder head had a crossflow design, the block used sleeved cylinder liners and a main bearing girdle to add rigidity to the block and the crankshaft rode on five main bearings.

The CVCC engine is unique in that its cylinder head is equipped with three valves per cylinder, instead of the usual two. This design employs the usual intake and exhaust valves, and beside each intake valve is an auxiliary intake valve which is much smaller than its counterpart. This auxiliary intake valve has its own separate precombustion chamber (adjacent to the main chamber with a crossover passage), its own intake manifold passages and carburetor circuit.

Briefly, the CVCC engine operates as follows: at the beginning of the intake stroke, a small but very rich mixture is inducted into the precombustion chamber, while next door in the main combustion chamber, a large but very lean mixture is inducted. (A rich mixture has a high proportion of fuel in the air/fuel ratio, while a lean mixture has a low proportion of fuel.) At the end of the compression stroke, ignition occurs. The spark plug, located in the precombustion chamber, easily ignites the rich auxiliary mixture and this ignition spreads out into the main combustion chamber, where the large lean mixture is ignited. This two-stage combustion process allows the engine to operate efficiently with a much leaner overall air/fuel ratio. So, whereas the 1975 and later non-CVCC engines require a belt-driven air injection system to control pollutants, the CVCC engines accomplish this internally and gets better gas mileage to boot.

On the 1984–88 models, Honda decided to improve engine breathing by replacing the single large main intake valve with a pair of smaller ones. This allows a much greater total intake valve area than a single valve and it also permits intake valve timing to be staggered slightly. This gave Honda engineers a unique opportunity to design air swirl into the combustion process. Such swirl (turbulence in the combustion chamber) not only tends to reduce engine knock but improves combustion speed and therefore engine efficiency, especially at low speeds. These engines retain the auxiliary intake valve on the exhaust side of the head.

All of the 1988 and later Honda engines have substituted fuel injection for carburetion, and the CVCC system is no more. In the interest of good fuel efficiency, increased power and low emissions, the two intake valves were retained. Fuel injection now makes CVCC obsolete because fuel injection provides the charge directly to the port of each cylinder under greater pressure than a carburetor is capable and permits the engine to run on leaner mixtures.

Another interesting feature of Honda engines used on 1988 and later Civic series models is the use of a single camshaft—instead of the more common two camshafts—to drive sixteen valves. The single camshaft consists of eight exhaust and eight intake lobes, actuating sixteen rocker arms. The only exception to date is the del Sol VTEC, which does use two camshafts.

Engine Overhaul Tips

Most engine overhaul procedures are fairly standard. In addition to specific parts replacement procedures and complete specifications for your individual engine, this section is also a guide to accepted rebuilding procedures. Examples of standard rebuilding practices are shown and should be used along with specific details concerning your particular engine.

Competent and accurate machine shop services will ensure maximum performance, reliability and engine life.

In most instances, it is more profitable for the do-it-yourself mechanic to remove, clean and inspect the component, buy the necessary parts and deliver these to a shop for actual machine work.

On the other hand, much of the rebuilding work (crankshaft, block, bearings, pistons, rods and other components) is well within the scope of the do-it-yourself mechanic.

TOOLS

The tools required for an engine overhaul or parts replacement will depend on the depth of your involvement. With a few exceptions, they will be the tools found in a mechanic's tool kit (see Section 1). More in depth work will require any or all of the following:

- a dial indicator (reading in thousandths) mounted on a universal base
- micrometers and telescope gauges
- jaw and screw-type pullers
- scraper
- valve spring compressor
- ring groove cleaner
- piston ring expander and compressor
- ridge reamer
- cylinder hone or glaze breaker
- Plastigage®
- engine stand

Use of most of these tools is illustrated in this section. Many can be rented for a one-time use from a local parts jobber or tool supply house specializing in automotive work.

Occasionally, the use of special tools is called for. See the information on Special Tools and the Safety Notice in the front of this book substituting another tool.

INSPECTION TECHNIQUES

Procedures and specifications are given in this section for inspecting, cleaning and assessing the wear limits of most major components. Other procedures such as Magnaflux® and Zyglo® can be used to locate material flaws and stress cracks. Magnaflux® is a magnetic process applicable only to ferrous materials. The Zyglo® process coats the material with a fluorescent dye penetrant and can be used on any material. Check for suspected surface cracks can be more readily made using spot check dye. The dye is sprayed onto the suspected area, wiped off and the area sprayed with a developer. Cracks will show up brightly.

OVERHAUL TIPS

Aluminum has become extremely popular for use in engines, due to its low weight. Observe the following precautions when handling aluminum parts:

- Never hot tank aluminum parts (the caustic hot-tank solution will eat the aluminum).
- Remove all aluminum parts (identification tag, etc.) from engine parts prior to hot-tanking.
- Always coat threads lightly with engine oil or anti-seize compound before installation, to prevent seizure.
- Never over-torque bolts or spark plugs, especially in aluminum threads. Stripped threads in any component can be repaired using any of several commercial repair kits (Heli-Coil®, Microdot®, Keenserts®, etc.).

When assembling the engine, any parts that will be in frictional contact must be prelubed to provide lubrication at initial start-up. Any product specifically formulated for this purpose can be used but engine oil is not recommended as a prelube.

When semi-permanent (locked, but removable) installation of bolts or nuts is desired, threads should be cleaned and coated with Loctite® or another similar commercial non-hardening sealant.

REPAIRING DAMAGED THREADS

▶ See Figures 15, 16, 17, 18 and 19

Several methods of repairing damaged threads are available. Heli-Coil® (shown here), Keenserts® and Microdot® are among the most widely used. All involve basically the same principle—drilling out stripped threads, tapping the

3-6 ENGINE AND ENGINE REBUILDING

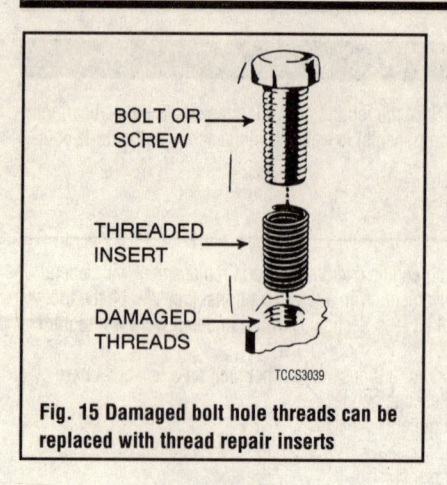

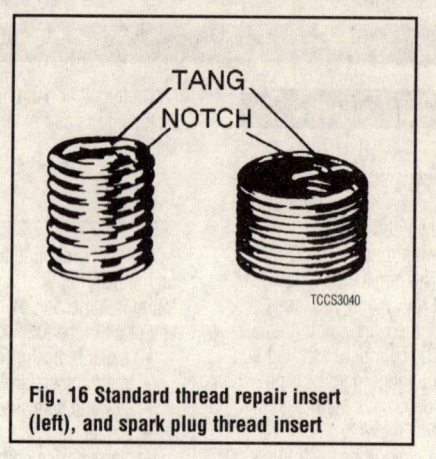

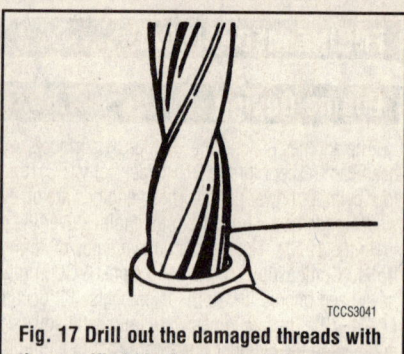

Fig. 15 Damaged bolt hole threads can be replaced with thread repair inserts

Fig. 16 Standard thread repair insert (left), and spark plug thread insert

Fig. 17 Drill out the damaged threads with the specified bit. Be sure to drill completely through the hole or to the bottom of a blind hole

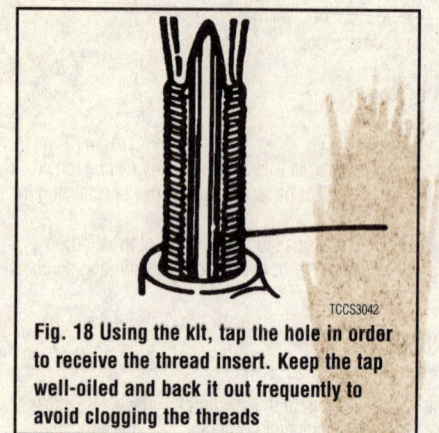

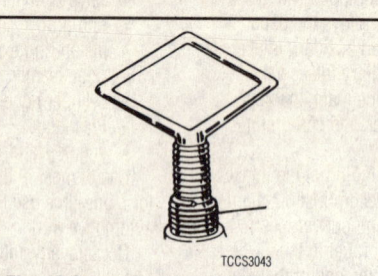

Fig. 18 Using the kit, tap the hole in order to receive the thread insert. Keep the tap well-oiled and back it out frequently to avoid clogging the threads

Fig. 19 Screw the threaded insert onto the installer tool until the tang engages the slot. Thread the insert into the hole until it is 1/4 to 1/2 turn below the top surface, then remove the tool and break off the tang using a hammer and punch

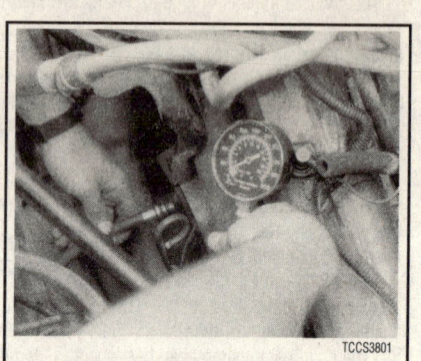

Fig. 20 The screw-in type gauge is more accurate than gauges that are merely held against the spark plug hole

hole and installing a pre-wound insert—making welding, plugging and oversize fasteners unnecessary.

Two types of thread repair inserts are usually supplied: a standard type for most Inch Coarse, Inch Fine, Metric Coarse and Metric Fine thread sizes and a spark plug type to fit most spark plug port sizes. Consult the individual manufacturer's catalog to determine exact applications. Typical thread repair kits will contain a selection of prewound threaded inserts, a tap (corresponding to the outside diameter threads of the insert) and an installation tool. Spark plug inserts usually differ because they require a tap equipped with pilot threads and a combined reamer/tap section. Most manufacturers also supply blister-packed thread repair inserts separately in addition to a master kit containing a variety of taps and inserts plus installation tools.

Before effecting a repair to a threaded hole, remove any snapped, broken or damaged bolts or studs. Penetrating oil can be used to free frozen threads; the offending item can be removed with locking pliers or with a screw or stud extractor. After the hole is clear, the thread can be repaired, as follows:

Checking Engine Compression

▶ See Figure 20

A noticeable lack of engine power, excessive oil consumption and/or poor fuel mileage measured over an extended period are all indicators of internal engine wear. Worn piston rings, scored or worn cylinder bores, blown head gaskets, sticking or burnt valves and worn valve seats are all possible culprits here. A check of each cylinder's compression will help you locate the problems.

As mentioned in the "Tools and Equipment" portion of Section 1, a screw-in type compression gauge is more accurate than the type you simply hold against the spark plug hole, although it takes slightly longer to use. It is worth the effort to obtain a more accurate reading.

1. Warm the engine to normal operating temperatures.
2. Remove all spark plugs.
3. Disconnect the high tension lead from the ignition coil.
4. On carbureted vehicles, fully open the throttle either by operating the carburetor throttle linkage by hand or by having an assistant "floor" the accelerator pedal. On the fuel injected vehicles, disconnect the cold start valve and all injector connections.
5. Screw the compression gauge into the No. 1 spark plug hole until the fitting is snug.

➡ Be careful not to cross thread the plug hole or over-tighten, as aluminum threads are easily stripped.

6. Ask an assistant to depress the accelerator pedal fully on both carbureted and fuel injected vehicles. Then, while you read the compression gauge, ask the assistant to crank the engine two or three times in short bursts using the ignition switch.
7. Read the compression gauge at the end of each series of cranks and record the highest of these readings. Repeat this procedure for each of the engine's cylinders. Compare the highest reading of each cylinder to the compression pressure specifications in the "Tune-Up Specifications" chart in Section 2. The specs in this chart are maximum values.

➡ A cylinder's compression pressure is usually acceptable if it is not less than 75 percent of maximum. For example, if the highest cylinder is 134 psi, the lowest should be 101 psi. The difference between each cylinder should be no more than 12-14 pounds.

8. If a cylinder is unusually low, pour a tablespoon of clean engine oil into the cylinder through the spark plug hole and repeat the compression test. If the compression comes up after adding the oil, it appears that the cylinder's piston rings or bore are damaged or worn. If the pressure remains low, the valves may not be seating properly (a valve job is needed) or the head gasket may be blown near that cylinder. If compression in any two adjacent cylinders is low and if the addition of oil doesn't help the compression, there is leakage past the head gasket. Oil and coolant water in the combustion chamber can result from this problem. There may be evidence of water droplets on the engine dipstick when a head gasket has blown.

ENGINE AND ENGINE REBUILDING 3-7

GENERAL ENGINE SPECIFICATIONS

Year	Engine ID/VIN	Engine Displacement Liters (cc)	Fuel System Type	Net Horsepower @ rpm	Net Torque @ rpm (ft. lbs.)	Bore x Stroke (in.)	Compression Ratio	Oil Pressure @ rpm
1984	EV1	1.3 (1342)	3 bbl	60 @ 5500	73 @ 3500	2.91 x 3.02	10.0:1	50 @ 2000
	EW1	1.5 (1488)	3 bbl	76 @ 6000	84 @ 3500	2.91 x 3.42	9.2:1	55 @ 2000
1985	EW1	1.5 (1488)	3 bbl	65 @ 5500	81 @ 3500	2.91 x 3.42	9.2:1	50 @ 2000
	EV1	1.3 (1342)	3 bbl	60 @ 5500	73 @ 3500	2.91 x 3.02	10.0:1	50 @ 2000
	EW1	1.5 (1488)	3 bbl	76 @ 6000	84 @ 3500	2.91 x 3.42	9.2:1	55 @ 2000
	EW3	1.5 (1488)	EFI	65 @ 5500	81 @ 3500	2.91 x 3.42	9.2:1	50 @ 2000
1986	EV1	1.3 (1342)	3 bbl	91 @ 5500	93 @ 4500	2.91 x 3.42	8.7:1	50 @ 2000
	EW1	1.5 (1488)	3 bbl	60 @ 5500	73 @ 3500	2.91 x 3.02	10.0:1	50 @ 2000
	EW3	1.5 (1488)	3 bbl	70 @ 6000	84 @ 3500	2.91 x 3.41	9.2:1	50 @ 2000
1987	EV1	1.3 (1342)	3 bbl	58 @ 4500	80 @ 2500	3.91 x 3.41	8.7:1	50 @ 2000
	EW1	1.5 (1488)	EFI	91 @ 5500	93 @ 4500	2.91 x 3.41	9.2:1	50 @ 2000
	EW3	1.5 (1488)	3 bbl	76 @ 6000	84 @ 3500	2.91 x 3.41	9.2:1	50 @ 2000
1988	D15A2	1.3 (1342)	3 bbl	58 @ 4500	80 @ 2500	2.91 x 3.07	10.0:1	50 @ 2000
	D13A2	1.3 (1342)	EFI	91 @ 5500	93 @ 4500	2.91 x 3.41	9.6:1	50 @ 2000
	D13A3	1.5 (1493)	DP-FI	70 @ 5500	83 @ 3000	2.95 x 3.33	8.7:1	50 @ 2000
	D15B1	1.5 (1493)	DP-FI	92 @ 6000	89 @ 4500	2.95 x 3.33	9.6:1	50 @ 2000
	D15B2	1.5 (1493)	MPFI	62 @ 4500	90 @ 2000	2.95 x 3.33	9.2:1	50 @ 2000
	D15B6	1.6 (1590)	DP-FI	105 @ 6000	98 @ 5000	2.95 x 3.54	9.2:1	74 @ 3000
1989	D16A6	1.6 (1590)	DP-FI	92 @ 6000	89 @ 4500	2.95 x 3.54	9.2:1	74 @ 3000
	D15B1	1.5 (1493)	MPFI	62 @ 4500	90 @ 2000	2.95 x 3.33	9.2:1	50 @ 3000
	D15B2	1.5 (1493)	DP-FI	70 @ 5500	83 @ 3000	2.95 x 3.33	9.2:1	50 @ 3000
	D15B6	1.6 (1590)	DP-FI	92 @ 6000	100 @ 5000	2.95 x 3.54	9.1:1	50 @ 3000
1990	D16A6	1.6 (1590)	MPFI	70 @ 5500	83 @ 3000	2.95 x 3.33	9.2:1	50 @ 3000
	D15B1	1.5 (1493)	MPFI	62 @ 4500	89 @ 4500	2.95 x 3.33	9.2:1	50 @ 3000
	D15B2	1.5 (1493)	MPFI	92 @ 6000	90 @ 2000	2.95 x 3.33	9.6:1	50 @ 3000
	D15B6	1.6 (1590)	MPFI	102 @ 5900	89 @ 4500	2.95 x 3.54	9.2:1	50 @ 3000
1991	D16A6	1.6 (1590)	MPFI	62 @ 4500	90 @ 2000	2.95 x 3.33	9.2:1	50 @ 3000
	D15B1	1.5 (1493)	MPFI	108 @ 6000	98 @ 5000	2.95 x 3.33	9.1:1	50 @ 3000
	D15B7	1.5 (1493)	DP-FI	70 @ 5000	91 @ 2000	2.95 x 3.33	9.1:1	50 @ 3000
	D15B8	1.5 (1493)	MPFI	102 @ 5900	98 @ 5000	2.95 x 3.33	9.2:1	50 @ 3000
1992	D1521	1.5 (1493)	MPFI	92 @ 5500	97 @ 4500	2.95 x 3.33	9.3:1	50 @ 3000
	D1626	1.6 (1590)	MPFI	125 @ 6600	106 @ 5200	2.95 x 3.54	9.2:1	50 @ 3000
	D15B7	1.5 (1493)	MPFI	102 @ 5900	98 @ 5000	2.95 x 3.33	9.2:1	50 @ 3000
1993	D1521	1.5 (1493)	MPFI	70 @ 5000	91 @ 2000	2.95 x 3.33	9.2:1	50 @ 3000
	D15B7	1.5 (1493)	MPFI	92 @ 5500	97 @ 4500	2.95 x 3.33	9.3:1	50 @ 3000
	D1626	1.6 (1590)	MPFI	125 @ 6600	106 @ 5200	2.95 x 3.54	9.2:1	50 @ 3000
1994	D1521	1.5 (1493)	MPFI	70 @ 5000	91 @ 2000	2.95 x 3.33	9.3:1	50 @ 3000
	D15B7	1.5 (1493)	MPFI	92 @ 5500	97 @ 4500	2.95 x 3.33	9.1:1	50 @ 3000
	D1626	1.6 (1590)	MPFI	102 @ 5900	98 @ 5000	2.95 x 3.33	9.3:1	50 @ 3000
	D15B8	1.5 (1493)	MPFI	92 @ 5500	97 @ 4500	2.95 x 3.33	9.1:1	50 @ 3000
	B16A3	1.6 (1590)	MPFI	160 @ 7600	111 @ 7000	3.19 x 3.05	10.2:1	50 @ 3000
1995	D15B7	1.5 (1493)	MPFI	102 @ 5900	98 @ 5000	3.19 x 3.05	9.2:1	50 @ 3000
	D15B8	1.5 (1493)	MPFI	70 @ 5000	91 @ 2000	2.95 x 3.33	9.1:1	50 @ 3000
	D1521	1.5 (1493)	MPFI	92 @ 5500	97 @ 4500	2.95 x 3.54	9.3:1	50 @ 3000
	D1626	1.6 (1590)	MPFI	125 @ 6600	106 @ 5200	2.95 x 3.54	9.2:1	50 @ 3000
	B16A3	1.6 (1595)	MPFI	160 @ 7600	111 @ 7000	3.19 x 3.05	10.2:1	50 @ 3000

EFI: Electronic Fuel Injection
DP-FI: Dual Point Fuel Injection
MPFI: Multi Port Fuel Injection

VALVE SPECIFICATIONS

Year	Engine ID/VIN	Engine Displacement Liters (cc)	Seat Angle (deg.)	Face Angle (deg.)	Spring Test Pressure (lbs. @ in.)	Spring Installed Height (in.)	Stem-to-Guide Clearance (in.) Intake	Stem-to-Guide Clearance (in.) Exhaust	Stem Diameter (in.) Intake	Stem Diameter (in.) Exhaust
1984	EV1	1.3 (1342)	45	45	NA	1.896 [1]	0.001-0.0020	0.002-0.0030	0.2591-0.2594	0.2579-0.2583
	EW1	1.5 (1488)	45	45	NA		0.001-0.0020	0.002-0.0030	0.2591-0.2594	0.2732-0.2736
	EW3						0.001-0.0020	0.002-0.0030	0.2591-0.2594	0.2579-0.2583
1985	EV1	1.3 (1342)	45	45	NA	1.896 [1]	0.001-0.0020	0.002-0.0030	0.2591-0.2594	0.2679-0.2583
	EW1	1.5 (1488)	45	45	NA	1.896 [1]	0.001-0.0020	0.002-0.0030	0.2591-0.2594	0.2732-0.2736
	EW3						0.001-0.0020	0.002-0.0030	0.2591-0.2594	0.2579-0.2583
1986	EV1	1.3 (1342)	45	45	NA	1.896 [1]	0.001-0.0020	0.002-0.0030	0.2591-0.2594	0.2579-0.2583
1987	EW1	1.5 (1488)	45	45	NA	1.896 [1]	0.001-0.0020	0.002-0.0030	0.2591-0.2594	0.2579-0.2583
	EW3						0.001-0.0020	0.002-0.0030	0.2591-0.2594	0.2579-0.2583
	D15A2	1.3 (1342)	45	45	NA	[3]	0.001-0.0020	0.002-0.0030	0.2591-0.2594	0.2579-0.2583
1988	D13A2	1.5 (1488)	45	45	NA	[3]	0.001-0.0020	0.002-0.0030	0.2591-0.2594	0.2579-0.2583
	D13A3						0.001-0.0020	0.002-0.0030	0.2591-0.2594	0.2583
	D15B2	1.5 (1493)	45	45	NA	[2]	0.001-0.0020	0.002-0.0030	0.2157-0.2161	0.2147-0.2150
	D15B6									
	D16A6	1.6 (1590)	45	45	NA	[2]	0.001-0.0020	0.002-0.0030	0.2157-0.2161	0.2147-0.2150
1989	D15B1	1.5 (1493)	45	45	NA	[2]	0.001-0.0020	0.002-0.0030	0.2157-0.2161	0.2147-0.2150
	D15B2									
	D15B6									
	D16A6	1.6 (1590)	45	45	NA	[2]	0.001-0.0020	0.002-0.0030	0.2157-0.2161	0.2147-0.2150
1990	D15B1	1.5 (1493)	45	45	NA	NA	0.001-0.0020	0.002-0.0030	0.2157-0.2161	0.2147-0.2150
	D16A6	1.6 (1590)	45	45	NA	NA	0.001-0.0020	0.002-0.0030	0.2157-0.2161	0.2147-0.2150
1991	D15B1	1.5 (1493)	45	45	NA	NA	0.001-0.0020	0.002-0.0030	0.2157-0.2161	0.2147-0.2150
	D16A6	1.6 (1590)	45	45	NA	NA	0.001-0.0020	0.002-0.0030	0.2157-0.2161	0.2147-0.2150
1992	D15B7	1.5 (1493)	45	45	NA	NA	0.001-0.0020	0.002-0.0030	0.2157-0.2161	0.2146-0.2150
	D1626	1.6 (1590)	45	45	NA	NA	0.001-0.0020	0.002-0.0030	0.2157-0.2161	0.2146-0.2150
1993	D15B7	1.5 (1493)	45	45	NA	NA	0.0010-0.0030	0.0020-0.0050	0.2146-0.2161	0.2134-0.2150
	D15B8	1.5 (1493)	45	45	NA	NA	0.0010-0.0030	0.0020-0.0050	0.2146-0.2161	0.2134-0.2150
	D1521	1.5 (1493)	45	45	NA	NA	0.0010-0.0030	0.0020-0.0050	0.2146-0.2161	0.2134-0.2150
	D1626	1.6 (1590)	45	45	NA	NA	0.0010-0.0030	0.0020-0.0050	0.2146-0.2161	0.2134-0.2150

3-8 ENGINE AND ENGINE REBUILDING

CRANKSHAFT AND CONNECTING ROD SPECIFICATIONS
All measurements are given in inches.

Year	Engine ID/VIN	Engine Displacement Liters (cc)	Main Brg. Journal Dia.	Crankshaft Main Brg. Oil Clearance	Shaft End-play	Thrust on No.	Journal Diameter	Connecting Rod Oil Clearance	Side Clearance
1984	EV1	1.3 (1342)	1.7707-1.7717	0.0009-0.0017	0.004-0.014	3	1.4951-1.4961	0.0008-0.0015	0.006-0.0120
	EW1	1.5 (1488)	1.9676-1.9685	0.0009-0.0017	0.004-0.014	3	1.6626-1.6635	0.0008-0.0015	0.006-0.0120
1985	EV1	1.3 (1342)	1.7707-1.7717	0.0009-0.0017	0.004-0.014	3	1.4951-1.4961	0.0008-0.0015	0.006-0.0120
	EW1 EW3	1.5 (1488)	1.9676-1.9685	0.0009-0.0017	0.004-0.014	3	1.6626-1.6635	0.0008-0.0015	0.006-0.0120
1986	EV1	1.3 (1342)	1.7707-1.7717	0.0009-0.0017	0.0040-0.0140	3	1.4951-1.4961	0.0008-0.0015	0.006-0.012
	EW1 EW3	1.5 (1488)	1.9676-1.9685	0.0009-0.0017	0.004-0.0140	3	1.6626-1.6535	0.0008-0.0015	0.006-0.012
1987	D15A2	1.3 (1342)	1.7707-1.7717	0.0009-0.0017	0.004-0.014	3	1.4951-1.4961	0.0008-0.0015	0.006-0.012
	D13A2 D15A3	1.5 (1488)	1.9676-1.9685	0.0009-0.0017	0.004-0.014	3	1.6526-1.6535	0.0008-0.0015	0.006-0.012
1988	D15B2	1.5 (1493)	1.7707-1.7718	0.0010-0.0017	0.004-0.0140	4	1.7707-1.7717	0.0008-0.0015	0.006-0.012
	D16A6	1.6 (1590)	2.1644-2.1654	0.0010-0.0017	0.004-0.0140	4	1.6626-1.6635	0.0008-0.0015	0.006-0.012
1989	D15B1 D15B2	1.5 (1493)	1.7707-1.7718	0.0010-0.0017	0.004-0.0140	4	1.6626-1.6635	0.0008-0.0015	0.006-0.012
	D16A6	1.6 (1590)	2.1644-2.1654	0.0010-0.0017	0.004-0.0140	4	1.7707-1.7717	0.0008-0.0015	0.006-0.012
1990	D15B1	1.5 (1493)	1.7707-1.7718	5	0.004-0.014	4	1.6526-1.6535	0.0008-0.0015	0.008-0.014
	D16A6	1.6 (1590)	2.1644-2.1654	6	0.004-0.0140	4	1.7707-1.7717	0.0008-0.0015	0.008-0.014
1991	D15B1	1.5 (1493)	1.7707-1.7718	5	0.004-0.014	4	1.6526-1.6535	0.0008-0.0015	0.006-0.012
	D16A6	1.6 (1590)	2.1644-2.1654	6	0.004-0.014	4	1.7707-1.7717	0.0008-0.0015	0.006-0.012
1992	D15B7	1.5 (1493)	1.7707-1.7717	5	0.004-0.014	4	1.6526-1.6535	0.0010-0.0017	0.008-0.014
	D16Z6	1.6 (1590)	2.1644-2.1654	6	0.004-0.014	4	1.7707-1.7717	0.0010-0.0017	0.008-0.014
1993	D15B7	1.5 (1493)	1.7707-1.7717	7	0.0040-0.0180	4	1.7707-1.7717	0.0008-0.0020	0.0060-0.0160
	D15B8	1.5 (1493)	1.7707-1.7717	7	0.0040-0.0180	4	1.7707-1.7717	0.0008-0.0020	0.0060-0.0160
	D15Z1	1.5 (1493)	1.7707-1.7717	7	0.0040-0.0180	4	1.7707-1.7717	0.0010-0.0020	0.0060-0.0160
	D16Z6	1.6 (1590)	2.1644-2.1654	7	0.0040-0.0180	4	1.6626-1.6535	0.0008-0.0020	0.0060-0.0160

VALVE SPECIFICATIONS

Year	Engine ID/VIN	Engine Displacement Liters (cc)	Seat Angle (deg.)	Face Angle (deg.)	Spring Test Pressure (lbs. @ in.)	Spring Installed Height (in.)	Stem-to-Guide Clearance (in.) Intake	Stem-to-Guide Clearance (in.) Exhaust	Stem Diameter (in.) Intake	Stem Diameter (in.) Exhaust
1994	D15B7	1.5 (1493)	45	45	NA	NA	0.0010-0.0030	0.0020-0.0050	0.2146-0.2161	0.2134-0.2150
	D15B8	1.5 (1493)	45	45	NA	NA	0.0010-0.0030	0.0020-0.0050	0.2146-0.2161	0.2134-0.2150
	D15Z1	1.5 (1493)	45	45	NA	NA	0.0010-0.0030	0.0020-0.0050	0.2146-0.2161	0.2134-0.2150
	D16Z6	1.6 (1590)	45	45	NA	NA	0.0010-0.0030	0.0020-0.0050	0.2146-0.2161	0.2134-0.2150
	B16A3	1.6 (1595)	45	45	NA	NA	0.0010-0.0030	0.0020-0.0040	0.2144-0.2459	0.2134-0.2150
1995	D15B7	1.5 (1493)	45	45	NA	NA	0.0010-0.0030	0.0020-0.0040	0.2150-0.2160	0.2130-0.2150
	D15B8	1.5 (1493)	45	45	NA	NA	0.0010-0.0030	0.0020-0.0040	0.2150-0.2160	0.2130-0.2150
	D15Z1	1.5 (1493)	45	45	NA	NA	0.0010-0.0030	0.0020-0.0050	0.2150-0.2160	0.2130-0.2150
	D16Z6	1.6 (1590)	45	45	NA	NA	0.0010-0.0030	0.0020-0.0050	0.2150-0.2160	0.2130-0.2150
	B16A3	1.6 (1595)	45	45	NA	NA	0.0010-0.0030	0.0020-0.0040	0.2144-0.2459	0.2134-0.2150

NA - Not Available
1. Auxiliary: 1.311
2. 1493cc and 1590cc:
   Intake: 1.8498-1.8683
   Exhaust: 1.9278-1.9463
3. 1342cc and 1488cc:
   Intake: 1.660
   Exhaust: 1.690
   Auxiliary: 0.098 (carbureted)

ENGINE AND ENGINE REBUILDING 3-9

PISTON AND RING SPECIFICATIONS
All measurements are given in inches.

Year	Engine ID/VIN	Engine Displacement Liters (cc)	Piston Clearance	Ring Gap Top Compression	Ring Gap Bottom Compression	Ring Gap Oil Control	Ring Side Clearance Top Compression	Ring Side Clearance Bottom Compression	Ring Side Clearance Oil Control
1984	EV1	1.3 (1342)	0.00040-0.0020	0.0060-0.0014	0.0060-0.0140	0.0060-0.0140	0.0012-0.0024	0.0012-0.0024	Snug
1985	EW1	1.5 (1488)	0.0040-0.0020	0.0060-0.0140	0.0060-0.0140	0.0060-0.0140	0.0012-0.0024	0.0012-0.0024	Snug
1986	EV1	1.3 (1342)	0.00040-0.0020	0.0060-0.0014	0.0060-0.0140	0.0060-0.0140	0.0012-0.0024	0.0012-0.0024	Snug
	EW1 EW3	1.5 (1488)	0.00040-0.0020	0.0060-0.0014	0.0060-0.0140	0.0060-0.0140	0.0012-0.0024	0.0012-0.0024	Snug
	EV1	1.3 (1342)	0.0040-0.0020	0.0060-0.0014	0.0060-0.0140	0.0080-0.0240	0.0012-0.0024	0.0012-0.0022	Snug
1987	EW1 EW3	1.5 (1488)	0.00040-0.0020	0.0060-0.0014	0.0060-0.0140	0.0080-0.0240	0.0012-0.0024	0.0012-0.0022	Snug
	D13A2	1.3 (1342)	0.0040-0.0020	0.0060-0.0014	0.0060-0.0140	0.0080-0.0240	0.0012-0.0024	0.0012-0.0022	Snug
1988	D15A2	1.5 (1488)	0.00040-0.0020	0.0060-0.0014	0.0060-0.0140	0.0080-0.0240	0.0012-0.0024	0.0012-0.0022	Snug
	D15B1 D15B2 D15B6	1.5 (1493)	0.00040-0.0016	0.0060-0.0014	0.0060-0.0140	0.0080-0.0240	0.0012-0.0024	0.0012-0.0022	Snug
	D16A6	1.6 (1590)	0.00040-0.0016	0.0060-0.0140	0.0060-0.0140	0.0080-0.0240	0.0012-0.0024	0.0012-0.0022	Snug
1989	D15B1 D15B2 D15B6	1.5 (1493)	0.0040-0.0016	0.0060-0.0120	0.012-0.0180	0.0080-0.0240	0.0012-0.0024	0.0012-0.0022	Snug
	D16A6	1.6 (1590)	0.00040-0.0016	0.0060-0.0014	0.0120-0.0180	0.0120-0.0180	0.0012-0.0024	0.0012-0.0024	Snug
1990	D15B1	1.5 (1493)	0.00040-0.0016	0.0060-0.0014	0.0120-0.0180	0.0120-0.0180	0.0012-0.0024	0.0012-0.0024	Snug
	D16A6	1.6 (1590)	0.00040-0.0016	0.0060-0.0014	0.0120-0.0180	0.0080-0.0310	0.0012-0.0024	0.0012-0.0024	Snug
1991	D15B1	1.5 (1493)	0.00040-0.0020	0.0060-0.0140	0.0120-0.0180	0.0080-0.0310	0.0012-0.0024	0.0012-0.0024	Snug
	D16A6	1.6 (1590)	0.00040-0.0020	0.0060-0.0140	0.0120-0.0180	0.0080-0.0310	0.0012-0.0024	0.0012-0.0024	Snug
1992	D15B1	1.5 (1493)	0.00040-0.0016	0.0060-0.0014	0.0120-0.0180	0.0080-0.0280	0.0012-0.0024	0.0012-0.0024	Snug
	D16Z6	1.6 (1590)	0.00040-0.0016	0.0060-0.0140	0.0120-0.0180	0.0080-0.0240	0.0012-0.0024	0.0012-0.0024	Snug
1993	D15B7	1.5 (1493)	0.00040-0.0020	0.0060-0.0240	0.0120-0.0280	0.0080-0.0310	0.0014-0.0050	0.0012-0.0050	Snug
	D15B8	1.5 (1493)	0.00040-0.0020	0.0060-0.0240	0.0120-0.0280	0.0080-0.0310	0.0014-0.0050	0.0012-0.0050	Snug
	D15Z1	1.5 (1493)	0.00040-0.0020	0.0060-0.0240	0.0120-0.0280	0.0080-0.0310	0.0014-0.0050	1	Snug
	D16Z6	1.6 (1590)	0.00040-0.0020	0.0060-0.0240	0.0120-0.0280	0.0080-0.0310	0.0012-0.0050	0.0012-0.0050	Snug

CRANKSHAFT AND CONNECTING ROD SPECIFICATIONS
All measurements are given in inches.

Year	Engine ID/VIN	Engine Displacement Liters (cc)	Crankshaft Main Brg. Journal Dia.	Crankshaft Main Brg. Oil Clearance	Shaft End-play	Thrust on No.	Connecting Rod Journal Diameter	Connecting Rod Oil Clearance	Connecting Rod Side Clearance
1994	D15B7	1.5 (1493)	1.7707-1.7717	7	0.0040-0.0180	4	1.6526-1.6535	0.0008-0.0020	0.0060-0.0160
	D15B8	1.5 (1493)	1.7707-1.7717	7	0.0040-0.0180	4	1.6526-1.6535	0.0008-0.0020	0.0060-0.0160
	D15Z1	1.5 (1493)	1.7707-1.7717	7	0.0040-0.0180	4	1.6526-1.6535	0.0008-0.0020	0.0060-0.0160
	D16Z6	1.6 (1590)	2.1644-2.1654	7	0.0040-0.0180	4	1.7707-1.7717	0.0008-0.0020	0.0060-0.0160
	B16A3	1.6 (1595)	8	2	0.0040-0.0180	4	1.7707-1.7717	0.0013-0.0024	0.0060-0.0160
1995	D15B7	1.5 (1493)	1.7707-1.7717	1	0.0040-0.0180	4	1.6526-1.6535	0.0008-0.0020	0.0060-0.0160
	D15B8	1.5 (1493)	1.7707-1.7717	1	0.0040-0.0180	4	1.6526-1.6535	0.0008-0.0020	0.0060-0.0160
	D15Z1	1.5 (1493)	1.7707-1.7717	1	0.0040-0.0180	4	1.6526-1.6535	0.0008-0.0020	0.0060-0.0160
	D16Z6	1.6 (1590)	2.1644-2.1654	1	0.0040-0.0180	4	1.7707-1.7717	0.0008-0.0020	0.0060-0.0160
	B16A3	1.6 (1595)	8	2	0.0040-0.0180	4	1.7707-1.7717	0.0013-0.0024	0.0060-0.0160

1. No. 3: 0.0013-0.0024
2. No. 1-2, 4-5: 0.0009-0.0020
   No. 3: 0.0012-0.0024
3. No. 2, 3, 4: 0.0010-0.0017
   No. 1, 5: 0.00007-0.0014
4. No. 2, 4: 0.0010-0.0017
5. No. 2, 3, 4: 0.0010-0.0017
   No. 1, 5: 0.00007-0.0014
6. No. 1, 5: 0.0010-0.0017
   No. 2, 4: 0.0010-0.0017
   No. 3: 0.0012-0.0019
7. No. 2-4: 0.0009-0.0020
   No. 1, 5: 0.00007-0.0020
8. No. 1-2, 4-5: 2.1644-2.1654
   No. 3: 2.1642-2.1651

3-10 ENGINE AND ENGINE REBUILDING

TORQUE SPECIFICATIONS
All readings in ft. lbs.

Year	Engine ID/VIN	Engine Displacement Liters (cc)	Cylinder Head Bolts	Main Bearing Bolts	Rod Bearing Bolts	Crankshaft Damper Bolts	Flywheel Bolts	Manifold Intake	Manifold Exhaust	Spark Plugs	Lug Nut
1984	EV1	1.3 (1342)	43	33	20	83	76 [1]	16	23	13	80
	EV1	1.5 (1488)	43	33	20	83	76 [1]	16	23	13	80
1985	EV1	1.3 (1342)	43	33	20	83	76 [1]	16	23	13	80
	EW1	1.5 (1488)	43	33	20	83	76 [1]	16	23	13	80
	EW3										
1986	EV1	1.3 (1342)	43	33	20	83	76 [1]	16	23	13	80
	EW1	1.5 (1488)	43	33	20	83	76 [1]	16	23	13	80
	EW3										
1987	D15A2	1.3 (1342)	43	33	20	83	76 [1]	16	23	13	80
	D13A2	1.5 (1488)	43	33	20	83	76 [1]	16	23	13	80
1988	D15A3	1.5 (1493)	49	48	23	83	87 [1]	25	25	13	80
	D15B2										
	D15B6										
1989	D16A6	1.6 (1590)	49	48	23	83	87 [1]	25	25	13	
	D15B1	1.5 (1493)	47	33 [2]	23	119 [2]	87 [1]	16	23	13	80
	D15B2										
	D15B6										
1990	D16A6	1.6 (1590)	47	33 [2]	23	119 [2]	87 [1]	16	23	13	80
	D15B1	1.5 (1493)	47	47 [2]	23	119 [2]	87 [1]	16	23	13	80
1991	D16A6	1.6 (1590)	47	33 [2]	23	119 [2]	87 [1]	16	23	13	80
	D15B1	1.5 (1493)	47	47 [2]	23	119 [2]	87 [1]	16	23	13	80
1992	D16A6	1.6 (1590)	47	33 [2]	23	119 [2]	87 [1]	16	23	13	80
	D16Z6	1.6 (1590)	47	38 [2]	23	119 [2]	87 [1]	16	23	13	80
1993	D15B7	1.5 (1493)	47	33 [2]	23	134 [2]	87 [1]	17	23	13	80
	D15B8	1.5 (1493)	47	33 [2]	23	134 [2]	87 [1]	17	23	13	80
	D15Z1	1.5 (1493)	53	33 [2]	23	134 [2]	87 [1]	17	23	13	80
	D16Z6	1.6 (1590)	53	38 [2]	23	134 [2]	87 [1]	17	23	13	80
1994	D15B7	1.5 (1493)		33 [3]	23	134	87 [1]	17	23	13	80
	D15B8	1.5 (1493)		33 [3]	23	134	87 [1]	17	23	13	80
	D15Z1	1.5 (1493)		33 [6]	23	134	87 [1]	17	23	13	80
	D16Z6	1.6 (1590)		38 [6]	23	134	87 [1]	17	23	13	80
	B16A3	1.6 (1595)		58 [6]	30	130	87 [1]	17	23	13	80
1995	D15B7	1.5 (1493)		33 [3]	23	134	87 [1]	17	23	13	80
	D15B8	1.5 (1493)		33 [6]	23	134	87 [1]	17	23	13	80
	D15Z1	1.5 (1493)	4	33 [6]	23	134	87 [1]	17	23	13	80
	D16Z6	1.6 (1590)	4	38 [6]	23	134	87 [1]	17	23	13	80
	B16A3	1.6 (1595)	5	58 [6]	30	130	87	17	23	13	80

1. Automatic transaxle: 54 ft. lbs.
2. Dip bolts in clean engine oil
3. Step 1: 22 ft. lbs. Step 2: 47 ft. lbs.
4. Step 1: 22 ft. lbs. Step 2: 53 ft. lbs.
5. Step 1: 22 ft. lbs. Step 2: 61 ft. lbs.
6. 18 ft. lbs.

PISTON AND RING SPECIFICATIONS
All measurements are given in inches.

Year	Engine ID/VIN	Engine Displacement Liters (cc)	Piston Clearance	Ring Gap Top Compression	Ring Gap Bottom Compression	Oil Control	Ring Side Clearance Top Compression	Ring Side Clearance Bottom Compression	Oil Control
1994	D15B7	1.5 (1493)	0.00040-0.0020	0.0060-0.0240	0.0120-0.0280	0.0080-0.0310	0.0012-0.0050	0.0012-0.0050	Snug
	D15B8	1.5 (1493)	0.00040-0.0016	0.0060-0.0240	0.0120-0.0280	0.0080-0.0310	0.0012-0.0050	0.0012-0.0050	Snug
	D15Z1	1.5 (1493)	0.00040-0.0016	0.0060-0.0240	0.0120-0.0280	0.0080-0.0310	0.0014-0.0050		Snug [1]
	D16Z6	1.6 (1590)	0.00040-0.0016	0.0060-0.0240	0.0120-0.0280	0.0080-0.0310	0.0012-0.0050	0.0012-0.0050	Snug
	B16A3	1.6 (1595)	0.00040-0.0016	0.0080-0.0240	0.0160-0.0280	0.0080-0.0310	0.0018-0.0051		Snug [2]
1995	D15B7	1.5 (1493)	0.00040-0.0020	0.0060-0.0240	0.0120-0.0280	0.0080-0.0350	0.0012-0.0050	0.0012-0.0050	Snug
	D15B8	1.5 (1493)	0.00040-0.0016	0.0060-0.0240	0.0120-0.0280	0.0080-0.0360	0.0012-0.0050	0.0012-0.0050	Snug
	D15Z1	1.5 (1493)	0.00040-0.0016	0.0060-0.0240	0.0120-0.0280	0.0080-0.0310	0.0014-0.0050		Snug [1]
	D16Z6	1.6 (1590)	0.00040-0.0016	0.0060-0.0240	0.0120-0.0280	0.0080-0.0310	0.0012-0.0050	0.0012-0.0050	Snug
	B16A3	1.6 (1595)	0.00040-0.0020	0.0080-0.0240	0.0160-0.0280	0.0080-0.0310	0.0018-0.0051		Snug [2]

1. TEIKOKU manufactured piston ring: 0.0014-0.0050
   RIKEN manufactured piston ring: 0.0012-0.0050
2. TEIKOKU manufactured piston ring: 0.0015-0.0051
   RIKEN manufactured piston ring: 0.0018-0.0051

ENGINE AND ENGINE REBUILDING 3-11

Engine

REMOVAL & INSTALLATION

✽✽ CAUTION

If the vehicle you are working on has air conditioning and the repair operation requires removal of a component of the air conditioning system, do not disconnect the refrigerant lines. If it is impossible to move the component out of the way with the lines attached, the air conditioning system must be evacuated. (Please refer to Section 1 before you decide to evacuate the system.) The air conditioning system contains R-12 or R134a refrigerant gas under pressure. This gas can be very dangerous to your health, as well as the environment, and damage to the air conditioning system could result. Consult your local laws concerning refrigerant discharge and recycling. In many areas, it may be illegal for anyone but a certified technician to service the A/C system. Always use an approved recovery station when discharging the air conditioning.

1984–87 Models

▶ See Figures 21 thru 30

1. Apply the parking brake and place blocks behind the rear wheels. Raise the front of the vehicle and support it on jackstands.
2. Disconnect both battery cables from the battery. Remove the battery and the battery tray from the engine compartment.
3. Using a scratch awl or a permanent marking pen, enscribe a line where the hood brackets meet the inside of the hood; this will help realign the hood during the installation. Remove the hood-to-bracket bolts and the hood.

➡Use of an instant camera to record component placement and hook-up could prove invaluable to a trouble-free re-installation. Take time to carefully remove parts and have a designated "safe" place to store them where things will not get lost. Tag and label fasteners (if necessary) or loosely thread them back for secure keeping. Tag all wires, connectors, hoses and belts (as necessary) that might be confusing when it is time to put it all back together.

4. Remove the engine and wheel well splash shields.

✽✽ CAUTION

Use care when removing radiator cap and when draining oil to avoid scalding by hot engine coolant, steam or oil. The EPA warns that prolonged contact with used engine oil may cause a number of skin disorders, including cancer. You should make every effort to minimize your exposure to used engine oil. Protective gloves should be worn when changing the oil. Wash your hands and any other exposed skin areas as soon as possible after exposure to used engine oil. Soap and water, or waterless hand cleaner should be used.

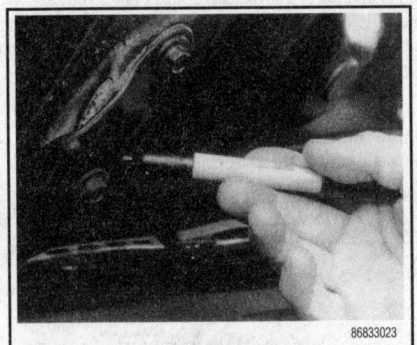

Fig. 21 Removing the hood makes engine removal easier. Here, the bracket-to-hood alignment is being marked

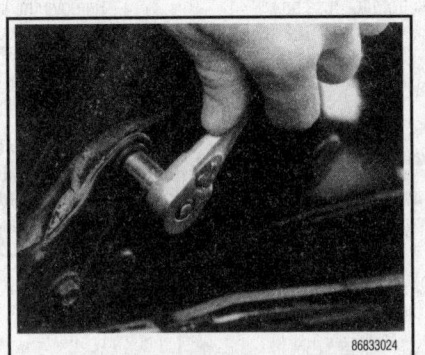

Fig. 22 You should have an assistant hold the hood as it is detached to keep it from falling into the cowling

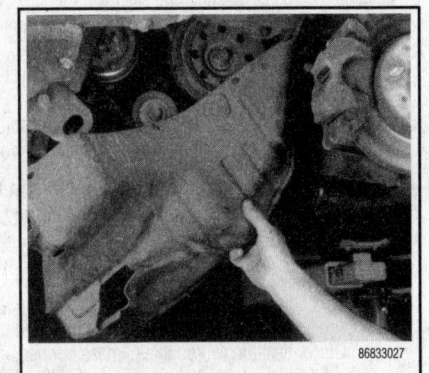

Fig. 23 Removing a driver's side wheel well splash shield

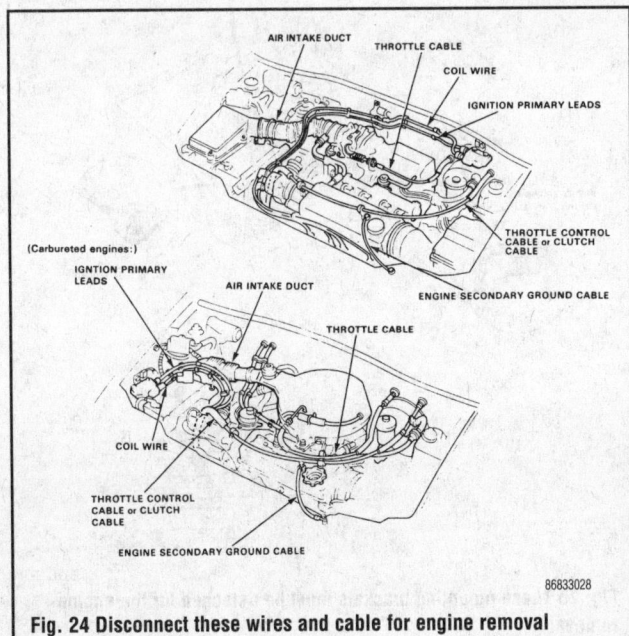

Fig. 24 Disconnect these wires and cable for engine removal

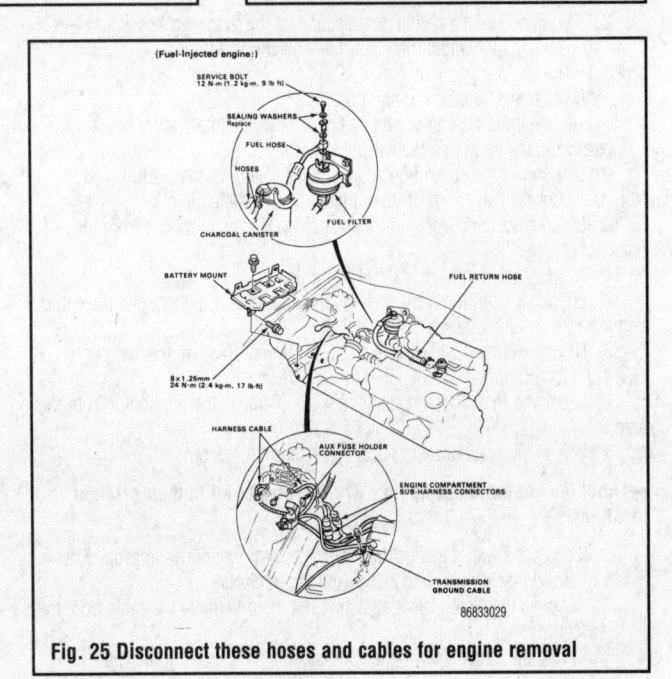

Fig. 25 Disconnect these hoses and cables for engine removal

3-12 ENGINE AND ENGINE REBUILDING

➡Always dispose of waste engine fluids properly.

   5. Drain the engine oil, coolant and transaxle fluid.

✷✷ CAUTION

When draining coolant, keep in mind that cats and dogs are attracted by ethylene glycol antifreeze, and are quite likely to drink any that is left in an uncovered container or in puddles on the ground. This will prove fatal in sufficient quantity. Always drain the coolant into a sealable container. Coolant should be reused unless it is contaminated or several years old.

   6. If equipped with a carburetor, perform the following procedures:
     a. Disconnect and label all hoses leading to the air cleaner.
     b. Remove the air cleaner cover and filter.
     c. Remove the air cleaner hold-down bolts. Lift the air cleaner, then, disconnect the temperature sensor wire, the remaining two hoses and the air cleaner.
   7. If equipped with fuel injection, perform the following procedures:
     a. Disconnect the air intake duct and vacuum hose.
     b. While holding the top of the fuel filter stationary with an open-end wrench, loosen the center service bolt with a box wrench; place a shop rag over the area so that any fuel spray will be absorbed. Carefully and slowly loosen the top service bolt, of the fuel filter, one turn.
     c. Disconnect the fuel return hose from the pressure regulator. Remove the fuel return hose bolt and the hose.
     d. From directly under the air cleaner outlet, disconnect the electrical harness connector. Disconnect the electrical harness cable from the fuse box.
     e. Remove the front panel-to-valve cover ground cable.
     f. Disconnect the brake booster vacuum hose.
     g. If equipped with air conditioning, disconnect the idle control solenoid valve.
     h. Loosen the throttle cable locknut and the adjusting nut, then, slip the end of the cable from the bracket and the accelerator linkage.

➡When removing the throttle cable, be sure not to use pliers and perform the work carefully so not to bend the cable. If you bend the cable, be sure to replace it.

     i. Disconnect the No. 1 control box connector, (this control box is in front of the shock tower on the driver's side), then, lift the control box from its retaining bracket and support it next to the engine.
     j. Disconnect the two engine harness connectors. These are made of white plastic and are located near the brake fluid reservoir.
     k. Remove the distributor. Label and disconnect the radiator and heater hoses from the engine, leaving the heater valve cable in position.
   8. If equipped with a carburetor, disconnect the following hoses and wires:
     a. The engine compartment sub-harness connector.
     b. The engine secondary cable.
     c. The brake booster vacuum hose.
     d. If equipped with air conditioning, remove the idle control solenoid hoses from the valve and the valve.
   9. Disconnect the control box connector(s). Remove the control box(s) from the bracket(s) and allow it (them) hang next to the engine.
   10. Disconnect the purge control solenoid valve vacuum hose from the charcoal canister.
   11. If equipped, remove the air jet controller.
   12. Loosen the throttle cable locknut and adjusting nut, then, slip the cable end from the throttle bracket.
   13. Disconnect the fuel hose from the fuel pump. Remove the fuel pump cover and the pump.
   14. Label and remove the spark plug wires. Remove the distributor from the engine.
   15. Remove the radiator and heater hoses from the engine.

➡Label the heater hoses so they will be reinstalled in their original locations.

   16. If equipped with a manual transaxle, perform the following procedures:
     a. Disconnect the ground cable from the transaxle.
     b. Loosen the clutch cable adjusting nut, then, remove the cable from the release arm.
     c. From the clutch housing, disconnect the shift lever torque rod.
     d. Slide the shift rod pin retainer out of the way. Using a pin punch, drive the pin out and remove the shift rod.
   17. If equipped with an automatic transaxle, perform the following procedures:
     a. Remove the oil cooler hoses from the transaxle, allow the fluid to drain from the hoses, then, secure the hoses out of the way near the radiator.
     b. From the inside of the vehicle, remove the center console.
     c. Position the shift lever in Reverse, then remove the lock pin from the end of the shift cable.
     d. Unbolt and remove the shift cable holder.
     e. Disconnect the throttle control cable from the throttle lever. Loosen the lower locknut on the throttle cable bracket and remove the cable from the bracket.

➡Do not move the upper locknut as it will change the transaxle shift points.

   18. Remove the speedometer cable clip and pull the cable from the holder.

➡Do not remove the holder from the transaxle, as doing so may allow the speedometer gear to fall into the transaxle.

   19. Using penetrating oil, squirt it on the exhaust header pipe nuts, then, remove the nuts and pipe.
   20. If working on a 4WD Wagon, perform the following procedures:
     a. Using a crayon or paint, mark the drive shaft-to-transaxle flange.
     b. Remove the drive shaft-to-transaxle flange bolts and disconnect the drive shaft from the transaxle.
   21. To remove the halfshafts on all models, perform the following procedures:
     a. Remove the jackstands and lower the vehicle. Loosen the 32mm spindle nuts with a socket. Raise the vehicle and resupport on jackstands.
     b. Remove the front wheel spindle nut.
     c. Place a floor jack under the lower control arm, then, remove the ball joint cotter pin and nut.

➡Be certain the lower control arm is positioned securely on top of the floor jack so that it doesn't suddenly jump or spring off when the ball joint remover is used.

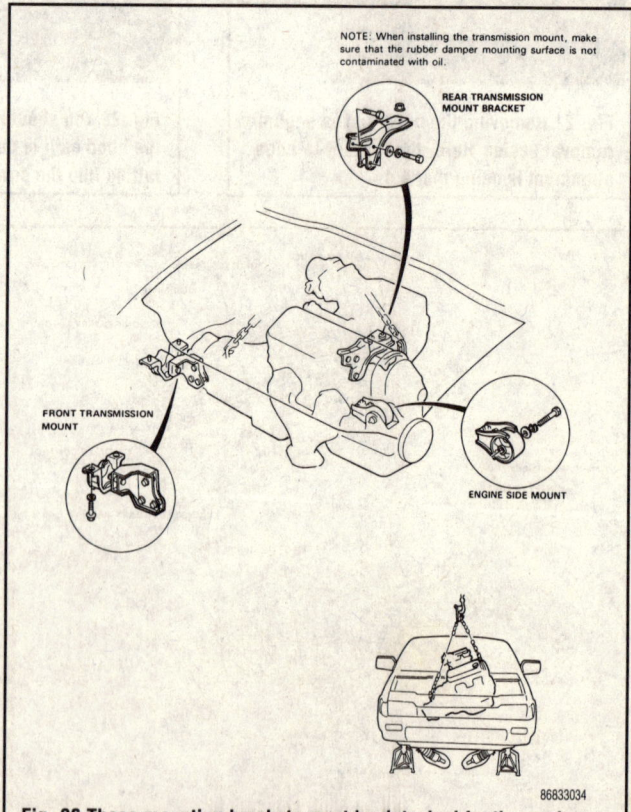

Fig. 26 These mounting brackets must be detached for the engine removal

ENGINE AND ENGINE REBUILDING 3-13

d. Using a ball joint remover tool or equivalent, separate the ball joint from the front hub.

e. Slowly, lower the floor jack to lower the control arm. Pull the hub outward and off the halfshaft.

f. Using a small prybar, pry out the inboard CV-joint approximately ½ in. (13mm) to release the spring clip from the groove in the transaxle.

g. Pull the halfshaft out of the transaxle case.

22. Using a lifting sling, attach it to the engine block, then raise the hoist to remove the slack from the chain.

23. Remove the rear transaxle mount bracket. Remove bolts from the front transaxle mount and the engine side mount.

24. If equipped with air conditioning, perform the following procedures:

a. Loosen the belt adjusting bolts and remove the belt.

b. Remove the air conditioning compressor bolts, then wire it up out of the way on the front beam.

➡ Do not disconnect the air conditioning refrigerant lines; the compressor can be moved without discharging the system.

c. Remove the lower compressor mounting bracket.

25. Disconnect the alternator wiring harness connectors. Remove the alternator belt. Remove the alternator mounting bolts and the alternator.

26. Check that the engine and transaxle are free of any hoses or electrical connectors.

27. Slowly raise and remove the engine from the vehicle.

To install:

28. Installation is the reverse order of the removal procedure. Use the following steps to aid in the installation procedure.

29. Tighten the engine mount bolts in the following sequence; be sure to replace the rear transaxle bolt and the front transaxle bolt with new bolts:

a. Rear transaxle mount bracket—33 ft. lbs. (45 Nm).

b. Front transaxle mount—33 ft. lbs. (45 Nm).

c. Engine side mount—33 ft. lbs. (45 Nm).

➡ Failure to tighten the bolts in the proper sequence can cause excessive noise and vibration, as well as reduce bushing life. Be sure to check that the bushings are not twisted or offset.

30. Check that the spring clip on the end of each drive shaft clicks into place. Be sure to use new spring clips on installation.

Fig. 27 Removing the exhaust header from the underside

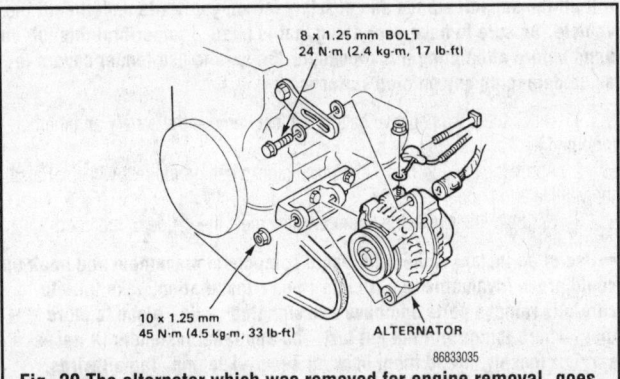

Fig. 29 The alternator which was removed for engine removal, goes back in place as shown

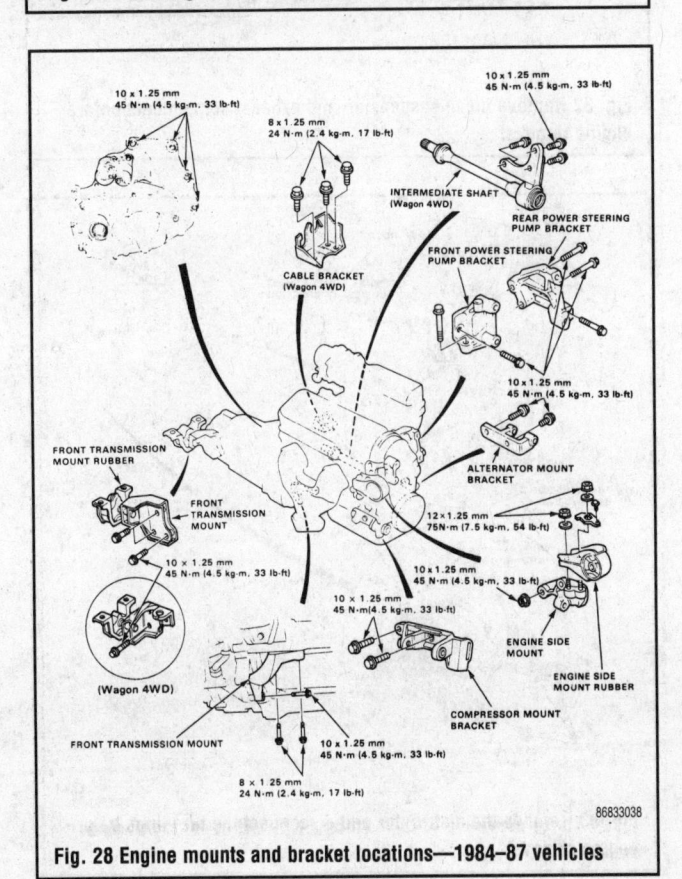

Fig. 28 Engine mounts and bracket locations—1984–87 vehicles

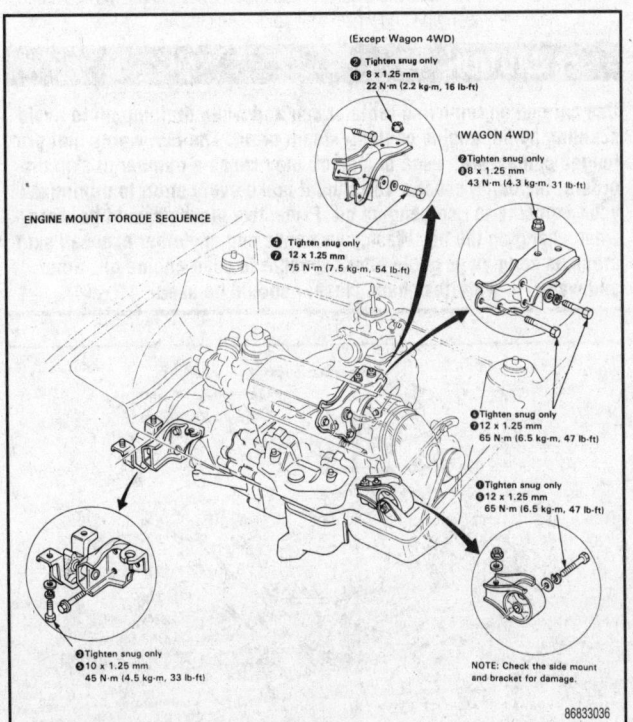

Fig. 30 Reinstallation requires tightening the engine mounts in the sequence shown. Failure to tighten the bolts in the proper order may cause excessive noise and vibration, as well as reduce bushing life. Check that the bushings are not twisted or offset

3-14 ENGINE AND ENGINE REBUILDING

31. After assembling the fuel line parts, turn the ignition switch (do not operate the starter) to the **ON** position so the fuel pump is operated for approximately two seconds so as to pressurize the fuel system. Repeat this procedure 2–3 times and check for a possible fuel leak.

32. Bleed the air from the cooling system at the bleed bolt with the heater valve open.

33. Adjust the throttle cable tension, install the air conditioning compressor and belt and adjust all belt tensions. Adjust the clutch cable free-play and check that the transaxle shifts into gear smoothly.

34. Check the ignition timing.

35. Install the speedometer cable, be sure to align the tab on the cable end with the slot holder. Install the clip so the bent leg is on the groove side. After installing, pull the speedometer cable to make sure it is secure.

1988–91 Models

▶ See Figures 31, 32, 33, 34 and 35

➡Make sure that all jacks and jackstands are placed properly and hoist brackets are attached to the correct positions on the engine. Always apply the parking brake and block the rear wheels so the vehicle will not roll off the support stands and possibly fall on you while underneath the vehicle. Be sure to have everything that is needed to perform this job on hand before attempting this procedure. Be sure to use fender covers to avoid damaging any painted surfaces.

1. Disconnect the negative battery cables. Remove the battery and the battery tray.
2. Apply the parking brake and place blocks behind the rear wheels. Raise the vehicle and support it safely.
3. Scribe a line where the hood brackets meet the inside of the hood.

➡Use of an instant camera to record component placement and hook-up could prove invaluable to a trouble-free re-installation. Take time to carefully remove parts and have a designated "safe" place to store them where things will not get lost. Tag and label fasteners (if necessary) or loosely thread them back for secure keeping. Tag all wires, connectors, hoses and belts (as necessary) that might be confusing when it is time to put it all back together.

4. Disconnect the windshield washer fluid tubes. Unbolt and remove the hood.
5. Remove the engine and wheel well splash shields.

※※ CAUTION

Use care when removing radiator cap and when draining oil to avoid scalding by hot engine coolant, steam or oil. The EPA warns that prolonged contact with used engine oil may cause a number of skin disorders, including cancer. You should make every effort to minimize your exposure to used engine oil. Protective gloves should be worn when changing the oil. Wash your hands and any other exposed skin areas as soon as possible after exposure to used engine oil. Soap and water, or waterless hand cleaner should be used.

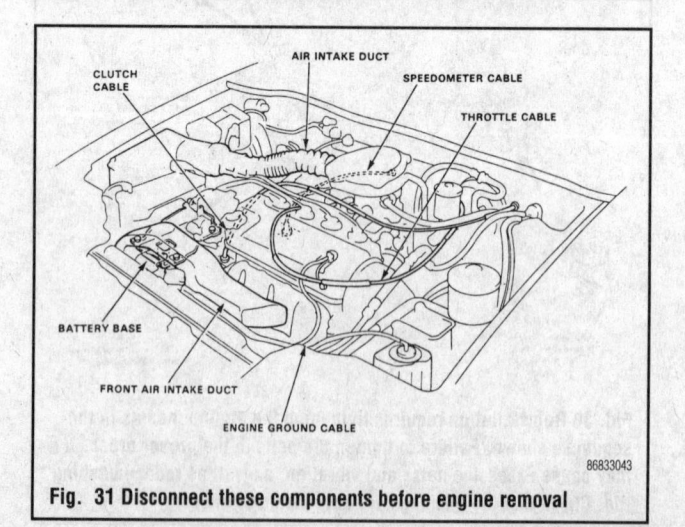

6. Drain the oil from the engine, the coolant from the radiator and the transaxle oil from the transaxle.

※※ CAUTION

When draining coolant, keep in mind that cats and dogs are attracted by ethylene glycol antifreeze, and are quite likely to drink any that is left in an uncovered container or in puddles on the ground. This will prove fatal in sufficient quantity. Always drain the coolant into a sealable container. Coolant should be reused unless it is contaminated or several years old.

7. Remove the air intake duct and the front air intake duct.
8. Relieve the fuel pressure from the fuel system, by slowly loosening the banjo bolt on the fuel filler approximately one turn.

➡Keep any and all open flames away from the work area. Before disconnecting any fuel lines, the fuel pressure should be relieved. Place a suitable shop towel over the fuel filler to prevent the pressurized fuel from spraying over the engine.

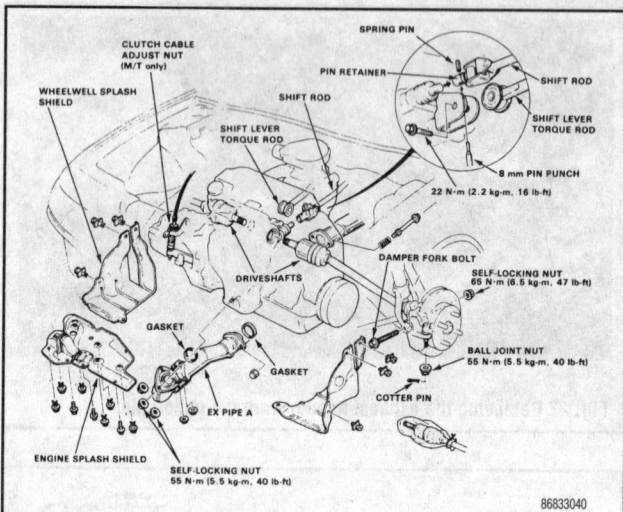

Fig. 32 Remove these suspension and exhaust components before engine removal

Fig. 31 Disconnect these components before engine removal

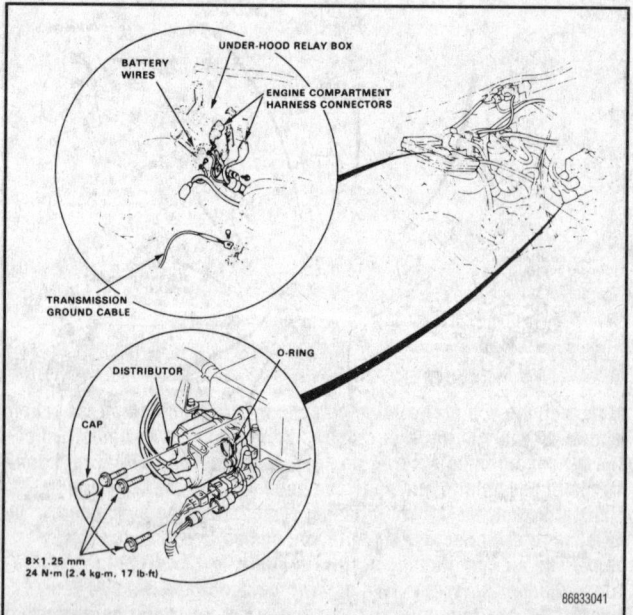

Fig. 33 Remove the distributor and disconnect the fuel lines before engine removal

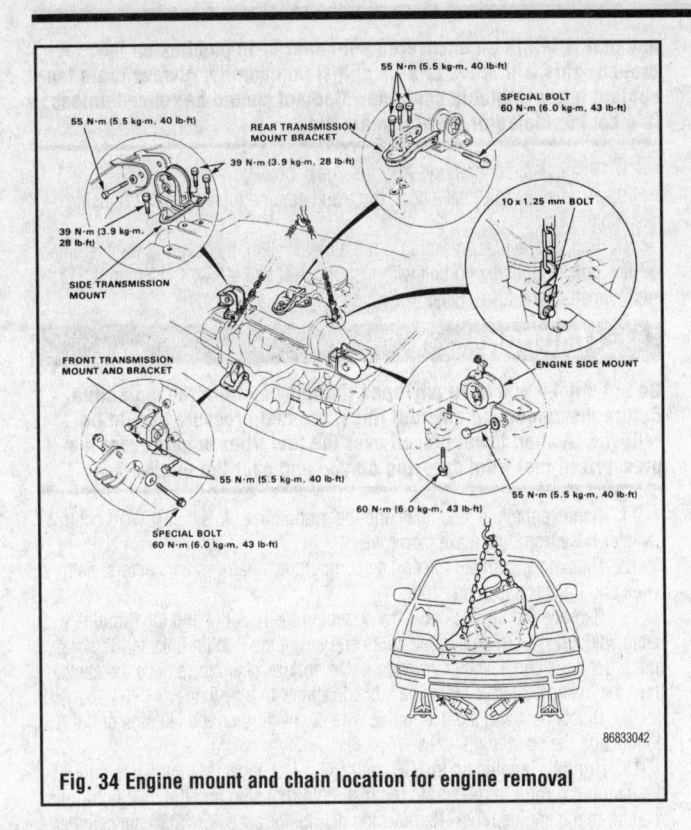

ENGINE AND ENGINE REBUILDING 3-15

Fig. 34 Engine mount and chain location for engine removal

9. Disconnect and tag the engine compartment harness connectors, battery wires and transaxle ground cable.
10. Remove the throttle cable by loosening the lock nut and the throttle cable adjust nut, then slip the throttle cable end out of the throttle bracket and accelerator linkage. Be sure not to bend the cable when removing it. Do not use pliers to remove the cable from the linkage. Always replace a kinked cable with a new one.
11. Disconnect and tag the engine wire connectors and spark plug wires. Bring the engine up to TDC on the No. 1 cylinder. Mark the position of the distributor rotor in relation to the distributor housing and the distributor in relation to the engine block. Remove the distributor assembly from the cylinder head.
12. Disconnect the radiator hoses and heater hoses. Disconnect the transaxle fluid cooler lines. Remove the speedometer cable.

➡ Do not remove the speedometer cable holder, as doing so may allow the speedometer gear to fall into the transaxle housing.

13. If equipped with power steering, remove the mounting bolts and power steering belt, then without disconnecting the hoses, pull the pump away from its mounting bracket and lay aside.
14. Disconnect and tag the alternator wiring, remove the alternator adjusting bolts, mounting bolts and belt. Remove the alternator from the vehicle.
15. Loosen the air conditioning belt adjust bolt and the idler puller nut. Remove the compressor mounting bolts. Disconnect the air conditioning suction and discharge lines, only if it is necessary. Lift the compressor out of the bracket with the air conditioning hoses attached and wire the compressor to the front beam of the vehicle.

✱✱ CAUTION

If the vehicle you are working on has air conditioning and the repair operation you will perform requires the removal of a component of the air conditioning system, do not disconnect the refrigerant lines. If it is impossible to move the component out of the way with the lines attached, the air conditioning system must be evacuated. (Please refer to Section 1 before you decide to evacuate the system.) The air conditioning system contains R-12 or R134a refrigerant gas under pressure. This gas can be very dangerous to your health, the environment, and damage to the air conditioning system could result. Consult your local laws concerning refrigerant discharge and recycling. In many areas it may be illegal for anyone but a certified technician to service the A/C system. Always use an approved recovery station when discharging the air conditioning.

16. If equipped with an automatic transaxle, proceed as follows:
    a. Remove the header pipe, header pipe bracket, torque converter cover and shift control cable holder.
    b. Remove the shift control cable by removing the cotter pin, control pin and control lever roller from the control lever.
17. If equipped with manual transaxles, remove the shift lever torque rod, shift rod and clutch cable. On reassembly, slide the retainer back into place after driving in the spring pin.
18. Remove the wheel well splash shields and engine splash shields. Remove the right and left halfshafts from the transaxle and cover the shafts with a plastic bag so as to prevent the oil from spilling over the work area. Be sure to coat all precision finished surfaces with clean engine oil or grease.
19. On 4WD vehicles equipped with automatic transaxles, remove the cable clip and the control pin. Loosen the shift control cable nut and then remove the control cable.
20. On 4WD vehicles equipped with manual transaxles, remove the cotter pins and the 3 cable bracket mounting bolts. Remove the cable bracket from the rear of the transaxle mount bracket.
21. Attach a suitable chain hoist to the engine block hoist brackets and raise the hoist just enough to remove the slack from the chain. To attach the rear engine chain, remove the plastic radiator hose bracket and hook the chain to the top of the clutch cable bracket.
22. Remove the rear transaxle mount bracket. Remove the bolts from the front transaxle bolt mount. Remove the bolts from the engine side mount. Remove the bolts from the engine side transaxle mounts.
23. Check that the engine/transaxle assembly are completely free of vacuum, fuel, coolant hoses and electrical wires.
24. Slowly raise the engine approximately 6 in. (152mm) and stop. Check again that the engine/transaxle assembly are completely free of vacuum, fuel, coolant hoses and electrical wires.
25. Raise the engine/transaxle assembly all the way up and out of the vehicle, once it is clear from the vehicle, lower the assembly into a suitable engine stand.

To install:

26. Installation is the reverse order of the removal procedure. Use the following steps to aid in the installation procedure.
27. Tighten the engine mount bolts in the following sequence; be sure to replace the rear transaxle bolt and the front transaxle bolt with new bolts:
    a. Side transaxle mount—40 ft. lbs. (54 Nm).
    b. Rear transaxle mount bracket—43 ft. lbs. (58 Nm).
    c. Front transaxle mount—43 ft. lbs. (58 Nm).
    d. Engine side mount—40 ft. lbs. (54 Nm).

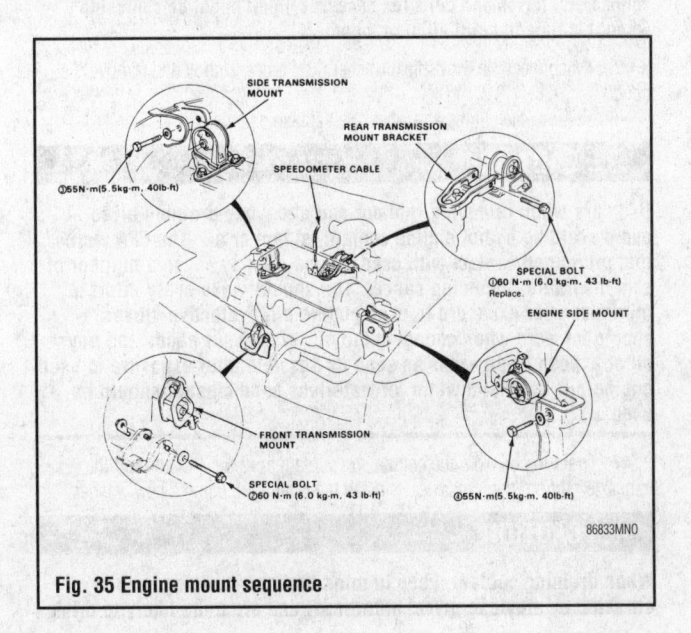

Fig. 35 Engine mount sequence

3-16 ENGINE AND ENGINE REBUILDING

➡Failure to tighten the bolts in the proper sequence can cause excessive noise and vibration, as well as reduce bushing life. Be sure to check that the bushings are not twisted or offset.

28. Check that the spring clip on the end of each drive shaft clicks into place. Be sure to use new spring clips on installation.

29. After assembling the fuel line parts, turn the ignition switch (do not operate the starter) to the **ON** position so the fuel pump is operated for approximately two seconds so as to pressurize the fuel system. Repeat this procedure 2–3 times and check for a possible fuel leak.

30. Bleed the air from the cooling system at the bleed bolt with the heater valve open.

31. Adjust the throttle cable tension, install the air conditioning compressor and belt and adjust all belt tensions. Adjust the clutch cable free-play and check that the transaxle shifts into gear smoothly.

32. Check the ignition timing.

33. Install the speedometer cable, be sure to align the tab on the cable end with the slot holder. Install the clip so the bent leg is on the groove side. After installing, pull the speedometer cable to make sure it is secure.

1992–95 Engines

◆ See Figures 36 thru 46

➡Make sure that all jacks and jackstands are placed properly and hoist brackets are attached to the correct positions on the engine. Always apply the parking brake and block the rear wheels so the vehicle will not roll off the support stands and possibly fall on you while underneath the vehicle. Be sure to have everything that is needed to perform this job on hand before attempting this procedure. Be sure to use fender covers to avoid damaging any painted surfaces.

1. Disconnect the negative battery cables. Remove the battery and the battery tray.

2. Apply the parking brake and place blocks behind the rear wheels. Raise the vehicle and support it safely.

➡Be sure to have the 5-digit code for the original radio. When electical power is reconnected, the word "CODE" will be displayed on the radio screen. This must be entered to resume radio operation.

3. Remove the radiator cap.

4. Scribe a line or use a permanent felt tip pen to mark where the hood brackets meet the inside of the hood.

➡Use of an instant camera to record component placement and hook-up could prove invaluable to a trouble-free re-installation. Take time to carefully remove parts and have a designated "safe" place to store them where things will not get lost. Tag and label fasteners (if necessary) or loosely thread them back for secure keeping. Tag all wires, connectors, hoses and belts (as necessary) that might be confusing when it is time to put it all back together.

5. Disconnect the windshield washer fluid tubes. Unbolt and remove the hood.

6. Remove the engine and wheel well splash shields.

✸✸ CAUTION

Use care when removing radiator cap and when draining oil to avoid scalding by hot engine coolant, steam or oil. The EPA warns that prolonged contact with used engine oil may cause a number of skin disorders, including cancer. You should make every effort to minimize your exposure to used engine oil. Protective gloves should be worn when changing the oil. Wash your hands and any other exposed skin areas as soon as possible after exposure to used engine oil. Soap and water, or waterless hand cleaner should be used.

7. Drain the oil from the engine, the coolant from the radiator and the transaxle oil from the transaxle. Reinstall the drain plug using a new washer.

✸✸ CAUTION

When draining coolant, keep in mind that cats and dogs are attracted by ethylene glycol antifreeze, and are quite likely to drink any that is left in an uncovered container or in puddles on the ground. This will prove fatal in sufficient quantity. Always drain the coolant into a sealable container. Coolant should be reused unless it is contaminated or several years old.

8. Remove the underhood ABS fuse/relay box.

9. Remove the air intake duct, the resonator and the air cleaner housing assembly.

10. Relieve the fuel pressure from the fuel system, by covering the fuel filter service bolt atop the banjo bolt with a shop towel and slowly loosening it approximately one turn. (Refer to Section 1 for more detail).

✸✸ CAUTION

Do not smoke and keep any open flame away from the work area. Before disconnecting any fuel lines, the fuel pressure should be relieved. A shop towel placed over the fuel filler is to prevent the pressurized fuel from spraying on you and over the engine.

11. Remove the fuel feed hose and the evaporative emission (EVAP) control canister hose from the intake manifold.

12. Disconnect and tag the engine compartment harness connectors, battery wires and transaxle ground cable.

13. Remove the throttle cable by loosening the lock nut and the throttle cable adjusting nut, then slip the cable end out of the throttle linkage. Be sure not to bend the cable when removing it. Do not use pliers to remove the cable from the linkage. Always replace a kinked cable with a new one.

14. Disconnect and tag the engine wire connectors on the left side of the engine and the spark plug wires.

15. Bring the engine up to TDC on the No. 1 cylinder. Mark the position of the distributor rotor in relation to the distributor housing and the distributor in relation to the engine block. Remove the distributor assembly from the cylinder head.

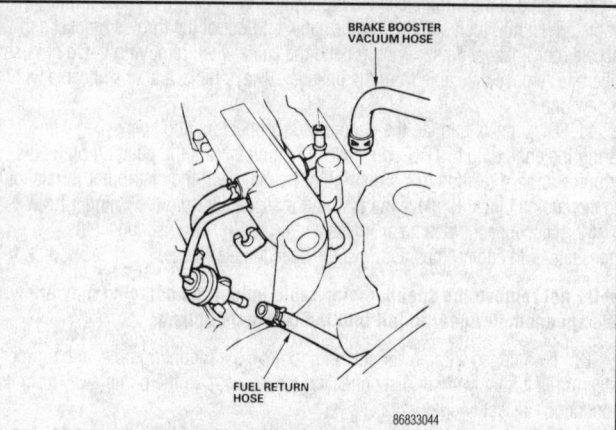

Fig. 36 Disconnect and remove the brake booster vacuum hose as well as the fuel return hose

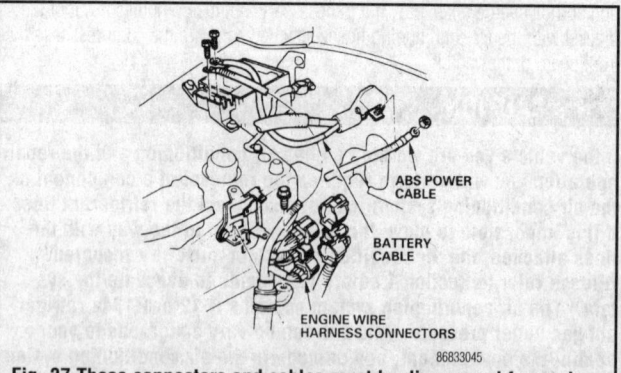

Fig. 37 These connectors and cables must be disengaged for engine removal

ENGINE AND ENGINE REBUILDING   3-17

16. Remove the fuel return hose and brake booster vacuum hose.

17. Remove the engine wire harness connectors, terminal and clamps on the right side of engine compartment.

18. Remove the battery cable and starter cable from the underhood fuse/relay box and ABS power cable from the battery terminal, and the engine ground cable from the body side.

19. Disconnect the radiator hoses and heater hoses. Disconnect the transaxle fluid cooler lines. Remove the speedometer cable.

➡ Do not remove the speedometer cable holder, as doing so may allow the speedometer gear to fall into the transaxle housing.

20. Remove power steering belt and pump, then without disconnecting the hoses, pull the pump away from its mounting bracket and lay aside.

21. Remove the air conditioning (A/C) belt and compressor and connector but do **not** disconnect the A/C hoses.

※※ CAUTION

If the work you are performing requires the removal of a component of the air conditioning system, do not disconnect the refrigerant lines. If it is impossible to move the component out of the way with the lines attached, the air conditioning system must be evacuated. (Please refer to Section 1 before you decide to evacuate the system.) The air conditioning system contains R-12 or R134a refrigerant gas under pressure. This gas can be very dangerous to your health, the environment, and damage to the air conditioning system could result. Consult your local laws concerning refrigerant discharge and recycling. In many areas it may be illegal for anyone but a certified technician to service the A/C system. Always use an approved recovery station when discharging the air conditioning.

22. If equipped with an automatic transaxle, proceed as follows:
    a. Remove the transmission ground cable and automatic transmission fluid (ATF) cooler hoses.
    b. Remove the upper and lower radiator hoses and heater hoses that were disconnected in Step 19.
    c. Remove exhaust pipe "A" and its support bracket.
    d. Remove the shift cable by removing the shift cable cover and control lever.

➡ Raise the vehicle as high on jackstands as is safely possible to make it less awkward working under the car.

23. If equipped with a manual transaxle:
    a. Remove the the clutch slave cylinder and pipe/hose assembly but do **not** disconnect the pipe/hose assembly.
    b. Remove the shift shift rod and extension rod.

24. Remove the damper fork.

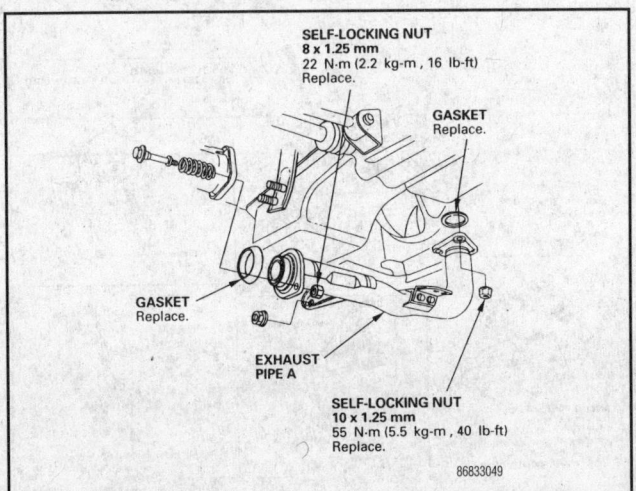

Fig. 38 Exhaust pipe "A" (also known in Honda terminology as a "header pipe") must be removed completely. Make sure to have the correct gasket for re-assembly

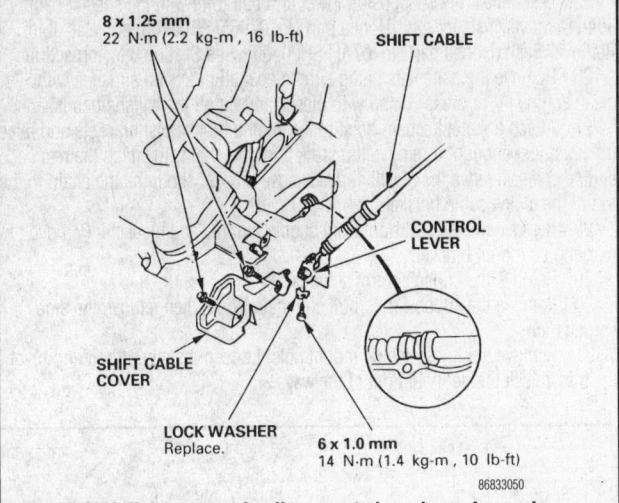

Fig. 39 Shift linkage must be disconnected as shown for engine removal

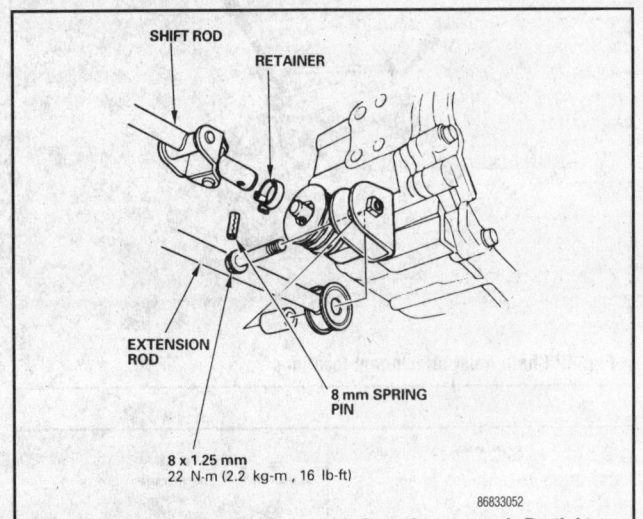

Fig. 40 Location of damper fork and bolts to be removed. Re-tighten to 16 ft. lbs. (22 Nm)

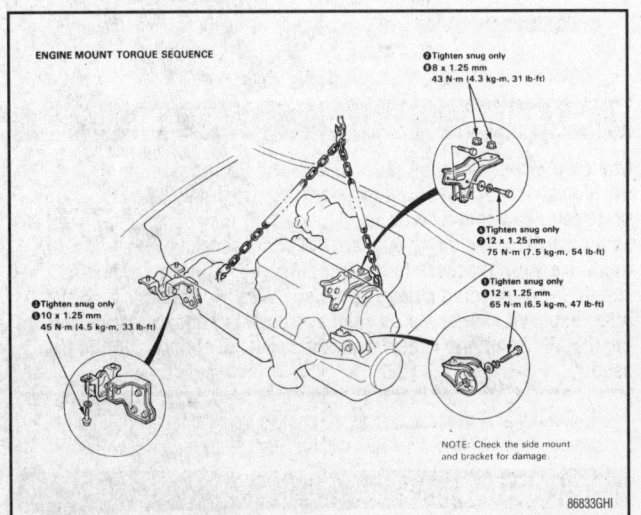

Fig. 41 During engine installation, tighten the mounts to their specified torque values

3-18 ENGINE AND ENGINE REBUILDING

25. Disconnect the suspension lower arm ball joint with the special ball Joint Remover/Installer tool (Honda part No. 07965-SB00100) and Ball Joint Remover Base (Honda part No. 07AF-SH20200). See Section 8 for procedure.

26. Remove the halfshafts being sure to coat all precision surfaces with clean engine oil or grease, wrap with plastic and tie off the halfshaft ends.

27. Attach a suitable chain hoist to the engine block hoist brackets and raise the hoist just enough to remove the slack from the chain. To attach the rear engine chain, remove the plastic radiator hose bracket and hook the chain to the top of the clutch cable bracket.

28. Remove the left and right front engine mounts and brackets. Remove the rear engine mount bracket.

29. Remove the cruise control actuator.

30. Remove the support nuts/bolt and mount bolt, then remove the side engine mount.

31. Remove the transmission mount nuts. Loosen the mount bolt and pivot the transmission side mount out of the way.

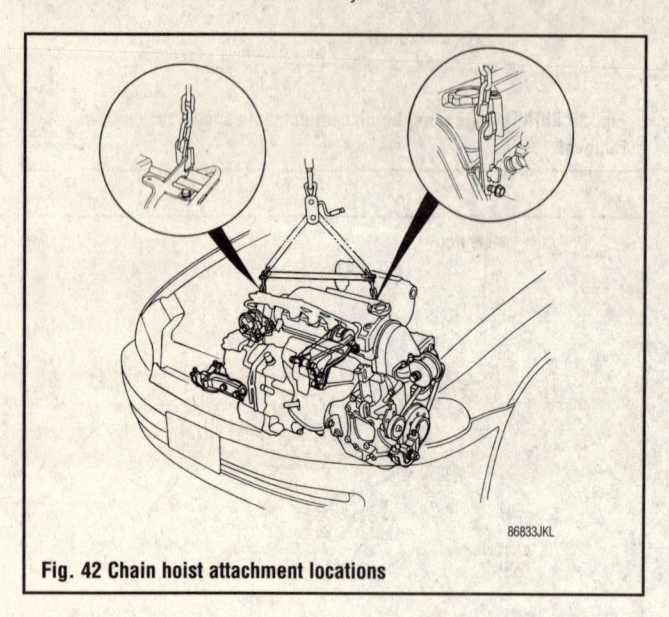

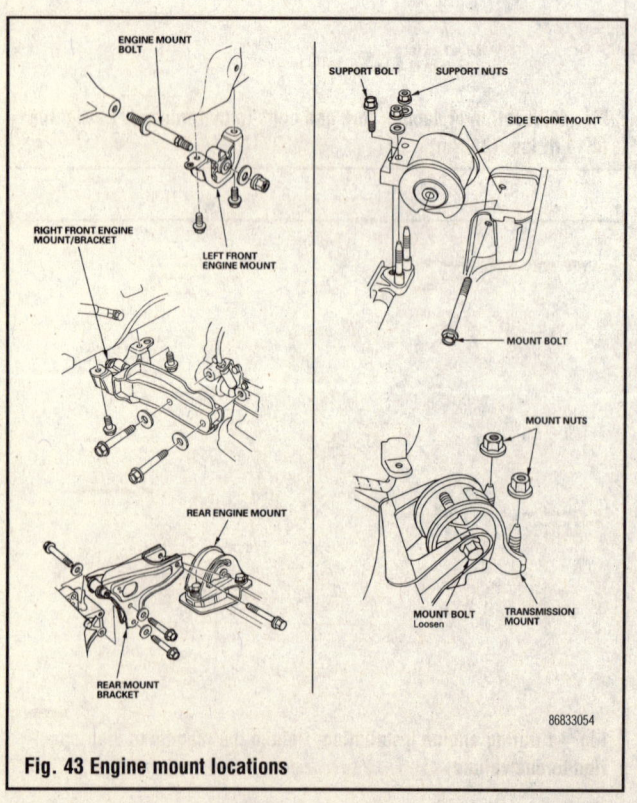

32. Check that the engine/transaxle assembly are completely free of vacuum, fuel, coolant hoses and electrical wires.

33. Slowly raise the engine approximately 6 in. (15 cm) and stop. Check again that the engine/transaxle assembly are completely free of vacuum, fuel, coolant hoses and electrical wires.

34. Raise the engine/transaxle assembly all the way up and out of the vehicle, once it is clear from the vehicle, lower the assembly into a suitable engine stand.

To install:

➡ Civic series models not equipped with A/C have an additional mount bracket. Tighten 10 x 1.25mm bolt to 40 ft. lbs. (55 Nm).

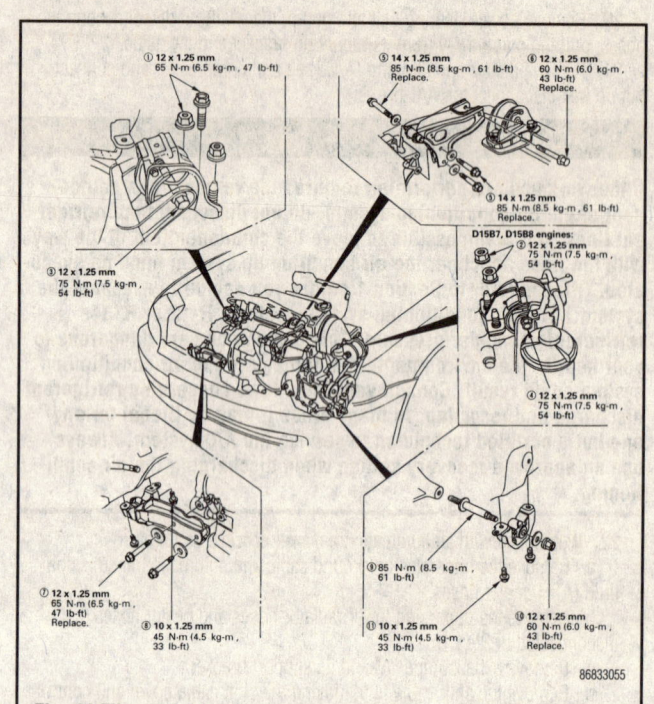

Fig. 42 Chain hoist attachment locations

Fig. 43 Engine mount locations

Fig. 44 Finger-tighten the engine mount bolts until all are in place. Then, tighten the mounts in proper order to the torque figures shown

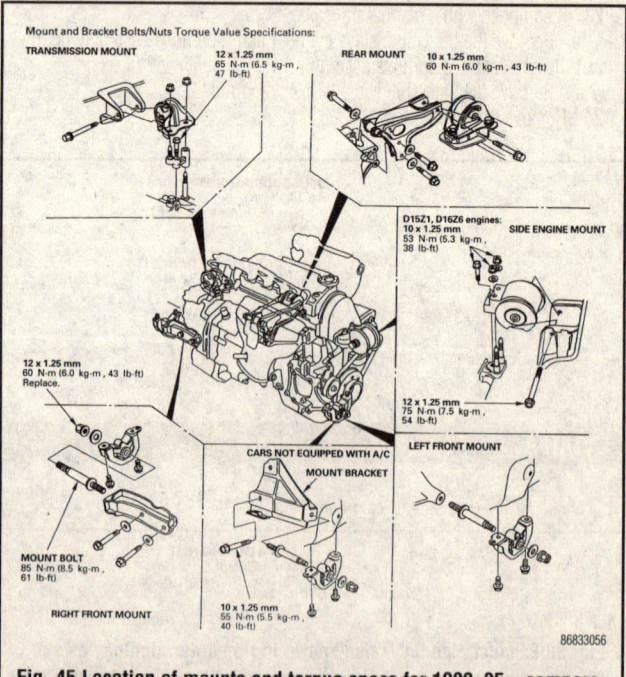

Fig. 45 Location of mounts and torque specs for 1992–95—compare to previous drawing for engine re-installation

ENGINE AND ENGINE REBUILDING 3-19

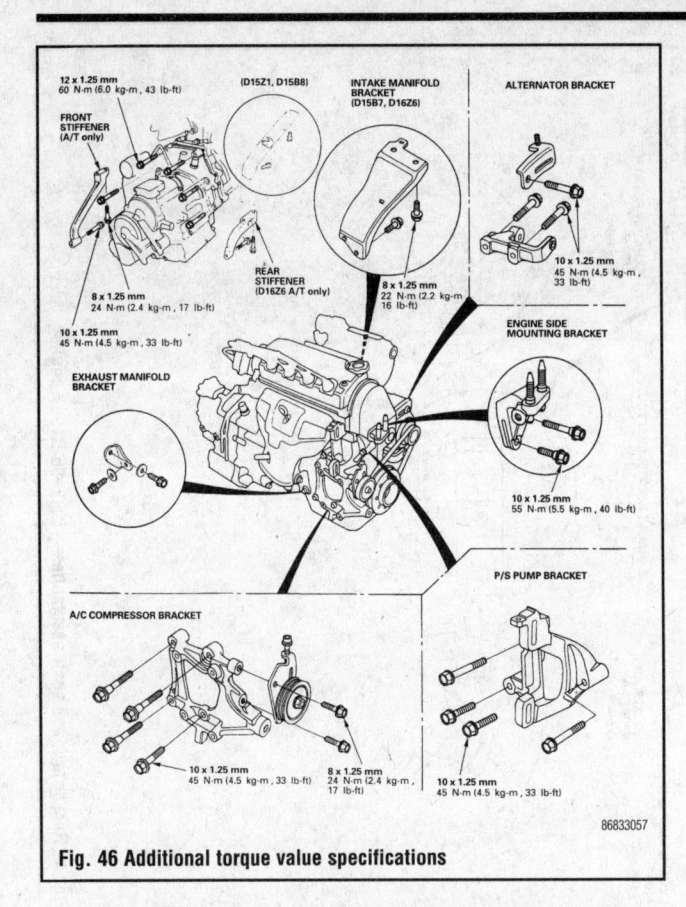

Fig. 46 Additional torque value specifications

35. Installation is the reverse order of the removal procedure. Use the following steps to aid in the installation procedure.

36. Tighten the engine mount bolts in the sequence shown in accompanying drawing. Be sure to replace the rear transaxle bolt and the front transaxle bolt with new bolts.

➡ **Failure to tighten the bolts in the proper sequence can cause excessive noise and vibration, as well as reduce bushing life. Be sure to check that the bushings are not twisted or offset.**

37. Check that the spring clip on the end of each drive shaft clicks into place. Be sure to use new spring clips on installation.

38. After assembling the fuel line parts, turn the ignition switch (do not operate the starter) to the **ON** position so the fuel pump is operated for approximately two seconds so as to pressurize the fuel system. Repeat this procedure 2–3 times and check for a possible fuel leak.

39. Bleed the air from the cooling system at the bleed bolt with the heater valve open.

40. Adjust the throttle cable tension, install the air conditioning compressor and belt and adjust all belt tensions. Adjust the clutch cable free-play and check that the transaxle shifts into gear smoothly.

41. Check the ignition timing.

42. Install the speedometer cable, be sure to align the tab on the cable end with the slot holder. Install the clip so the bent leg is on the groove side. After installing, pull the speedometer cable to make sure it is secure.

Cylinder Head Cover

REMOVAL & INSTALLATION

1. On some models, it will be necessary to remove the air cleaner assembly.
2. Disconnect the negative battery cable. Remove the ground cable, the spark plug wires (this will be necessary on the 1988–95 models) and the throttle cable (if necessary) from the cylinder head cover.
3. If equipped, remove the PCV hose from the cylinder head cover.
4. Remove the cylinder head cover-to-cylinder head nuts, the washer/grommet assemblies and the cylinder head cover. On the 1984–87 models there are 2 or 4 cap screws that hold on the cylinder head cover, on 1988–91 models there are 4 or 6 cap screws, and on 1992–95 models, there are 3 cap screws that hold on the cylinder head cover.

➡ **If the cover is difficult to remove, use a rubber mallet or equivalent soft-head hammer to bump it loose.**

5. Using a gasket scraper or putty knife, clean the gasket mounting surfaces.
6. To install, use a new gasket, sealant (if necessary) and reverse the removal procedures. Tighten the rocker arm cover-to-cylinder head nuts to 7 ft. lbs. (10 Nm).
7. Connect the spark plug wires and reconnect anything removed from the cylinder head arm cover.

Rocker Arms/Shafts

The rocker arms and shafts are an assembly; they must be removed from the engine as a unit. Always follow the torque sequence carefully when installing the rocker shaft assembly.

REMOVAL & INSTALLATION

▶ See Figures 47 thru 58

1. Disconnect the negative battery cable.
2. Remove the cylinder head cover and bring the No. 1 cylinder to TDC.
3. Loosen the adjusting screws two turns at a time, in a crisscross pattern, to prevent damaging the valves or rocker arm assembly.
4. Unscrew the camshaft holder bolts and remove the rocker arm/shaft assemblies. Leave the rocker arm bolts in place as the shafts are removed to keep the bearing caps, springs and rocker arms in place on the shafts.

➡ **If the rocker arms or shafts are to be replaced, identify the parts as they are removed from the shafts to ensure reinstallation in the original location.**

To install:

5. Lubricate the camshaft journals and lobes.
6. Set the rocker arm assembly in place and loosely install the bolts. Tighten each bolt two turns at a time in the proper sequence to ensure that the

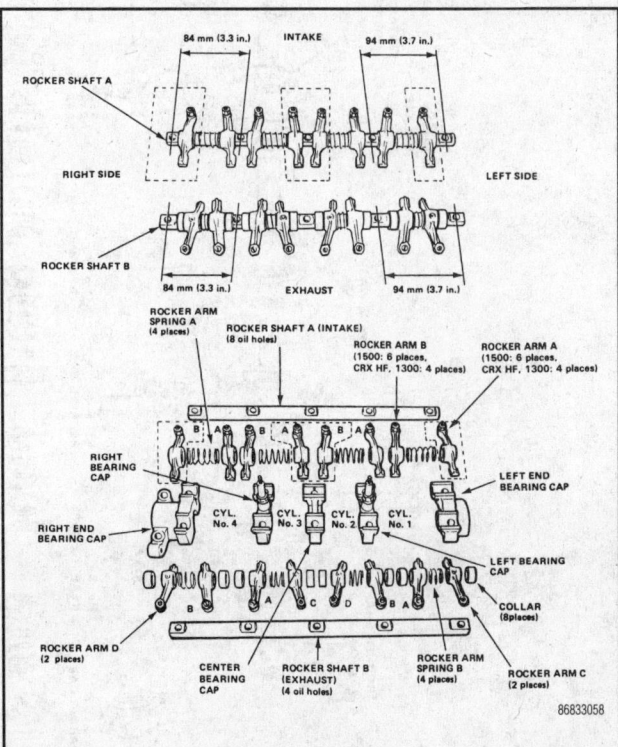

Fig. 47 Rocker arm and shaft assemblies—1984–87 vehicles (all)

3-20 ENGINE AND ENGINE REBUILDING

Fig. 49 Rocker arm and shaft assemblies—1992-95 D15B7 engine

Fig. 48 Rocker arm and shaft assemblies—1988-91 vehicles

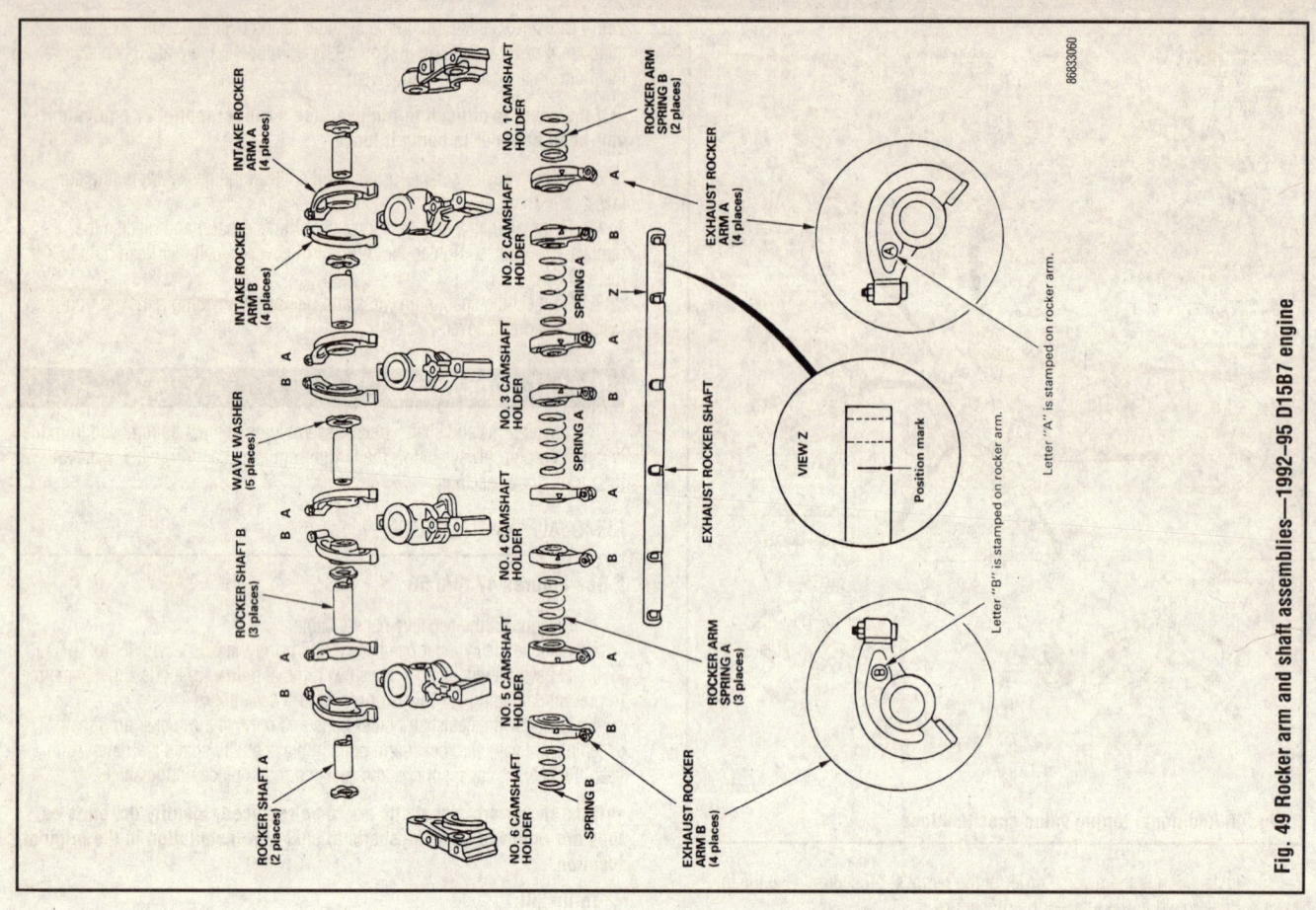

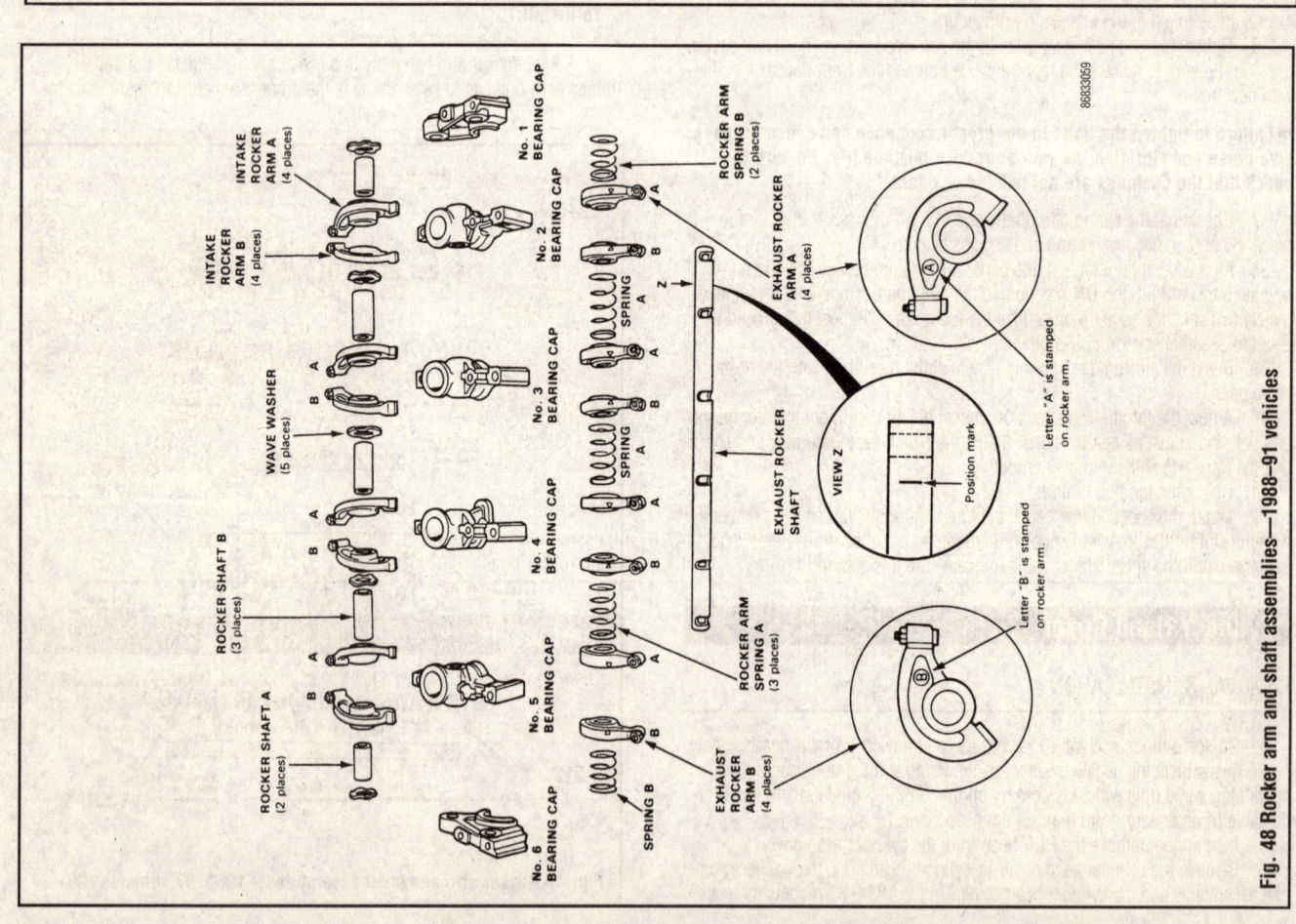

ENGINE AND ENGINE REBUILDING   3-21

Fig. 50 Rocker assembly torque sequence—1984–87 engines

Fig. 51 Rocker assembly torque sequence—1988–91 engines

Fig. 52 Rocker assembly torque sequence—1992–95 D15B8 engines

Fig. 53 Rocker assembly torque sequence—1992–95 D15B7 engines

Fig. 54 Rocker assembly torque sequence—1992–95 D15Z1 engines

Fig. 55 Rocker assembly torque sequence—1992–95 D16Z6 engines

Fig. 56 To remove the rocker arm and shaft assembly loosen, but do not remove the camshaft holder bolts

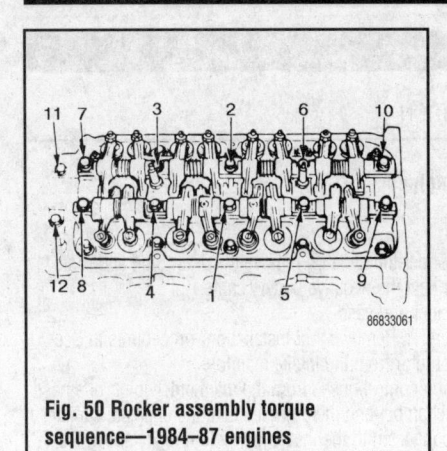

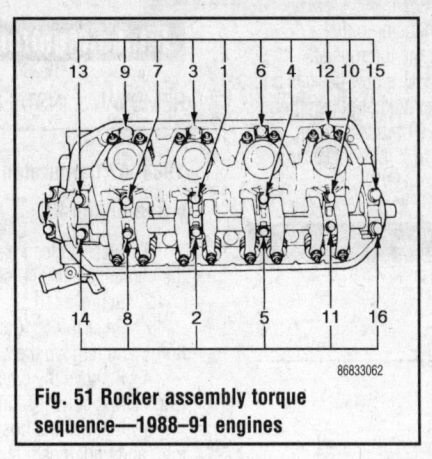

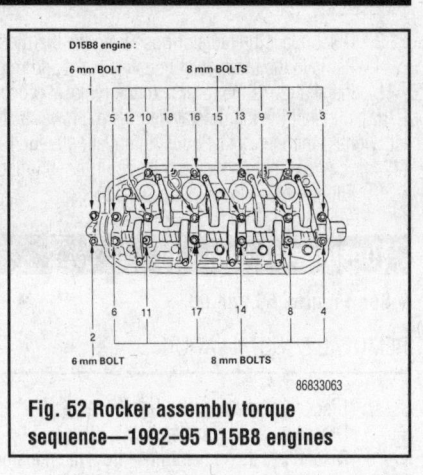

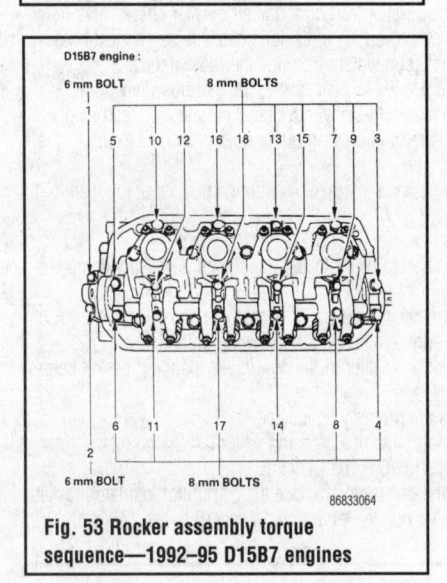

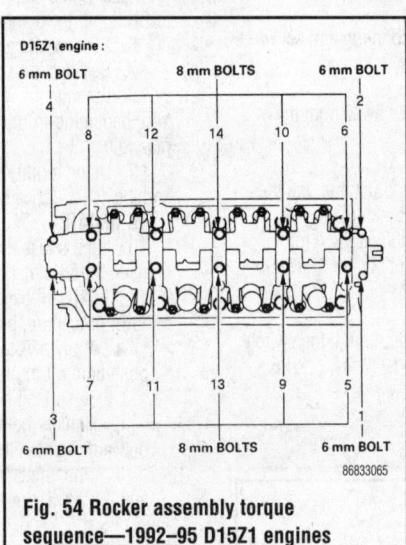

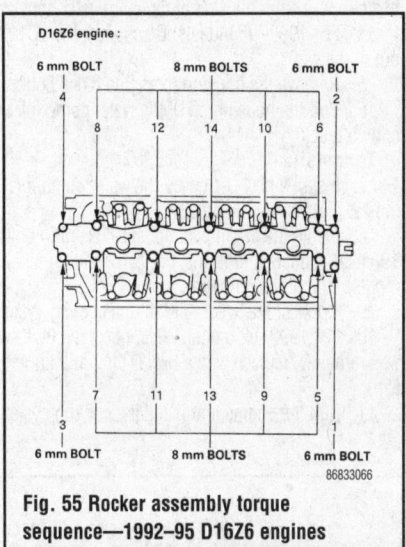

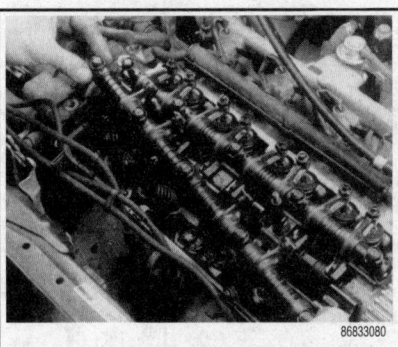

Fig. 57 The exhaust side rocker arm and shaft assembly is removed here intact, held together by the camshaft holder bolts

Fig. 58 The intake side rocker arms are removed as an assembly—1987 CRX Si shown

rockers do not bind on the valves. Tighten the camshaft holder bolts for 1984–91 to 16 ft. lbs. (22 Nm). For 1992–95 engines, tighten to the torque specified:

8mm bolts:
D16Z6, D15Z1 engines: 14 ft. lbs. (20 Nm)
D15B7, D15B8 engines: 16 ft. lbs. (22 Nm)
6mm bolts:
All engines: 9 ft. lbs. (12 Nm)

7. Replace the cylinder head cover and reconnect the negative battery cable.

Thermostat

The thermostat is located in a housing at the end of the water pump inlet pipe. The housing is on the fire wall side of the engine. The thermostat can be removed without removing the housing from the engine.

REMOVAL & INSTALLATION

1. Disconnect the negative battery cable.
2. Drain the cooling system.

3-22 ENGINE AND ENGINE REBUILDING

3. Disconnect the radiator hose from the thermostat housing outlet.
4. Remove the thermostat housing outlet and remove the thermostat.
5. Installation is the reverse of the removal procedure. Use new gaskets and O-rings. Install the thermostat with the pin towards the thermostat housing outlet. Tighten the thermostat housing outlet bolts for 1984–91 models to 9 ft. lbs. (12 Nm), and tighten housing bolts for 1992–95 models to 7 ft. lbs. (10 Nm).
6. Fill and bleed the cooling system.

Thermostat Housing

▶ See Figures 59 and 60

REMOVAL & INSTALLATION

1. Disconnect the negative battery cable.
2. Drain the cooling system.
3. Disconnect the radiator hose from the housing neck. Disconnect the heater hoses from the side of the thermostat housing.
4. For 1992–95 models: Disconnect the ECT switch coupling and set wires out of the way.
5. Remove the housing to engine block bolts.
6. Pull the housing off of the water pump inlet tube and away from the engine block.

To install:

7. Install a new O-ring on the water pump inlet pipe and position the thermostat housing on the pipe.
8. Install the thermostat housing retaining bolts and tighten them for 1984–91 models to 9 ft. lbs. (12 Nm), and tighten housing bolts for 1992–95 models to 7 ft. lbs. (10 Nm).
9. Connect the hoses and fill the cooling system.
10. For 1992–95 models: Reconnect the ECT switch coupling. If the switch was removed, replace with a new O-ring and tighten to 16-19 ft. lbs. (22-26 Nm).
11. Run the engine and bleed the cooling system.

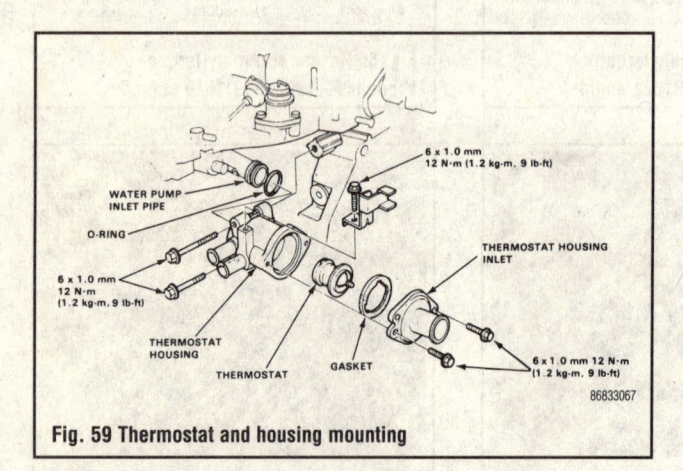

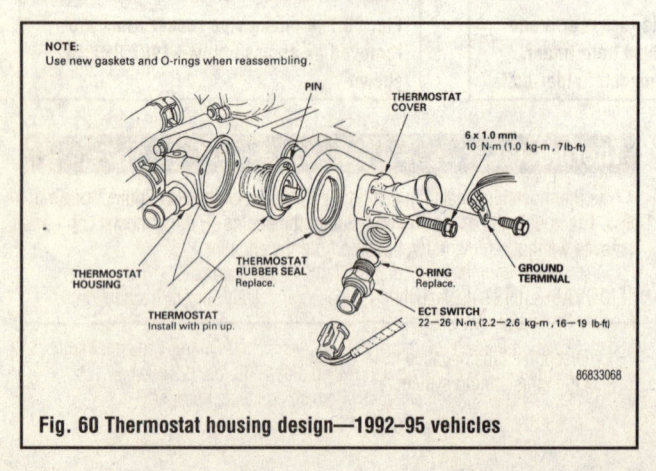

Intake Manifold

REMOVAL & INSTALLATION

1984–87 Carbureted Engines

▶ See Figure 61

1. Place a clean pan under the radiator, remove the drain plug and drain the cooling system. Disconnect the negative battery cable.
2. Remove the air cleaner and case.
3. Refer to the "Carburetor, Removal and Installation" procedures in Section 5 and remove the carburetor from the intake manifold.
4. Remove the emission control hoses from the manifold T-joint; one hose leads to the condensation chamber and the other leads to the charcoal canister.
5. Remove the hose connected to the intake manifold directly above the T-joint and underneath the carburetor, leading to the air cleaner check valve (refer to Section 5 for diagrams of the various emission control hose connections).
6. Remove the electrical connectors from the thermoswitches.
7. Remove the solenoid valve located next to the thermoswitch.
8. Remove the intake manifold-to-cylinder head nuts in a crisscross pattern, beginning from the center and moving out to both ends, then, remove the manifold.
9. Using a putty knife, clean the gasket mounting surfaces of the manifold and the cylinder head.

To install:

10. If the intake manifold is to be replaced, transfer all necessary components to the new manifold.
11. To install, use new gaskets, sealant (if necessary) and reverse the removal procedure, being sure to observe the following points:
   a. Apply a water-resistant sealer to the new intake manifold gasket before positioning it in place.
   b. Be sure all hoses are properly connected.
   c. Tighten the manifold nuts in a crossing pattern, from the inside nuts outward. Tighten the nuts to 16 ft. lbs (22 Nm).
   d. When installing the carburetor, replace the carburetor insulator gasket and O-ring. Tighten the carburetor mounting nuts to 16 ft. lbs. (22 Nm).

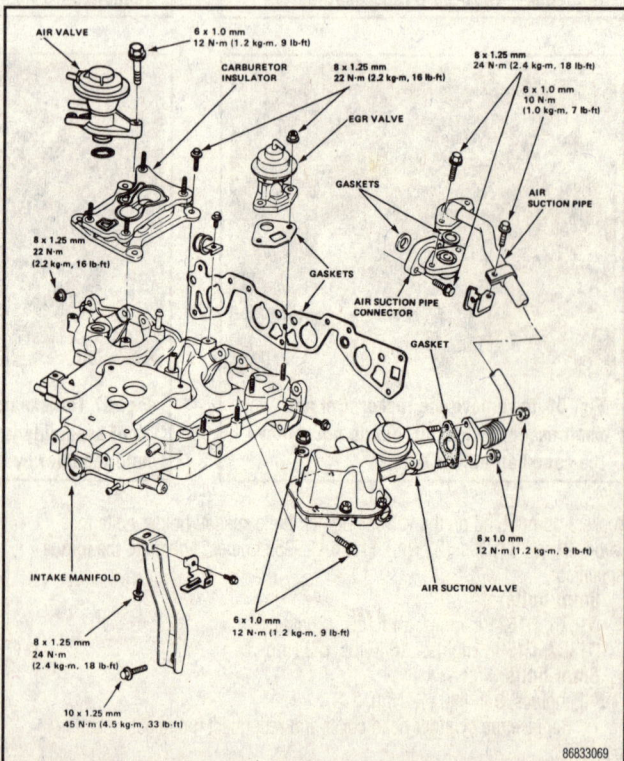

Fig. 59 Thermostat and housing mounting

Fig. 60 Thermostat housing design—1992–95 vehicles

Fig. 61 Intake manifold and components—1984–87 carbureted engines

ENGINE AND ENGINE REBUILDING 3-23

1985-91 Fuel Injected Models (All)

▶ See Figures 62, 63 and 64

1. Disconnect the negative battery cable. Drain the cooling system, into a suitable container.
2. Label and disconnect all related electrical connectors and vacuum lines.

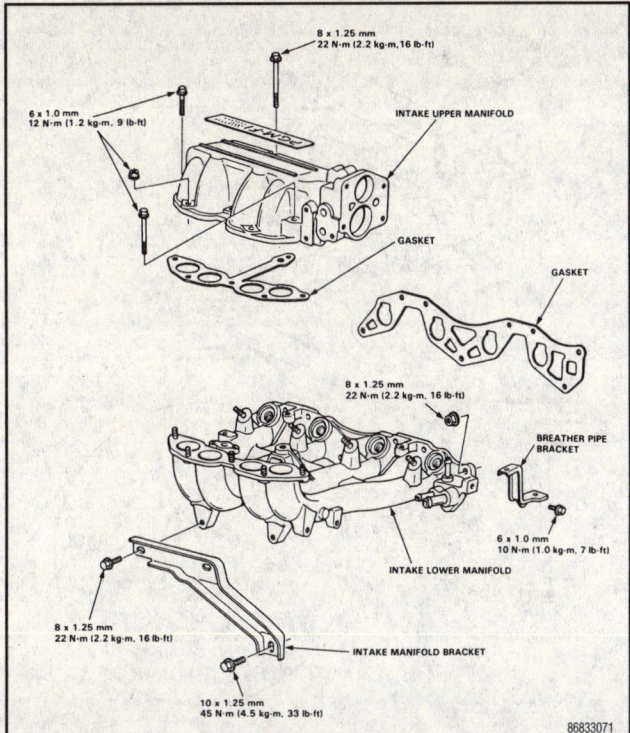

Fig. 62 Intake manifold and components—1984–87 multi-port fuel injected engines

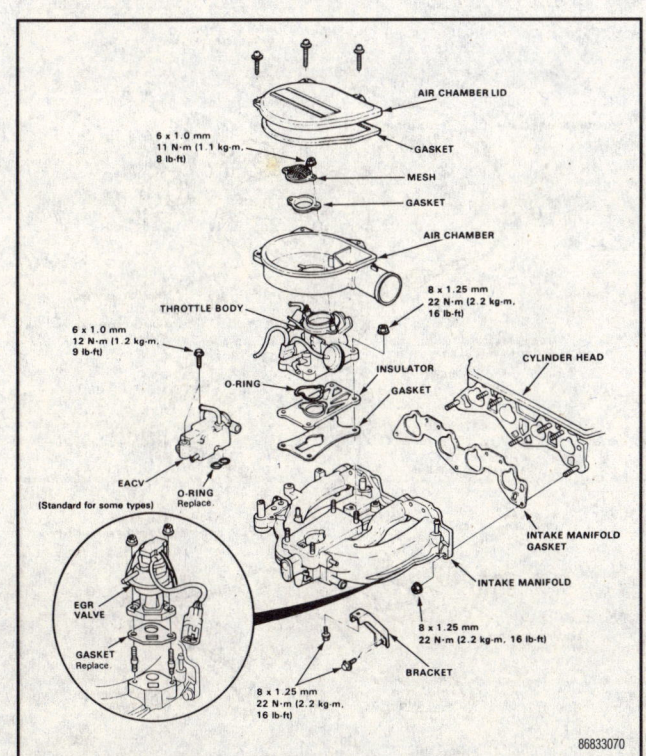

Fig. 63 Intake manifold, throttle body and related components—1988–91 DP-FI engines

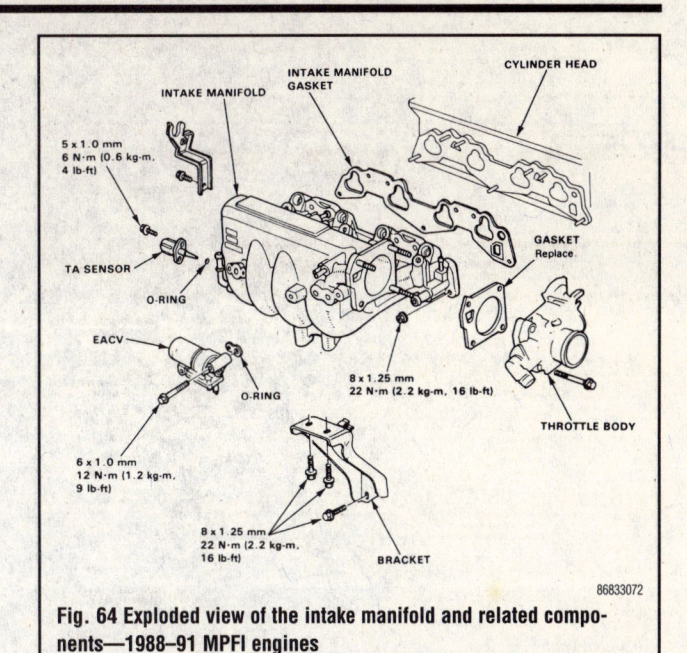

Fig. 64 Exploded view of the intake manifold and related components—1988–91 MPFI engines

3. Properly relieve the fuel system pressure (Refer to Section 5 for the correct procedure.)
4. Remove the throttle body assembly on 1.5L engines with dual-point fuel injection. On all other engines, remove the fuel injector manifold and fuel injectors.
5. As required, remove the fast idle control valve, the air bleed valve, the EGR valve and all related brackets.
6. Remove the intake manifold retaining bolts. Remove the intake manifold assembly from the vehicle. Discard the gaskets.

➡1984–87 multi-port fuel injected engines have a bolt-together upper and lower manifold chamber. Separate the upper chamber from the lower manifold before removing the assembly from the vehicle.

To install:

7. Installation is the reverse of the removal procedure.
8. Make sure all gasket mating surfaces are clean prior to installation.
9. Always use new gaskets.
10. Tighten the intake manifold nuts in 2–3 steps in a crisscross pattern starting with the inside nuts. Tighten the nuts to 16 ft. lbs. (22 Nm).
11. On 1984–87 models attach the upper manifold to the lower manifold and tighten the 6mm bolts and nut to 9 ft. lbs. (12 Nm) and the 8mm bolt to 16 ft. lbs. (22 Nm).
12. Make sure that all hoses and wires are connected properly.
13. Refill the cooling system.

1992–95 Engines

▶ See Figures 65 and 66

1. Disconnect the negative battery cable. Drain the cooling system, into a suitable container.
2. Label and disconnect all related electrical connectors and vacuum lines.
3. Properly relieve the fuel system pressure (Refer to Section 5 for the correct procedure.)
4. Remove the fuel injector manifold and four individual fuel injectors.
5. As required, remove the fast idle control valve, the air bleed valve, the EGR valve and all related brackets.
6. Remove the intake manifold retaining bolts. Remove the intake manifold assembly from the vehicle. Discard the gaskets.

To install:

7. Installation is the reverse of the removal procedure.
8. Make sure all gasket mating surfaces are clean prior to installation.
9. Always use new gaskets.
10. Tighten the intake manifold nuts in 2–3 steps in a crisscross pattern starting with the inside nuts. Tighten the nuts to 17 ft. lbs. (23 Nm).
11. Make sure that all hoses and wires are connected properly.
12. Refill the cooling system.

3-24 ENGINE AND ENGINE REBUILDING

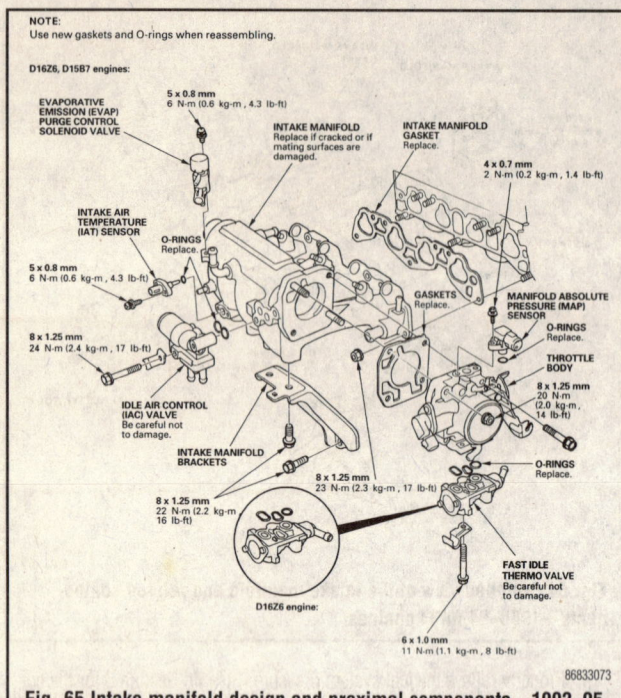

Fig. 65 Intake manifold design and proximal components—1992–95 D16Z6 and D15B7 engines

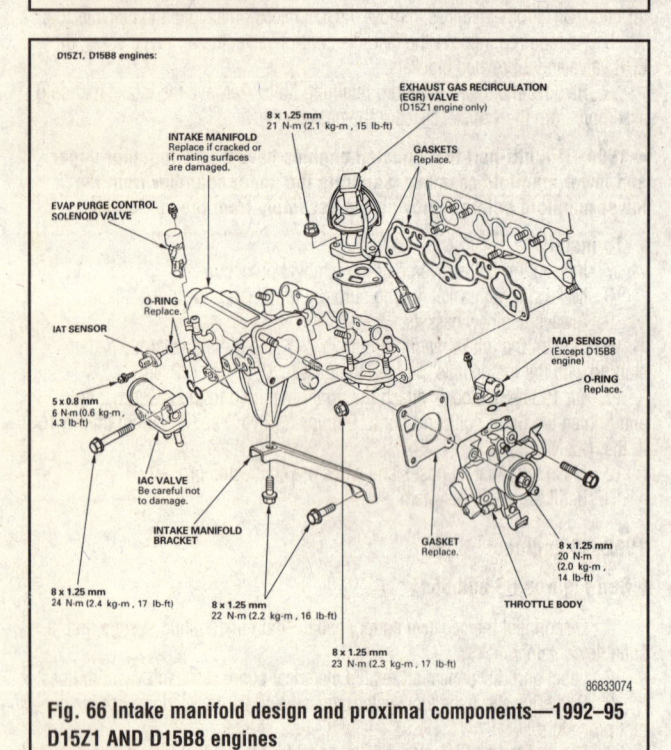

Fig. 66 Intake manifold design and proximal components—1992–95 D15Z1 AND D15B8 engines

Exhaust Manifold

REMOVAL & INSTALLATION

▸ See Figures 67 thru 72

✲✲ CAUTION

Do not perform this operation on a warm or hot engine or serious burns could result.

1. Disconnect the negative battery cable. Raise and safely support the front of the vehicle. Remove the header pipe nuts or catalytic converter-to-exhaust manifold nuts and separate the pipe from the manifold.
2. Disconnect and remove the oxygen sensor (if equipped).
3. If equipped, remove the EGR and the air suction tubes from the exhaust manifold.

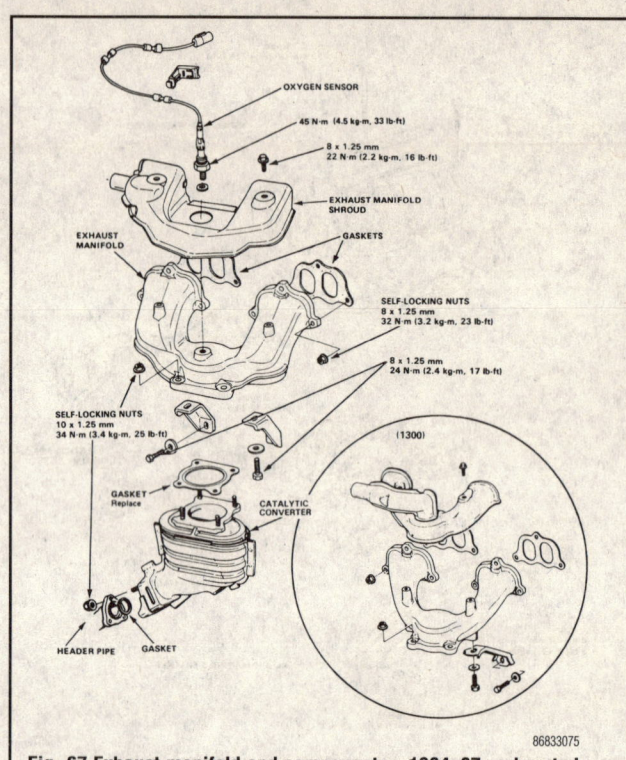

Fig. 67 Exhaust manifold and components—1984–87 carbureted engine

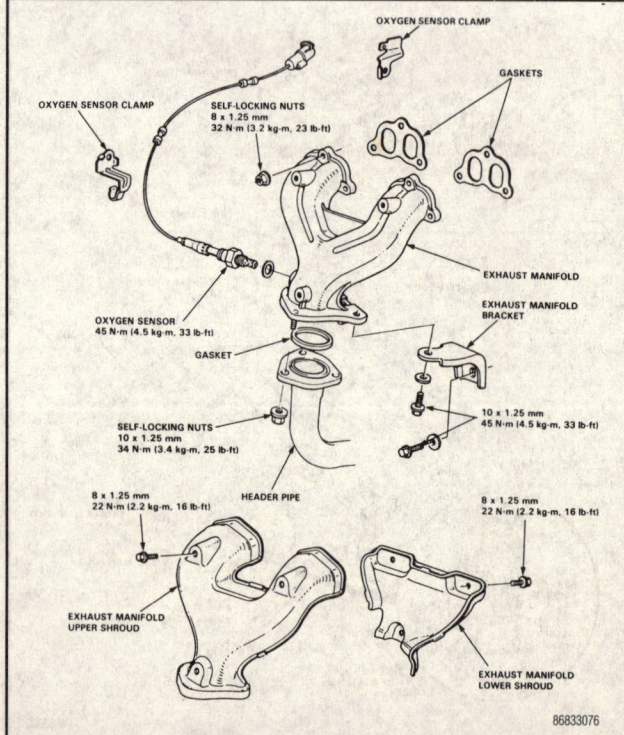

Fig. 68 Exhaust manifold and components—1985–87 fuel injected engine

ENGINE AND ENGINE REBUILDING 3-25

4. Remove the exhaust manifold shroud.
5. Remove the exhaust manifold bracket bolts.
6. Using a crisscross pattern and starting from the center, remove the exhaust manifold-to-cylinder head nuts, the manifold, the gaskets and discard them.

To install:

7. Using a gasket scraper or sharp putty knife, clean the gasket mounting surfaces.

➡ Always use new self locking bolts when working on the exhaust manifold.

8. To install, use new gaskets and mount the manifold in position. Tighten the exhaust manifold-to-cylinder head nuts (using a crisscross pattern, starting from the center) to 23 ft. lbs. (32 Nm) and the header pipe-to-exhaust manifold nuts or converter-to-manifold bolts to 23 ft. lbs. (32 Nm), and the oxygen sensor-to-exhaust manifold to 33 ft. lbs. (45 Nm). Tighten the manifold bracket bolts to 16 ft. lbs. (22 Nm) and the manifold shroud bolts to 16 ft. lbs. (22 Nm). Tighten the converter shroud bolts to 9 ft. lbs. (12 Nm).

9. Start the engine and check for exhaust leaks.

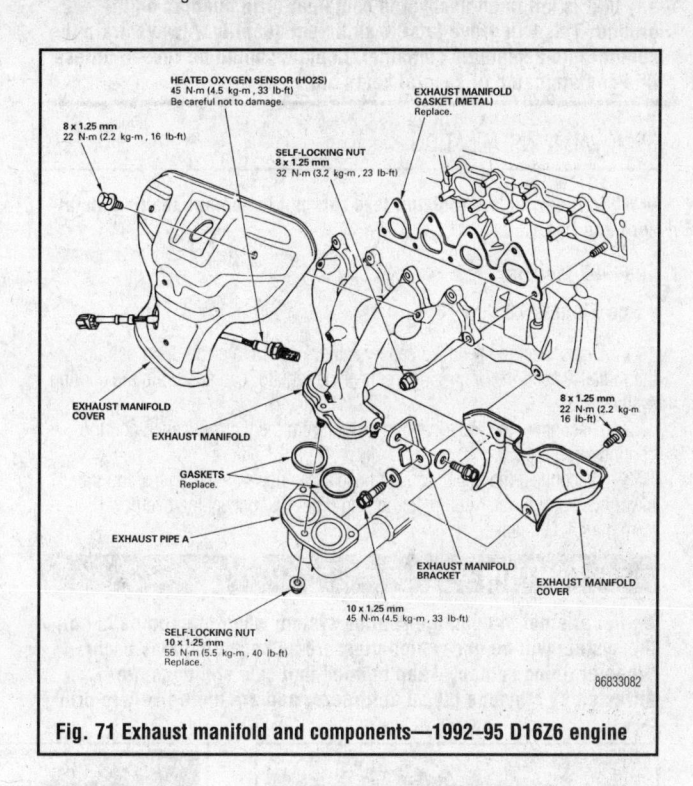

Fig. 69 Exhaust manifold and components—1988–91 Canadian engine

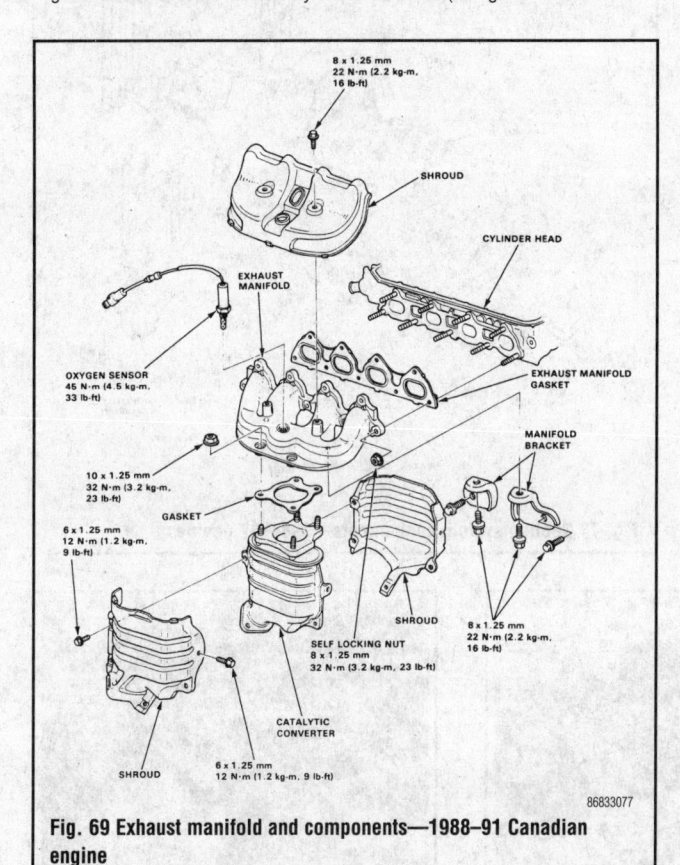

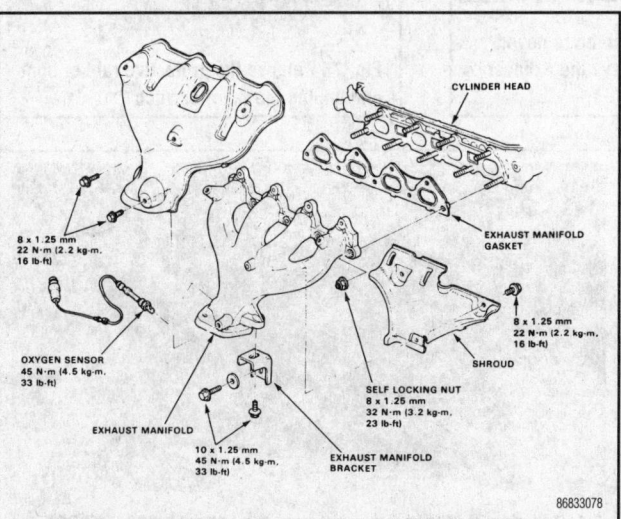

Fig. 70 Exhaust manifold and components—1988–91 vehicles (Except Canada)

Fig. 71 Exhaust manifold and components—1992–95 D16Z6 engine

Fig. 72 Exhaust manifold and components—all D15Z1 and California D15B8 engines

3-26 ENGINE AND ENGINE REBUILDING

Radiator

⚠️ CAUTION

When draining coolant, keep in mind that cats and dogs are attracted by ethylene glycol antifreeze, and are quite likely to drink any that is left in an uncovered container or in puddles on the ground. This will prove fatal in sufficient quantity. Always drain the coolant into a sealable container. Coolant should be reused unless it is contaminated or several years old.

REMOVAL & INSTALLATION

→ When removing the radiator, take care not to damage the fine metal core and fins.

1984–87 Models

▶ See Figures 73 thru 78

1. Disconnect the negative battery cable. Position a clean drain pan under the radiator, open the drain plug, remove the radiator cap and drain the cooling system.
2. Disconnect the electrical connectors from the thermo-switch and the cooling fan motor.
3. Disconnect the upper coolant hose at the upper radiator tank and the lower hose at the water pump connecting pipe. Disconnect the overflow hose from the coolant tank.

⚠️ CAUTION

Do not attempt to open the cooling system when the engine is hot; the system will be under high pressure and scalding may occur. When draining coolant, keep in mind that cats and dogs are attracted by ethylene glycol antifreeze, and are quite likely to drink any that is left in an uncovered container or in puddles on the ground. This will prove fatal in sufficient quantity. Always drain the coolant into a sealable container. Coolant may be reused unless it is contaminated or several years old.

4. Remove the radiator-to-chassis bolts, lower supports and the radiator with the fan attached. The fan can be easily unbolted from the back of the radiator.

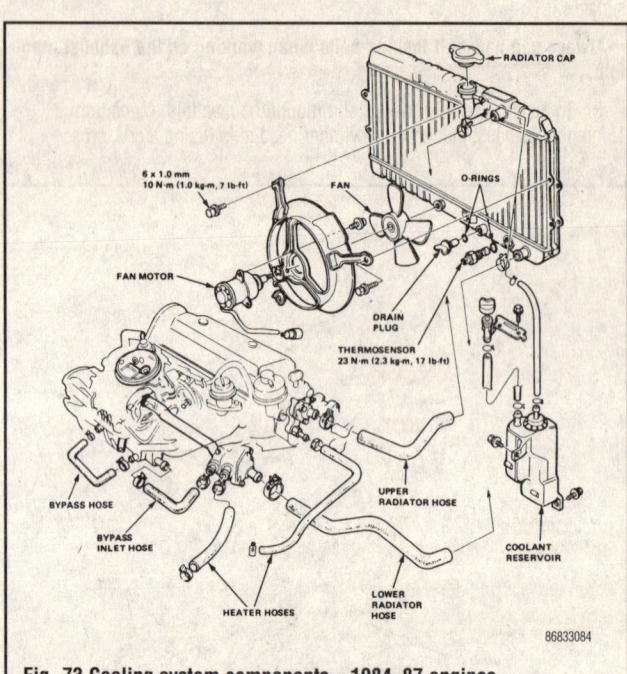

Fig. 73 Cooling system components—1984–87 engines

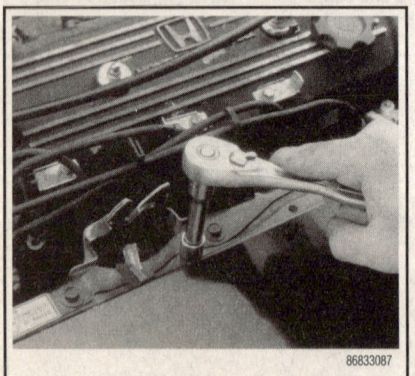

Fig. 74 Unbolting the radiator from the radiator support

Fig. 75 The radiator has to be removed from the underside. Remove the exhaust header for clearance

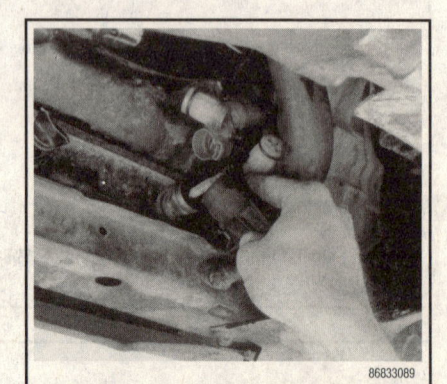

Fig. 76 Remove the protective rubber boot and unplug the thermoswitch

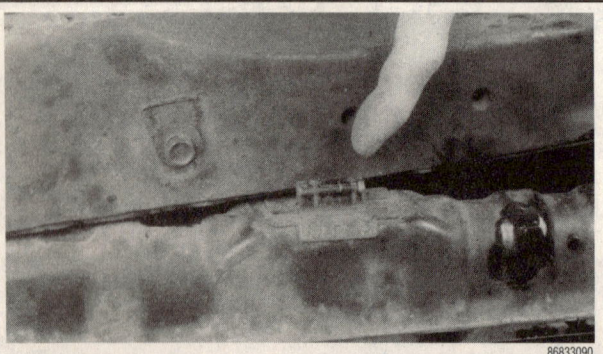

Fig. 77 The notch on the underside of the radiator fits into a slot to hold it into place

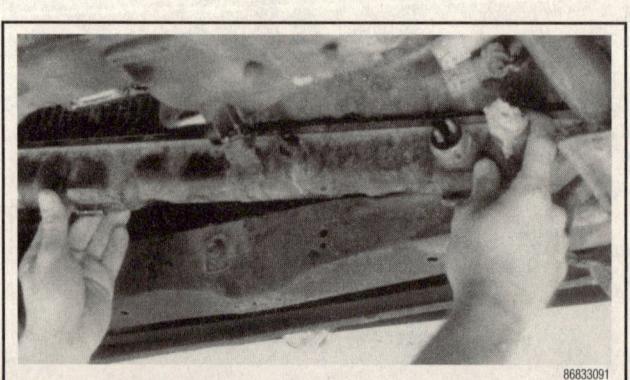

Fig. 78 Lowering the radiator out from the bottom—1984–87 CRX shown

ENGINE AND ENGINE REBUILDING   3-27

5. Inspect the hoses for damage, leaks and/or deterioration; if necessary, replace them. If the radiator fins are clogged, wash off any insects or dirt with low pressure water.

➡ For some models, it may be necessary to place the car on jackstands, remove the exhaust header and any obstruction, then remove the radiator from the underside of the car.

To install:

6. To install, use new O-rings (if used) and reverse the removal procedure. Tighten the radiator-to-chassis bolts to 7 ft. lbs. (9.5 Nm).

7. Refill and bleed the cooling system; if equipped with a coolant reservoir, be sure to fill it to proper level. Start the engine, allow it to reach normal operating temperatures and check for leaks.

➡ Check all cooling system hoses for any signs of damage, leaks or deterioration and replace if necessary.

1988–91 Models

♦ See Figure 79

1. Disconnect the negative battery cable.
2. Properly drain the cooling system into a suitable container.
3. Disconnect the fan motor and motor connector.
4. Disconnect the upper and lower radiator hoses.
5. Disconnect and plug the automatic transaxle cooling lines at the radiator, if equipped.
6. Disconnect the coolant reservoir overflow hose.
7. Remove the radiator attaching bolts and brackets.
8. Remove the radiator with the cooling fan attached.
9. Remove the cooling fan and shroud from the radiator.

➡ Check all cooling system hoses for any signs of damage, leaks or deterioration and replace if necessary.

To install:

10. Attach the cooling fan and shroud to the radiator and install the assembly.
11. Attach the radiator bolts and brackets. Tighten the radiator bracket bolts to 7 ft. lbs. (10 Nm).
12. Connect the coolant reservoir overflow hose.
13. Connect the automatic transaxle cooling lines to the radiator, if equipped.
14. Connect the upper and lower radiator hoses.
15. Connect the fan motor and thermo-switch wire connector.
16. Refill the system with the proper type and quantity of coolant, check for leaks and bleed the cooling system.
17. Reconnect the negative battery cable.

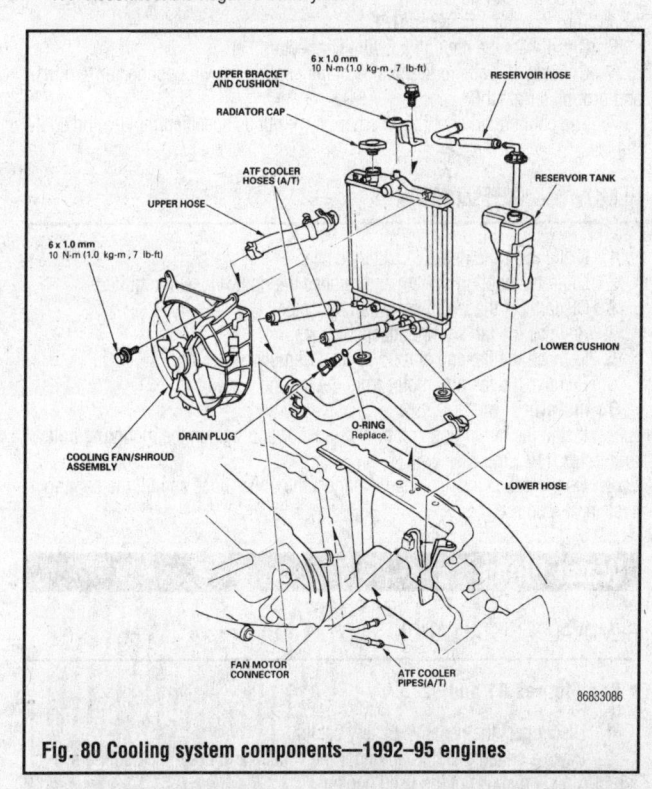

1992–95 Models

♦ See Figure 80

1. Disconnect the negative battery cable.
2. Properly drain the cooling system into a suitable container.
3. Disconnect the fan motor and motor connector.
4. Disconnect the upper and lower radiator hoses.
5. Disconnect and plug the automatic transaxle cooling lines at the radiator, if equipped.
6. Disconnect the coolant reservoir overflow hose.
7. Remove the radiator attaching bolts and brackets.
8. Remove the radiator with the cooling fan attached.
9. Remove the cooling fan and shroud from the radiator.

➡ Check all cooling system hoses for any signs of damage, leaks or deterioration and replace if necessary.

To install:

10. Attach the cooling fan and shroud to the radiator and install the assembly.
11. Attach the radiator bolts and brackets. Tighten the radiator bracket bolts to 7 ft. lbs. (10 Nm).
12. Connect the coolant reservoir overflow hose.
13. Connect the automatic transaxle cooling lines to the radiator, if equipped.
14. Connect the upper and lower radiator hoses.
15. Connect the fan motor and thermo-switch wire connector.
16. Refill the system with the proper type and quantity of coolant, check for leaks and bleed the cooling system.
17. Reconnect the negative battery cable.

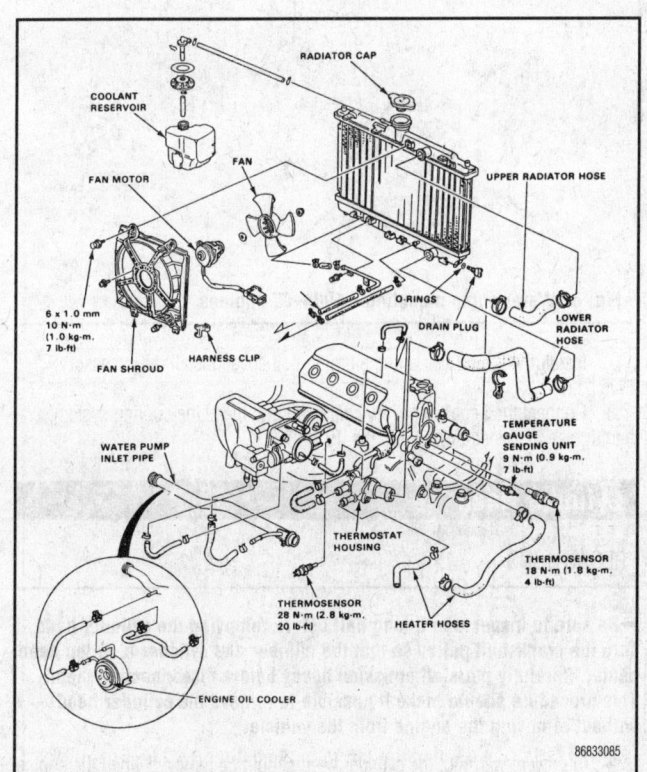

Fig. 79 Cooling system components—1988–91 engines

Fig. 80 Cooling system components—1992–95 engines

Condenser Fan

TESTING

1. Turn the ignition, air conditioner and blower speed switch **ON**.
2. Check for voltage at the condenser fan motor using a voltmeter.

3-28 ENGINE AND ENGINE REBUILDING

3. There should be approximately 12 volts, if not as specified check the fuses, relay, wiring and control unit.
4. Turn the ignition switch **OFF**.
5. Disconnect the condenser fan wire connector.
6. Connect 12 volts to terminal **A** of the cooling fan wire connector terminal and ground terminal **B**.
7. The cooling fan should run smoothly, if not as specified replace the condenser fan.

REMOVAL & INSTALLATION

1. Disconnect the negative battery cable.
2. Drain the cooling system and remove the upper hose, if required.
3. Disconnect the fan motor wire connector.
4. Remove the fan shroud attaching bolts.
5. Remove the fan and shroud as an assembly.
6. Remove the fan and motor from the shroud.

To install:

7. Install the fan in position on the radiator and tighten the mounting bolts to 7 ft. lbs. (10 Nm).
8. Connect the electrical lead. Connect the upper hose and fill the cooling system, if required.

Cooling Fan

TESTING

1. Turn the ignition switch **ON**.
2. Check for ground at the thermo-switch, when the engine reaches the proper operating temperature the switch should close and complete the ground.
3. Check the wiring, the relays, the control unit, if these are within specification, replace the thermo-switch.
4. Turn the ignition switch **OFF**.
5. Disconnect the cooling fan wire connector.
6. Connect 12 volts to terminal **A** of the cooling fan wire connector terminal and ground terminal **B**.
7. The cooling fan should run smoothly, if not as specified replace the cooling fan.

REMOVAL & INSTALLATION

1. Disconnect the negative battery cable.
2. Drain the cooling system and remove the upper hose, if required.
3. Disconnect the fan motor wire connector.
4. Remove the fan shroud attaching bolts.
5. Remove the fan and shroud as an assembly.
6. Remove the fan and motor from the shroud.

To install:

7. Install the fan in position on the radiator and tighten the mounting bolts to 7 ft. lbs. (10 Nm).
8. Connect the electrical lead. Connect the upper hose and fill the cooling system, if required.

Water Pump

REMOVAL & INSTALLATION

♦ See Figures 81 and 92

1. Disconnect the negative battery cable.
2. Place a suitable drain pan under the vehicle and drain the cooling system. Properly dispose of the used coolant.
3. Remove the accessory drive belts.
4. Refer to the timing belt removal procedure in this section and remove the timing belt cover and the timing belt.
5. Remove the water pump mounting bolts and the water pump.

To install:

6. Use a new O-ring seal and install the water pump on the block. Tighten the water pump mounting bolts to 9 ft. lbs. (12 Nm).

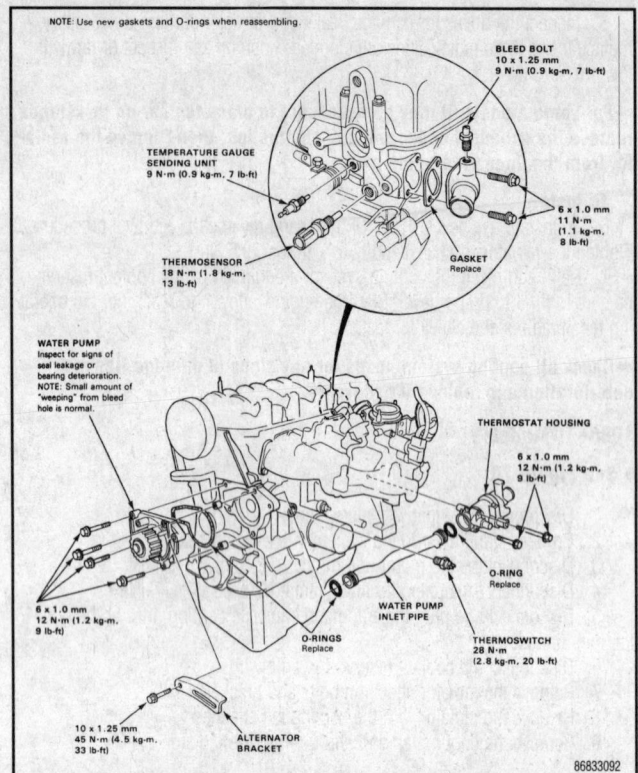

Fig. 81 Exploded view of common water pump and peripheral accessories—1988–91 vehicle shown

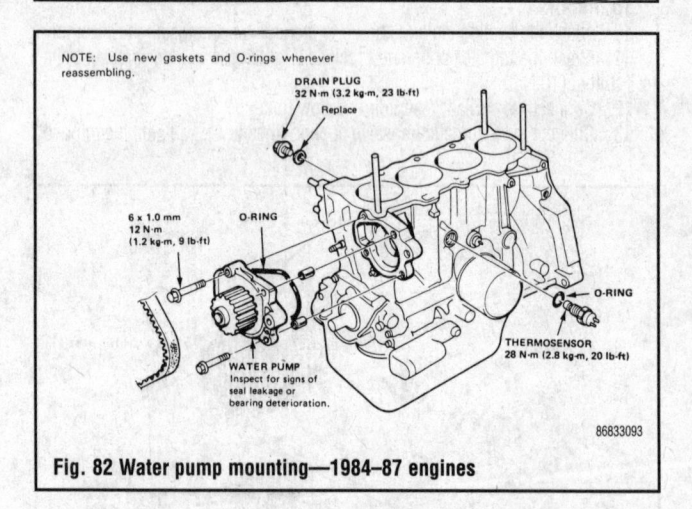

Fig. 82 Water pump mounting—1984–87 engines

7. Install the timing belt and the timing belt cover. Install the accessory drive belts.
8. Connect the negative battery cable. Fill and bleed the cooling system, then tighten the bleed bolt to 7 ft. lbs. (9 Nm).

Cylinder Head

PRECAUTIONS

➡ Be sure to inspect the timing belt before removing the cylinder head. Turn the crankshaft pulley so that the number one cylinder is at top dead center. Carefully mark all emission hoses before disconnecting them. This procedure should make it possible to remove the cylinder head without removing the engine from the vehicle.

• To prevent warping, the cylinder head should be removed when the engine is cold (below 100°F (38°C).

ENGINE AND ENGINE REBUILDING   3-29

- Remove oil, scale or carbon deposits accumulated from each part. When de-carbonizing take care not to score or scratch the mating surfaces.
- After washing the oil holes or orifices in each part, make sure they are not restricted by blowing out with compressed air.
- If the parts will not be reinstalled immediately after washing, spray the parts with a rust preventive to protect from corrosion.

➡ Taking a few pictures with an instant camera before disassembly could prove to be extremely helpful during assembly.

REMOVAL & INSTALLATION

✱✱ CAUTION

When draining coolant, keep in mind that cats and dogs are attracted by ethylene glycol antifreeze, and are quite likely to drink any that is left in an uncovered container or in puddles on the ground. This will prove fatal in sufficient quantity. Always drain the coolant into a sealable container. Coolant may be reused unless it is contaminated or several years old.

1984–87 Engines

◆ See Figures 83 thru 97

1. Disconnect the negative battery terminal.
2. Place a clean drain pan under the radiator, remove the radiator cap, the drain plug (from the front side of the engine block, if equipped) and drain the cooling system.
3. To remove the air cleaner, perform the following procedures:
    a. Remove the air cleaner cover and filter.
    b. Disconnect the hot and cold air intake ducts, then, remove the air chamber hose.
    c. Remove the three bolts holding the air cleaner.
    d. Lift the air cleaner housing, then, remove the remaining hoses and the air temperature sensor wire.
    e. Remove the air cleaner.
4. On fuel injected models, relieve the fuel pressure using the following procedures:

    a. Slowly loosing the service bolt on top of the fuel filter about one turn.

➡ Place a rag under the filter during this procedure to prevent fuel from spilling onto the engine.

   b. Disconnect the fuel return hose from the pressure regulator. Remove the special nut and then, remove the fuel hose.
5. Remove the brake booster vacuum tube from the intake manifold.
6. Remove the engine ground wire from the valve cover and disconnect the wires from the fuel cut-off solenoid valve, the automatic choke and the thermosensor.
7. Disconnect the fuel lines.
8. Label and disconnect the spark plug wires from the spark plugs, then, remove the distributor assembly.
9. Disconnect the throttle cable from the carburetor or throttle body.
10. Disconnect the hoses from the charcoal canister and from the No. 1 control box at the tubing manifold.
11. For California or high altitude models, disconnect the air jet controller.
12. If equipped with air conditioning, disconnect the idle control solenoid hoses.
13. Disconnect the upper radiator heater and bypass hoses.
14. On fuel injected models, disconnect the engine sub harness connectors and the following couplers from the head and the intake manifold.
- The four injector couplers
- The TA sensor connector
- The ground connector
- The TW sensor connector
- The throttle sensor connector
- The crankshaft angle sensor coupler
15. Remove the thermostat housing-to-intake manifold hose.
16. Disconnect the oxygen sensor coupler.
17. Remove the exhaust manifold bracket and manifold bolts, then, remove the manifold.
18. Remove the bolts from the intake manifold and bracket.
19. Disconnect the breather chamber-to-intake manifold hose.
20. Remove the valve and timing belt covers.
21. Loosen the timing belt tensioner adjustment bolt, then, remove the belt.
22. Remove the cylinder head bolts in the reverse order given in the head bolt torque sequence.

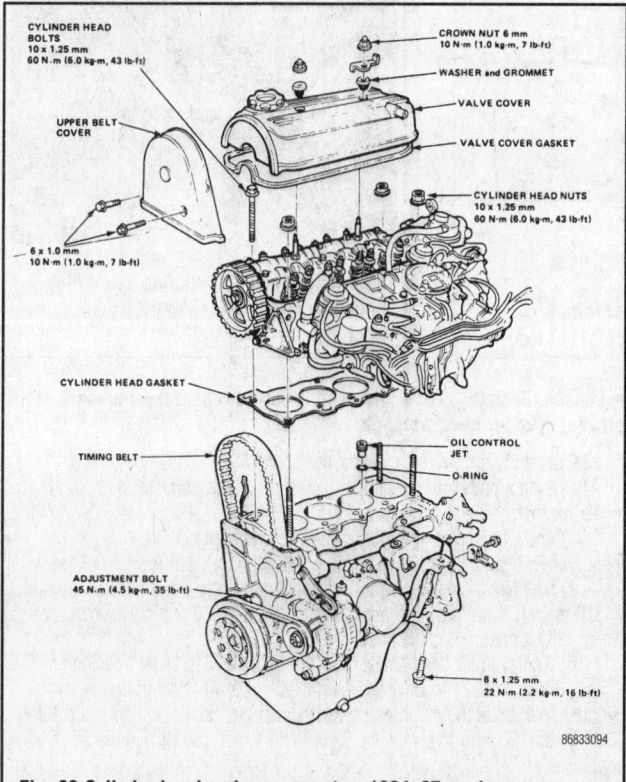

Fig. 83 Cylinder head and components—1984–87 engines

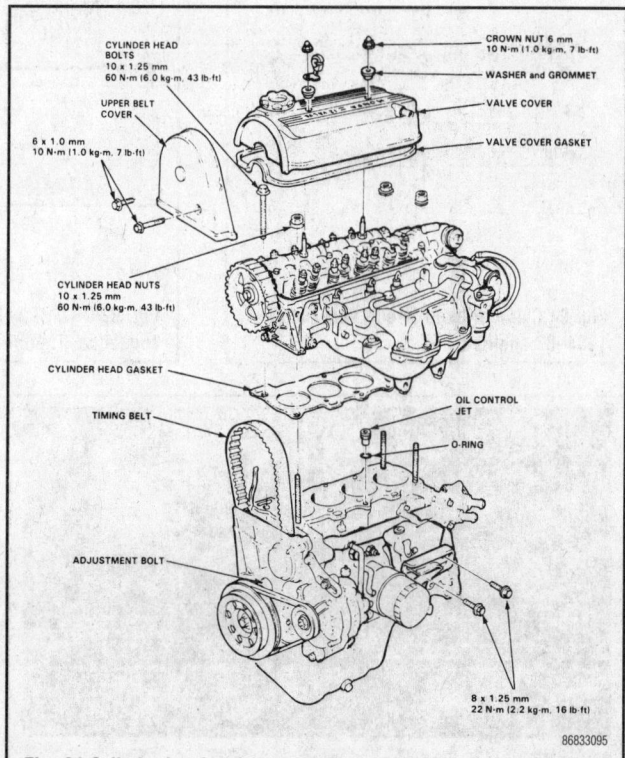

Fig. 84 Cylinder head and components—1985–87 fuel injected models

3-30 ENGINE AND ENGINE REBUILDING

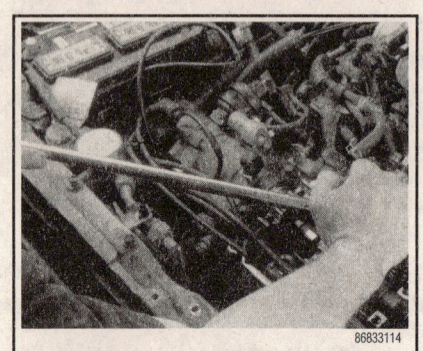

Fig. 85 Loosening the cylinder head bolts using a breaker bar—once they are cracked loose, a ratchet will work best

Fig. 86 Use a gasket scraper to thouroughly clean the remaining gasket material from the mating surfaces.

Fig. 87 Using a breaker bar to loosen the head bolts—1985–87 fuel injected engine shown

Fig. 88 Notice special oil flow grooves spiraling down the shaft

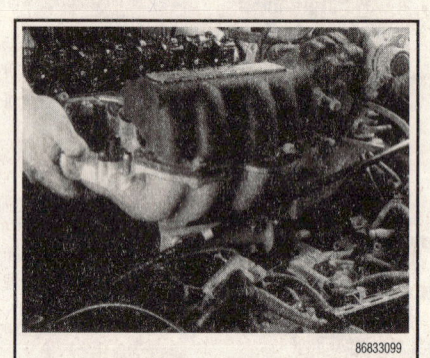

Fig. 89 The cylinder head is here lifted from the engine with its two-piece intake manifold and throttle body still attached.

Fig. 90 Top-side view of the entire cylinder head assembly

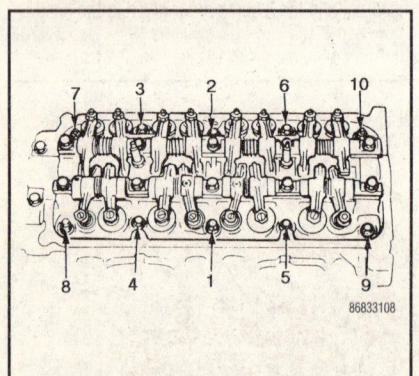

Fig. 91 Cylinder head torque sequence—1984–87 engines

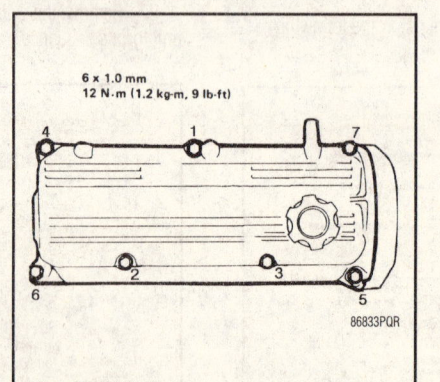

Fig. 92 CRX HF valve cover torque sequence. Tighten to 9 ft. lbs. (12 Nm)

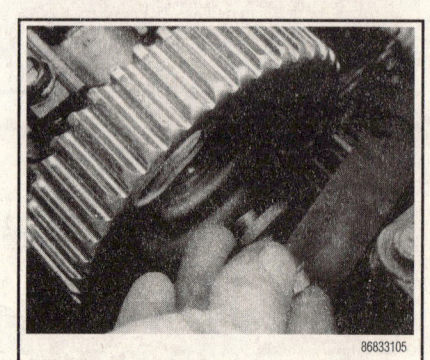

Fig. 93 Replacing the timing belt pulley bolt. Be sure to tighten to the proper torque

Fig. 94 Replace the belt making sure the teeth line up between the belt and pulley

➡ Loosen the bolts 1/3 of a turn each time and repeat the sequence to prevent cylinder head warpage.

23. Carefully remove the cylinder head from the engine.
24. Using a putty knife, clean the gasket mounting surfaces.

To install:

25. Scrape any remaining gasket material from the mating surfaces and install a new head gasket. Be sure to pay attention to the following points:

    a. The No. 1 cylinder must be at top dead center with the camshaft pulley **UP** mark (if applicable, see also Section 1 for timing belt replacement) on top before positioning the head in place.

    b. The cylinder head dowel pins and oil control jet must be aligned.

26. Replace the cylinder head using clean or new bolts and tighten the cylinder head bolts in two progressive steps and in the proper sequence. First, to 22 ft. lbs. (30 Nm) in sequence, then, to 43 ft. lbs. (60 Nm) in the same sequence.

27. Replace the timing belt and water pump pulleys and replace belt.

ENGINE AND ENGINE REBUILDING   3-31

Fig. 95 The timing belt properly installed. Note that there is no "UP" mark on the top of this particular pulley, although there are opposed horizontal alignment marks enscribed on either side of the center bolt. These marks are sight lines that should be even with the top of the cylinder head as shown

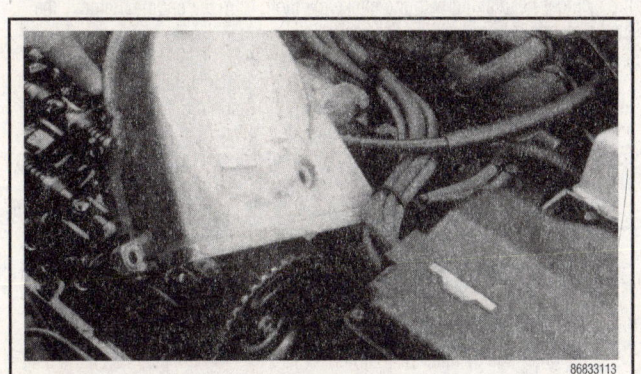

Fig. 96 Replacing the timing belt pulley cover—tighten the 6 x 1.0mm bolt to 7 ft. lbs. (10 Nm)

Fig. 97 Using a torque wrench to tighten cylinder head bolts to the proper specification

28. Replace the valve cover and tighten to 9 ft. lbs. (12 Nm) starting with the inner bolts then working out progressively. (Tighten CRX HF valve cover in proper two-step sequence whereby you apply low torque on the first round and apply final torque on second round—see drawing).
29. Replace the breather chamber to intake manifold hose. Replace the thermostat housing-to-intake manifold hose.
30. Replace the exhaust manifold using a new gasket and clean or new bolts coated lightly with anti-seize compound.
31. Reconnect the header pipe to the exhaust manifold using a new gasket.
32. Install the distributor aligning it to the mating end of the end of the camshaft.
33. On fuel injected models, connect the engine sub-harness connectors and the following couplers from the cylinder head and the intake manifold.
- The four injector couplers
- The TA sensor connector
- The ground connector
- The TW sensor connector
- The throttle sensor connector
- The crankshaft angle sensor coupler
- The oxygen sensor coupler

34. Replace all other related hoses including: radiator, charcoal canister-to-control box, fuel lines, brake booster vacuum tube.
35. If equipped with A/C, replace the idle control solenoid hoses.
36. If a California or HI ALT model, replace the air jet controller.
37. Reconnect the throttle cable to carbureter or fuel injector throttle body (as applicable).
38. Replace all remaining wires, including: spark plug wires, engine ground wires, wires to the fuel cut-off solenoid valve, automatic choke, and the thermosensor.
39. Tighten the fuel system connections to the proper torque in order to accomodate fuel pressure when running.
40. Replace the air cleaner, refill the radiator and reconnect the negative battery terminal.
41. After installation, check to see that all hoses and wires are installed correctly and nothing was overlooked.

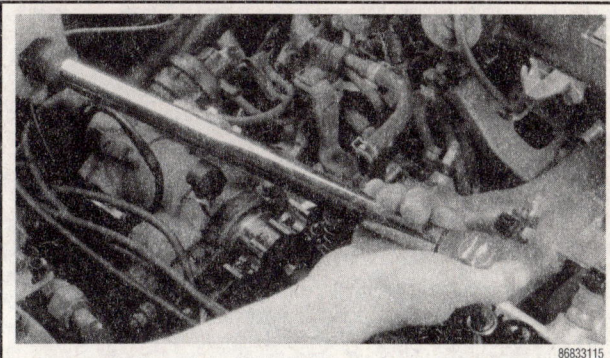

➡ If you took photographs before starting, now compare the before to after results if necessary.

42. Run the engine and check the ignition timing. Check for coolant or oil leaks.

1988–91 Engines

◆ See Figures 98, 99, 100 and 101

1. Be sure the engine is cold. Disconnect the negative battery cable. Drain the coolant.
2. Remove the brake booster vacuum hose from the brake master cylinder power booster. Remove the engine secondary ground cable from the valve cover.
3. Remove the air intake hose and the air chamber. Relieve the fuel pressure. Disconnect the fuel hoses and fuel return hose.

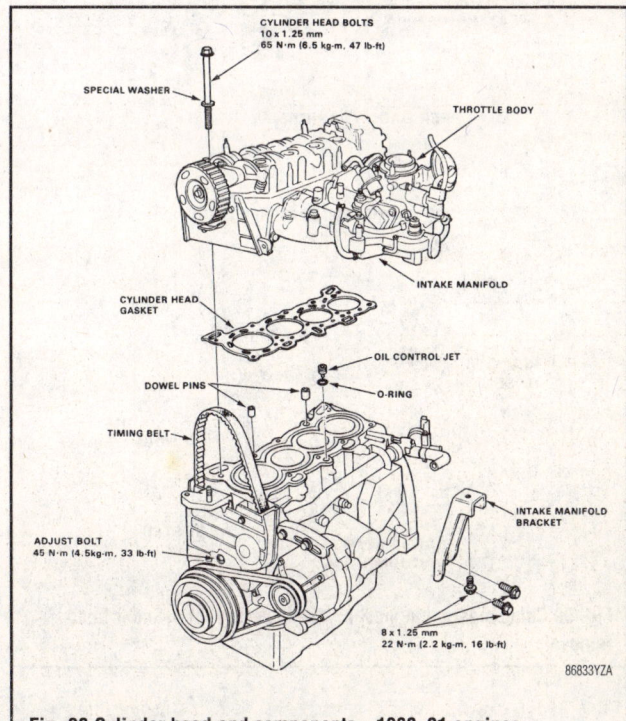

Fig. 98 Cylinder head and components—1988–91 engines

3-32 ENGINE AND ENGINE REBUILDING

4. Remove the air intake hose and resonator hose. Disconnect the throttle cable at the throttle body. On vehicles equipped with automatic transaxles, disconnect the throttle control cable at the throttle body.

5. Disconnect the charcoal canister hose at the throttle valve.

6. Disconnect the following engine wire connectors from the cylinder head and the intake manifold:
   a. 14-prong connector from the main wiring harness.
   b. EACV connector.
   c. Intake air temperature sensor connector.
   d. Throttle angle sensor connector.
   e. Injector connectors.
   f. Ignition coil from the distributor.
   g. Top dead center/crank sensor connector from the distributor.
   h. Coolant temperature gauge sender connector.
   i. Coolant temperature sensor connector.
   j. Oxygen sensor.

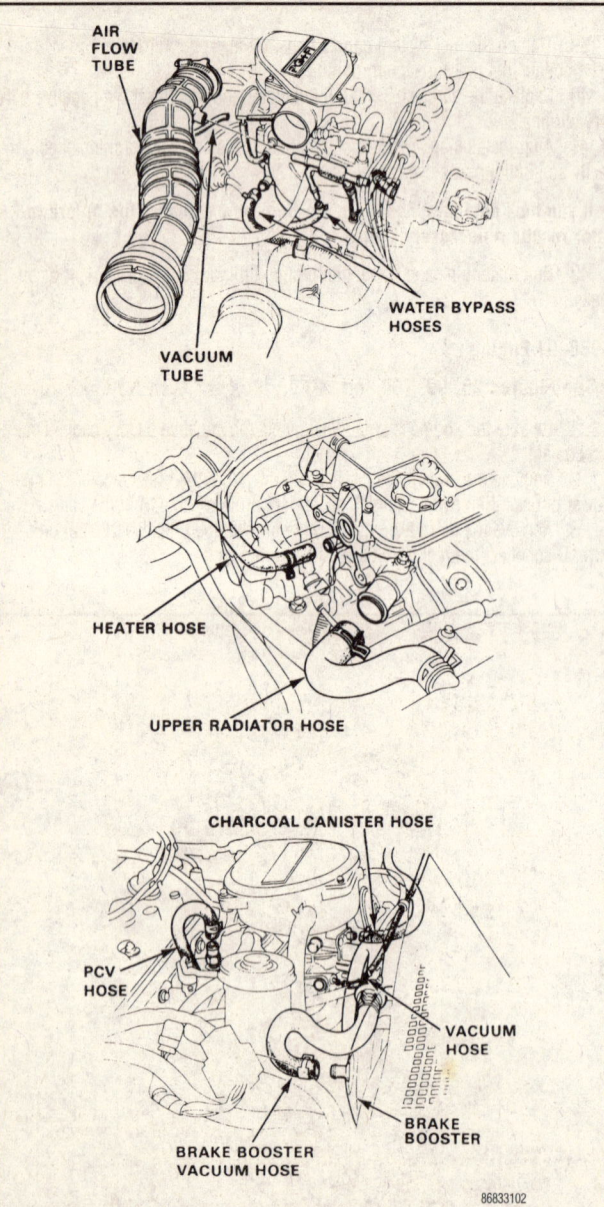

Fig. 99 Components that must be disconnected for cylinder head removal

7. Disconnect the vacuum hoses and the water bypass hoses from the intake manifold and throttle body.

8. Remove the upper radiator hose and the heater hoses from the cylinder head.

9. Remove the PCV hose, charcoal canister hose and vacuum hose from the intake manifold, and remove the vacuum hose from the brake master cylinder power booster.

10. Loosen the air conditioning idler pulley and remove the air conditioning belt. Remove the alternator belt. If equipped with power steering, remove the power steering belt and pump bracket.

11. Remove the intake manifold bracket. Remove the exhaust manifold bracket, then remove the header pipe.

12. Remove the exhaust manifold shroud, then remove the exhaust manifold.

13. Mark the position of the distributor in relation to the engine block, remove and tag the spark plug wires and remove the distributor assembly.

14. Remove the valve cover. Remove the timing belt cover.

15. Mark the direction of rotation on the timing belt. Loosen the timing belt adjuster bolt, then remove the timing belt from the camshaft pulley.

➡Do not crimp or bend the timing belt more than 90 degrees or smaller than 1 in. (25mm) in diameter.

16. Remove the cylinder head bolts. Once the bolts are all removed, remove the cylinder head along with the intake manifold from the engine. Remove the intake manifold from the cylinder head.

To install:

17. Be sure the cylinder head and the engine block surfaces are clean, level and straight and place a new head gasket on the mounting surface.

18. Be sure the **UP** mark on the timing belt pulley is at the top or the horizontal alignment marks are even with the top of the cylinder head when sighting with your eye looking straight-on.

19. Install the intake manifold making sure the cylinder head dowel pins and control jet are aligned. Starting with the inner nuts, tighten the nuts in a crisscross pattern in 2–3 steps to 17 ft. lbs. (23 Nm).

20. Install the bolts that secure the intake manifold to its bracket but do not tighten them at this point.

21. Position the cam correctly and install the cylinder head bolts.

22. Tighten the cylinder head bolts in two steps. On the first step tighten all the bolts, in sequence, to 22 ft. lbs. (30 Nm). On the final step, using the same sequence, tighten the bolts to 47 ft. lbs. (65 Nm).

23. On the Standard and Si vehicles, install the exhaust manifold and tighten the nuts in a crisscross pattern in two or three steps to 25 ft. lbs. (34 Nm) starting with the inner nuts.

24. On the CRX HF vehicles, install the catalytic converter to the exhaust manifold, then install the exhaust manifold assembly and tighten the bolts to 25 ft. lbs. (34 Nm).

25. Install the header pipe onto the exhaust manifold. Tighten the bolts to the intake manifold bracket. Install the header pipe on to its bracket.

26. After the installation procedure is complete, check that all tubes, hoses and connectors are installed correctly. Adjust the valve timing.

➡If you took photographs before starting, now compare the instalation with the "before" pictures; correct any unintended discrepancies.

1992–95 Engines

◆ See Figures 102 thru 109

➡Be sure to inspect the timing belt before removing the cylinder head. Turn the crankshaft pulley so that the number one cylinder is at top dead center. Mark all emission hoses before disconnecting them. Be sure you have the original radio code number before disconnecting the battery (if applicable). This procedure does not require removing the engine from the vehicle.

1. Be sure the engine is cold to avoid warping the cylinder head (wait until the coolant temperature drops below 100°F (38°C) before loosening the retaining bolts). Disconnect the negative battery cable. Drain the cooling system and relieve the system fuel pressure.

ENGINE AND ENGINE REBUILDING 3-33

Fig. 100 Cylinder head torque sequence— 1988–91 vehicles except CRX HF

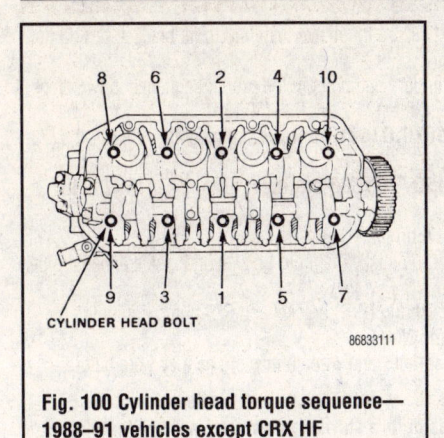

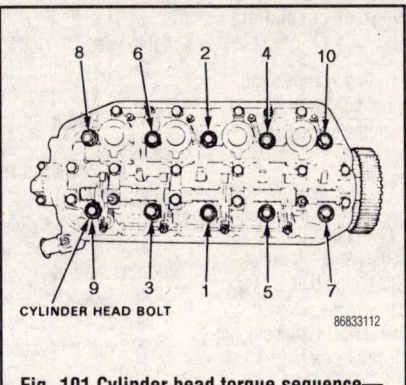

Fig. 101 Cylinder head torque sequence— 1988–91 vehicles CRX HF

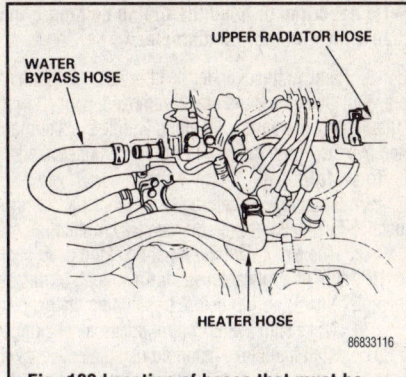

Fig. 102 Location of hoses that must be removed for cylinder head removal

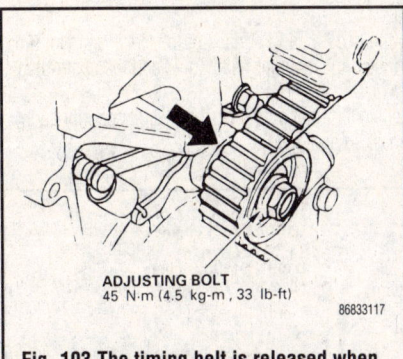

Fig. 103 The timing belt is released when the tensioner is loosened slightly and pushed to allow slack for the belt

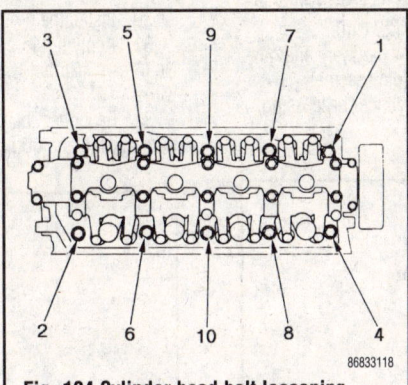

Fig. 104 Cylinder head bolt loosening sequence—1992–95 vehicles

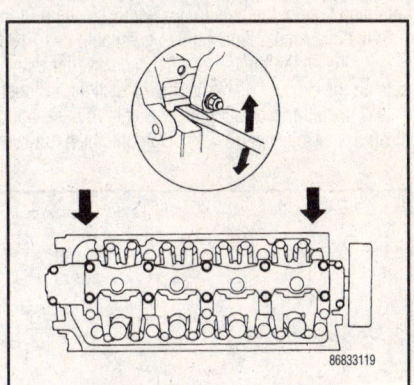

Fig. 105 Pry the head from the block using a suitable prytool

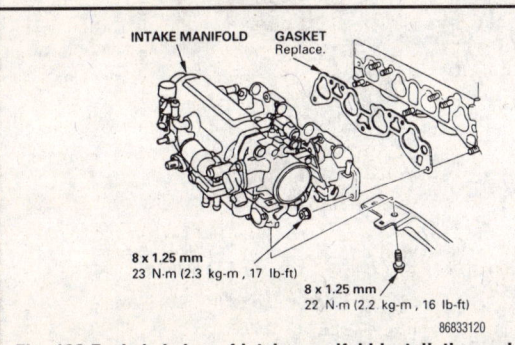

Fig. 106 Exploded view of intake manifold installation and proper torque figures. Tighten the bolts in the sequence described in step No. 20—1992–95 vehicles

** CAUTION

Never smoke or allow open flame or spark near the car when working on the fuel system. Only drain fuel into an approved container.

2. Remove the brake booster vacuum hose from the brake master cylinder power booster. Remove the engine secondary ground cable from the valve cover.

➡An instant camera (if available) to take a few pictures of everything in place could prove invaluable when it is time for re-installation.

3. Remove the air intake hose and the air chamber. Relieve the fuel pressure. Disconnect the fuel feed hose and evaporative emission control canister (EVAP control) hose from the intake manifold.

4. Remove the throttle cable at the throttle body (A/T only).

➡Do not bend the throttle cable when removing it. If it gets kinked, replace with a new one.

5. Remove the fuel return hose and brake booster vacuum hose.
6. Disconnect the charcoal canister hose at the throttle valve.
7. Disconnect the following engine wire connectors from the cylinder head and the intake manifold:
   a. Four injector connectors
   b. Intake air temperature (IAT) sensor connector
   c. Throttle position sensor connector
   d. Manifold absolute pressure (MAP) sensor connector
   e. Ground terminal (at thermostat cover)
   f. Heated oxygen sensor (HO2s) connector
   g. Engine coolant temperature (ECT) sensor connector
   h. ECT switch connector
   i. ECT gauge sending unit connector
   j. VTEC solenoid valve connector (D16Z6, D15Z1 engines)
   k. VTEC pressure switch connector (D16Z6, D15Z1 engines)
   l. Exhaust gas recirculation (EGR) valve lift sensor connector (D15Z1 engine)
   m. TDC/CKP/CYP sensor connector

8. Remove and tag the spark plug wires and remove the distributor assembly. Remove the engine ground cable on the cylinder head cover.
9. Remove the power steering (P/S) belt and pump but do not disconnect the P/S hose. Remove the P/S bracket.
10. Remove the emission vacuum hoses and water bypass hoses from the intake manifold assembly.
11. Remove the upper radiator hose and heater hose from the cylinder head and the water bypass hose from the thermostat housing.
12. Remove the intake manifold bracket.
13. Remove self-locking nuts and disconnect the the exhaust manifold bracket, then remove exhaust pipe A (formerly known as the "header pipe"). Then remove the exhaust manifold bracket.
14. Remove the PCV hose, remove the cylinder head cover, then remove the upper cover.
15. Loosen the adjusting bolt 1/2 turn.
16. Push the tensioner to release tension from the timing belt, then retighten the adjusting bolt.
17. Remove the timing belt.

3-34 ENGINE AND ENGINE REBUILDING

→Do not crimp or bend the timing belt more than 90 degrees or smaller than 1 inch (25mm) in diameter.

18. Remove the cylinder head bolts in sequence as shown in accompanying drawing. Once the bolts are all removed, gently pry the cylinder head with a suitable tool and remove it along with the intake manifold from the engine. Remove the intake manifold from the cylinder head.

To install:

19. Scrape any remaining gasket material from the mating surfaces and install a new head gasket. Be sure to pay attention to the following points:

    a. The No. 1 cylinder must be at top dead center with the camshaft pulley **UP** mark (if applicable, see further along in this section for timing belt replacement) on top before positioning the head in place.

    b. Make sure the oil control orifice is not clogged (D15B7, D15B8 engines).

20. Install the intake manifold using a new gasket and clean or new hardware and tighten the nuts in a crisscross pattern in two or three steps beginning with the inner nuts. Final tightening should be to 17 ft. lbs. (23 Nm).

21. Install the exhaust manifold using a new gasket and clean or new hardware and tighten the nuts in a crisscross pattern in two or three steps beginning with the inner nuts. Final tightening should be to 16 ft. lbs. (22 Nm).

22. Install the three-way catalytic converter using all new hardware and gaskets on the exhaust manifold, then install the exhaust manifold assembly

23. Install two dowel pins, a new head gasket and the cylinder head using clean or new oil-dipped bolts. Do not tighten fully yet.

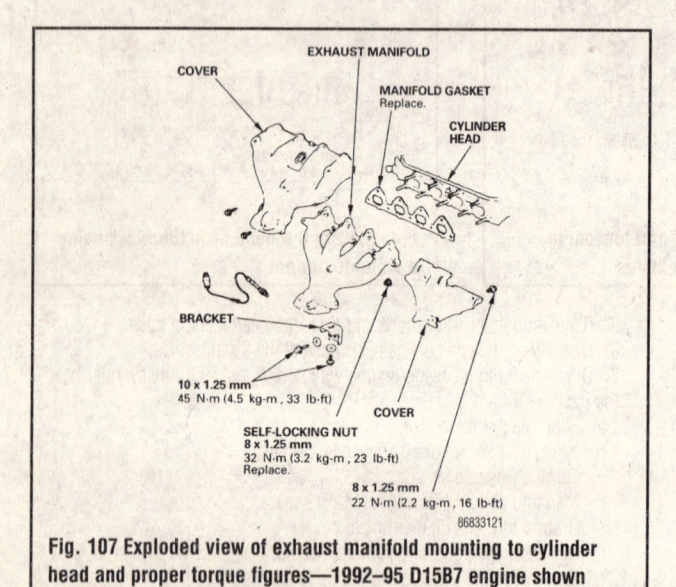

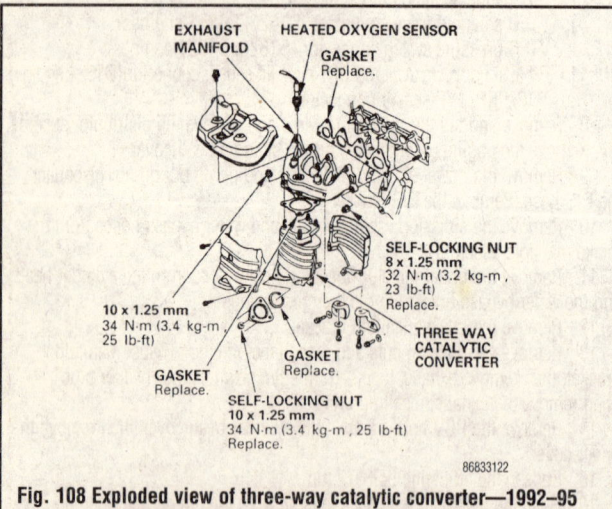

24. Install the bolts that secure the intake manifold to its bracket, but do not tighten yet.

25. Now tighten the cylinder head bolts in the proper sequence in two steps:

    a. **1st step:** initially tighten to 22 ft. lbs. (30 Nm)

    b. **2nd step for D16Z6, D15Z1 engines:** fully tighten to 53 ft. lbs (73 Nm), or

    c. **2nd step for D15B7, D15B8 engines:** fully tighten to 47 ft. lbs. (65 Nm)

26. Install exhaust pipe A on the exhaust manifold.

27. Tighten the bolts for the intake manifold bracket and install exhaust pipe A on its bracket.

28. Replace the power steering (P/S) bracket and P/S pump tightening the bracket bolts to 33 ft. lbs. (45 Nm).

29. Reconnect all hoses, tubes and wire connectors as they were before removal.

→If you took photographs before starting, now compare the installation with the "before" pictures; correct any unintended discrepancies.

30. Install the timing belt.

31. Apply liquid gasket to the head mating surface of the No. 1 and No. 5 or No. 6 camshaft holders, then install the cylinder head cover and the spark plug wires.

32. Reconnect battery negative terminal and enter the 5-digit security code to re-activate the original radio (if so equipped).

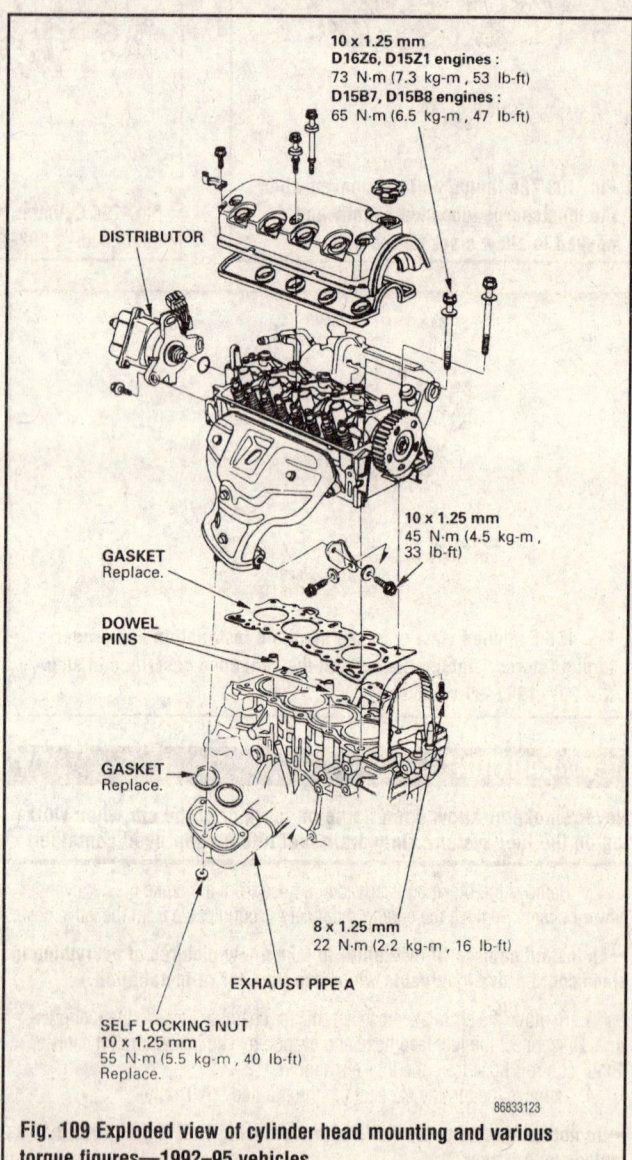

Fig. 107 Exploded view of exhaust manifold mounting to cylinder head and proper torque figures—1992–95 D15B7 engine shown

Fig. 108 Exploded view of three-way catalytic converter—1992–95 D15Z1 and D15B8 engines for 49 states. The similar California version for same engines uses the same torque figures

Fig. 109 Exploded view of cylinder head mounting and various torque figures—1992–95 vehicles

ENGINE AND ENGINE REBUILDING 3-35

CLEANING AND INSPECTION

♦ See Figures 110, 111 and 112

1. Refer to the Valves, Removal and Installation procedure in this section and remove the valve assemblies from the cylinder head.
2. Using a small wire power brush, clean the carbon from the combustion chambers and the valve ports.
3. Inspect the cylinder head for cracks in the exhaust ports, combustion chambers or external cracks in the water chamber.
4. Thoroughly clean the valve guides using a suitable wire bore brush or expanding wire-type cleaner.

➟Excessive valve stem-to-bore clearance will cause excessive oil consumption and may cause valve breakage. Insufficient clearance will result in noisy and sticky functioning of the valve and will disturb engine smoothness.

5. Measure the valve stem clearance as follows:
   a. Clamp a dial indicator on one side of the cylinder head rocker arm cover gasket rail.
   b. Locate the indicator so movement of the valve stem from side-to-side (crosswise to the head) will cause a direct movement of the indicator stem. The indicator stem must contact the side of the valve stem just above the valve guide.
   c. Prop the valve head about 1/16 in. (1.6mm) off the valve seat.
   d. Move the stem of the valve from side-to-side using light pressure to obtain a clearance reading. If the clearance exceeds specifications, it will be necessary to ream (for original valves) the valve guides.
6. Inspect the rocker arm studs for wear or damage.
7. Install a dial micrometer into the valve guide and check the valve seat for concentricity.

RESURFACING

♦ See Figures 113 and 114

1. Using a straightedge, check the cylinder head for warpage.

2. If the warpage exceeds 0.076mm in a 152mm span or 0.152mm over the total length, the cylinder head must be resurfaced. Resurfacing can be performed at most machine shops.

➟If the camshaft-to-holder oil clearances are not within specification, the cylinder head cannot be resurfaced.

Valves

REMOVAL & INSTALLATION

♦ See Figures 115, 116 and 117

1. Refer to the cylinder head and camshaft removal procedures elsewhere in this section, then remove the cylinder head from the engine and the camshaft from the cylinder head.
2. Using a plastic mallet, tap each valve stem to loosen the valve keepers.
3. Using a valve spring compressor tool, compress the valve springs, then, remove the valve keepers, retainers and springs. Remove the valves from the opposite side of the cylinder head.

➟When removing the valves and components, keep them in exact order for reinstallation purposes.

4. To replace the valve seals, simply pull the seal from the valve guide. If the valve seals are being reused, it is a good idea to replace the springs around the seal's neck.

➟The exhaust valve seal uses a black spring, while the intake valve seal uses a white spring.

5. Inspect the valves for wear, damage and/or cracks. If necessary, reface the valves on a valve grinding machine.

➟When replacing the valve springs, place the closely wound end toward the cylinder head.

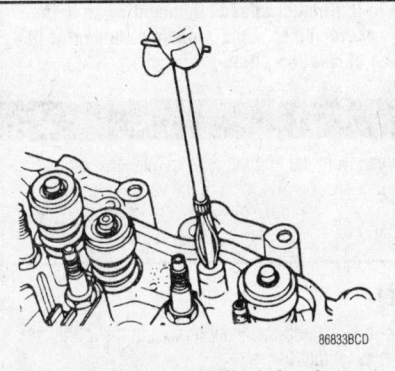

Fig. 110 Using an expanding wire cleaner to clean the valve guides

Fig. 111 Using a dial indicator to measure the valve stem clearance

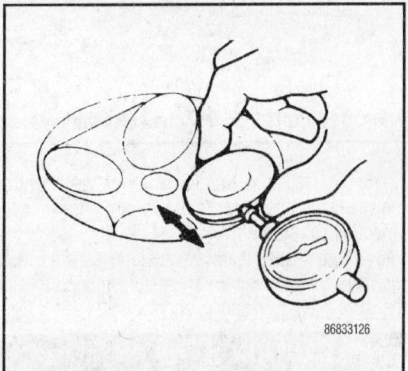

Fig. 112 Measure the valve movement in the head using a dial indicator

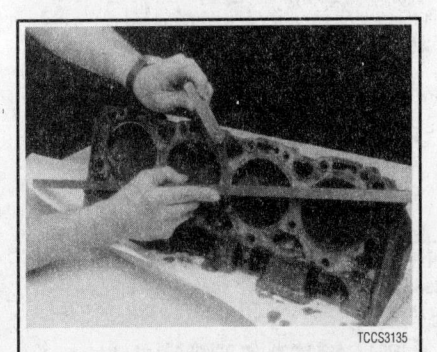

Fig. 113 Use a straightedge and feeler gauge to check for possible cylinder head warpage

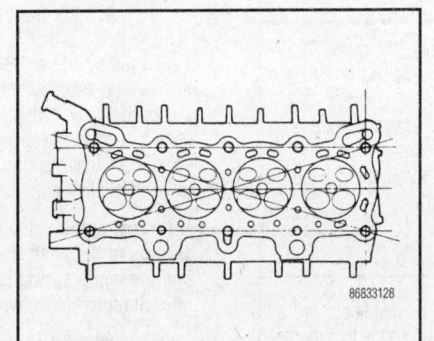

Fig. 114 Use this pattern to check the cylinder head height

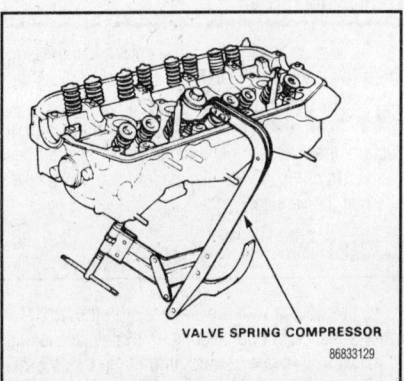

Fig. 115 Using a valve spring compressor to remove the valves

3-36 ENGINE AND ENGINE REBUILDING

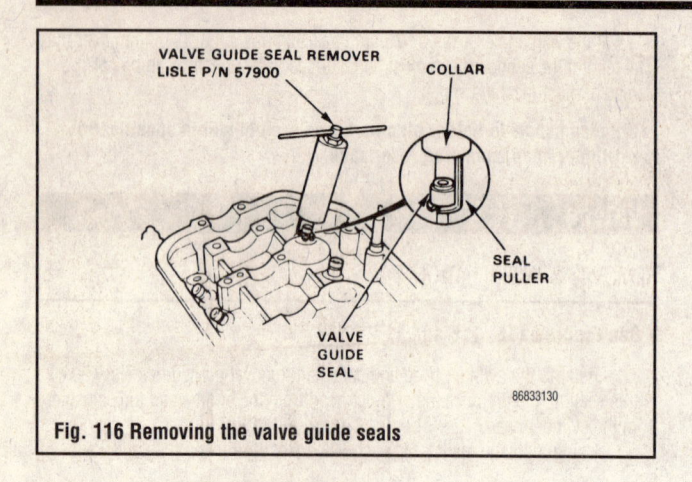

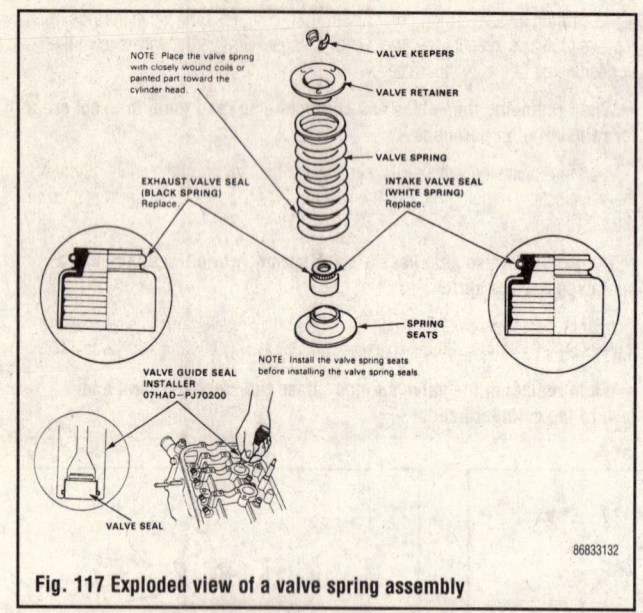

Fig. 116 Removing the valve guide seals

Fig. 117 Exploded view of a valve spring assembly

6. To assemble, first lubricate the valves with clean engine oil and slide the valves up and down in the valve guides to make sure they move smoothly.
7. Then install the seals and lubricate the parts with clean engine oil. Use the valve spring compressor tool to compress the springs and carefully install the valve keepers, retainers and springs.

✲✲ CAUTION

When removing the valve spring compressor tool, remove it slowly and make sure the valve keepers are fully seated; otherwise, the springs may fly off.

8. After removing the valve spring compressor, tap the valve stems 2–3 times to make sure the valve keepers and valves are fully seated. Only tap each valve stem along its axis so you do not bend it.
9. Refer to rocker arms/shafts installation procedures elsewhere in this section, then reassemble the rest of the cylinder head.
10. To complete the installation, use new gaskets and install the cylinder head on the vehicle.

INSPECTION

Inspect the valve faces and seats (in the head) for pits, burned spots and other evidence of poor seating. If a valve face is in such bad shape that the head of the valve must be ground, in order to true the face, discard the valve, because the sharp edge will run too hot. The correct angle for valve faces is 45 degrees. Refacing is best performed by a reputable machine shop.

Check the valve stem for scoring and burned spots. If not noticeably scored or damaged, clean the valve stem with solvent to remove all gum and varnish. Clean the valve guides using an expanding wire-type valve guide cleaner. If you have access to a dial indicator for measuring valve stem-to-guide clearance, mount it so the stem of the indicator is at 90 degrees to the valve stem and as closer to the valve guide as possible. Move the valve off its seat, then measure the valve guide-to-stem clearance by rocking the stem back and for the actuate the dial indicator. Measure the valve stem diameter using a micrometer and compare it to specifications to determine whether the stem or guide wear is responsible for the excess clearance. If a dial indicator and micrometer are not available, take the cylinder head and valves to a reputable machine shop for inspection.

REFACING

➡ **All valve grinding operations should be performed by a qualified machine shop; only the valve lapping operation may be performed by the inexperienced mechanic.**

Valve Lapping

When valve faces and seats have been refaced and/or recut, or if they are determined to be in good condition, the valves must be lapped in to ensure efficient sealing when the valve closes against the seat.

1. Invert the cylinder head so the combustion chambers are facing upward.
2. Lightly lubricate the valve stems with clean engine oil and coat the valve seats with valve grinding compound. Install the valves in the cylinder head as numbered.
3. Using a valve lapping tool, attach the suction cup to a valve head. You will probably have to moisten the suction cup to securely attach the tool to the valve.
4. Rotate the tool between the palms, changing position and lifting the tool often to prevent grooving. Lap the valve until a smooth polished seat is evident (you may have to add a bit more compound after some lapping is done).
5. Remove the valve and tool, then, remove all traces of the grinding compound with a solvent-soaked rag or rinse the head with solvent.

➡ **Valve lapping can also be done by fastening a suction to a piece of drill rod in a egg-beater type hand drill. Proceed as above, using the drill as a lapping tool. Due to the higher speeds involved when using the hand drill, care must be exercised to avoid grooving the seat. Lift the tool and change direction of rotation often.**

Valve Springs

If the cylinder head is removed from the engine, refer to the valve removal and installation procedures in this section and remove the valve springs.

REMOVAL & INSTALLATION

▶ See Figures 118 thru 123

1. Refer to the rocker arm removal and installation procedures in this section and remove the rocker arm assembly.
2. Using an old set of rocker arm shafts, install them in place of the rocker arm assembly; this will prevent bending or scratching the original set of shafts.
3. Remove the spark plug(s) from the cylinder head.
4. Rotate the crankshaft to position the cylinder (being worked on) to the TDC of its compression stroke.
5. Using a Spark Plug Air Hold tool, install it into the spark plug hole and inject compressed air into the cylinder to hold the valves in place.
6. Using a valve spring compressor tool, compress the spring(s) and remove the valve keepers. Relax the spring compressor tool, then remove the valve retainer, the spring(s) and the valve seal; keep the parts in the exact order for reinstallation purposes.

➡ **If reusing the valve seals, it is recommended to replace the neck springs of the seal; the black spring is used on the exhaust valve seal and the white spring is used on the intake valve seal. The seals are not interchangeable between the intake and the exhaust valves.**

7. Inspect the springs for fatigue, cracks and/or uniformity; if necessary, replace them.
8. To assemble, first lubricate the valves with clean engine oil and slide the valves up and down in the valve guides to make sure they move smoothly.

ENGINE AND ENGINE REBUILDING   3-37

Fig. 118 Use a valve spring compressor to relieve the pressure from the spring

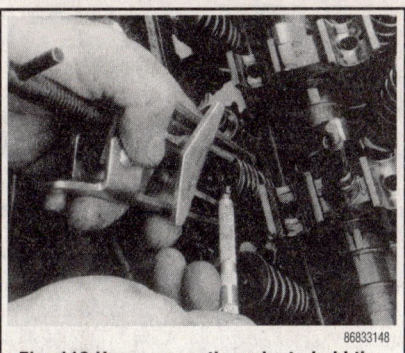

Fig. 119 Use a magnetic probe to hold the valve keepers and prevent them from falling as they are removed

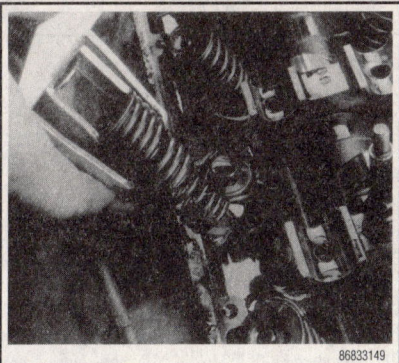

Fig. 120 Slide the springs off the valve guide leaving the stem, seal and seat

Fig. 121 Remove the valve spring seal and seat with pliers, then move on to the next spring assembly

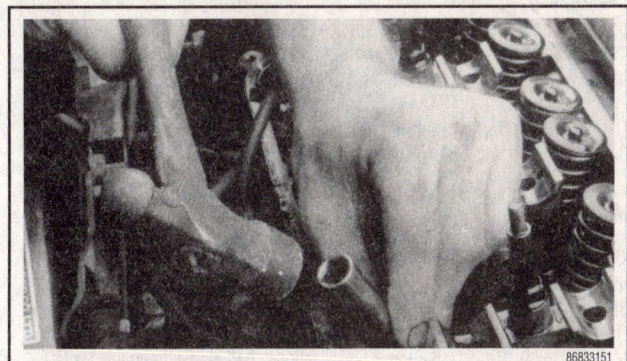

Fig. 122 Install the valve guide seals using a deep socket or suitable tool

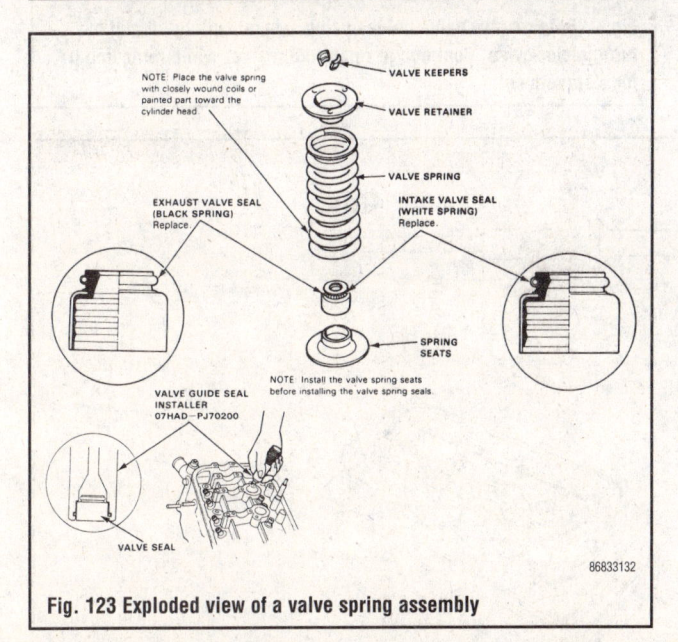

Fig. 123 Exploded view of a valve spring assembly

9. Install the valve seals using a valve guide seal installer or suitable tool.
10. Then install the seals and lubricate the parts with clean engine oil. Use the valve spring compressor tool to compress the springs and carefully install the valve keepers, retainers and springs.

※ CAUTION

When removing the valve spring compressor tool, remove it slowly and make sure the valve keepers are fully seated; otherwise, the springs may fly off.

11. After removing the valve spring compressor, tap the valve stems with a plastic mallet 2–3 times to make sure the valve keepers and valves are fully seated. Only tap each valve stem along its axis so you do not bend it.
12. Refer to the rocker arms/shafts and camshaft installation procedures in this section, then reassemble the rest of the cylinder head.

➡ **When installing the springs, be sure to position the closely wound coil ends or painted part toward the cylinder head.**

13. To complete the installation, use new gaskets and install the cylinder head on the vehicle.
14. Refill the cooling system, connect the negative battery terminal, start the engine and check for normal running and leaks.

INSPECTION

1. Position the valve spring on a flat, clean surface next to a square.
2. Rotate the spring against the square to measure the distortion (out-of-roundness). If the spring height varies (between similar springs) by more than 1/16 in. (1.6mm), replace the spring.

Valve Seats

The valve seats can be machined during a valve job to provide optimum sealing between the valve and the seat.

The seating services should be performed by a professional machine shop which has the specialized knowledge and tools necessary to perform the service.

Valve Guides

If replacement valve guides are not available or you determine the replacement procedure is too involved, the guides can be reconditioned using a procedure known as knurling (machining the inner surface).

3-38 ENGINE AND ENGINE REBUILDING

REMOVAL & INSTALLATION

▸ See Figures 124 and 125

1. Refer to the valve removal/installation procedures in this section and remove the valves.
2. Place the cylinder head in an oven and heat it to 300°F using an accurately positioned cooking thermometer to determine temperature; this procedure will loosen the valve guides enough to drive them out.
3. Place the new valve guides in the freezer section of a refrigerator for about an hour; this will caus them to contract making them easier to install.
4. Using a hammer and the Valve Guide Driver tool No. 07942–SA50000, 07942–8230000 or equivalent (for exhaust side) or 07942–6570100, 07942–6110000 or equivalent (for intake side), drive the valve guide(s) toward the camshaft side of the cylinder head.
5. Using a hammer and the Valve Guide Driver tool No. 07942–SA50000, 07942–8230000 or equivalent (for exhaust side) or 07942–6570100, 07942–6110000 or equivalent (for intake side), remove the new guide from the freezer and drive it into the cylinder head from the camshaft side until the guide projects above the cylinder head surface. Valve guide installed height for 1984–87 engines is 0.8 in (20.0mm) (intake), and 0.76 in (19.0mm) (exhaust); for 1988–91 engines it is 0.648 in. (16.2mm); for 1992–95 D16Z6, D15Z1 engines it is 0.714–0.734 in (17.85–18.35mm) (intake), and 0.746–0.766 in. (18.65–19.15mm) (exhaust); for 1992–95 D15B7, D15B8 engines it is 0.638–0.658 in. (15.95–16.45) (intake and exhaust).

➡ If installing a whole set of guides (8 or 16 depending on engine), it may be necessary to re-heat the cylinder head more than once.

6. Using the Valve Guide Reamer tool No. 07984–SA50000, 07984–689010A or equivalent (exhaust) or 07984–6110000, 07984–657010A or equivalent, coat the reamer with cutting oil and ream the valve guides to the proper valve stem fit. Use the reamer with an in-out motion while rotating it. For the finished dimension of the valve guide, check the Valve Specifications chart elsewhere in this section.

➡ When replacing the valve springs, place the closely wound end toward the cylinder head.

7. To assemble the cylinder head, first lubricate the valves with clean engine oil and slide the valves up and down in the valve guides to make sure they move smoothly.

8. Then install the valve guide seals and lubricate the parts with clean engine oil. Use a valve spring compressor to compress the valve springs and carefully install the valve keepers, retainers and springs. (The exhaust valve seal uses a black spring, while the intake valve seal uses a white spring.)

✽✽ CAUTION

When removing the valve spring compressor tool, do it slowly and make sure the valve keepers are fully seated; otherwise, the springs may fly off suddenly.

9. After removing the valve spring compressor, tap the valve stems 2–3 times to make sure the valve keepers and valves are fully seated. Only tap each valve stem along its axis so you do not bend it.
10. Refer to "Rocker Arms/Shafts, Installation" and "Camshaft, Installation" in this section for procedures to assemble the rest of the cylinder head.
11. To complete the installation, use new gaskets and install the cylinder head on the vehicle.

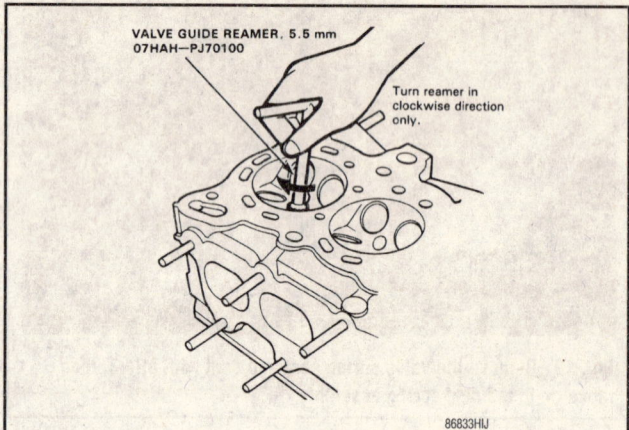

Fig. 125 Ream new valve guides using special cutting oil and by turning clockwise. Continue to rotate the reamer while removing it for a smooth cut

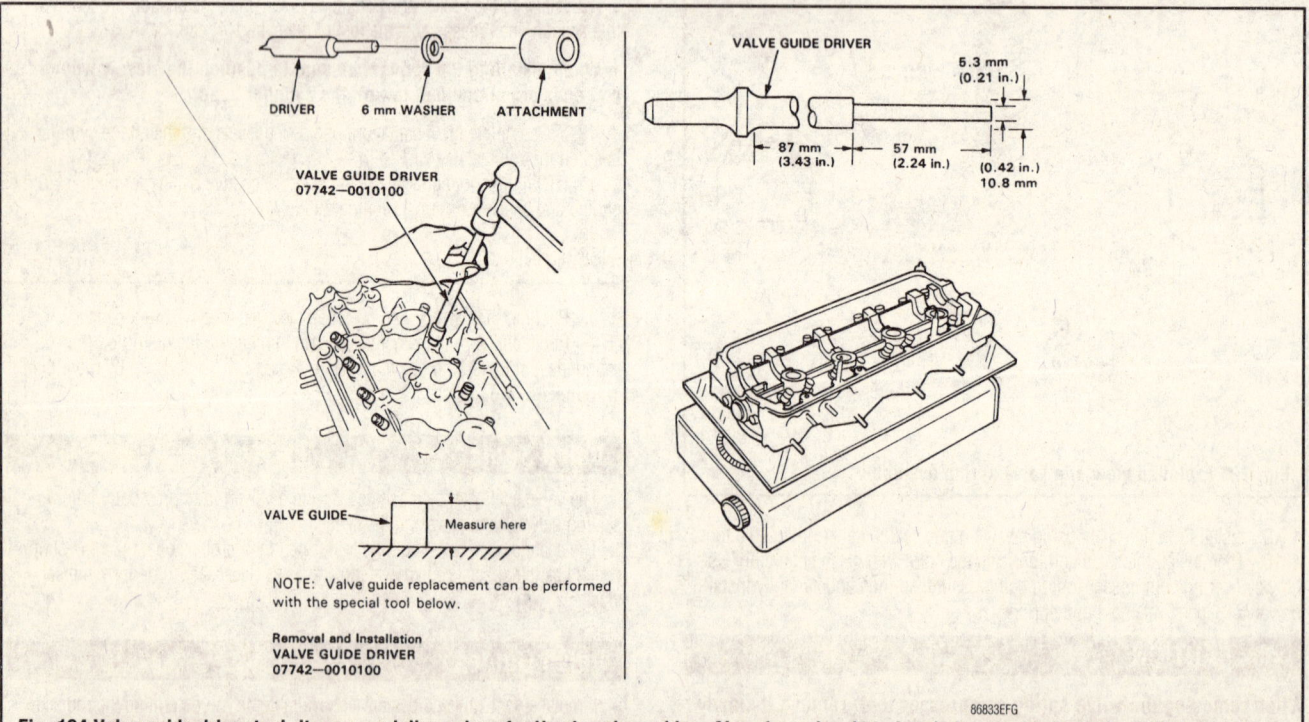

Fig. 124 Valve guide driver tool, its use, and dimensions for Honda valve guides. Also shown is a hot plate being used to heat the cylinder head to enable removal/replacement of guides

ENGINE AND ENGINE REBUILDING   3-39

12. Refill the cooling system. Start the engine and check for normal operation and leaks.

KNURLING

Knurling is a process in which the metal on the valve guide bore is displaced and raised, thereby reducing the clearance. It also provides excellent oil control. The option of knurling rather than reaming valve guides should be discussed with a reputable machinist or engine specialist.

Oil Pan

REMOVAL & INSTALLATION

Except 4WD Civic Wagon

▶ See Figures 126 thru 132

1. Disconnect the negative battery cable.
2. Raise and safely support the vehicle.
3. Remove the engine splash shield, if equipped.
4. Drain the engine oil.

➡ Be sure to properly dispose of used engine oil, it is an environmental hazard.

5. Remove the exhaust header pipe, if necessary.
6. Remove the oil pan bolts and nuts and the oil pan.

※※ CAUTION

Wear safety goggles when doing this and related procedures. Oil in eyes is hazardous, and being incompatible with water, it does not wash out easily.

7. Installation is the reverse of the removal procedure. Make sure all gasket mating surfaces are clean prior to installation.

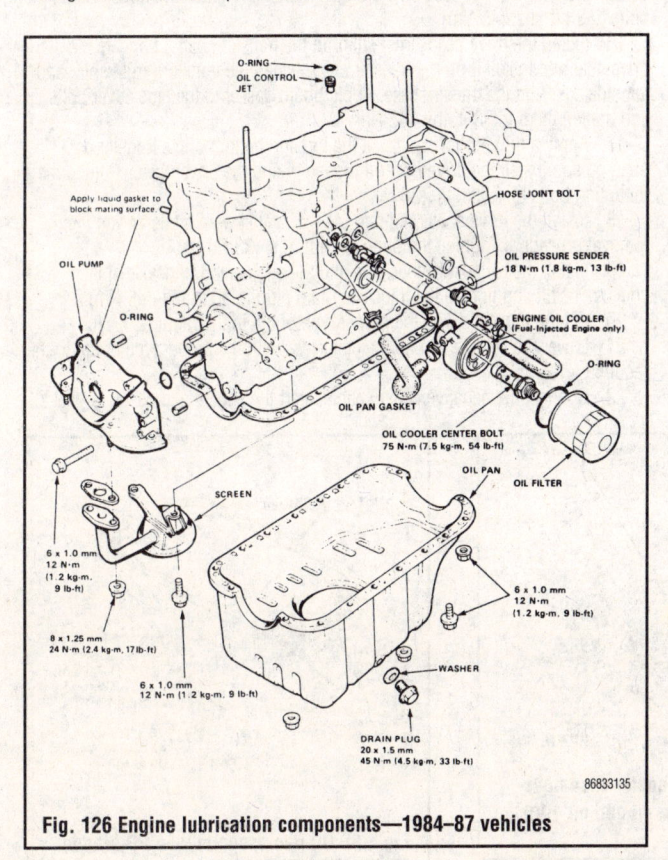

8. Apply a coat of sealant to both sides of the oil pan gasket. Tighten the bolts and nuts in 2 steps in a crisscross pattern beginning at the center of the pan. The final torque should be 9 ft. lbs. (12 Nm). Tighten the oil pan drain plug to 33 ft. lbs. (45 Nm).

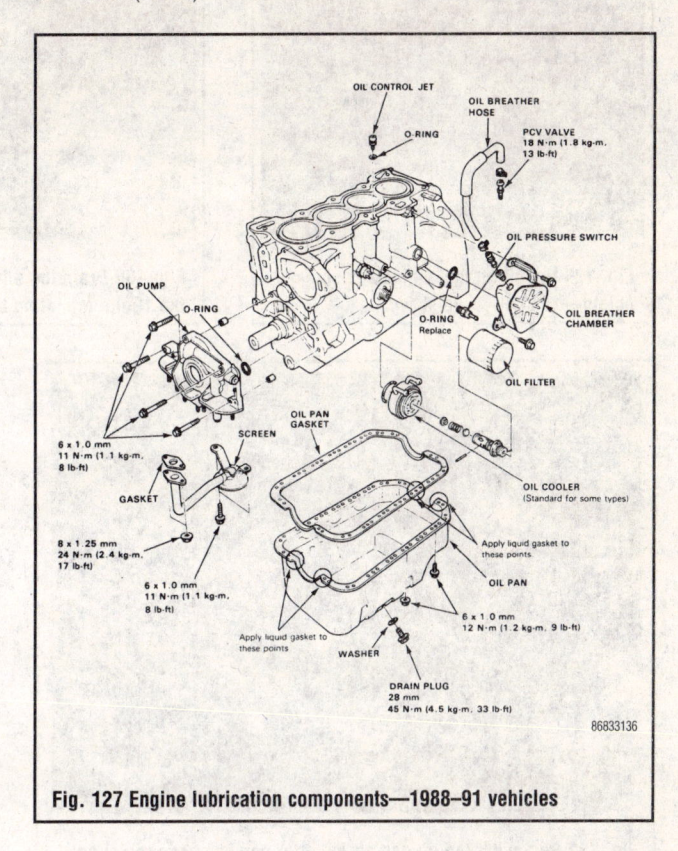

Fig. 127 Engine lubrication components—1988–91 vehicles

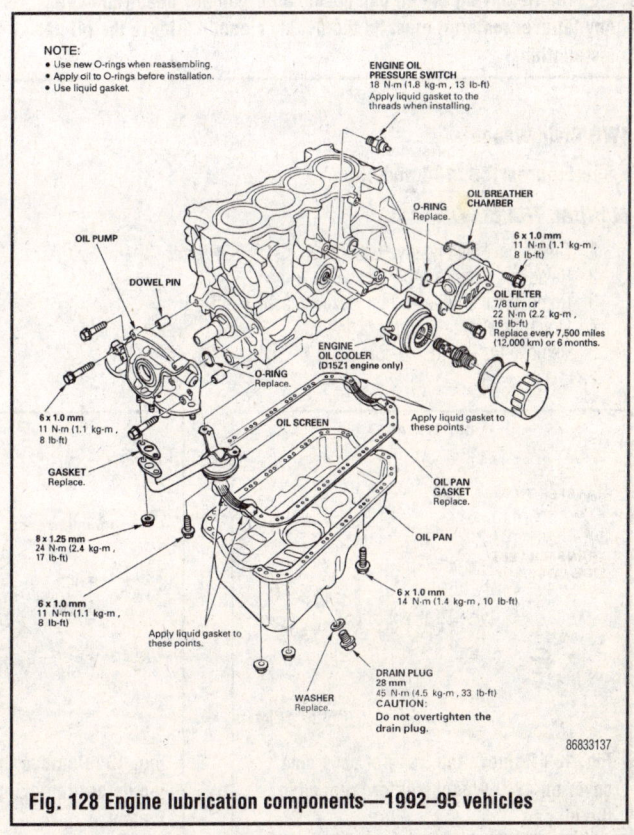

Fig. 126 Engine lubrication components—1984–87 vehicles

Fig. 128 Engine lubrication components—1992–95 vehicles

3-40 ENGINE AND ENGINE REBUILDING

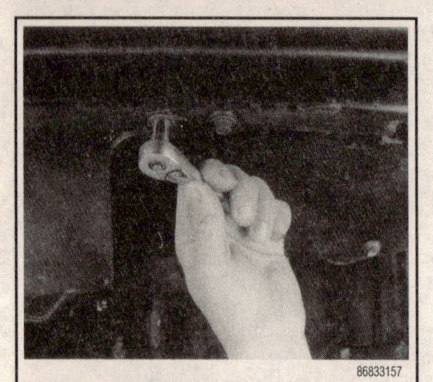

Fig. 129 Unbolting the engine splash shield—1984–87 style shown

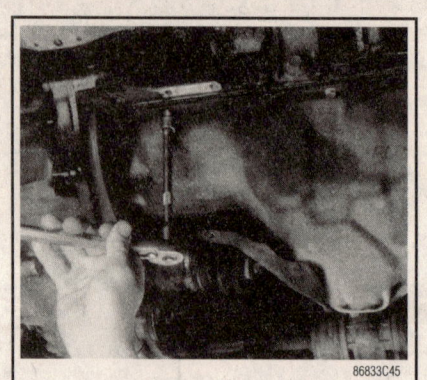

Fig. 130 Drain the engine oil, then remove the 10mm mounting bolts . . .

Fig. 131 . . . and carefully lower the pan

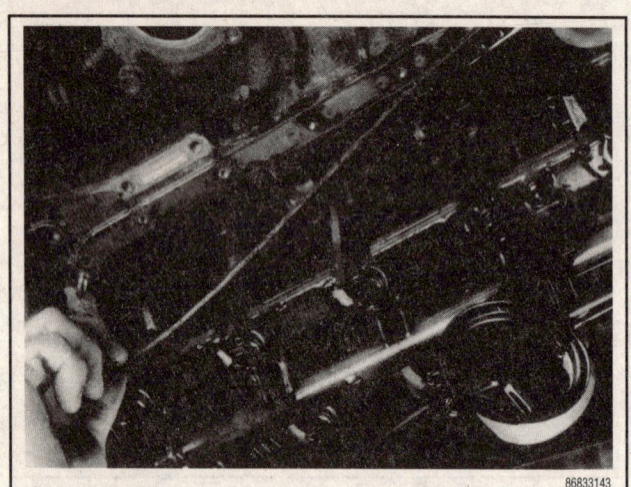

Fig. 132 Removing the oil pan gasket after pan has been removed. Any leftover material must be thoroughly cleaned prior to the oil pan installation

4WD Civic Wagon

▶ See Figures 133, 134 and 135

MANUAL TRANSAXLE

1. Disconnect the negative battery cable.
2. Raise and safely support the vehicle.
3. Remove the engine and transaxle splash shield.
4. Drain the engine and transaxle oil.
5. Remove the exhaust header pipe.
6. Mark the position of the drive shaft flange in relation to the companion flange and remove the drive shaft.
7. Remove the left side cover from the transfer case.

➥Be careful not to damage the thrust shim and mating surface.

8. Remove the driven gear from the transfer case.

➥Be careful not to damage the thrust shim and mating surface.

9. Remove the transfer case from the clutch housing.
10. Remove the clutch case cover.
11. Remove the oil pan by removing the bolts and nuts.

To install:

12. Clean all gasket mating surfaces.
13. Apply sealant to both sides of a new oil pan gasket and install the gasket and the oil pan. Tighten the bolts and nuts in two steps in a crisscross pattern starting at the center of the pan. The final torque should be 9 ft. lbs. (12 Nm).
14. Install and tighten the oil drain plug to 33 ft. lbs. (45 Nm).
15. Apply liquid gasket to the clutch housing mating surface of the transfer case. Install the transfer case on the clutch housing. Tighten the transfer case bolts to 33 ft. lbs. (45 Nm).
16. Install the drive gear thrust shim on the transfer shaft. Lubricate the drive gear and install it on the transfer shaft. Install the transfer thrust shim and left side cover on the transfer case. Apply liquid gasket to the side cover bolts and tighten them to 33 ft. lbs. (45 Nm).
17. Apply a thin film of sealant at the top and bottom of the transfer case opening and install the driven gear thrust shim and the driven gear. Tighten the mounting bolts to 19 ft. lbs. (26 Nm).
18. Install the drive shaft, aligning the marks that were made during the removal procedure. Tighten the bolts to 24 ft. lbs. (33 Nm).
19. Install the exhaust header pipe and the engine and transaxle splash shields. Install and tighten the transaxle drain plug to 30 ft. lbs. (40 Nm).
20. Fill the transaxle with the proper type of oil, to the required level.
21. Lower the vehicle and fill the crankcase with the proper type of oil, to the required level.
22. Connect the negative battery cable, start the engine and check for leaks.

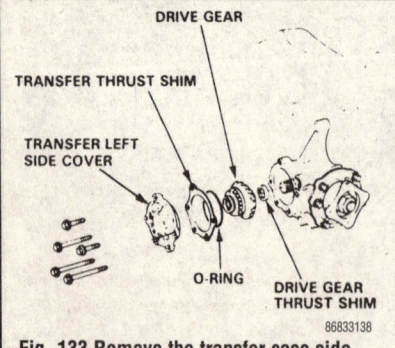

Fig. 133 Remove the transfer case side cover on 4WD Wagons before removing the oil pan

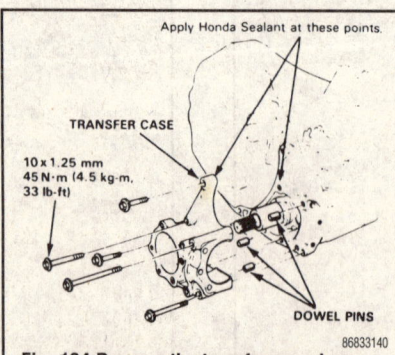

Fig. 134 Remove the transfer case housing before removing the oil pan on 4WD Wagon

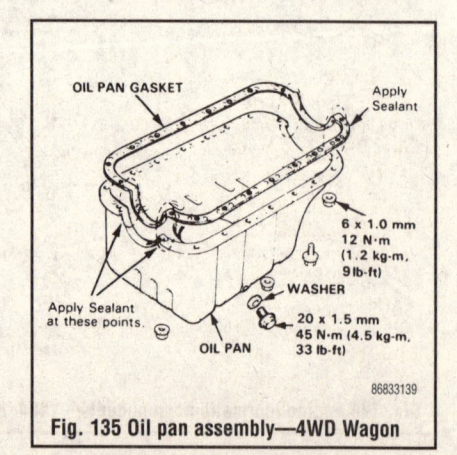

Fig. 135 Oil pan assembly—4WD Wagon

ENGINE AND ENGINE REBUILDING 3-41

AUTOMATIC TRANSAXLE

1. Disconnect the negative battery cable.
2. Raise and safely support the vehicle.
3. Remove the engine and transaxle splash shield.
4. Drain the engine and transaxle oil.
5. Remove the exhaust header pipe.
6. Mark the position of the drive shaft flange in relation to the companion flange and remove the drive shaft.
7. Remove the driven gear assembly from the transfer case.
8. Remove the left side cover and then the drive gear from the transfer case. Rotate the cover using the bolt closest to the front of the vehicle as the axis. This bolt is not removed from the cover.

➡Be careful not to damage the thrust shim and mating surface.

9. Remove the transfer case from the clutch housing.
10. Remove the clutch case cover.
11. Remove the oil pan by removing the bolts and nuts.

To install:
12. Clean all gasket mating surfaces.
13. Apply sealant to both sides of a new oil pan gasket and install the gasket and the oil pan. Tighten the bolts and nuts in 2 steps in a crisscross pattern starting at the center of the pan. The final torque should be 9 ft. lbs. (12 Nm).
14. Install and tighten the oil drain plug to 33 ft. lbs. (45 Nm).
15. Apply liquid gasket to the clutch housing mating surface of the transfer case. Attach a new O-ring to the groove in the transfer left side cover.
16. Install the transfer case on the clutch housing. Install the bolt that remained in the left side cover, in the transfer case before installing the case on the clutch housing. Tighten the transfer case bolts to 33 ft. lbs. (45 Nm).
17. Install the drive gear thrust shim on the transfer shaft. Lubricate the drive gear and install it on the transfer shaft. Install the transfer thrust shim and left side cover on the transfer case. Apply liquid gasket to the side cover bolts and tighten them to 33 ft. lbs. (45 Nm).
18. Apply a thin film of sealant at the top and bottom of the transfer case opening and install the driven gear thrust shim and the driven gear. Tighten the mounting bolts to 19 ft. lbs. (26 Nm).
19. Install the drive shaft, aligning the marks that were made during the removal procedure. Tighten the bolts to 24 ft. lbs. (33 Nm).
20. Install the exhaust header pipe and the engine and transaxle splash shields. Install and tighten the transaxle drain plug to 29 ft. lbs. (40 Nm).
21. Lower the vehicle and fill the crankcase with the proper type of oil, to the required level. Fill the transaxle with the proper type and quantity of oil.
22. Connect the negative battery cable, start the engine and check for leaks.

Oil Pump

REMOVAL & INSTALLATION

▶ See Figures 136, 137 and 138

1. Disconnect the negative battery cable.
2. Raise and safely support the vehicle.
3. Drain the engine oil.
4. Bring the No. 1 cylinder to TDC. The mark on the crankshaft pulley should align with the index mark on the timing cover.
5. Remove the necessary accessory drive belts and the crankshaft pulley.
6. Remove the valve cover and the timing belt covers.
7. Remove the following components:
   a. Timing belt tensioner.
   b. Timing belt.
   c. Timing belt drive pulley.
8. Remove the oil pan and oil screen.
9. Remove the oil pump mount bolts and the oil pump assembly.

To install:
10. Apply a light coat of oil to the seal lip.
11. Install the two dowel pins and new O-ring on the cylinder block.

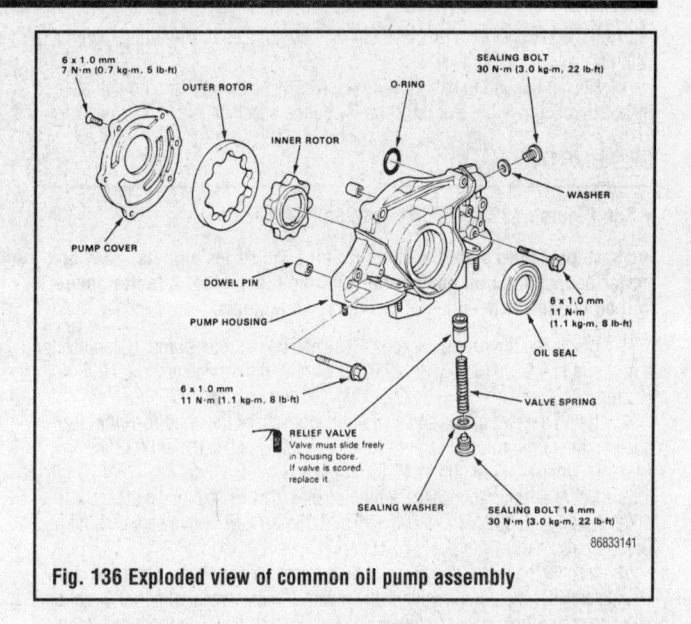

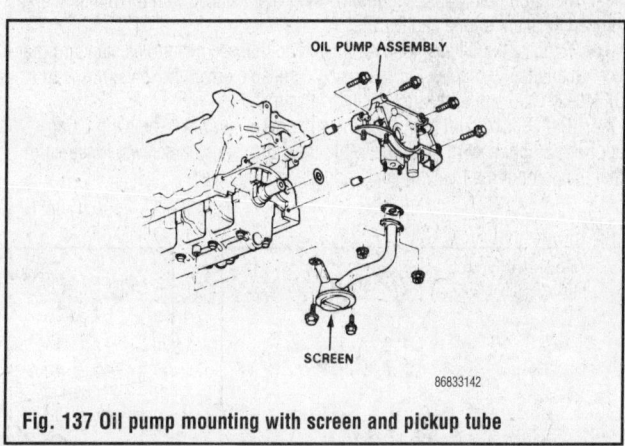

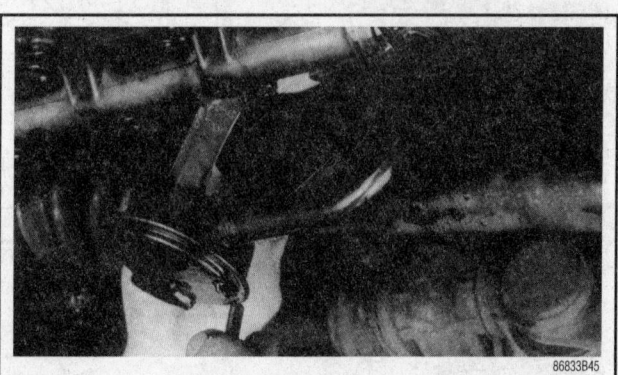

Fig. 136 Exploded view of common oil pump assembly

Fig. 137 Oil pump mounting with screen and pickup tube

Fig. 138 Removing the oil pump pickup tube and screen—1984–87 vehicles

12. Make sure the mating surfaces are clean and dry and apply liquid gasket to the cylinder block mating surfaces of the oil pump. Do NOT apply liquid gasket to the O-ring grooves.

➡If more than five minutes elapse after applying liquid gasket without installing the pump, peel off the liquid gasket and reapply a new coating, then bolt together within the five minute time frame.

13. Install and bolt on the pump within five minutes of applying liquid gasket. Tighten the bolts to 9 ft. lbs. (12 Nm). Install the oil screen, oil pan and

3-42 ENGINE AND ENGINE REBUILDING

remaining components. After assembly, wait 30 minutes before filling the engine with oil.

14. Install the timing belt. Be sure to properly tension the timing belt after installation. Tighten the crankshaft pulley bolt to specification.

OIL PUMP OVERHAUL

♦ See Figures 139, 140, 141, 142 and 143

➡ These procedures are the same for all Civic series models. They are performed with the oil pump removed from the engine. A feeler gauge will be needed to take the following measurements.

1. Check the inner-to-outer rotor clearance on the rotor pump. Clearance for a new unit is 0.0008–0.0056 in. (0.02–0.14mm), the service limit is 0.008 in. (0.2mm).
2. Using a feeler gauge and a straight edge, check the body-to-rotor clearance on the pump rotor. Clearance for a new unit is is 0.0012–0.0032 in. (0.03–0.08mm), with 0.006 in. (0.15mm) as the service limit.
3. Using a feeler gauge, check the housing to outer rotor clearance. Clearance should be 0.0004–0.0072 in. (0.10–0.18mm) new, with a service limit of 0.008 in. (0.2mm).
4. Inspect the pump housing and both rotors for scoring or other damage. Decide whether the pump requires any repair or replacement of parts. (Inner and outer rotors on both upper and lower halves are installed with the punch marks aligned adjacent to one another.)
5. Replace the oil seal (special tools required, see picture) as standard practice. Gently tap in the new oil seal using a suitable driver or Honda tool part No. 07749-001000 and a 52×55mm attachment.
6. Using petroleum jelly, pack the pump cavity and reassemble it using applying a light coat of liquid gasket to the pump housing screws. Make sure that the pump turns freely in your hand when assembled.

Timing Belt Cover and Seal

REMOVAL & INSTALLATION

1. Disconnect the negative battery cable.
2. Raise and safely support the vehicle.
3. Remove the left front wheel and tire assembly.
4. Remove the left front wheel well splash shield.
5. If equipped, remove the power steering belt and pump.
6. If equipped with air conditioning, remove the adjust pulley with bracket and the belt.
7. Remove the power steering bracket, loosen the alternator adjust-bolt and through-bolt and remove the alternator belt.
8. Use a suitable device to support the engine. Remove the engine side support bolts and nut and then remove the side mount rubber.
9. Remove the valve cover, the crankshaft pulley bolt and the crankshaft pulley.
10. Remove the timing belt upper and lower cover.
11. Installation is the reverse of the removal procedure. Apply engine oil to the crankshaft pulley. On 1984–88 models, tighten the pulley bolt to 83 ft. lbs. (115 Nm). On 1988–91 models, tighten the pulley bolt to 119 ft. lbs. (165 Nm). On 1992–95 models, tighten the pulley bolt to 134 ft. lbs. (134 Nm).

OIL SEAL REPLACEMENT

1. Disconnect the negative battery cable.
2. Remove the timing belt cover and the timing belt.
3. Remove the crankshaft timing sprocket.
4. Using a suitable seal removal tool, remove the seal from the front of the engine.

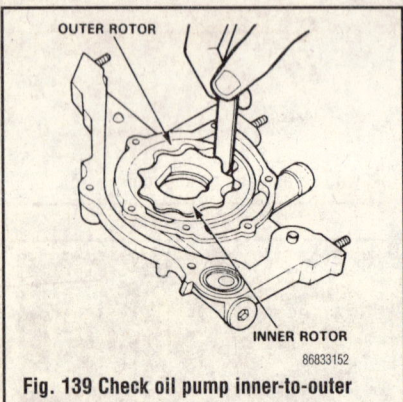

Fig. 139 Check oil pump inner-to-outer rotor clearance

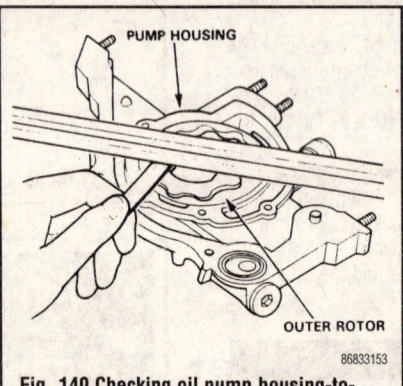

Fig. 140 Checking oil pump housing-to-rotor radial clearance

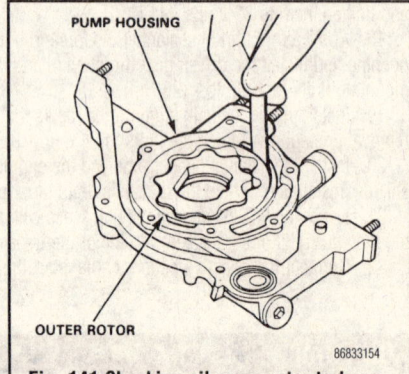

Fig. 141 Checking oil pump rotor-to-housing axial clearance

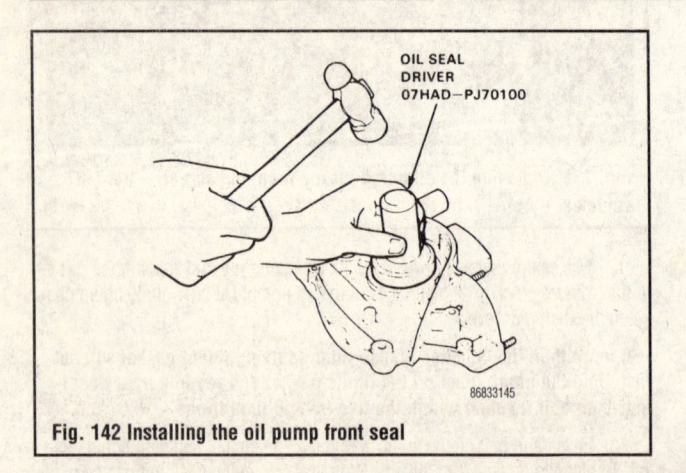

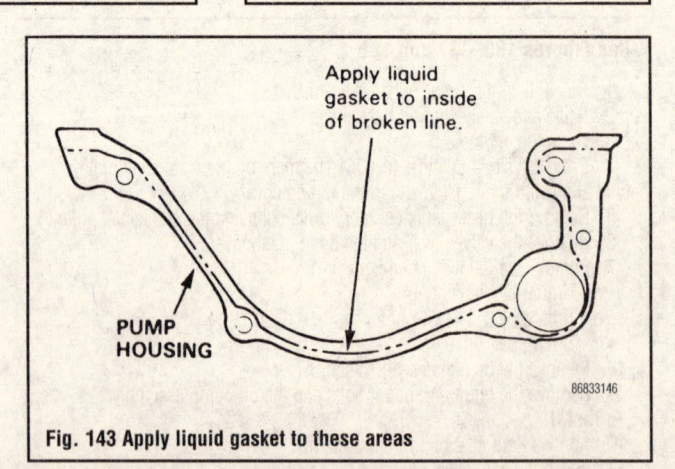

Fig. 142 Installing the oil pump front seal

Fig. 143 Apply liquid gasket to these areas

ENGINE AND ENGINE REBUILDING

5. Installation is the reverse of the removal procedure. Place a thin coat of oil on the seal lip prior to installation. Use a suitable seal driver to install the seal. Be sure to install the seal with the open (spring) side facing the inside of the engine.

Timing Belt and Tensioner

REMOVAL & INSTALLATION

1984–87 Engines

▶ See Figures 144, 145 and 146

1. Turn the crankshaft pulley until it is at Top Dead Center.
2. Remove the pulley belt, water pump pulley, crankshaft pulley, and timing gear cover. Mark the direction of timing belt rotation.
3. Loosen, but do not remove, the tensioner adjusting bolt and pivot bolt.
4. Slide the timing belt off of the camshaft timing gear and the crankshaft pulley gear and remove it from the engine.
5. To remove the camshaft timing gear pulley, first remove the center bolt and then remove the pulley with a pulley remover or a brass hammer. This can be accomplished by simply removing the timing belt upper cover, loosening the tensioner bolts, and sliding the timing belt off of the gear to expose the gear for removal.

➡If you remove the timing gear with the timing belt cover in place, be sure not to let the woodruff key fall inside the timing cover when removing the gear from the camshaft. Inspect the timing belt. Replace if over 90,000 miles old, if oil soaked (find source of oil leak also), or if worn on leading edges of belt teeth.

To install:

6. Be sure to install the crankshaft pulley and the camshaft timing gear pulley in the top dead center position. Align the marks on the camshaft timing gear so they are parallel with the top of the cylinder head and the woodruff key is facing up.

➡When installing the timing belt, do not allow oil to come in contact with the belt. Oil will cause the rubber to swell. Be careful not to twist the belt unnecessarily or bend in a 90° angle, since it is made with fiberglass; nor should you use tools having sharp edges when installing or removing the belt. Be sure to install the belt with the arrow facing in the same direction it was facing during removal.

7. After installing the belt, adjust the belt tension by first rotating the crankshaft counterclockwise 1/4 turn. Then, re-tighten the adjusting bolt and finally the tensioner pivot bolt.
8. Replace the other components removed for belt access.
9. Start the car, check timing and adjust if necessary

✳✳ WARNING

Do not remove the adjusting or pivot bolts, only loosen them. When adjusting, do not use any force other than the adjuster spring. If the belt is too tight, it will result in a shortened belt life.

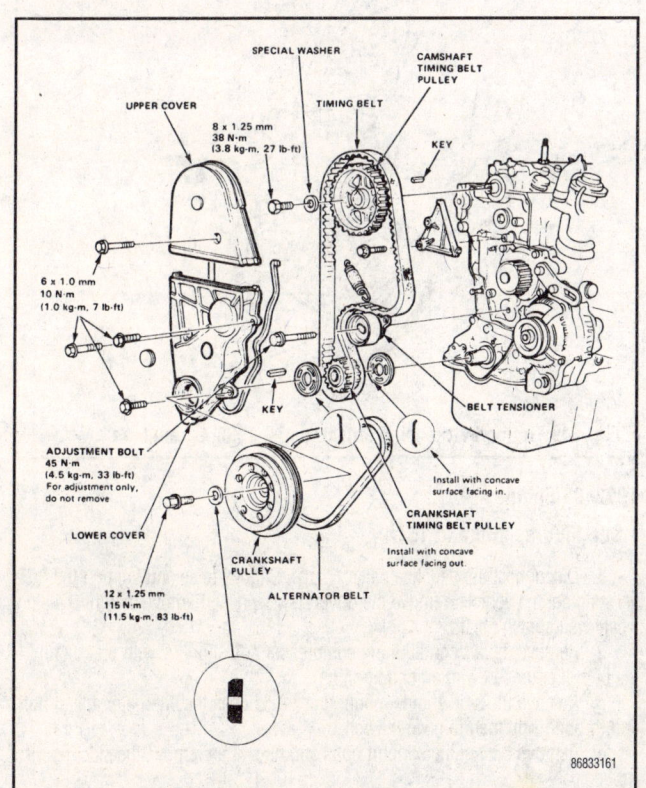

Fig. 144 Timing belt and front cover assembly—1984–87 engines

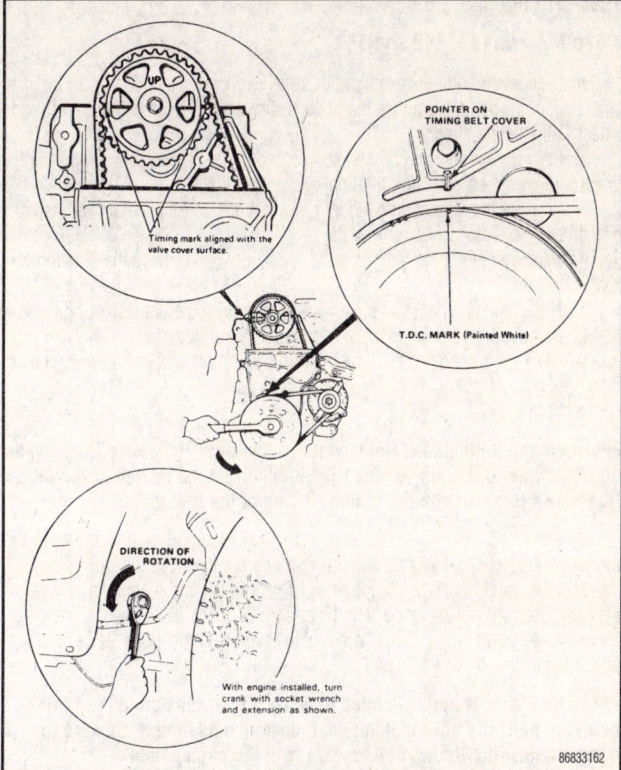

Fig. 145 Positioning the crankshaft and camshaft pulleys before installing the timing belt—1984–87 engines

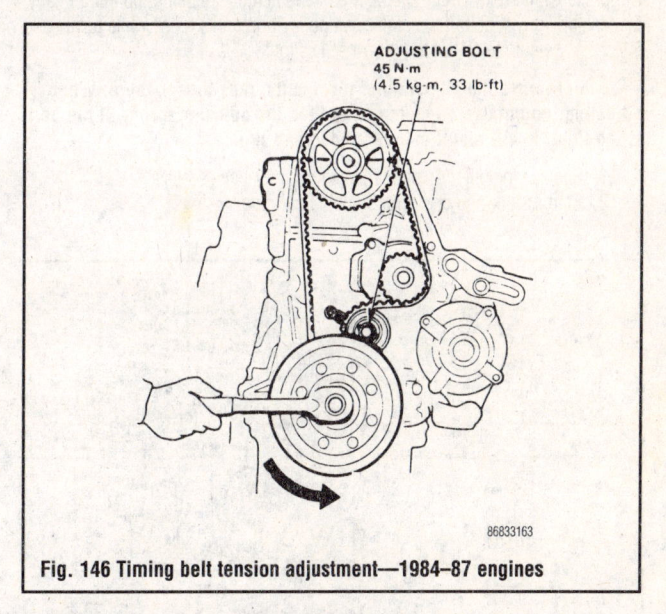

Fig. 146 Timing belt tension adjustment—1984–87 engines

3-44 ENGINE AND ENGINE REBUILDING

1988–91 Engines

◆ See Figures 147, 148 and 149

1. Disconnect the negative battery cable. Slightly raise and support the driver's side of the vehicle, remove the driver's side wheel. Remove the driver's side wheel well splash shield.
2. Remove the air conditioning compressor adjust pulley with bracket and the belt (if equipped with air conditioning).
3. Remove the side engine mount bracket. Loosen the alternator adjusting bolt and through-bolt, then remove the belt.
4. Remove the engine support bolts and nuts, then remove the side mount rubber.
5. Unfasten the crankshaft pulley bolt and remove the crankshaft pulley.
6. Remove the timing belt upper cover and lower cover.
7. Mark the rotation of the timing belt (for easy installation). Loosen the tensioner adjusting bolt, then remove the timing belt.

To install:

➡ Inspect the timing belt. Replace it if it has been in service longer than 90,000 miles, if it is oil soaked (find and correct the source of the oil leak), or if it is worn on the leading edges of the belt teeth.

8. Before installing the belt, double check the position of the engine. The timing pointer must be aligned with the mark on the flywheel; the cam pulley must be set so the small dot or the word UP is vertical and the marks on the edges of the pulley are aligned with the surface of the cylinder head.
9. Installation is the reverse order of the removal procedure. Be sure to adjust the timing belt as follows:

➡ The tensioner is spring-loaded to apply proper tension to the timing belt automatically after making the following adjustment. Be sure to always adjust the timing belt tension with the engine cold.

10. After installing the timing belt, adjust the belt tension by first rotating the crankshaft counterclockwise 1/4 turn or three teeth on the camshaft pulley (this will put tension on the timing belt). Then, retighten the adjusting bolt and finally the tensioner pivot bolt. If the crankshaft pulley bolt broke loose while turning the crank, be sure to re-tighten it to specifications.

➡ Do not remove the adjusting or pivot bolts, only loosen them. When adjusting, do not use any force other than the adjuster spring. If the belt is too tight, it will result in a shortened belt life.

11. Replace the other components removed for belt access.
12. Start the car and check for smooth operation.

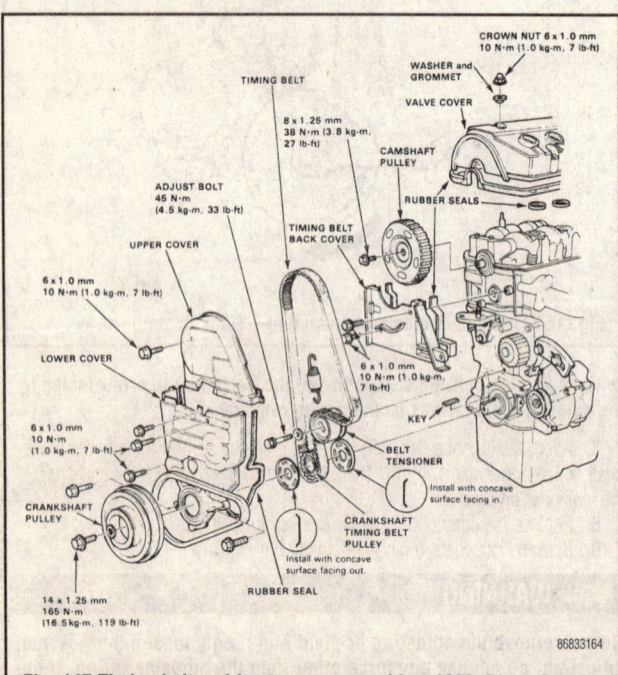

Fig. 147 Timing belt and front cover assembly—1988–91 engines

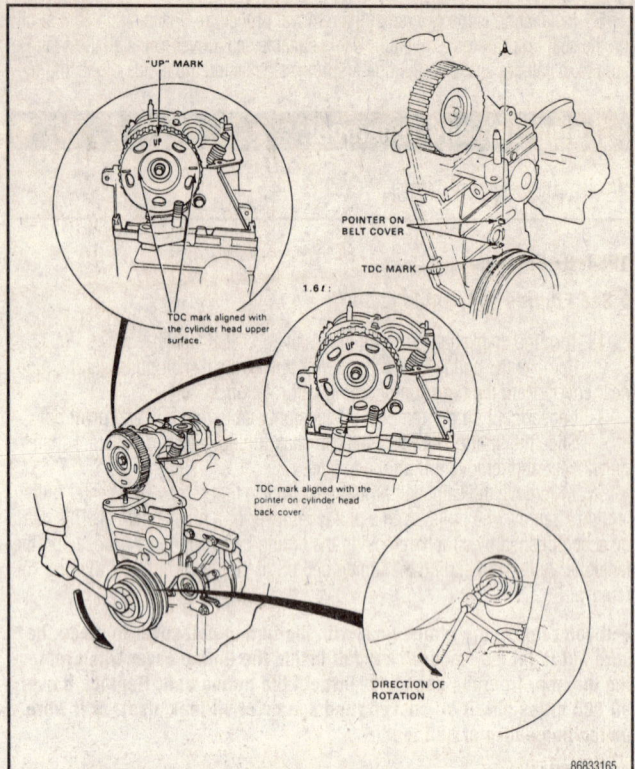

Fig. 148 Positioning the crankshaft and camshaft pulleys before installing the timing belt—1988–91 engines

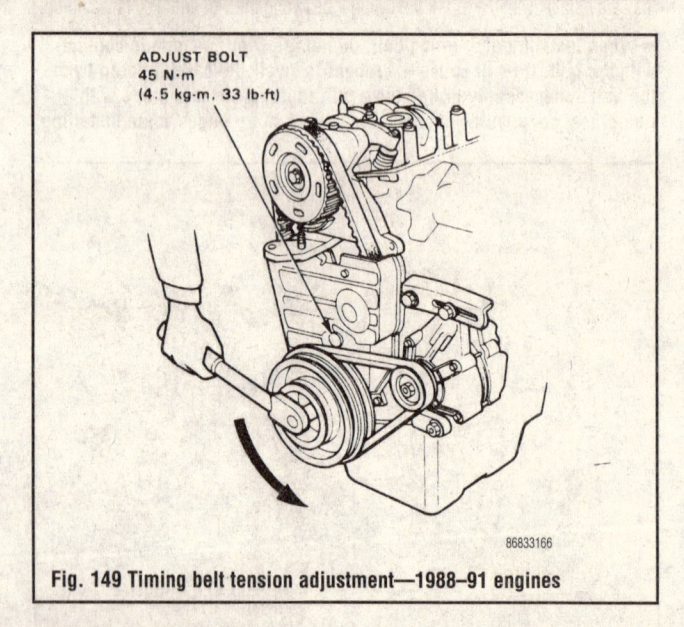

Fig. 149 Timing belt tension adjustment—1988–91 engines

1992–95 Engines

◆ See Figures 150 and 151

1. Disconnect the negative battery cable. Slightly raise and support the driver's side of the vehicle, remove the driver's side wheel. Remove the driver's side wheel well splash shield.
2. Remove the air conditioning compressor adjust pulley with bracket and the belt (if equipped with air conditioning).
3. Remove the side engine mount bracket. Loosen the alternator adjust bolt and through-bolt then remove the belt.
4. Remove the engine support bolts and nuts, then remove the side mount rubber.
5. Unfasten the crankshaft pulley bolt and remove the crankshaft pulley.

ENGINE AND ENGINE REBUILDING   3-45

6. Remove the timing belt upper cover and lower cover.
7. Mark the rotation of the timing belt (for easy installation). Loosen the adjusting bolt, then remove the timing belt.

➡ Inspect the timing belt. Replace it if it has been in service longer than 90,000 miles, if it is oil soaked (find and correct the source of the oil leak), or if it is worn on the leading edges of the belt teeth.

To install:
8. Double check the position of the engine. The timing pointer must be aligned with the mark on the flywheel; the cam pulley must be set so the small dot or the word UP is vertical and the marks on the edges of the pulley are aligned with the surface of the head.
9. Install the belt on the pulleys.
10. Be sure to adjust the timing belt as follows:

➡ The tensioner is spring-loaded to apply proper tension to the timing belt automatically after making the following adjustment. Be sure to always adjust the timing belt tension with the engine cold.

11. After installing the timing belt, adjust the belt tension by first rotating the crankshaft counterclockwise 1/4 turn or three teeth on the camshaft pulley (this will put tension on the timing belt). Then, re-tighten the adjusting bolt and finally the tensioner pivot bolt. After adjustments are complete, be sure to re-tighten the crankshaft pulley bolt to specifications.
12. Start the car and check for smooth operation.

➡ Do not remove the adjusting or pivot bolts, only loosen them. When adjusting, do not use any force other than the adjuster spring. If the belt is too tight, it will result in a shortened belt life.

Camshaft Sprockets

REMOVAL & INSTALLATION

➡ Be sure to use the timing cover and timing belt removal and installation procedures to aid in the removal of the sprockets.

1. Turn the crankshaft pulley until No. 1 piston is at Top Dead Center of the compression stroke. This can be determined by observing the valves (all closed) or by feeling for pressure in the spark plug hole (with your thumb or a compression gauge) as the engine is turned.
2. Remove the pulley belt, water pump pulley (if so equipped), crankshaft pulley, and timing gear cover. Mark the direction of timing belt rotation.
3. Loosen, but do not remove, the tensioner adjusting bolt and pivot bolt.

4. Slide the timing belt off the camshaft sprocket, crankshaft sprocket and the water pump sprocket (if so equipped), then remove it from the engine.
5. To remove either the camshaft or crankshaft timing sprocket, first remove the center bolt and then remove the sprocket with a pulley remover or a brass hammer.

To install:
6. Be sure to position the crankshaft and camshaft timing sprockets in the Top Dead Center position. Tighten the camshaft sprocket bolt to 27 ft. lbs. (38 Nm)

When installing the timing belt, do not allow oil to come in contact with the belt. Oil will cause the rubber to swell. Be careful not to bend or twist the belt unnecessarily, since it is made with fiberglass; nor should you use tools having sharp edges when installing or removing the belt. Be sure to install the belt with the arrow facing in the same direction it was facing during removal.

After installing the timing belt, adjust the belt tension by first rotating the crankshaft counterclockwise 1/4 turn or three teeth on the camshaft pulley. Then, retighten the adjusting bolt and finally the tensioner pivot bolt.

➡ Do not remove the adjusting or pivot bolts, only loosen them. When adjusting, do not use any force other than the adjuster spring. If the belt is too tight, it will result in a shortened belt life.

Camshaft

REMOVAL & INSTALLATION

▶ See Figure 152

➡ To facilitate the installation, make sure that No. 1 piston is at Top Dead Center before removing the camshaft. Be sure to use the timing cover and timing belt removal and installation procedures to aid in the removal of the sprockets.

1. Disconnect the negative battery cable.
2. Bring the piston in the No. 1 cylinder to TDC on the compression stroke.
3. Remove the cylinder head cover, timing belt front covers and the timing belt.
4. Remove the camshaft sprocket.
5. Remove the rocker arm/shaft assembly.
6. Remove the camshaft and camshaft seal.

To install:
7. Lubricate the lobes and journals of the camshaft prior to installation.
8. Install the camshaft with the keyway facing up.

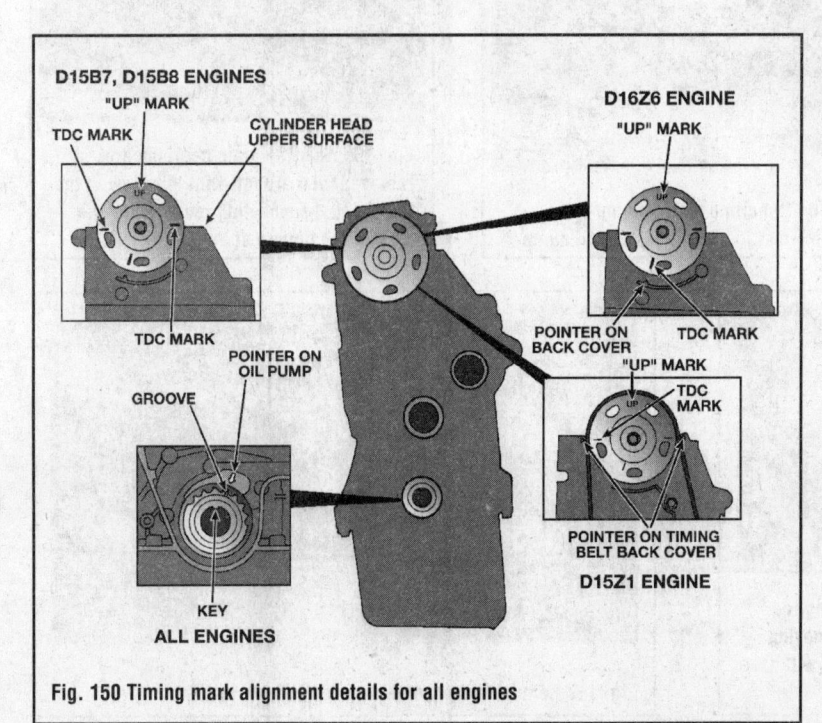

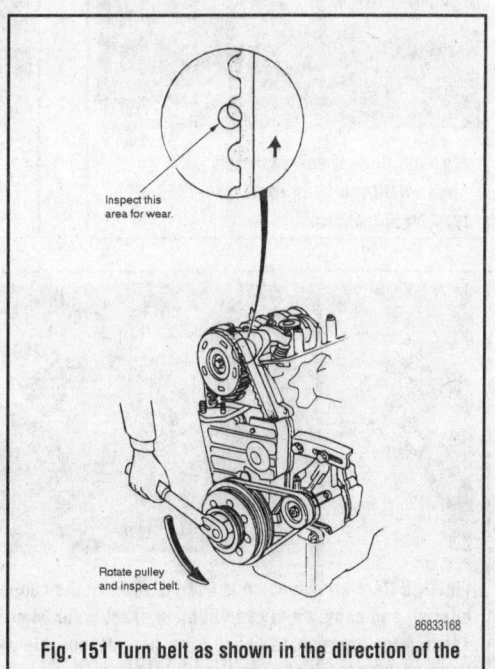

Fig. 150 Timing mark alignment details for all engines

Fig. 151 Turn belt as shown in the direction of the arrow. Be sure to check the belt's teeth for wear

3-46 ENGINE AND ENGINE REBUILDING

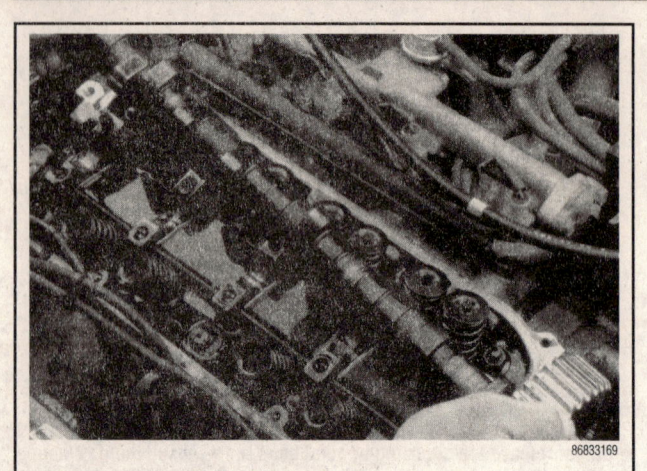

Fig. 152 Removing the camshaft—1987 CRX Si shown

9. Install the rocker arm/shaft assembly as follows:
   a. Loosen the rocker arm locknuts and back off the adjust screws.
   b. Set the rocker arm/shaft assembly in place and loosely install the bolts.
   c. Tighten each bolt two turns at a time in the proper sequence to ensure that the rockers do not bind on the valves.
   d. Tighten the camshaft holder bolts for 1984–91 to 16 ft. lbs. (22 Nm). For 1992–95 engines tighten to the following specifications:

8mm bolts:
D16Z6, D15Z1 engines: 14 ft. lbs. (20 Nm)
D15B7, D15B8 engines: 16 ft. lbs. (22 Nm)

6mm bolts:
All engines: 9 ft. lbs. (12 Nm)

10. Lubricate a new camshaft seal and install using a suitable tool.
11. Properly set the tension of the timing belt after installation.
12. Tighten the crankshaft pulley bolt to specification. Adjust the valve lash.

INSPECTION

Degrease the camshaft using safe solvent, clean all oil grooves. Visually inspect the cam lobes and bearing journals for excessive wear. If a lobe is questionable, check all lobes and journals with a micrometer.

Measure the lobes from nose to base and again at 90°. The lift is determined by subtracting the second measurement from the first. If all exhaust lobes and all intake lobes are not identical, the camshaft must be reground or replaced. Measure the bearing journals and compare to specifications. If a journal is worn there is a good chance that the cam bearings are worn too, requiring replacement.

If the lobes and journals appear intact, place the front and rear cam journals in V-blocks and rest a dial indicator on the center journal. Rotate the camshaft to check for straightness, if deviation exceeds 0.025mm, replace the camshaft.

Pistons and Connecting Rods

REMOVAL & INSTALLATION

▶ See Figures 153 thru 161

✱✱ CAUTION

When draining coolant, keep in mind that cats and dogs are attracted by ethylene glycol antifreeze, and are quite likely to drink any that is left in an uncovered container or in puddles on the ground. This will prove fatal in sufficient quantity. Always drain the coolant into a sealable container. Coolant should be reused unless it is contaminated or several years old.

➡ Refer to the appropriate sections in this manual to assist you in performing the following procedure.

For removal with the engine out of the vehicle, begin with Step 8.
1. Disconnect the negative battery cable. Remove the hood.

Fig. 153 Bottom end assembly as it appears after oil pan removal—1984–87 type engine shown

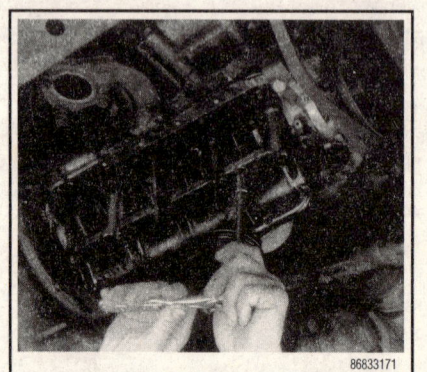

Fig. 154 Unbolting the oil pump screen and pick-up tube that feeds the oil pump

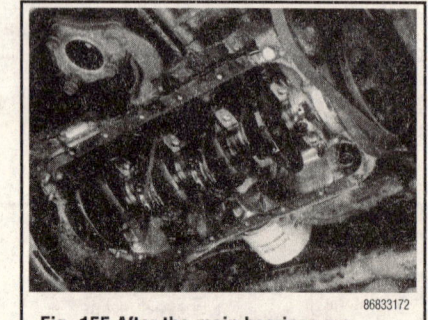

Fig. 155 After the main bearings are removed, you are left with this view of the No. 2 and 3 connecting rod bearing end caps at the bottom of their strokes

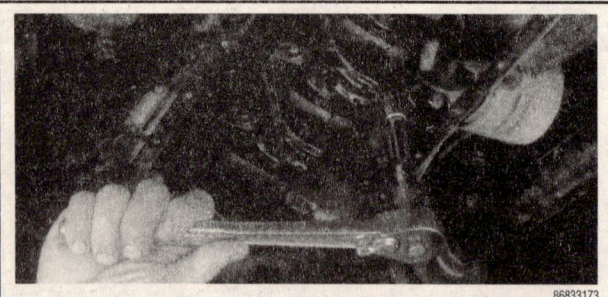

Fig. 156 Here an extension is used to remove the connecting rod bearing end caps. Keep everything in exact order when removing. These parts are mated to each other and all reusable parts will have to go back as originally installed

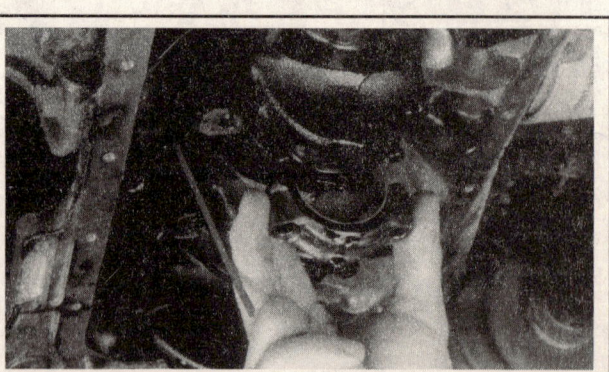

Fig. 157 Removing the connecting rod bearing end cap

ENGINE AND ENGINE REBUILDING 3-47

2. Drain the radiator.
3. Drain the engine oil.

❋❋ CAUTION

The EPA warns that prolonged contact with used engine oil may cause a number of skin disorders, including cancer! You should make every effort to minimize your exposure to used engine oil. Protective gloves should be worn when changing the oil. Wash your hands and any other exposed skin areas as soon as possible after exposure to used engine oil. Soap and water, or waterless hand cleaner should be used.

4. Raise the front of the vehicle and support it with safety stands.
5. Attach a chain to the clutch cable bracket on the transaxle case and raise just enough to take the load off of the center mount.

➡ Do not remove the left engine mount.

6. Remove the center beam and engine lower mount.
7. Remove the cylinder head.
8. Loosen the oil pan bolts and remove the oil pan and flywheel dust shield. Loosen the oil pan bolts in a crisscross pattern beginning with the outside bolt. To remove the oil pan, lightly tap the corners of the oil pan with a mallet. It is not necessary to remove the gasket unless it is damaged.

❋❋ WARNING

Do not pry the oil pan off with the tip of a screwdriver.

9. Remove the oil passage block, the oil pump screen/pick-up tube and the oil pump assembly.

❋❋ WARNING

As soon as the oil passage block bolts are loosened, the oil in the oil line may flow out.

10. Working from the underside of the vehicle, unbolt the main bearing caps (a one piece casting) and remove keeping everything in the exact order as when installed.
11. Remove the connecting rod bearing caps.

Fig. 158 Piston and connecting rod assembly with engine block—1984–87 engines

➡ The connecting rod and main bearings incorporated in the Civic engines are "plain" bearings as opposed to some form of roller bearing (such as "needle" or "ball" bearings). Plain bearings are hydraulic bearings that used a micro-thin film of engine oil to suspend and separate the fast moving metal parts. Precisely machined tolerances, sufficient oil pressure and oil quality are all the more critical for proper

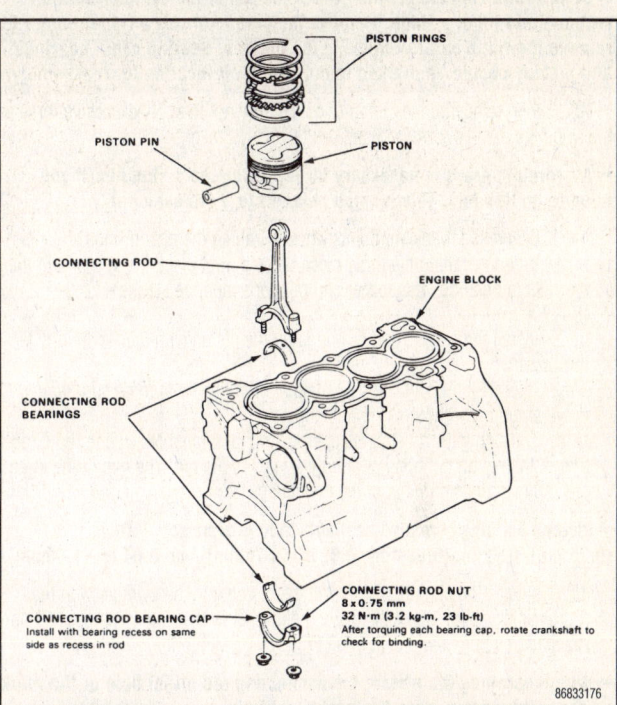

Fig. 159 Piston and connecting rod assembly with engine block—1988–91 engines

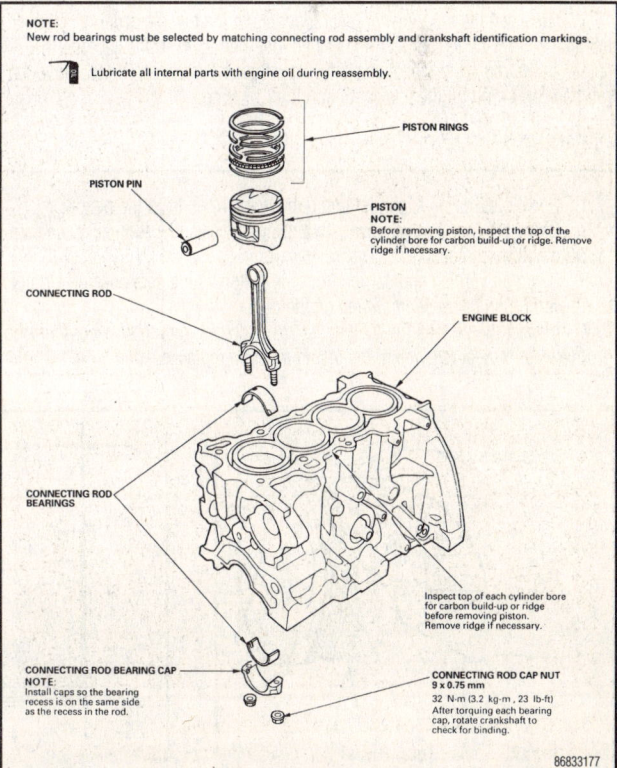

NOTE:
New rod bearings must be selected by matching connecting rod assembly and crankshaft identification markings.

Lubricate all internal parts with engine oil during reassembly.

Fig. 160 Piston and connecting rod assembly with engine block—1992–95 engines

3-48 ENGINE AND ENGINE REBUILDING

operation and longevity. Failure of one of these factors (such as a marginal bearing that is re-used and then fails) could result in the demise of a plain bearing in minutes or seconds.

12. Using the wooden handle of a hammer, push the pistons and connecting rods out of the cylinders.

➡ Before removing the pistons, check the top of the cylinder bore for carbon build-up or a ridge. Remove the carbon or use a ridge reamer to remove the ridge before removing the pistons. Bearing caps, bearings, and pistons should be marked to indicate their location for reassembly.

13. When removing the piston rings, be sure not to apply excessive force as the rings are made of cast iron and can be easily broken.

➡ A hydraulic press is necessary for removing the piston pin (if you need to go this far). This is a job best left to a professional.

14. Observe the following points when installing the piston rings:
   a. When installing the three-piece oil ring, first place the spacer and then the rails in position. The spacer and rail gaps must be staggered 0.8–1.2 in. (20–30mm).
   b. Install the second and top rings on the piston with their markings facing upward.
   c. After installing all rings on the piston, rotate them to be sure they move smoothly without signs of binding.
   d. The ring gaps must be staggered 120° and must NOT be in the direction of the piston pin boss or at right angles to the pin. The gap of the three-piece oil ring refers to that of the middle spacer.

➡ Pistons and rings are also available in four oversizes, 0.01 in. (0.25mm), 0.02 in. (0.50mm), 0.03 in. (0.75mm), and 0.04 in. (1.00mm).

15. Using a ring compressor, install the piston into the cylinder with the skirt protruding about 1/3 of the piston height below the ring compressor. Prior to installation, apply a thin coat of oil to the rings and to the cylinder wall.

➡ When installing the piston, the connecting rod oil jet hole or the mark on the piston crown faces the intake manifold.

16. Using the wooden handle of a hammer, slowly press the piston into the cylinder. Guide the connecting rod so it does not damage the crankshaft journals.

17. Reassemble the remaining components in the reverse order of removal. Install the connecting rod bearing caps so that the recess in the cap and the recess in the rod are on the same side. After tightening the cap bolts, move the rod back and forth on the journal to check for binding.

CLEANING AND INSPECTION

1. Use a piston ring expander and remove the rings from the piston.
2. Clean the ring grooves using an appropriate cleaning tool, exercise care to avoid cutting too deeply.
3. Clean all varnish and carbon from the piston with a safe solvent. Do not use a wire brush or caustic solution on the pistons.
4. Inspect the pistons for scuffing, scoring, cracks, pitting or excessive ring groove wear. If wear is evident, the piston must be replaced.

5. Have the piston and connecting rod assembly checked by a machine shop for correct alignment, piston pin wear and piston diameter. If the piston has collapsed it will have to be replaced or knurled to restore original diameter. Connecting rod bushing replacement, piston pin fitting and piston changing can be handled by the machine shop.

CYLINDER BORE INSPECTION

◆ See Figure 162

Check the cylinder bore for wear using a telescope gauge and a micrometer, measure the cylinder bore diameter perpendicular to the piston pin at a point 2.54 in. (63.5mm) below the top of the engine block. Measure the piston skirt perpendicular to the piston pin. The difference between the two measurements is the piston clearance. If the clearance is within specifications, 0.003 in. (0.08mm), finish honing or glaze breaking is all that is required. If clearance is excessive a slightly oversize piston may be required. If greatly oversize, the engine will have to be bored and 0.01 in. (0.25mm) or larger oversized pistons installed.

FITTING AND POSITIONING PISTON RINGS

◆ See Figures 163 and 164

1. Take the new piston rings and compress them, one at a time into the cylinder that they will be used in. Press the ring about 1 in. (25mm) below the top of the cylinder block using an inverted piston.
2. Use a feeler gauge and measure the distance between the ends of the ring. This is called measuring the ring end gap. Compare the reading to the one called for in the specifications table. File the ends of the ring with a fine file to obtain necessary clearance.

➡ If inadequate ring end gap is utilized, ring breakage will result.

3. Inspect the ring grooves on the piston for excessive wear or taper. If necessary have the grooves recut for use with a standard ring and spacer. A machine shop can handle this job for you.
4. Check the ring grooves by rolling the new piston ring around the groove to check for burrs or carbon deposits. If any are found, remove with a fine file. Hold the ring in the groove and measure side clearance with a feeler gauge. If clearance is excessive, spacer(s) will have to be added.

➡ Always add spacers above the piston ring.

5. Install the ring on the piston, lower oil ring first. Use a ring installing tool on the compression rings. Consult the instruction sheet that comes with the rings to be sure they are installed with the correct side up. A mark on the ring usually faces upward.
6. When installing oil rings, first, install the expanding ring in the groove. Hold the ends of the ring butted together (they must not overlap) and install the bottom rail (scraper) with the end about 1 in. (25mm) away from the butted end of the control ring. Install the top rail about 1 in. (25mm) away from the butted end of the control but on the opposite side from the lower rail.
7. Install the two compression rings.
8. Consult the illustration for ring positioning, arrange the rings as shown, install a ring compressor and insert the piston and rod assembly into the engine.

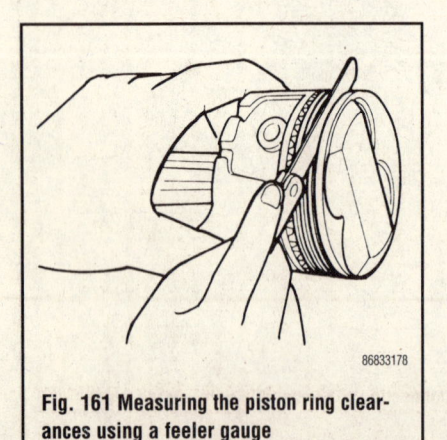

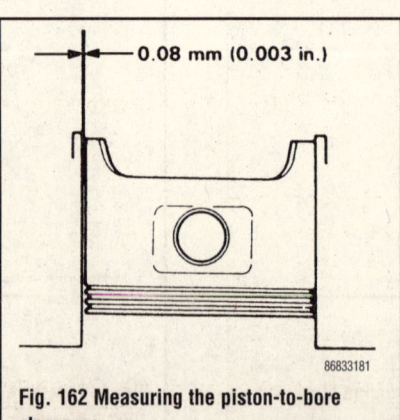

Fig. 161 Measuring the piston ring clearances using a feeler gauge

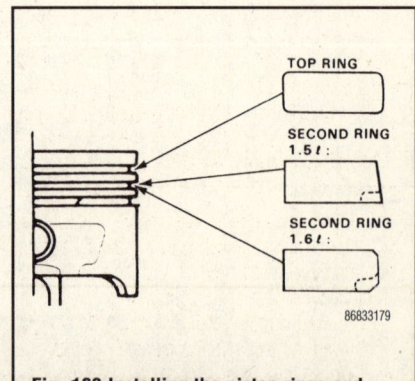

Fig. 162 Measuring the piston-to-bore clearance

Fig. 163 Installing the piston rings and positioning them on the piston

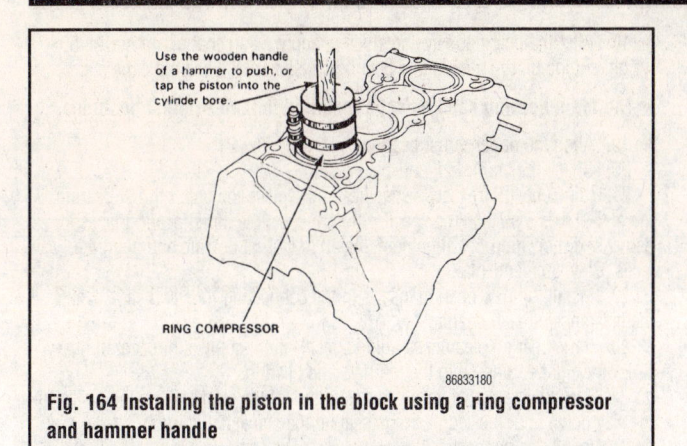

ENGINE AND ENGINE REBUILDING 3-49

Fig. 164 Installing the piston in the block using a ring compressor and hammer handle

Rear Main Seal

REMOVAL & INSTALLATION

♦ See Figure 165

The rear main oil seal is installed in a separate housing on the rear of the block. Replacement of the seal requires the removal of the transaxle, flywheel and clutch housing, as well as the oil pan. Refer to the appropriate sections for the removal and installation of the above components. Both the front and rear main seal are installed after the crankshaft has been tightened, in the event it was removed. Special drivers are used.

Fig. 165 Installing the rear main oil seal in the housing

Crankshaft and Main Bearings

♦ See Figures 166, 167 and 168

The crankshaft main bearings on the Honda engines are retained by a large bridge type main bearing assembly. This is essentially one main bearing assembly. The 1988-91 1.6L (1590cc) engines use a main bearing cap and separate bridge assembly while the 1992-95 1.6L (1590cc and 1595cc) have returned to the one piece mainbearing cap/bridge design (see illustrations). The removal procedures are the same for both.

REMOVAL & INSTALLATION

1. Remove the engine from the vehicle and place it on a work stand.
2. Remove the crankshaft pulley attaching bolts and washer.
3. Remove the front cover and the air conditioning idler pulley assembly, if so equipped. Remove cover assembly.
4. Check the timing belt deflection. Remove the timing belt and sprockets.
5. Invert the engine on work stand. Remove the flywheel and the rear seal cover. Remove the oil pan and gasket. Remove the oil pump inlet and the oil pump assembly.

6. Ensure all bearing caps (main and connecting rod) are marked so they can be installed in their original positions. Turn the crankshaft until the connecting rod from which cap is being removed is up. Remove the connecting rod

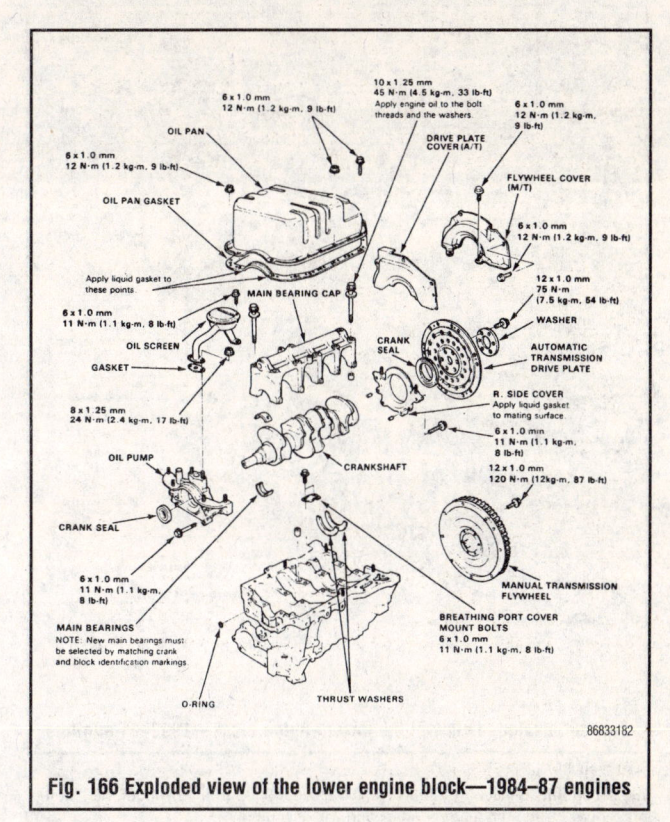

Fig. 166 Exploded view of the lower engine block—1984-87 engines

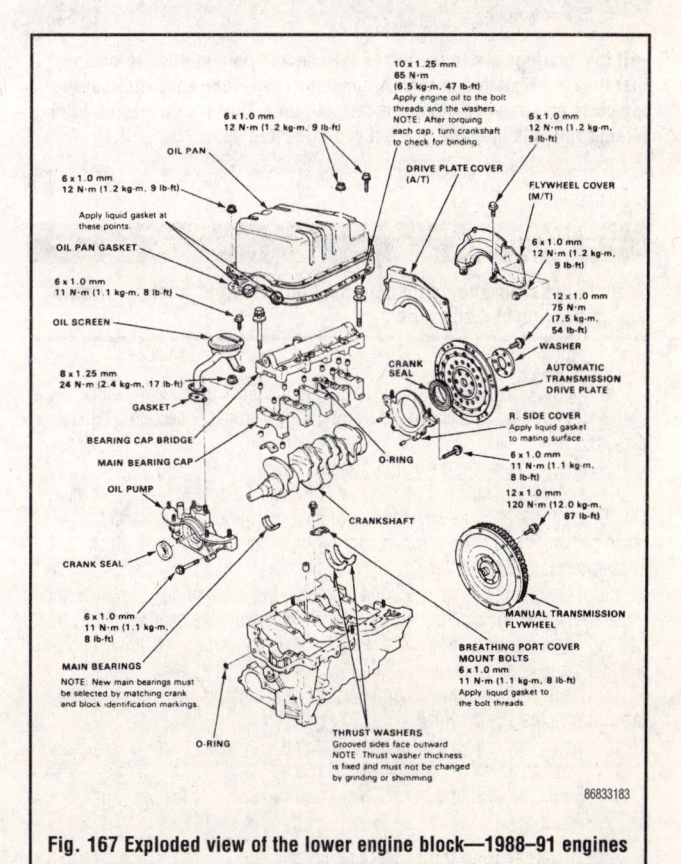

Fig. 167 Exploded view of the lower engine block—1988-91 engines

3-50 ENGINE AND ENGINE REBUILDING

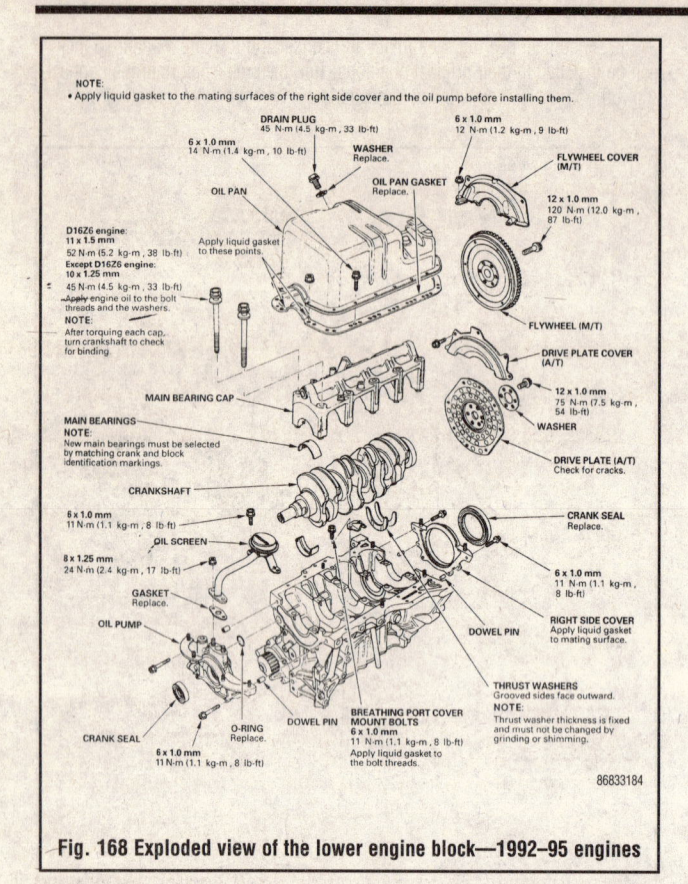

Fig. 168 Exploded view of the lower engine block—1992–95 engines

cap. Push the connecting rod and piston assembly up in the cylinder. Repeat the procedure for the remaining connecting rod assemblies.

7. Remove the main bearing caps.

➡ If the bearings are to be reused they should be identified to ensure that they are installed in their original position. Take care while aranging parts that must go back in a certain order. There is no way of determining which is the right piece if you mix them up.

8. Carefully lift crankshaft out of block so upper thrust bearing surfaces are not damaged.

✱✱ WARNING

Handle the crankshaft with care to avoid possible fracture or damage to the finished surfaces.

To install:

9. Remove the main bearing inserts from the block and bearing caps.
10. Remove the connecting rod bearing inserts from connecting rods and caps.
11. Install a new rear oil seal in rear seal cover.
12. Apply a thin coat of Polyethylene Grease to the rear crankshaft surface. Do not apply sealer to the area forward of oil sealer groove. Inspect all the machined surfaces on the crankshaft for nicks, scratches or scores which could cause premature bearing wear.
13. If the crankshaft main bearing journals have been refinished to a definite undersize, install the correct undersize bearings. Ensure the bearing inserts and bearing bores are clean. Foreign material under the inserts will distort the bearing and cause a failure.
14. Place the upper main bearing inserts in position in the bores with the tang fitted in the slot provided.
15. Install the lower main bearings inserts in the bearing caps.
16. Carefully lower the crankshaft into place.
17. Check the clearance of each main bearing. Select fit the bearings for proper clearance.
18. After the bearings have been fitted, apply a thin coat of heavy SAE 50 weight engine oil to journals and bearings. Install all the bearing caps.

➡ **The main bearing caps must be installed in their original positions.**

19. Align the upper thrust bearing.
20. Check the crankshaft end-play.
21. If the end-play exceeds specification, replace the upper thrust bearing. If the end-play is less than the specification, inspect the thrust bearing faces for damage, dirt or improper alignment. Install the thrust bearing and align the faces. Check the end-play.
22. Install the new bearing inserts in the connecting rods and caps. Check the clearance of each bearing.
23. If the bearing clearances are to specification, apply a thin coat of heavy SAE 50 weight engine oil to the journals and bearings.
24. Turn the crankshaft throw to the bottom of the stroke. Push the piston all the way down until the rod bearings seat on the crankshaft journal.
25. Install the connecting rod cap.
26. After the piston and connecting rod assemblies have been installed, check the connecting rod crankshaft journal.
27. Turn the engine on the work stand so the front end is up. Install the timing belt, sprockets, front cover, oil seal and the crankshaft pulley.
28. Clean the oil pan, oil pump and the oil pump screen assembly.
29. Prime the oil pump by filling the inlet opening with oil and rotating the pump shaft until oil emerges from the outlet opening. Install the oil pump.
30. Position the flywheel on the crankshaft. Install the attaching bolts. Tighten to specification.

➡ **On the flywheel (manual transaxle only) locate clutch disc and install pressure plate.**

31. Turn the engine on the work stand so the engine is in the normal upright position. Install the oil level dipstick. Install the accessory drive pulley, if so equipped. Install and adjust the drive belt and the accessory belts to specification.
32. Install either the clutch assembly or the torque converter.
33. Install the oil pan.
34. Remove the engine from work stand. Install the engine in the vehicle.

CLEANING AND INSPECTION BEARING OIL CLEARANCE

◆ See Figures 169, 170, 171, 172 and 173

Remove cap from the bearing to be checked. Using a clean, dry rag, thoroughly clean all oil from crankshaft journal and bearing insert.

➡ **Plastigage® is soluble in oil, therefore, oil on the journal or bearing could result in erroneous readings.**

Place a piece of Plastigage® along the full width of the bearing insert, reinstall cap, and tighten to specifications.

➡ **Torque specifications are given in the engine specifications earlier in this section.**

Remove bearing cap, and determine bearing clearance by comparing width of Plastigage® to the scale on Plastigage® envelope. Journal taper is determined by comparing width of the bearing insert, reinstall cap, and tighten to specifications.

➡ **Do not rotate crankshaft with Plastigage® installed. If bearing insert and journal appear intact, and are within tolerances, no further main bearing service is required. If bearing or journal appear defective, cause of failure should be determined before replacement.**

CRANKSHAFT END-PLAY/CONNECTING ROD SIDE PLAY

Place a prybar between a main bearing cap and crankshaft casting taking care not to damage any journals. Pry backward and forward, measure the distance between the thrust bearing and crankshaft with a feeler gauge. Compare the reading with specifications. If too great a clearance is determined, a main bearing with a larger thrust surface or crank machining may be required. Check with an automotive machine shop for their advice.

ENGINE AND ENGINE REBUILDING 3-51

Fig. 169 Measuring the connecting rod end-play using a feeler gauge

Fig. 170 Measuring the crankshaft end-play using a dial indicator

Fig. 171 Plastigage® in place on a lower main bearing shell

Connecting rod clearance between the rod and crankthrow casting can be checked with a feeler gauge. Pry the rod carefully on one side as far as possible and measure the distance on the other side of the rod.

CRANKSHAFT REPAIRS

If a journal is damaged on the crankshaft, repair is possible by having the crankshaft machined to a standard undersize.

In most cases, however, since the engine must be removed from the car and disassembled, some thought should be given to replacing the damaged crankshaft with a reground shaft kit. A reground crankshaft kit contains the necessary main and rod bearings for installation. The shaft has been ground and polished to undersize specifications and will usually hold up well if installed correctly.

Flywheel/Flex Plate

REMOVAL & INSTALLATION

♦ See Figures 178 and 179

1. Remove the transmission.
2. Remove the flywheel/flex plate attaching bolts and the flywheel.
3. The rear cover plate can be removed (manual transmission only).

To install:

All major rotating components including the flex plate/flywheel are individually balance to zero. Engine assembly balancing is not required. Balance weights should not be installed on new flywheels.

4. Install the rear cover plate, if removed.
5. Position the flywheel on the crankshaft and install the attaching bolts. Tighten the attaching bolts to specification shown in the Torque specification chart using the illustrated sequence (figs. 201 and 202).

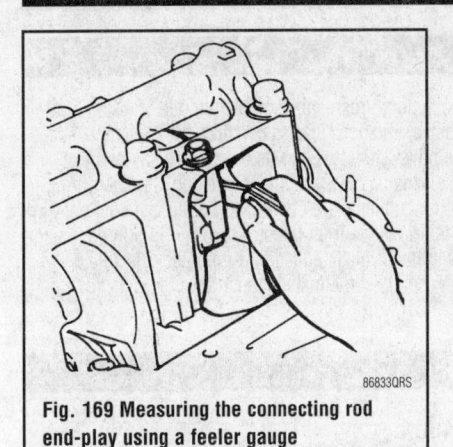

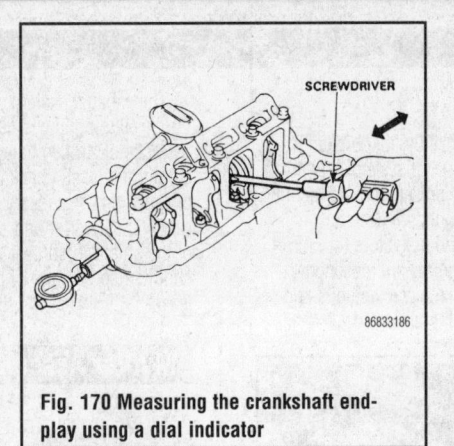

Fig. 172 Apply a strip of gauging material to the bearing journal, then install and torque the cap

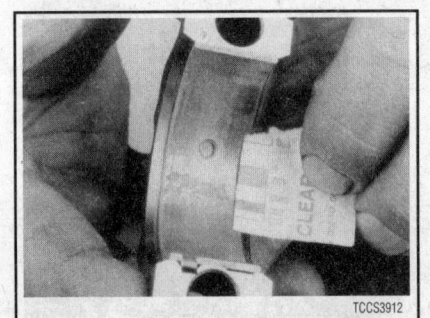

Fig. 173 After the bearing cap has been removed, use the gauge supplied with the gauge material to check bearing clearances

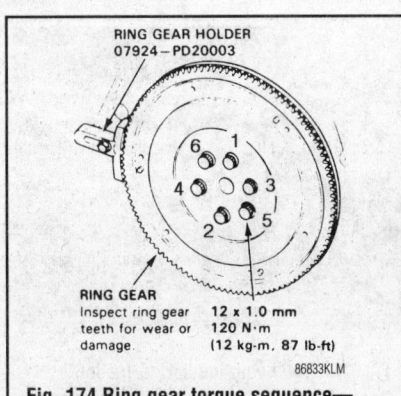

Fig. 174 Ring gear torque sequence—manual transaxle equipped vehicles

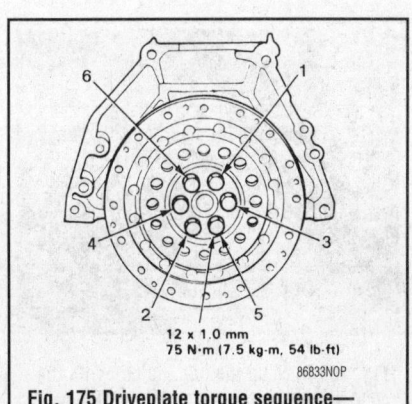

Fig. 175 Driveplate torque sequence—automatic transaxle equipped vehicles

3-52 ENGINE AND ENGINE REBUILDING

EXHAUST SYSTEM

▶ See Figures 176 thru 187

Inspect inlet pipes, outlet pipes and mufflers for cracked joints, broken welds and corrosion damage that would result in a leaking exhaust system. It is normal for a certain amount of moisture and staining to be present around the muffler seams. The presence of soot, light surface rust or moisture does not indicate a faulty muffler. Inspect the clamps, brackets and insulators for cracks and stripped or badly corroded bolt threads. When flat joints are loosened and/or disconnected to replace a shield pipe or muffler, replace the bolts and flange nuts if there is reasonable doubt that its service life is limited.

The exhaust system, including brush shields, must be free of leaks, binding, grounding and excessive vibrations. These conditions are usually caused by loose or broken flange bolts, shields, brackets or pipes. If any of these conditions exist, check the exhaust system components and alignment. Align or replace as necessary. Brush shields are positioned on the underside of the catalytic converter and should be free from bends which would bring any part of the shield in contact with the catalytic converter or muffler. The shield should also be clear of any combustible material such as dried grass or leaves.

Muffler Assembly

REMOVAL & INSTALLATION

1. Raise the vehicle and support on jackstands.
2. Remove the U-bolt assembly and the rubber insulators from the hanger brackets and remove the muffler assembly. Slide the muffler assembly toward the rear of the car to disconnect it from the exhaust pipe.
3. Replace parts as needed.
4. Position the muffler assembly under the car and slide it forward onto the exhaust pipe. Check that the slot in the muffler and the tab on the exhaust pipe are fully engaged. Make sure that the gaskets are positioned properly. Tighten the muffler-to-exhaust pipe bolts to 16 ft. lbs. (22 Nm).
5. Install the rubber insulators on the hanger assemblies. Install the U-bolt and tighten
6. Check the system for leaks. Lower the vehicle.

Catalytic Converter and/or Pipe Assembly

Some models have the catalytic converter in line under the vehicle and some models have the converter bolted directly to the header pipe.

Fig. 176 Undercar view of the front-half of the exhaust system with splash shields in place—1987 CRX Si shown

Fig. 177 The engine splash shield prior to removal. If yours is bent or missing bolts, now is a good time to repair or replace it

Fig. 178 After unbolting the shield, remove it for access to the header pipe (Exhaust Pipe A) flange

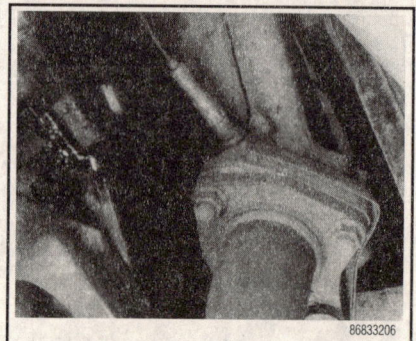

Fig. 179 The exhaust header pipe flange. Note the exhaust gas sensor screwed in to the lower portion of the exhaust manifold

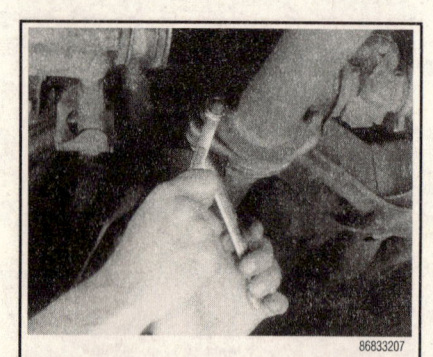

Fig. 180 Back up the nut and bolt with the right size wrench (now is not the time to try to substitute SAE for metric tools!)

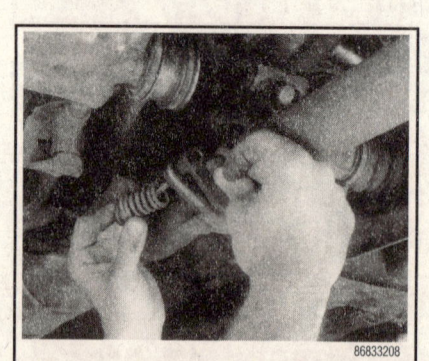

Fig. 181 Removing the spring-loaded flange bolt and nut from the rear portion of the header (pipe A)

Fig. 182 Undercar view of the exhaust pipe, catalytic converter and muffler. Inspect especially for severity of rust or holes

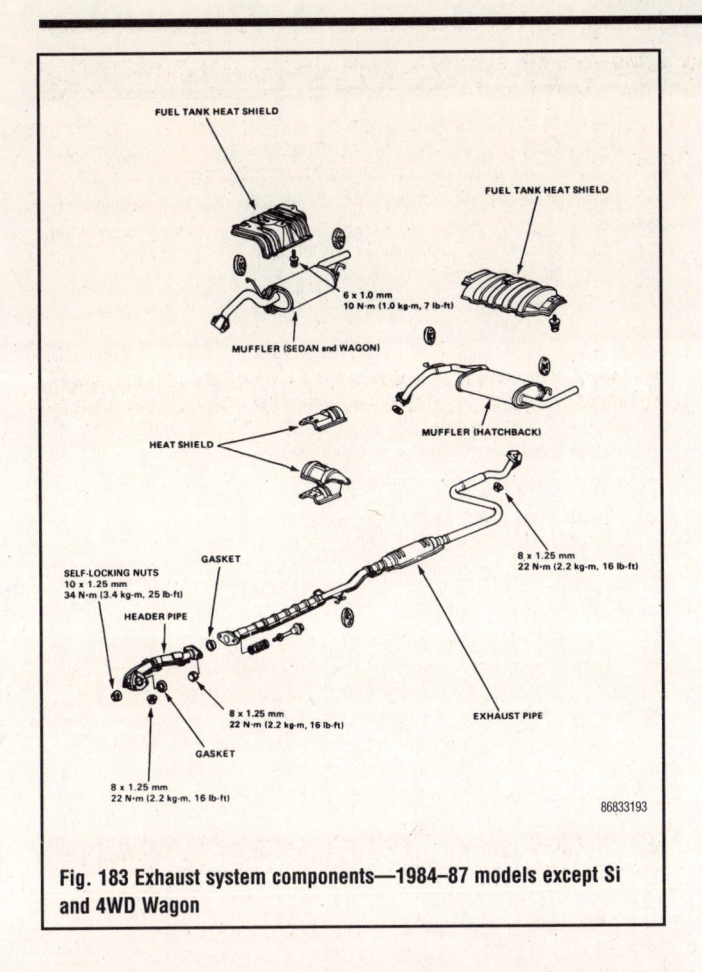

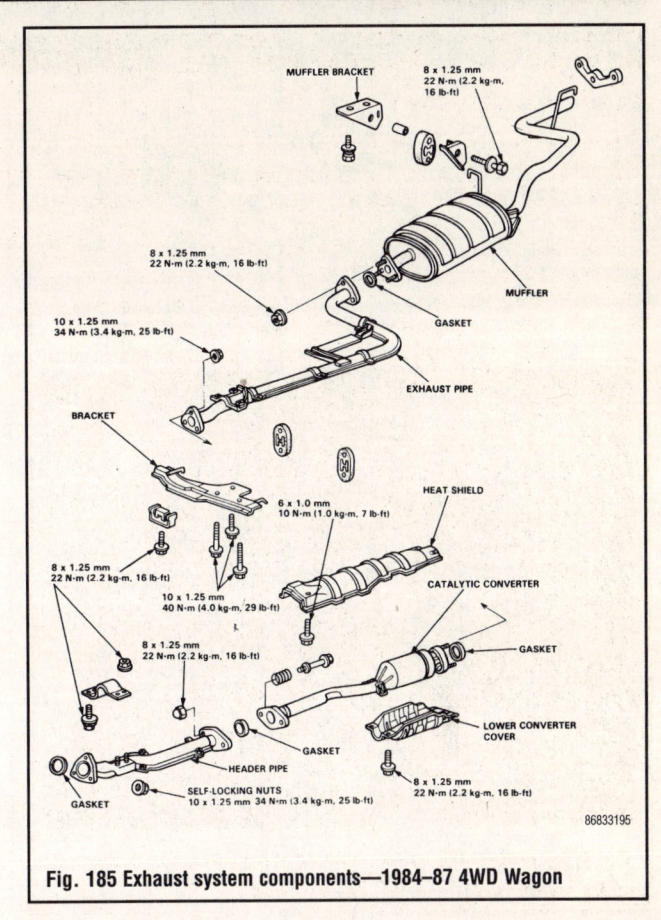

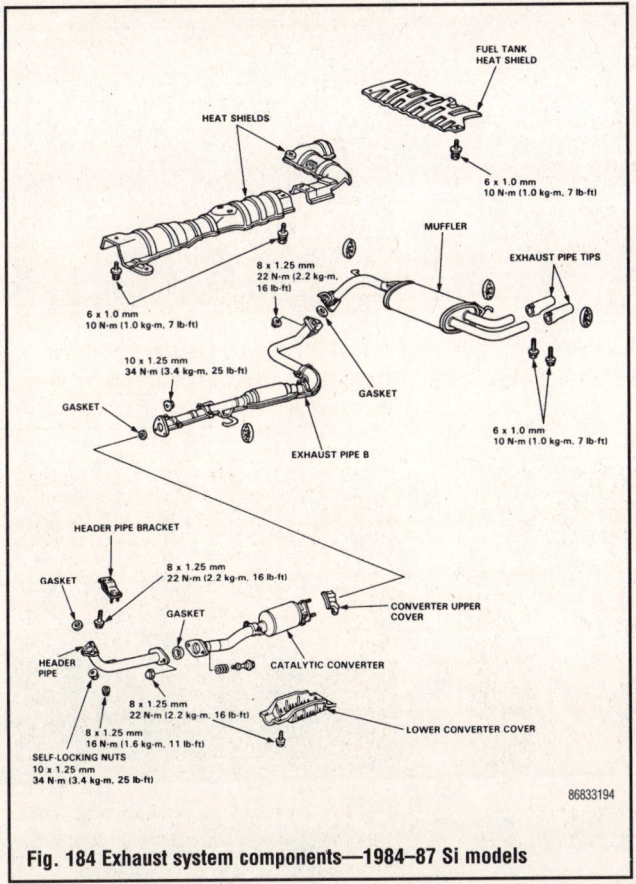

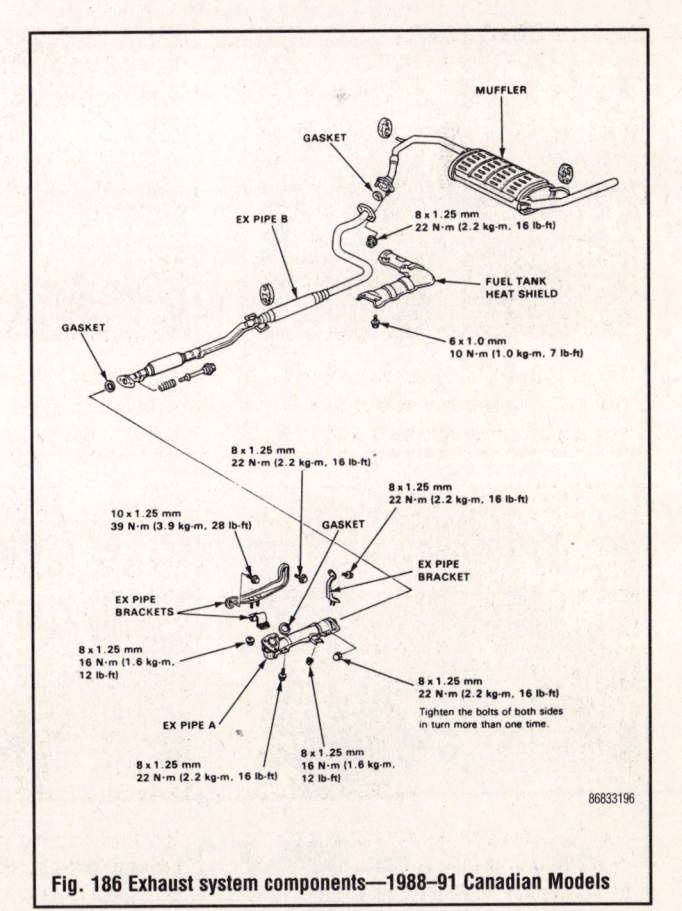

ENGINE AND ENGINE REBUILDING   3-53

Fig. 183 Exhaust system components—1984–87 models except Si and 4WD Wagon

Fig. 185 Exhaust system components—1984–87 4WD Wagon

Fig. 184 Exhaust system components—1984–87 Si models

Fig. 186 Exhaust system components—1988–91 Canadian Models

3-54 ENGINE AND ENGINE REBUILDING

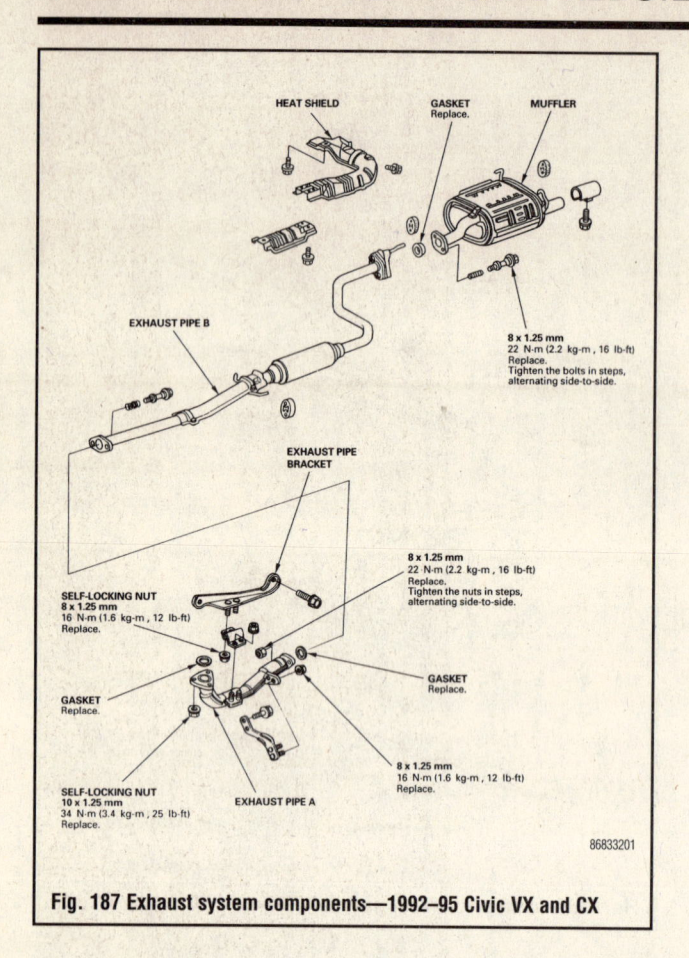

Fig. 187 Exhaust system components—1992–95 Civic VX and CX

REMOVAL & INSTALLATION

1. Raise the vehicle and support on jackstands.
2. Remove the front catalytic converter flange fasteners at the flex joint and discard the flex joint gasket, remove the rear U-bolt connection.
3. Separate the catalytic converter inlet and outlet connections. Remove the converter.
4. Install the converter to the muffler. Install a new flex joint gasket.
5. Install the converter and muffler assembly to the inlet pipe/flex joint. Connect the air hoses and position the U-bolt.
6. Align the exhaust system into position and, starting at the front of the system, tighten all the nuts and bolts. All pipe connecting bolts are tightened to 16 ft. lbs. (22 Nm) and all pipe-to-converter bolts are tightened to 25 ft. lbs. (34 Nm). Header pipe-to-exhaust manifold bolts are tightened to 25 ft. lbs. (34 Nm). Converter-to-exhaust manifold bolts are tightened to 25 ft. lbs (34 Nm).
7. Check the system for leaks. Lower the vehicle.

EMISSION CONTROLS 4-2
POSITIVE CRANKCASE VENTILATION
  (PCV) SYSTEM 4-2
    OPERATION 4-2
    SERVICE 4-2
FUEL EVAPORATIVE EMISSION
  CONTROL SYSTEM 4-2
    OPERATION 4-2
    SERVICE 4-2
EXHAUST GAS RECIRCULATION (EGR)
  SYSTEM 4-4
    CARBURETED ENGINES 4-4
    FUEL INJECTED ENGINES 4-4
CATALYTIC CONVERTER 4-5
    DESCRIPTION 4-5
ADDITIONAL EXHAUST EMISSION
  CONTROLS 4-6
    CARBURETED VEHICLES 4-6
    COMPONENT TESTING 4-7
**ELECTRONIC ENGINE
CONTROLS 4-9**
PROGRAMMED FUEL INJECTION
  (PGM-FI) FAULT MEMORY 4-9
    DESCRIPTION 4-9
    ENTERING SELF-DIAGNOSTICS AND
      DIAGNOSTIC CODE DISPLAY 4-9
    CLEARING TROUBLE CODES 4-11
PROGRAMMED FUEL INJECTION
  (PGM-FI) SYSTEM 4-13
    GENERAL DESCRIPTION 4-13
    SERVICE PRECAUTIONS 4-14
DIAGNOSTIC CHARTS 4-15
VACUUM DIAGRAMS 4-21

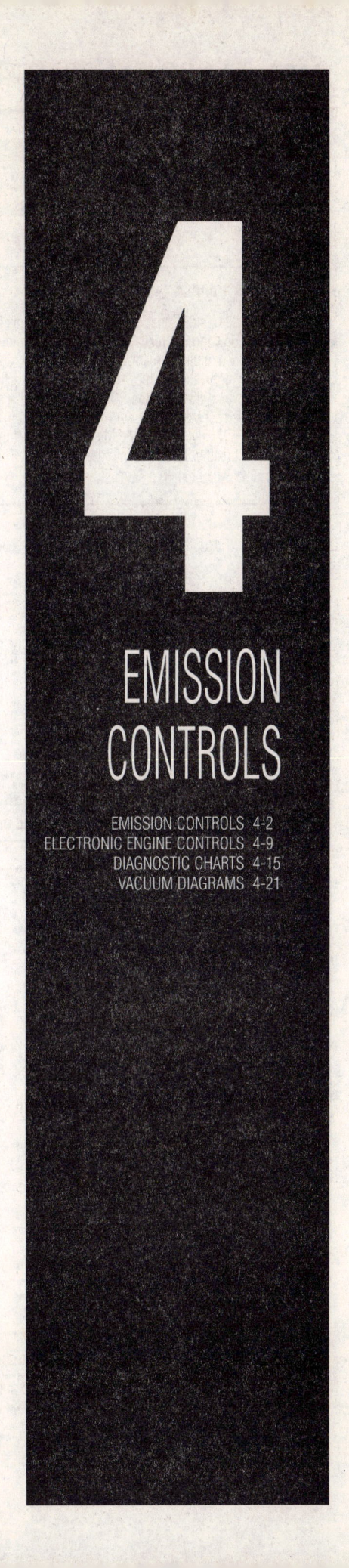

4

EMISSION CONTROLS

EMISSION CONTROLS 4-2
ELECTRONIC ENGINE CONTROLS 4-9
DIAGNOSTIC CHARTS 4-15
VACUUM DIAGRAMS 4-21

4-2 EMISSION CONTROLS

EMISSION CONTROLS

Positive Crankcase Ventilation (PCV) System

OPERATION

♦ See Figures 1 and 2

The PCV system is designed to prevent blow-by gas from escaping to the atmosphere by drawing it directly into the intake manifold. The PCV valve contains a spring-loaded plunger which is lifted in proportion to intake manifold vacuum. Under high vacuum, blow-by is returned directly to the manifold through the breather chamber and PCV valve with fresh air. When intake manifold vacuum decreases (wide open throttle) and vacuum in the air cleaner increases, the PCV valve closes and the blow-by is returned through the cylinder head cover breather hose and into the air cleaner. The PCV valve also acts as a check valve when the engine backfires to prevent crankcase explosion.

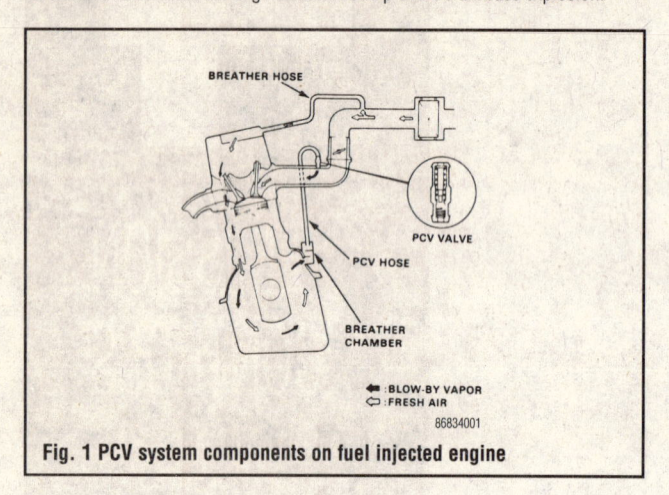

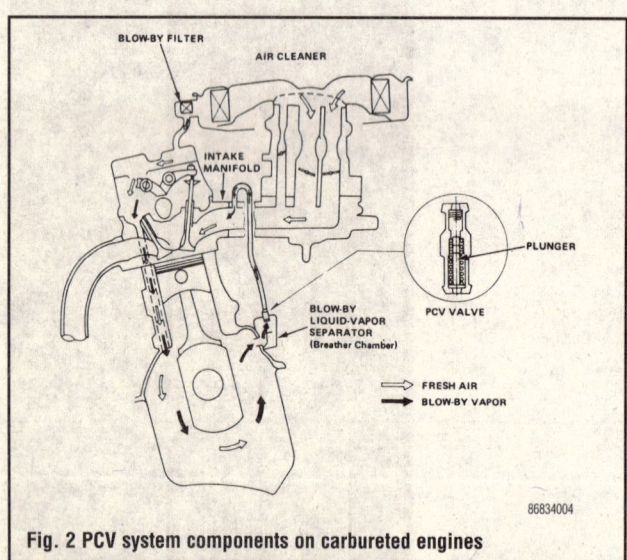

Fig. 1 PCV system components on fuel injected engine

Fig. 2 PCV system components on carbureted engines

SERVICE

➥For PCV valve removal, refer to Section 1 of this manual.

1. Locate the PCV valve. With the engine running at idle, use pliers to pinch the PCV hose between the valve and the intake manifold. It should be possible to hear the spring loaded valve click shut.
2. If no click is heard, remove the valve from the engine but leave it connected to the intake manifold. A hissing sound should be heard and vacuum should be felt at the valve inlet.
3. If no vacuum is felt, remove the valve from the intake manifold hose. The engine should stall or almost stall. This will indicate a stuck valve or clogged hose. Check the breather hose for clogging or leaks and replace any defective parts.

Fuel Evaporative Emission Control System

OPERATION

♦ See Figure 3

The function of this system is to minimize the amount of fuel vapors escaping into the air. The system consists of a charcoal canister, a canister vapor purge system, tank vapor control system, and on the carbureted engine, a carburetor vapor control system. When fuel evaporates in the tank or carburetors, the vapors pass through the vent hoses to the charcoal canister where they are stored. The purge control system uses solenoid valves operated by the ECU to draw fresh air into the canister and carry the vapors into the intake manifold when the engine is running.

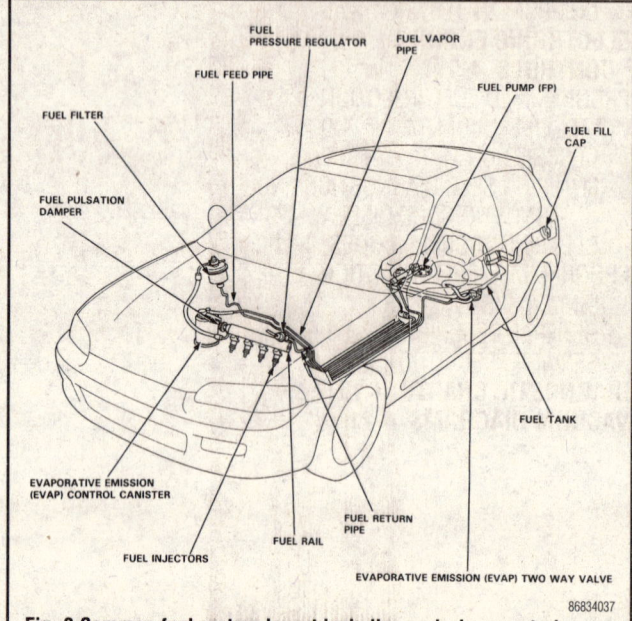

Fig. 3 Common fuel system layout including emissions control devices—1992–95 vehicles

Charcoal Canister

A sealed, maintenance-free charcoal canister is used on all vehicles and is located in the engine compartment. Fuel tank vents lead to the canister. Fuel vapors are temporarily held in the canister's activated charcoal until they can be drawn into the intake manifold and burned in the combustion chamber.

SERVICE

Vapor Purge Control System

This system consists of a purge control diaphragm valve on top of the canister and a solenoid valve controlling the vacuum signal. In the vapor line between the tank and the charcoal canister is a 2-way valve which provides some resistance to flow in either direction. It allows air to flow into the tank as fuel is drawn out. It also allows fuel vapor to travel the other way to the canister when the pressure in the fuel tank is above the valve opening pressure.

➥Complete testing of the purge control system on a vehicle with an automatic transmission requires running the engine with the transmission in gear and the drive wheels turning.

EMISSION CONTROLS 4-3

FUEL INJECTED ENGINE

▶ See Figures 4, 5, 6 and 7

1. With the engine cold, disconnect the vacuum hose from the top of the purge control diaphragm and connect a vacuum gauge to it. The gauge on a hand vacuum pump can be used.
2. With the engine cold and at idle, the solenoid valve should be closed. The gauge should show no vacuum when the engine is first started. As the engine warms up, the solenoid valve will open to open the purge control valve. On CRX SI, the valve should open in about 10 seconds.
3. If there is no vacuum by the time the radiator fan has run, check the purge cut-off solenoid and the wiring to it. The solenoid should have 12 volts at the black/yellow wire when the ignition switch is **ON**. The ECU completes the ground circuit.
4. Reconnect the vacuum line to the valve and connect the gauge to the bottom of the canister where fresh air is admitted. Run the engine at about 3500 rpm. If no vacuum appears when the solenoid valve is open, the canister may be broken. Test the 2-way valve and make sure the fuel filler cap is not leaking.
5. Stop the engine. To test the 2-way valve, remove the fuel filler cap, disconnect the fuel tank vent line, connect a hand vacuum pump to the line and draw a vacuum. It should momentarily stabilize at 0.2–0.6 in. Hg (5–15mm Hg) of vacuum.
6. At the same line, apply low air pressure. The pressure should momentarily stabilize at 0.4–1.4 in. Hg (10–35mm Hg) of vacuum. Correct vacuum and pressure readings mean the 2-way valve is functioning and the tank is venting properly.

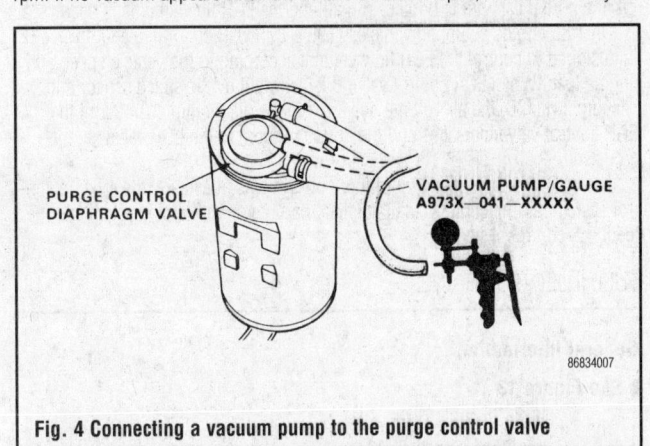

CARBURETED ENGINE

▶ See Figures 8, 9 and 10

1. Raise and safely support the front of the vehicle so the drive wheels are free to turn. With the engine cold, disconnect vacuum hose No. 19 from the top of the purge control diaphragm and connect a vacuum gauge to it. The gauge on a hand vacuum pump can be used.
2. Start the engine and run at idle. There should be no vacuum. If there is vacuum, check for 12 volts to the purge cut-off solenoid valve at the orange wire. If there is voltage, the ECU is getting an incorrect input signal or the ECU is faulty. If there is no voltage, the solenoid valve is probably stuck open.
3. Warm the engine until the radiator fan runs. Put the shift selector in **2** and run the engine at 3500 rpm. If the vacuum lines are routed correctly and not leaking, there should be vacuum. If there is no vacuum and there is power to the purge cut-off solenoid at the orange wire when the drive wheels are turning at about 20 mph, the solenoid ground or solenoid is probably faulty. If there is no power to the solenoid, go to the ECU self-diagnostic test.
4. Reconnect the vacuum line to the valve and connect the gauge to the bottom of the canister where fresh air is admitted. Run the engine at about 3500 rpm. If no vacuum appears when the solenoid valve is open, the canister may be broken. Test the 2-way valve and make sure the fuel filler cap is not leaking.
5. Stop the engine. To test the 2-way valve, remove the fuel filler cap, disconnect the fuel tank vent line, connect a hand vacuum pump to the line and draw a vacuum. It should momentarily stabilize at 0.2–0.6 in. Hg (5–15mm Hg) of vacuum.

Fig. 4 Connecting a vacuum pump to the purge control valve

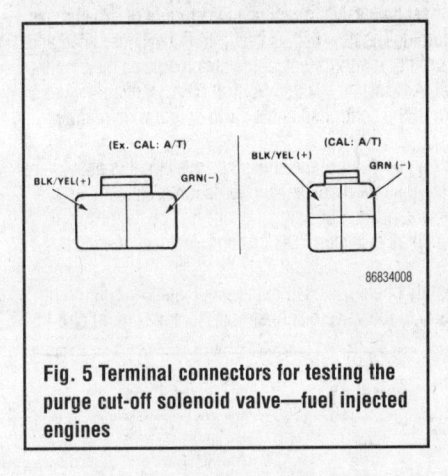

Fig. 5 Terminal connectors for testing the purge cut-off solenoid valve—fuel injected engines

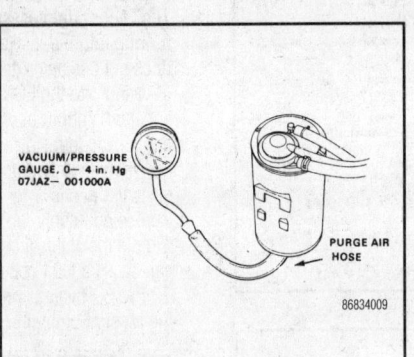

Fig. 6 Testing the purge fresh air hose using a vacuum gauge—fuel injected engines

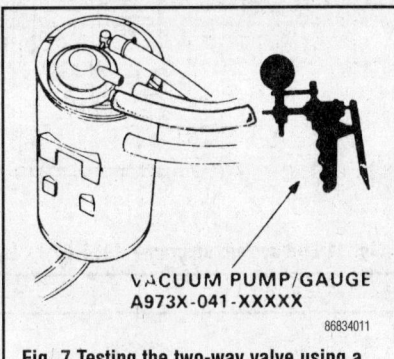

Fig. 7 Testing the two-way valve using a vacuum pump/gauge—fuel injected engines

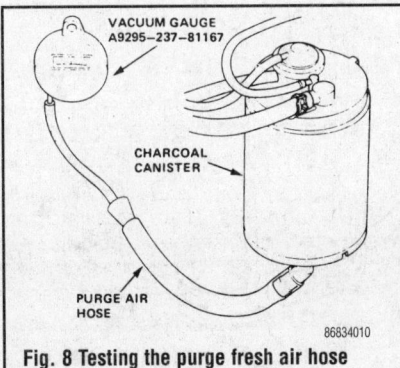

Fig. 8 Testing the purge fresh air hose using a vacuum gauge—carbureted engines

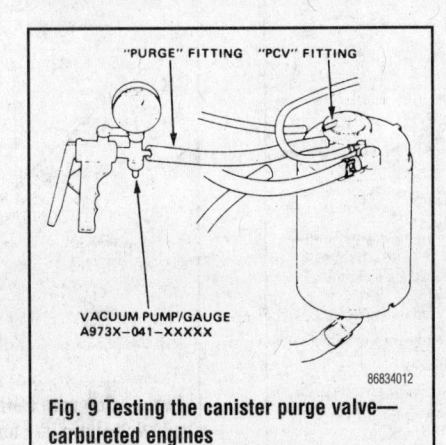

Fig. 9 Testing the canister purge valve—carbureted engines

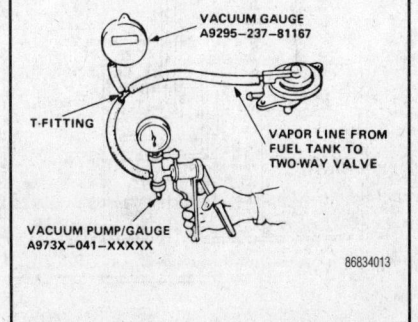

Fig. 10 Testing the two-way valve using a vacuum pump/gauge—carbureted engines

4-4 EMISSION CONTROLS

6. At the same line, apply low air pressure. The pressure should momentarily stabilize at 0.4–1.4 in. Hg (10–35mm Hg) of vacuum. Correct vacuum and pressure readings mean the 2-way valve is functioning and the tank is venting properly.

Exhaust Gas Recirculation (EGR) System

CARBURETED ENGINES

General Information

♦ See Figures 11 and 12

The EGR system is designed to reduce oxides of nitrogen emission, by recirculating exhaust gas through the EGR valve and the intake manifold into the combustion chambers.

The EGR valve is operated by vacuum from the carburetor port and provides EGR volume proportional to engine loads (intake air volume) by the operation of the EGR control valves A and B.

The EGR system recirculates exhaust gas only after the engine is up to normal operating temperature and then only while accelerating or cruising: these two conditions create the majority of NOx emissions. EGR flow is cut off during idle, deceleration and cold engine operation to assure good combustion during these conditions:

• The vacuum signal is ported above the idle throttle valve position to eliminate EGR at idle.
• The EGR control solenoid valve A is designed to cut off the EGR flow. When EGR control solenoid valve A is off and ported vacuum is introduced into the EGR valve and EGR control valve A, the EGR valve is opened, allowing exhaust into the intake.
• With a cold engine, when the thermosensor B is on, EGR control solenoid valve A is energized and the EGR valve is closed.
• When the intake manifold vacuum rises during deceleration, the control switch is turned on to activate EGR control solenoid valve A and the EGR valve flow is cut off.

On 1.5L engine equipped vehicles (1984–87) with automatic transaxle and on all Wagons (1984–87), the EGR control solenoid valve C is designed to control the EGR flow volume according to the driving conditions. When the EGR control solenoid valve C is activated by the control unit venturi vacuum is cut off, the ported vacuum is diluted with air and the EGR flow volume is decreased.

On 1.3L engine equipped vehicles (1984–87), EGR control solenoid valve B is designed to control the EGR flow volume according to the vehicle speed and engine load. When EGR control valve B is activated by the speed sensor and vacuum switch B, the air volume flowing into the diaphragm chamber of the EGR control valve increases and the ported vacuum to the EGR valve is increased.

The EGR system is made up of four sub-systems, these are; the mixture control system, air injection system, auxiliary carburetor air supply system and the feedback control system.

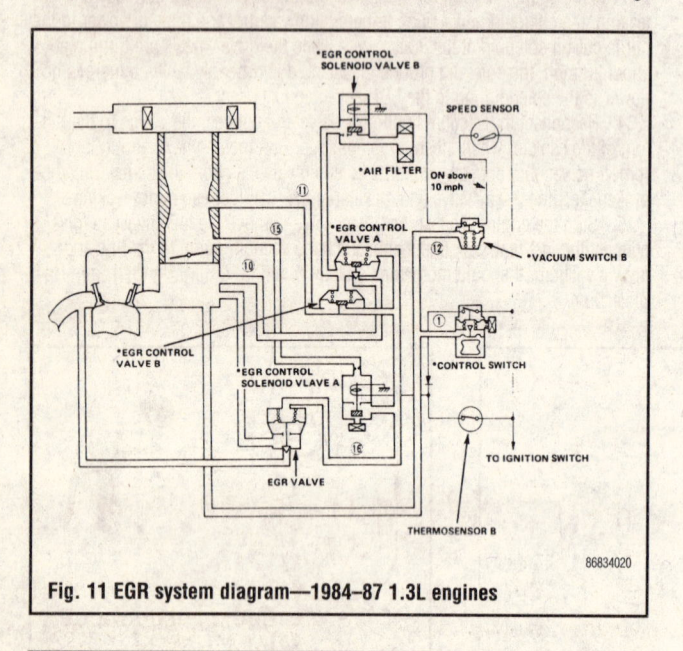

FUEL INJECTED ENGINES

General Information

♦ See Figure 13

The Exhaust Gas Recirculation (EGR) system is used to control emissions measured at the tailpipe. Components included are the EGR valve and controls. The EGR system uses a poppet type valve to regulate the amount of exhaust gas flowing into the intake manifold. The flow path is cast into the cylinder head and does not use any external tubing. A control solenoid valve regulates the amount of vacuum to the EGR valve. A Constant Volume Control (CVC) valve provides a constant amount of vacuum to the control solenoid valve over a wide range of engine vacuum.

When servicing or repairing any part of the emissions system, it is absolutely essential to check for any obvious mechanical faults or failures. Remember, a trouble code only indicates which sensor or circuit is affected by the problem. Simple mechanical faults such as a vacuum leak or poor electrical connection can cause a fault code.

The EGR valve is not simply open or closed but is modulated by controlling the amount of engine vacuum to the valve diaphragm, up to a maximum of 8 in.

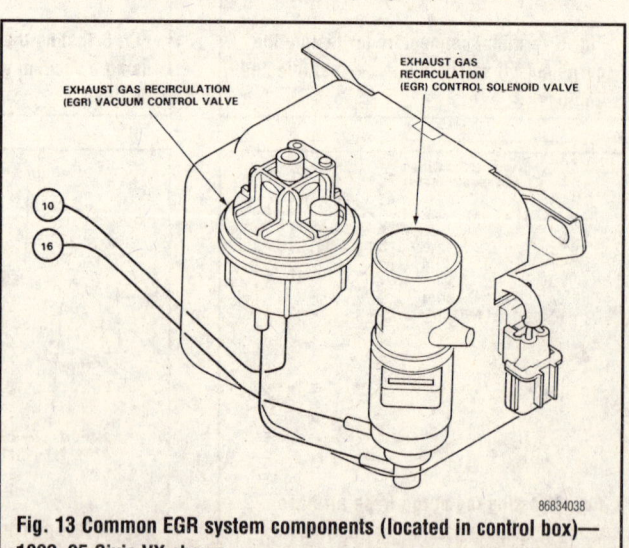

Fig. 11 EGR system diagram—1984–87 1.3L engines

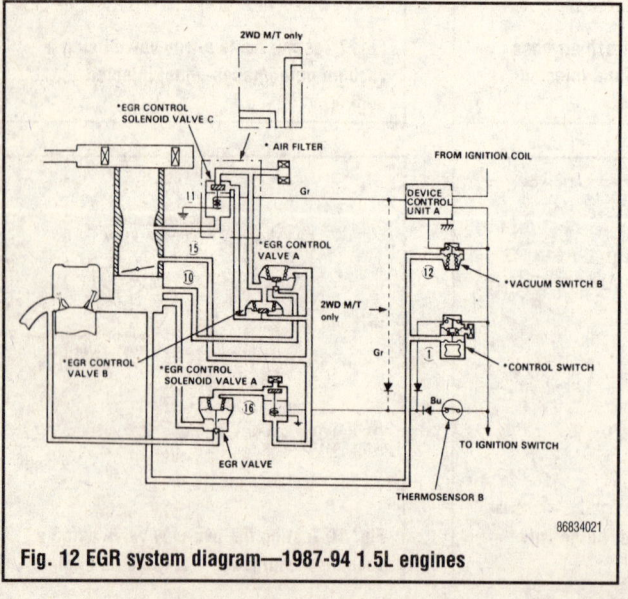

Fig. 12 EGR system diagram—1987–94 1.5L engines

Fig. 13 Common EGR system components (located in control box)—1992–95 Civic VX shown

EMISSION CONTROLS 4-5

Hg of vacuum. The EGR valve has a sensor in the top of the valve that reports valve lift to the ECU (same as ECM, 1992–95). The ECU modulates the position of the control solenoid valve to control the EGR valve lift according to an internal program. Upstream of the control solenoid valve, the CVC provides a constant supply of vacuum so EGR control is precise under all manifold vacuum conditions. An air chamber in-line between the CVC and control solenoid acts as an expansion chamber to dampen any vacuum pulses. The control solenoid valve, CVC and air chamber are in the control box on the firewall.

Service

♦ See Figures 14, 15, 16 and 17

➥Only one Civic model manufactured after 1992 is equipped with the EGR system: the Civic VX.

1. First check that all vacuum lines and electrical connections are in good condition. If the engine will not run at idle, disconnect the vacuum hose to the EGR valve and plug it. If the engine will now idle, the problem is in the EGR control system. Run the engine until the radiator fan comes on. Disconnect the vacuum supply hose No. 16 to the EGR valve and connect a hand vacuum pump with a gauge to the valve.
2. With the engine at idle, draw a vacuum on the EGR valve. The engine should stall and the valve should hold vacuum. If not, replace the EGR valve.
3. Connect the hand pump with vacuum gauge to the No. 16 vacuum hose from the control solenoid valve and restart the engine; there should be no vacuum at idle.
4. If there is vacuum to the EGR valve at idle, check the wiring for the control solenoid. One wire should have 12 volts any time the engine is running. The ground wire goes to the ECU, which modulates the control solenoid opening by controlling the ground circuit. Turn the ignition switch **OFF** and use an ohmmeter to see if the wire between the ECU and control solenoid is shorted to ground. If the wiring is correct, the ECU is getting an incorrect input signal or the ECU is faulty.
5. The vacuum going to the control solenoid valve should be about 8 in. Hg at idle. Connect the pump with gauge to the hose coming from the air chamber. If the vacuum is not correct, read the vacuum at the CVC valve outlet. Full manifold vacuum should be available at the CVC inlet. If the air chamber or CVC valve leak or are not functioning properly, the units cannot be repaired.
6. To test the EGR valve lift sensor, turn the ignition **OFF** and disconnect the wiring to the EGR valve. Turn the ignition switch **ON** and check for 5 volts to the sensor at one of the connector terminals. Switch the ignition **OFF**, connect an ohmmeter across the center and either of the end terminals on the EGR valve lift sensor connector and operate the EGR valve with the vacuum pump. The resistance should change as the valve opens and closes.

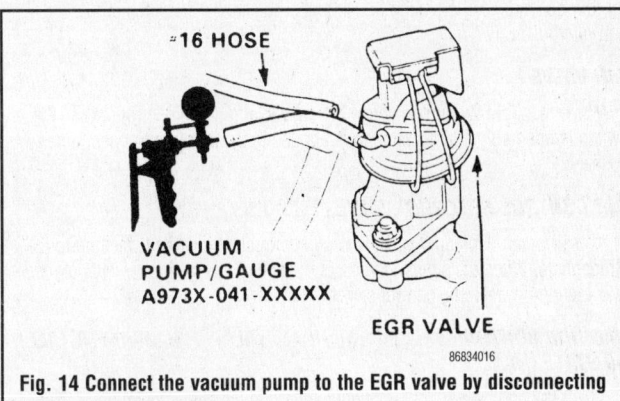

Fig. 14 Connect the vacuum pump to the EGR valve by disconnecting vacuum hose No. 16

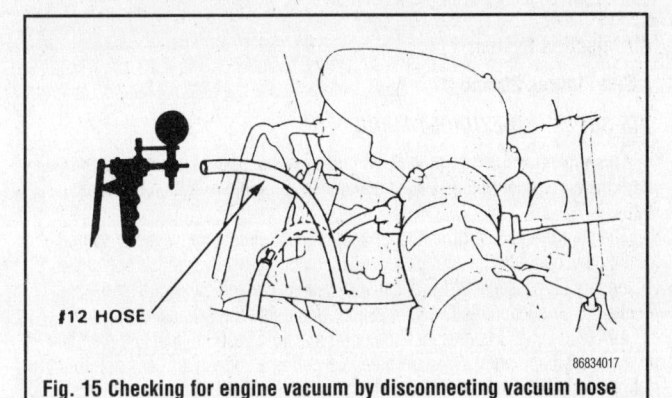

Fig. 15 Checking for engine vacuum by disconnecting vacuum hose No. 12

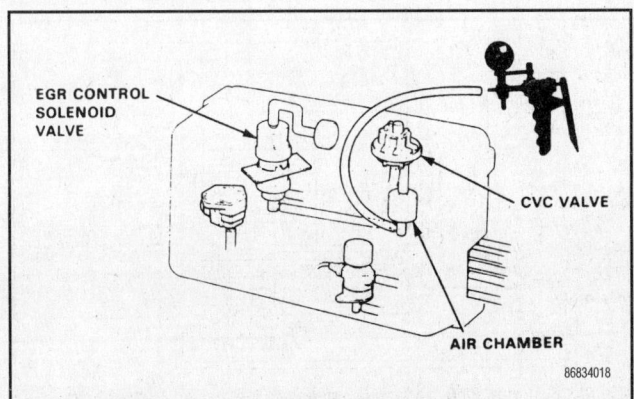

Fig. 16 Connecting the vacuum pump to the CVC valve

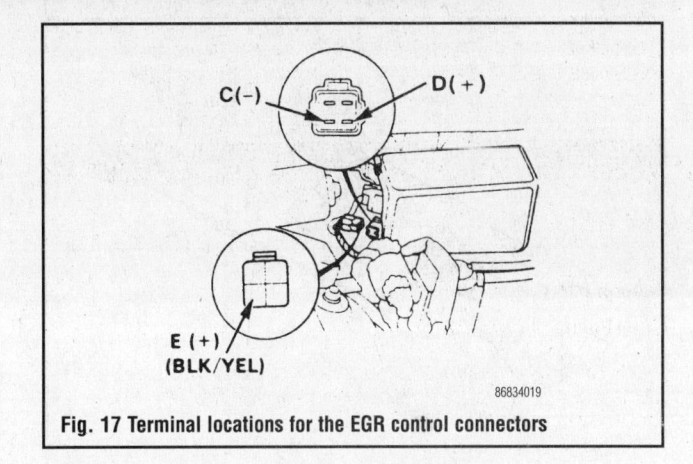

Fig. 17 Terminal locations for the EGR control connectors

Catalytic Converter

DESCRIPTION

♦ See Figure 18

The most conspicuous emission control is the catalytic converter. Its function is to combine unburned hydrocarbons (HC) and carbon monoxide (CO) with oxygen to produce carbon dioxide (CO_2). It also breaks down oxides of nitrogen (NOx). The catalyst operates in a very narrow range of air/fuel concentration. The oxygen sensor in the exhaust system converts the unused oxygen concentration to a voltage which the ECU reads. The ECU constantly adjusts engine air/fuel ratio to provide the proper gas feed to the catalyst.

This is a totally passive device, in that there are no actuators or sensors. It has no influence on the operation of the engine, unless it has melted and causes high exhaust back pressure. If this is suspected, the catalyst can be removed for inspection. By looking into the catalyst inlet, it should be possible to see through almost all of the passages in the honey comb pattern ceramic blocks. If there is any melting which may cause high back pressure, it will be quite obvious. On some vehicles, the catalyst is mounted directly to the exhaust manifold, on others it is farther down stream. On all vehicles, the catalytic converter is before the muffler.

4-6 EMISSION CONTROLS

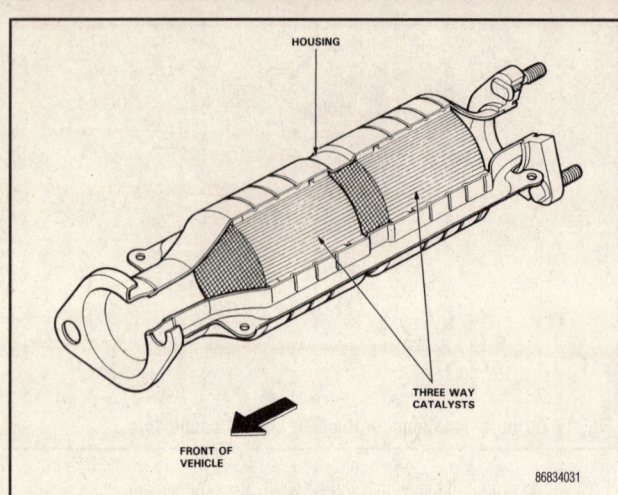

Fig. 18 Cutaway view of the Three Way Catalytic Converter (TWC). This passive exhaust system device reduces emissions by converting hydrocarbons (HC), carbon monoxide (CO), and oxides of nitrogen (NOx) in the exhaust gas to carbon dioxide (CO_2), dinitrogen (N_2) and water vapor

Additional Exhaust Emission Controls

CARBURETED VEHICLES

Mixture Control System

▶ See Figure 19

The anti-afterburn valve and air valve let fresh air into the intake manifold when the manifold vacuum suddenly increases; such as during deceleration.

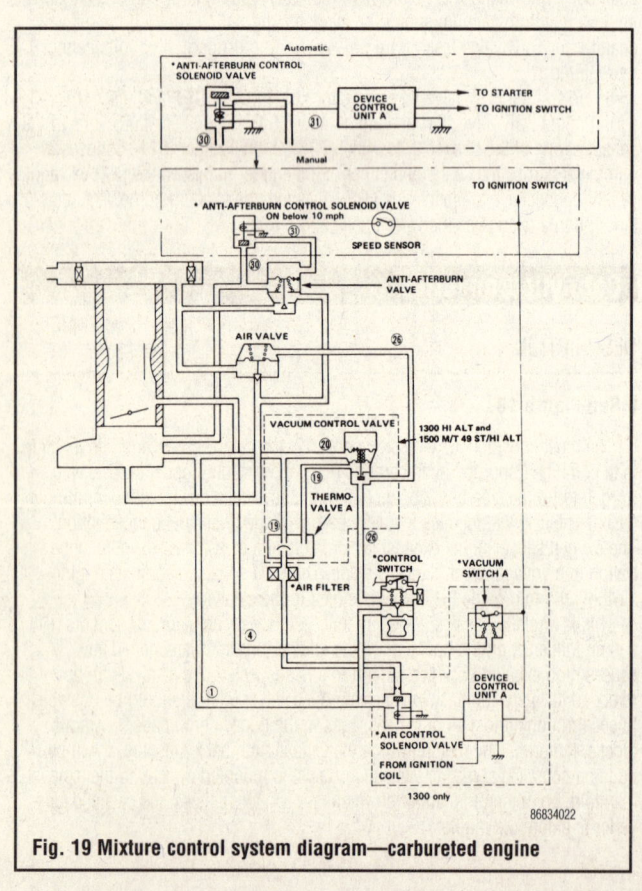

ANTI-AFTERBURN VALVE

When manifold vacuum suddenly increases (i.e. sudden deceleration), the diaphragm/valve unit is pulled downward. Air flow from the air chamber to the sensing chamber is restricted by an orifice, creating a pressure differential on the diaphragm, which holds the valve open. This unbalanced condition lasts for a few seconds until the pressure in both chambers is equalized by air entering through the orifice, then the spring pushes the diaphragm up, closing the valve.

ANTI-AFTERBURN CONTROL SOLENOID VALVE

The anti-afterburn control solenoid valve is designed to cut off manifold vacuum flow to the top side of the anti-afterburn valve diaphragm.

On Manual transaxle cars, when the vehicle speed is below 10 mph, the anti-afterburn control solenoid valve opens and the anti-afterburn valve does not open because there is no vacuum difference on both sides of the diaphragm.

On automatic transaxle cars, the anti-afterburn control solenoid valve is off for four seconds after cranking the engine. Thereafter, it is on to activate the anti-afterburn valve.

AIR VALVE

When manifold vacuum is above the set vacuum of the control switch, the switch is open so manifold vacuum is applied to the air valve to open the air passage.

AIR CONTROL SOLENOID VALVE (1.3L ONLY)

The air control solenoid valve is designed to bleed manifold vacuum to the atmosphere. When the solenoid valve is opened, it takes a higher level of manifold vacuum to open the control switch.

VACUUM CONTROL VALVE (1.3L HI ALT (ALL); 1.5L 2WD M/T (ALL); 49 STATE (HI ALT)

When the engine coolant temperature exceeds the set temperature of thermovalve A and manifold vacuum suddenly increases, the diaphragm/valve is opened, which allows manifold vacuum into the air valve.

Air Injection System

▶ See Figures 20 and 21

AIR SUPPLY-TO-EXHAUST MANIFOLD

This system makes use of vacuum pulses in the exhaust manifold to draw air from the air cleaner to the exhaust manifold to promote oxidation of hydrocarbons.

The control unit energizes the air suction control solenoid valve to apply manifold vacuum to the air suction cut-off diaphragm valve, which opens the secondary air passage. When negative pressure, created by exhaust pulses, opens the air suction reed valve, fresh air pours into the exhaust manifold.

Air bleed valve B is designed to cut off manifold vacuum flow to the air suction cut-off diaphragm. When air bleed valve B opens fresh air does not pour into the exhaust manifold because the air suction cut-off diaphragm valve closes the secondary air passage. The air chamber acts as silencer to reduce exhaust noise.

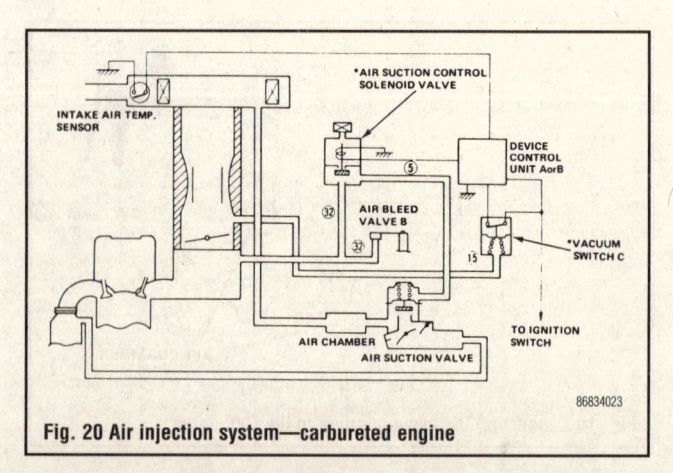

Fig. 19 Mixture control system diagram—carbureted engine

Fig. 20 Air injection system—carbureted engine

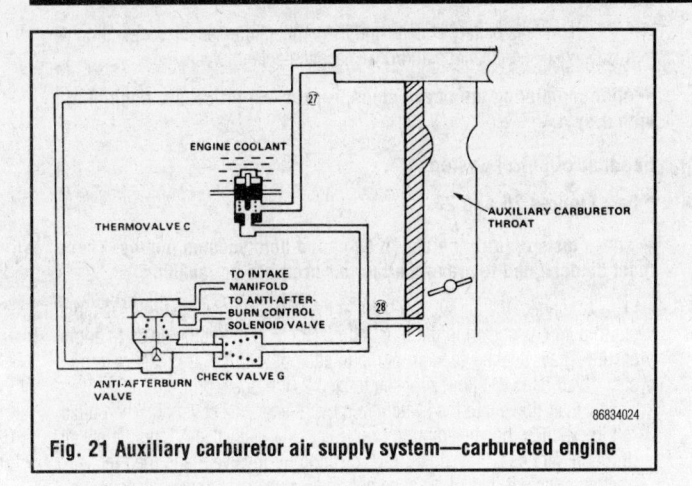

EMISSION CONTROLS   4-7

Fig. 21 Auxiliary carburetor air supply system—carbureted engine

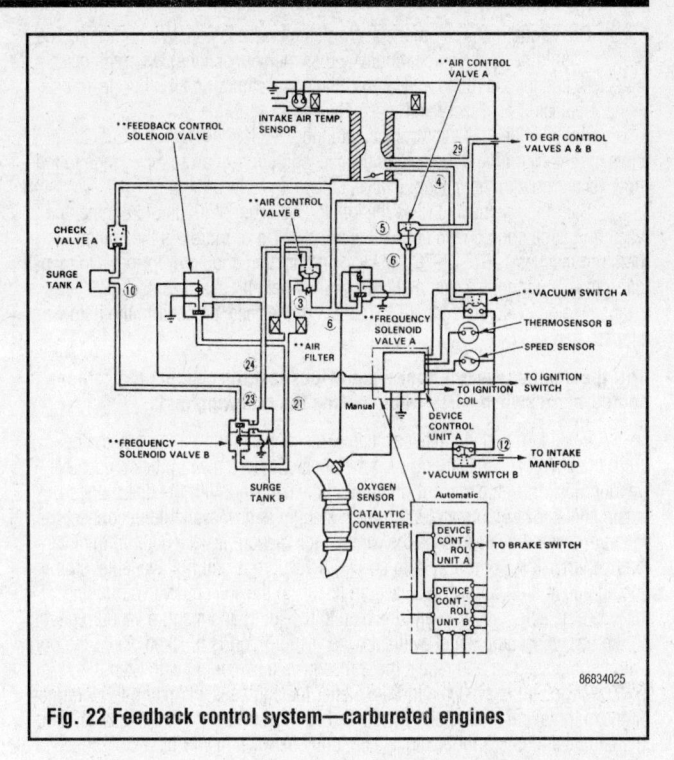

Fig. 22 Feedback control system—carbureted engines

Auxiliary Carburetor Air Supply System

When the engine coolant temperature is below the set temperature of thermovalve C, fresh air pours into the auxiliary intake manifold to prevent an overly-rich mixture.

During gear shifting or deceleration, the anti-afterburn valve senses the change in intake manifold vacuum, and provides fresh air into the intake manifold.

Under such conditions, fresh air is also supplied from the anti-afterburn valve into the intake ports of the auxiliary carburetor throat through check valve G in order to prevent an incombustible mixture in the auxiliary combustion chamber.

Feedback Control System

The Feedback Control System maintains the proper air/fuel mixture ratio by allowing air into the intake manifold, as is necessary, to adjust a temporarily fuel-rich condition. This system is made up of two subsystems; the X-system and the M-system.

The X-system consists of air control valve B, frequency solenoid valve B, feedback control solenoid valve, check valve B, surge tanks A and B, and control unit. When frequency solenoid valve B and feedback control solenoid valve are activated by the control unit, manifold vacuum is applied to air control valve B. When air control valve B opens, it allows the correct amount of air to be fed into the intake manifold.

Surge tank A acts as a vacuum reservoir while surge tank B dampens the vacuum pulses so the relatively steady vacuum is applied to air control valve B.

The M-system consists of air control valve A, frequency solenoid valve A and control unit. When frequency solenoid valve A is opened by control unit, air control valve A, which has been already opened by vacuum from the carburetor port, feeds the correct amount of air into the intake manifold. The amount of air is proportional to intake air volume.

COMPONENT TESTING

Intake Air Temperature Control System (Engine Cold)

♦ See Figure 22

1. Inspect for loose, disconnected, or deteriorated vacuum hoses and replace as necessary.
2. Remove the air cleaner cover and element.
3. With the transmission in Neutral and the blue distributor disconnected, engage the starter motor for approximately two (2) seconds. Manifold vacuum to the vacuum motor should completely raise the air control valve door. Once opened, the valve door should stay open unless there is a leak in the system.
4. If the valve door does not open, check the intake manifold port by passing a No. 78 — 0.016 in. (0.4mm) diameter — drill or compressed air through the orifice in the manifold.
5. If the valve door still does not open, proceed to the following steps:
   a. Vacuum Motor Test: Disconnect the vacuum line from the vacuum motor inlet pipe. Fully open the air control valve door; block the vacuum motor inlet pipe, then release the door. If the door does not remain open, the vacuum motor is defective. Replace as necessary and repeat Steps 1–3.
   b. Air Bleed Valve Test: Unblock the inlet pipe and make sure that the valve door fully closes without sticking or binding. Reconnect the vacuum line to the vacuum motor inlet pipe. Connect a vacuum source (for example, hand vacuum pump) to the manifold vacuum line (disconnect at the intake manifold fixed orifice) and draw enough vacuum to fully open the valve door. If the valve door closes with the manifold vacuum line plugged (by the vacuum pump), then vacuum is leaking through the air bleed valve. Replace as necessary and repeat Steps 1–3.

✲✲ WARNING

Never force the air bleed valve (bi-metal strip) on or off its valve seat. The bimetallic strip and the valve seat may be damaged.

   c. Check Valve Test: Again draw a vacuum (at the manifold vacuum line) until the valve door opens. Unplug the line by disconnecting the pump from the manifold vacuum line. If the valve door closes, vacuum is leaking past the check valve. Replace as necessary and repeat Steps 1–3.
6. After completing the above steps, replace the air cleaner element and cover and fit a vacuum gauge into the line leading to the vacuum motor.
7. Start the engine and raise the idle to 1500–2000 rpm. As the engine warms, the vacuum gauge reading should drop to zero.

➥Allow sufficient time for the engine to reach normal operating temperature — when the cooling fan cycles on and off.

If the reading does not drop to zero before the engine reaches normal operating temperature, the air bleed valve is defective and must be replaced. Repeat Step 3 as a final check.

Temperature and Transaxle Controlled Spark Advance (Engine Cold)

♦ See Figures 23 and 24

1. Check for loose, disconnected or deteriorated vacuum hoses and replace (if necessary).
2. Check the coolant temperature sensor switch for proper operation with an ohmmeter or 12V light. The switch should normally be open (no continuity across the switch terminals) when the coolant temperature is below approximately 120°F (49°C). If the switch is closed (continuity across the terminals), replace the switch and repeat the check.

4-8 EMISSION CONTROLS

3. On manual transaxle models, check the transmission sensor switch. The switch should be open (no continuity across the connections) when 4th gear is selected, and closed (continuity across the connections) in all other gear positions. Replace if necessary and repeat the check.

4. Remove the spark control vacuum tube, leading between the spark advance/retard unit and the solenoid valve, and connect a vacuum gauge to the now vacant hole in the solenoid valve, according to the diagram.

5. Start the engine and raise the idle to 2000 rpm. With a cold engine, the vacuum gauge should read approximately 3 in. Hg or more. As the coolant temperature reaches 120°F (49°C), the vacuum reading should drop to 0. On manual transaxle models, vacuum should return when 4th gear is selected (transmission switch is opened). If this is not the case, proceed to the following steps:

➥If the engine is warm from the previous test, disconnect the coolant temperature switch wires when making the following tests.

6. If vacuum is not initially available, disconnect the vacuum signal line from the charcoal canister and plug the open end, which will block a possible vacuum leak from the idle cut-off valve of the canister. With the line plugged, again check for vacuum at 2000 rpm. If vacuum is now available, reconnect the vacuum signal line and check the canister for vacuum leaks. (Refer to the Evaporative Emission Control System check.) If vacuum is still not available, stop the engine and disconnect the vacuum line from the solenoid valve (the line between the solenoid valve and the manifold T-joint) and insert a vacuum gauge in the line. If vacuum is not available, the vacuum port is blocked. Clear the port with compressed air and repeat the test sequence beginning with Step 3.

7. If vacuum is available in Step 5 after the engine is warm and in all ranges of the automatic transaxle and in 1st, 2nd and 3rd of the manual transaxle, stop the engine and check for electrical continuity between the terminals of the coolant temperature sensor:

   a. If there is no continuity (engine is warm), replace the temperature sensor switch and recheck for continuity.

   b. If there is continuity, check the battery voltage to the vacuum solenoid. If no voltage is available (ignition switch **ON**), check the wiring, fuses and connections.

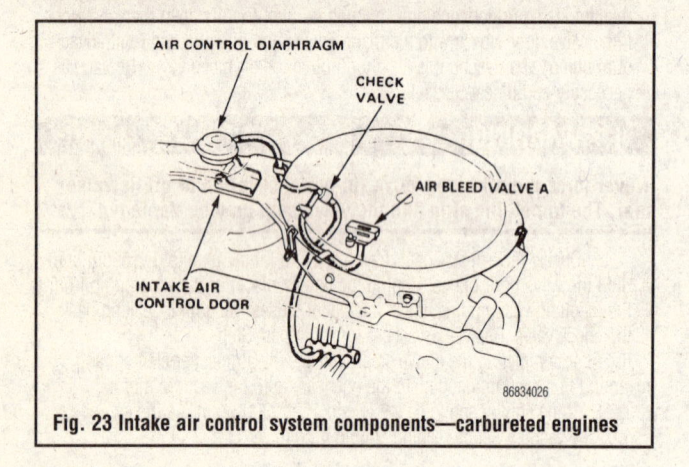

   c. If there is battery voltage and the temperature sensor is operating correctly, check connections and/or replace the solenoid valve.

➥After completing the above steps, repeat test procedure beginning with Step 4.

Feedback Control System

▶ See Figures 25 and 26

➥These tests require the use of two hand held vacuum pumps. These must be designed to provide either air pressure or vacuum.

1. Disconnect the air suction hose at the vacuum hose manifold. Tee in a vacuum pump as shown in the illustration. Hook up the pump so as to apply pressure. Then, use the vacuum pump to attempt to force air pressure into the system. If air does not flow, proceed to step 2. If it does, proceed to 2a.

   a. If air flows into the system, remove the air box cover and then pinch off the vacuum hose leading to frequency solenoid valve A. Then, blow air through the hose. If air flows, replace air control valve B and then recheck that air does not flow. If air does not flow, replace the frequency solenoid valve A and retest as in Step 1.

2. Disconnect vacuum hose No. 24 at surge tank B. Connect the vacuum pump to the end of this hose so as to apply pressure. Then, disconnect the hose leading from Air Control Valve B and frequency solenoid valve A. Connect the hand pump so as to apply pressure. Apply vacuum at hose #24 and blow air into the other hose (the B-valve side of the Air Control Valve). Air should flow. If it does, proceed to Step 3. If not, proceed to 2a.

   a. Replace air control valve B and then retest the system to make sure air does flow.

3. Jumper battery voltage (from the + terminal) to the green/white connector of connector box #2 (see illustration). Blow air into the vacuum hose going to frequency valve A. If air flows, the valve is okay and you should proceed to Step 4. If not, proceed to Step 3a.

   a. If air does not flow, disconnect the hose leading from frequency solenoid valve A to air control valve A and check for air flow with the pump in operation. If air flows, replace control valve A and then retest to make sure the system now works. If air does not flow, replace frequency solenoid valve A and re-test to make sure the system now works.

4. Disconnect vacuum hose #29 at the vacuum hose manifold and connect the vacuum pump to the open end. Apply vacuum to the open end and blow air into the air suction hose leading to Control Valve A. When there is vacuum applied via the pump, air should flow without resistance. When vacuum is released, air should flow into the system only with significant resistance; that is, pressure should build up on the gauge of the pump supplying pressure and then be released only gradually. If this occurs, reconnect the air suction hose and hose #29 and then go on to Step 5. If there is no such change in resistance, replace air control valve A and then repeat the test.

5. Jumper battery voltage from the battery (+) terminal to the frequency solenoid valve B. Disconnect the vacuum hose #23 from surge tank B and connect a vacuum pump to it. Then, apply vacuum. Vacuum should build up on the pressure gauge. If there is vacuum, go to Step 6. If not, pinch the hose leading from the frequency solenoid valve B to the constant vacuum valve. Then, repeat the test. If there is now vacuum, replace the constant vacuum valve and then retest to make sure the system is repaired. If there is still no vacuum, replace

Fig. 23 Intake air control system components—carbureted engines

Fig. 24 Testing the vacuum advance diaphragm—disconnect vacuum hose No. 2 to connect the vacuum pump

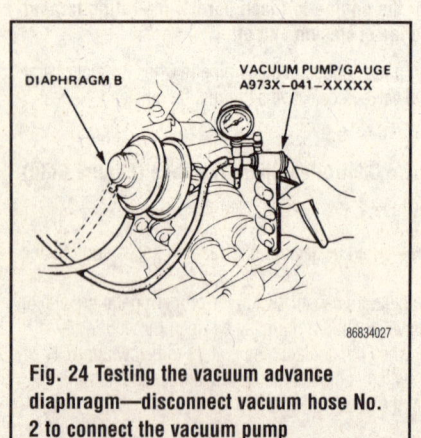

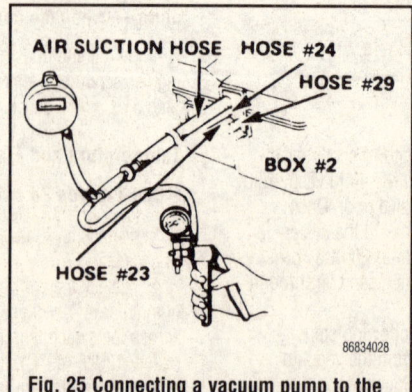

Fig. 25 Connecting a vacuum pump to the air control valve

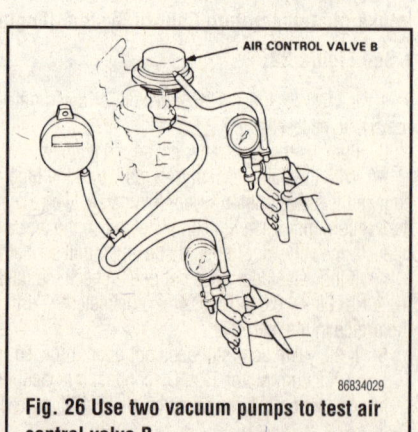

Fig. 26 Use two vacuum pumps to test air control valve B

EMISSION CONTROLS 4-9

the frequency solenoid valve B and repeat the test to make sure the system is repaired.

6. Leaving the vacuum pump hooked up, disconnect the jumper wire leading to frequency solenoid valve B. Make sure there is still vacuum in the system. Vacuum should disappear when the wire is disconnected. If it does disappear, reconnect vacuum hose #23 and go on to Step 7. If it does not disappear, replace solenoid valve B and retest to make sure it now disappears.

7. Remove the cover of control box #2 and disconnect the hose connecting the constant vacuum valve and frequency solenoid valve B from its connection within the box. Then, connect a vacuum pump to the open end of the hose. Now, start the engine, allow it to idle, and measure the vacuum with a gauge. It should stabilize in the range 6–11 in.Hg (manual transaxle) vehicles and 2–7 in.Hg (automatic transaxle). If vacuum stabilizes in this range, the constant vacuum valve is okay, and you should reconnect the hose. If not, replace the constant vacuum valve, and then retest to make sure you have corrected the malfunction.

ELECTRONIC ENGINE CONTROLS

➥The electronic engine control system is extremely complex, diagnosis and repair of this system requires that care be taken and all precautions followed. Failure to follow all directions carefully, will result in expensive computer damage. If unfamiliar with the use of testing equipment and the operation of he fuel system, take the vehicle to an authorized dealer for service.

Programmed Fuel Injection (PGM-FI) Fault Memory

DESCRIPTION

Programmed Fuel Injection (PGM-FI) System consists of three sub-systems: air intake, electronic control, and fuel. The Electronic Control System, in order to get fuel into the cylinders at the correct instant and in the correct amount must perform various interrelated functions. The Electronic Control Unit (ECU) on pre-1992 vehicles or Electronic Control Module (ECM) on 1992-95 vehicles is the computer "brain" of the PGM-FI system. It uses an eight-bit microcomputer and consists of a CPU (Central Processing Unit), memories, and I/O (Input/Output) ports. Basic data stored in memory is compensated by the ECU or ECM in response to signals sent from various sensors to provide the correct air/fuel mixture for all engine needs.

The unit contains memories for the basic discharge duration at various engine speeds and manifold pressures. The basic discharge duration, after being read out from the memory, is further modified by signals sent from the various sensors to obtain the final discharge duration. Other functions also include:

- Starting Control — The fuel system must vary the air/fuel ratio to suit different operating requirements. For example, the mixture must be rich for starting. The memories also contain the basic discharge durations to be read out by signals from the starter switch, and engine speed and coolant temperature sensors, thereby providing extra fuel needed for starting.
- Injector Control — The ECU controls the discharge durations at various engine speeds and loads.
- Electronic Air Control — The ECU controls the EACV to maintain correct idle speed based on engine and accessories demand.
- Ignition Timing Control — The ECU controls the basic ignition timing based on engine load, engine rpm, vehicle speed and coolant temperature.
- Fuel Pump Control — When the speed of the engine falls below the prescribed limit, electric current to the fuel pump is cut off, preventing the injectors from discharging fuel.
- Fuel Cut-Off Control — During deceleration with the throttle valve nearly closed, electric current to the injectors is cut off at speeds over 900 rpm, contributing to improved fuel economy. Fuel cut-off action also takes place when engine speed exceeds approximately 7500 rpm regardless of the position of the throttle valve.
- Safety — A fail-safe system monitors the sensors and detects any abnormality in the ECU, ensuring safe driving even if one or more sensors are faulty, or if the ECU malfunctions.
- Self-Diagnosis — When an abnormality occurs, the ECU lights the engine warning light on the dashboard to alert the driver and stores the failure code in erasable memory to be read by the servicing technician. The ECU LED will display the code any time the ignition is turned **ON**.

ENTERING SELF-DIAGNOSTICS AND DIAGNOSTIC CODE DISPLAY

Self-Diagnosis Indicators

1984–87 VEHICLES

▶ See Figures 27 and 28

The quick reference chart covers the most common failure modes for the PGM-FI. The probable causes are listed in order of most easily checked first, then progressing to more difficult fixes. Run through all the causes listed. If problem is still unsolved, go on to the more detailed troubleshooting. Troubleshooting is divided into different LED displays. Find the correct light display and begin again.

For all the conditions listed, the PGM-FI warning light on the dashboard must be on (comes on and stays on). This indicates a problem in the electrical portion of the fuel injection system. At that time, check the LED display (self-diagnosis system) in the ECU.

On all Civic models, there are 4 LED displays. They are part of the ECU, which is located under the passenger's seat. They are numbered 1, 2, 4 and 8, as counted from right to left.

Sometimes the dash warning light and/or ECU LED(s) will come on, indicating a system problem, when, in fact, there is only a bad or intermittent electrical connection. To troubleshoot a bad connection, note the ECU LED pattern that is lit, refer to the diagnosis chart and check the connectors associated with the items mentioned in the "Possible Cause" column for that LED pattern (disconnect, clean or repair if necessary and reconnect those connections). Then, reset the ECU memory as described, restart the car and drive it for a few minutes and then recheck the car and drive it for a few minutes and then recheck the LED(s). If the same pattern lights up, begin system troubleshooting; if it does not light up, the problem was only a bad connection.

The memory for the PGM-FI warning light on the dashboard will be erased when the ignition switch is turned off; however, the memory for the LED display(s) will not be canceled. Thus, the warning light will not come on when the ignition switch is again turned on unless the trouble is once more detected. Troubleshooting should be done according to the LED display(s) even if the warning light is off.

Other ECU information:

- After making repairs, disconnect the battery negative cable from the battery negative terminal for at least 10 seconds and reset the ECU memory. After reconnecting the cable, check that the LED display is turned off.
- Turn the ignition switch on. The PGM-FI warning light should come on for about 2 seconds. If the warning light won't come on, check for:

1. Blown warning light bulb
2. Blown No. 3 fuse (causing faulty back up light, seat belt alarm, clock, memory function of the car radio)
3. Open circuit in Yellow wire between No. # 3 fuse and combination meter
4. Open circuit in Green/Orange wire between combination meter and control unit

- After the PGM-FI warning light and self-diagnosis indicators have been turned on, turn the ignition switch off. If the LED display fails to come on when the ignition switch is turned on again, check for:

5. Blown No. 10 fuse
6. Open circuit in White/Green wire between ECU A17 terminal and No. 10 fuse

4-10 EMISSION CONTROLS

LED Display	Symptom	Possible Cause
○ ○ ○ ○ (Dash warning light off)	• Engine will not start.	• Loose or poorly connected power line to ECU • Disconnected control unit ground wire • Faulty ECU
○ ○ ○ ○ (Dash warning light on)	• Engine will not start. • No particular symptom shown	• Disconnected control unit ground wire • Short circuit in combination meter or warning light wire • Faulty ECU
○ ○ ○ ● (1)	• No particular symptom shown • Erratic idling • Erratic injector, coupler and wiring • Insufficient fuel	• Disconnected oxygen sensor coupler • Spark plug mis-fire • Short or open circuit in oxygen sensor circuit • Faulty oxygen sensor
○ ○ ● ○ (2)	• No particular symptom shown or system does not operate	• Faulty ECU
○ ○ ● ● (2,1)	• Fuel fouled plug • Frequent engine stalling • Hesitation	• Disconnected manifold absolute pressure sensor coupler • Short or open circuit in manifold air pressure sensor wire • Faulty manifold absolute pressure sensor
○ ● ○ ○ (4)	• No particular symptom shown or system does not operate	• Faulty ECU
○ ● ○ ● (4,1)	• Hesitation • Fuel fouled plug • Frequent engine stalling	• Disconnected manifold air pressure sensor piping
○ ● ● ○ (4,2)	• High idle speed during warm-up • High idle speed • Hard starting at low temperature	• Disconnected coolant temperature sensor coupler • Open circuit in coolant temperature sensor wire • Faulty coolant temperature sensor (thermostat housing)
○ ● ● ● (4,2,1)	• Poor engine response to opening throttle rapidly • High idle speed • Engine does not rev up when cold	• Disconnected throttle angle sensor coupler • Open or short circuit in throttle angle sensor wire • Faulty throttle angle sensor
● ○ ○ ○ (8)	• Engine does not rev up • High idle speed • Erratic idling	• Short or open circuit in crank angle sensor wire • Crank angle sensor wire interfering with high spark plug wires • Faulty crank angle sensor
● ○ ○ ● (8,1)	Same as above	Same as above
● ○ ● ○ (8,2)	• High idle speed • Erratic idling when very cold	• Disconnected intake air temperature sensor • Open circuit in intake air temperature sensor wire • Faulty intake air temperature sensor
● ○ ● ● (8,2,1)	• No particular symptom shown • High idle speed	• Disconnected idle mixture adjuster sensor coupler • Open or short circuit in idle mixture adjuster sensor wire • Faulty idle mixture adjuster sensor
● ● ○ ○ (8,4)	• No particular symptom shown or system does not operate at all.	• Faulty ECU
● ● ○ ● (8,4,1)	• Poor acceleration at high altitude • Hard starting at high altitude when cold	• Disconnected atmospheric pressure sensor coupler • Open or short circuit in atmospheric pressure sensor wire • Faulty atmospheric pressure sensor
● ● ● ○ (8,4,2)	• No particular symptom shown or system does not operate at all	• Faulty ECU
● ● ● ● (8,4,2,1)	Same as above	Same as above

NOTE: Some failure indications (such as when only the No. 1 indication is lit) require the full test procedures on the following pages to confirm that the failure has or has not been eliminated.

Fig. 27 Self-diagnostic fault code displays—1984–87 fuel injected vehicles

EMISSION CONTROLS   4-11

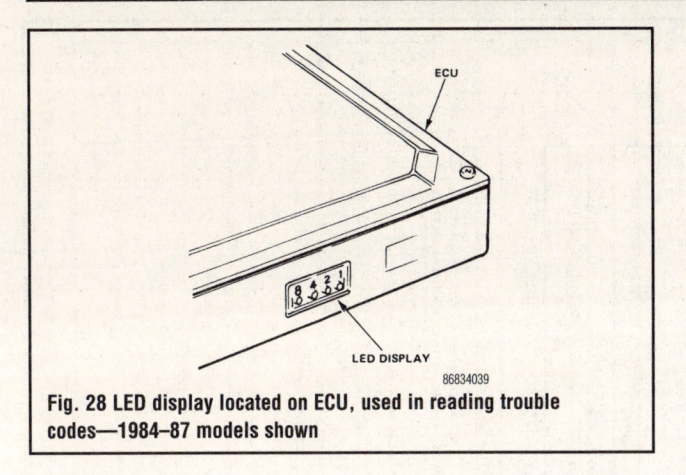

Fig. 28 LED display located on ECU, used in reading trouble codes—1984–87 models shown

• Replace the ECU only after making sure that all couplers and connectors are connected securely.

1988–95 VEHICLES

▶ See Figures 29, 30, 31 and 32

The quick reference troubleshooting charts (located at the front of the diagnosis part of the section) covers the most common failure modes for the PGM-FI. The probable causes are listed in order of most-easily-checked first, then progressing to more difficult fixes. Run through all the causes listed. If problem is still unsolved, go on to the more detailed troubleshooting. Troubleshooting is divided into different LED displays. Find the correct light display and begin again.

For all the conditions listed, the PGM-FI warning light on the dashboard must be on (comes on and stays on). This indicates a problem in the electrical portion of the fuel injection system. At that time, check the LED display (self-diagnosis system) in the ECU/ECM.

There is only one LED display. The LED will blink consecutively to indicate the trouble code. The ECU/ECM is located beneath the access panel, under the carpet on the passengers side of the vehicle.

Sometimes the dash warning light and/or ECU/ECM LED will come on, indicating a system problem, when, in fact, there is only a bad or intermittent electrical connection. To troubleshoot a bad connection, note the ECU/ECM LED pattern that is lit, refer to the diagnosis chart and check the connectors associated with the items mentioned in the "Possible Cause" column for that LED pattern (disconnect, clean or repair if necessary and reconnect those connections). Then, reset the ECU/ECM memory as described, restart the car and drive it for a few minutes and then recheck the car and drive it for a few minutes and then recheck the LED. If the same pattern lights up, begin system troubleshooting; if it does not light up, the problem was only a bad connection.

The memory for the PGM-FI warning light on the dashboard will be erased when the ignition switch is turned **OFF**; however, the memory for the LED display will not be canceled. Thus, the warning light will not come on when the ignition switch is again turned **ON** unless the trouble is once more detected. Troubleshooting should be done according to the LED display even if the warning light is OFF.

If the LED blinks codes that don't exist, clear the ECU, road test the vehicle and check them again. If the same codes repeat, replace the ECU.

CLEARING TROUBLE CODES

The memory for the PGM-FI warning light on the dashboard may be erased when the ignition switch is turned **OFF**; however, the memory for the LED display will not be canceled. Thus, the warning light may not come ON when the ignition switch is again turned **ON** unless the trouble is once more detected. Troubleshooting should be done according to the LED display even if the warning light is OFF.

To clear the ECU/ECM trouble code memory remove the ECU/ECM memory power fuse for at least 10 seconds. The correct fuse to remove is the No. 10 fuse for 1984–87 models, the HAZARD fuse at the main fuse box for the 1988–91 models, and the BACK UP fuse (7.5 A) in the uinder-hood fuse/relay box.

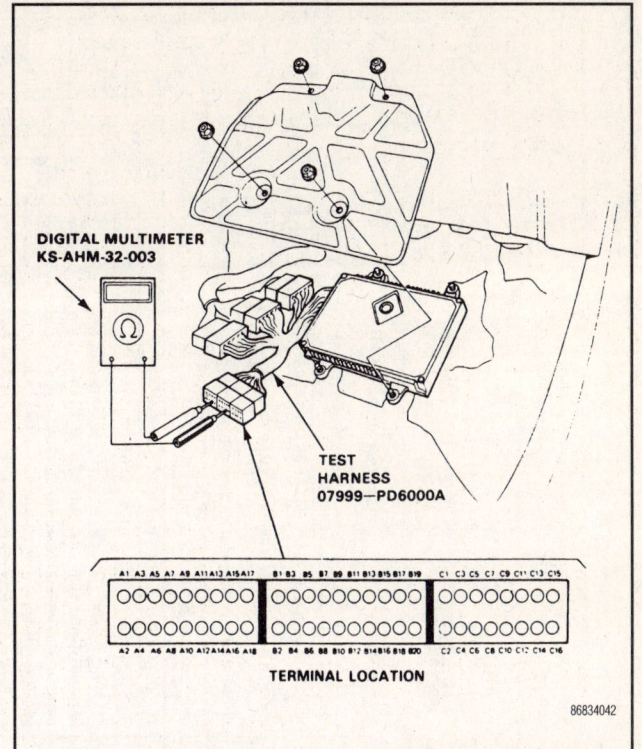

Fig. 29 The ECU is located under the carpet and protective panel—1988–91 fuel injected vehicles

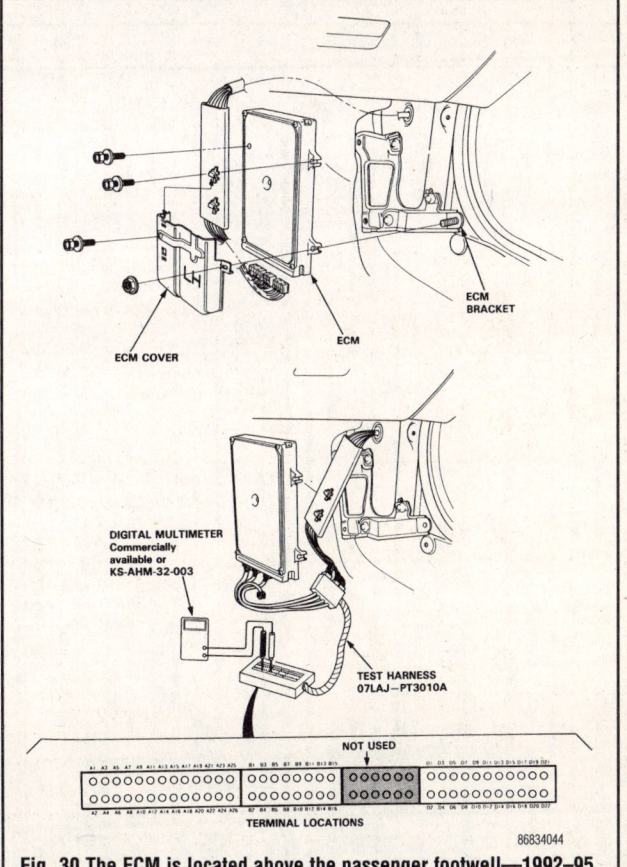

Fig. 30 The ECM is located above the passenger footwell—1992–95 vehicles

4-12 EMISSION CONTROLS

DIAGNOSTIC TROUBLE CODE (DTC)	SYSTEM INDICATED
0	ENGINE CONTROL MODULE (ECM)
1	HEATED OXYGEN SENSOR (HO2S) OR OXYGEN SENSOR (O2S)
3	MANIFOLD ABSOLUTE PRESSURE (MAP SENSOR)
4	CRANKSHAFT POSITION (CKP SENSOR)
6	ENGINE COOLANT TEMPERATURE (ECT SENSOR)
7	THROTTLE POSITION (TP SENSOR)
8	TOP DEAD CENTER POSITION (TDC SENSOR)
9	No. 1 CYLINDER POSITION (CYP SENSOR)
10	INTAKE AIR TEMPERATURE (IAT SENSOR)
12	EXHAUST GAS RECIRCULATION (EGR) SYSTEM
13	BAROMETRIC PRESSURE (BARO SENSOR)
14	IDLE AIR CONTROL (IAC VALVE)
15	IGNITION OUTPUT SIGNAL
16	FUEL INJECTOR
17	VEHICLE SPEED SENSOR (VSS)
19	LOCK-UP CONTROL SOLENOID VALVE A/B (A/T)
20	ELECTRICAL LOAD DETECTOR (ELD)
21	VARIABLE VALVE TIMING & VALVE LIFT ELECTRONIC CONTROL SOLENOID VALVE (VTEC SOLENOID VALVE)
22	VARIABLE VALVE TIMING & VALVE LIFT ELECTRONIC CONTROL PRESSURE SWITCH (VTEC PRESSURE SWITCH)
41	HEATED OXYGEN SENSOR (HO2S) HEATER
43	FUEL SUPPLY SYSTEM
48	HEATED OXYGEN SENSOR (HO2S) [D15Z1 engine (except California model)]

- If codes other than those listed above are indicated, verify the code. If the code indicated is not listed above, replace the ECM.
- The MIL may come on, indicating a system problem when, in fact, there is a poor or intermittent electrical connection. First, check the electrical connections, clean or repair connections if necessary.
- The MIL does not come on when there is a malfunction in the Electrical Load Detector (ELD) circuit. However, it will indicate the code when the Service Check Connector is jumped.

Fig. 32 Self-diagnostic code chart —1992–95 vehicles

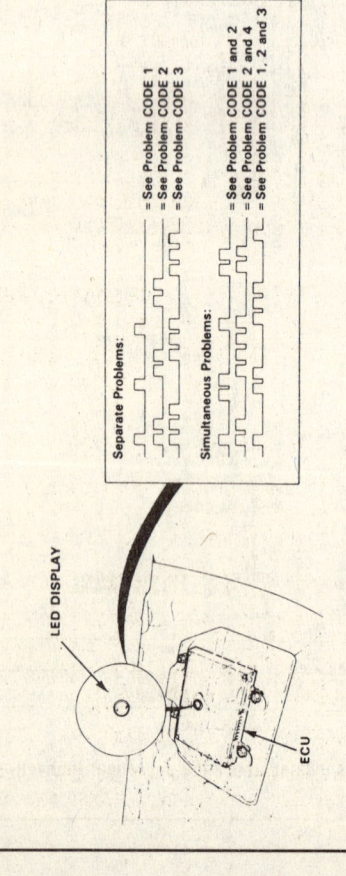

Separate Problems:
= See Problem CODE 1
= See Problem CODE 2
= See Problem CODE 3

Simultaneous Problems:
= See Problem CODE 1 and 2
= See Problem CODE 2 and 4
= See Problem CODE 1, 2 and 3

SELF-DIAGNOSIS INDICATOR BLINKS	SYSTEM INDICATED	PAGE
0	ECU	11-27
1	OXYGEN CONTENT	11-32
3	MANIFOLD ABSOLUTE PRESSURE	11-34
5		11-38
4	CRANK ANGLE	11-40, 44
6	COOLANT TEMPERATURE	11-50
7	THROTTLE ANGLE	11-52
8	TDC POSITION	11-42, 46
9	No. 1 CYLINDER POSITION (1.6ℓ)	11-48
10	INTAKE AIR TEMPERATURE	11-54
12	EXHAUST GAS RECIRCULATION SYSTEM (1.5ℓ CAL: A/T)	11-132
13	ATMOSPHERIC PRESSURE	11-56
14	ELECTRONIC AIR CONTROL	11-71
15	IGNITION OUTPUT SIGNAL	11-58
16	FUEL INJECTOR	11-90, 95
17	VEHICLE SPEED SENSOR	11-60
19	LOCK-UP CONTROL SOLENOID VALVE (A/T)	11-62
20	ELECTRIC LOAD	11-64

If codes other than those listed above are indicated, count the number of blinks again. If the indicator is in fact blinking these codes, substitute a known-good ECU and recheck. If the indication goes away, replace the original ECU.
The Check Engine light and ECU LED may come on, indicating a system problem, when, in fact, there is a poor or intermittent electrical connection. First, check the electrical connections, clean or repair connections if necessary.
If the Check Engine light is on and LED stays on, replace the ECU.

Fig. 31 Self-diagnostic code display sample and fault code chart —1988–91 vehicles

EMISSION CONTROLS 4-13

➡ Some vehicle computers have a learning ability. If a change is noted in vehicle performance after clearing codes, it may be due to the computer's learning ability. To restore performance, warm vehicle to normal operating temperature and drive at part throttle with moderate acceleration, before performing any additional diagnosis.

Programmed Fuel Injection (PGM-FI) System

GENERAL DESCRIPTION

Programmed Fuel Injection (PGM-FI) System consists of three sub-systems: air intake, electronic control, and fuel.

Air Intake System

The system supplies air for all engine needs. It consists of the air cleaner, air intake pipe, throttle body, idle control system, fast idle mechanism, and intake manifold. A resonator in the air intake pipe provides additional silencing as air is drawn into the system.

THROTTLE BODY

The throttle body, either a side-draft type or down-draft type, depending on the engine, with the primary air horn at the top. To prevent icing of the throttle valves and air horn walls, under certain atmospheric conditions, the lower portion of the throttle body is heated by engine coolant. A throttle sensor is attached to the primary throttle valve to sense changes in throttle opening. A dash pot is used to slow the throttle as it approaches the closed position.

IDLE CONTROL SYSTEM

The air/fuel ratio during idling is controlled by the electronic control unit and various solenoid valves such as Electronic Air Control Valve (EACV), fast idle valve and air boost valves. With the exception of the A/C idle control solenoid valve, these change the amounts of air bypassing into the air intake manifold. The A/C control solenoid valve opens the throttle when the air conditioner is turned on by signals sent from the ECU.

BYPASS CONTROL SOLENOID VALVE (BPCSV)

When engine rpm is below 5000 rpm, the BPCSV directs air flow through the long intake path for higher torque. At higher speeds, intake air flow is through the shorter path to reduce resistance to air flow.

IDLE ADJUSTER (BYPASS CIRCUIT)

Fuel cut-off takes place at a set position or angle of the throttle valve. If the throttle valve is moved to adjust idle speed, this position or angle will be changed and the system may not cut off fuel supply. To solve this problem, the throttle body contains an adjustable bypass circuit. This circuit is designed to control the amount of air bypassing into the intake manifold without changing the position of throttle valve. The idle speed usually does not require adjustment. When the idle control system is in operation, the idle adjustment screw has no effect on the idle speed.

FAST IDLE MECHANISM

To prevent erratic running when the engine is warming up, it is necessary to raise the idle speed. The air bypass valve is controlled by a thermowax plunger. When the thermowax is cold, the valve is open. When the thermowax is heated, the valve is closed. With the engine cold and the thermowax consequently cold, additional air bypasses into the intake manifold so that the engine idles faster than normal. When the engine reaches operating temperature, the valve begins to close, reducing the amount of air bypassing into the manifold.

Electronic Control System

CONTROL SYSTEM

In order to get fuel into the cylinders at the correct instant and in correct amount, the control system must perform various separate functions. The Electronic Control Unit (ECU, or on 1992-later cars, Engine Control Module, ECM), the heart of the PGM-FI, uses an eight-bit microcomputer and consists of a CPU (Central Processing Unit), memories, and I/O (Input/Output) ports. Basic data stored in the memories are compensated by the signals sent from the various sensors to provide the correct air/fuel mixture for all engine needs.

ELECTRONIC CONTROL UNIT (ECU)

The unit contains memories for the basic discharge duration at various engine speeds and manifold pressures. The basic discharge duration, after being read out from the memory, is further modified by signals sent from various sensors to obtain the final discharge duration. Other functions also include:

- Starting Control — The fuel system must vary the air/fuel ratio to suit different operating requirements. For example, the mixture must be rich for starting. The memories also contain the basic discharge durations to be read out by signals from the starter switch, and engine speed and coolant temperature sensors, thereby providing extra fuel needed for starting.
- Injector Control — The ECU/ECM controls the discharge durations at various engine speeds and loads.
- Electronic Air Control — The ECU/ECM controls the EACV to maintain correct idle speed based on engine and accessories demand.
- Ignition Timing Control — The ECU/ECM controls the basic ignition timing based on engine load, engine rpm, vehicle speed and coolant temperature.
- Fuel Pump Control — When the speed of the engine falls below the prescribed limit, electric current to the fuel pump is cut off, preventing the injectors from discharging fuel.
- Fuel Cut-Off Control — During deceleration with the throttle valve nearly closed, electric current to the injectors is cut off at speeds over 900 rpm, contributing to improved fuel economy. Fuel cut-off action also takes place when engine speed exceeds approximately 7500 rpm regardless of the position of the throttle valve.
- Safety — A fail-safe system monitors the sensors and detects any abnormality in the ECU/ECM, ensuring safe driving even if one or more sensors are faulty, or if the ECU/ECM malfunctions.
- Self-Diagnosis — When a abnormality occurs, the ECU/ECM lights the engine warning light and stores the failure code in erasable memory. The ECU/ECM LED will display the code any time the ignition is turned **ON**.

CRANK ANGLE SENSOR (TDC/CYL SENSORS)

The sensors and distributor are designed as an assembly to save space and weight. The entire unit consist of a pair of rotors, TDC and CYL, and a pickup for each rotor. Since the rotors are coupled to the camshaft, they turn together as a unit as the camshaft rotates. The CYL sensor detects the position of the No. 1 cylinder as the base for the Sequential Injection whereas the TDC sensor serves to determine the injection timing for each cylinder. The TDC sensor is also used to detect engine speed to read out the basic discharge duration for different operating conditions.

Most 1990 vehicles, incorporate the TDC sensor, CYL sensor and Crank sensor in a single assembly. The function of these components does not change, but if any single sensor is defective, the entire distributor must be replaced as an assembly.

MANIFOLD ABSOLUTE PRESSURE SENSOR (MAP SENSOR)

The sensor converts manifold air pressure readings into electrical voltage signals and sends them to the ECU/ECM. This information with signals from the crank angle sensor is then used to read out the basic discharge duration from the memory.

ATMOSPHERIC PRESSURE SENSOR (PA SENSOR)

Like the MAP sensor, the unit converts atmospheric pressures into voltage signals and sends them to the ECU. The signals then modify the basic discharge duration to compensate for changes in the atmospheric pressure.

COOLANT TEMPERATURE SENSOR (TW SENSOR)

The sensor uses a temperature-dependent diode (thermistor) to measure differences in the coolant temperature. The basic discharge duration is read out by the signals sent from this sensor through the ECU. The resistance of the thermister decreases with a rise in coolant temperature.

INTAKE AIR TEMPERATURE SENSOR (TA SENSOR)

This device is also a thermistor and is placed in the intake manifold. It acts much like the water temperature sensor but with a reduced thermal capacity for

4-14 EMISSION CONTROLS

quicker response. The basic discharge duration read out from the memory is again compensated for different operating conditions by the signals sent from this sensor through the ECU/ECM.

THROTTLE ANGLE SENSOR

This sensor is essentially a variable resistor. In construction, the rotor shaft is connected to the throttle valve shaft such that, as the throttle valve is moved, the resistance varies, altering the output voltage to the control unit.

OXYGEN SENSOR

The oxygen sensor, by detecting the oxygen content in the exhaust gas, maintains the stoichiometric air/fuel ratio. In operation, the ECU/ECM receives the signals from the sensor and changes the duration during which fuel is injected. The oxygen sensor is located in the exhaust manifold.

The sensor is a hollow shaft of zirconia with a closed end. The inner and outer surfaces are plated with platinum, thus forming a platinum electrode. The inner surface or chamber is open to the atmosphere whereas the outer surface is exposed to the exhaust gas flow through the manifold.

Voltage is induced at the platinum electrode when there is any difference in oxygen concentration between the two layers of air over the surfaces. Operation of the device is dependent upon the fact that voltage induced changes sharply as the stoichiometric air/fuel ratio is exceeded when the electrode is heated above a certain temperature.

Some 1990 models, use a heated oxygen sensor. The heater stabilizes the sensor's outputs and allow the sensor to heat quicker after the engine as been started.

STARTER SWITCH

The air/fuel mixture must be rich for starting. During cranking, the ECU/ECM detects signal from the starter switch and increases the amount of fuel injected into the manifold according to the engine temperature. The amount of fuel injected is gradually reduced when the starter switch is turned **OFF**.

Fuel System

FUEL PUMP

The fuel pump is a compact impeller design and is installed inside the fuel tank, thereby saving space and simplifying the fuel line system.

The fuel pump is comprised of a DC motor, a circumference flow pump, a relief valve for protecting the fuel line systems, a check valve for retaining residual pressure, an inlet port, and a discharge port. The pump assembly consists of the impeller (driven by the motor), the pump casing (which forms the pumping chamber), and cover of the pump.

PRESSURE REGULATOR

The fuel pressure regulator maintains a constant fuel pressure to the injectors. The spring chamber of the pressure regulator is connected to the intake manifold to constantly maintain the fuel pressure at 36 psi (248 kpa) higher than the pressure in the manifold. When the difference between the fuel pressure and manifold pressure exceeds 36 psi (248 kpa), the diaphragm is pushed upward, and the excess fuel is fed back into the fuel tank through the return line.

FUEL INJECTOR

The injector is of the solenoid-actuated constant-stroke pintle type consisting of a solenoid, plunger, needle valve and housing. When current is applied to the solenoid coil, the valve lifts up and pressurized fuel fills the inside of the injector and is injected close to the intake valve. Because the needle valve lifts and the fuel pressure are constant, the injection quantity is determined by the length of time that the valve is open, i.e., the duration the current is supplied to the solenoid coil. The injector is sealed by an O-ring and seal ring at the top and bottom. These seals also reduce operating noise.

RESISTOR

The injector timing, which controls the opening and closing intervals, must be very accurate since it dictates the air/fuel mixture ratio. The injector must also be durable. For the best possible injector response, it is necessary to shorten the current rise time when voltage is applied to the injector coil. Therefore, the number of windings of the injector coil is reduced to reduce the inductance in the coil. This, however, makes low resistance in the coil, allowing a large amount of current to flow through the coil. As a result, the amount of heat generated is high, which compromises the durability of the coil. Flow of current in the coil is therefore restricted by a resistor installed in series between the electric power source and the injector coil.

MAIN RELAY

The main relay is a direct coupler type which contains the relays for the electronic control unit power supply and the fuel pump power supply. This relay is installed at the back of the fuse box.

SERVICE PRECAUTIONS

- Do not operate the fuel pump when the fuel lines are empty.
- Do not operate the fuel pump when removed from the fuel tank.
- Do not reuse fuel hose clamps.
- Make sure all ECU harness connectors are fastened securely. A poor connection can cause an extremely high surge voltage in the coil and condenser and result in damage to integrated circuits.
- Keep ECU all parts and harnesses dry during service.
- Before attempting to remove any parts, turn **OFF** the ignition switch and disconnect the battery ground cable.
- Always use a 12 volt battery as a power source.
- Do not attempt to disconnect the battery cables with the engine running.
- Do not disconnect and wiring connector with the engine running, unless instructed to do so.
- Do not depress the accelerator pedal when starting.
- Do not rev up the engine immediately after starting or just prior to shutdown.
- Do not apply battery power directly to injectors.

EMISSION CONTROLS 4-15

DIAGNOSTIC CHARTS

4-16 EMISSION CONTROLS

TROUBLESHOOTING GUIDE — 1988–91

NOTE: Across each row in the chart, the systems that could be sources of a symptom are ranked in the order they should be inspected starting with ①. Find the symptom in the left column, read across to the most likely source, then refer to the page listed at the top of that column. If inspection shows the system is OK, try the next most likely system ②, etc.

SYMPTOM	ECU	OXYGEN SENSOR	MANIFOLD ABSOLUTE PRESSURE SENSOR	TDC/CRANK SENSOR	COOLANT TEMPERATURE SENSOR	THROTTLE ANGLE SENSOR	INTAKE AIR TEMPERATURE SENSOR	ATMOSPHERIC PRESSURE SENSOR
CHECK ENGINE WARNING LIGHT TURNS ON	□ or ☼	⚠	⚠	⚠	⚠	⚠	⚠	⚠
SELF-DIAGNOSIS INDICATOR (LED) BLINKS	◎ or ☼	①	④ or ⑤	④ or ⑤	⑥	⑦	⑩	⑪
ENGINE WON'T START	③							
DIFFICULT TO START ENGINE WHEN COLD	(BU)				①			
IRREGULAR IDLING — WHEN COLD FAST IDLE OUT OF SPEC	(BU)				③			
IRREGULAR IDLING — ROUGH IDLE	(BU)		③					
IRREGULAR IDLING — WHEN WARM RPM TOO HIGH	(BU)							
IRREGULAR IDLING — WHEN WARM RPM TOO LOW	(BU)		③					
FREQUENT STALLING — WHILE WARMING UP	(BU)							
FREQUENT STALLING — AFTER WARMING UP	(BU)							
POOR PERFORMANCE — MISFIRE OR ROUGH RUNNING	(BU)	③	②					
POOR PERFORMANCE — FAILS EMISSION TEST	(BU)					②		
POOR PERFORMANCE — LOSS OF POWER	(BU)							

SYMPTOM	PGM-FI IGNITION OUTPUT SIGNAL	VEHICLE SPEED SENSOR	LOCK-UP CONTROL SOLENOID VALVE	ELECTRIC LOAD DETECTOR	IDLE CONTROL ELECTRONIC AIR CONTROL VALVE	OTHER IDLE CONTROLS	FUEL SUPPLY FUEL INJECTOR	OTHER FUEL SUPPLY	AIR INTAKE	EMISSION CONTROL EGR CONTROL SYSTEM (CAL. A/T only)	OTHER EMISSION CONTROLS
CHECK ENGINE WARNING LIGHT TURNS ON	⚠	⚠	⚠	□	⚠		⚠			⚠	
SELF-DIAGNOSIS INDICATOR (LED) BLINKS	⑫	⑧	⑲	⑨	⑭		⑯			⑫	
ENGINE WON'T START					②		②	①			
DIFFICULT TO START ENGINE WHEN COLD					①	②					
IRREGULAR IDLING — ROUGH IDLE					②	①	③				③
IRREGULAR IDLING — WHEN WARM RPM TOO HIGH				③	②						
IRREGULAR IDLING — WHEN WARM RPM TOO LOW				②	①		③				
FREQUENT STALLING — WHILE WARMING UP	②										
FREQUENT STALLING — AFTER WARMING UP					①		②	③			③
POOR PERFORMANCE — MISFIRE OR ROUGH RUNNING					③		①	②		③	
POOR PERFORMANCE — FAILS EMISSION TEST								①			
POOR PERFORMANCE — LOSS OF POWER							③	①			

If codes other than those listed above are indicated, count the number of blinks again. If the indicator is in fact blinking these codes, substitute a known-good ECU and recheck. If the indication goes away, replace the original ECU.

(BU): When the Check Engine warning light and the self-diagnosis indicator are on, the back-up system is in operation. Substitute a known-good ECU and recheck. If the indication goes away, replace the original ECU.

EMISSION CONTROLS 4-17

TROUBLESHOOTING GUIDE — 1988–91

NOTE: Across each row in the chart, the systems that could be sources of a symptom are ranked in the order they should be inspected starting with 1. Find the symptom in the left column, read across to the most likely source, then refer to the page listed at the top of that column. If inspection shows the system is OK, try the next most likely system 2, etc.

SYMPTOM	ECU	OXYGEN SENSOR	MANIFOLD ABSOLUTE PRESSURE SENSOR	TDC CRANK CYL SENSOR	COOLANT TEMPERATURE SENSOR	THROTTLE ANGLE SENSOR	INTAKE AIR TEMPERATURE SENSOR	ATMOSPHERIC PRESSURE SENSOR	IGNITION OUTPUT SIGNAL	VEHICLE SPEED SENSOR	ELECTRIC LOAD DETECTOR	ELECTRONIC AIR CONTROL VALVE	OTHER IDLE CONTROLS	FUEL INJECTOR	OTHER FUEL SUPPLY	AIR INTAKE	EMISSION CONTROL
CHECK ENGINE WARNING LIGHT TURNS ON	☐ or ☀																
SELF-DIAGNOSIS INDICATOR (LED) BLINKS	☐ or ✱	①	③ or ⑤	④ or ⑧ or ⑨	⑥	⑦	⑩	⑬	⑮	⑰	⑳	⑭		⑯			
ENGINE WON'T START	③																
DIFFICULT TO START ENGINE WHEN COLD	BU		③		①									②			
WHEN COLD FAST IDLE OUT OF SPEC	BU				②												
ROUGH IDLE	BU		③									①	②	②			
WHEN WARM RPM TOO HIGH	BU											①	②				
WHEN WARM RPM TOO LOW	BU										③	②	①	②			
WHILE WARMING UP	BU										③	①		②	③		
AFTER WARMING UP	BU											①	②	②	③		
MISFIRE OR ROUGH RUNNING	BU	③	②									②	②	①			
FAILS EMISSION TEST	BU	③	②											①			①
LOSS OF POWER	BU		③			②											

If codes other than those listed above are indicated, count the number of blinks. If the indicator is in fact blinking these codes, substitute a known-good ECU and recheck. If the indication goes away, replace the original ECU.

(BU): When the Check Engine warning light and the self-diagnosis indicator are on, the back-up system is in operation. Substitute a known-good ECU and recheck. If the indication goes away, replace the original ECU.

4-18 EMISSION CONTROLS

Fig. 34 Troubleshooting Guide—1992–95 Civics, except del Sol

Fig. 33 Troubleshooting Guide—1992–95 Civics (all models)

PGM-FI: LOCK-UP CONTROL SOLENOID VALVE (A/T)	PGM-FI: ELECTRICAL LOAD DETECTOR (Except Canada)	PGM-FI: KNOCK SENSOR (B16A3 engine)	PGM-FI: *** VARIABLE VALVE TIMING & VALVE LIFT ELECTRONIC CONTROL SOLENOID VALVE	PGM-FI: *** VARIABLE VALVE TIMING & VALVE LIFT ELECTRONIC CONTROL PRESSURE SWITCH	IDLE CONTROL: IDLE AIR CONTROL VALVE	IDLE CONTROL: OTHER IDLE CONTROLS	FUEL SUPPLY: FUEL INJECTOR	FUEL SUPPLY: OTHER FUEL SUPPLY	INTAKE AIR	EMISSION CONTROLS
CHECK		CHECK	CHECK	CHECK	CHECK		CHECK			
19	20	23	21	22	14		16			
								②		
								②		
					①	②				
					①	②	②			
					①	②				
	③				①	②	②			
					①	②		③		
					②			①		
							①			
										①
			③	③			③	①	③	

86834112

Fig. 35 Troubleshooting Guide—1992–95 del Sol

NOTE:

- Across each row in the chart, the sub-systems that could be sources of a symptom are ranked in the order they should be inspected, starting with ①. Find the symptom in the left column, read across to the most likely source, then refer to the page listed at the top of that column. If inspection shows the system is OK, try the next system ②, etc.
- If the idle speed is out of specification and the Malfunction Indicator Lamp (MIL) does not blink Diagnostic Trouble Code (DTC) 14, go to inspection described on the following page.

SYMPTOM		IDLE ADJUSTING SCREW	IDLE AIR CONTROL VALVE	AIR CONDITIONING SIGNAL	ALTERNATOR FR SIGNAL	AUTOMATIC TRANSAXLE GEAR POSITION SWITCH SIGNAL	M/T CLUTCH SWITCH SIGNAL	STARTER SWITCH SIGNAL	BRAKE SWITCH SIGNAL	POWER STEERING PRESSURE SWITCH SIGNAL	FAST IDLE THERMO VALVE	HOSES AND CONNECTIONS
DIFFICULT TO START ENGINE WHEN COLD											①	
WHEN COLD FAST IDLE OUT OF SPEC (1,000–2,000 rpm)		③	②								①	
ROUGH IDLE			②									①
WHEN WARM RPM TOO HIGH		③	①							③	②	③
WHEN WARM RPM TOO LOW	Idle speed is below specified rpm (no load)	②	①									
	Idle speed does not increase after initial start up.		①									
	On models with automatic transmission, the idle speed drops in gear		②			①						
	Idle speeds drops when air conditioner in ON		②	①								
	Idle speed drops when steering wheel is turning		②							①		
	Idle speed fluctuates with electrical load		②		③							①
FREQUENT STALLING	WHILE WARMING UP	②	①									
	AFTER WARMING UP	①	②									
FAILS EMISSION TEST												①

86834157

Fig. 36 Idle Control System—1992–95 System Troubleshooting Guide

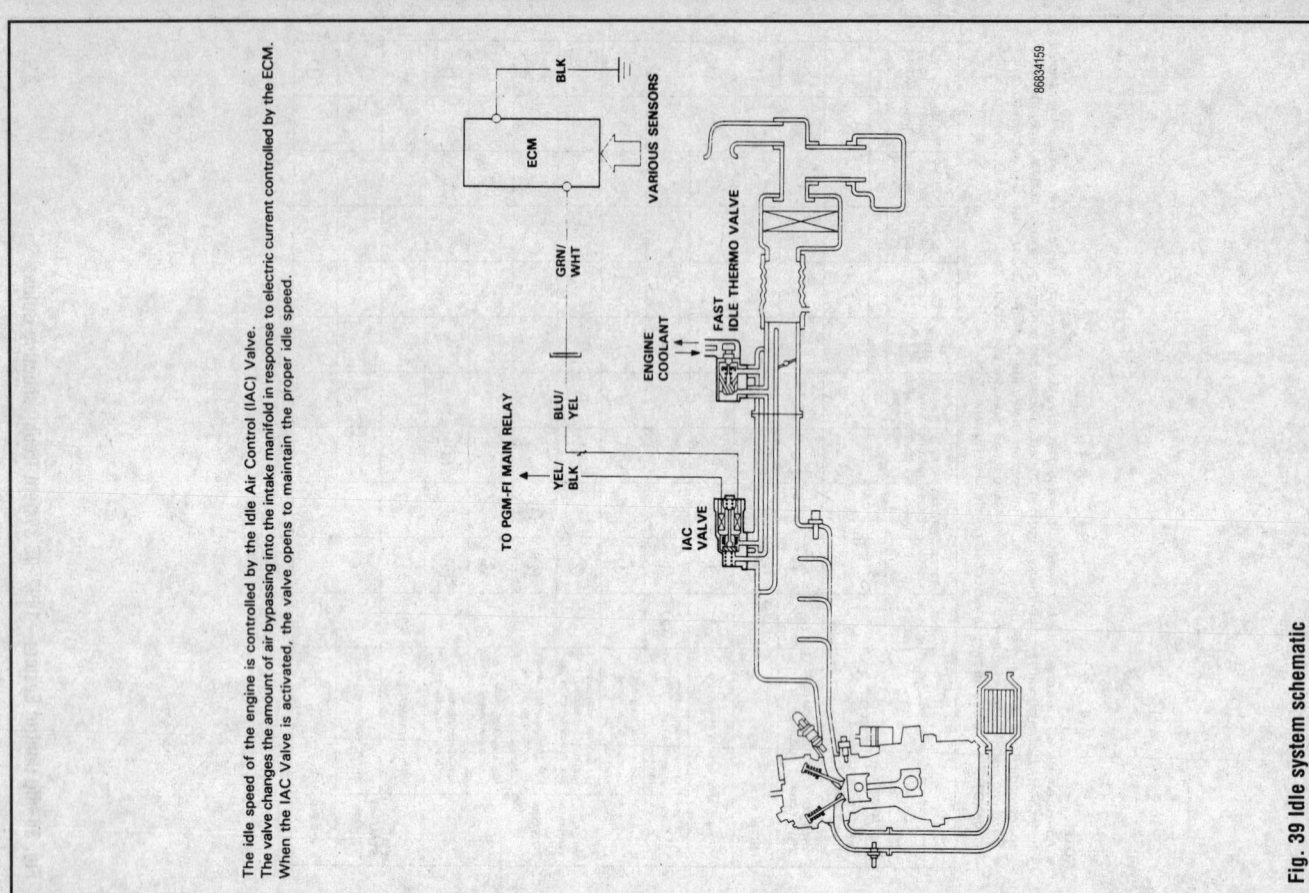

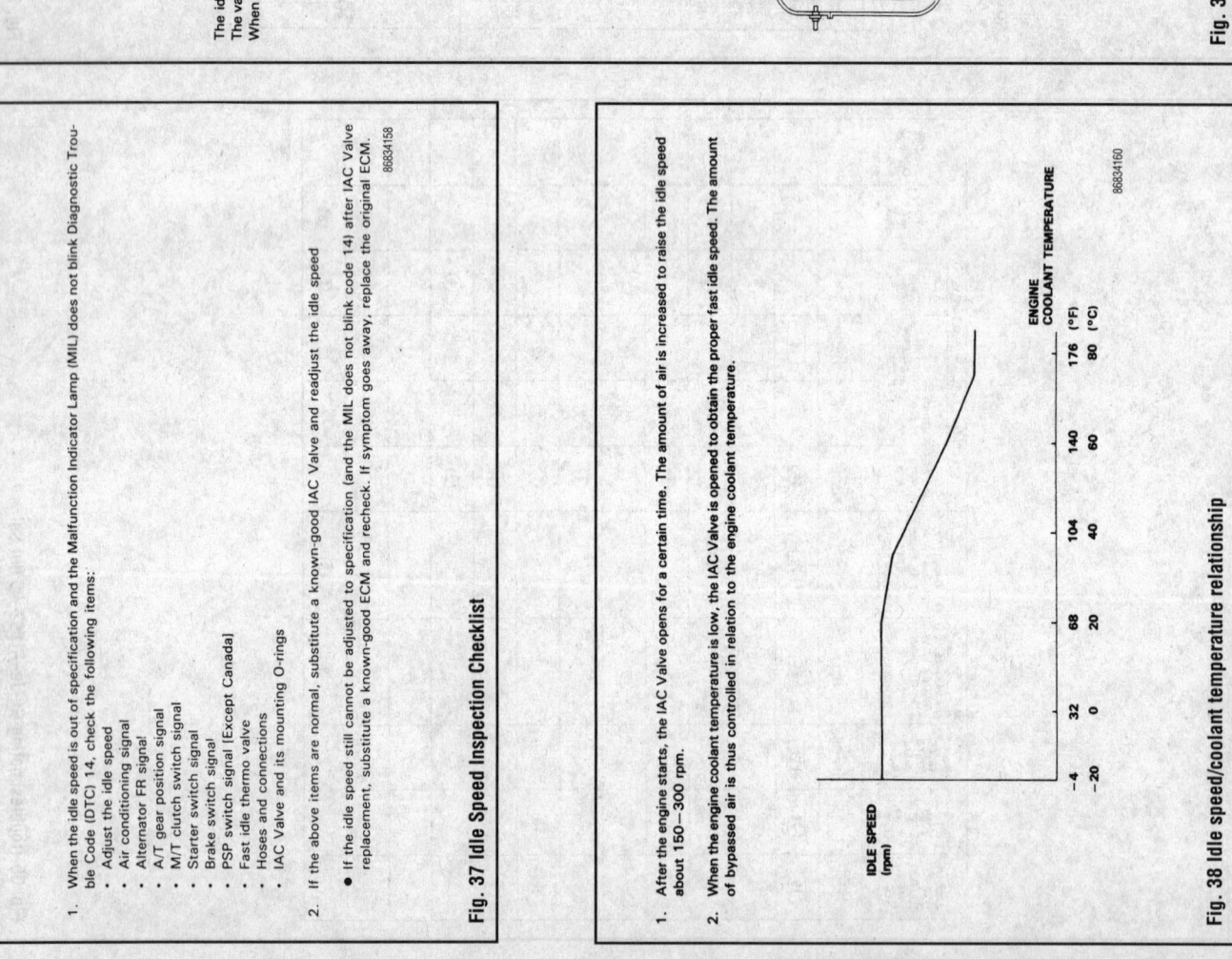

4-20 EMISSION CONTROLS

The idle speed of the engine is controlled by the Idle Air Control (IAC) Valve. The valve changes the amount of air bypassing into the intake manifold in response to electric current controlled by the ECM. When the IAC Valve is activated, the valve opens to maintain the proper idle speed.

Fig. 39 Idle system schematic

1. When the idle speed is out of specification and the Malfunction Indicator Lamp (MIL) does not blink Diagnostic Trouble Code (DTC) 14, check the following items:
   - Adjust the idle speed
   - Air conditioning signal
   - Alternator FR signal
   - A/T gear position signal
   - M/T clutch switch signal
   - Starter switch signal
   - Brake switch signal
   - PSP switch signal (Except Canada)
   - Fast idle thermo valve
   - Hoses and connections
   - IAC Valve and its mounting O-rings

2. If the above items are normal, substitute a known-good IAC Valve and readjust the idle speed

   • If the idle speed still cannot be adjusted to specification (and the MIL does not blink code 14) after IAC Valve replacement, substitute a known-good ECM and recheck. If symptom goes away, replace the original ECM.

Fig. 37 Idle Speed Inspection Checklist

1. After the engine starts, the IAC Valve opens for a certain time. The amount of air is increased to raise the idle speed about 150—300 rpm.
2. When the engine coolant temperature is low, the IAC Valve is opened to obtain the proper fast idle speed. The amount of bypassed air is thus controlled in relation to the engine coolant temperature.

Fig. 38 Idle speed/coolant temperature relationship

EMISSION CONTROLS 4-21

VACUUM DIAGRAMS

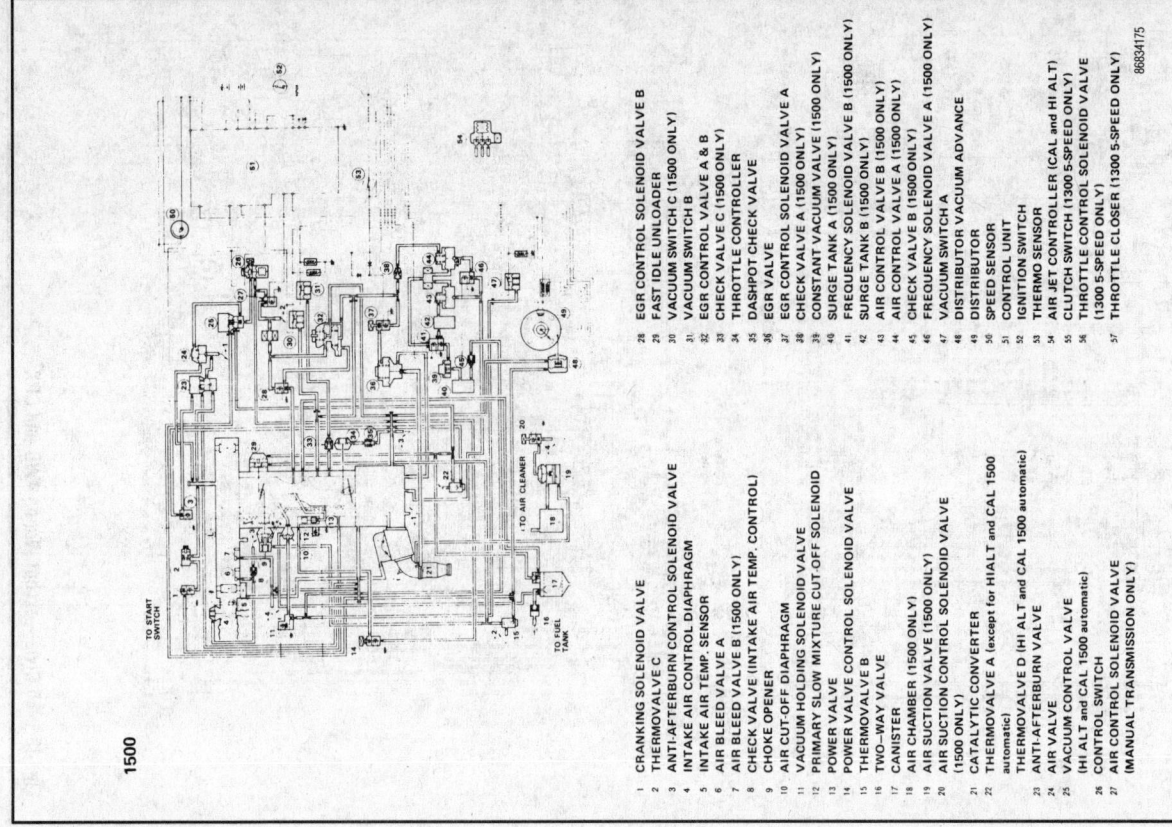

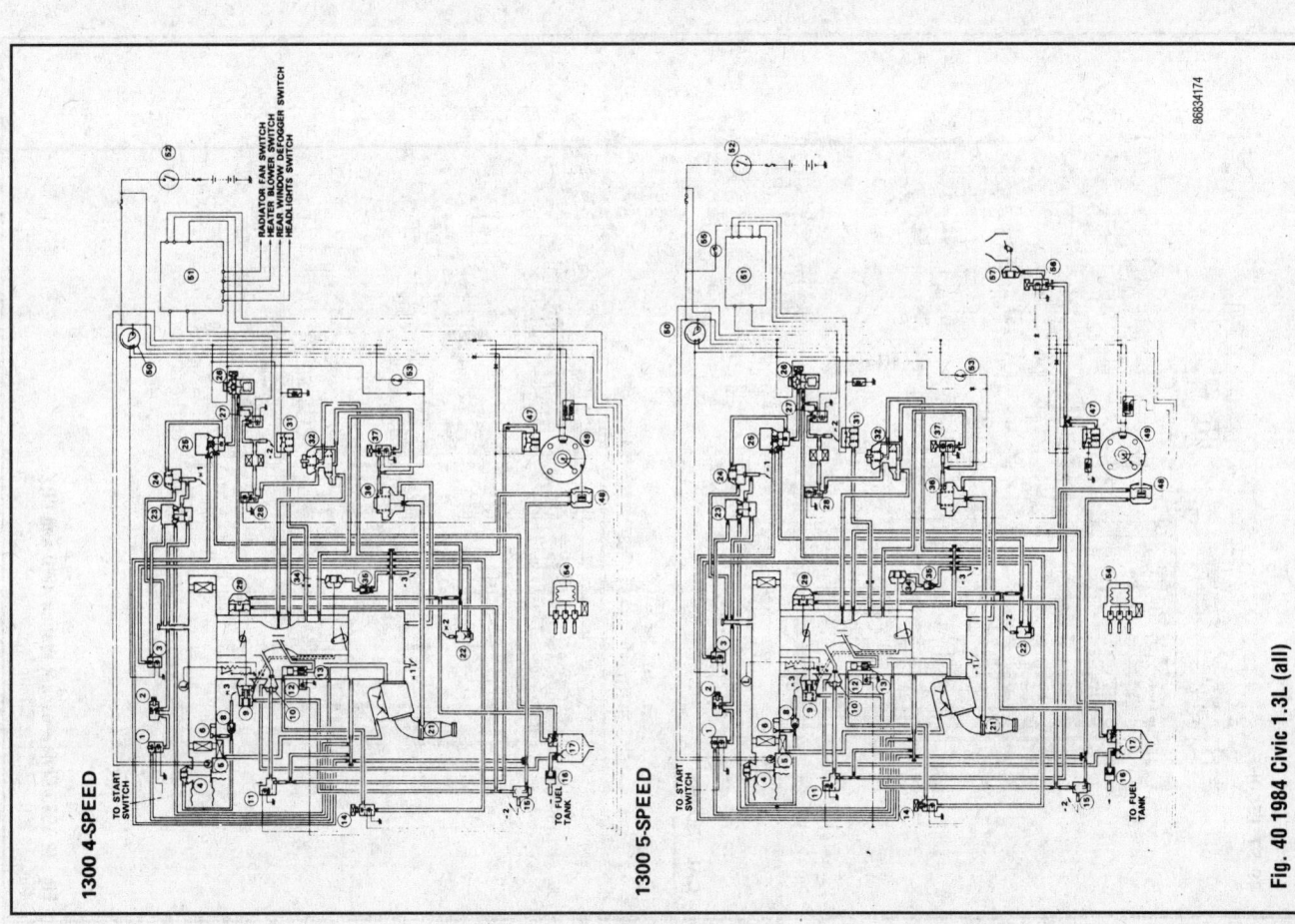

- 1 CRANKING SOLENOID VALVE
- 2 THERMOVALVE C
- 3 ANTI-AFTERBURN CONTROL SOLENOID VALVE
- 4 INTAKE AIR CONTROL DIAPHRAGM
- 5 INTAKE AIR TEMP. SENSOR
- 6 AIR BLEED VALVE A
- 7 AIR BLEED VALVE B (1500 ONLY)
- 8 CHECK VALVE (INTAKE AIR TEMP. CONTROL)
- 9 CHOKE OPENER
- 10 AIR CUT-OFF DIAPHRAGM
- 11 VACUUM HOLDING SOLENOID VALVE
- 12 PRIMARY SLOW MIXTURE CUT-OFF SOLENOID
- 13 POWER VALVE
- 14 POWER VALVE CONTROL SOLENOID VALVE
- 15 THERMOVALVE B
- 16 TWO-WAY VALVE
- 17 CANISTER
- 18 AIR CHAMBER (1500 ONLY)
- 19 AIR SUCTION VALVE (1500 ONLY)
- 20 AIR SUCTION CONTROL SOLENOID VALVE (1500 ONLY)
- 21 CATALYTIC CONVERTER
- 22 THERMOVALVE A (except for HIALT and CAL 1500 automatic)
- 23 ANTI-AFTERBURN VALVE
- 24 AIR VALVE
- 25 VACUUM CONTROL VALVE (HI ALT and CAL 1500 automatic)
- 26 AIR CONTROL SOLENOID VALVE (MANUAL TRANSMISSION ONLY)
- 27
- 28 EGR CONTROL SOLENOID VALVE B
- 29 FAST IDLE UNLOADER
- 30 VACUUM SWITCH C (1500 ONLY)
- 31 VACUUM SWITCH B
- 32 EGR CONTROL VALVE A & B
- 33 CHECK VALVE C (1500 ONLY)
- 34 DASHPOT CHECK VALVE
- 35 EGR VALVE
- 37 EGR CONTROL SOLENOID VALVE A
- 38 CHECK VALVE A (1500 ONLY)
- 39 CONSTANT VACUUM VALVE (1500 ONLY)
- 40 SURGE TANK A (1500 ONLY)
- 41 FREQUENCY SOLENOID VALVE B (1500 ONLY)
- 42 SURGE TANK B (1500 ONLY)
- 43 AIR CONTROL VALVE B (1500 ONLY)
- 44 AIR CONTROL VALVE A (1500 ONLY)
- 45 CHECK VALVE B (1500 ONLY)
- 46 FREQUENCY SOLENOID VALVE A (1500 ONLY)
- 47 VACUUM SWITCH A
- 48 DISTRIBUTOR VACUUM ADVANCE
- 49 SPEED SENSOR
- 50 CONTROL UNIT
- 51 IGNITION SWITCH
- 53 THERMO SENSOR
- 54 AIR JET CONTROLLER (CAL and HI ALT)
- 55 CLUTCH SWITCH (1300 5-SPEED ONLY)
- 56 THROTTLE CONTROL SOLENOID VALVE (1300 5-SPEED ONLY)
- 57 THROTTLE CLOSER (1300 5-SPEED ONLY)

Fig. 41 1984 Civic/CRX (all)

Fig. 40 1984 Civic 1.3L (all)

4-22 EMISSION CONTROLS

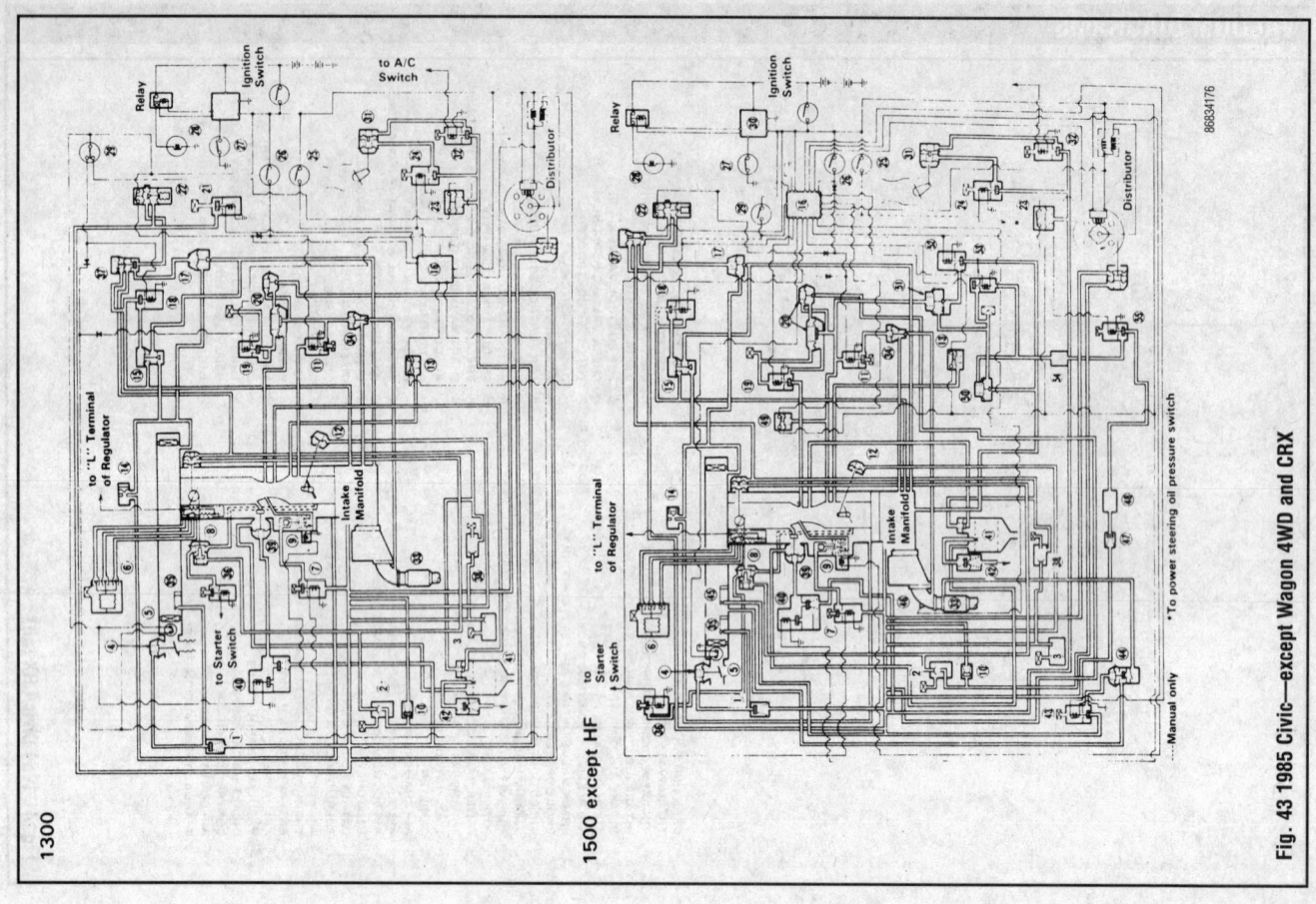

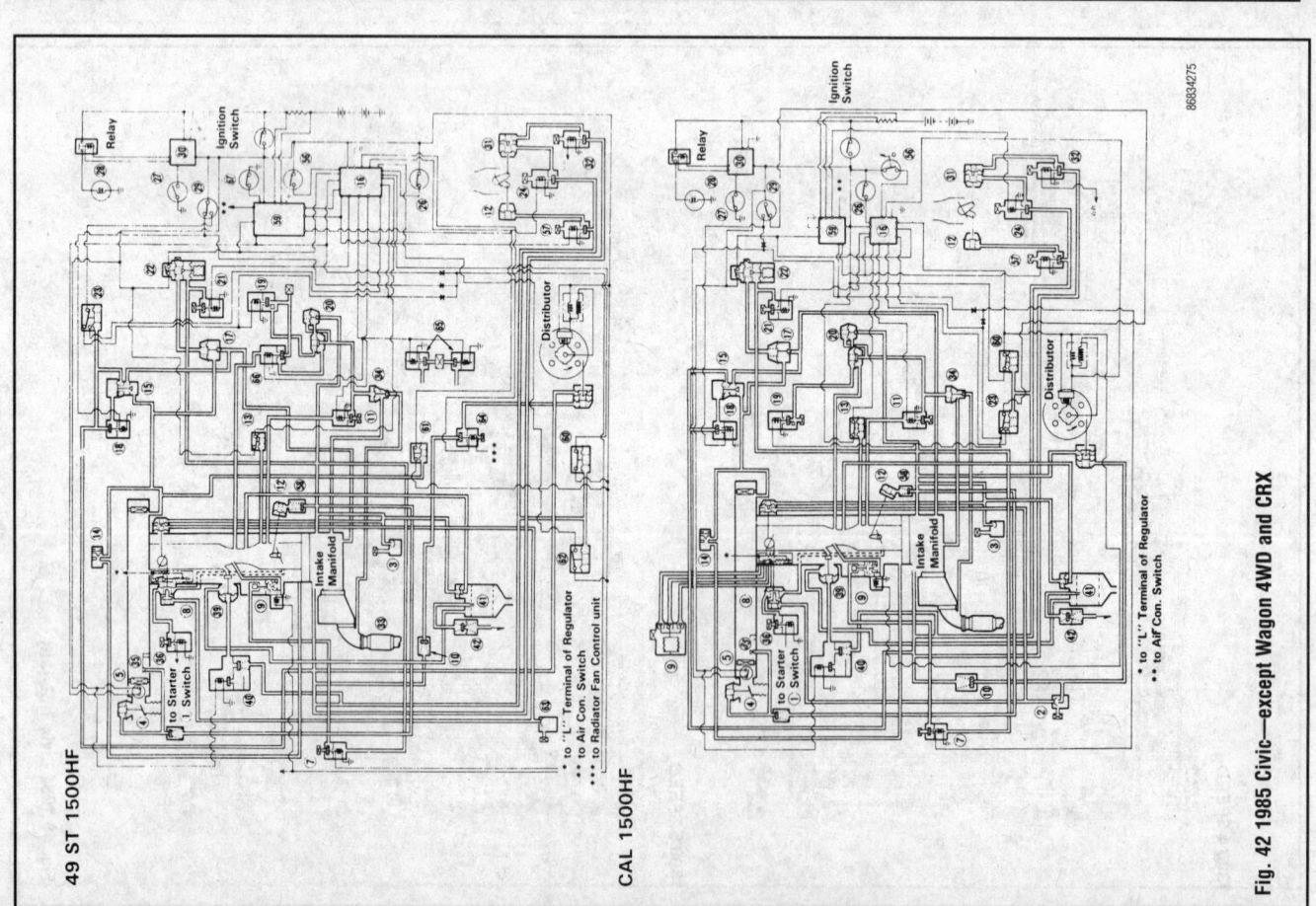

Fig. 43 1985 Civic—except Wagon 4WD and CRX

Fig. 42 1985 Civic—except Wagon 4WD and CRX

EMISSION CONTROLS 4-23

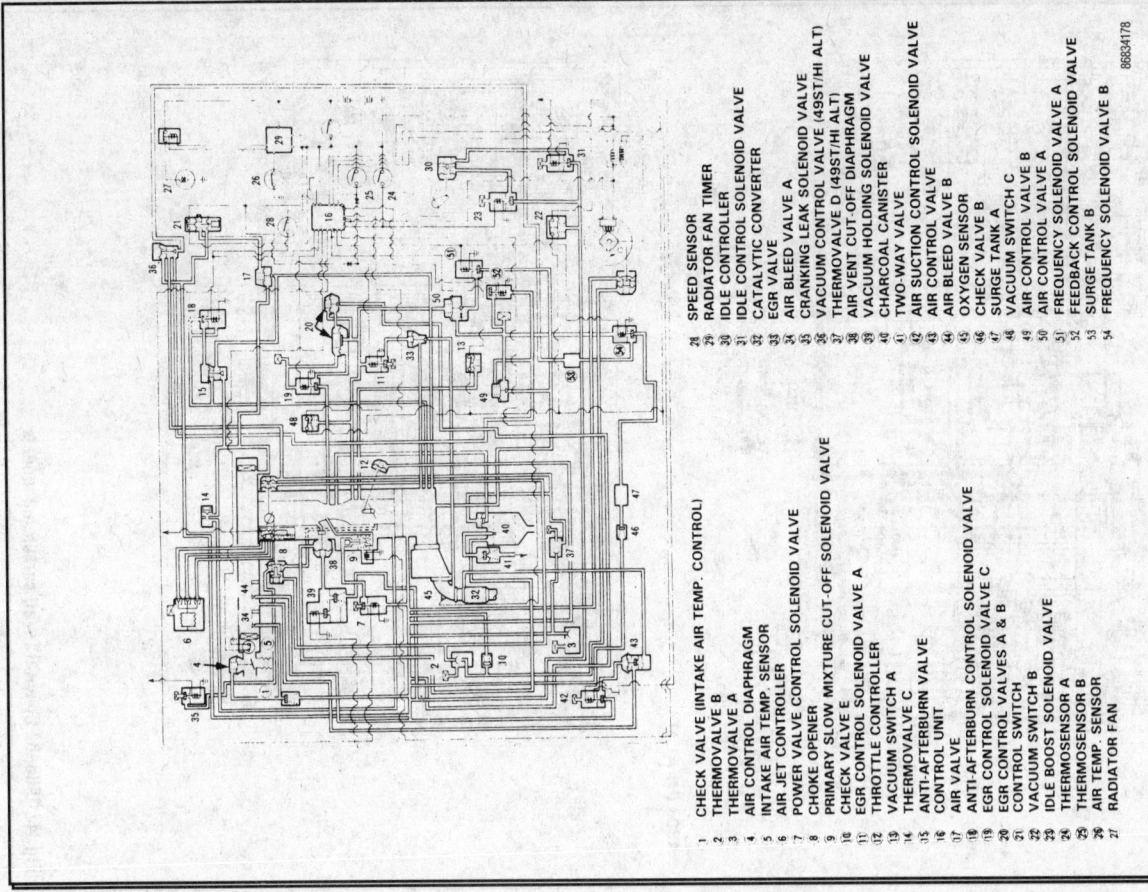

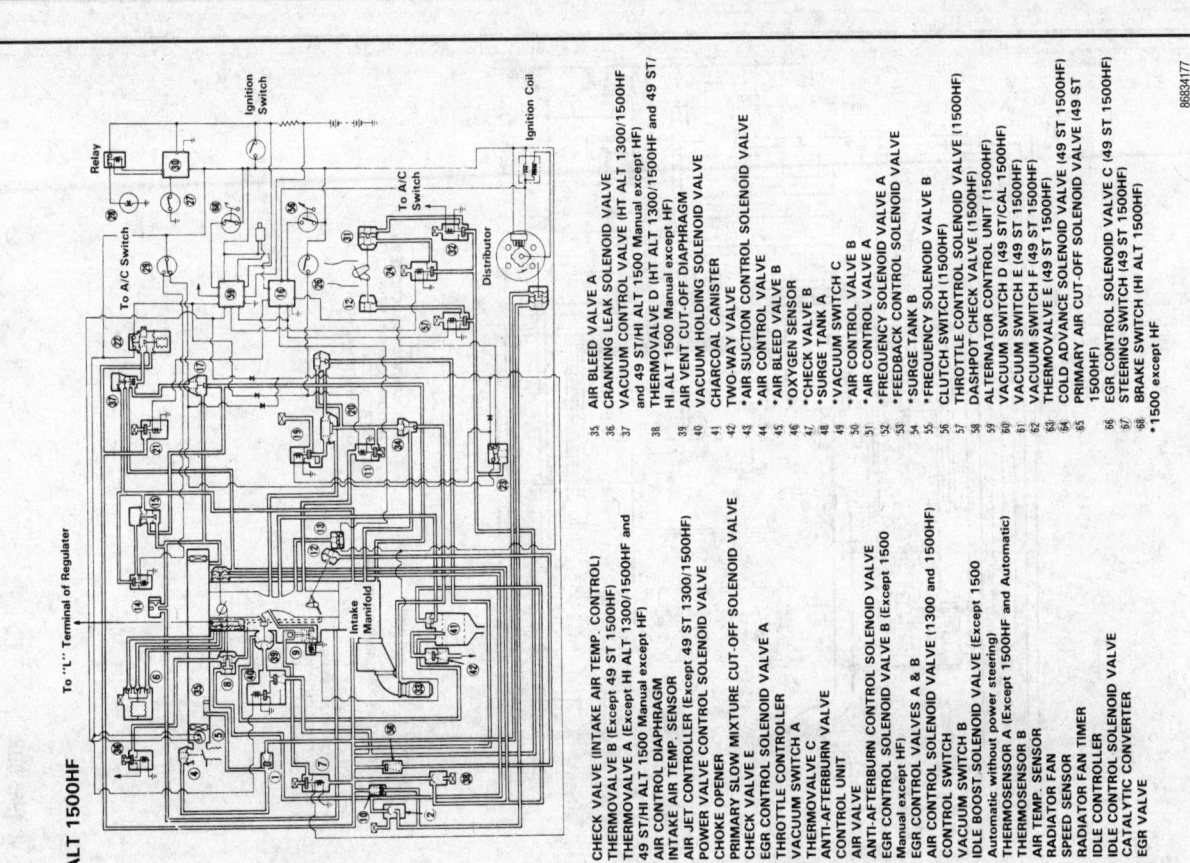

28 SPEED SENSOR
29 RADIATOR FAN TIMER
30 IDLE CONTROLLER
31 IDLE CONTROL SOLENOID VALVE
32 CATALYTIC CONVERTER
33 EGR VALVE
34 AIR BLEED VALVE A
35 CRANKING LEAK SOLENOID VALVE (49ST/HI ALT)
36 VACUUM CONTROL VALVE (49ST/HI ALT)
37 THERMOVALVE D (49ST/HI ALT)
38 AIR VENT CUT-OFF DIAPHRAGM
39 VACUUM HOLDING SOLENOID VALVE
40 CHARCOAL CANISTER
41 TWO-WAY VALVE
42 AIR SUCTION CONTROL SOLENOID VALVE
43 AIR BLEED VALVE B
44 CHECK VALVE B
45 OXYGEN SENSOR
46 SURGE TANK A
47 VACUUM SWITCH C
48 AIR CONTROL VALVE B
49 AIR CONTROL VALVE B
50 FREQUENCY SOLENOID VALVE A
51 FEEDBACK CONTROL SOLENOID VALVE
52 SURGE TANK B
53 FREQUENCY SOLENOID VALVE B

1 CHECK VALVE (INTAKE AIR TEMP. CONTROL)
2 THERMOVALVE B
3 THERMOVALVE A
4 AIR CONTROL DIAPHRAGM
5 INTAKE AIR TEMP. SENSOR
6 AIR JET CONTROLLER
7 POWER VALVE CONTROL SOLENOID VALVE
8 CHOKE OPENER
9 PRIMARY SLOW MIXTURE CUT-OFF SOLENOID VALVE
10 CHECK VALVE A
11 EGR CONTROL SOLENOID VALVE A
12 THROTTLE CONTROLLER
13 VACUUM SWITCH A
14 THERMOVALVE C
15 ANTI-AFTERBURN VALVE
16 CONTROL UNIT
17 AIR VALVE
18 ANTI-AFTERBURN CONTROL SOLENOID VALVE
19 EGR CONTROL SOLENOID VALVES A & B
20 EGR CONTROL VALVES A & B
21 CONTROL SWITCH
22 VACUUM SWITCH B
23 IDLE BOOST SOLENOID VALVE
24 THERMOSENSOR B
25 AIR TEMP. SENSOR
26 RADIATOR FAN
27 RADIATOR FAN

Fig. 45 1985 Civic Wagon 4WD

35 AIR BLEED VALVE A
36 CRANKING LEAK SOLENOID VALVE
37 VACUUM CONTROL VALVE (HT ALT 1300/1500HF and 49 ST/HI ALT 1500 Manual except HF)
38 THERMOVALVE D (HT ALT 1300/1500HF and 49 ST HI ALT 1500 Manual except HF)
39 AIR VENT CUT-OFF DIAPHRAGM
40 VACUUM HOLDING SOLENOID VALVE
41 CHARCOAL CANISTER
42 TWO-WAY VALVE
43 AIR SUCTION CONTROL SOLENOID VALVE
44 AIR BLEED VALVE B
45 CHECK VALVE B
46 OXYGEN SENSOR
47 SURGE TANK A
48 VACUUM SWITCH C
49 AIR CONTROL VALVE A
50 FREQUENCY SOLENOID VALVE A
51 FEEDBACK CONTROL SOLENOID VALVE
52 SURGE TANK B
53 FREQUENCY SOLENOID VALVE B
54 SURGE TANK B
55 FREQUENCY SOLENOID VALVE B
56 CLUTCH SWITCH (1500HF)
57 THROTTLE CONTROL SOLENOID VALVE (1500HF)
58 DASHPOT CHECK VALVE (1500HF)
59 ALTERNATOR CONTROL UNIT (1500HF)
60 VACUUM SWITCH D (49 ST/CAL 1500HF)
61 VACUUM SWITCH F (49 ST 1500HF)
62 THERMOVALVE E (49 ST 1500HF)
63 COLD ADVANCE SOLENOID VALVE (49 ST 1500HF)
64 PRIMARY AIR CUT-OFF SOLENOID VALVE (49 ST 1500HF)
65 STEERING SWITCH (49 ST 1500HF)
66 EGR CONTROL SOLENOID VALVE C (49 ST 1500HF)
67 BRAKE SWITCH (HI ALT 1500HF)
68 *1500 except HF

1 CHECK VALVE (INTAKE AIR TEMP. CONTROL)
2 THERMOVALVE B (Except 49 ST 1500HF)
3 THERMOVALVE A (Except HI ALT 1300/1500HF and 49 ST/HI ALT 1500 Manual except HF)
4 AIR CONTROL DIAPHRAGM
5 INTAKE AIR TEMP. SENSOR
6 AIR JET CONTROLLER (Except 49 ST 1300/1500HF)
7 POWER VALVE CONTROL SOLENOID VALVE
8 CHOKE OPENER
9 PRIMARY SLOW MIXTURE CUT-OFF SOLENOID VALVE
10 CHECK VALVE E
11 EGR CONTROL SOLENOID VALVE A
12 THROTTLE CONTROLLER
13 VACUUM SWITCH A
14 THERMOVALVE C
15 ANTI-AFTERBURN VALVE
16 CONTROL UNIT
17 AIR VALVE
18 ANTI-AFTERBURN CONTROL SOLENOID VALVE
19 EGR CONTROL SOLENOID VALVE B (Except 1500 Manual except HF)
20 EGR CONTROL VALVES A & B (1300 and 1500HF)
21 CONTROL SWITCH
22 VACUUM SWITCH B
23 IDLE BOOST SOLENOID VALVE (Except 1500)
24 THERMOSENSOR A (Except 1500HF and Automatic without power steering)
25 THERMOSENSOR B
26 AIR TEMP. SENSOR
27 RADIATOR FAN
28 RADIATOR FAN TIMER
29 SPEED SENSOR
30 IDLE CONTROLLER
31 IDLE CONTROL SOLENOID VALVE
32 CATALYTIC CONVERTER
33 EGR VALVE

Fig. 44 1985 Civic—except Wagon 4WD and CRX

4-24 EMISSION CONTROLS

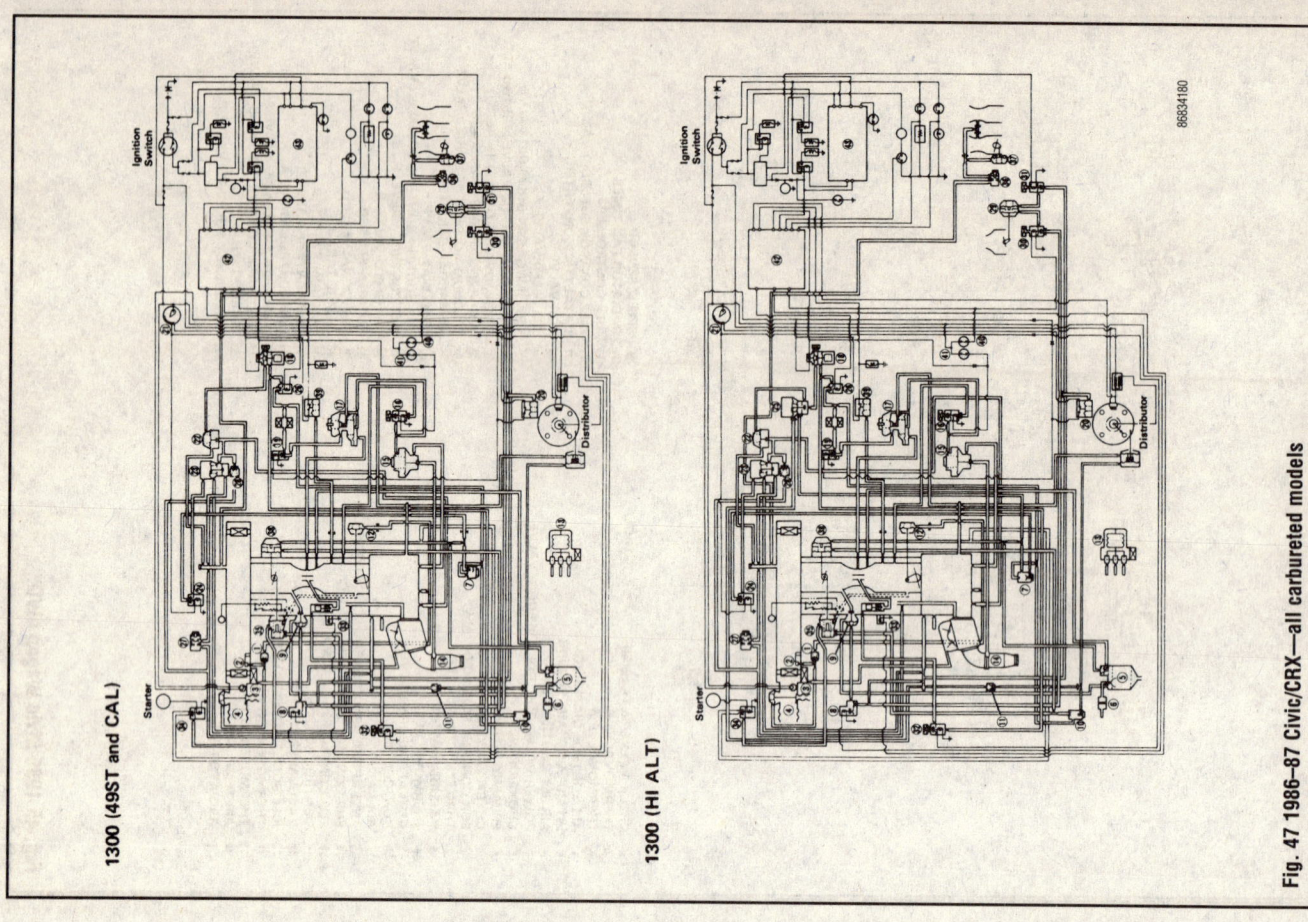

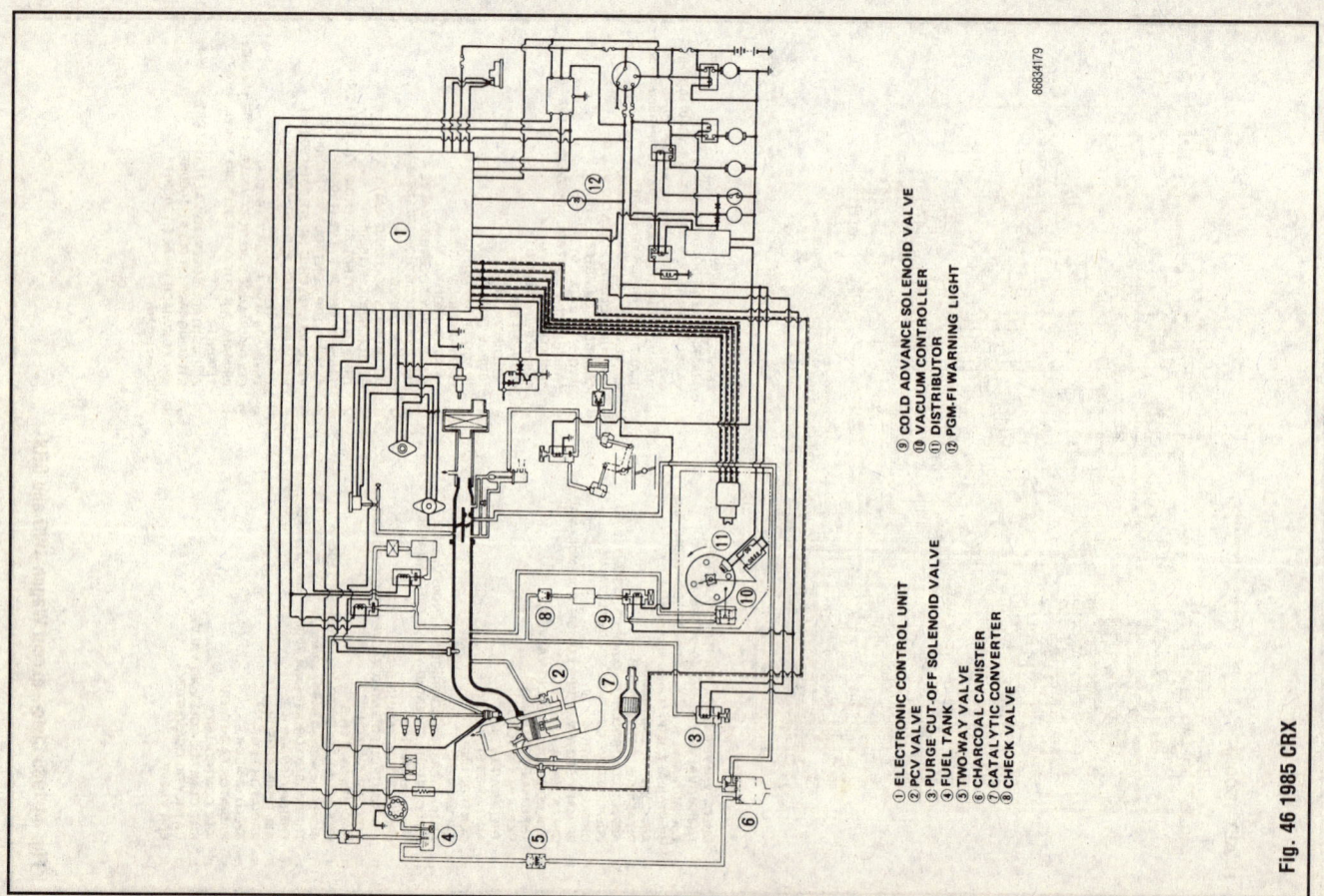

Fig. 47 1986–87 Civic/CRX—all carbureted models

1300 (49ST and CAL)

1300 (HI ALT)

Fig. 46 1985 CRX

1. ELECTRONIC CONTROL UNIT
2. PCV VALVE
3. PURGE CUT-OFF SOLENOID VALVE
4. FUEL TANK
5. TWO-WAY VALVE
6. CHARCOAL CANISTER
7. CATALYTIC CONVERTER
8. CHECK VALVE
9. COLD ADVANCE SOLENOID VALVE
10. VACUUM CONTROLLER
11. DISTRIBUTOR
12. PGM-FI WARNING LIGHT

EMISSION CONTROLS 4-25

1 CHECK VALVE (INTAKE AIR TEMP.)	33 OXYGEN SENSOR
2 AIR BLEED VALVE A	34 CHECK VALVE A
3 AIR BLEED VALVE B	35 ACCUMULATOR
4 INTAKE AIR TEMP. SENSOR	36 FREQUENCY SOLENOID VALVE B
5 AIR CONTROL DIAPHRAGM	37 PULSE RECTIFIER
6 CHARCOAL CANISTER	38 FEEDBACK CONTROL SOLENOID VALVE
7 TWO-WAY VALVE	39 FREQUENCY SOLENOID VALVE A
8 THERMOVALVE A	40 AIR CONTROL VALVE A
9 VACUUM HOLDING SOLENOID VALVE	41 AIR CONTROL VALVE B
10 AIR VENT CUT-OFF DIAPHRAGM	42 IDLE CONTROLLER
11 THERMOVALVE B	43 IDLE BOOST SOLENOID VALVE (M/T and A/T with power steering)
12 CHECK VALVE E	44 A/C IDLE BOOST SOLENOID VALVE
13 COLD ADVANCE SOLENOID VALVE (A/T)	45 POWER VALVE CONTROL SOLENOID VALVE
14 THROTTLE CONTROLLER	46 PRIMARY SLOW MIXTURE CUT-OFF SOLENOID VALVE
15 AIR JET CONTROLLER	47 CRANKING LEAK SOLENOID VALVE
16 CATALYTIC CONVERTER	48 CHOKE OPENER
17 EGR VALVE	49 THERMOVALVE D
18 EGR CONTROL SOLENOID VALVE A	50 SECONDARY DIAPHRAGM
19 EGR CONTROL SOLENOID VALVE A & B	51 FAST IDLE UNLOADER
20 CONTROL SWITCH	52 VACUUM SWITCH C
21 EGR CONTROL SOLENOID VALVE C (A/T & 4WD)	53 VACUUM SWITCH A
22 VACUUM SWITCH B	54 AIR TEMP. SENSOR
23 SPEED SENSOR	55 THERMOSENSOR A
24 AIR VALVE	56 THERMOSENSOR B
25 ANTI-AFTERBURN VALVE	57 NEUTRAL SWITCH (A/T)
26 ANTI-AFTERBURN CONTROL SOLENOID VALVE	58 PARKING SWITCH (A/T)
27 VACUUM CONTROL VALVE (49 ST-M/T)	59 DEVICE CONTROL UNIT A
28 AIR SUCTION CONTROL SOLENOID VALVE	60 DEVICE CONTROL UNIT B (A/T)
29 SILENCER	61 RADIATOR FAN TIMER
30 AIR SUCTION VALVE	62 POWER STEERING OIL PRESSURE SWITCH
31 THERMOVALVE C	
32 CHECK VALVE G	

Fig. 49 1986–87 Civic/CRX—all carbureted models (legend applies to all 1986–87 1.5L carbureted models)

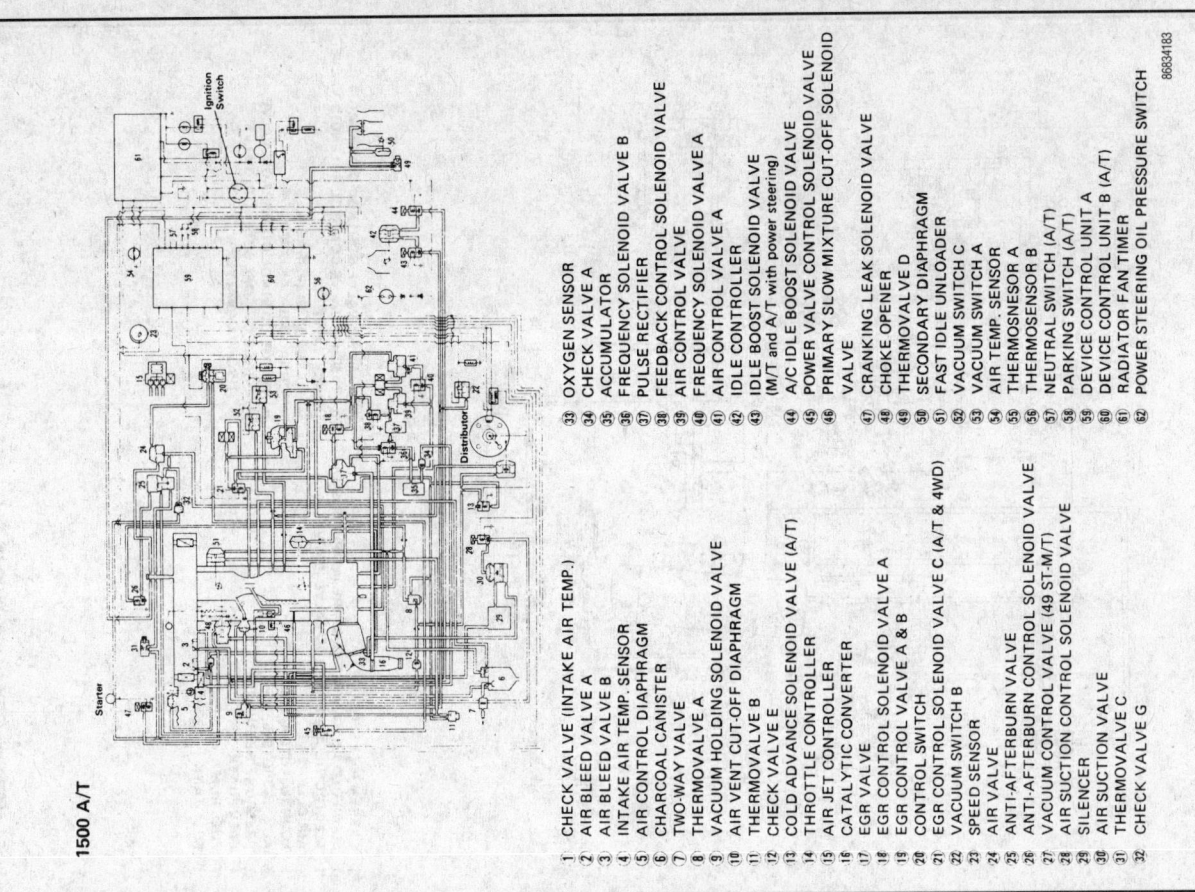

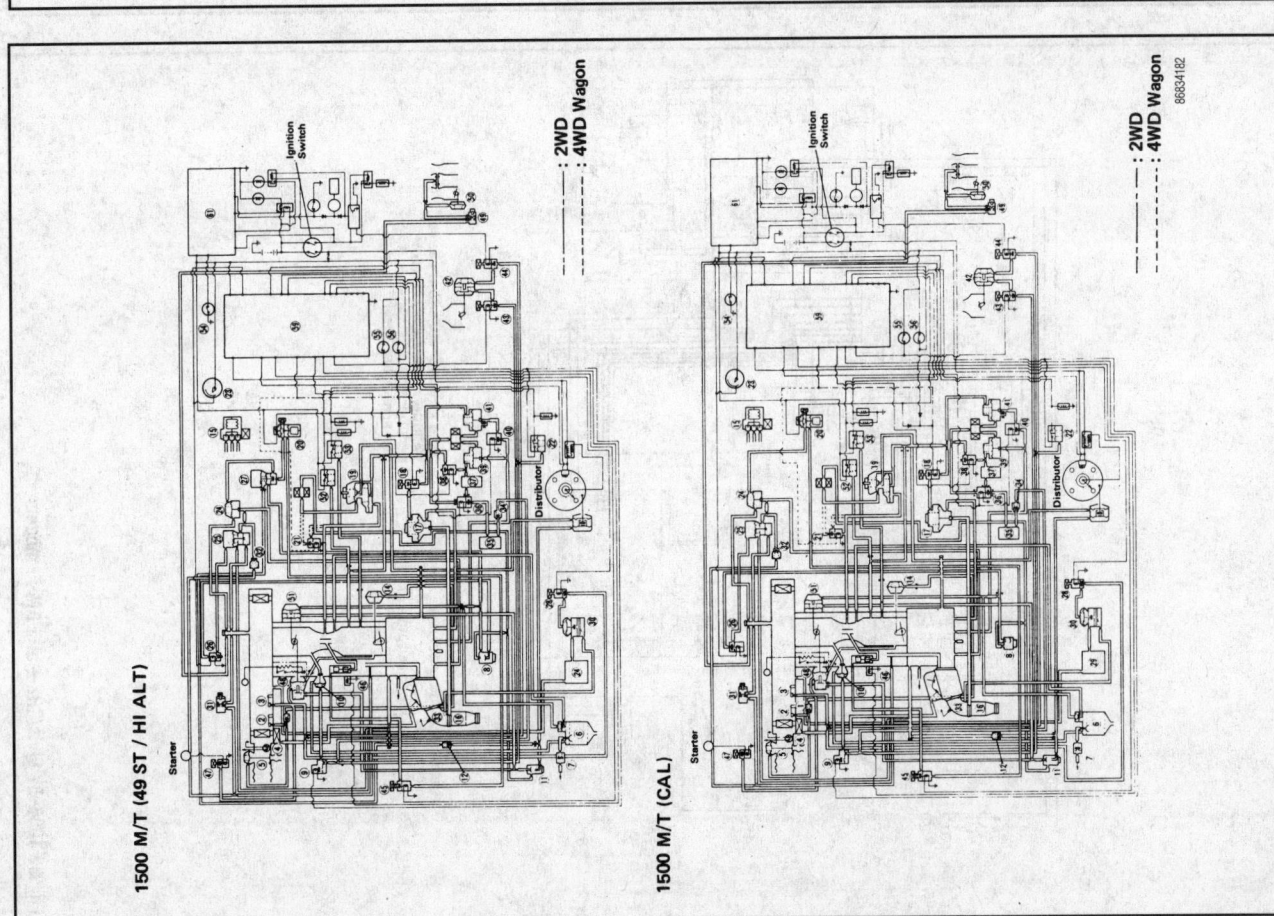

Fig. 48 1986–87 Civic/CRX—all carbureted models

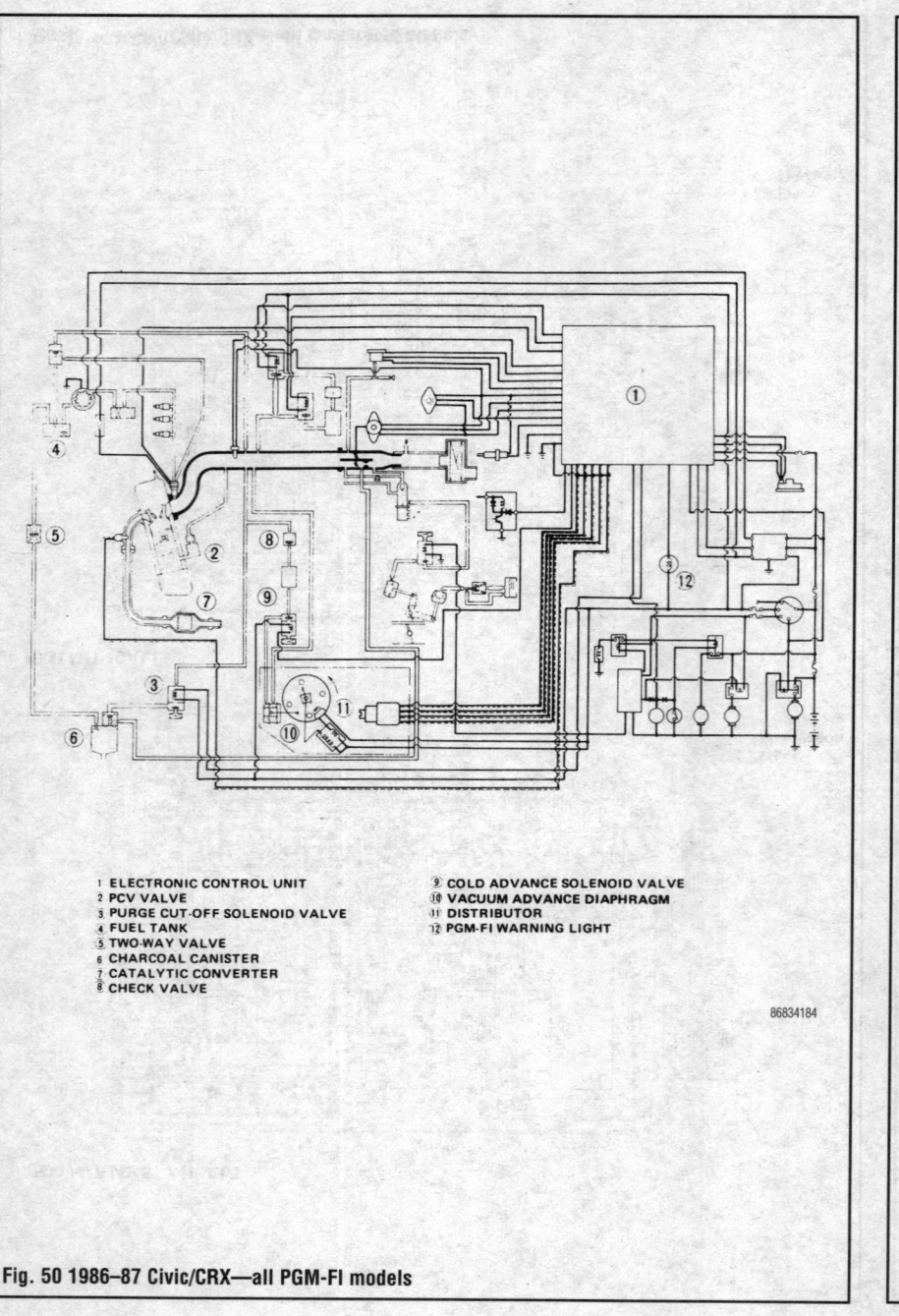

4-26 EMISSION CONTROLS

① OXYGEN (O$_2$) SENSOR
② MANIFOLD ABSOLUTE PRESSURE (MAP) SENSOR
③ ELECTRONIC AIR CONTROL VALVE (EACV)
④ AIR CLEANER
⑤ MAIN INJECTOR
⑥ AUX. INJECTOR
⑦ PRESSURE REGULATOR
⑧ FUEL FILTER
⑨ FUEL PUMP
⑩ FUEL TANK
⑪ TANDEM VALVE CONTROL DIAPHRAGM
⑫ TANDEM VALVE CONTROL SOLENOID VALVE
⑬ PCV VALVE
⑭ DASHPOT DIAPHRAGM
⑮ CHARCOAL CANISTER
⑯ PURGE CONTROL DIAPHRAGM VALVE
⑰ PURGE CUT-OFF SOLENOID VALVE
⑱ TWO-WAY VALVE

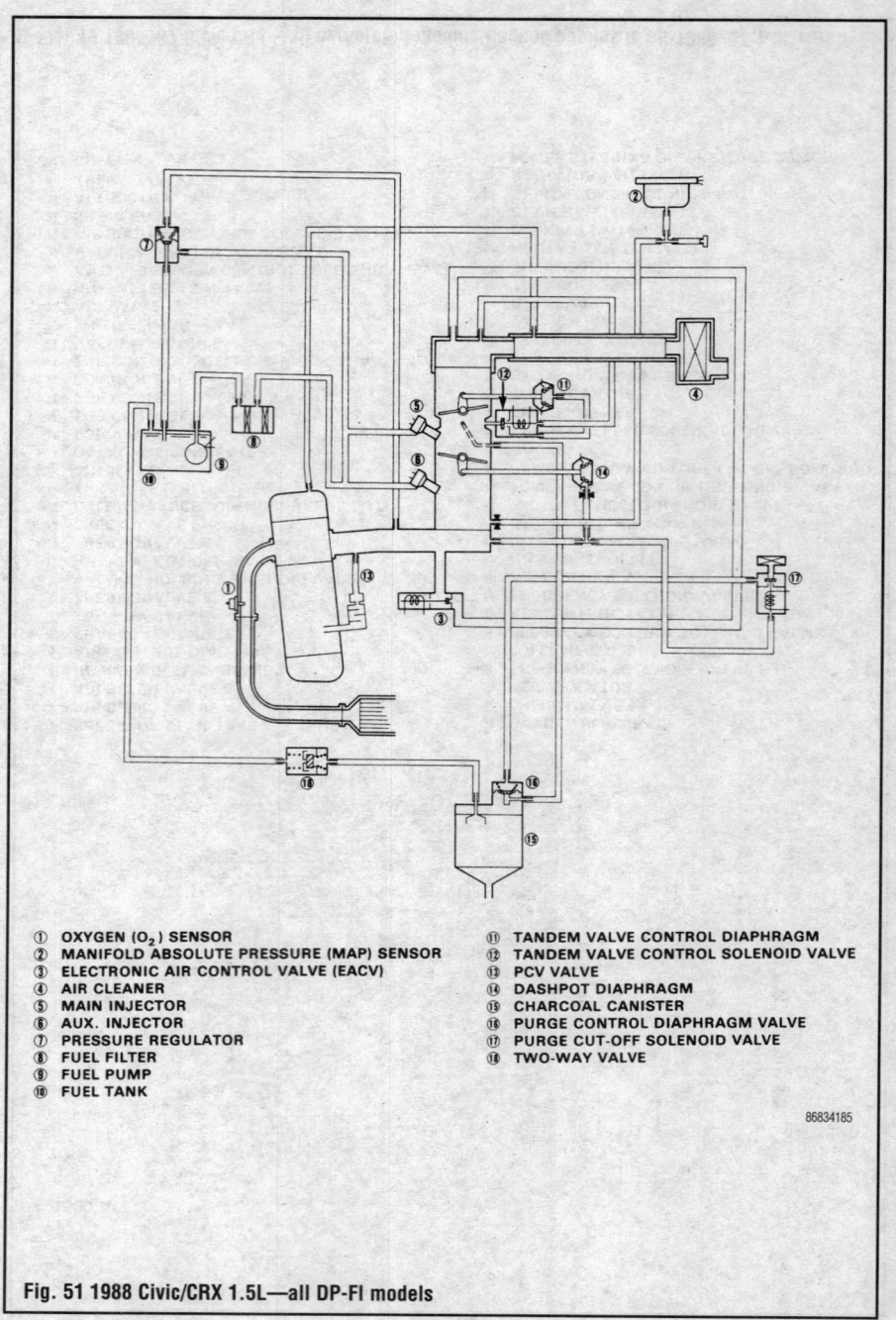

Fig. 51 1988 Civic/CRX 1.5L—all DP-FI models

1 ELECTRONIC CONTROL UNIT
2 PCV VALVE
3 PURGE CUT-OFF SOLENOID VALVE
4 FUEL TANK
5 TWO-WAY VALVE
6 CHARCOAL CANISTER
7 CATALYTIC CONVERTER
8 CHECK VALVE
9 COLD ADVANCE SOLENOID VALVE
10 VACUUM ADVANCE DIAPHRAGM
11 DISTRIBUTOR
12 PGM-FI WARNING LIGHT

Fig. 50 1986–87 Civic/CRX—all PGM-FI models

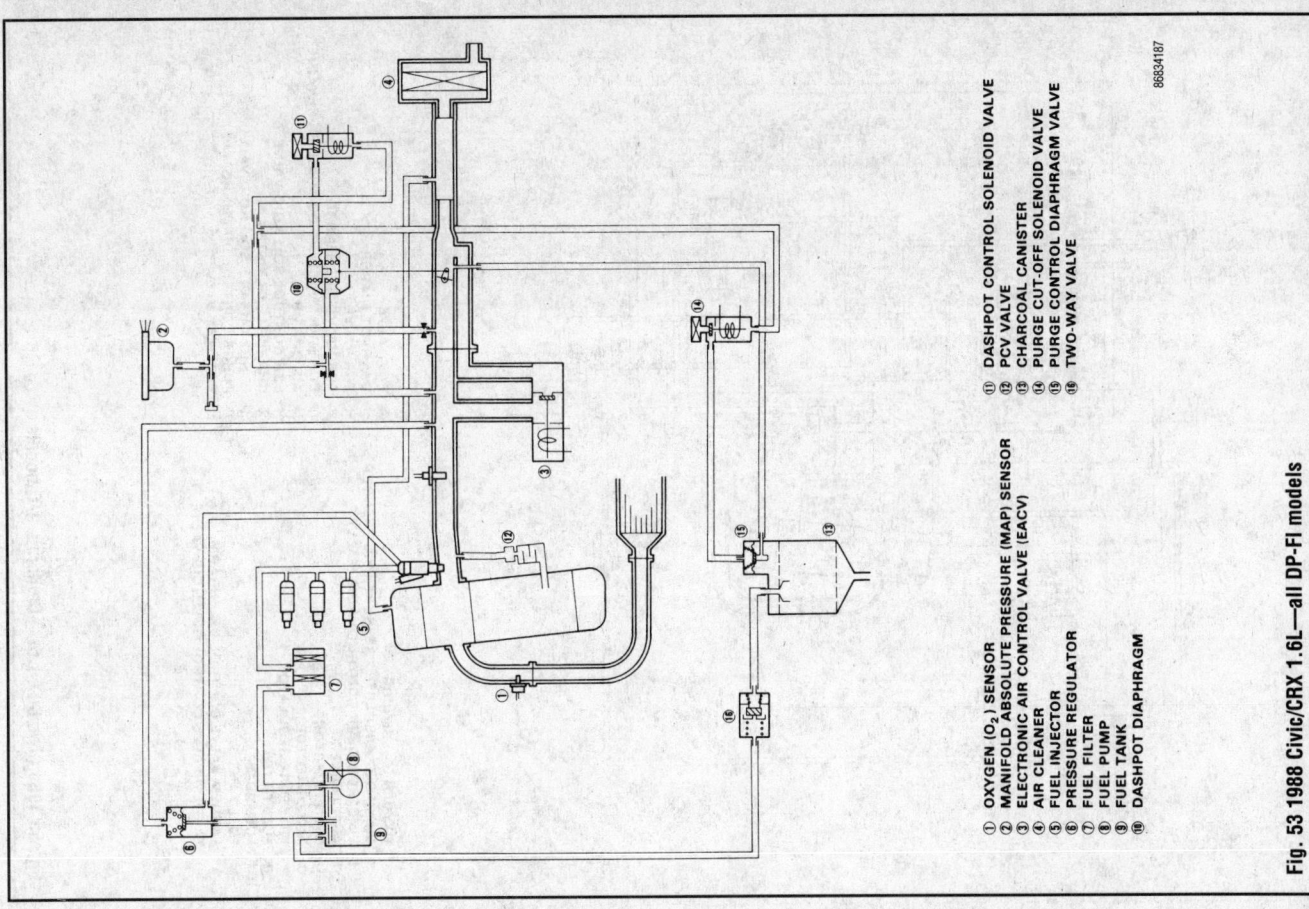

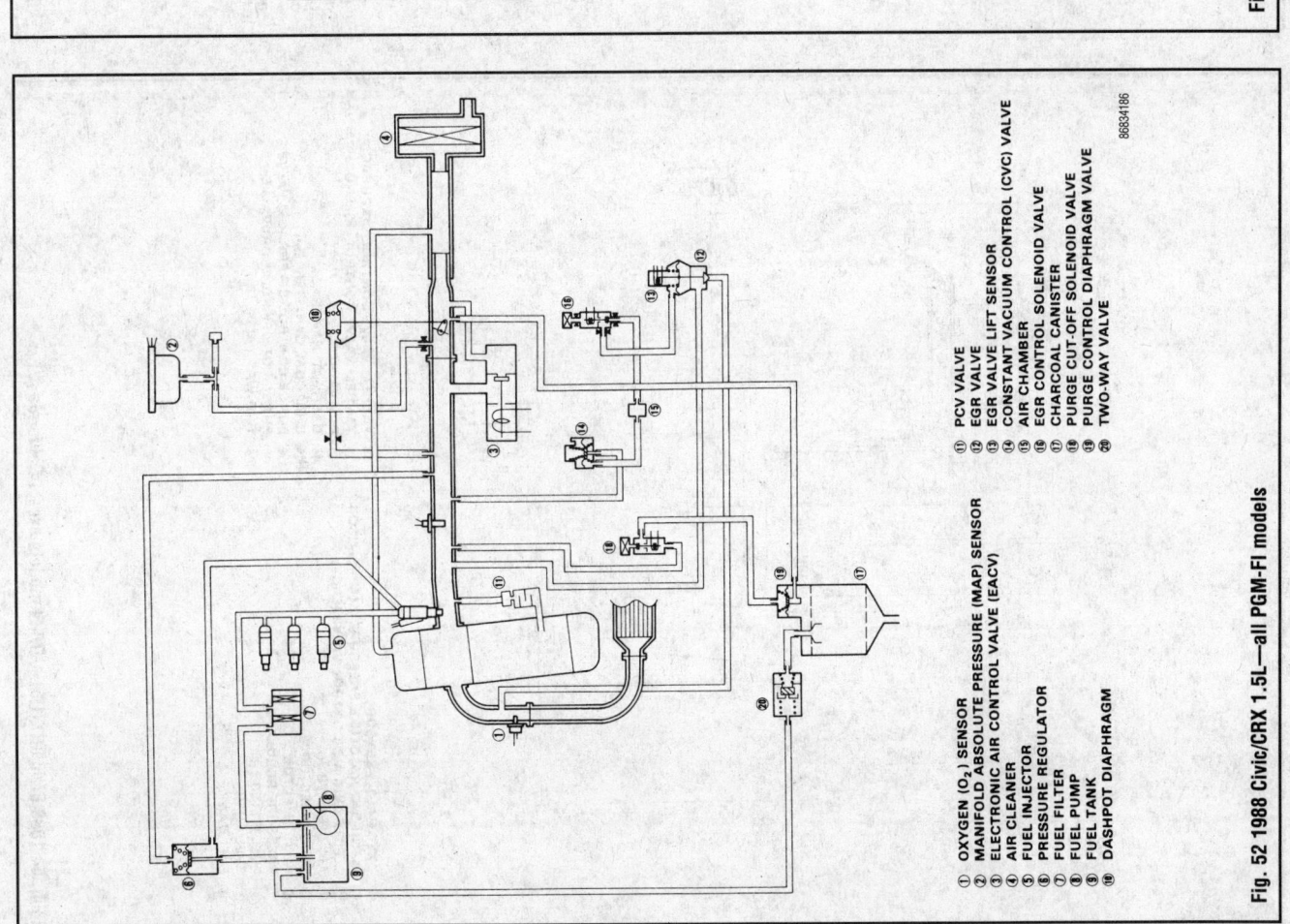

EMISSION CONTROLS 4-27

① OXYGEN (O₂) SENSOR
② MANIFOLD ABSOLUTE PRESSURE (MAP) SENSOR
③ ELECTRONIC AIR CONTROL VALVE (EACV)
④ AIR CLEANER
⑤ FUEL INJECTOR
⑥ PRESSURE REGULATOR
⑦ FUEL FILTER
⑧ FUEL PUMP
⑨ FUEL TANK
⑩ DASHPOT DIAPHRAGM

⑪ DASHPOT CONTROL SOLENOID VALVE
⑫ PCV VALVE
⑬ CHARCOAL CANISTER
⑭ PURGE CUT-OFF SOLENOID VALVE
⑮ PURGE CONTROL DIAPHRAGM VALVE
⑯ TWO-WAY VALVE

Fig. 53 1988 Civic/CRX 1.6L—all DP-FI models

① OXYGEN (O₂) SENSOR
② MANIFOLD ABSOLUTE PRESSURE (MAP) SENSOR
③ ELECTRONIC AIR CONTROL VALVE (EACV)
④ AIR CLEANER
⑤ FUEL INJECTOR
⑥ PRESSURE REGULATOR
⑦ FUEL FILTER
⑧ FUEL PUMP
⑨ FUEL TANK
⑩ DASHPOT DIAPHRAGM

⑪ PCV VALVE
⑫ EGR VALVE
⑬ EGR VALVE LIFT SENSOR
⑭ CONSTANT VACUUM CONTROL (CVC) VALVE
⑮ AIR CHAMBER
⑯ EGR CONTROL SOLENOID VALVE
⑰ CHARCOAL CANISTER
⑱ PURGE CUT-OFF SOLENOID VALVE
⑲ PURGE CONTROL DIAPHRAGM VALVE
⑳ TWO-WAY VALVE

Fig. 52 1988 Civic/CRX 1.5L—all PGM-FI models

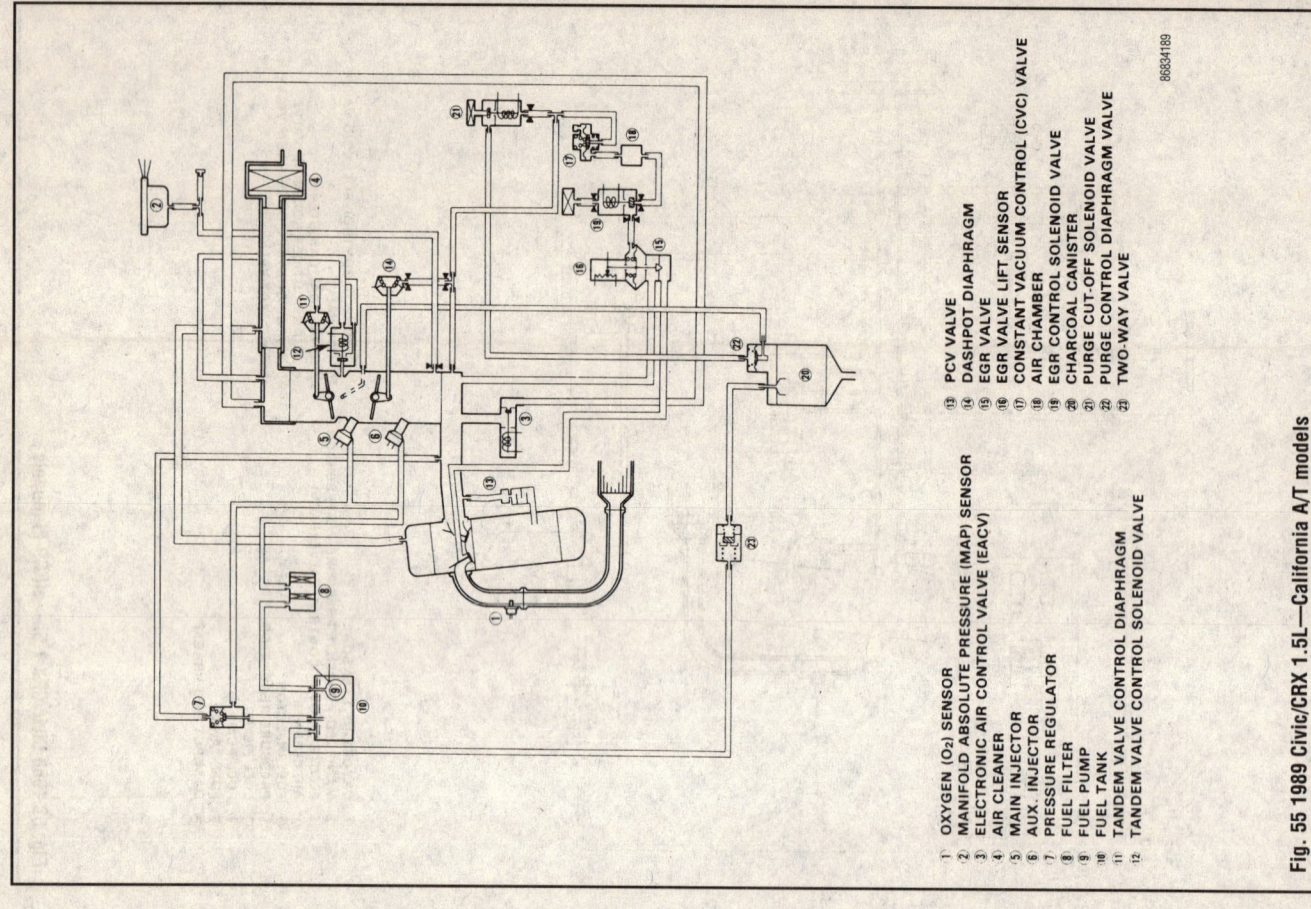

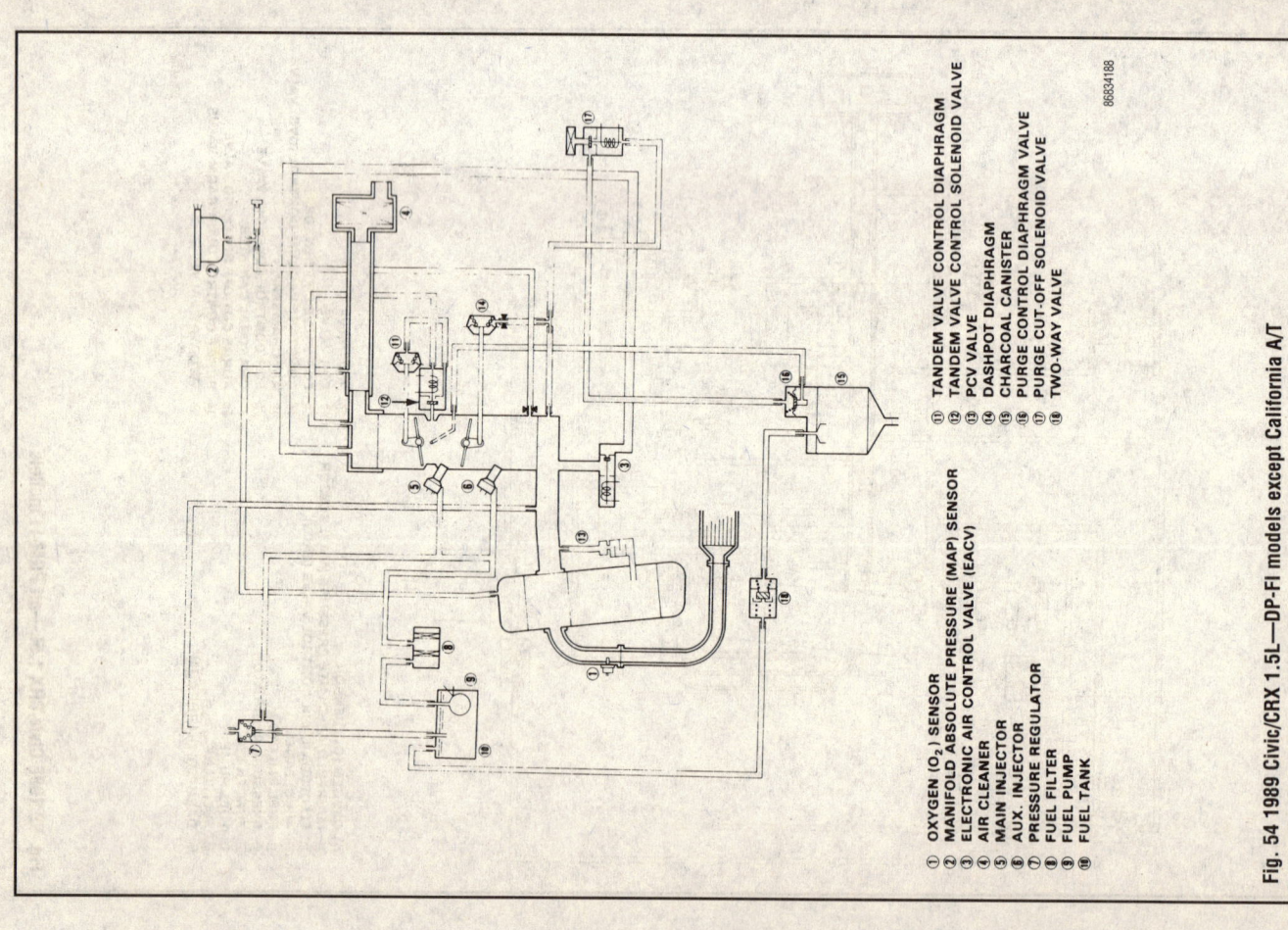

4-28 EMISSION CONTROLS

① OXYGEN (O₂) SENSOR
② MANIFOLD ABSOLUTE PRESSURE (MAP) SENSOR
③ ELECTRONIC AIR CONTROL VALVE (EACV)
④ AIR CLEANER
⑤ MAIN INJECTOR
⑥ AUX. INJECTOR
⑦ PRESSURE REGULATOR
⑧ FUEL FILTER
⑨ FUEL PUMP
⑩ FUEL TANK
⑪ TANDEM VALVE CONTROL DIAPHRAGM
⑫ TANDEM VALVE CONTROL SOLENOID VALVE
⑬ PCV VALVE
⑭ DASHPOT DIAPHRAGM
⑮ EGR VALVE
⑯ EGR VALVE LIFT SENSOR
⑰ CONSTANT VACUUM CONTROL (CVC) VALVE
⑱ AIR CHAMBER
⑲ EGR CONTROL SOLENOID VALVE
⑳ CHARCOAL CANISTER
㉑ PURGE CUT-OFF SOLENOID VALVE
㉒ PURGE CONTROL DIAPHRAGM VALVE
㉓ TWO-WAY VALVE

Fig. 55 1989 Civic/CRX 1.5L—California A/T models

① OXYGEN (O₂) SENSOR
② MANIFOLD ABSOLUTE PRESSURE (MAP) SENSOR
③ ELECTRONIC AIR CONTROL VALVE (EACV)
④ AIR CLEANER
⑤ MAIN INJECTOR
⑥ AUX. INJECTOR
⑦ PRESSURE REGULATOR
⑧ FUEL FILTER
⑨ FUEL PUMP
⑩ FUEL TANK
⑪ TANDEM VALVE CONTROL DIAPHRAGM
⑫ TANDEM VALVE CONTROL SOLENOID VALVE
⑬ PCV VALVE
⑭ DASHPOT DIAPHRAGM
⑮ CHARCOAL CANISTER
⑯ PURGE CONTROL DIAPHRAGM VALVE
⑰ PURGE CUT-OFF SOLENOID VALVE
⑱ TWO-WAY VALVE

Fig. 54 1989 Civic/CRX 1.5L—DP-Fi models except California A/T

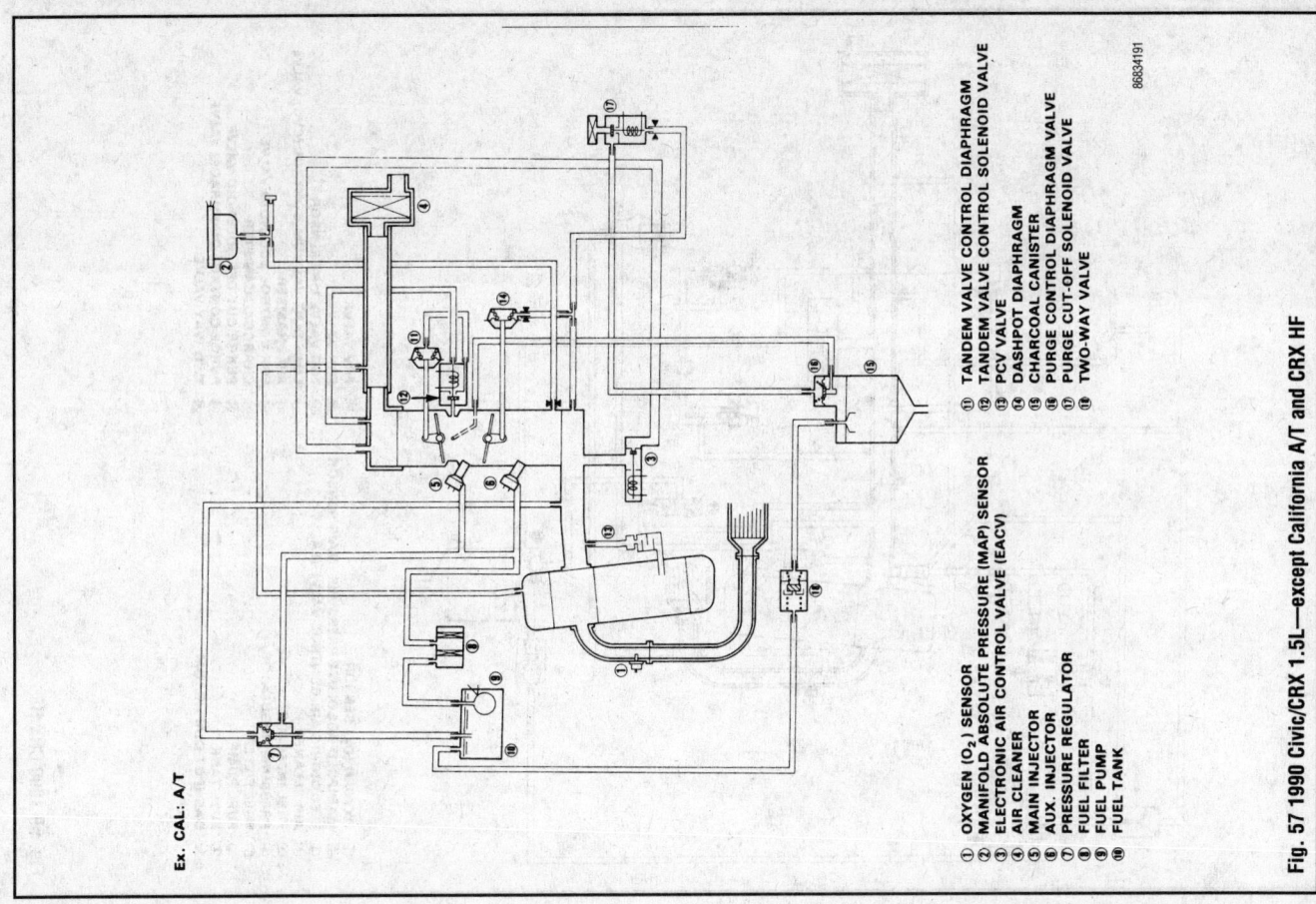

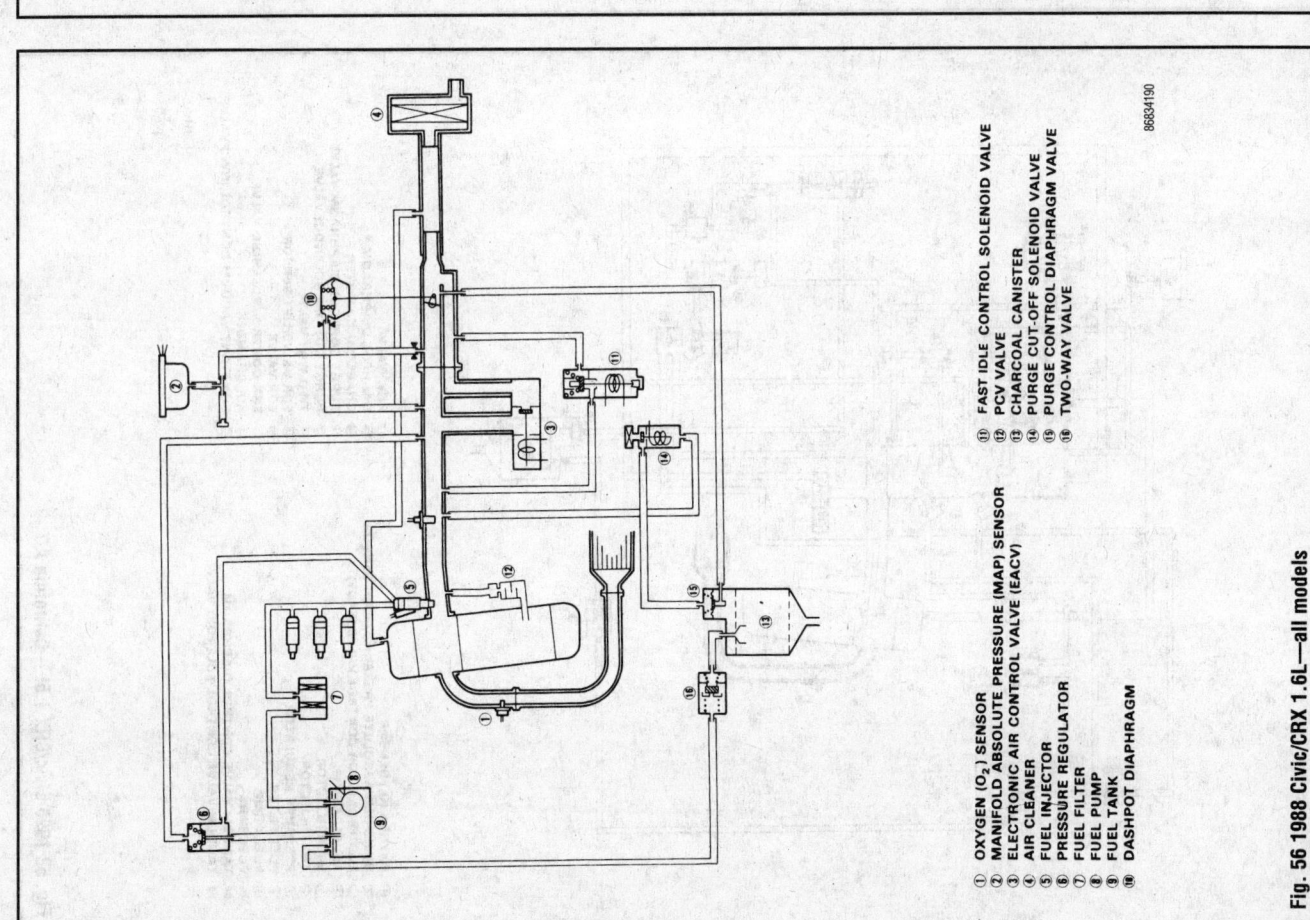

EMISSION CONTROLS 4-29

① OXYGEN (O₂) SENSOR
② MANIFOLD ABSOLUTE PRESSURE (MAP) SENSOR
③ ELECTRONIC AIR CONTROL VALVE (EACV)
④ AIR CLEANER
⑤ MAIN INJECTOR
⑥ AUX. INJECTOR
⑦ PRESSURE REGULATOR
⑧ FUEL FILTER
⑨ FUEL PUMP
⑩ FUEL TANK
⑪ TANDEM VALVE CONTROL DIAPHRAGM
⑫ TANDEM VALVE CONTROL SOLENOID VALVE
⑬ PCV VALVE
⑭ DASHPOT DIAPHRAGM
⑮ CHARCOAL CANISTER
⑯ PURGE CONTROL DIAPHRAGM VALVE
⑰ PURGE CUT-OFF SOLENOID VALVE
⑱ TWO-WAY VALVE

Fig. 57 1990 Civic/CRX 1.5L—except California A/T and CRX HF

① OXYGEN (O₂) SENSOR
② MANIFOLD ABSOLUTE PRESSURE (MAP) SENSOR
③ ELECTRONIC AIR CONTROL VALVE (EACV)
④ AIR CLEANER
⑤ FUEL INJECTOR
⑥ PRESSURE REGULATOR
⑦ FUEL FILTER
⑧ FUEL PUMP
⑨ FUEL TANK
⑩ DASHPOT DIAPHRAGM
⑪ FAST IDLE CONTROL SOLENOID VALVE
⑫ PCV VALVE
⑬ CHARCOAL CANISTER
⑭ PURGE CUT-OFF SOLENOID VALVE
⑮ PURGE CONTROL DIAPHRAGM VALVE
⑯ TWO-WAY VALVE

Fig. 56 1988 Civic/CRX 1.6L—all models

4-30 EMISSION CONTROLS

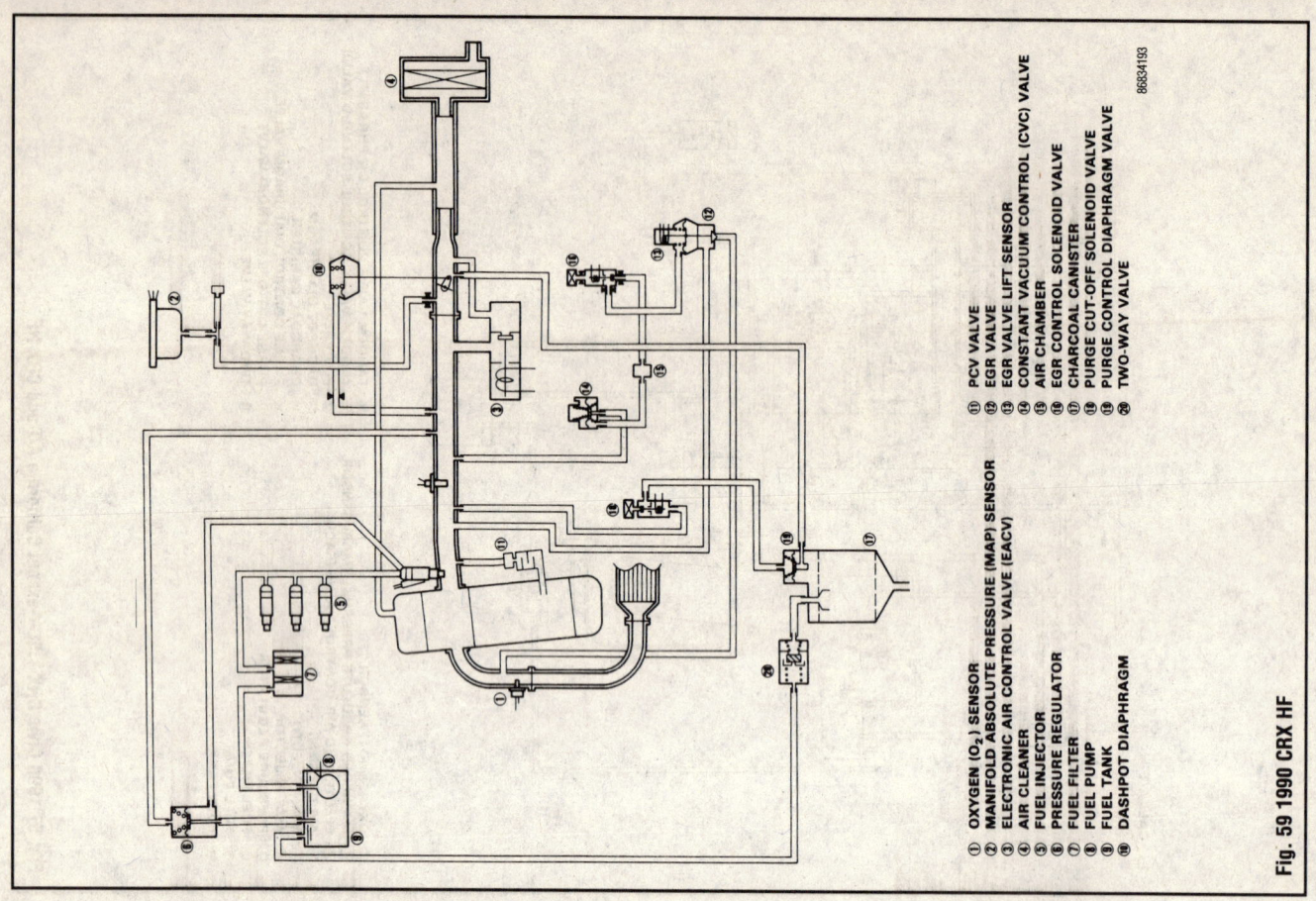

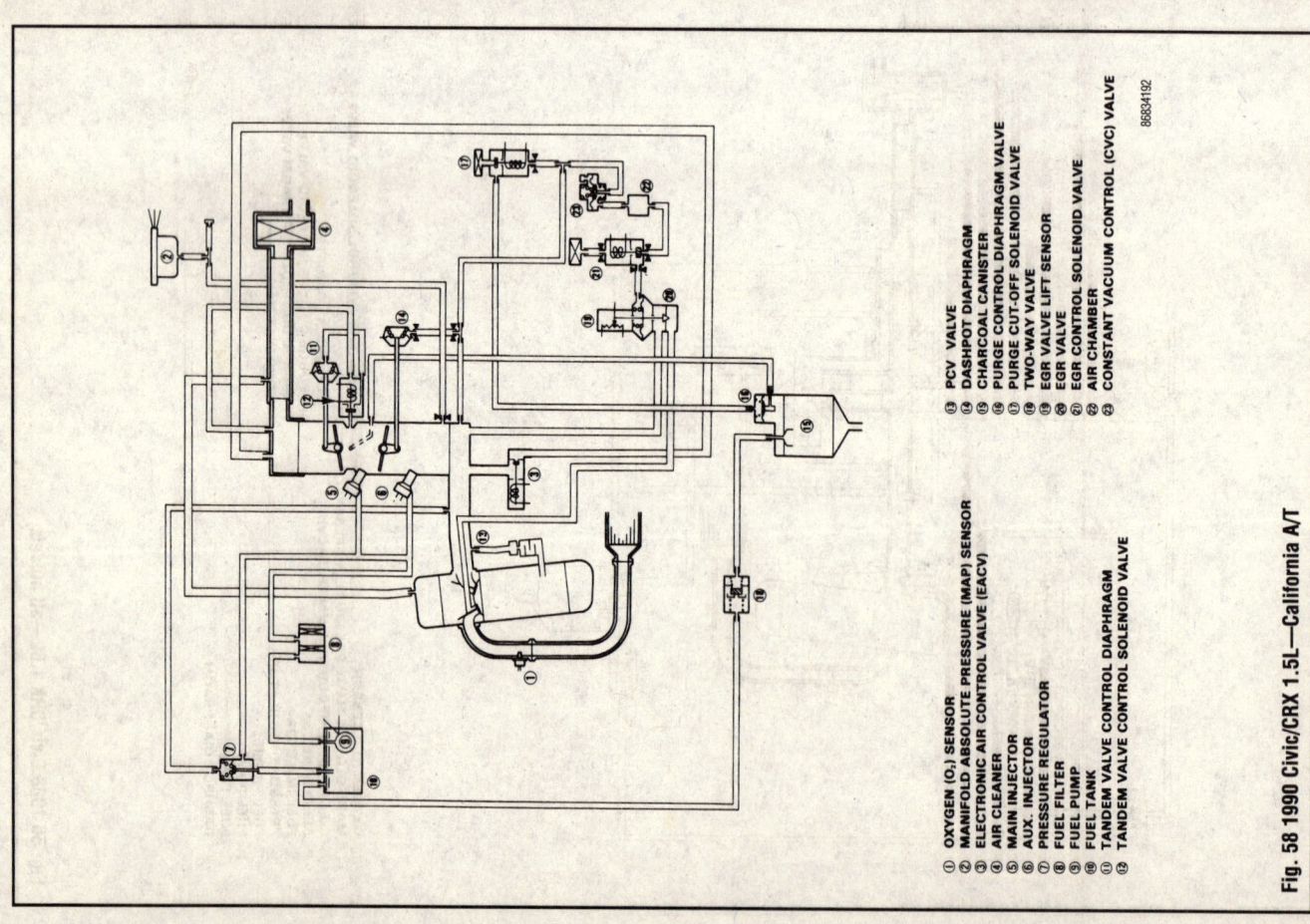

Fig. 59 1990 CRX HF

1. OXYGEN (O₂) SENSOR
2. MANIFOLD ABSOLUTE PRESSURE (MAP) SENSOR
3. ELECTRONIC AIR CONTROL VALVE (EACV)
4. AIR CLEANER
5. FUEL INJECTOR
6. PRESSURE REGULATOR
7. FUEL FILTER
8. FUEL PUMP
9. FUEL TANK
10. DASHPOT DIAPHRAGM
11. PCV VALVE
12. EGR VALVE
13. EGR VALVE LIFT SENSOR
14. CONSTANT VACUUM CONTROL (CVC) VALVE
15. AIR CHAMBER
16. EGR CONTROL SOLENOID VALVE
17. CHARCOAL CANISTER
18. PURGE CUT-OFF SOLENOID VALVE
19. PURGE CONTROL DIAPHRAGM VALVE
20. TWO-WAY VALVE

Fig. 58 1990 Civic/CRX 1.5L—California A/T

1. OXYGEN (O₂) SENSOR
2. MANIFOLD ABSOLUTE PRESSURE (MAP) SENSOR
3. ELECTRONIC AIR CONTROL VALVE (EACV)
4. AIR CLEANER
5. MAIN INJECTOR
6. AUX. INJECTOR
7. PRESSURE REGULATOR
8. FUEL FILTER
9. FUEL PUMP
10. FUEL TANK
11. TANDEM VALVE CONTROL DIAPHRAGM
12. TANDEM VALVE CONTROL SOLENOID VALVE
13. PCV VALVE
14. DASHPOT DIAPHRAGM
15. CHARCOAL CANISTER
16. PURGE CONTROL DIAPHRAGM VALVE
17. PURGE CUT-OFF SOLENOID VALVE
18. TWO-WAY VALVE
19. EGR VALVE LIFT SENSOR
20. EGR VALVE
21. EGR CONTROL SOLENOID VALVE
22. AIR CHAMBER
23. CONSTANT VACUUM CONTROL (CVC) VALVE

EMISSION CONTROLS 4-31

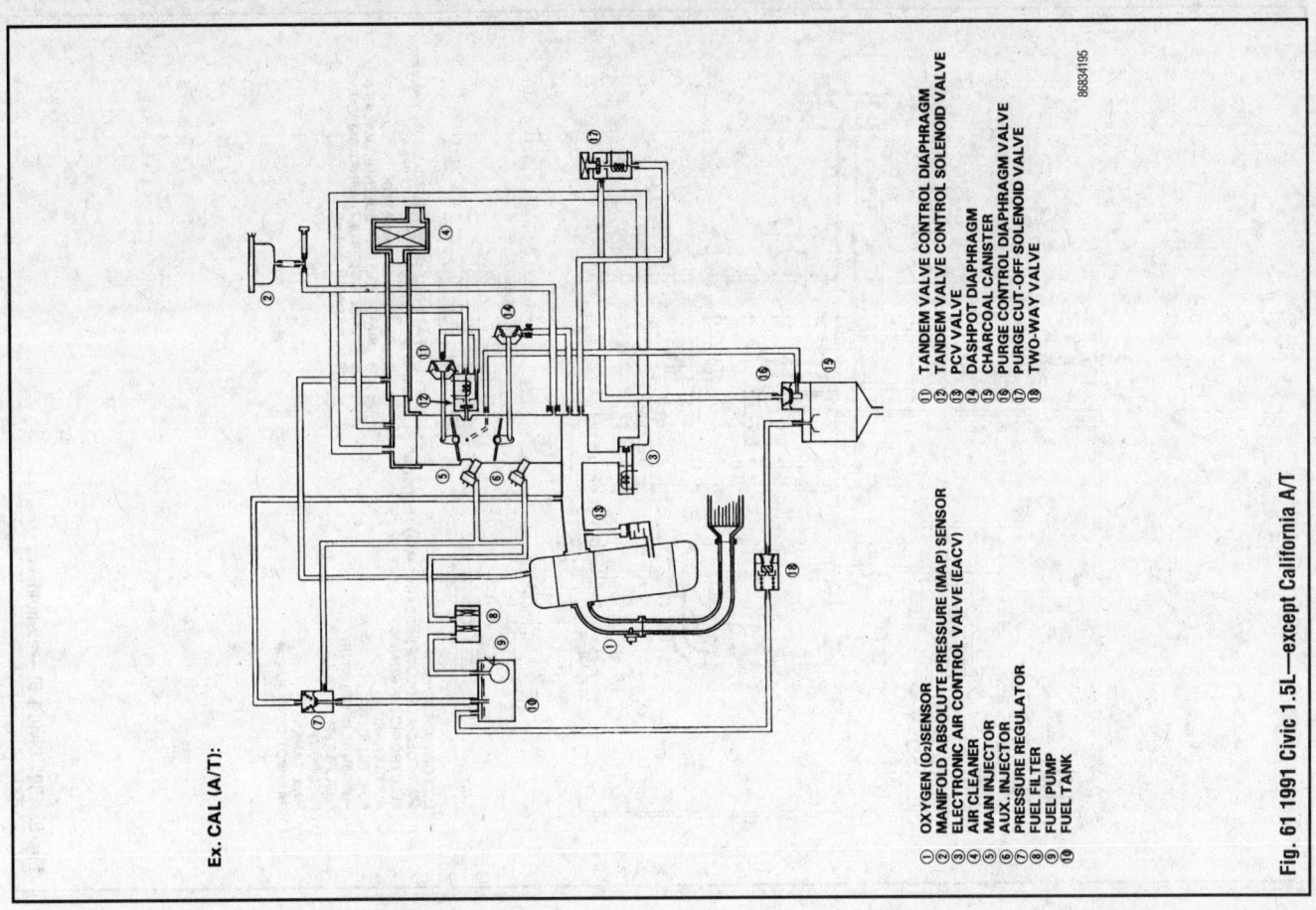

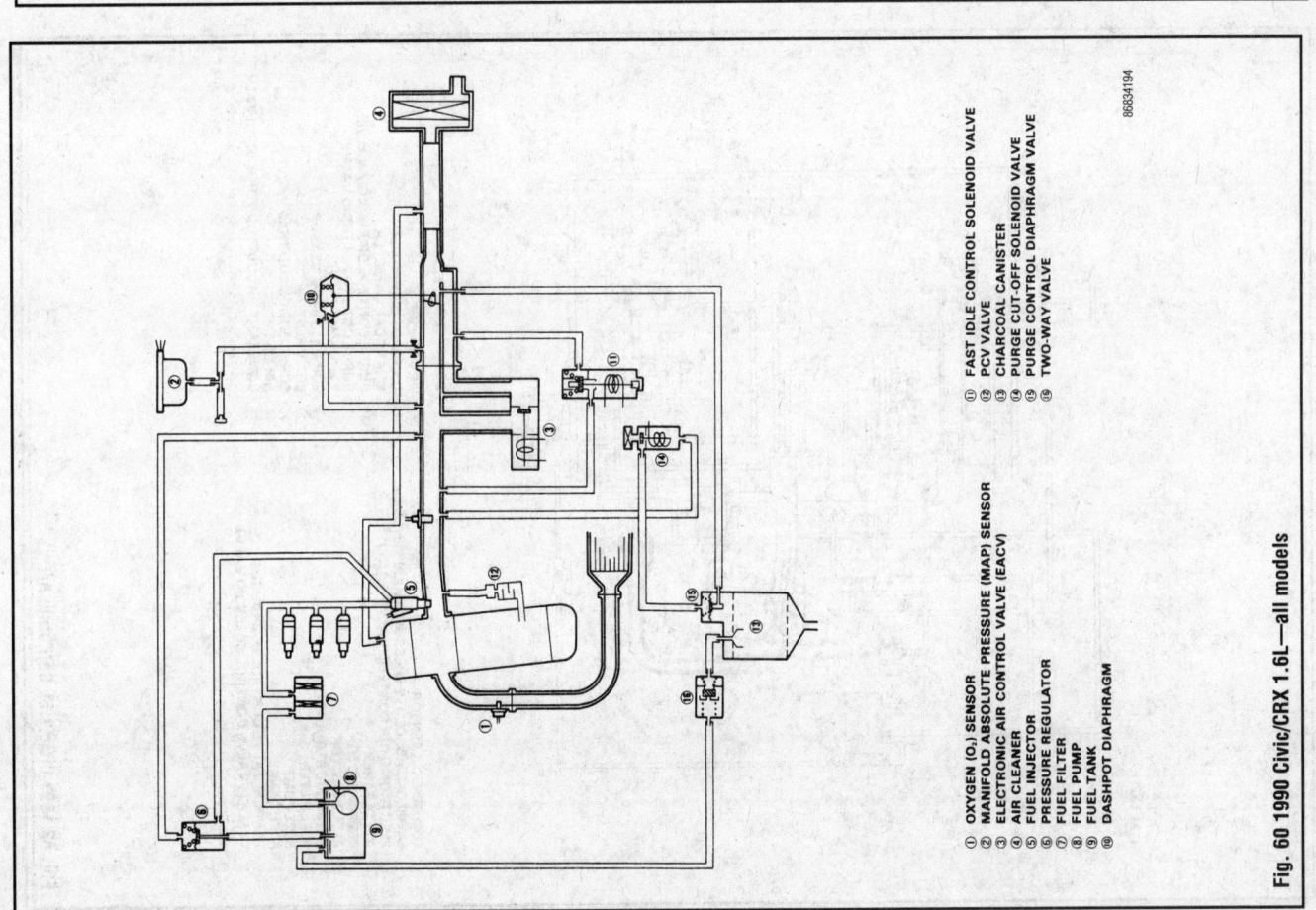

Ex. CAL (A/T):

① OXYGEN (O₂) SENSOR
② MANIFOLD ABSOLUTE PRESSURE (MAP) SENSOR
③ ELECTRONIC AIR CONTROL VALVE (EACV)
④ AIR CLEANER
⑤ MAIN INJECTOR
⑥ AUX. INJECTOR
⑦ PRESSURE REGULATOR
⑧ FUEL FILTER
⑨ FUEL PUMP
⑩ FUEL TANK

⑪ TANDEM VALVE CONTROL DIAPHRAGM
⑫ TANDEM VALVE CONTROL SOLENOID VALVE
⑬ PCV VALVE
⑭ DASHPOT DIAPHRAGM
⑮ CHARCOAL CANISTER
⑯ PURGE CONTROL DIAPHRAGM VALVE
⑰ PURGE CUT-OFF SOLENOID VALVE
⑱ TWO-WAY VALVE

Fig. 61 1991 Civic 1.5L—except California A/T

① OXYGEN (O₂) SENSOR
② MANIFOLD ABSOLUTE PRESSURE (MAP) SENSOR
③ ELECTRONIC AIR CONTROL VALVE (EACV)
④ AIR CLEANER
⑤ FUEL INJECTOR
⑥ PRESSURE REGULATOR
⑦ FUEL FILTER
⑧ FUEL PUMP
⑨ FUEL TANK
⑩ DASHPOT DIAPHRAGM

⑪ FAST IDLE CONTROL SOLENOID VALVE
⑫ PCV VALVE
⑬ CHARCOAL CANISTER
⑭ PURGE CUT-OFF SOLENOID VALVE
⑮ PURGE CONTROL DIAPHRAGM VALVE
⑯ TWO-WAY VALVE

Fig. 60 1990 Civic/CRX 1.6L—all models

4-32 EMISSION CONTROLS

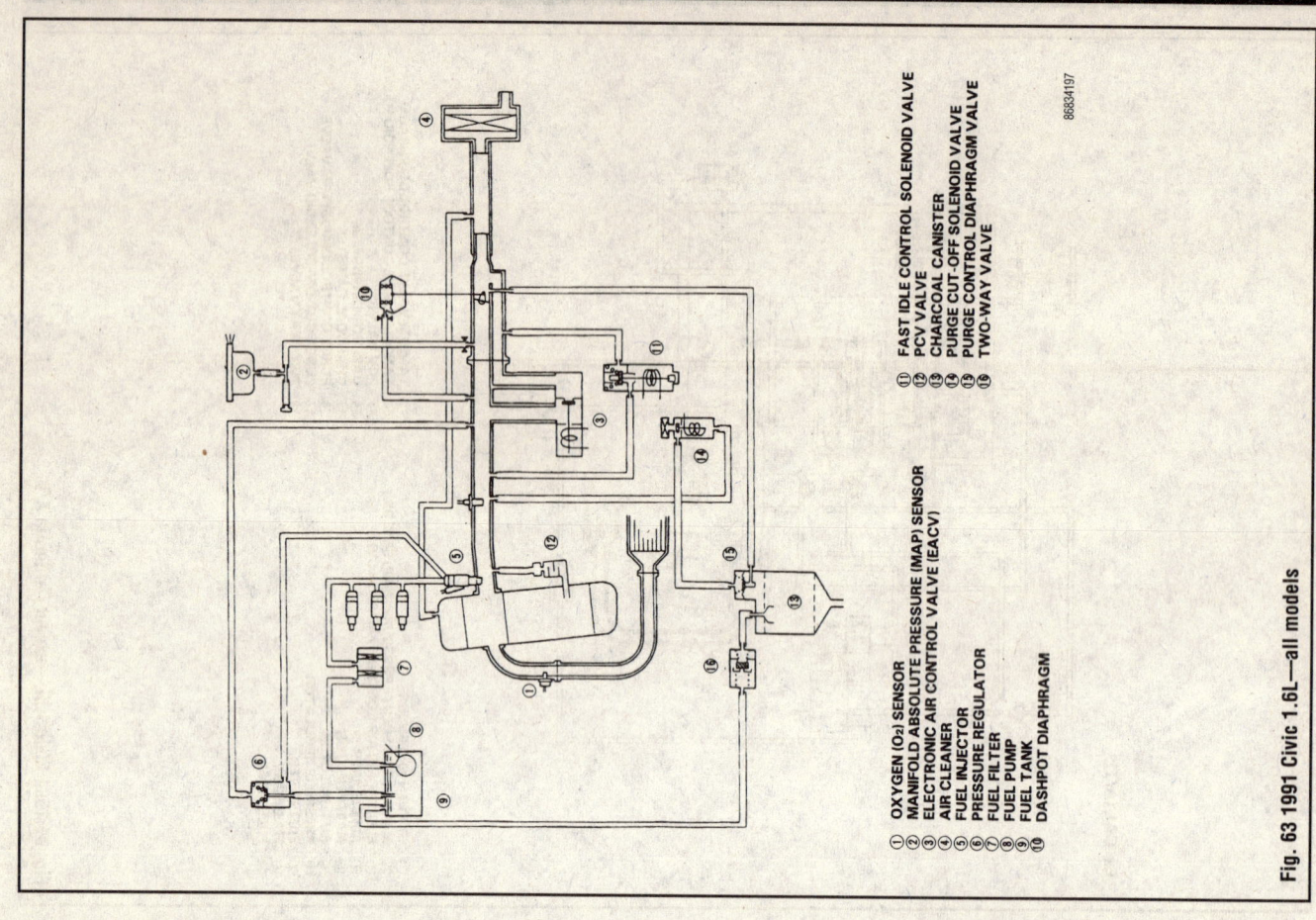

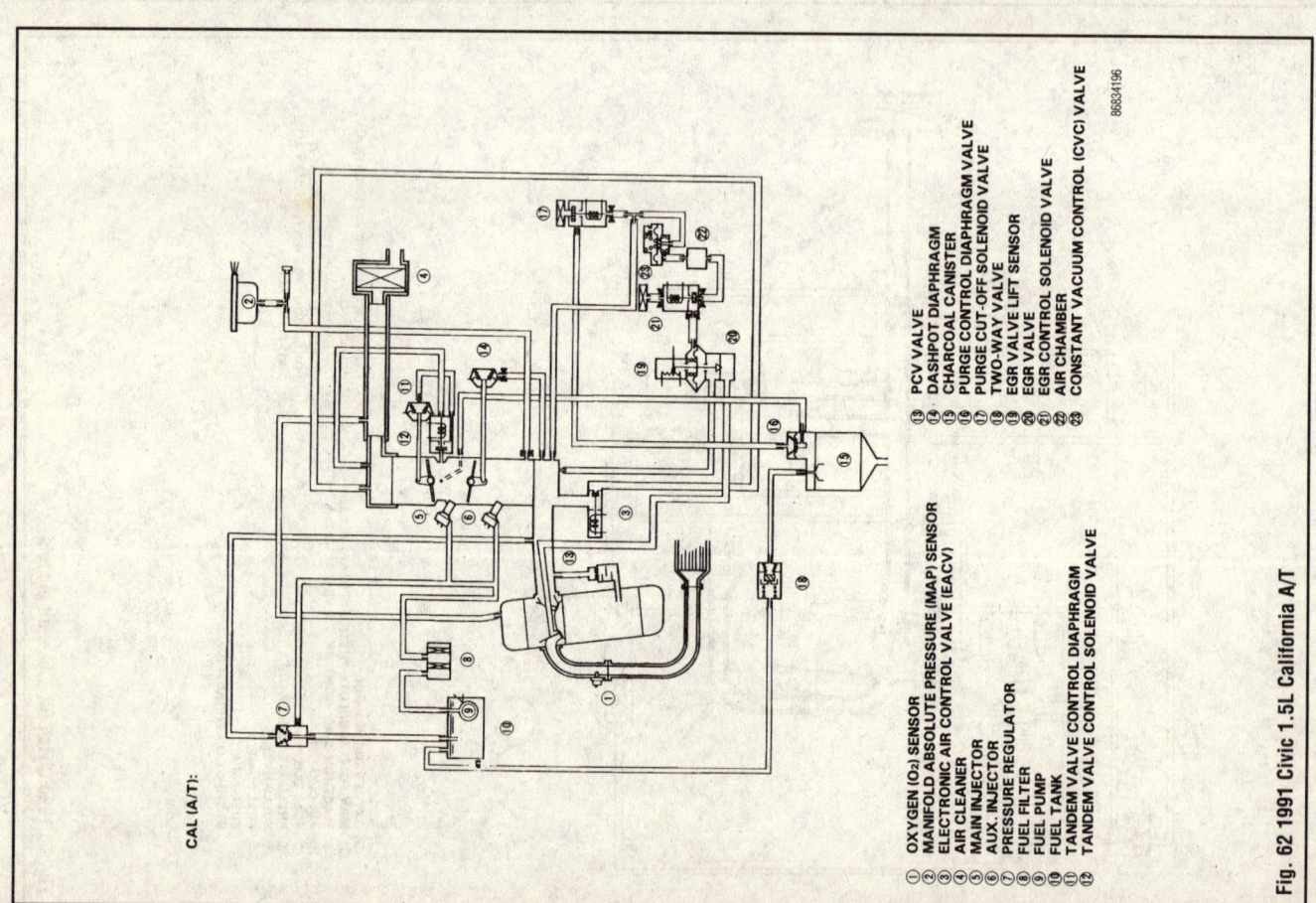

① OXYGEN (O₂) SENSOR
② MANIFOLD ABSOLUTE PRESSURE (MAP) SENSOR
③ ELECTRONIC AIR CONTROL VALVE (EACV)
④ AIR CLEANER
⑤ FUEL INJECTOR
⑥ PRESSURE REGULATOR
⑦ FUEL FILTER
⑧ FUEL PUMP
⑨ FUEL TANK
⑩ DASHPOT DIAPHRAGM
⑪ FAST IDLE CONTROL SOLENOID VALVE
⑫ PCV VALVE
⑬ CHARCOAL CANISTER
⑭ PURGE CUT-OFF SOLENOID VALVE
⑮ PURGE CONTROL DIAPHRAGM VALVE
⑯ TWO-WAY VALVE

Fig. 63 1991 Civic 1.6L—all models

CAL (A/T):

① OXYGEN (O₂) SENSOR
② MANIFOLD ABSOLUTE PRESSURE (MAP) SENSOR
③ ELECTRONIC AIR CONTROL VALVE (EACV)
④ AIR CLEANER
⑤ MAIN INJECTOR
⑥ AUX. INJECTOR
⑦ PRESSURE REGULATOR
⑧ FUEL FILTER
⑨ FUEL PUMP
⑩ FUEL TANK
⑪ TANDEM VALVE CONTROL DIAPHRAGM
⑫ TANDEM VALVE CONTROL SOLENOID VALVE
⑬ PCV VALVE
⑭ DASHPOT DIAPHRAGM
⑮ CHARCOAL CANISTER
⑯ PURGE CONTROL DIAPHRAGM VALVE
⑰ PURGE CUT-OFF SOLENOID VALVE
⑱ TWO-WAY VALVE
⑲ EGR VALVE LIFT SENSOR
⑳ EGR VALVE
㉑ EGR CONTROL SOLENOID VALVE
㉒ AIR CHAMBER
㉓ CONSTANT VACUUM CONTROL (CVC) VALVE

Fig. 62 1991 Civic 1.5L California A/T

EMISSION CONTROLS 4-33

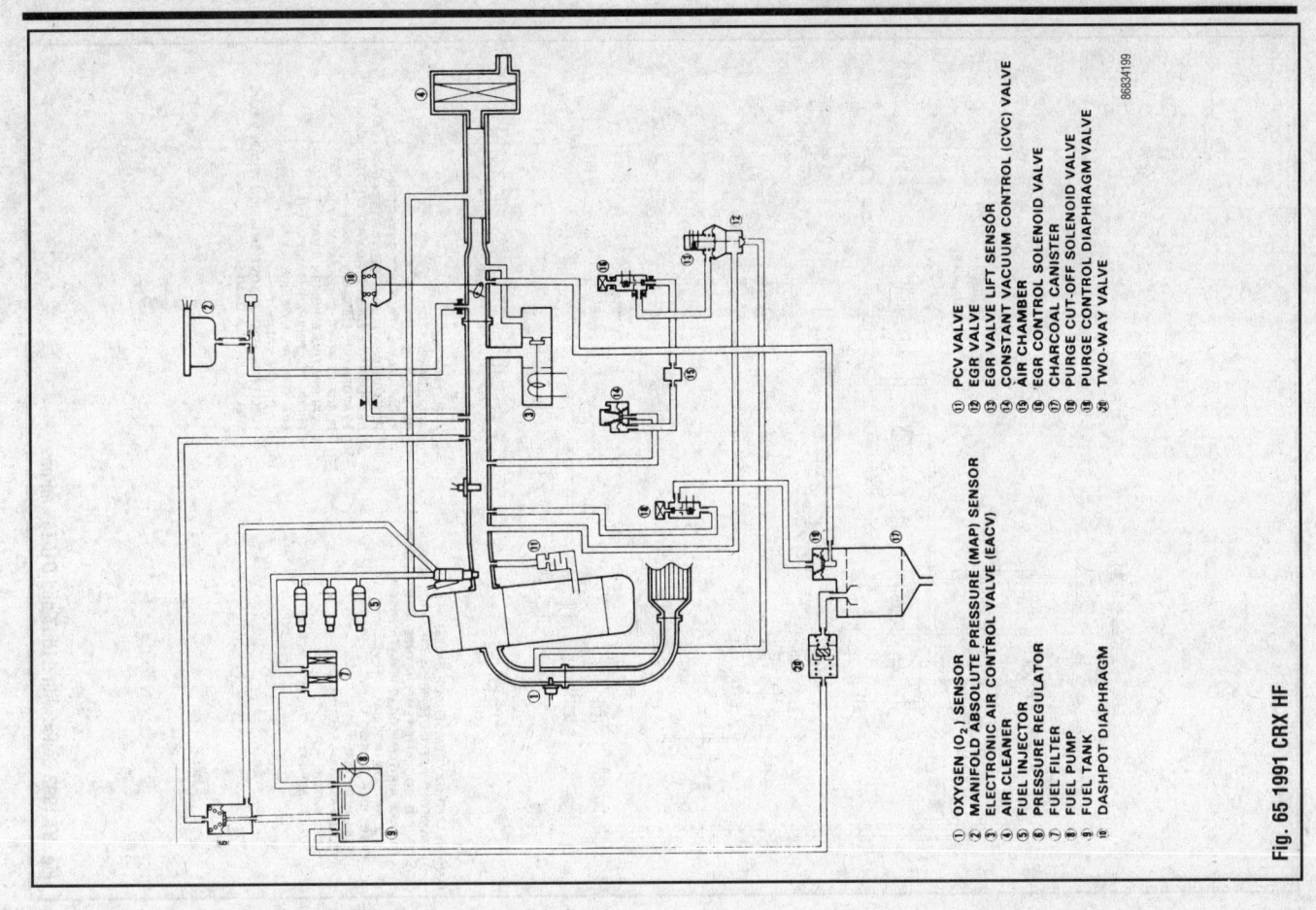

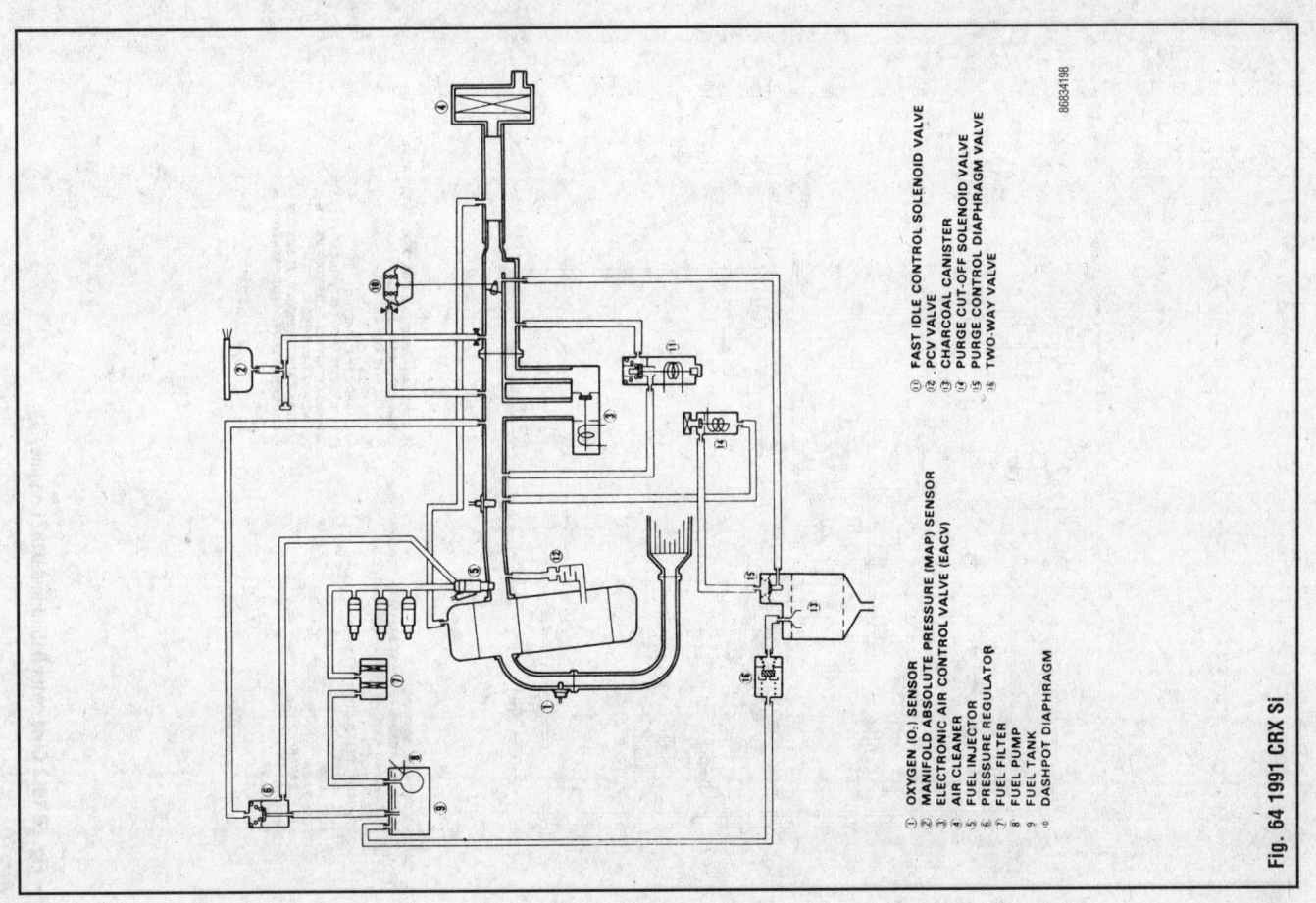

Fig. 65 1991 CRX HF

① OXYGEN (O₂) SENSOR
② MANIFOLD ABSOLUTE PRESSURE (MAP) SENSOR
③ ELECTRONIC AIR CONTROL VALVE (EACV)
④ AIR CLEANER
⑤ FUEL INJECTOR
⑥ PRESSURE REGULATOR
⑦ FUEL FILTER
⑧ FUEL PUMP
⑨ FUEL TANK
⑩ DASHPOT DIAPHRAGM
⑪ PCV VALVE
⑫ EGR VALVE
⑬ EGR VALVE LIFT SENSOR
⑭ CONSTANT VACUUM CONTROL (CVC) VALVE
⑮ AIR CHAMBER
⑯ EGR CONTROL SOLENOID VALVE
⑰ CHARCOAL CANISTER
⑱ PURGE CUT-OFF SOLENOID VALVE
⑲ PURGE CONTROL DIAPHRAGM VALVE
⑳ TWO-WAY VALVE

Fig. 64 1991 CRX Si

① OXYGEN (O₂) SENSOR
② MANIFOLD ABSOLUTE PRESSURE (MAP) SENSOR
③ ELECTRONIC AIR CONTROL VALVE (EACV)
④ AIR CLEANER
⑤ FUEL INJECTOR
⑥ PRESSURE REGULATOR
⑦ FUEL FILTER
⑧ FUEL PUMP
⑨ FUEL TANK
⑩ DASHPOT DIAPHRAGM
⑪ FAST IDLE CONTROL SOLENOID VALVE
⑫ PCV VALVE
⑬ CHARCOAL CANISTER
⑭ PURGE CUT-OFF SOLENOID VALVE
⑮ PURGE CONTROL DIAPHRAGM VALVE
⑯ TWO-WAY VALVE

4-34 EMISSION CONTROLS

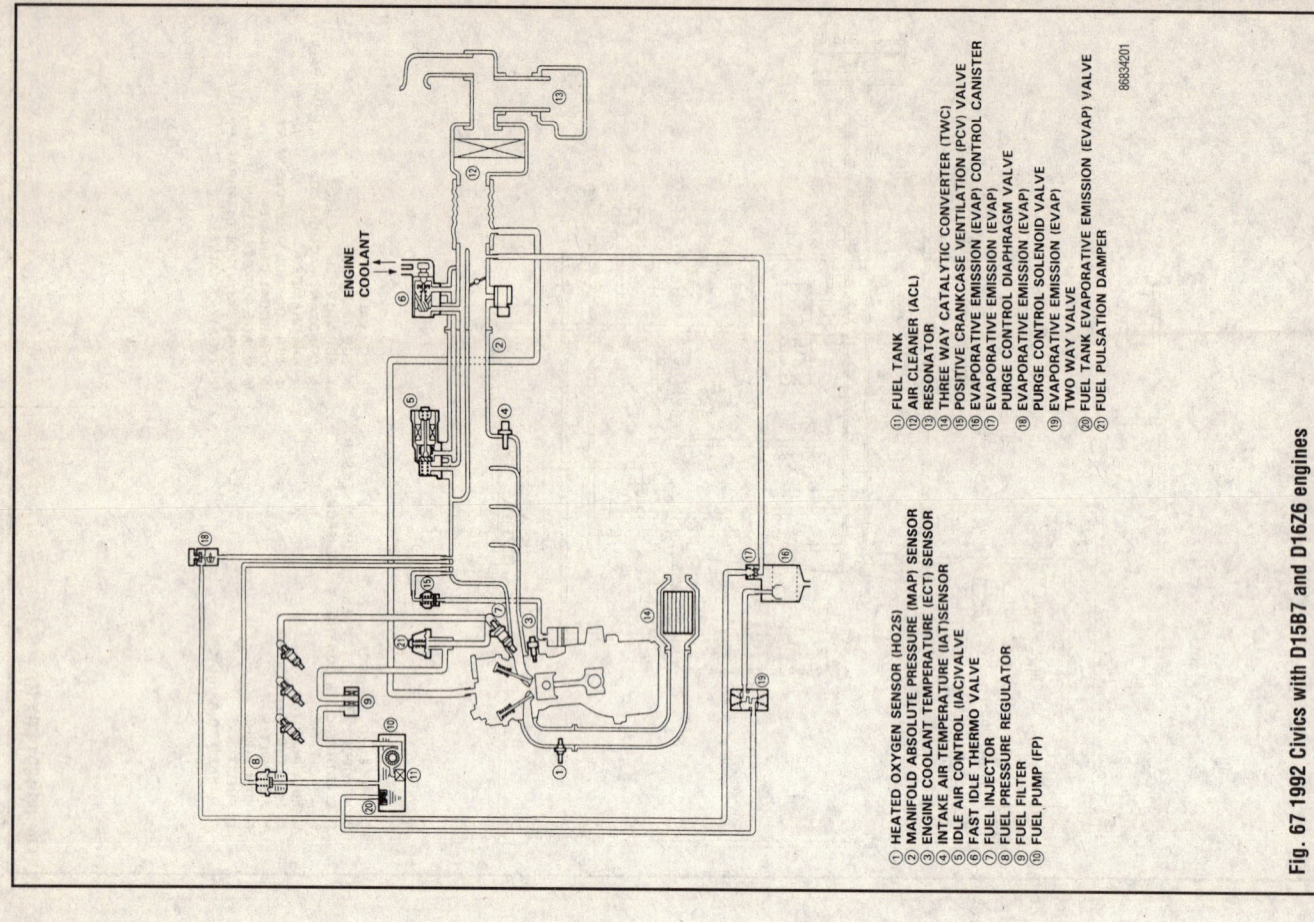

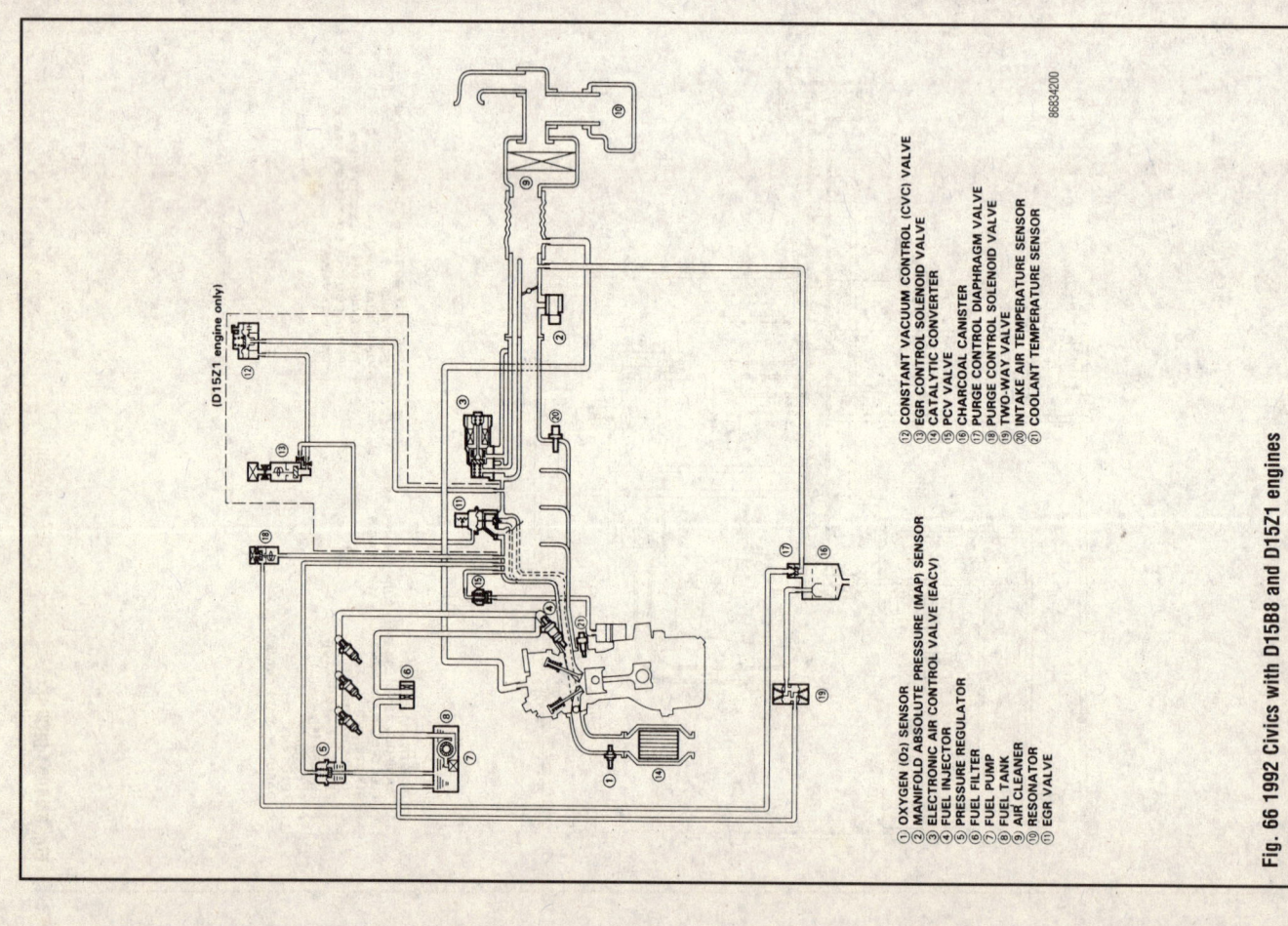

① HEATED OXYGEN SENSOR (HO2S)
② MANIFOLD ABSOLUTE PRESSURE (MAP) SENSOR
③ ENGINE COOLANT TEMPERATURE (ECT) SENSOR
④ INTAKE AIR TEMPERATURE (IAT) SENSOR
⑤ IDLE AIR CONTROL (IAC) VALVE
⑥ FAST IDLE THERMO VALVE
⑦ FUEL INJECTOR
⑧ FUEL PRESSURE REGULATOR
⑨ FUEL FILTER
⑩ FUEL PUMP (FP)
⑪ FUEL TANK
⑫ AIR CLEANER (ACL)
⑬ RESONATOR
⑭ THREE WAY CATALYTIC CONVERTER (TWC)
⑮ POSITIVE CRANKCASE VENTILATION (PCV) VALVE
⑯ EVAPORATIVE EMISSION (EVAP) CONTROL CANISTER
⑰ EVAPORATIVE EMISSION (EVAP)
⑱ PURGE CONTROL DIAPHRAGM VALVE
⑲ EVAPORATIVE EMISSION (EVAP)
⑳ PURGE CONTROL SOLENOID VALVE
㉑ EVAPORATIVE EMISSION (EVAP)
㉒ TWO WAY VALVE
㉓ FUEL TANK EVAPORATIVE EMISSION (EVAP) VALVE
㉔ FUEL PULSATION DAMPER

Fig. 67 1992 Civics with D15B7 and D16Z6 engines

① OXYGEN (O2) SENSOR
② MANIFOLD ABSOLUTE PRESSURE (MAP) SENSOR
③ ELECTRONIC AIR CONTROL VALVE (EACV)
④ FUEL INJECTOR
⑤ PRESSURE REGULATOR
⑥ FUEL FILTER
⑦ FUEL PUMP
⑧ FUEL TANK
⑨ AIR CLEANER
⑩ RESONATOR
⑪ EGR VALVE
⑫ CONSTANT VACUUM CONTROL (CVC) VALVE
⑬ EGR CONTROL SOLENOID VALVE
⑭ CATALYTIC CONVERTER
⑮ PCV VALVE
⑯ CHARCOAL CANISTER
⑰ PURGE CONTROL DIAPHRAGM VALVE
⑱ PURGE CONTROL SOLENOID VALVE
⑲ TWO-WAY VALVE
⑳ INTAKE AIR TEMPERATURE SENSOR
㉑ COOLANT TEMPERATURE SENSOR

Fig. 66 1992 Civics with D15B8 and D15Z1 engines

EMISSION CONTROLS 4-35

① OXYGEN SENSOR(O2S)
   [D15B8 engine (except California model)]
① HEATED OXYGEN SENSOR (HO2S)
② MANIFOLD ABSOLUTE PRESSURE (MAP) SENSOR
③ ENGINE COOLANT TEMPERATURE (ECT) SENSOR
④ INTAKE AIR TEMPERATURE (IAT) SENSOR
⑤ IDLE AIR CONTROL (IAC) VALVE
⑥ FUEL INJECTOR
⑦ FUEL PRESSURE REGULATOR
⑧ FUEL FILTER
⑨ FUEL PUMP (FP)
⑩ FUEL TANK
⑪ AIR CLEANER (ACL)
⑫ RESONATOR
⑬ EXHAUST GAS RECIRCULATION (EGRI) VALVE
   [D15Z1 engine]
⑭ EXHAUST GAS RECIRCULATION (EGR)
   VACUUM CONTROL VALVE [D15Z1 engine]
⑮ EXHAUST GAS RECIRCULATION (EGR)
   CONTROL SOLENOID VALVE [D15Z1 engine]
⑯ THREE WAY CATALYTIC CONVERTER (TWC)
⑰ POSITIVE CRANKCASE VENTILATION (PCV) VALVE
⑱ EVAPORATIVE EMISSION (EVAP) CONTROL CANISTER
⑲ EVAPORATIVE EMISSION (EVAP)
⑳ PURGE CONTROL DIAPHRAGM VALVE
㉑ PURGE CONTROL SOLENOID VALVE
   EVAPORATIVE EMISSION (EVAP)
㉒ TWO WAY VALVE
   FUEL TANK EVAPORATIVE EMISSION (EVAP) VALVE

Fig. 69 1993 Civics with D15B8 and D15Z1 engines

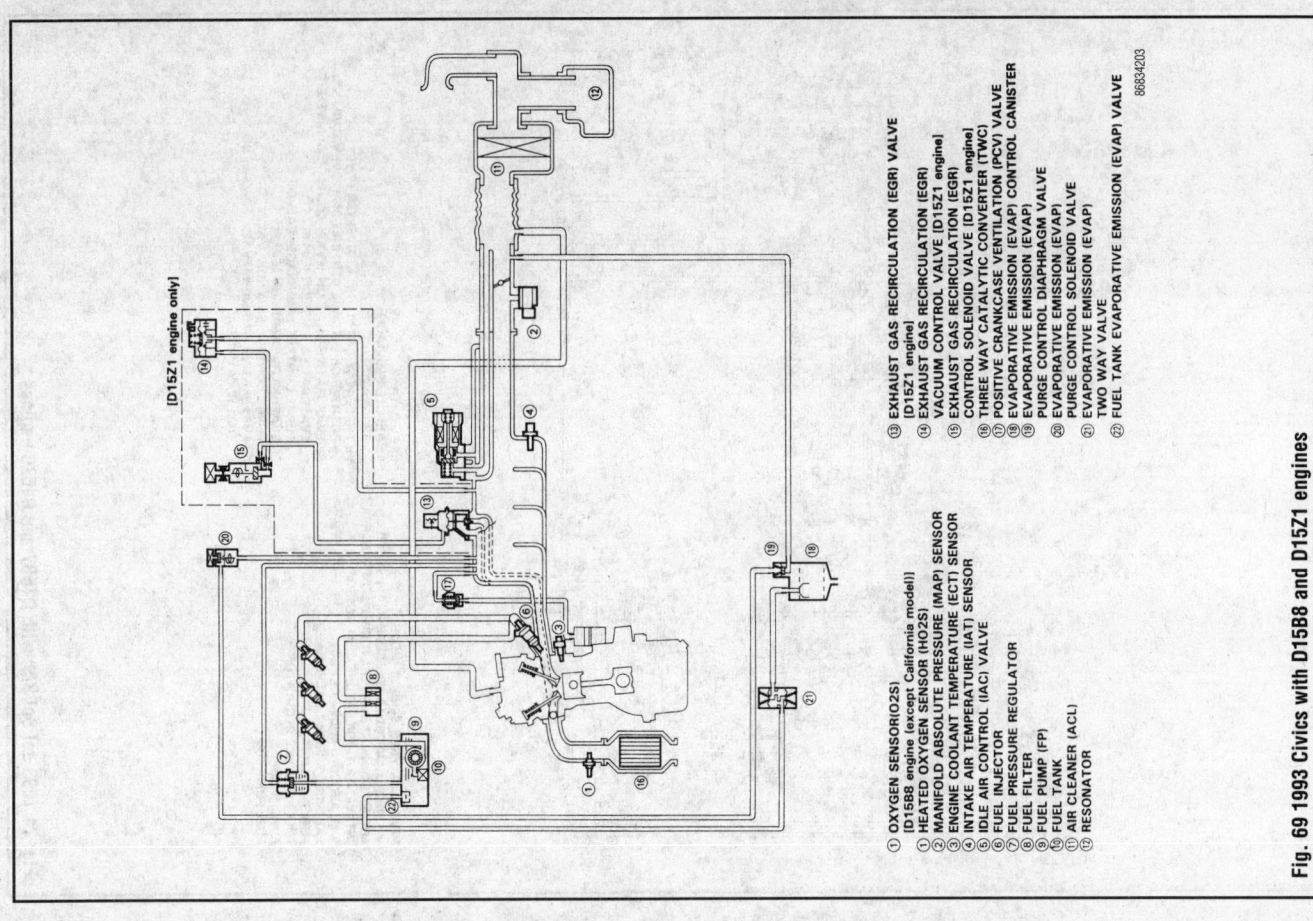

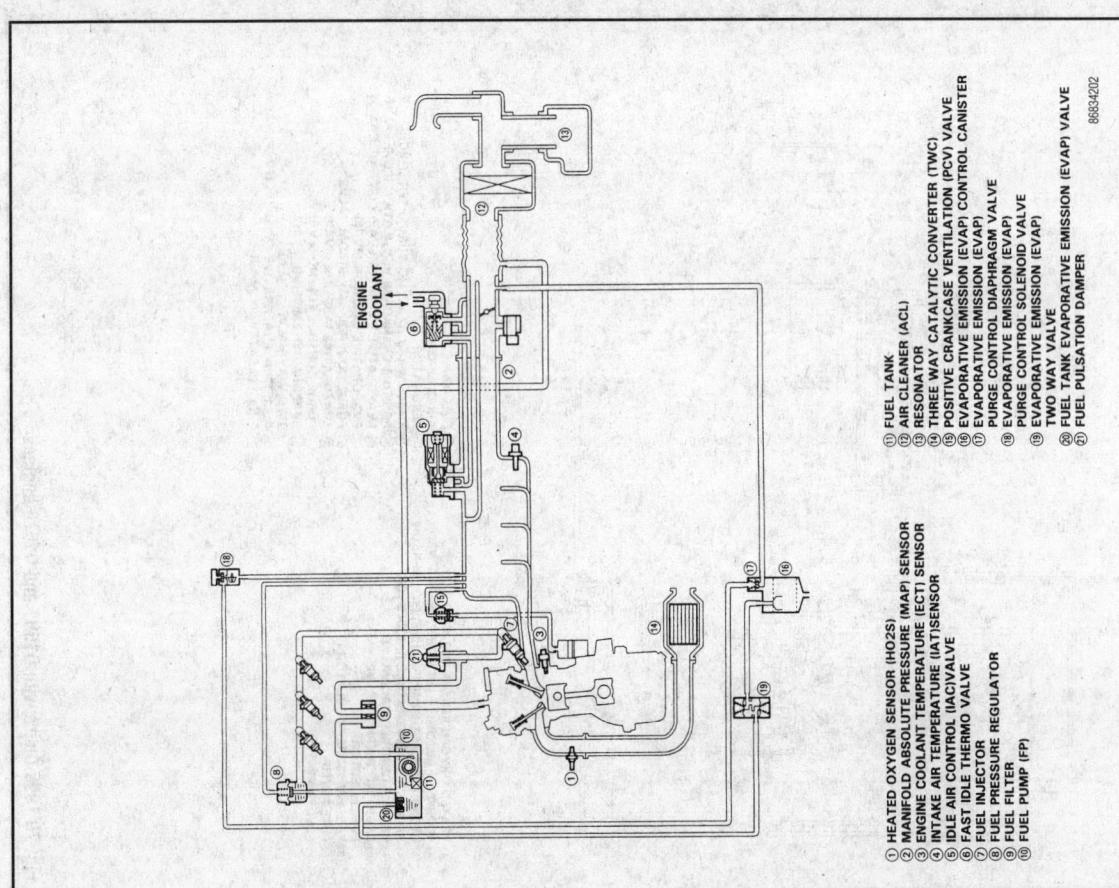

① HEATED OXYGEN SENSOR (HO2S)
② MANIFOLD ABSOLUTE PRESSURE (MAP) SENSOR
③ ENGINE COOLANT TEMPERATURE (ECT) SENSOR
④ INTAKE AIR TEMPERATURE (IAT) SENSOR
⑤ IDLE AIR CONTROL (IAC) VALVE
⑥ FAST IDLE THERMO VALVE
⑦ FUEL INJECTOR
⑧ FUEL PRESSURE REGULATOR
⑨ FUEL FILTER
⑩ FUEL PUMP (FP)
⑪ FUEL TANK
⑫ AIR CLEANER (ACL)
⑬ RESONATOR
⑭ THREE WAY CATALYTIC CONVERTER (TWC)
⑮ POSITIVE CRANKCASE VENTILATION (PCV) VALVE
⑯ EVAPORATIVE EMISSION (EVAP) CONTROL CANISTER
⑰ EVAPORATIVE EMISSION (EVAP)
   PURGE CONTROL DIAPHRAGM VALVE
⑱ EVAPORATIVE EMISSION (EVAP)
   PURGE CONTROL SOLENOID VALVE
⑲ EVAPORATIVE EMISSION (EVAP)
⑳ TWO WAY VALVE
㉑ FUEL TANK EVAPORATIVE EMISSION (EVAP) VALVE
   FUEL PULSATION DAMPER

Fig. 68 1993 Civics with D15B7 and D16Z6 engines

4-36 EMISSION CONTROLS

① HEATED OXYGEN SENSOR (HO2S)
② MANIFOLD ABSOLUTE PRESSURE (MAP) SENSOR
③ ENGINE COOLANT TEMPERATURE (ECT) SENSOR
④ INTAKE AIR TEMPERATURE (IAT) SENSOR
⑤ IDLE AIR CONTROL (IAC) VALVE
⑥ FAST IDLE THERMO VALVE
⑦ FUEL INJECTOR
⑧ FUEL PRESSURE REGULATOR
⑨ FUEL FILTER
⑩ FUEL PUMP (FP)
⑪ FUEL TANK
⑫ AIR CLEANER (ACL)
⑬ RESONATOR
⑭ THREE WAY CATALYTIC CONVERTER (TWC)
⑮ POSITIVE CRANKCASE VENTILATION (PCV) VALVE
⑯ EVAPORATIVE EMISSION (EVAP) CONTROL CANISTER
⑰ EVAPORATIVE EMISSION (EVAP)
⑱ PURGE CONTROL DIAPHRAGM VALVE
⑲ PURGE CONTROL SOLENOID VALVE
⑳ EVAPORATIVE EMISSION (EVAP)
TWO WAY VALVE

Fig. 71 1993 del Sol S/Si with D15B7 and D16Z6 engines

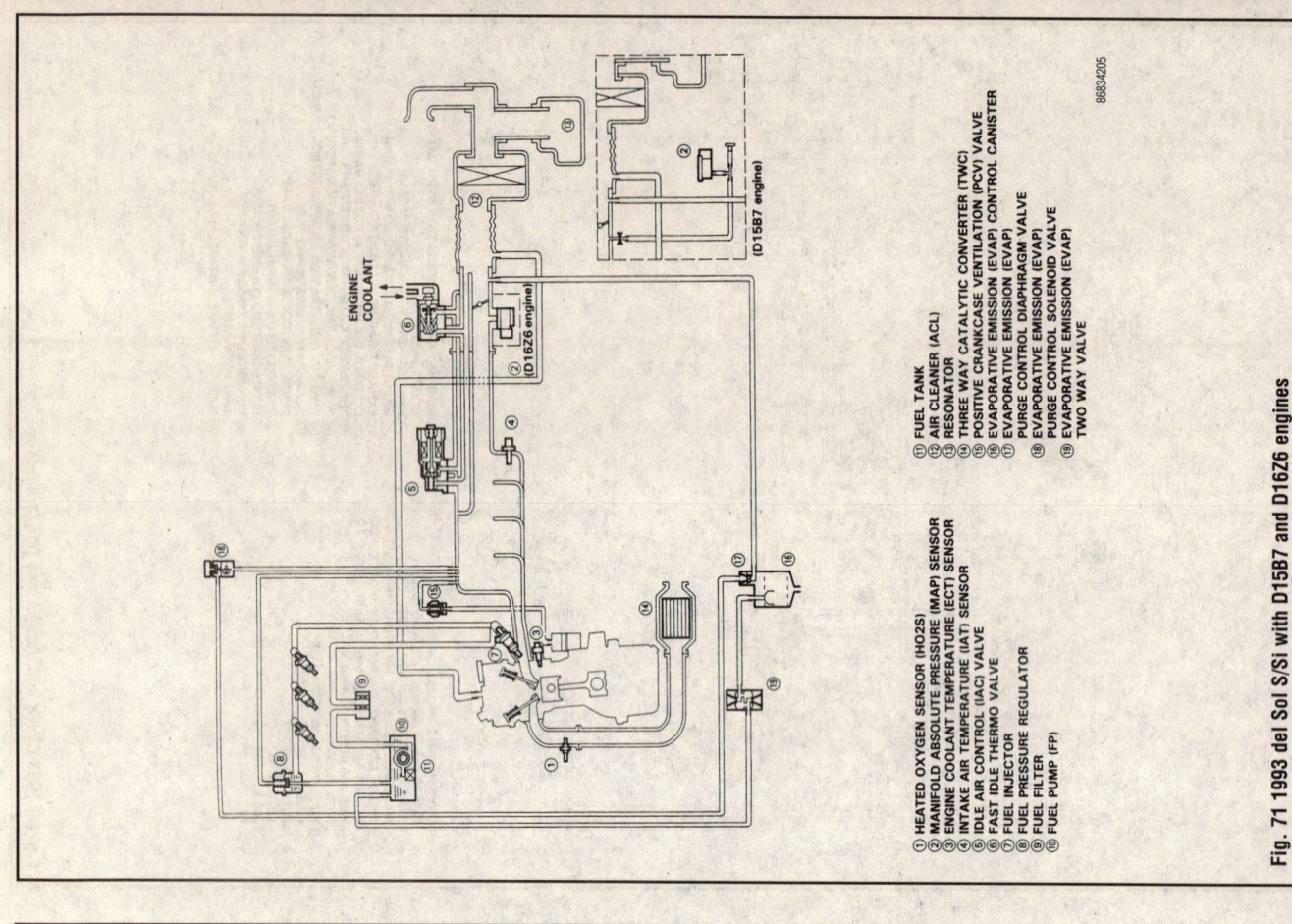

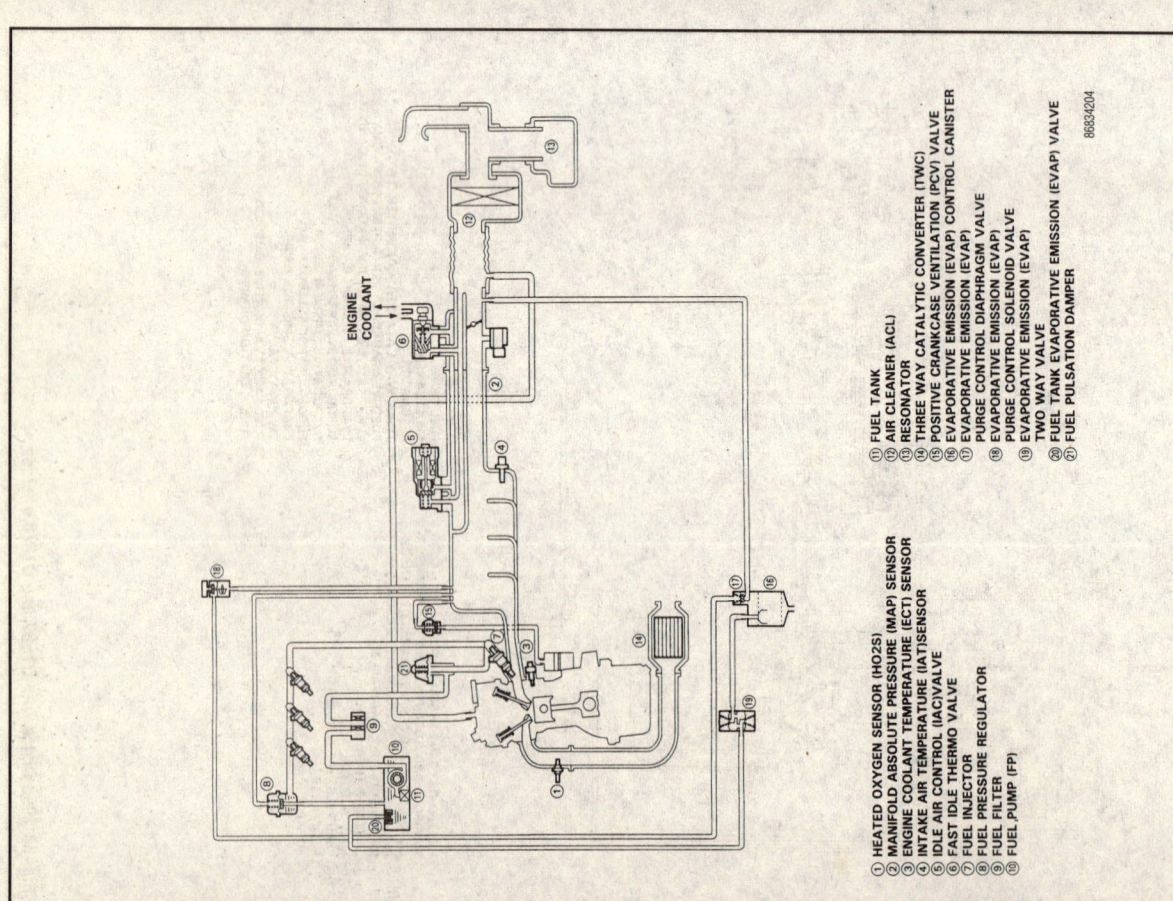

① HEATED OXYGEN SENSOR (HO2S)
② MANIFOLD ABSOLUTE PRESSURE (MAP) SENSOR
③ ENGINE COOLANT TEMPERATURE (ECT) SENSOR
④ INTAKE AIR TEMPERATURE (IAT) VALVE
⑤ IDLE AIR CONTROL (IAC) VALVE
⑥ FAST IDLE THERMO VALVE
⑦ FUEL INJECTOR
⑧ FUEL PRESSURE REGULATOR
⑨ FUEL FILTER
⑩ FUEL PUMP (FP)
⑪ FUEL TANK
⑫ AIR CLEANER (ACL)
⑬ RESONATOR
⑭ THREE WAY CATALYTIC CONVERTER (TWC)
⑮ POSITIVE CRANKCASE VENTILATION (PCV) VALVE
⑯ EVAPORATIVE EMISSION (EVAP) CONTROL CANISTER
⑰ EVAPORATIVE EMISSION (EVAP)
⑱ PURGE CONTROL DIAPHRAGM VALVE
⑲ PURGE CONTROL SOLENOID VALVE
⑳ EVAPORATIVE EMISSION (EVAP)
TWO WAY VALVE
㉑ FUEL TANK EVAPORATIVE EMISSION (EVAP) VALVE
㉒ FUEL PULSATION DAMPER

Fig. 70 1993 Civics with D15B7 and D16Z6 engines

EMISSION CONTROLS 4-37

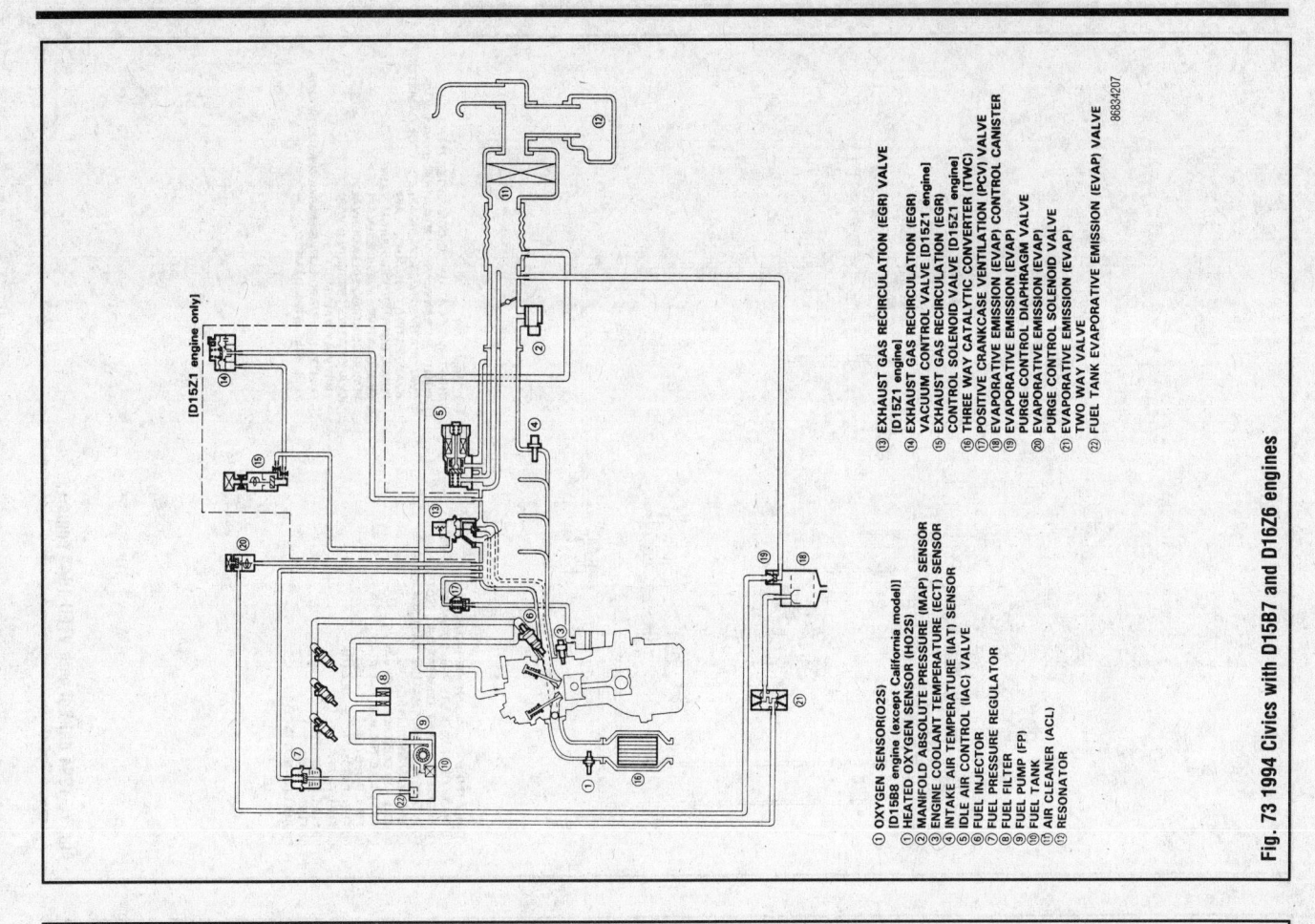

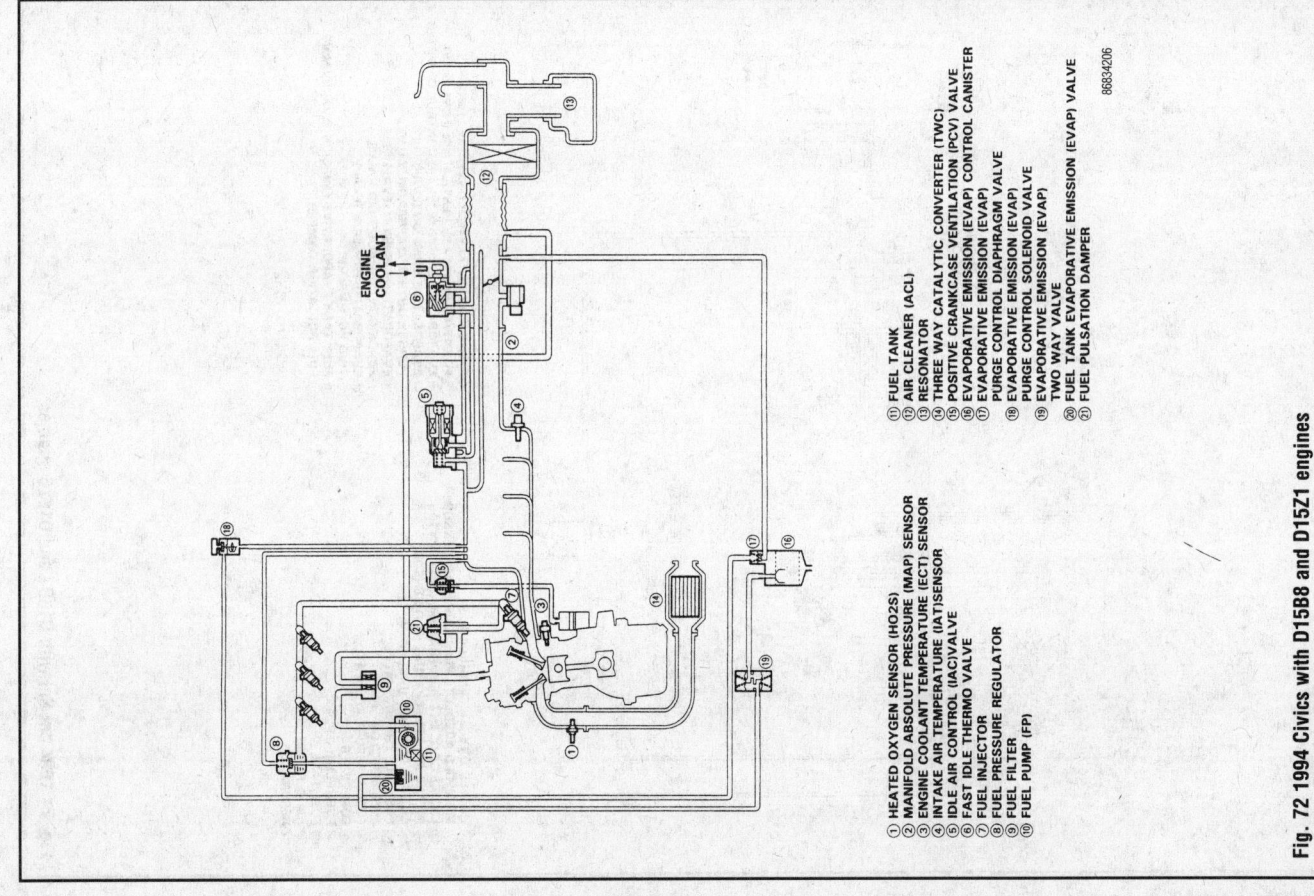

① OXYGEN SENSOR (O2S) [D15B8 engine (except California model)]
① HEATED OXYGEN SENSOR (HO2S)
② MANIFOLD ABSOLUTE PRESSURE (MAP) SENSOR
③ ENGINE COOLANT TEMPERATURE (ECT) SENSOR
④ INTAKE AIR TEMPERATURE (IAT) SENSOR
⑤ IDLE AIR CONTROL (IAC) VALVE
⑥ FUEL INJECTOR
⑦ FUEL PRESSURE REGULATOR
⑧ FUEL FILTER
⑨ FUEL PUMP (FP)
⑩ FUEL TANK
⑪ AIR CLEANER (ACL)
⑫ RESONATOR
⑬ EXHAUST GAS RECIRCULATION (EGR) VALVE [D15Z1 engine]
⑭ EXHAUST GAS RECIRCULATION (EGR) VACUUM CONTROL VALVE [D15Z1 engine]
⑮ EXHAUST GAS RECIRCULATION (EGR) CONTROL SOLENOID VALVE [D15Z1 engine]
⑯ THREE WAY CATALYTIC CONVERTER (TWC)
⑰ POSITIVE CRANKCASE VENTILATION (PCV) VALVE
⑱ EVAPORATIVE EMISSION (EVAP) CONTROL CANISTER
⑲ PURGE CONTROL DIAPHRAGM VALVE
⑳ PURGE CONTROL SOLENOID VALVE
㉑ EVAPORATIVE EMISSION (EVAP) TWO WAY VALVE
㉒ FUEL TANK EVAPORATIVE EMISSION (EVAP) VALVE

Fig. 73 1994 Civics with D15B7 and D1626 engines

① HEATED OXYGEN SENSOR (HO2S)
② MANIFOLD ABSOLUTE PRESSURE (MAP) SENSOR
③ ENGINE COOLANT TEMPERATURE (ECT) SENSOR
④ INTAKE AIR TEMPERATURE (IAT) SENSOR
⑤ IDLE AIR CONTROL (IAC) VALVE
⑥ FAST IDLE THERMO VALVE
⑦ FUEL INJECTOR
⑧ FUEL PRESSURE REGULATOR
⑨ FUEL FILTER
⑩ FUEL PUMP (FP)
⑪ FUEL TANK
⑫ AIR CLEANER (ACL)
⑬ RESONATOR
⑭ THREE WAY CATALYTIC CONVERTER (TWC)
⑮ POSITIVE CRANKCASE VENTILATION (PCV) VALVE
⑯ EVAPORATIVE EMISSION (EVAP) CONTROL CANISTER
⑰ EVAPORATIVE EMISSION (EVAP)
⑱ PURGE CONTROL DIAPHRAGM VALVE
⑲ PURGE CONTROL SOLENOID VALVE
⑳ TWO WAY VALVE
㉑ FUEL TANK EVAPORATIVE EMISSION (EVAP) VALVE
㉒ FUEL PULSATION DAMPER

Fig. 72 1994 Civics with D15B8 and D15Z1 engines

4-38 EMISSION CONTROLS

① HEATED OXYGEN SENSOR (HO2S)
② MANIFOLD ABSOLUTE PRESSURE (MAP) SENSOR
③ ENGINE COOLANT TEMPERATURE (ECT) SENSOR
④ INTAKE AIR TEMPERATURE (IAT) SENSOR
⑤ KNOCK SENSOR (KS)
⑥ IDLE AIR CONTROL (IAC) VALVE
⑦ FAST IDLE THERMO VALVE
⑧ FUEL INJECTOR
⑨ FUEL PRESSURE REGULATOR
⑩ FUEL FILTER
⑪ FUEL PUMP (FP)
⑫ FUEL TANK
⑬ AIR CLEANER (ACL)
⑭ RESONATOR
⑮ THREE WAY CATALYTIC CONVERTER (TWC)
⑯ POSITIVE CRANKCASE VENTILATION (PCV) VALVE
⑰ EVAPORATIVE EMISSION (EVAP) CONTROL CANISTER
⑱ EVAPORATIVE EMISSION (EVAP)
⑲ PURGE CONTROL DIAPHRAGM VALVE
⑳ PURGE CONTROL SOLENOID VALVE
㉑ EVAPORATIVE EMISSION (EVAP)
㉒ TWO WAY VALVE
㉓ FUEL TANK EVAPORATIVE EMISSION (EVAP) VALVE
㉔ FUEL PULSATION DAMPER

Fig. 75 1994 del Sol with VTEC 16A3 engine

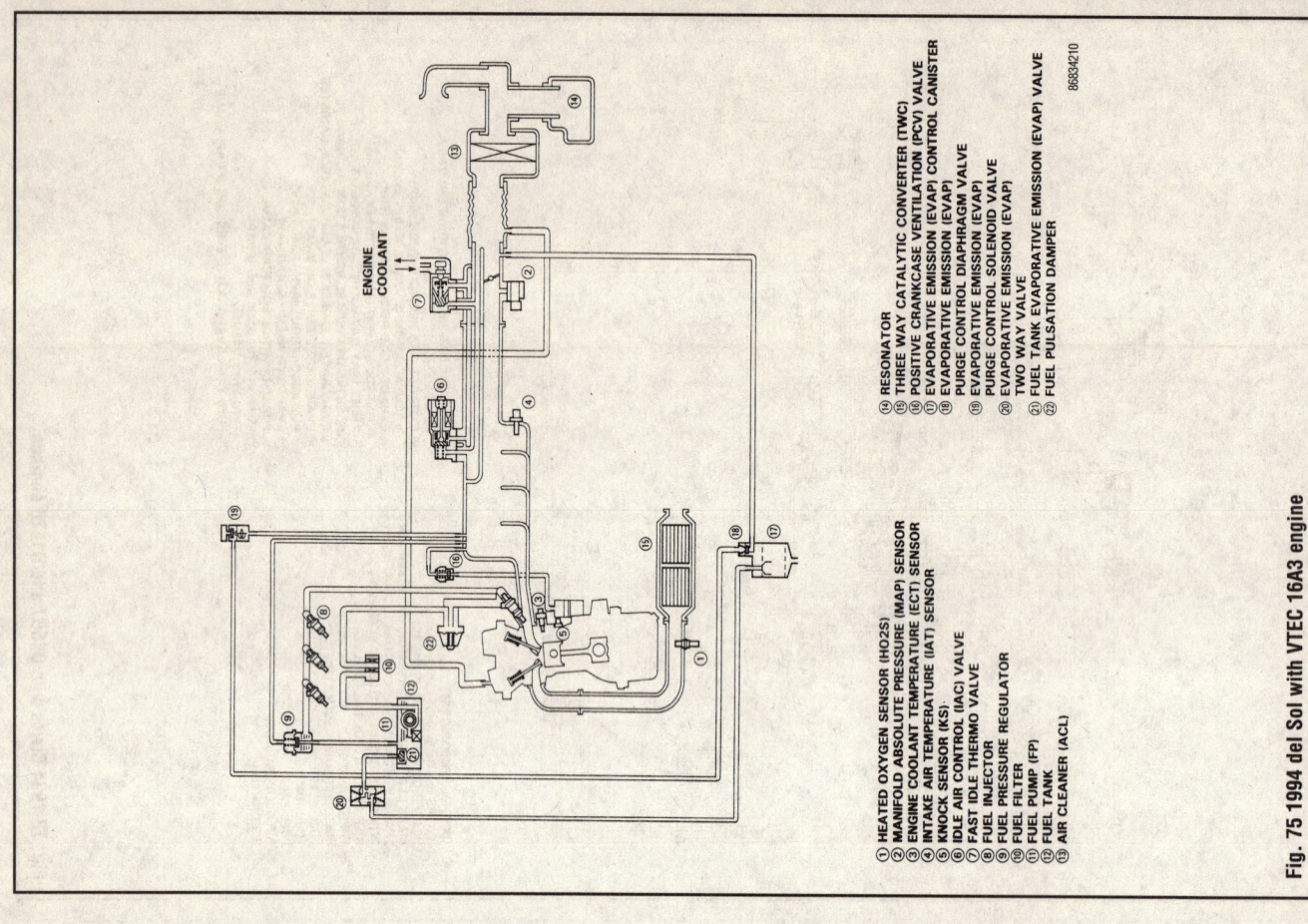

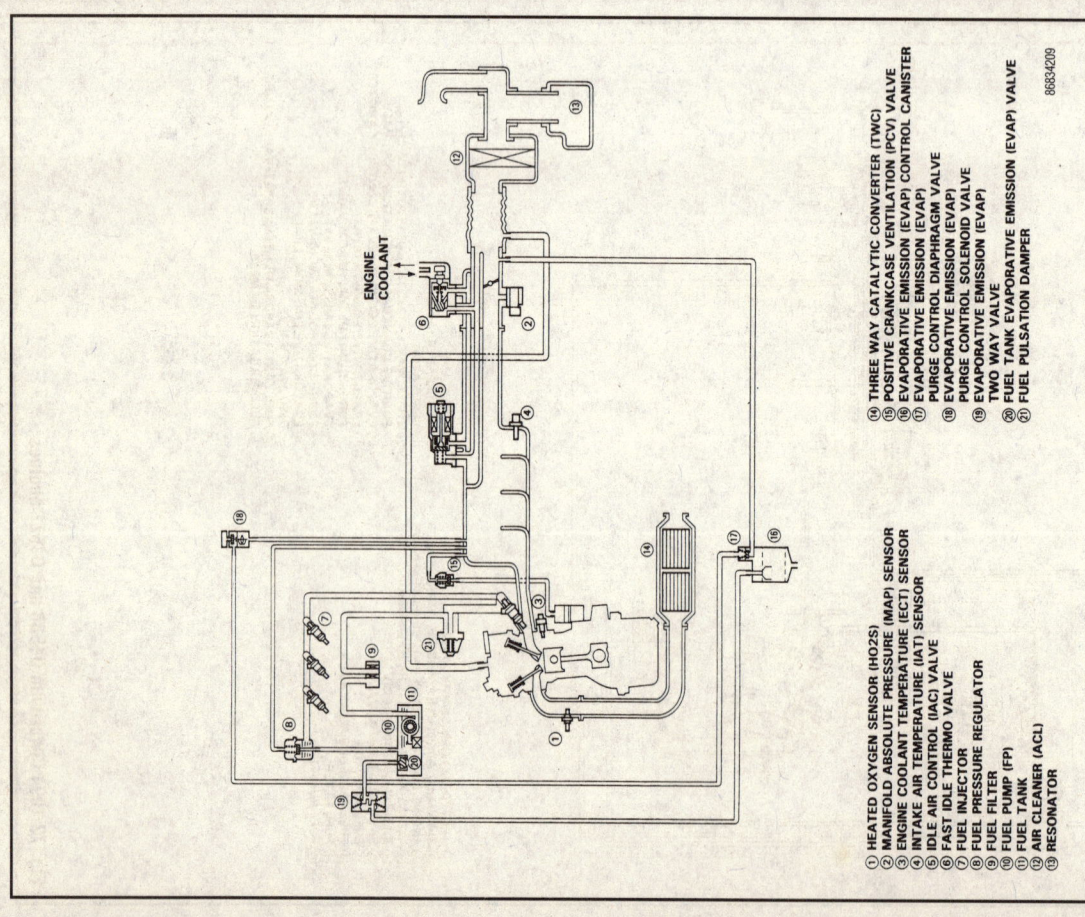

① HEATED OXYGEN SENSOR (HO2S)
② MANIFOLD ABSOLUTE PRESSURE (MAP) SENSOR
③ ENGINE COOLANT TEMPERATURE (ECT) SENSOR
④ INTAKE AIR TEMPERATURE (IAT) SENSOR
⑤ IDLE AIR CONTROL (IAC) VALVE
⑥ FAST IDLE THERMO VALVE
⑦ FUEL INJECTOR
⑧ FUEL PRESSURE REGULATOR
⑨ FUEL FILTER
⑩ FUEL PUMP (FP)
⑪ FUEL TANK
⑫ AIR CLEANER (ACL)
⑬ RESONATOR
⑭ THREE WAY CATALYTIC CONVERTER (TWC)
⑮ POSITIVE CRANKCASE VENTILATION (PCV) VALVE
⑯ EVAPORATIVE EMISSION (EVAP) CONTROL CANISTER
⑰ EVAPORATIVE EMISSION (EVAP)
⑱ PURGE CONTROL DIAPHRAGM VALVE
⑲ PURGE CONTROL SOLENOID VALVE
⑳ EVAPORATIVE EMISSION (EVAP)
㉑ TWO WAY VALVE
㉒ FUEL TANK EVAPORATIVE EMISSION (EVAP) VALVE
㉓ FUEL PULSATION DAMPER

Fig. 74 1994 del Sol with D15B7 and D16Z6 engines

EMISSION CONTROLS 4-39

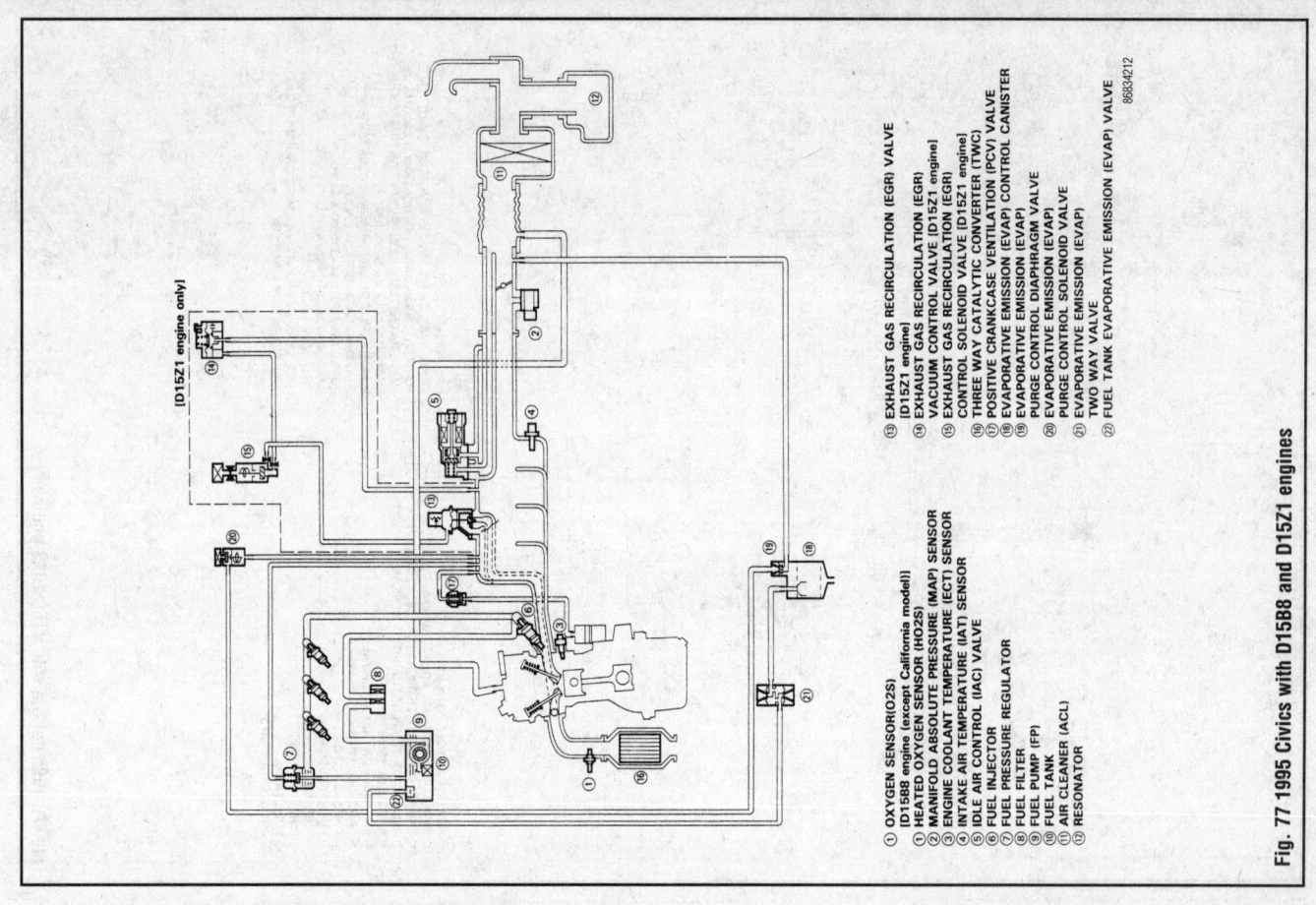

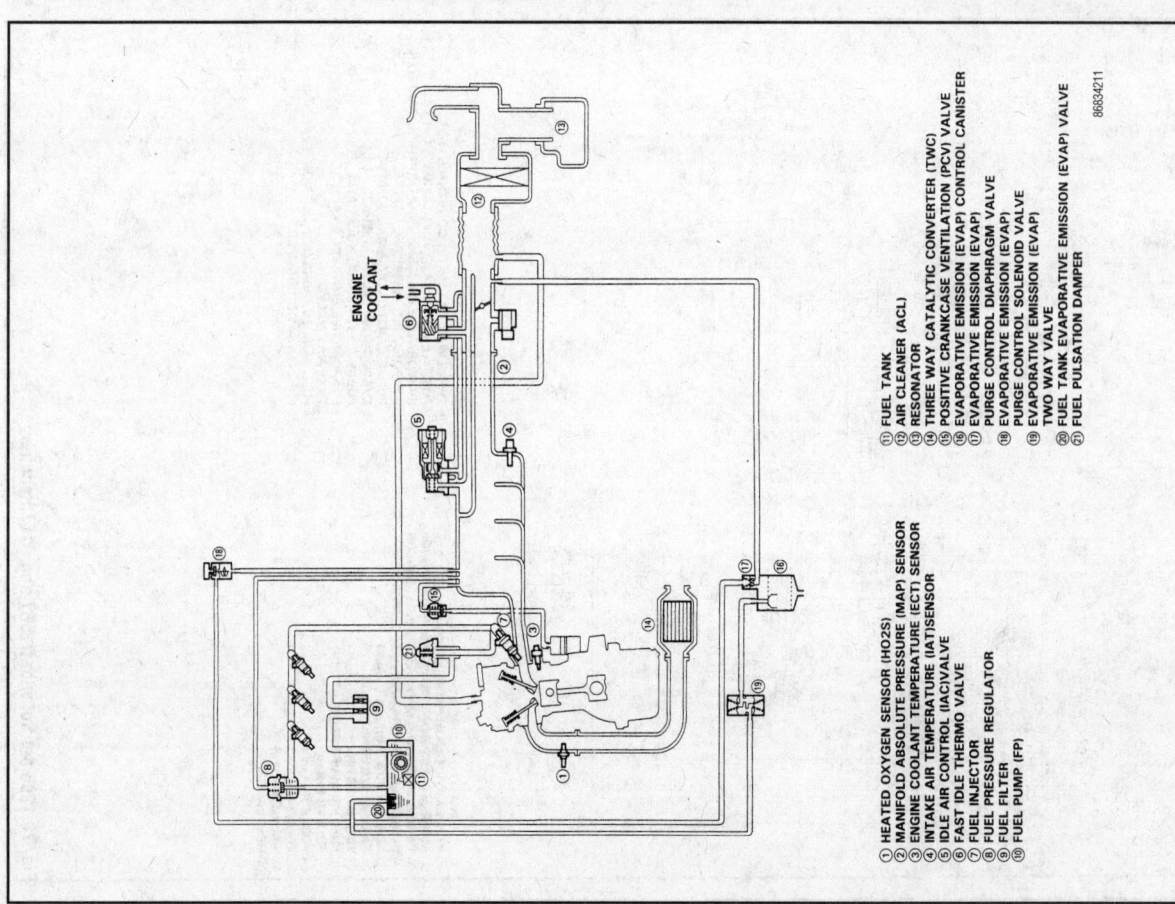

① OXYGEN SENSOR (O2S)
  (D15B8 engine (except California model))
② HEATED OXYGEN SENSOR (HO2S)
  (D15Z1 engine)
③ MANIFOLD ABSOLUTE PRESSURE (MAP) SENSOR
④ ENGINE COOLANT TEMPERATURE (ECT) SENSOR
⑤ INTAKE AIR TEMPERATURE (IAT) SENSOR
⑥ IDLE AIR CONTROL (IAC) VALVE
⑦ FUEL INJECTOR
⑧ FUEL PRESSURE REGULATOR
⑨ FUEL FILTER
⑩ FUEL PUMP (FP)
⑪ FUEL TANK
⑫ AIR CLEANER (ACL)
⑬ RESONATOR
⑭ EXHAUST GAS RECIRCULATION (EGR) VALVE
  (D15Z1 engine)
⑮ EXHAUST GAS RECIRCULATION (EGR)
  VACUUM CONTROL VALVE (D15Z1 engine)
⑯ EXHAUST GAS RECIRCULATION (EGR)
  CONTROL SOLENOID VALVE (D15Z1 engine)
⑰ THREE WAY CATALYTIC CONVERTER (TWC)
⑱ POSITIVE CRANKCASE VENTILATION (PCV) VALVE
⑲ EVAPORATIVE EMISSION (EVAP) CONTROL CANISTER
⑳ EVAPORATIVE EMISSION DIAPHRAGM VALVE
㉑ PURGE CONTROL SOLENOID VALVE
㉒ EVAPORATIVE EMISSION (EVAP)
  TWO WAY VALVE
㉓ FUEL TANK EVAPORATIVE EMISSION (EVAP) VALVE

Fig. 77 1995 Civics with D15B8 and D15Z1 engines

① HEATED OXYGEN SENSOR (HO2S)
② MANIFOLD ABSOLUTE PRESSURE (MAP) SENSOR
③ ENGINE COOLANT TEMPERATURE (ECT) SENSOR
④ INTAKE AIR TEMPERATURE (IAT) SENSOR
⑤ FAST IDLE THERMO VALVE
⑥ IDLE AIR CONTROL (IAC) VALVE
⑦ FUEL INJECTOR
⑧ FUEL PRESSURE REGULATOR
⑨ FUEL FILTER
⑩ FUEL PUMP (FP)
⑪ FUEL TANK
⑫ AIR CLEANER (ACL)
⑬ RESONATOR
⑭ THREE WAY CATALYTIC CONVERTER (TWC)
⑮ POSITIVE CRANKCASE VENTILATION (PCV) VALVE
⑯ EVAPORATIVE EMISSION (EVAP) CONTROL CANISTER
⑰ EVAPORATIVE EMISSION DIAPHRAGM VALVE
⑱ PURGE CONTROL SOLENOID VALVE
⑲ EVAPORATIVE EMISSION (EVAP)
⑳ TWO WAY VALVE
㉑ FUEL TANK EVAPORATIVE EMISSION (EVAP) VALVE
㉒ FUEL PULSATION DAMPER

Fig. 76 1995 Civics with D15B7 and D16Z6 engines

4-40 EMISSION CONTROLS

① HEATED OXYGEN SENSOR (HO2S)
② MANIFOLD ABSOLUTE PRESSURE (MAP) SENSOR
③ ENGINE COOLANT TEMPERATURE (ECT) SENSOR
④ INTAKE AIR TEMPERATURE (IAT) SENSOR
⑤ KNOCK SENSOR (KS)
⑥ IDLE AIR CONTROL (IAC) VALVE
⑦ FAST IDLE THERMO VALVE
⑧ FUEL INJECTOR
⑨ FUEL PRESSURE REGULATOR
⑩ FUEL FILTER
⑪ FUEL PUMP (FP)
⑫ FUEL TANK
⑬ AIR CLEANER (ACL)
⑭ RESONATOR
⑮ THREE WAY CATALYTIC CONVERTER (TWC)
⑯ POSITIVE CRANKCASE VENTILATION (PCV) VALVE
⑰ EVAPORATIVE EMISSION (EVAP) CONTROL CANISTER
⑱ PURGE CONTROL DIAPHRAGM VALVE
⑲ PURGE CONTROL SOLENOID VALVE
⑳ EVAPORATIVE EMISSION (EVAP)
㉑ TWO WAY VALVE
㉒ FUEL TANK EVAPORATIVE EMISSION (EVAP) VALVE
㉓ FUEL PULSATION DAMPER

Fig. 79 1995 del Sol with VTEC 16A3 engine

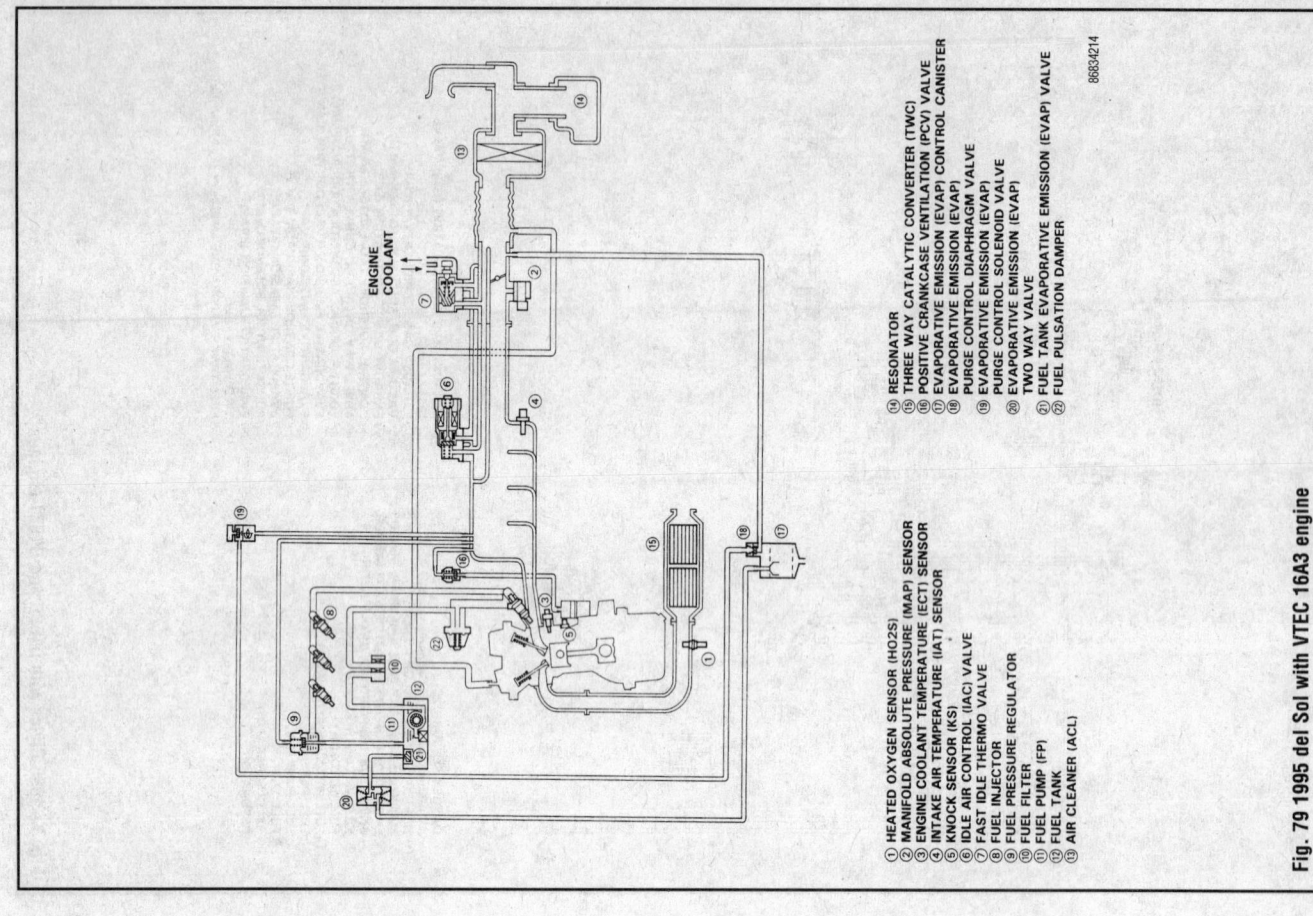

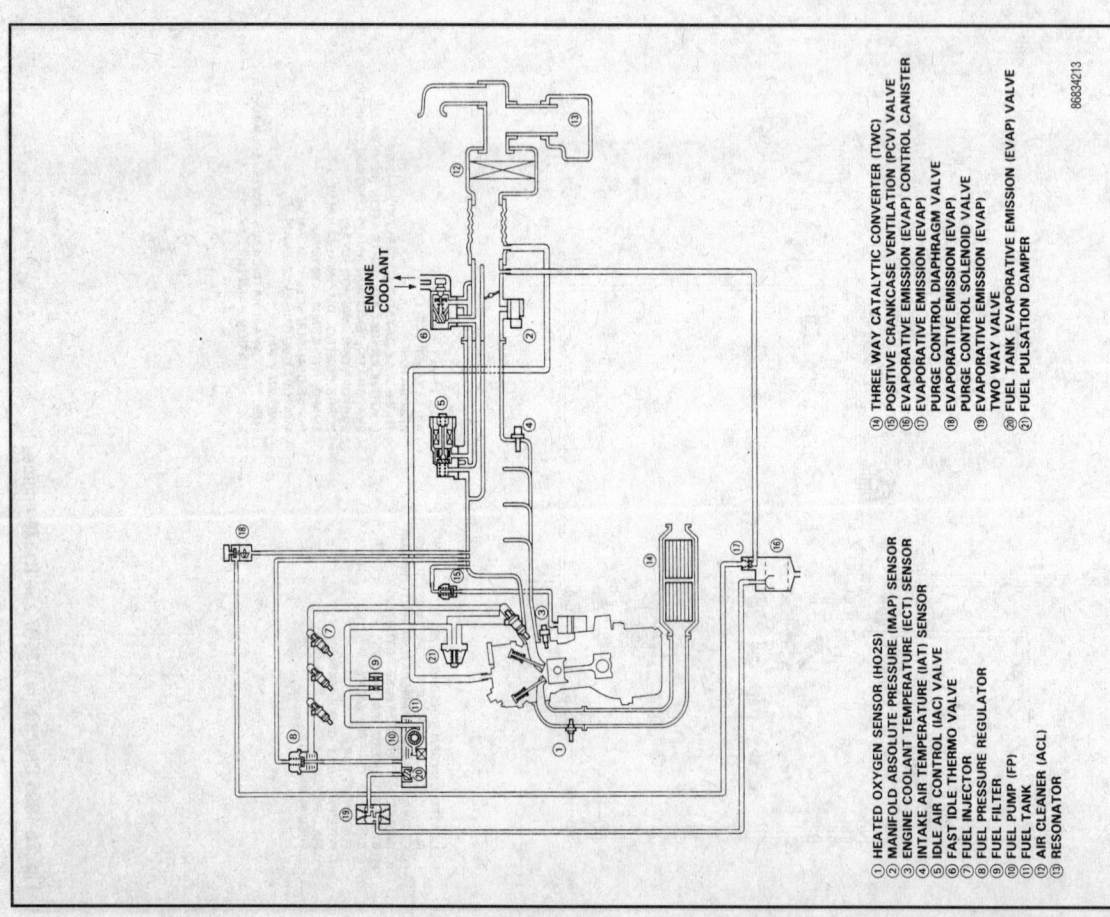

① HEATED OXYGEN SENSOR (HO2S)
② MANIFOLD ABSOLUTE PRESSURE (MAP) SENSOR
③ ENGINE COOLANT TEMPERATURE (ECT) SENSOR
④ INTAKE AIR TEMPERATURE (IAT) SENSOR
⑤ IDLE AIR CONTROL (IAC) VALVE
⑥ FAST IDLE THERMO VALVE
⑦ FUEL INJECTOR
⑧ FUEL PRESSURE REGULATOR
⑨ FUEL FILTER
⑩ FUEL PUMP (FP)
⑪ FUEL TANK
⑫ AIR CLEANER (ACL)
⑬ RESONATOR
⑭ THREE WAY CATALYTIC CONVERTER (TWC)
⑮ POSITIVE CRANKCASE VENTILATION (PCV) VALVE
⑯ EVAPORATIVE EMISSION (EVAP) CONTROL CANISTER
⑰ PURGE CONTROL DIAPHRAGM VALVE
⑱ PURGE CONTROL SOLENOID VALVE
⑲ EVAPORATIVE EMISSION (EVAP)
⑳ TWO WAY VALVE
㉑ FUEL TANK EVAPORATIVE EMISSION (EVAP) VALVE
㉒ FUEL PULSATION DAMPER

Fig. 78 1995 del Sol with D15B7 and D16Z6 engines

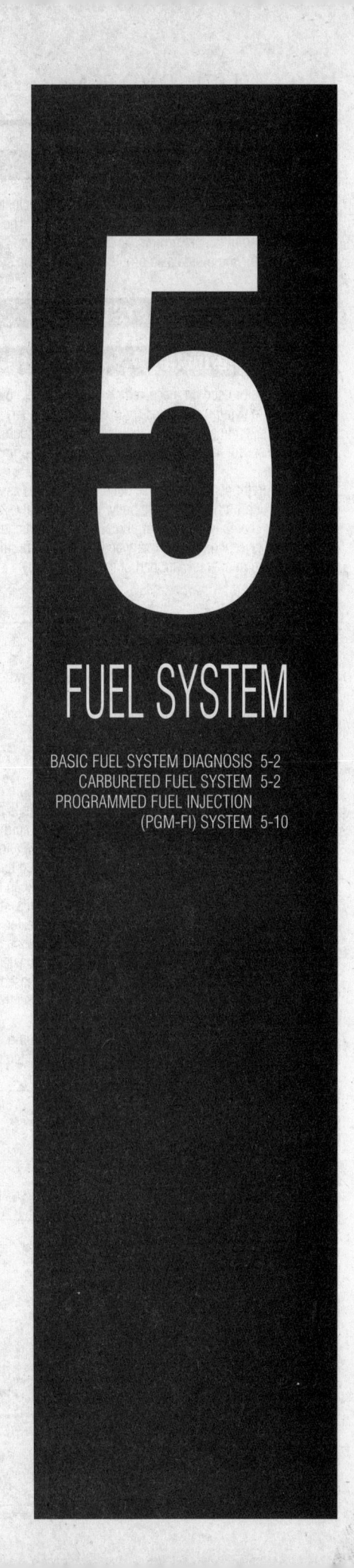

**BASIC FUEL SYSTEM
 DIAGNOSIS 5-2
CARBURETED FUEL SYSTEM 5-2**
GENERAL INFORMATION 5-2
    AUTOMATIC CHOKE 5-2
    CHOKE OPENER SYSTEM 5-2
    ACCELERATOR PUMP 5-2
    CARBURETOR HEAT RISER 5-3
    VACUUM CONTROLLED
      SECONDARY 5-3
    POWER VALVE 5-3
FUEL SYSTEM SERVICE
 PRECAUTIONS 5-3
FUEL TANK 5-4
    REMOVAL & INSTALLATION 5-4
MECHANICAL FUEL PUMP 5-4
    REMOVAL & INSTALLATION 5-4
    TESTING 5-5
CARBURETOR ADJUSTMENTS 5-5
    TROUBLESHOOTING 5-5
    ADJUSTMENTS 5-5
CARBURETOR ASSEMBLY 5-5
    REMOVAL & INSTALLATION 5-9
    OVERHAUL 5-9
**PROGRAMMED FUEL INJECTION
 (PGM-FI) SYSTEM 5-10**
GENERAL DESCRIPTION 5-10
    AIR INTAKE SYSTEM 5-10
    ELECTRONIC CONTROL
      SYSTEM 5-11
    FUEL SYSTEM 5-12
FUEL SYSTEM SERVICE
 PRECAUTIONS 5-12
RELIEVING FUEL SYSTEM
 PRESSURE 5-13
    COMPONENT TEST
      PROCEDURES 5-13
ELECTRIC FUEL PUMP 5-17
    PRESSURE TESTING 5-17
    REMOVAL & INSTALLATION 5-18
FUEL INJECTOR 5-18
    REMOVAL & INSTALLATION 5-18
FUEL PRESSURE REGULATOR 5-20
    REMOVAL & INSTALLATION 5-20
THROTTLE BODY 5-21
    REMOVAL & INSTALLATION 5-21
FUEL TANK 5-22
    REMOVAL & INSTALLATION 5-22

5

FUEL SYSTEM

BASIC FUEL SYSTEM DIAGNOSIS 5-2
CARBURETED FUEL SYSTEM 5-2
PROGRAMMED FUEL INJECTION
(PGM-FI) SYSTEM 5-10

5-2 FUEL SYSTEM

BASIC FUEL SYSTEM DIAGNOSIS

When there is a problem starting or driving a vehicle, two of the most important checks involve the ignition and the fuel systems. The questions most mechanics attempt to answer first, "is there spark?" and "is there fuel?" will often lead to solving most basic problems. For ignition system diagnosis and testing, please refer to the information on engine electrical components and ignition systems found earlier in this manual. If the ignition system checks out (there is spark), then you must determine if the fuel system is operating properly (is there fuel?).

CARBURETED FUEL SYSTEM

General Information

The carburetor used on these vehicles is a 2-barrel down-draft fixed venturi type. There are two throttle bores (primary and secondary) in the carburetor. During normal engine operation, fuel is supplied through the primary bore; at high engine speed, fuel is supplied through both the primary and secondary throttle bores.

The carburetor consists of three major components assembled in layers: the air horn, mixing body and throttle body. The choke valve is in the air horn. The venturi, float chamber, accelerator pump, actuator and choke housing are in the mixing body layer, while the primary and secondary throttle valves, secondary diaphragm and pilot screw are part of the throttle body.

AUTOMATIC CHOKE

♦ See Figure 1

This system provides easy engine starting under a wide range of air temperatures. The system consists of the following:
- Choke valve and linkage
- Choke heater
- Choke opener
- Fast idle and fast idle unloader system

The choke valve is located in the primary throttle bore of the carburetor. When the engine is not running, the choke valve angle is determined by the bimetallic coil spring acting against the choke return spring. When the engine is running, the choke opener also affects the choke valve angle.

When the engine is started, electric current supplied to the main choke heater causes the bimetallic coil spring to gradually open the choke valve. As the air temperature in the choke cap rises, the thermal switch turns on and electric current is also directed to the secondary choke heater. This speeds the opening of the choke valve during its final stages. The combination of heater and thermistor keeps the bimetallic coil spring at a constant high temperature.

The choke opener adjusts the choke valve for increased air flow once the engine begins to fire. It operates in two steps according to coolant temperature and it operates independently of the fast idle setting.

The fast idle unloader operates in two steps according to the characteristics of two thermovalves, which sensor the engine coolant temperature differently.

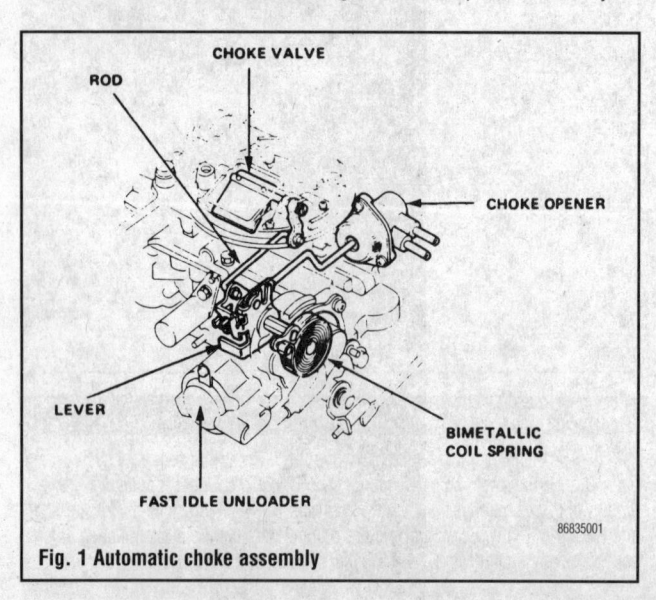

Fig. 1 Automatic choke assembly

CHOKE OPENER SYSTEM

♦ See Figure 2

The fast idle cam is engaged and disengaged by depressing the accelerator pedal and is also disengaged by the fast idle unloading mechanism.

The unloading mechanism consists of a fast idle unloader and thermovalves **A** and **B**. The unloader has two diaphragms to release the throttle valve in two steps. When the coolant temperature reaches the first thermovalve temperature setting, the valve closes to shut off the vacuum bleed. This allows the inside diaphragm of the unloader to retract to the first step by manifold vacuum. Then, as the coolant temperature further rises and reaches the second thermovalve temperature setting, the second valve closes and manifold vacuum is applied onto the outside diaphragm of the unloader. This allows the unloader to operate on the second step.

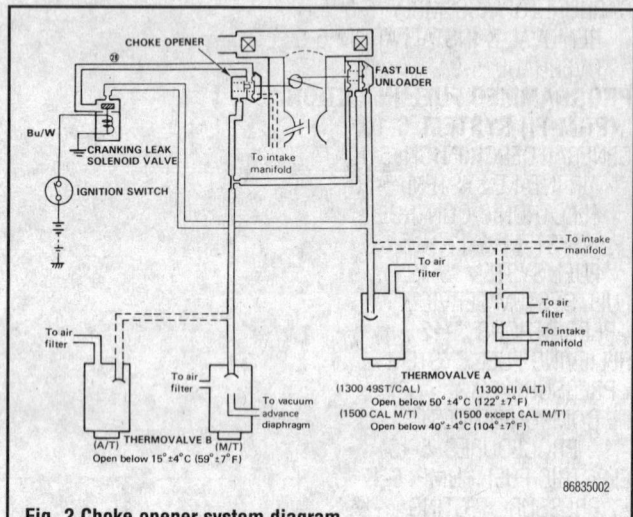

Fig. 2 Choke opener system diagram

Choke Opener

This system is designed to promote easy starting. During engine start-up, manifold vacuum is transmitted to the choke opener; thus, the choke valve is opened a fixed amount.

The first thermovalve works to open the choke valve in response to engine coolant temperature. When the engine coolant temperature is below the set temperature of the thermovalve, it opens and manifold vacuum is bled from the valve. In this situation the choke opener diaphragm is retracted to an intermediate position because of the balance between the vacuum and the spring force of the choke opener.

When the engine coolant temperature exceeds the set temperature of the thermovalve, it closes to shut off the vacuum bleed and this allows the choke opener to fully retract and pull the choke valve open.

ACCELERATOR PUMP

♦ See Figure 3

The accelerator pump supplies the extra fuel necessary to maintain the best possible air/fuel mixture when the accelerator pedal is depressed suddenly at low engine speeds.

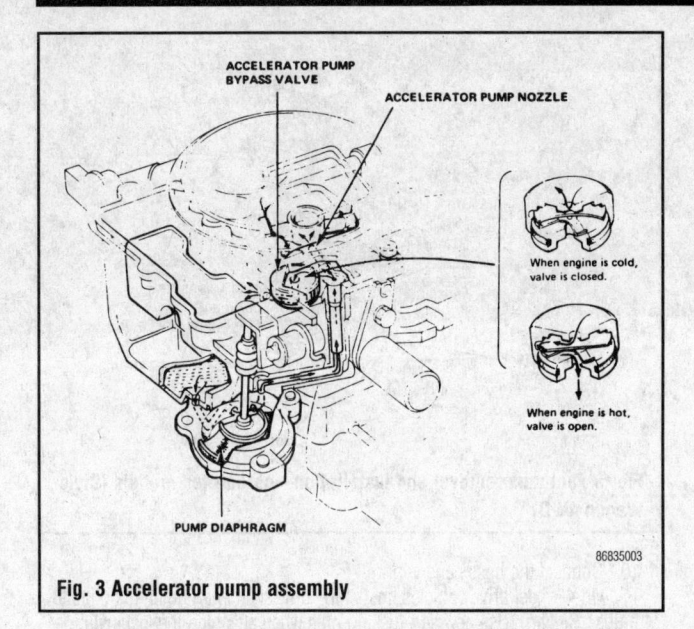

FUEL SYSTEM 5-3

Fig. 3 Accelerator pump assembly

When the accelerator pedal is depressed, the pump rod, which is connected to the throttle lever, pushes down on the accelerator pump diaphragm. This opens the outlet check valve and allows fuel to be pumped up to the accelerator pump nozzle, where it is sprayed into the carburetor's primary throat.

The fuel flow from the accelerator pump is further regulated by the temperature-sensitive bypass valve in the carburetor. When the engine is cold, the bypass portion of the valve is closed so that the accelerator pump nozzle will get the maximum of fuel available. When the engine warms up to its normal operating temperature, the bypass valve opens and this allows some of the fuel, which would ordinarily be routed to the nozzle, to be bled back into the carburetor float chamber.

CARBURETOR HEAT RISER

▶ See Figure 4

A heat riser, utilizing engine coolant, is built into the intake manifold to prevent carburetor icing and to improve cold driveability. This improves air/fuel mixture atomization. Since excessive heat can cause fuel percolation, a bypass valve with a thermostat shuts off coolant flow at normal temperatures.

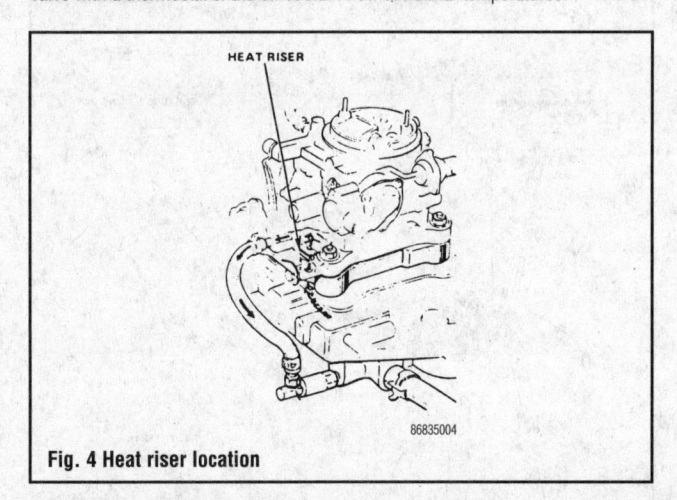

Fig. 4 Heat riser location

VACUUM CONTROLLED SECONDARY

▶ See Figure 5

The secondary throttle valve is controlled by venturi vacuum and thermovalve D. When thermovalve D is below its specified temperature, vacuum to the secondary diaphragm is bled to the air cleaner. Above that temperature, thermovalve D gradually closes the air bleed and supplies venturi vacuum to the secondary diaphragm.

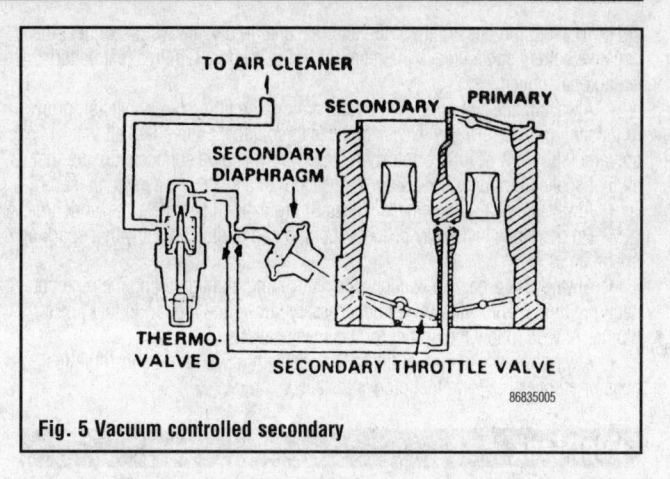

Fig. 5 Vacuum controlled secondary

POWER VALVE

▶ See Figure 6

This system provides supplementary fuel to the primary main fuel passage when the vehicle is run at high load or in the power mode.

1.3L Engines

When the intake air temperature is below 60°F (16°C), the power valve is open because the power valve control solenoid does not allow vacuum to the power valve.

When the vehicle is below 10 mph, with the intake air temperature above 60°F (16°C), the power valve is still open because the power valve solenoid will not allow vacuum to the power valve.

When the vehicle speed exceeds 10 mph (16 km), the power valve is then controlled by the signal from the vacuum switch B to the power valve control solenoid valve. The power valve control solenoid valve is energized by low manifold vacuum to the vacuum switch.

1.5L Engine

The control unit energizes the power valve control solenoid to cut off manifold vacuum to the power valve which supplies fuel to the primary main fuel passage.

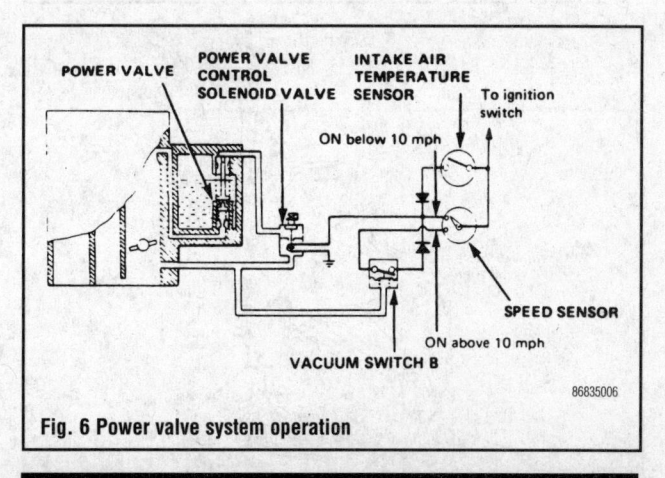

Fig. 6 Power valve system operation

Fuel System Service Precautions

Safety is the most important factor when performing not only fuel system maintenance but any type of maintenance. Failure to conduct maintenance and repairs in a safe manner may result in serious personal injury or death. Service and testing of the vehicle's fuel system components can be accomplished safely and effectively by adhering to the following rules and guidelines.

5-4 FUEL SYSTEM

• To avoid the possibility of fire and personal injury, always disconnect the negative battery cable unless the repair or test procedure requires that battery voltage be applied.
• Always place a shop towel or cloth around the fitting or connection prior to loosening to absorb any excess fuel due to spillage. Ensure that all fuel spillage (should it occur) is quickly removed from engine surfaces. Ensure that all fuel soaked cloths or towels are deposited into a suitable waste container.
• Always keep a dry chemical (Class B) fire extinguisher near the work area.
• Do not allow fuel spray or fuel vapors to come into contact with a spark or open flame.
• Always use a backup wrench when loosening and tightening fuel line connection fittings. This will prevent unnecessary stress and torsion to fuel line piping. Always follow the proper torque specifications.
• Always replace worn fuel fitting O-rings with new. Do not substitute fuel hose or equivalent where fuel pipe is installed.

Fuel Tank

REMOVAL & INSTALLATION

▶ See Figures 7 and 8

1. Raise and safely support the vehicle, then remove the rear wheels.
2. Remove the drain bolt and drain the fuel from the tank. Be sure to take the appropriate fire safety precautions.
3. In the luggage area on 2-wheel drive vehicles, remove the access covers and disconnect the fuel pump and gauge unit wiring. On 4-wheel drive vehicles, the access cover is under the rear seat.
4. To disconnect the hoses, loosen the clamps and slide them back. Carefully twist the hose while pulling it off the fitting to avoid damage to the hose or the flared fittings.
5. On 4-wheel drive vehicles, remove the exhaust pipe and muffler, then remove the rear driveshaft.
6. Place a jack under the tank and remove the nuts to allow the tank straps to fall free. The tank may stick to the vehicle's undercoating but it can be pried out of the mounts. Be careful not to damage the fittings.

To install:

7. Position the tank under the vehicle and install the straps, washers and nuts onto the hooks.
8. Wiggle the tank while tightening the nuts a few turns at a time to make sure the tank seats properly into the mount. Tighten the strap nuts to 16 ft. lbs. (22 Nm).
9. Make sure the drain plug in installed with a new gasket and tighten to 36 ft. lbs. (50 Nm).
10. Connect the hoses and wiring.
11. On 4-wheel drive vehicles, install the driveshaft and tighten the bolts to 24 ft. lbs. (33 Nm). Use new self-locking nuts when attaching the exhaust pipe to the catalytic converter and tighten to 25 ft. lbs. (34 Nm).

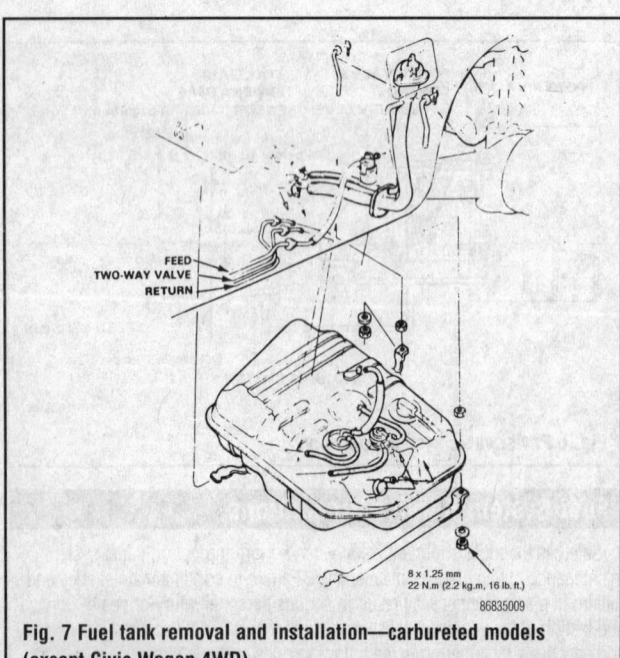

Mechanical Fuel Pump

▶ See Figures 9 and 10

REMOVAL & INSTALLATION

The fuel pump is located on the passengers side of the engine towards the front of the car.

1. Disconnect the negative battery cable. Remove the fuel pump cover.
2. Clamp the fuel lines closed with fuel line clamps or equivalent, to prevent fuel from running out. Wrap a rag around the fuel lines and disconnect them from the fuel pump.

➡ **When disconnecting the fuel lines, twist and slide them off to avoid damage.**

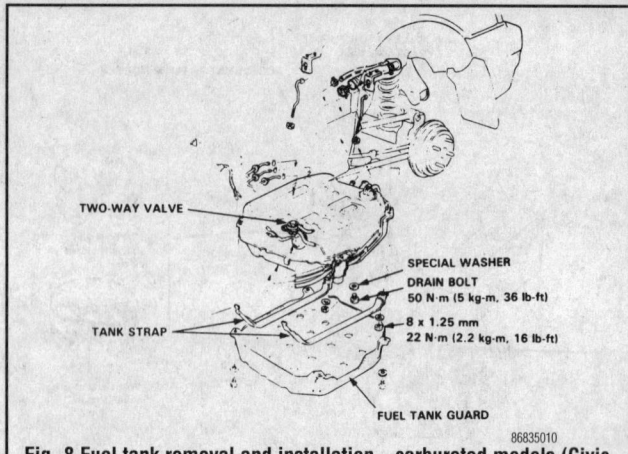

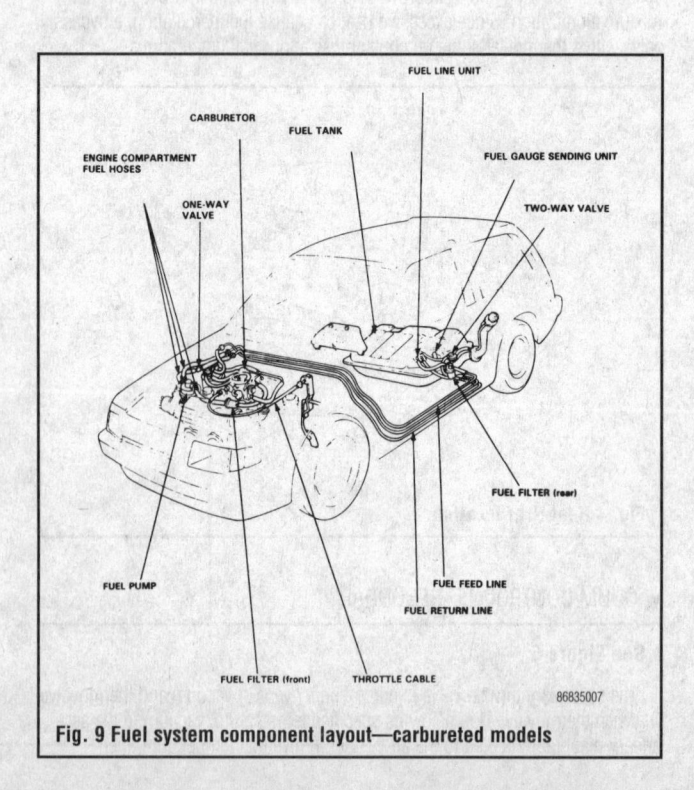

Fig. 8 Fuel tank removal and installation—carbureted models (Civic Wagon 4WD)

Fig. 7 Fuel tank removal and installation—carbureted models (except Civic Wagon 4WD)

Fig. 9 Fuel system component layout—carbureted models

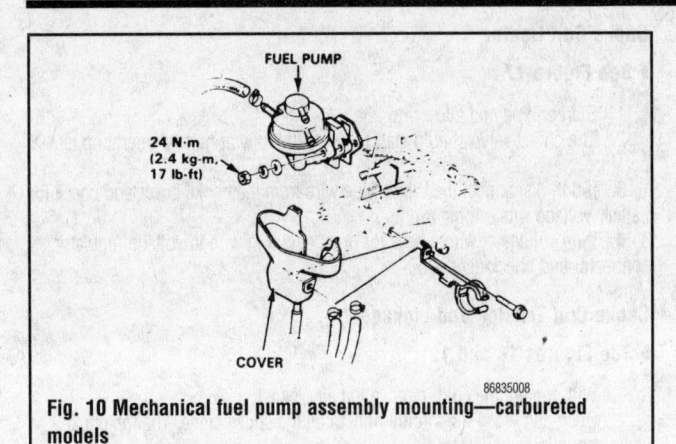

FUEL SYSTEM 5-5

Fig. 10 Mechanical fuel pump assembly mounting—carbureted models

3. Remove the fuel pump retaining bolts and remove the pump from the engine.

To Install:
4. Clean the gasket mating surfaces and using a new gasket, install the fuel pump in position on the block.
5. Tighten the fuel pump retaining bolts to 17 ft. lbs. (24 Nm).
6. Reconnect the fuel lines and make sure that they are tightly clamped. Do not over tighten the clamps or you will damage the hoses.
7. Connect the negative battery cable and start the vehicle. Check for leaks at the fuel pump.

TESTING

♦ See Figures 11 and 12

1. Check the following items:
   a. Looseness of the pump connector.
   b. Looseness of the upper or lower body and cover screws.
   c. Looseness of the rocker arm pin.
   d. Contamination or clogging of the air hole.
   e. Improper operation of the pump.
2. Check to see if there are signs of oil or fuel around the air hole. If so, the diaphragm is damaged and you must replace the pump.
3. Disconnect the fuel line to the fuel filter, in the engine compartment and connect a fuel pressure gauge to it.
4. Disconnect and plug the fuel return line at the fuel pump.
5. Start the engine and allow it to idle until the fuel pressure stabilizes, then shut the engine **OFF**.
6. The fuel pressure gauge should read 2.7–3.8 psi. (17.7–26.5 kPa) at idle. If the gauge reads at least 2.7 psi (17.7 kPa), then go on to Step 7, if the gauge reads less than 2.7 psi. (17.7 kPa), the fuel pump must be replaced.
7. Remove the pressure gauge and place a graduated glass container under the fuel hose.
8. Start the engine and allow the car to idle for exactly 60 seconds, then shut the engine **OFF**.

9. The fuel volume pumped into the container should be 5.7 oz. (170cc) or more, if the volume was less than 5.7 oz. (170cc), replace the pump.

➥Check for a clogged fuel filter or fuel line and for a pinched fuel line before replacing the fuel pump.

Carburetor Adjustments

TROUBLESHOOTING

Carburetor problems are among the most difficult internal combustion engine malfunctions to diagnose. If you have a carburetor problem, read the description of carburetor systems in the beginning of this section. Consider which system or combination of systems are in operation when the problem occurs. Some troubleshooting tips are given in the system operation descriptions.

The most reliable way for a nonprofessional to diagnose a bad carburetor is to eliminate all other possible sources of the problem. If you suspect the carburetor is the problem, perform the adjustments given in this section. Check the ignition system to ensure that the spark plugs, contact points, and condenser are in good shape and adjusted properly. Check the emission control equipment following the instructions given in Section 4 of this manual. Check the ignition timing adjustment. Check all vacuum hoses on the engine for loose connections, splits or breaks. Make sure the carburetor and intake manifold attaching bolts are tightened to the proper torque.

If you do determine that the carburetor is malfunctioning, and the adjustments in this section don't help, you have three alternatives: you can take it to a professional shop and let them fix it, you can buy a new or rebuilt carburetor to replace the one now on your vehicle, or you can buy a carburetor rebuilding kit and overhaul your carburetor.

ADJUSTMENTS

♦ See Figure 13

Fast Idle

During cold engine starting and the engine warm-up period, a specially enriched fuel mixture is required. If the engine fails to run properly or if the engine over-revs with the choke closed in cold weather, the fast idle system should be checked and adjusted. This is accomplished with the carburetor installed.

1. Run the engine to normal operating temperature.
2. Connect a tachometer according to the tool manufacturer's instructions.
3. Disconnect and plug the hose from the fast idle unloader.
4. Shut the engine **OFF**, hold the choke valve closed, and open and close the throttle to engage the fast idle cam.
5. Start the engine, run it for one minute. Fast idle speed should be 2300–3300 rpm (manual transaxle) or 2200–3200 rpm (automatic transaxle).
6. Adjust the idle by turning the fast idle screw.

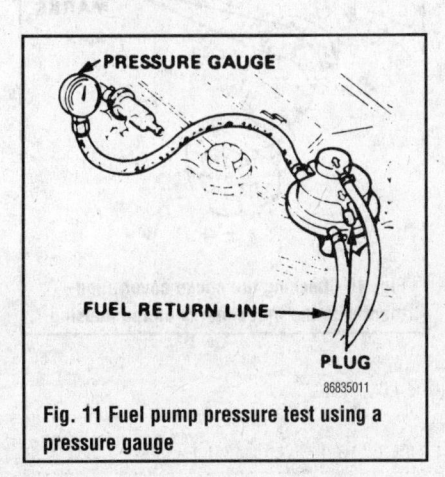

Fig. 11 Fuel pump pressure test using a pressure gauge

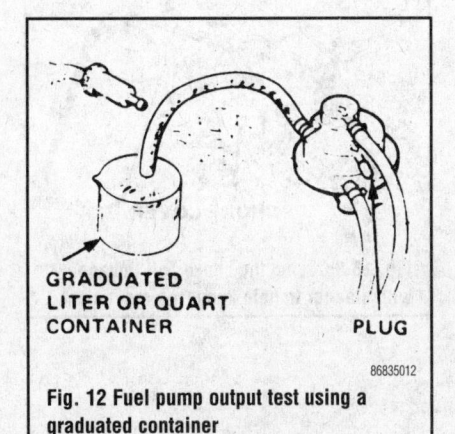

Fig. 12 Fuel pump output test using a graduated container

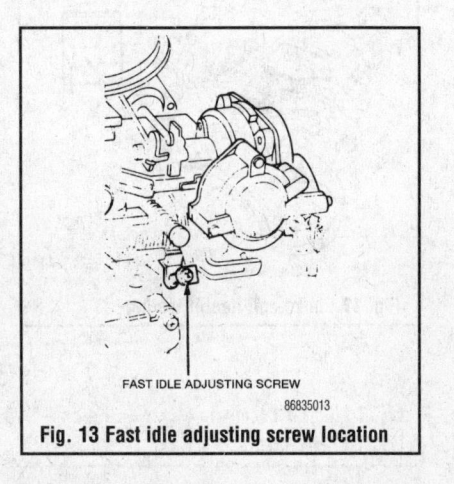

Fig. 13 Fast idle adjusting screw location

5-6 FUEL SYSTEM

Float and Fuel Level

♦ See Figures 14 and 15

Poor fuel combustion, black sooty exhaust, and fuel overflow are indications of improper float level. Lean running may also be a symptom, although you should also check for such causes as jets blocked by dirt and vacuum leaks.

1. Make sure the vehicle is level. Start the engine and run it until it is warm. Remove the air cleaner and move it aside, leaving as many hoses connected as possible. Snap the throttle open so the engine accelerates to 3000 rpm several times, then allow the engine to idle.
2. Wait until the fuel level stabilizes, and then check its location in the inspection window. If it is not centered, use a small screwdriver to adjust the fuel level, turning the screw no more than ⅛ turn every 15 seconds (so the fuel level will stabilize). When fuel level is correct, it is a good idea to put a dab of white paint near the screw to indicate that fuel level has been properly set. Stop the engine and replace the air cleaner.

Throttle Linkage

♦ See Figure 16

1. Remove the air cleaner assembly to provide access.
2. Check that the cable free-play (deflection) is 4.0–10.0mm (0.157–0.394 in.) This is measured right before the cable enters the throttle shaft bell crank.
3. If deflection is not to specifications, rotate the cable adjusting nuts in the required direction.
4. As a final check, have a friend press the gas pedal all the way to the floor, while you look down inside the throttle bore checking that the throttle plates reach the Wide Open Throttle (WOT) vertical position.
5. Install the air cleaner.

Choke Coil Heater

♦ See Figure 17

1. Start engine and allow it to idle.
2. The choke valve should start to open and fully at normal operating temperature.
3. If not, disconnect the White/Blue wire from the choke cover and check for battery voltage with engine running.
4. If no voltage at wire, check for open circuit in wire to voltage regulator connector and check alternator.

Choke Coil Tension and Linkage

♦ See Figures 18 and 19

1. With the engine cold, remove the air cleaner.
2. Open and close the throttle fully to let the choke close. The the choke valve should close completely.

→Above about 82°F (28°C), the choke will not close completely, but should still close to less than ⅛ in. (3mm).

3. If the choke closes properly, go on to the fast idle unloader test.
4. If the choke does not close properly, spray linkage with carburetor cleaner and check again (use a spray can with an extension on nozzle to reach linkage).

✱✱ CAUTION

Always wear safety goggles or a face shield when spraying carburetor cleaner.

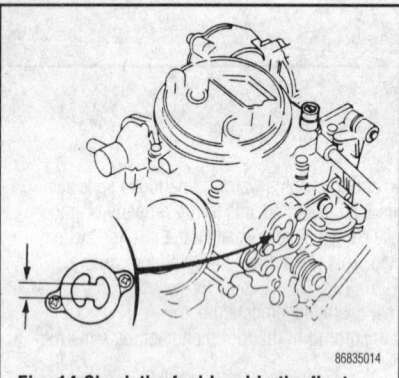

Fig. 14 Check the fuel level in the float bowl, through the level window

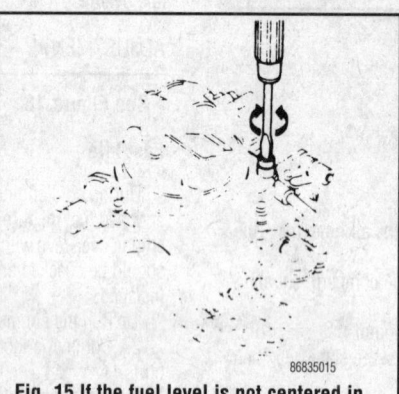

Fig. 15 If the fuel level is not centered in the level window, turn the adjusting screw

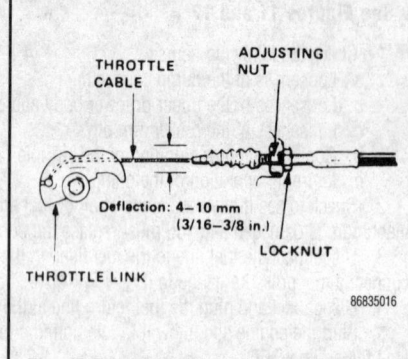

Fig. 16 Throttle cable end and adjusting nut

Fig. 17 Choke coil heater testing

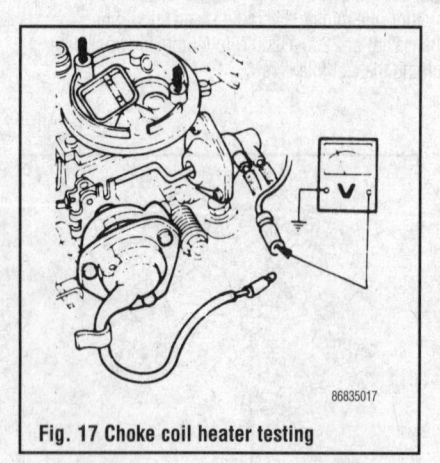

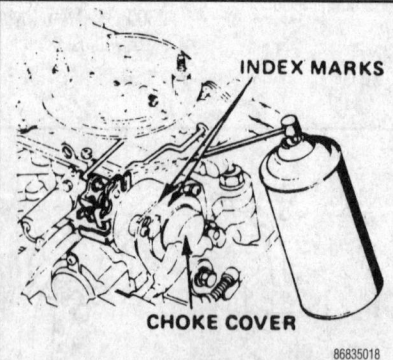

Fig. 18 Spraying the choke coil linkage with cleaner to help keep free movement

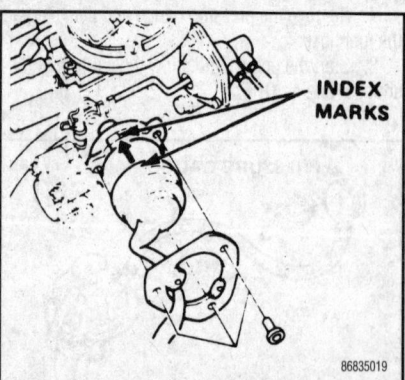

Fig. 19 Checking the choke cover alignment with the index marks on the housing

FUEL SYSTEM   5-7

5. If the choke still does not close properly, remove choke cover and inspect linkage for free movement. Repair or replace parts as necessary. Then, reinstall cover and adjust so index marks line up and retest.
6. If the choke still does not close properly, replace cover.

Fast Idle Unloader
▶ See Figure 20

1. With the engine cold, disconnect the two hoses from fast idle unloader.
2. Open and close throttle fully to engage fast idle cam.
3. Start engine. Engine should run at fast idle.
4. If engine does not run at fast idle, remove choke cover and check operation of fast idle cam.
5. Connect a vacuum pump to inside fitting of unloader and apply vacuum. Idle should drop.
6. If idle speed does not drop, check unloader for leaks, blockage or damaged diaphragm. Remove choke cover and check unloader rod for free movement. Repair or replace as necessary.
7. Reconnect hoses.
8. When engine warms, engine speed should drop below 1400 rpm as unloader pulls internal choke linkage off fast idle cam.
9. If fast idle does not drop below 1400 rpm, disconnect the two unloader hoses and check that vacuum is present.
10. If vacuum is present, check the unloader for leaks or blockage. Remove choke cover and check the unloader rod for free movement. Repair or replace parts as necessary. If there is no vacuum at inner fitting, check for vacuum at choke opener and thermovalve A. If there is no vacuum at outer fitting, check thermovalve A (for location, refer to the drawing accompanying the description of this subsystem.).

Choke Opener
▶ See Figures 21, 22, 23 and 24

1. With the engine cold, disconnect the choke heater wire.
2. Open and close the throttle fully to let the choke close.
3. Start the engine. The choke valve should partially open.
4. If the choke does not partially open, check the linkage for free movement, repair as necessary and retest.
5. If the choke valve still does not partially open, check the choke opener diaphragm by removing the two screws and attaching a hand vacuum pump to hose fitting. Hold finger over the orifices in the opener while applying vacuum to pull opener rod in completely, then stop. If the rod will not stay in, replace the opener. If the rod stays in check vacuum port in carburetor and clean out if necessary.
6. If the coolant temperature is below about 57°F (14°C), tab A on the choke opener lever should not be seated against the carburetor. If tab A is seated, disconnect the choke opener hose No. 18 from the upper fitting. If tab A comes off seat, check vacuum line No. 18 to the thermovalve for blockage and check that the thermovalve is open. If tab A does not come off its seat, press down on choke opener lever until it unseats. If it won't stay unseated, clean out the choke opener fitting with a 0.5mm drill bit and retest.
7. If coolant temperature is above about 77°F (25°C), tab A on choke opener lever should be seated against carburetor.
8. If tab A is not seated, check vacuum line No. 18 hose from upper fitting, for leaks and check that thermovalve A is closed. If tab A is seated, reconnect choke heater wires.
9. Disconnect the No. 26 vacuum hose from the choke opener and connect a vacuum pump to the hose and apply vacuum.
10. Vacuum should remain steady. Turn ignition switch to **START** with engine temperature below 104°F (40°C) vacuum should bleed off. If not go to the "Cranking Leak Test" testing procedure.

Vacuum Controlled Secondary
▶ See Figure 25

1. Disconnect the secondary diaphragm vacuum hose and attach a spare piece of hose between the diaphragm and a vacuum pump.
2. Open the throttle valve fully and apply a vacuum. Check that the diaphragm rod moves as vacuum is applied and that vacuum then remains steady.

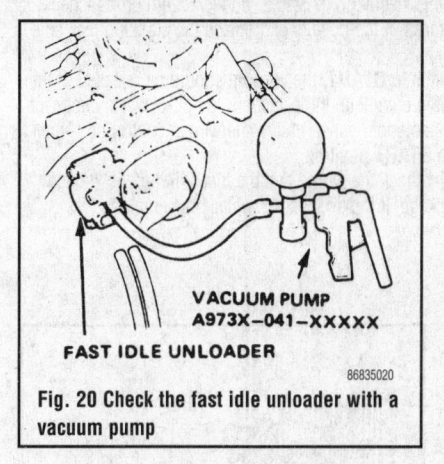

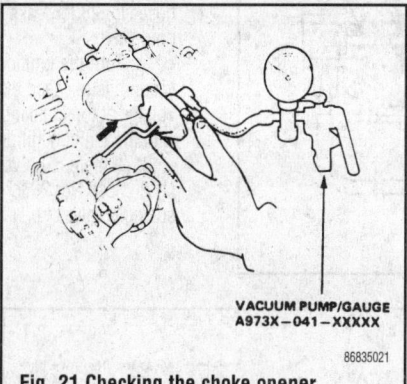

Fig. 20 Check the fast idle unloader with a vacuum pump

Fig. 21 Checking the choke opener diaphragm for movement

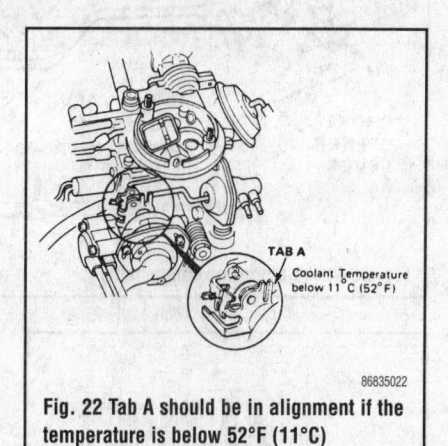

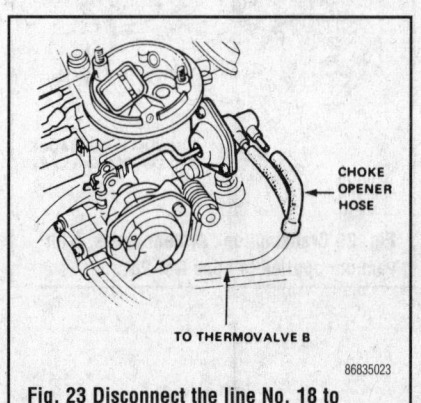

Fig. 22 Tab A should be in alignment if the temperature is below 52°F (11°C)

Fig. 23 Disconnect the line No. 18 to check if tab A moves from its seat

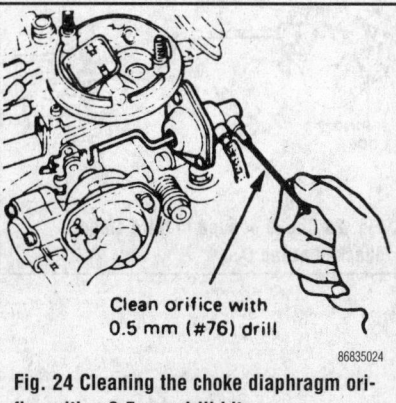

Fig. 24 Cleaning the choke diaphragm orifice with a 0.5 mm drill bit

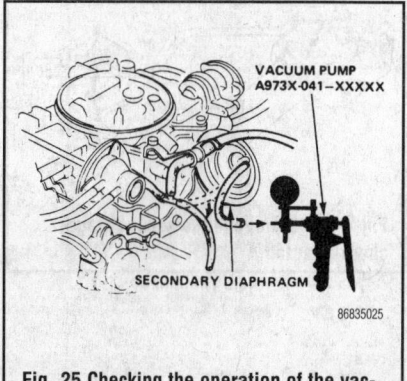

Fig. 25 Checking the operation of the vacuum controlled secondary

5-8 FUEL SYSTEM

3. If vacuum does not hold or rod does not move, first check hose for proper connection and condition, then replace diaphragm.

4. With the engine cold (water temperature below 122°F or 50°C), disconnect vacuum hose from 3-way joint, connect a vacuum pump and apply vacuum. Vacuum should not hold.

5. If vacuum holds, first make sure hose is not clogged, then replace thermovalve D.

6. After the engine has warmed up, disconnect vacuum hose from the 3-way joint, connect a vacuum pump and apply vacuum. Vacuum should hold.

7. If vacuum does not remain steady, check the hose for proper connection and that it is not clogged and replace thermovalve D.

8. Disconnect the vacuum hose from the 3-way joint and connect to a vacuum pump/gauge. Apply vacuum. Vacuum should not hold.

9. If vacuum is present, check hose, 3-way joint and clean vacuum port.

Choke Linkage

♦ See Figures 26, 27 and 28

➡ This procedure is not necessary unless the linkage has been bent, the choke opener was replaced or the car has poor cold starting. It can be performed while the engine is either hot or cold.

1. Remove choke cover.
2. While holding choke valve closed, open and close throttle fully to engage choke and fast idle linkage.
3. Disconnect the choke opener hose from the steel tubing manifold. Attach a check valve to the hose. Then pressurize the choke opener with compressed air, 15–85 psi (103–586 kPa), to hold bleed valve closed.
4. Push the choke opener lever towards opener until lever stops (opener rod seats against pressurized bleed valve), then pull choke drive lever down against opener lever (to take all free play out of linkage) and measure clearance between choke valve and casting.

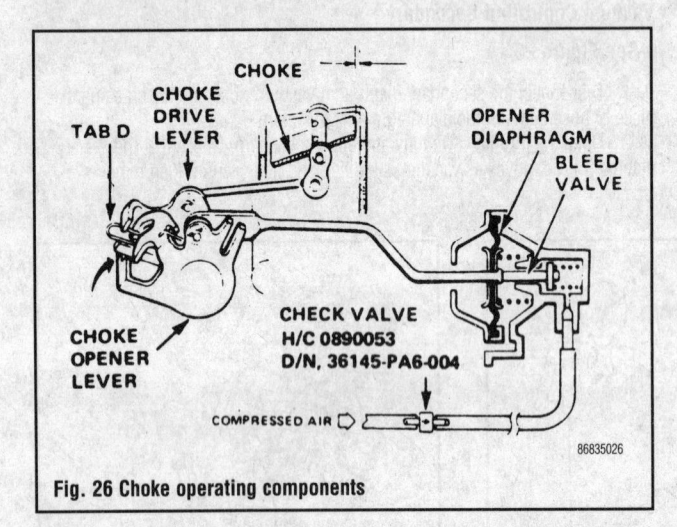

First stage clearance should be:
- 1.5L manual transaxle: 0.047–0.53 in. (1.20–1.34mm)
- 1.5L automatic transaxle: 0.40–0.46 in. (1.03–0.17mm)
- 1.3L 49 state model: 0.337–0.343 in. (0.79–0.93mm)
- 1.3L HI ALT: 0.037–0.043 in. (0.95–1.09mm)

5. Adjust clearance by bending tab D.
6. Remove the check valve, and reconnect the choke opener hose.
7. Hold both levers together, then push levers toward diaphragm again until lever stop (tab A on opener lever seats against carburetor). Measure the clearance at choke valve.

The second stage clearance should be:
- 1.5L manual transaxle: 0.067–0.074 in. (1.71–1.89mm)
- 1.3L and 1.5L automatic transaxle: 0.72–0.080 in. (1.84mm–2.02mm)

8. Adjust clearance by bending tab A.
9. While still holding opener lever tab A against the seat, release choke drive lever and measure clearance at choke valve (tab C on drive lever should stay seated against spring loop; if not, repeat Step 2 and recheck).

Third stage clearance should be:
- 1.5L models (all): 0.199 in.–0.215 in. (5.05mm–5.45mm)
- 1.3L models (all): 0.144 in.–0.160 in. (3.67mm–4.07mm)

10. Adjust clearance by bending tab C.

Cranking Leak System

♦ See Figures 29 and 30

1. Disconnect hose No. 28 from the lower fitting of the choke opener, attach a hand vacuum pump to the hose and draw vacuum. Vacuum should remain steady. If the vacuum remains steady go to Step 2. If the vacuum drops, check hose No. 28 for leaks or cracks, if the hose is ok, replace the cranking leak solenoid valve.

2. Turn the ignition switch to **START**, the vacuum should be released. If the vacuum is released the test is complete. If the vacuum is not released, check for voltage at the cranking leak solenoid valve (blue/white wire at control box No. 1) with the ignition still in the **START** position.

3. If there is no voltage, check the wiring and the fuses. If there is voltage check hose No. 28 for blockage. If the hose is ok, replace the cranking leak solenoid valve.

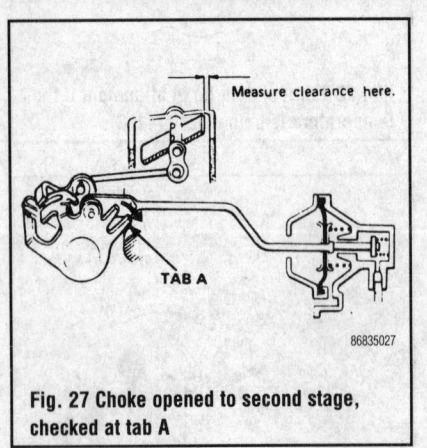

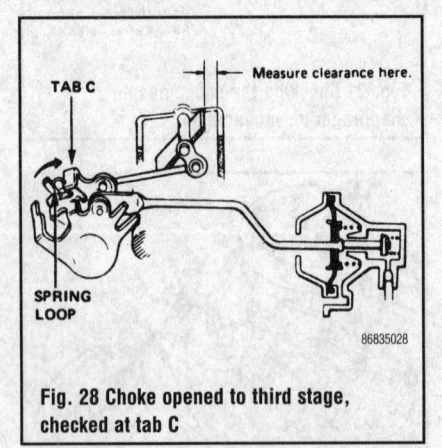

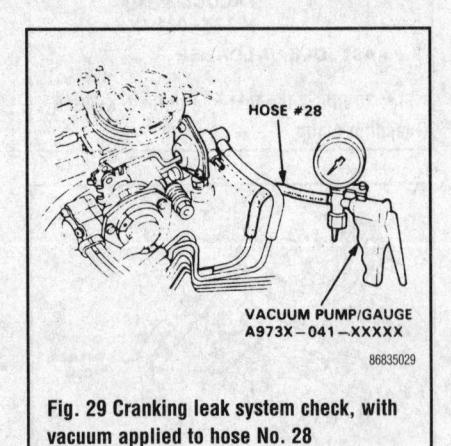

Fig. 26 Choke operating components

Fig. 27 Choke opened to second stage, checked at tab A

Fig. 28 Choke opened to third stage, checked at tab C

Fig. 29 Cranking leak system check, with vacuum applied to hose No. 28

FUEL SYSTEM 5-9

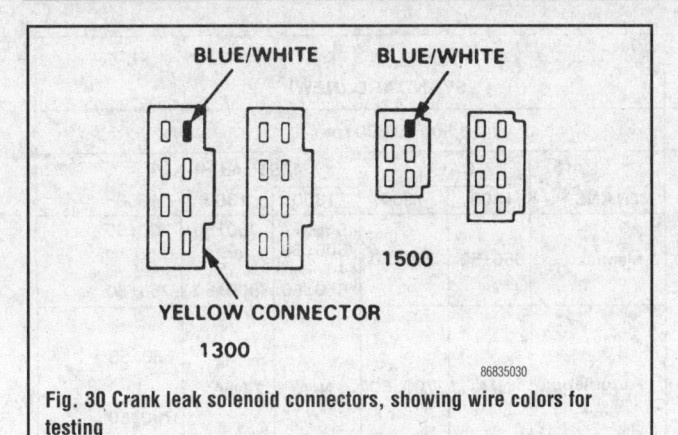

Fig. 30 Crank leak solenoid connectors, showing wire colors for testing

Carburetor Assembly

♦ See Figure 31

REMOVAL & INSTALLATION

1. Disconnect the negative battery cable.
2. Disconnect the following:
   a. Hot air tube.
   b. Vacuum hose between the one-way valve and the manifold at the manifold.
   c. Breather chamber (on air cleaner case) to intake manifold at the breather chamber.
   d. Hose from the air cleaner case to the valve cover.
   e. Hose from the carbon canister to the carburetor, at the carburetor.
   f. Throttle opener hose at the throttle opener.
3. Disconnect the fuel line at the carburetor. Plug the end of the fuel line to prevent dust entry.
4. Disconnect the choke and throttle control cables.
5. Disconnect the fuel shut-off solenoid wires.
6. Remove the carburetor retaining bolts and the carburetor. Leave the insulator on the manifold. Cover the intake manifold to prevent anything from falling into the manifold.

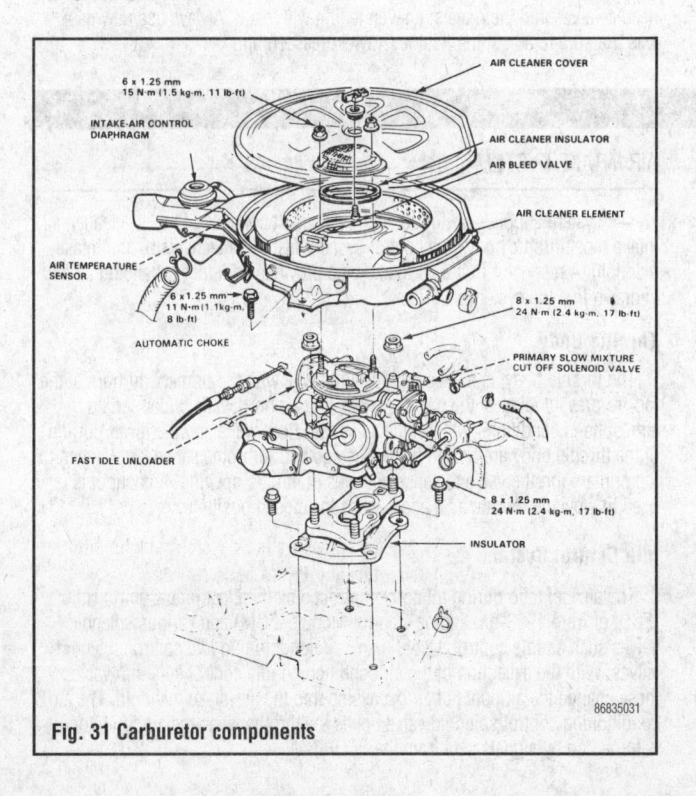

To install:

7. Install the carburetor assembly onto the insulator using a new gasket. Tighten the mounting nuts to 17 ft. lbs. (24 Nm).
8. Attach the vacuum and electrical connections.
9. Connect the fuel lines to the carburetor.
10. Install the air cleaner assembly and reconnect the intake air hoses.
11. Connect the negative battery cable. Check and adjust the idle as needed.

OVERHAUL

♦ See Figure 32

Honda does not supply overhaul information for their carburetors, outside of the instructions included in the overhaul kit. The following should be used along with the instructions in the kit.

Efficient carburetion depends greatly on careful cleaning and inspection during overhaul since dirt, gum, water or varnish in or on the carburetor parts are often responsible for poor performance.

Overhaul your carburetor in a clean, dust-free area. Carefully disassemble the carburetor, referring often to the exploded views. Keep all similar and look-alike parts segregated during disassembly and cleaning to avoid accidental interchange during assembly. Make a note of all jet sizes.

When the carburetor is disassembled, wash all parts (except diaphragms, electric choke units, pump plunger and any other plastic, leather, fiber or rubber parts) in clean carburetor solvent. Do not leave parts in the solvent any longer than is necessary to sufficiently loosen the deposits. Excessive cleaning may remove the special finish from the float bowl and choke valve bodies, leaving these parts unfit for service. Rinse all parts in clean solvent and blow them dry with compressed air or allow them to air dry. Wipe clean all cork, plastic, leather and fiber parts with a clean, lint-free cloth.

Blow out all passages and jets with compressed air and be sure that there are no restrictions or blockages. Never use wire or similar tools to clean jets, fuel passages or air bleeds. Clean all jets and valves separately to avoid accidental interchange.

Check all parts for wear or damage. If wear or damage is found, replace the defective parts. Especially check the following:

1. Check the float needle and seat for wear. If wear is found, replace the complete assembly.
2. Check the float hinge pin for wear and the float(s) for dents or distortion. Replace the float if fuel has leaked into it.
3. Check the throttle and choke shaft bores for wear or an out-of-round condition. Damage or wear to the throttle arm, shaft or shaft bore will often require replacement of the throttle body. These parts require a close tolerance of fit; wear may allow air leakage, which could affect starting and idling.

➡ Throttle shafts and bushings are not included in overhaul kits. They can be purchased separately.

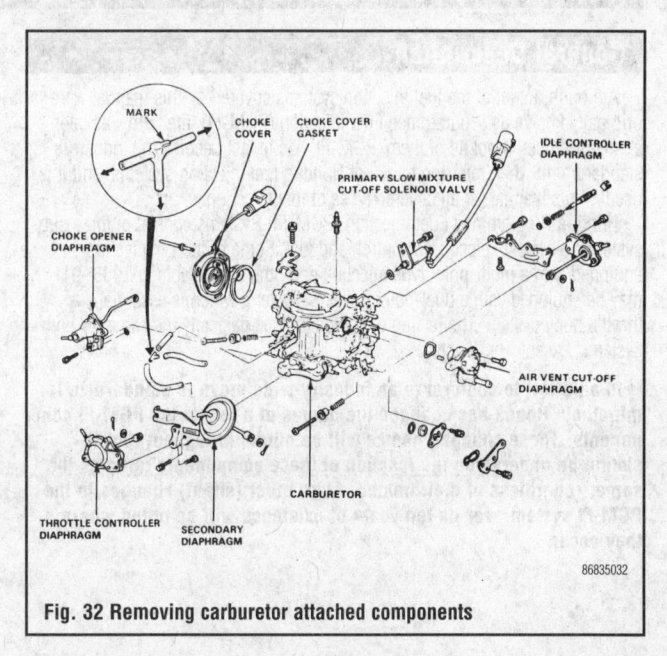

Fig. 31 Carburetor components

Fig. 32 Removing carburetor attached components

5-10 FUEL SYSTEM

Carburetor

MEASUREMENT		STANDARD (NEW)					
Carburetor	Choke fast idle	2,500–3,500 rpm					
	Idle speed (rpm) with headlights and cooling fan off	CAR TRANS.	CAL		49ST and HI ALT		
			1300	1500	1300	1300	1500
		Manual	650±50	700±50	5-spd: 600±50 4-spd: 650±50	750±50 ↕ 800±50	700±50 ↕ 750±50
		Automatic	N/A	700±50	N/A	N/A	650±50 ↕ 700±50
	*↕: The lower number is if idle is measured at high altitude; the higher number is if idle is measured at low altitude.						
	Idle CO	0.5–2.0					

4. Inspect the idle mixture adjusting needles for burrs or grooves. Any such condition requires replacement of the needle, since you will not be able to obtain a satisfactory idle.

5. Test the accelerator pump check valves. They should pass air one way but not the other. Test for proper seating by blowing and sucking on the valve. Replace the valve if necessary. If the valve is satisfactory, wash the valve again to remove breath moisture.

6. Check the bowl cover for warped surfaces with a straight edge.

7. Closely inspect the valves and seats for wear and damage, replacing as necessary.

8. After the carburetor is assembled, check the choke valve for freedom of operation.

Carburetor overhaul kits are recommended for each overhaul. These kits contain all gaskets and new parts to replace those that deteriorate most rapidly. Failure to replace all parts supplied with the kit (especially gaskets) can result in poor performance later.

Most carburetor manufacturers supply overhaul kits of three basic types: minor repair; major repair; and gasket kits. Basically, they contain the following:

Minor Repair Kits:
- All gaskets
- All diaphragms
- Float needle valve
- Volume control screw
- Spring for the pump diaphragm

Major Repair Kits:
- All jets and gaskets
- All diaphragms
- Float needle valve
- Volume control screw
- Pump ball valve
- Main jet carrier
- Float
- Complete intermediate rod
- Intermediate pump lever
- Complete injector tube
- Some cover hold down screws and washers

Gasket Kits:
- All gaskets

After cleaning and checking all components, reassemble the carburetor, using new parts and referring to the exploded view. When reassembling, make sure that all screws and jets are tight in their seats, but do not over tighten, as the tips will be distorted. Tighten all screws gradually, in rotation. Do not tighten needle valves into their seats; uneven jetting will result. Always use new gaskets. Be sure to adjust the float level when reassembling.

PROGRAMMED FUEL INJECTION (PGM-FI) SYSTEM

General Description

Although not all of the fuel injection systems covered by this manual were originally known as Programmed Fuel Injection (PGM-FI), the term was later applied to most, if not all of them. PGM-FI, has in fact become the industry's standard term when referring to any of Honda's fuel injection systems, and it is used in this manual for all covered types of fuel injection.

Honda's Programmed Fuel Injection (PGM-FI) Systems consist of three subsystems: air intake, electronic control, and fuel. Some of the vehicles are equipped with a multi-point type injection and others (available in 1988–91) may be equipped with a dual-point injection system, which is essentially a throttle body system. The testing of components is basically the same for both systems, except where noted.

→In an effort to conform to an industry-wide move to standardize terminology, Honda has changed the names of a few of the PGM-FI components. These changed names will be noted throughout, but it should be understood the function of these components remains the same, regardless of their names. Also, other (slight) changes to the PGM-FI system over its ten years of existence will be noted where they occur.

AIR INTAKE SYSTEM

The system supplies air for all engine needs. It consists of the air cleaner, air intake pipe, throttle body, idle control system, fast idle mechanism, and intake manifold. A resonator in the air intake pipe provides additional silencing as air is drawn into the system.

Throttle Body

The throttle body, is a 2-barrel side-draft type with the primary air horn at the top. To prevent icing of the throttle valves and air horn walls under certain atmospheric conditions, the throttle valve's air horn walls and the lower portion of the throttle body are heated by engine coolant. A throttle sensor is attached to the primary throttle valve to sense changes in throttle opening. A dashpot is used to slow the throttle as it approaches the closed position.

Idle Control System

The air/fuel ratio during idling is controlled by the Electronic Control Unit (ECU, or from 1993-on, Engine Control Module, ECM) and various solenoid valves such as idle control, fast idle and air conditioning idle control solenoid valves. With the exception of the air conditioning idle control solenoid valve, these change the amounts of air bypassing into the air intake manifold. The air conditioning control solenoid valve opens the throttle when the air conditioner is turned on by signals sent from the ECU/ECM.

FUEL SYSTEM 5-11

The Idle Control Solenoid Valve (ICSV) is used to compensate for idle speed reduction due to electrical, or other loads on the engine. The valve does this by bypassing additional air into the intake manifold. This additional air will allow the idle speed to increase to its normal speed (700–800 rpm). The operation depends upon changes in the voltage at the **FR** terminal of the alternator for quick response. The valve also lowers the fast idle speed in steps during warm-up, after the coolant temperature has reached 131°F (55°C). To prevent erratic running after the engine first fires, the valve is opened during cranking and immediately after starting to provide additional air into the intake manifold.

The Fast Idle Control Solenoid Valve prevents erratic running when the engine is warming up, and a higher idle speed is needed. When the atmospheric pressure is 26.4 in. Hg (660mm Hg) or less, the valve opens to bypass additional air into the intake manifold.

The air conditioning Idle Control Solenoid Valve maintains an idle speed of 700–800 rpm when the air conditioner is turned on. The valve causes the air conditioning idle control diaphragm to open the throttle valve which raises the idle speed. The valve is also opened when coolant temperature is low (immediately after starting) thereby ensuring stable idling regardless of position of the air conditioning switch.

Idle Adjuster (Bypass Circuit)

Fuel cut-off takes place at a set position or angle of the throttle valve. If the throttle valve is moved to adjust idle speed, this position or angle will be changed and the system may not cut off fuel supply. To solve this problem, the throttle body contains an adjustable bypass circuit. This circuit is designed to control the amount of air bypassing into the intake manifold without changing the position of the throttle valve. While the idle control system is in operation, the idle screw has no effect on the idle speed.

Usually idle speed adjustment does not require turning the idle adjustment screw since idle speed is adjusted automatically by the operation of the idle control system. Idle speed does not change by turning the idle adjust screw while the idle control system is in operation.

Fast Idle Mechanism

To prevent erratic running when the engine is warming up, it is necessary to raise the idle speed. The air bypass valve is controlled by a thermowax plunger. When the thermowax is cold, the valve is open. When the thermowax is heated, the valve is closed. With the engine cold and the thermowax consequently cold, additional air is bypassed into the intake manifold so that the engine idles faster than normal. When the engine reaches operating temperature, the valve begins to close, reducing the amount of air bypassing into the manifold.

ELECTRONIC CONTROL SYSTEM

In order to get fuel into the cylinders at the correct instant and in the correct amount, the control system must perform various separate functions. The Electronic Control Unit (ECU, or for 1993–95, Engine Control Module, ECM), the heart of the PGM-FI, uses an eight-bit microcomputer and consists of a Central Processing Unit (CPU), memory storage, and I/O (Input/Output) ports. Basic data stored in the memory are compensated by the signals sent from the various sensors to provide the correct air/fuel mixture for all engine needs.

Electronic Control Unit (ECU)

➡The ECU is called the Engine Control Module (ECM) on 1993–95 models.

The control unit contains (in memory) information stored for the basic duration ("on time") of the injectors at various engine speeds and manifold pressures. Depending on the conditions detected by the various sensors (coolant temperature, air temperature, etc.) the basic signal is then modified to deliver the correct air/fuel duration. Other related functions include:

• Starting Control: The fuel system must vary the air/fuel ratio to suit different operating requirements. For example, the mixture must be rich for starting. The information stored in memory also contains the basic duration (of the injectors) to be read out by signals from the starter switch, and engine speed and coolant temperature sensors, thereby providing extra fuel needed for starting.

• Fuel Pump Control: When the speed of the engine falls below the prescribed limit, electric current to the fuel pump is cut off, preventing the injectors from discharging fuel.

• Fuel Cut-Off Control: During deceleration with the throttle valve nearly closed, electric current to the injectors is cut off at speeds over 900 rpm, contributing to improved fuel economy. Fuel cut-off action also takes place when engine speed exceeds 7000 rpm regardless of the position of the throttle valve.

• Safety: A fail-safe system monitors the sensors and detects any abnormality in the ECU, ensuring engine operation driving even if one or more sensors are faulty, or if the ECU/ECM malfunctions.

Crank Angle (TDC/CKP/CYP) Sensors

➡1.5L models produced before 1993 do not utilize the cylinder position sensor (CYP).

The sensors and distributor are designed as an assembly to save space and weight. The entire unit on pre-1992 1.5L models consists of a pair of rotors, TDC and CRANK sensors, and a pickup for each rotor. All Civic models with a 1.6L engine and all 1.5L models produced 1993–95 and later are equipped with a third sensor, the CYP Sensor. Since the rotors are coupled to the camshaft, they turn together as a unit as the camshaft rotates. The CYP sensor detects the position of the No. 1 cylinder as the base for sequential fuel injection to each cylinder, whereas the TDC sensor serves to determine the injection timing for each cylinder. The CKP sensor determines the timing for fuel injection and ignition of each cylinder and also detects engine speed. The TDC sensor is also used to detect engine speed to read out the basic fuel injector duration for different operating conditions.

Manifold Absolute Pressure (MAP) Sensor

This sensor converts manifold air pressure readings into electrical voltage signals and sends them to the ECU/ECM. This information with signals from the crank angle sensor is then used to read out the basic fuel injector duration from the memory.

Atmospheric Pressure (PA) Sensor

Like the MAP sensor, this unit converts atmospheric pressures into voltage signals and sends them to the ECU/ECM. The signals then modify the basic fuel injector duration to compensate for changes in the atmospheric pressure.

Coolant Temperature (TW) Sensor

The sensor uses a temperature dependent diode (thermistor) to measure differences in the coolant temperature. The basic fuel injector duration is read out by the signals sent from this sensor through the ECU/ECM. The resistance of the thermister decreases with a rise in coolant temperature.

Intake Air Temperature (IAT) Sensor

This device is also a thermistor and is placed in the intake manifold. It acts much like the water temperature sensor but with a reduced thermal capacity for quicker response. The basic fuel injector duration is read from the memory and again modified to compensate for different operating conditions by the signals sent from this sensor through the ECU/ECM.

Throttle Position (TP) Sensor

This sensor is essentially a variable resistor (potentiometer). In construction, the rotor shaft is connected to the throttle valve shaft such that, as the throttle valve is moved, the resistance varies, altering the output voltage to the control unit.

Oxygen Sensor (O2S)

▶ See Figures 33 and 34

➡Models produced 1992–95 have a heater unit attached to the same type oxygen sensor, thereby making it a Heated Oxygen Sensor (HO2S). The heater was added when Honda found it stabilized the sensor's output. The description following for the two units is in every other respect the same.

The oxygen sensor, by detecting the oxygen content in the exhaust gas, helps the ECU/ECM decide how much fuel is necessary to maintain the stoichiometric air/fuel ratio. During operation, the ECU/ECM receives the signals from the sensor and changes the duration during which fuel is injected. The oxygen sensor is located in the exhaust manifold.

5-12 FUEL SYSTEM

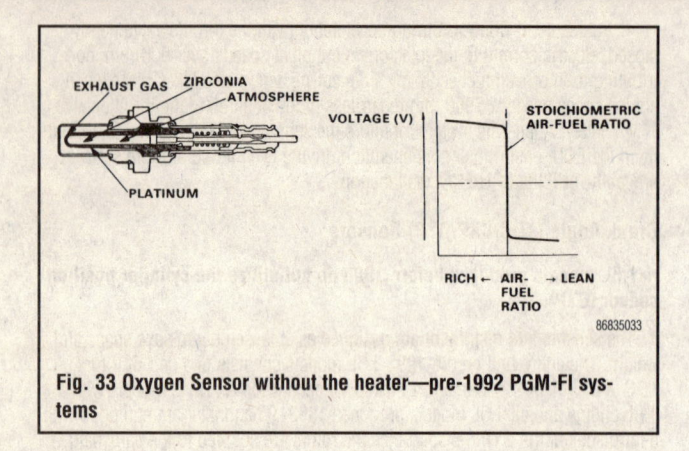

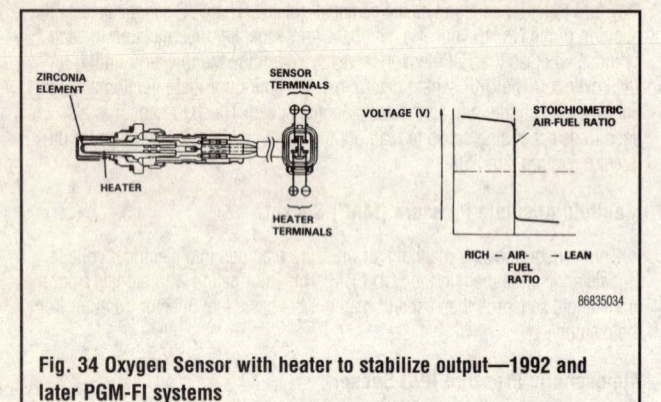

Fig. 33 Oxygen Sensor without the heater—pre-1992 PGM-FI systems

Fig. 34 Oxygen Sensor with heater to stabilize output—1992 and later PGM-FI systems

The sensor is a hollow shaft of zirconia with a closed end. The inner and outer surfaces are plated with platinum, thus forming a platinum electrode. The inner surface or chamber is open to the atmosphere whereas the outer surface is exposed to the exhaust gas flow through the manifold.

Voltage is induced at the platinum electrode when there is any difference in oxygen concentration between the two layers of air over the surfaces. Operation of the device is dependent upon the fact that voltage induced changes sharply as the stoichiometric air/fuel ratio is exceeded when the electrode is heated above a certain temperature.

Idle Mixture Adjuster (IMA) Sensor

The sensor is located in the control box. The primary objective of this unit is to maintain the correct air/fuel ratio at idling. No adjustment of the IMA sensor is necessary as the feedback control is performed by the oxygen sensor even during idling.

Starter Switch

The air/fuel mixture must be rich for starting. During cranking, the ECU/ECM detects the signal from the starter switch and increases the amount of fuel injected into the manifold according to the engine temperature. The amount of fuel injected is gradually reduced when the starter switch is turned off.

FUEL SYSTEM

Fuel Pump

The fuel pump is an inline, direct-drive type. Fuel is drawn into the pump through a filter, flows around the armature through the one-way valve and is delivered to the engine compartment. A baffle is provided to prevent fuel pulsation. The fuel pump has a relief valve to prevent excessive pressure. It opens if there is a blockage in the discharge side. When the relief valve opens, fuel flows from the high pressure to the low pressure side. A check valve is provided to maintain fuel pressure in the line after the pump is stopped. This is to ease restarting.

The pump section is composed of a rotor, rollers and pump spacer. When the rotor turns, the rollers turn and travel along the inner surface of the pump spacer by centrifugal force. The volume of the cavity enclosed by these three parts changes, drawing and pressurizing the fuel.

Pressure Regulator

The fuel pressure regulator maintains a constant fuel pressure to the injectors. The spring chamber of the pressure regulator is connected to the intake manifold to constantly maintain the fuel pressure at 36 psi (248 kPa) higher than the pressure in the manifold (40–47 psi or 280–330 kPa for 1992–95 models). When the difference between the fuel pressure and manifold pressure exceeds 36 psi (248 kPa) (or 40–47 psi / 280–330 kPa for 1992–95 models), the diaphragm is pushed upward, and the excess fuel is fed back into the fuel tank through the return line.

Injector

The injector is of the solenoid actuated constant stroke pintle type consisting of a solenoid, plunger, needle valve and housing. When current is applied to the solenoid coil, the valve lifts up and pressurized fuel fills the inside of the injector and is injected close to the intake valve. Because the needle valve lifts and the fuel pressure are constant, the injection quantity is determined by the length of time that the valve is open, i.e., the duration the current is supplied to the solenoid coil. The injector is sealed by an O-ring and seal ring at the top and bottom. These seals also reduce operating noise.

Resistor

The injector timing, which controls the opening and closing intervals, must be very accurate since it dictates the air/fuel mixture ratio. The injector must also be durable. For the best possible injector response, it is necessary to shorten the current rise time when voltage is applied to the injector coil. Therefore, the number of windings of the injector coil was reduced to lower the inductance in the coil. This, however, makes low resistance in the coil, allowing a large amount of current to flow through the coil. As a result, the amount of heat generated is high, which compromises the durability of the coil. Flow of current in the coil is therefore restricted by a resistor installed in series between the electric power source and the injector coil.

Main Relay

The main relay is a direct coupler type which contains the relays for the electronic control unit power supply and the fuel pump power supply. This relay is installed at the back of the fuse box.

Fuel System Service Precautions

Safety is the most important factor when performing not only fuel system maintenance but any type of maintenance. Failure to conduct maintenance and repairs in a safe manner may result in serious personal injury or death. Maintenance and testing of the vehicle's fuel system components can be accomplished safely and effectively by adhering to the following rules and guidelines.

- To avoid the possibility of fire and personal injury, always disconnect the negative battery cable unless the repair or test procedure requires that battery voltage be applied.
- Always relieve the fuel system pressure prior to disconnecting any fuel system component (injector, fuel rail, pressure regulator, etc.), fitting or fuel line connection. Exercise extreme caution whenever relieving fuel system pressure to avoid exposing skin, face and eyes to fuel spray. Please be advised that fuel under pressure may penetrate the skin or any part of the body that it contacts.
- Always place a shop towel or cloth around the fitting or connection prior to loosening to absorb any excess fuel due to spillage. Also be sure that all fuel spillage (should it occur) is quickly removed from engine surfaces. Ensure that all fuel soaked cloths or towels are deposited into a suitable waste container.
- Always keep a dry chemical (Class B) fire extinguisher near the work area.
- Do not allow fuel spray or fuel vapors to come into contact with a spark or open flame.
- Always use a backup wrench when loosening and tightening fuel line connection fittings. This will prevent unnecessary stress and torsion to fuel line piping. Always follow the proper torque specifications.
- Always replace worn fuel fitting O-rings with new. Do not substitute fuel hose or equivalent where fuel pipe is installed.

FUEL SYSTEM 5-13

Relieving Fuel System Pressure

It is important to release the pressure in the fuel system before performing any work on it. Fuel under pressure could be sprayed onto hot engine components and possibly ignite.

1. Make sure the engine is cold.
2. Disconnect the negative battery cable.
3. Remove the fuel filler cap.
4. Use a suitable box end wrench on the 6mm service bolt at the top of the fuel filter, while holding the special banjo bolt with another wrench.
5. Place a rag or shop towel over the 6mm service bolt.
6. Slowly loosen the 6mm service bolt one complete turn and allow the pressurized fuel to de-pressurize.
7. Unless the work you are performing requires removal of the service bolt, re-tighten the bolt, once the fuel pressure is relieved, to 11 ft. lbs. (15 Nm).

COMPONENT TEST PROCEDURES

➡ Detailed tests for 1988–95 models can be found in the Troubleshooting Flowcharts in Section 4.

Oxygen Sensor

▶ See Figure 35

1985–87 MODELS

1. Unplug the oxygen sensor connector.
2. Start engine and warm up for two minutes at 3000 rpm under no load. Raise engine speed to 4000 rpm and release throttle suddenly at least 5 times.
3. Within one minute after engine has been warmed up, measure voltage between connector terminal and body ground as described in Steps 4 and 5.

➡ If it takes more than one minute to complete checks, warm engine as in Step 2 before continuing.

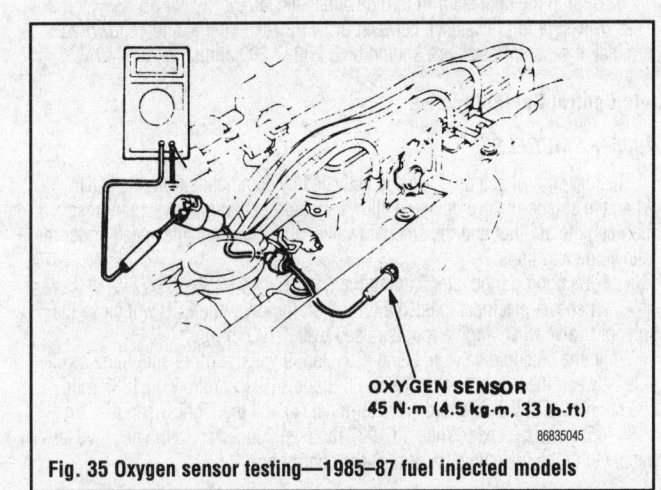

4. Raise engine speed to 5000 rpm, then lower to 2000 rpm by operating accelerator pedal, immediately turn ignition switch OFF. Voltage should be below 0.4 volts.
5. Disconnect the MAP sensor vacuum tube from the throttle body; plug opening in throttle body. Connect a hand vacuum pump to open end of vacuum tube and apply 12 in. Hg (300mm Hg), and raise engine speed to 4,000 rpm. Voltage should be above 0.6 volts.
6. Replace oxygen sensor if voltages are out of range.

Manifold Air Pressure (MAP) Sensor

▶ See Figure 36

1985–87 MODELS

1. Disconnect the MAP sensor vacuum hose from the throttle body; plug opening in throttle body. Connect a vacuum hand pump to open end of vacuum tube.
2. Unplug the connector from control unit. Install system checker harness (07999–PD6000A or equivalent) between control unit and wire harness connector.
3. Turn ignition switch ON. Connect a digital voltmeter positive probe to C11 terminal of system checker harness and negative probe to C14 terminal. Measure voltage between two terminals. Voltmeter should indicate between 0.5 volts at 4 in. Hg (100mm Hg) of vacuum and 4.5 volts at 45 in. Hg (1200mm Hg).
4. If voltage is incorrect, check vacuum tube for leakage, then check the wires between control unit and sensor for an open or short circuit. Replace sensor if wires are normal.

Atmospheric Pressure (PA) Sensor

1985–87 MODELS

▶ See Figure 37

1. Unplug the connector from the control unit. Install a system checker harness (07999–PD6000A or equivalent) between control unit and wire harness connector.
2. Turn ignition switch ON. Connect a digital voltmeter positive probe to C9 terminal of system checker harness and negative probe to C12 terminal. Measure voltage between two terminals. Voltmeter should indicate 2.76–2.96 volts.
3. If voltage is outside ranges, check for open or short circuit between ECU and PA sensor. Replace PA sensor with a new one if wires are in good condition.

Idle Mixture Adjuster (IMA) Sensor

1985–87 MODELS

▶ See Figure 38

1. Measure resistance between Green terminal and Yellow terminal (Yellow/White and Black terminal on 1987 models) of control box coupler. Resistance should be 250–6200 ohms.
2. Replace IMA sensor if resistance is out of range.

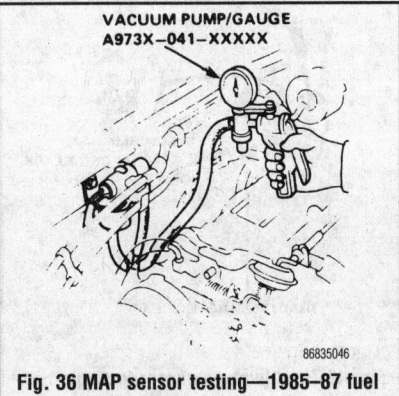

Fig. 35 Oxygen sensor testing—1985–87 fuel injected models

Fig. 36 MAP sensor testing—1985–87 fuel injected models

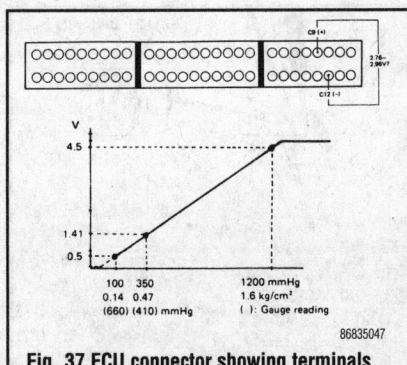

Fig. 37 ECU connector showing terminals for checking the Atmospheric Pressure (PA) Sensor—1985–87 fuel injected models

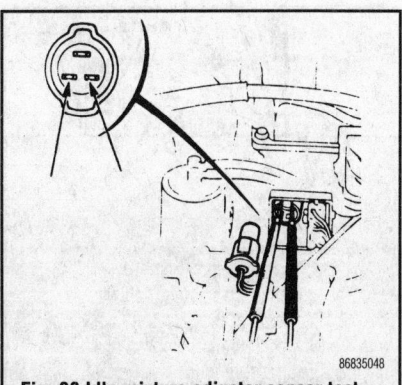

Fig. 38 Idle mixture adjuster sensor testing—1985–87 fuel injected models

5-14 FUEL SYSTEM

Intake Air Temperature (TA)/Coolant Temperature (TW) Sensor

1985–87 MODELS

▶ See Figures 39 and 40

1. Unplug the connector and remove TA/TW sensor from inlet manifold/cylinder head.
2. To test a sensor, suspend it in cold water and heat water slowly. Measure resistance between terminals. Measurements should be 9800–13,400 ohms at 95°F (40°C) and 2200–3500 ohms at 176°F (80°C).
3. Replace sensor if resistance is outside of range.

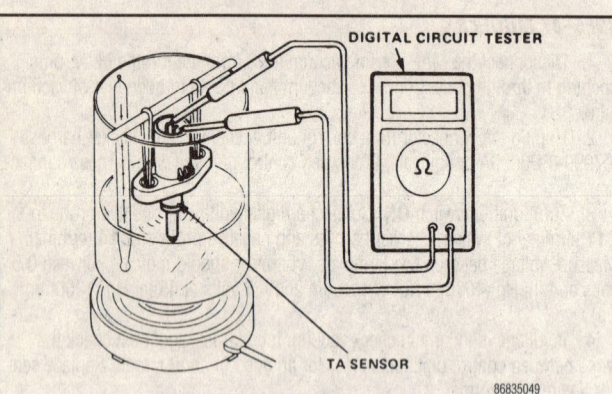

Fig. 39 Intake air temperature and coolant temperature sensor testing—1985–87 fuel injected models

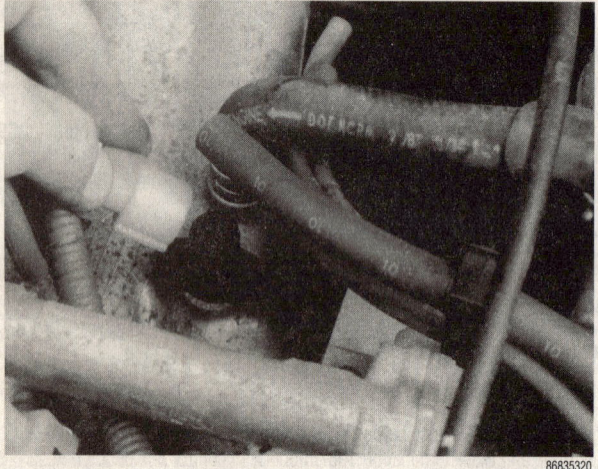

Fig. 40 Disengaging IAT sensor located on the intake manifold—1985–87-style CRX Si shown

Crank Angle Sensor (CYL Sensor)

1985–87 MODELS

▶ See Figure 41

→ If either CYL or TDC sensor tests bad, replace as an assembly.

1. Disconnect connector of crank angle sensor.
2. Measure resistance between the white and red wire at the terminal at the sensor. Resistance should be 6,500–8,500 ohms.
3. Measure insulation resistance between the white and red wires at the terminal of the sensor and crank angle sensor housing. Resistance should 100,000 ohms or more.

Crank Angle Sensor (TDC Sensor)

1985–87 MODELS

▶ See Figure 42

→ If either CYL or TDC sensor tests bad, replace as an assembly.

1. Unplug the connector of crank angle sensor.
2. Measure resistance between the brown and red wires at the sensor terminal. Resistance should be 6,500–8,500 ohms.
3. Measure insulation resistance between brown and red wires at the terminal of the sensor and crank angle sensor housing. Resistance should be 100,000 ohms or more.

Throttle Angle Sensor

1985–87 MODELS

▶ See Figure 43

→ Do not adjust the throttle valve stop screw since it is preset at factory.

1. Unplug the connector of throttle angle sensor.
2. Measure full resistance between Brown/Black terminal and Yellow/Red terminal at sensor. Resistance should be 3,200–7,200 ohms.

Idle Control Solenoid Valve

1985–87 MODELS

The idle control solenoid valve is activated by commands from the ECU. When the solenoid valve opens, it allows vacuum in the upper vacuum hose (between the air filter and the solenoid valve) and increases idle speed under the following conditions:

- For a short period after starting the engine.
- Whenever electrical loads are turned on (however the ECU will close the solenoid valve when engine rpm is raised over 1,700 rpm).

While the solenoid valve is being activated, 9 volts or more should be available between the Black wire on 1985–86 models (Black/Yellow on 1987 models) connected at the solenoid valve terminal (+) and the Light Green wire on 1985–86 models (Green/White on 1987 models) connected to the negative terminal (–) of the main wire harness at the control box.

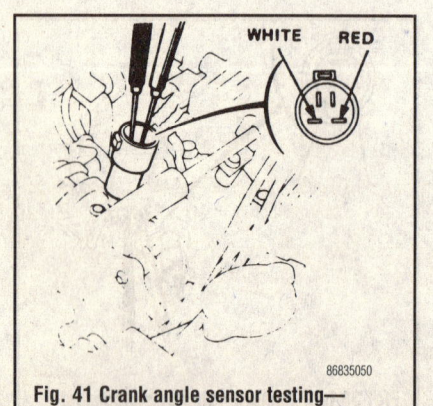

Fig. 41 Crank angle sensor testing—1985–87 fuel injected models

Fig. 42 TDC sensor testing—1985–87 fuel injected models

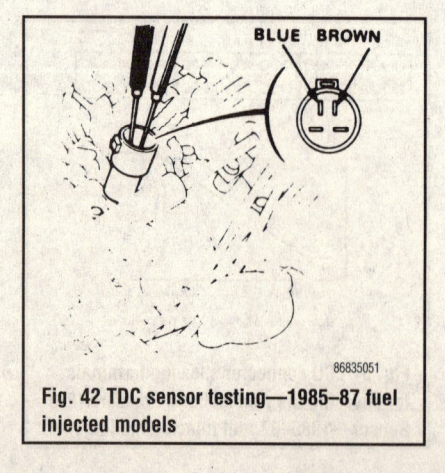

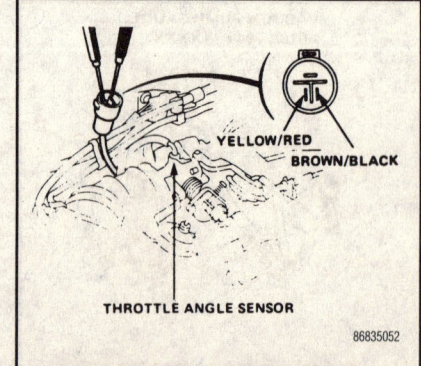

Fig. 43 Throttle angle sensor testing—1985–87 fuel injected models

FUEL SYSTEM 5-15

1. Open the control box lid and unplug the 8-pin connector at the wire harness from the control box.
2. Disconnect the vacuum hose between the idle control solenoid valve and the No. 10 vacuum hose (upper vacuum hose of idle control solenoid valve from 3-way joint on 1987–88 models) and connect the vacuum pump to the hose.
3. Apply vacuum to the hose. It should hold vacuum.
4. Connect battery positive terminal to Black terminal of control box and negative battery terminal to Light Green terminal.
5. Apply vacuum to hose. It should not hold vacuum. If vacuum holds, replace solenoid valve.

Fast Idle Control Solenoid Valve

1985–87 MODELS

The fast idle control solenoid valve is open when the atmospheric pressure is 26 in. Hg (660mm Hg) or less. Vacuum is produced in the vacuum hose (between the solenoid valve and air filter).

1. When the valve is open, 9 volts or more should be available for 1985–86 models between the Black terminal (+) and Yellow terminal (–) (Black/Yellow terminal (+) and Green/Yellow terminal (–) on 1987 models) of the main wire harness of the control box.
2. Open the control box lid and unplug the 8-pin connector.
3. Disconnect the lower vacuum hose of the fast idle control solenoid valve from the uppermost hose that connects to the 3-way joint and connect the vacuum pump for testing.
4. Apply vacuum to the hose. It should hold vacuum.
5. Connect the battery positive terminal to the Black terminal of control box coupler, and the battery negative terminal to Yellow terminal.
6. Apply vacuum to the hose. It should not hold vacuum. If the vacuum holds, replace the valve.

IDLE SPEED TOO HIGH AFTER WARMUP

▶ See Figures 44, 45, 46 and 47

➡The fast idle valve is factory adjusted, it should not be disassembled. Check PCV (engine breather) circuit tubing for breakage, disconnection, clogging, etc. Check that the throttle valves are fully closed.

1. Confirm that engine is adequately warmed.
2. Check whether the idling control function is normal.
3. Remove the cover of the fast idle valve.
4. Check that the valve is completely closed. If not, air is being sucked from valve seat area. It can be detected by putting your finger on valve seat area.
5. If any suction sound is felt, the valve is leaking. Replace the fast idle valve and adjust idle speed.

IDLE SPEED TOO LOW AFTER WARMUP

➡Fast idle valve is factory adjusted, it should not be disassembled. Check PCV (engine breather) circuit tubing for breakage, disconnection, clogging, etc. Check that throttle valve are fully closed.

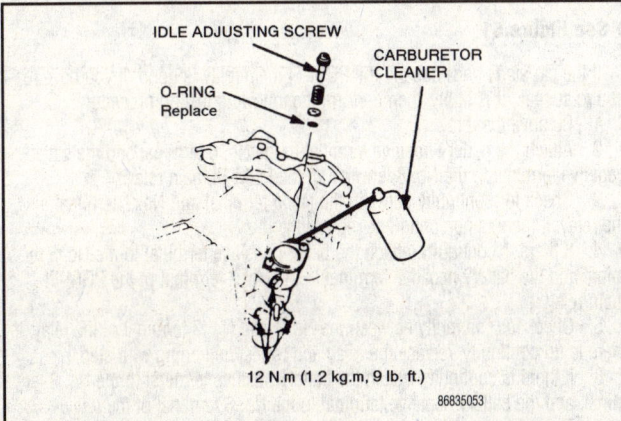

Fig. 44 Fast idle valve assembly—1985–87 fuel injected models

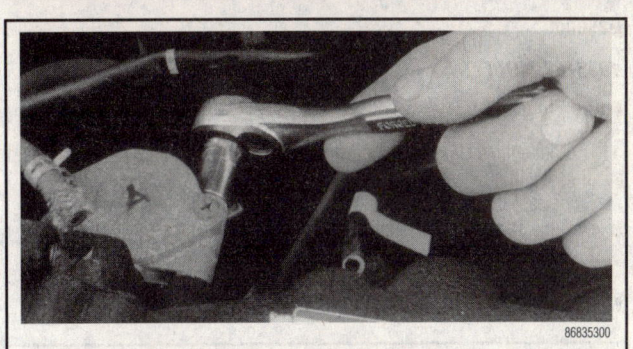

Fig. 45 If the fast idle valve is found faulty, unbolt the valve from its location on the back of the intake plenum—1985–87 multi-point fuel injected

Fig. 46 Once the faulty valve is unbolted, it pulls cleanly away from its mounting point—1985–87 multi-point fuel injected

Fig. 47 On all models, if fast idle valve replacement is required, do not re-use the old O-rings—1985–87 multi-point fuel injected shown

1. Remove idle adjusting screw.
2. Wash idle adjusting screw and air bypass channel with carburetor cleaner.
3. Readjust the idle speed after cleaning.

FAST IDLE SPEED IS LOW WHEN ENGINE IS COLD

➡The fast idle valve is factory adjusted, it should not be disassembled. Fast idle speed should be 1250–2250 rpm.

1. Remove fast idle valve assembly from throttle body.
2. Apply cold water and cool down wax part of fast idle valve to 41–86°F (5–30°C).
3. Blow through part **A** of the fast idle valve, and check that a fairly large amount of air flows without resistance.
4. If the air does not flow or resistance is great, replace the fast idle valve and adjust the idle speed.

5-16 FUEL SYSTEM

Dashpot

1985–87 MODELS

▶ See Figure 48

1. With engine **OFF**, slowly open throttle arm until dashpot rod is raised up as far as it will go.
2. Release throttle arm and measure time until throttle arm contacts stop screw. Time should be less than 2 seconds.
3. If time is over 2 seconds, replace dashpot check valve.
4. If rod does not operate, check for bound linkage, or for clogged check valve or vacuum line. If OK, replace dashpot.

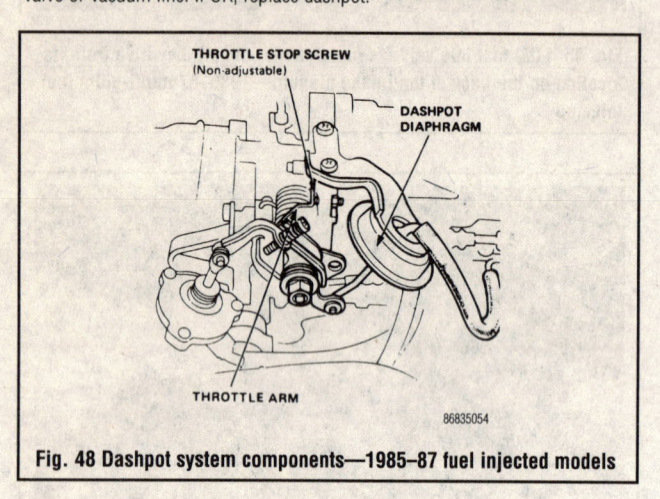

Fig. 48 Dashpot system components—1985–87 fuel injected models

Injector Test

1985–87 MODELS

➥Check following items before testing idle speed, ignition timing, valve clearance and idle CO%.

1. If engine will run, disconnect injector couplers with engine idling, and inspect change in idling speed. Idle should drop the same for each cylinder.
2. Check clicking sound of each injector by means of a stethoscope when engine is idling.
3. If any injector fails to make a clicking sound, check wiring between ECU and injector. Voltage at injector coupler should fluctuate 0–2 volts. If voltage is OK, replace injector.
4. If engine can not be started, remove coupler of injector, and measure resistance between terminals of injector. Resistance should be 1.5–2.5 ohms.
5. If resistance is not as specified, replace injector. If resistance is normal, check wiring between resistor and injector, wiring between resistors and control unit, and resistors.

Fuel System Resistor

1985–87 MODELS

▶ See Figure 49

1. Disconnect resistor coupler.
2. Check for resistance between each of resistor terminals **E, D, C and B** and power terminal **A**. Resistance should be 5–7 ohms.
3. Replace resistor with a new one if any of resistances are outside of specification.

Main Relay

▶ See Figure 50

1985–91 MODELS

1. Remove the main relay, near under-dash fuse box.
2. Connect battery positive terminal to No. 4 terminal and battery negative terminal to No. 8 terminal of main relay.
3. Check for continuity between No. 5 terminal and No. 7 terminal of main relay. If no continuity, replace main relay.

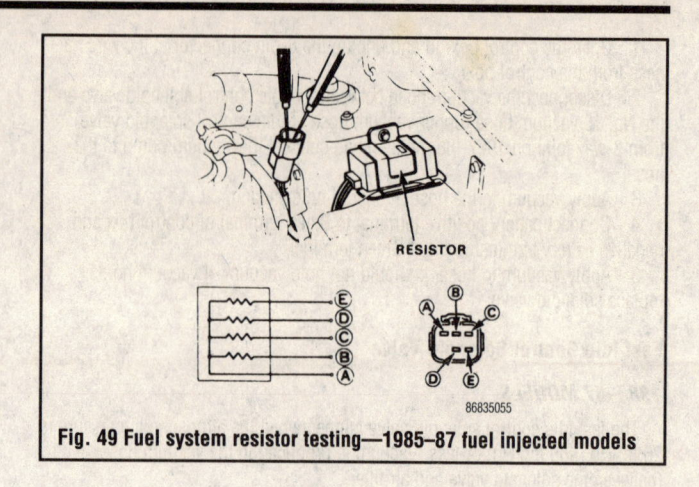

Fig. 49 Fuel system resistor testing—1985–87 fuel injected models

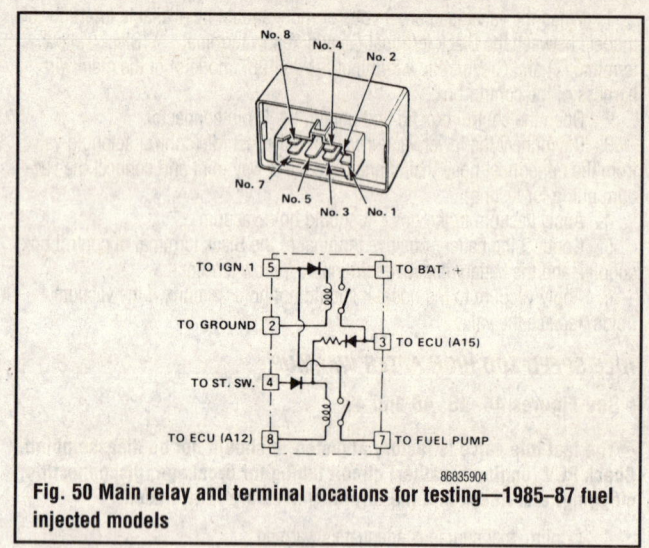

Fig. 50 Main relay and terminal locations for testing—1985–87 fuel injected models

4. Connect battery positive terminal to No. 5 terminal and battery negative terminal to No. 2 terminal of main relay.
5. Check that there is continuity between No. 1 terminal and No. 3 terminal of main relay. If there is no continuity, replace main relay.
6. Connect battery positive terminal to No. 3 terminal and battery negative terminal to No. 8 terminal of main relay.
7. Check that there is continuity between No. 5 terminal and No. 7 terminal of main relay. If there is no continuity, replace the PGM-FI main relay.

PGM-FI Main Relay

1992–95 MODELS

▶ See Figure 51

If the car starts and runs ok, the PGM-FI main relay is good. If you have reason to suspect it is faulty, then perform the following testing procedure:

1. Remove the relay.
2. Attach the battery positive terminal to the No. 6 terminal and the battery negative terminal to the No. 8 terminal of the PGM-FI main relay.
3. Check for continuity between the No. 5 terminal and No. 7 terminal on the relay. If there is no continuity, replace the relay.
4. If there is continuity, attach the battery positive terminal to the No. 5 terminal and the battery negative terminal to the No. 2 terminal of the PGM-FI main relay.
5. Check for continuity between the No. 1 and No. 3 terminal of the relay. If there is no continuity, replace the relay and re-test beginning with step 1.
6. If there is continuity, attach the battery positive terminal to the No. 3 terminal, and the battery negative terminal to the No. 8 terminal of the relay.
7. Then check that there is continuity between the No. 5 terminal and the No. 7 terminal. If there is no continuity, replace the relay and re-test beginning with step 1. If there is continuity, the relay is ok.

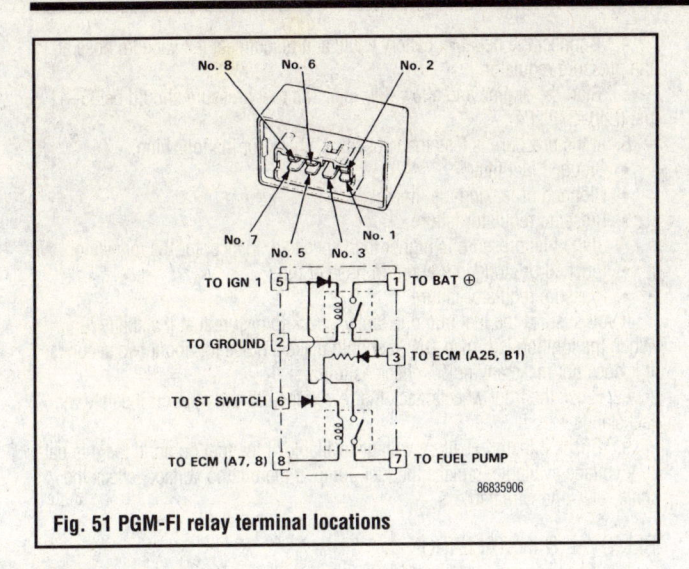

FUEL SYSTEM 5-17

Fig. 51 PGM-FI relay terminal locations

Harness

1985–87 MODELS

1. Keep ignition switch in the **OFF** position.
2. Unplug the main relay connector.
3. Connect positive probe of circuit tester to Yellow/White in coupler and ground the negative probe of the tester to the body ground. Tester should read battery voltage.
4. If there is no voltage, check wiring between battery and main relay as well as ECU fuse in engine compartment.
5. Connect positive terminal of tester to Black/Yellow wire of coupler and ground negative terminal of tester to body ground.
6. Turn ignition switch **ON**. Tester should read battery voltage.
7. If no voltage, check wiring from ignition switch and main relay as well as fuse No. 4.
8. Connect positive terminal of tester to Blue/White wire in coupler and ground negative terminal to body.
9. Turn the ignition switch to **START** position. The tester should read battery voltage.
10. If there is no voltage, check the wiring between the ignition switch and the main relay as well as starter fuse No. 1.
11. Connect a jumper wire between the Yellow/White wire and the Yellow wire for 1985–86 models (Black/Yellow wire on 1987 models) in coupler. The fuel pump should work.
12. If pump does not work, check wiring between battery and fuel pump and wiring from fuel pump to ground (Black wire).

1988–91 MODELS

1. Keep ignition switch in **OFF** position.
2. Disconnect main relay connector.
3. Check continuity between Black wire in connector and body ground.
4. Connect positive probe of circuit tester to Yellow/Black wire and negative probe of tester to Black wire. Tester should read battery voltage.
5. If there is no voltage, check wiring between battery and main relay as well as ECU (15A) fuse.
6. Connect positive probe of tester to Black/Yellow wire of connector and ground negative probe of tester to Black wire.
7. Turn ignition switch **ON**. Tester should read battery voltage.
8. If no voltage, check wiring from ignition switch and main relay as well as fuse No. 14.
9. Connect positive probe of tester to Blue/White wire and negative probe to Black wire.
10. Turn ignition switch to **START** position. Tester should read 10 volts or more.
11. If no voltage, check wiring between ignition switch and main relay as well as No. 2 (10A) fuse.
12. Connect a jumper wire between 2 Black/Yellow wires in connector. Turn ignition switch **ON**. Fuel pump should run.
13. If pump does not run, check wiring between battery and fuel pump and wiring from fuel pump to ground (Black wire).

Throttle Control Diaphragm

1988–91—EXCEPT 1988 CRX SI

▶ See Figure 52

1. Start the engine and allow to reach normal operating temperature.
2. Disconnect the No. 6 vacuum line from the dashpot diaphragm and check engine rpm.
3. If engine rpm is higher than 2000–3000 rpm, adjust by bending the tab.
4. If the engine speed does not change, connect a vacuum gauge to the No. 6 hose and check for vacuum.
5. If there is vacuum, replace the diaphragm. If there is no vacuum, check hose for blockage or leakage.
6. Reconnect vacuum hose and adjust idle as needed.

1988 CRX SI

1. Start the engine and allow to reach normal operating temperature.
2. Disconnect the No. 6 vacuum line from the dashpot diaphragm and check engine rpm.
3. If the engine speed does not change, connect a vacuum gauge to the No. 6 hose and check for vacuum.
4. If there is vacuum, replace the diaphragm. If there is no vacuum, check hose for blockage or leakage.
5. If engine rpm is out of range, inspect the throttle body.

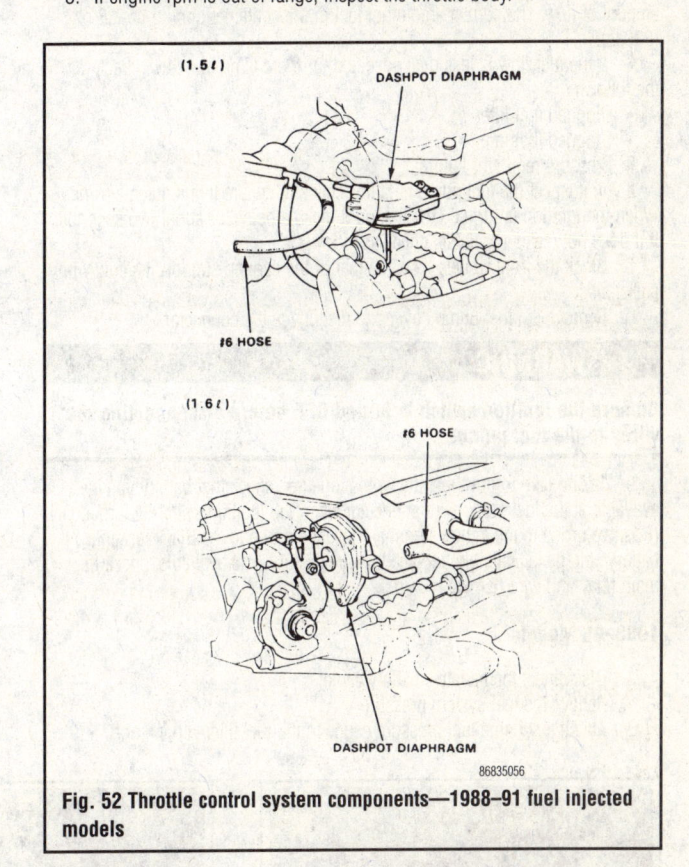

Fig. 52 Throttle control system components—1988–91 fuel injected models

Electric Fuel Pump

PRESSURE TESTING

1985–87 Models

▶ See Figure 53

1. With the ignition switch **OFF**, unplug the main relay connector from behind the fuse box.
2. Connect the Yellow wire to the Black/Yellow wire with a jumper.

5-18 FUEL SYSTEM

Fig. 53 Electric fuel pump mounting—1985–87

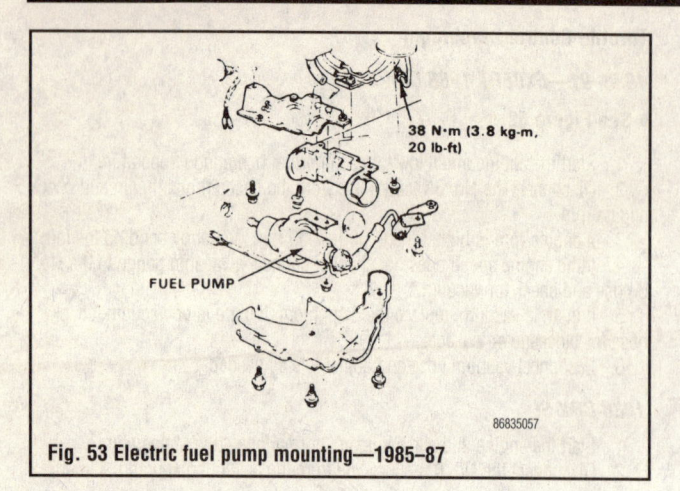

3. Relieve fuel pressure, then tighten the service bolt.
4. Disconnect the fuel return hose from the regulator and place the end of the hose in a graduated cylinder, or equivalent, in order to measure fuel flow by volume.
5. Turn the ignition switch **ON** for exactly 10 seconds, then measure the amount of flow. The correct amount of fuel in the container should be 7.8 oz. (230cc).
6. If the fuel flow is less than specified, or there is no fuel flow, check for the following:
- Clogged fuel filter
- Clogged fuel line
- Pressure regulator failure

If you suspect the fuel pump is faulty, check to ensure that it actually runs. When the ignition is turned **ON**, it should make a noise for about two seconds. If it does not make any noise, check as follows:

7. Block the front wheels, jack the rear of the car and support it safely with jackstands.
8. Remove the fuel pump cover and disengage the connector.

✳✳ WARNING

Be sure the ignition switch is turned OFF before disconnecting the wires to the fuel pump.

9. Taking care to not touch the wires together, check that battery voltage is available at the fuel pump wire connector when the ignition switch is turned ON. (Positive probe to the Yellow terminal, negative probe to the Black terminal). If battery voltage is available, replace the pump. If it is not available, check the main relay and wire harness.

1988–95 Models

1. Disconnect the negative battery cable.
2. Relieve the fuel system pressure.
3. Attach a suitable fuel pressure gauge to the fuel filter service port.

4. Connect the negative battery cable and disconnect the vacuum hose at the pressure regulator.
5. Start the engine and allow it to idle. The fuel pressure should be 35–41 psi (240–279 kPa).
6. If the pressure is less than specified, check for the following:
- Clogged fuel filter
- Clogged or leaking fuel line(s)
- Pressure regulator failure
7. If the fuel pressure is higher than specified, inspect the following:
- Pinched or clogged fuel return hose or pipe
- Pressure regulator failure

If you suspect the fuel pump is faulty, check to ensure that it actually runs. When the ignition is turned **ON**, it should make a noise for about two seconds. If it does not make any noise, check as follows:

8. Block the front wheels, jack the rear of the car and support it safely with jackstands.
9. Check if there is battery voltage available at the fuel pump. If there is battery voltage available, replace the fuel pump. If there is no voltage, check the main relay and wire harness.

REMOVAL & INSTALLATION

1985–87 Models

▶ See Figures 53, 54, 55, 56 and 57

The fuel pump is located under the left rear of the vehicle, it is bolted to the chassis and is covered by a 2-piece cover.

➡**Relieve the fuel system pressure before working on any part of the fuel system. Follow all safety precautions.**

1. Properly relieve the fuel system pressure.
2. Disconnect the negative battery cable and raise and safely support the vehicle.
3. Remove the left rear wheel. Remove the fuel pump cover assembly bolts and remove the cover.

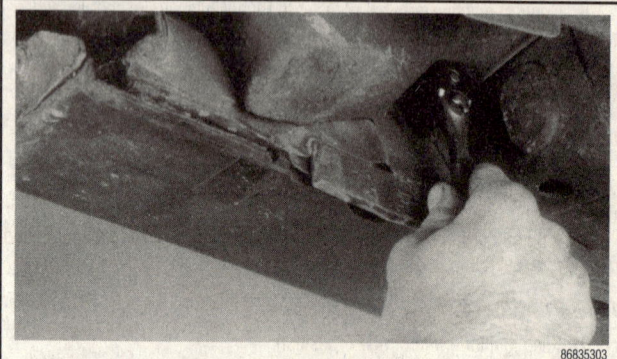

Fig. 54 The rear electric fuel pump as found on 1985–87 models is protected beneath this shroud—1985–87 style CRX shown

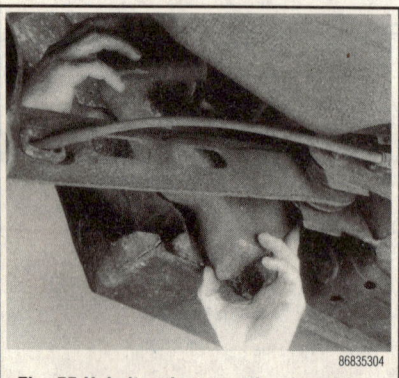

Fig. 55 Unbolt and remove the shroud for rear fuel pump access—1985–87

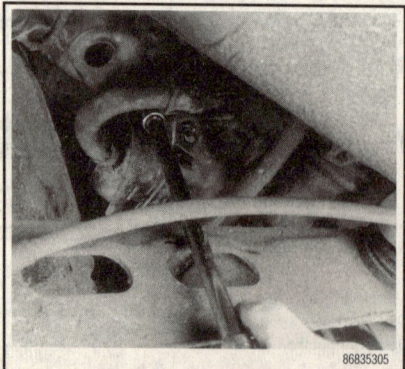

Fig. 56 Removing the bolts holding the fuel pump bracket—1985–87

Fig. 57 Removing the fuel line clamp from the rear fuel pump—1985–87

FUEL SYSTEM 5-19

4. Remove the three fuel pump retaining bolts and lower the pump slightly, disconnect the fuel lines from the pump.
5. Disconnect the electrical leads from the pump. Remove the clamp retaining the pump and slide the pump out of the mount. Remove the silencer from the fuel line.

To install:
6. Install the fuel pump on its mount and install the silencer.
7. Install the fuel lines on the pump. Tighten the silencer to 20 ft. lbs. (28 Nm).
8. Install the pump covers and the pump assembly in place.
9. Install the rear wheel. Connect the negative battery cable.
10. Turn the ignition key to the **ON** position and check the fuel connections for leaks. Lower the vehicle.

1988–91 Models

The fuel pump is located in the fuel tank.
1. Disconnect the negative battery cable.
2. Relieve the fuel system pressure.
3. Raise and safely support the vehicle.
4. Remove the fuel tank drain bolt and drain the fuel into a suitable container.
5. On Civic models (except CRX and Wagon) unplug the fuel pump electrical connector in the trunk. On CRX, remove the storage compartment and unplug the fuel pump electrical connector. On Civic Wagon, remove the rear seat and carefully disengage the fuel pump electrical harness.
6. On Civic Wagon with 4WD, remove the driveshaft from the rear differential and the exhaust pipe and muffler.
7. Remove the 2-way valve cover and fuel hose protector.
8. Unbolt the fuel lines.
9. Place a suitable support under the fuel tank.
10. Remove the fuel tank strap nuts and let the straps hang.
11. Remove the fuel tank.
12. Remove the fuel pump mounting nuts and the fuel pump.

To install:
13. Install the pump assembly into the tank and make sure the unit is properly sealed.
14. Install the tank into the vehicle and tighten the strap nuts. Connect the hoses and wiring.
15. On 4WD vehicles, connect the driveshaft and the exhaust system.
16. Put about three gallons of fuel into the tank and test the system.

1992–95 Models

The fuel pump is in the fuel tank and is reached through an access panel inside the car in the rear passenger or cargo area.
1. Relieve the fuel system pressure.
2. Remove the rear seat (if applicable) and remove the access panel.
3. Unbolt the fuel lines and unplug the electrical connector.
4. Remove the mounting nuts and carefully pull out the fuel pump from the fuel tank.

To install:
5. Place the fuel pump in the tank in the correct position and install the mounting nuts, tightening them to 4 ft. lbs. (6 Nm).
6. Plug in the wire connectors and bolt the banjo line fitting using a new washer. Tighten to 20 ft. lbs. (28 Nm).
7. When properly installed, briefly turn the ignition **ON**, to ensure proper operation. The pump should make a noise for about two seconds as the system pressurizes.
8. If ok, install the rear seat (if applicable).

Fuel Injector

REMOVAL & INSTALLATION

Dual Point Fuel Injection (DP-FI)

♦ See Figure 58

1. Disconnect the negative battery cable.
2. Relieve the fuel pressure.

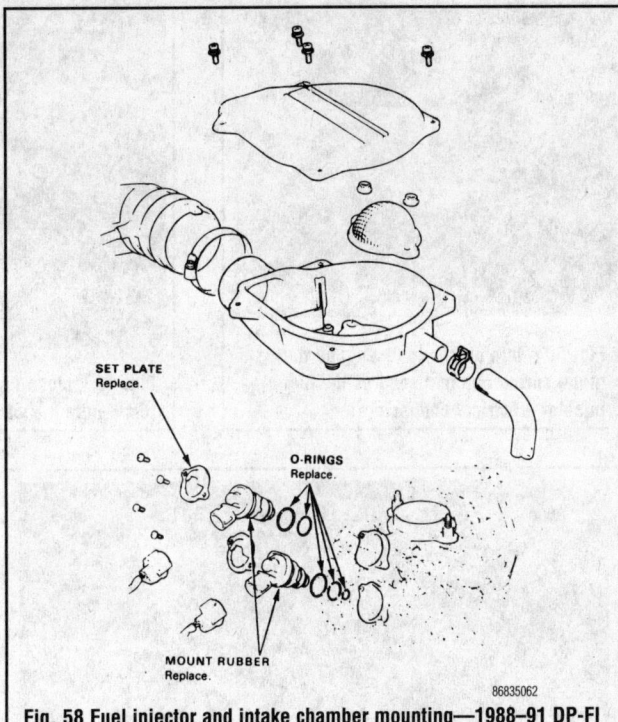

Fig. 58 Fuel injector and intake chamber mounting—1988–91 DP-FI equipped models

3. Remove the air intake chamber.
4. Disengage the electrical connector from the fuel injector.
5. Loosen the screws and remove the injector from the throttle body. Place a rag or shop towel over the throttle body after removal.
6. Installation is the reverse of the removal procedure. Use new O-rings and coat them with engine oil prior to installation. After the injector is inserted, make sure it turns smoothly approximately 30°.
7. Before installing the air intake chamber, connect the negative battery cable and turn the ignition switch **ON** for approximately two seconds. Repeat 2–3 times and check for fuel leaks.

Multi-Point Fuel Injection

♦ See Figures 59 thru 65

1. Disconnect the negative battery cable.
2. Relieve the fuel pressure.
3. Unplug the electrical connectors from the fuel injectors.
4. Disconnect the vacuum hose and fuel return hose from the fuel pressure regulator.

➡ Place a rag or shop towel over the hose and tube before disconnecting them.

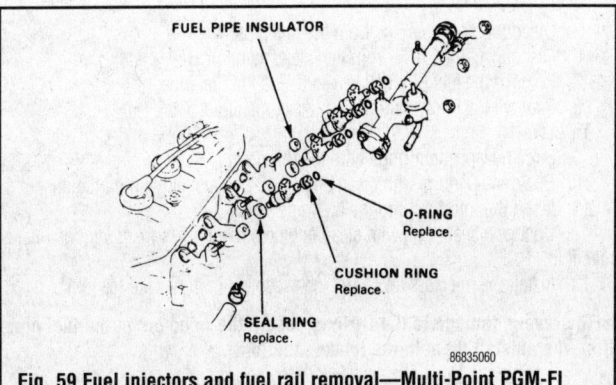

Fig. 59 Fuel injectors and fuel rail removal—Multi-Point PGM-FI equipped models

5-20 FUEL SYSTEM

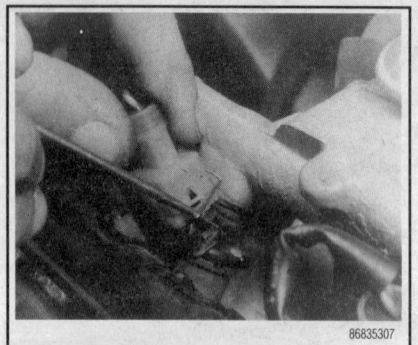

Fig. 60 A thin pick is to disengage the metal spring clip that secures the fuel injector electrical connectors

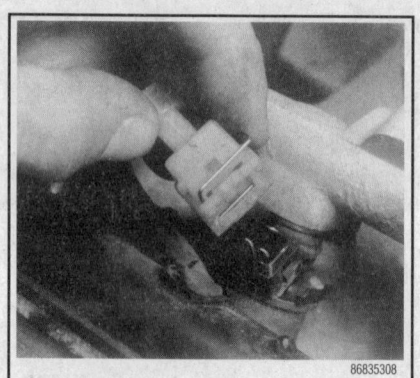

Fig. 61 A close-up view of the connectors—note the water-resistant design

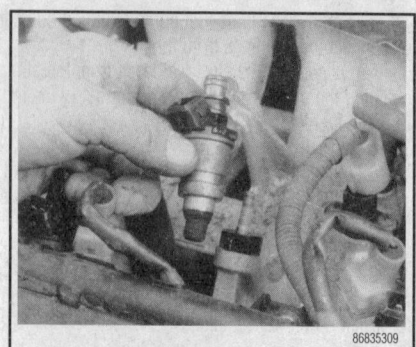

Fig. 62 Carefully pull out the fuel injector—notice carbon buildup from the combustion chamber—1985–87

Fig. 63 Remove the seal after the injector has been pulled out—1985–87

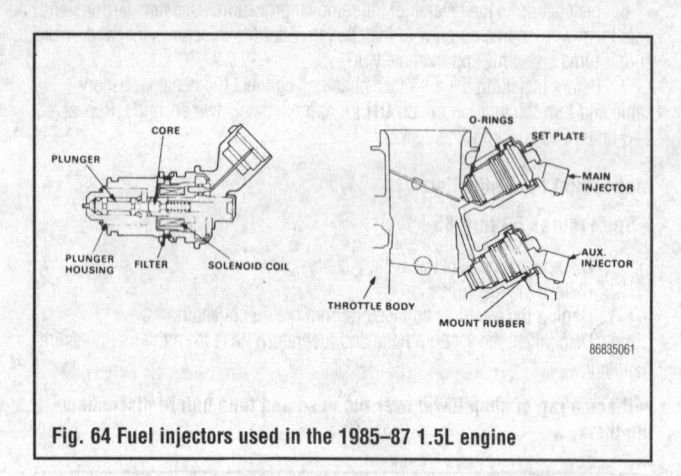

Fig. 64 Fuel injectors used in the 1985–87 1.5L engine

Fig. 65 Exploded view of fuel rail and related components—1992–95

5. Disconnect the fuel line from the fuel pipe.
6. Disconnect the EACV from the intake manifold.
7. Remove the fuel pipe retainer nuts and the fuel pipe.
8. Remove the injectors from the intake manifold.

To install:
9. Slide new cushion rings onto the injectors.
10. Coat new O-rings with clean engine oil and install them on the injectors.
11. Insert the injectors into the fuel pipe first.
12. Coat new seal rings with clean engine oil and press them into the intake manifold.
13. Install the injectors and fuel pipe assembly in the intake manifold.

➡ To prevent damage to the O-rings, install the injectors in the fuel pipe first, then install them in the intake manifold.

14. Align the center line marking on the fuel injector with the mark on the fuel pipe.

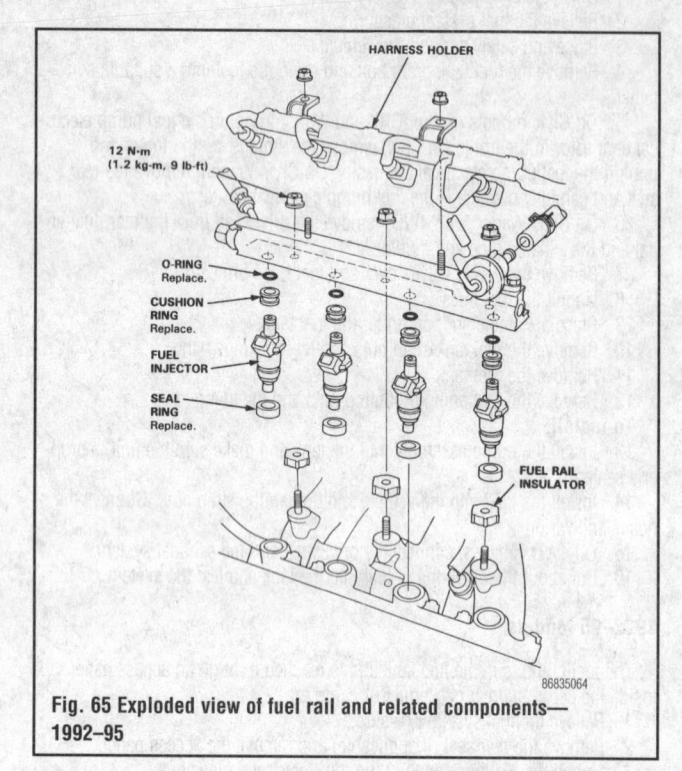

15. Install and tighten the fuel pipe retainer nuts to 9 ft. lbs (12 Nm).
16. Connect the fuel line to the fuel pipe, then the vacuum hose and fuel return line to the pressure regulator.
17. Connect the electrical connectors to the injectors.
18. Connect the negative battery cable and turn the ignition switch **ON** for 2 seconds, but do not operate the starter. Repeat 2–3 times and check for fuel leaks.

Fuel Pressure Regulator

REMOVAL & INSTALLATION

◆ See Figures 66, 67, 68 and 69

1. Disconnect the negative battery cable.
2. Relieve fuel pressure.
3. Remove the air intake chamber, if necessary to gain access.
4. Disconnect the vacuum hose and fuel hose.
5. Remove the 6mm retaining bolts.

To install:
6. Install a new O-ring.
7. Apply clean engine oil to the O-ring and mating surface of fuel pressure regulator.

FUEL SYSTEM 5-21

Fig. 66 Grasp and carefully pull the vacuum line from the fuel pressure regulator

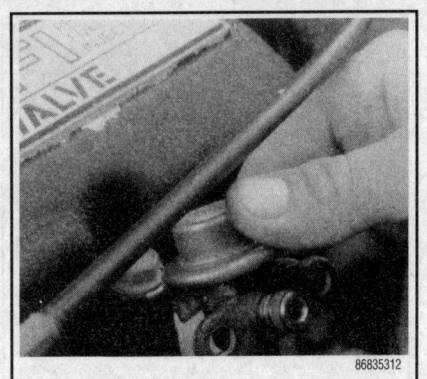

Fig. 67 After unbolting, carefully remove the fuel pressure regulator—1985–87

Fig. 68 Replace this O-ring before re-installation

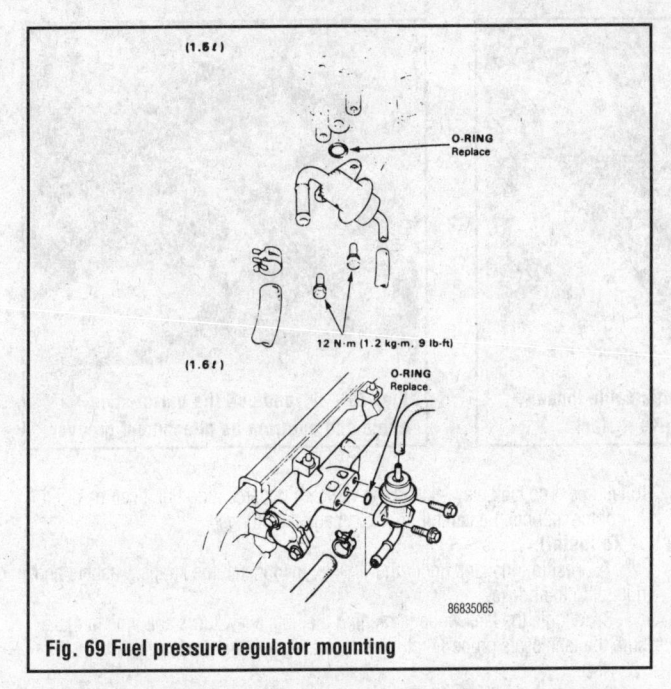

Fig. 69 Fuel pressure regulator mounting

8. Install the regulator. Tighten the mounting bolts to 9 ft. lbs. (12 Nm).
9. Connect the vacuum and fuel hoses.
10. Install the air chamber, if removed.
11. Connect the negative battery cable.
12. Turn ignition switch **ON** but do not operate starter. After fuel pump runs for approximately 2 seconds, fuel pressure in fuel line rises. Repeat this 3 times, then check whether there is any fuel leakage.

Throttle Body

♦ See Figure 70

REMOVAL & INSTALLATION

Dual Point Fuel Injection (DP-FI)

1. Properly relieve the fuel system pressure.
2. Disconnect the negative battery cable.
3. Remove the cover to the air intake chamber and the air intake screen.
4. Disconnect the air intake hose from the intake chamber by removing the large clamp.
5. Label and disconnect all vacuum hoses to the throttle body.
6. Disconnect the throttle cable from the throttle body.
7. Disconnect the fuel line banjo fitting from the side of the housing.

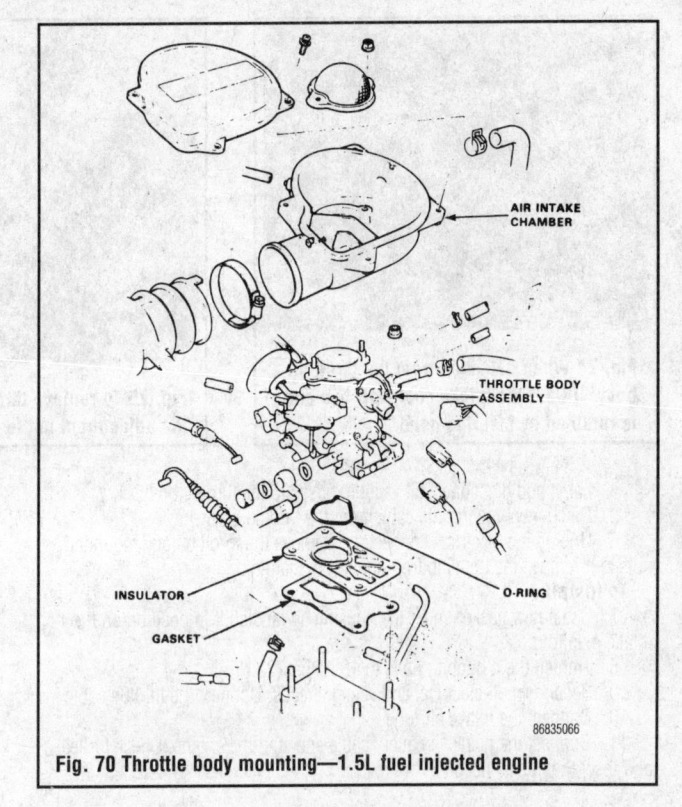

Fig. 70 Throttle body mounting—1.5L fuel injected engine

➡ Discard the washers from the banjo fitting, they should be replaced whenever the fitting is loosened.

8. Disengage the electrical connectors. Remove the throttle body mounting nuts and remove the housing from the intake manifold.

To install:

9. Use a new gasket under the insulator and a new O-ring under the housing and install the throttle body in position on the intake manifold.
10. Tighten the mounting bolts to 16 ft. lbs. (22 Nm).
11. Reconnect all electrical and vacuum leads. Connect the fuel line banjo fitting, using new washers. Connect the throttle cable.
12. Install the air intake housing and connect the intake air tube.
13. Connect the negative battery cable and start the engine. Check for leaks and reset the idle, as needed.

Multi-Point Fuel Injection

♦ See Figures 71 thru 76

1. Properly relieve the fuel system pressure.
2. Disconnect the negative battery cable.
3. Disconnect the air intake hose from the throttle body.

5-22 FUEL SYSTEM

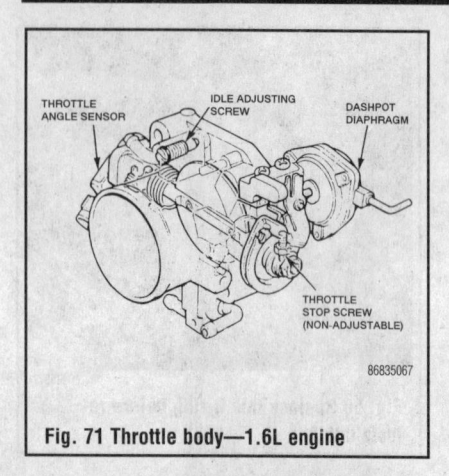

Fig. 71 Throttle body—1.6L engine

Fig. 72 Unclamp all hoses from air intake duct prior to removal

Fig. 73 Unclamp the air intake hose and pull it away from the throttle body

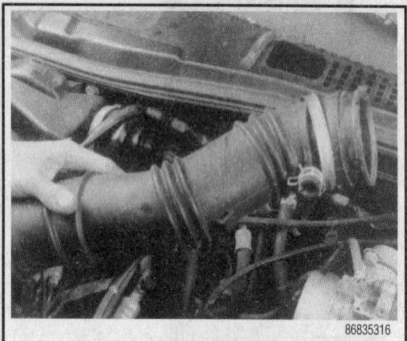

Fig. 74 When detached from the throttle body, lift out the intake duct—air box end is retained by a spring band

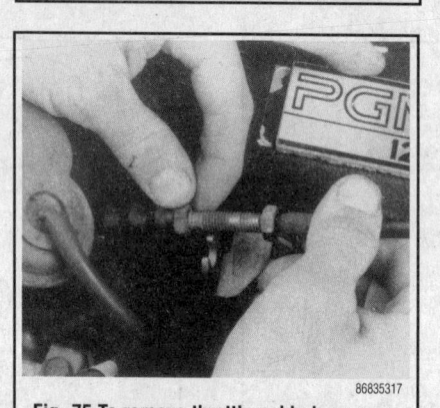

Fig. 75 To remove throttle cable loosen the adjustment nut to give it slack . . .

Fig. 76 . . . and pull the barrel-style cable end out from its attachment groove

4. Label and disconnect all vacuum hoses to the throttle body.
5. Disconnect the throttle cable from the throttle body.
6. Unplug the electrical connectors. Remove the throttle body mounting nuts and remove the housing from the intake manifold.

To install:

7. Use a new gasket under the housing and install it in position on the intake manifold.
8. Tighten the mounting bolts to 16 ft. lbs. (22 Nm).
9. Reconnect all electrical and vacuum leads. Connect the throttle cable.
10. Connect the intake air tube.
11. Connect the negative battery cable and start the engine. Check for leaks and reset the idle as needed.

Fuel Tank

REMOVAL & INSTALLATION

♦ See Figures 77, 78, 79, 80 and 81

1. Raise and safely support the vehicle then remove the rear wheels.
2. Remove the drain bolt and drain the fuel from the tank. Be sure to take the appropriate fire safety precautions.
3. In the luggage area on 2-wheel drive vehicles, remove the access covers and disconnect the fuel pump and gauge unit wiring. On 4-wheel drive vehicles, the access cover is under the rear seat.
4. To disconnect the hoses, loosen the clamps and slide them back. Carefully twist the hose while pulling it off the fitting to avoid damage to the hose or the flared fittings.
5. On 4-wheel drive vehicles, remove the exhaust pipe along with the muffler and the rear driveshaft.
6. Place a jack under the tank and remove the nuts to allow the tank straps to fall free. The tank may stick to the vehicle's undercoating but it can be pried out of the mounts. Be careful not to damage the fittings.

To install:

7. Position the tank under the vehicle, then install the straps, washers and nuts onto the hooks.
8. Wiggle the tank while tightening the nuts a few turns at a time to make sure the tank seats properly into the mount. tighten the strap nuts to 16 ft. lbs. (22 Nm).
9. Make sure the drain plug in installed with a new gasket and tighten to 36 ft. lbs. (50 Nm).
10. Connect the hoses and wiring.
11. On 4-wheel drive vehicles, install the driveshaft and tighten the bolts to 24 ft. lbs. (33 Nm). Use new self locking nuts when attaching the exhaust pipe to the catalytic converter and tighten to 25 ft. lbs. (34 Nm).

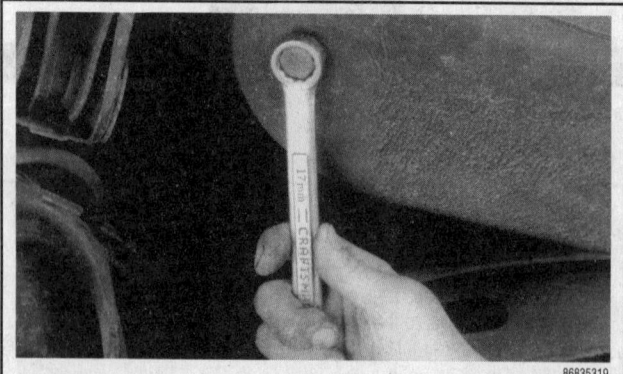

Fig. 77 Unbolt the fuel tank drain bolt with great care

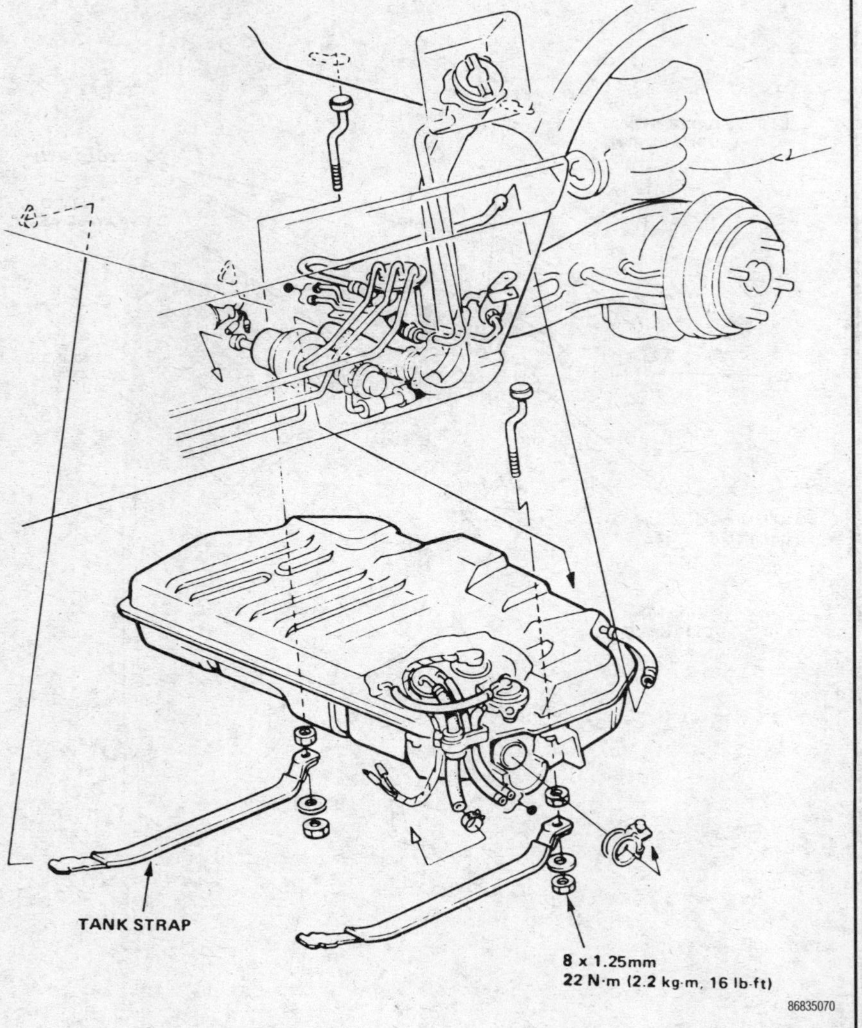

FUEL SYSTEM 5-23

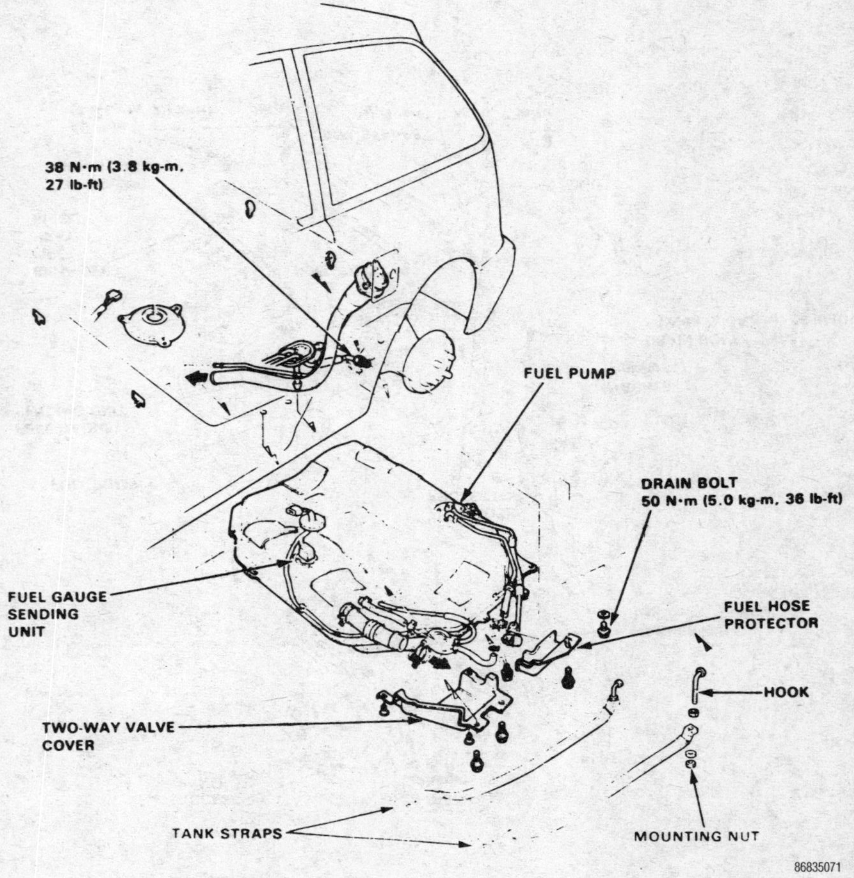

Fig. 79 Fuel tank mounting—1988-91 fuel injected models

Fig. 78 Fuel tank mounting—1985-87 fuel injected models

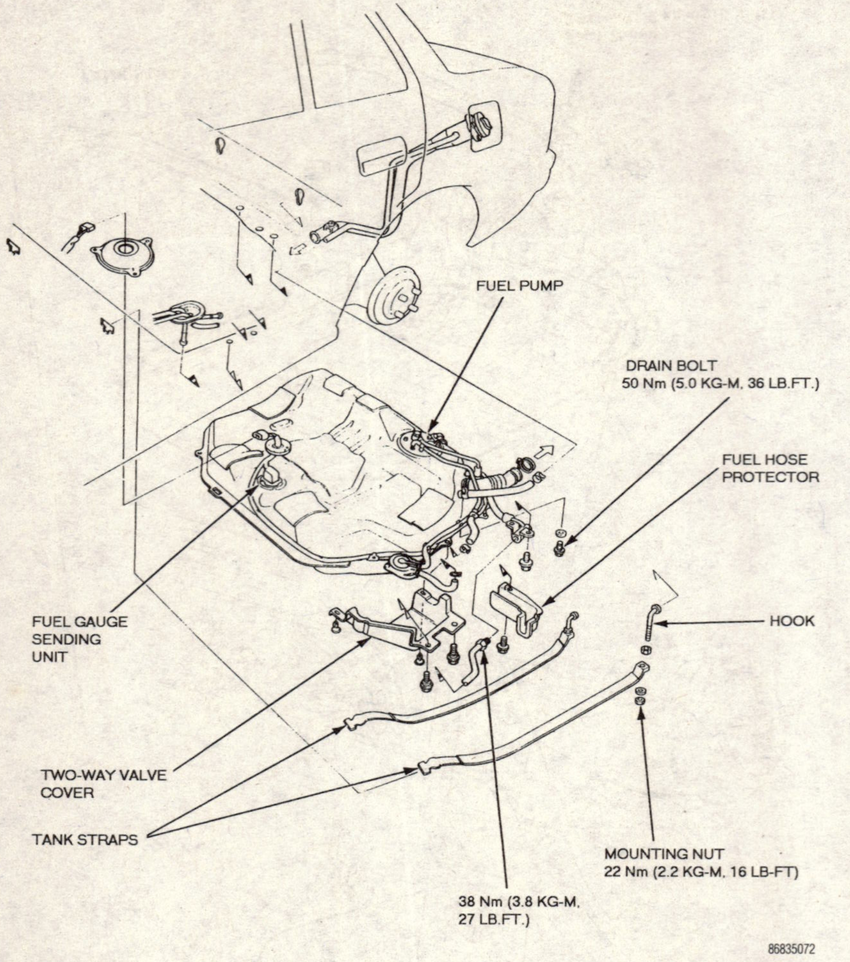

5-24 FUEL SYSTEM

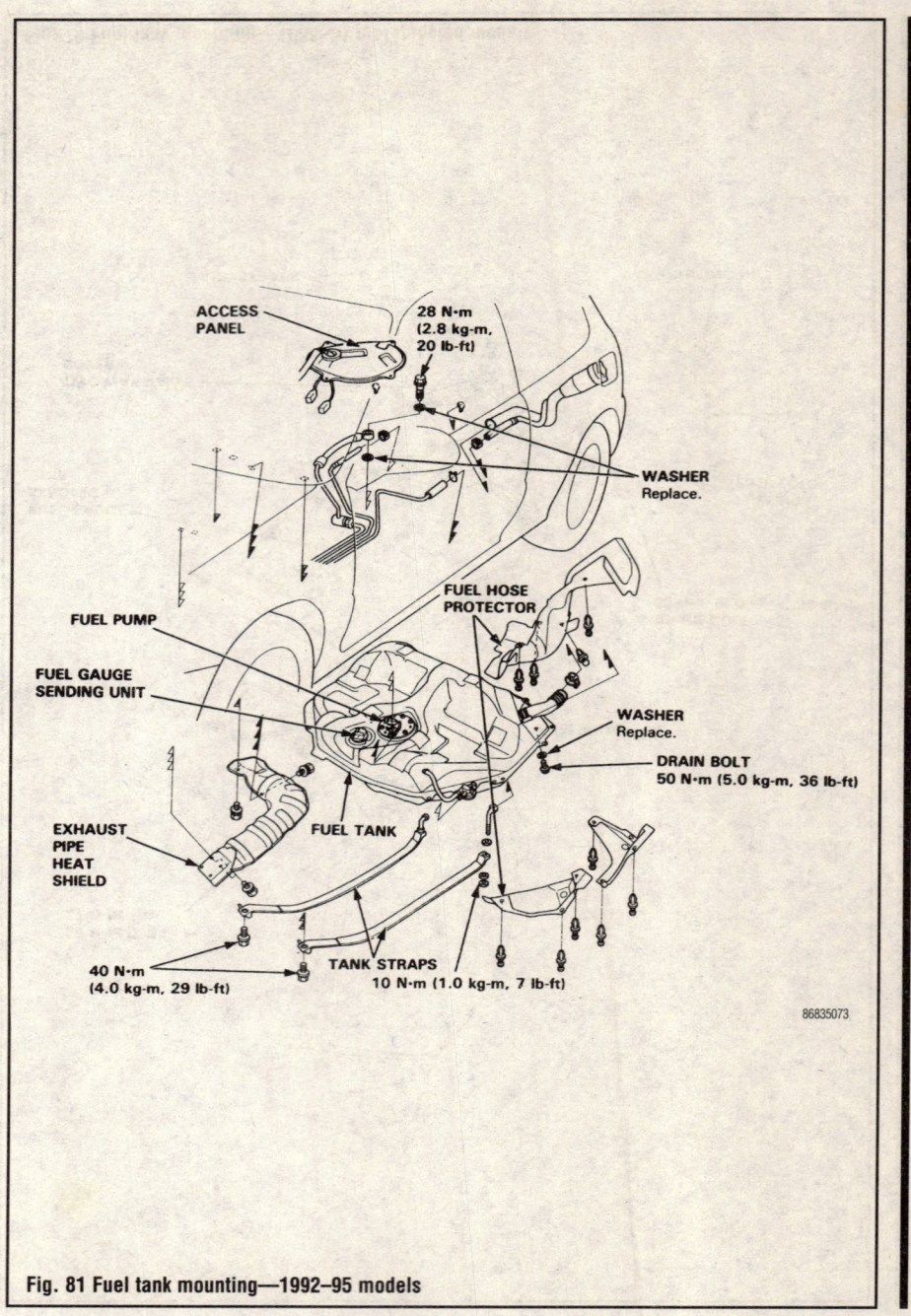

Fig. 81 Fuel tank mounting—1992–95 models

Fig. 80 Fuel tank mounting—1988–91 fuel injected models

6 CHASSIS ELECTRICAL

UNDERSTANDING AND TROUBLESHOOTING ELECTRICAL SYSTEMS 6-2
BASIC ELECTRICAL THEORY 6-2
　HOW DOES ELECTRICITY WORK: THE WATER ANALOGY 6-2
　OHM'S LAW 6-2
ELECTRICAL COMPONENTS 6-2
　POWER SOURCE 6-2
　GROUND 6-3
　PROTECTIVE DEVICES 6-3
　SWITCHES & RELAYS 6-3
　LOAD 6-4
　WIRING & HARNESSES 6-4
　CONNECTORS 6-4
TEST EQUIPMENT 6-5
　JUMPER WIRES 6-5
　TEST LIGHTS 6-5
　MULTIMETERS 6-5
TROUBLESHOOTING ELECTRICAL SYSTEMS 6-6
TESTING 6-6
　OPEN CIRCUITS 6-6
　SHORT CIRCUITS 6-6
　VOLTAGE 6-6
　VOLTAGE DROP 6-7
　RESISTANCE 6-7
WIRE AND CONNECTOR REPAIR 6-8
COMPONENT LOCATIONS 6-8
RELAY, SENSORS AND COMPUTER LOCATIONS 6-8
SUPPLEMENTAL RESTRAINT SYSTEM (SRS) 6-9
GENERAL INFORMATION 6-9
　SYSTEM OPERATION 6-9
　1992–95 SUPPLEMENTAL RESTRAINT SYSTEM (SRS) PRECAUTIONS 6-9
　DISARMING THE SYSTEM 6-11
　ENABLING THE SYSTEM 6-11
HEATING AND AIR CONDITIONING 6-11
GENERAL INFORMATION 6-11
AIR CONDITIONING COMPONENTS 6-11
　REMOVAL & INSTALLATION 6-11
BLOWER MOTOR 6-12
　REMOVAL & INSTALLATION 6-12
BLOWER MOTOR RESISTOR 6-12
　REMOVAL & INSTALLATION 6-12
HEATER CORE 6-12
　REMOVAL & INSTALLATION 6-12
MANUAL CONTROL HEAD 6-13
　REMOVAL & INSTALLATION 6-13
CONTROL CABLES 6-14
　ADJUSTMENT 6-14
　REMOVAL & INSTALLATION 6-15
FAN SWITCH 6-15
　TESTING 6-15
　REMOVAL & INSTALLATION 6-16
RECIRCULATION CONTROL MOTOR 6-16
　TESTING 6-16
　REMOVAL & INSTALLATION 6-17
CRUISE CONTROL 6-17
GENERAL DESCRIPTION 6-17
　SERVICE PRECAUTIONS 6-18
ENTERTAINMENT SYSTEM 6-18
RADIO 6-18
　REMOVAL & INSTALLATION 6-18
STEREO SPEAKERS 6-19
　REMOVAL & INSTALLATION 6-20
WINDSHIELD WIPERS AND WASHERS 6-21
WINDSHIELD WIPER ARMS 6-21
　REMOVAL & INSTALLATION 6-21
WINDSHIELD WIPER MOTOR 6-21
　REMOVAL & INSTALLATION 6-21
WASHER FLUID RESERVOIR AND WASHER PUMPS 6-23
　REMOVAL & INSTALLATION 6-23
INSTRUMENTS AND SWITCHES 6-23
INSTRUMENT CLUSTER 6-23
　REMOVAL & INSTALLATION 6-23
GAUGES 6-25
　REMOVAL & INSTALLATION 6-25
COMBINATION SWITCH 6-26
　REMOVAL & INSTALLATION 6-26
WINDSHIELD WIPER/WASHER SWITCH 6-26
　REMOVAL & INSTALLATION 6-26
HEADLIGHT SWITCH 6-26
　REMOVAL & INSTALLATION 6-26
CLOCK 6-26
　REMOVAL & INSTALLATION 6-26
REAR WINDOW DEFOGGER SYSTEM 6-27
　REMOVAL & INSTALLATION 6-27
　COMPONENT TESTING 6-27
LIGHTING 6-28
HEADLIGHTS 6-28
　REMOVAL & INSTALLATION 6-28
TAIL LIGHTS, REVERSE LIGHTS, REAR MARKER AND SIGNAL LIGHTS 6-29
　REMOVAL & INSTALLATION 6-29
FRONT TURN SIGNAL, SIDE MARKER AND LICENSE PLATE LIGHTS 6-29
　REMOVAL & INSTALLATION 6-29
CIRCUIT PROTECTION 6-30
FUSES 6-30
　REPLACMENT 6-30
FUSIBLE LINKS 6-31
　REPLACEMENT 6-31
RELAYS 6-31
　REPLACEMENT 6-31
WIRING DIAGRAMS 6-32
TROUBLESHOOTING CHART
　CRUISE CONTROL 6-18

UNDERSTANDING AND TROUBLESHOOTING ELECTRICAL SYSTEMS 6-2
COMPONENT LOCATIONS 6-8
SUPPLEMENTAL RESTRAINT SYSTEM (SRS) 6-9
HEATING AND AIR CONDITIONING 6-11
CRUISE CONTROL 6-17
ENTERTAINMENT SYSTEM 6-18
WINDSHIELD WIPERS AND WASHERS 6-21
INSTRUMENTS AND SWITCHES 6-23
LIGHTING 6-28
CIRCUIT PROTECTION 6-30
WIRING DIAGRAMS 6-32

6-2 CHASSIS ELECTRICAL

UNDERSTANDING AND TROUBLESHOOTING ELECTRICAL SYSTEMS

Basic Electrical Theory

♦ See Figure 1

For any 12 volt, negative ground, electrical system to operate, the electricity must travel in a complete circuit. This simply means that current (power) from the positive (+) terminal of the battery must eventually return to the negative (-) terminal of the battery. Along the way, this current will travel through wires, fuses, switches and components. If, for any reason, the flow of current through the circuit is interrupted, the component fed by that circuit will cease to function properly.

Perhaps the easiest way to visualize a circuit is to think of connecting a light bulb (with two wires attached to it) to the battery—one wire attached to the negative (-) terminal of the battery and the other wire to the positive (+) terminal. With the two wires touching the battery terminals, the circuit would be complete and the light bulb would illuminate. Electricity would follow a path from the battery to the bulb and back to the battery. It's easy to see that with longer wires on our light bulb, it could be mounted anywhere. Further, one wire could be fitted with a switch so that the light could be turned on and off.

The normal automotive circuit differs from this simple example in two ways. First, instead of having a return wire from the bulb to the battery, the current travels through the frame of the vehicle. Since the negative (-) battery cable is attached to the frame (made of electrically conductive metal), the frame of the vehicle can serve as a ground wire to complete the circuit. Secondly, most automotive circuits contain multiple components which receive power from a single circuit. This lessens the amount of wire needed to power components on the vehicle.

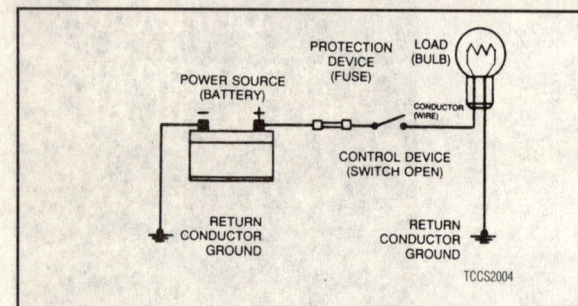

Fig. 1 This example illustrates a simple circuit. When the switch is closed, power from the positive (+) battery terminal flows through the fuse and the switch, and then to the light bulb. The light illuminates and the circuit is completed through the ground wire back to the negative (-) battery terminal. In reality, the two ground points shown in the illustration are attached to the metal frame of the vehicle, which completes the circuit back to the battery

HOW DOES ELECTRICITY WORK: THE WATER ANALOGY

Electricity is the flow of electrons—the subatomic particles that constitute the outer shell of an atom. Electrons spin in an orbit around the center core of an atom. The center core is comprised of protons (positive charge) and neutrons (neutral charge). Electrons have a negative charge and balance out the positive charge of the protons. When an outside force causes the number of electrons to unbalance the charge of the protons, the electrons will split off the atom and look for another atom to balance out. If this imbalance is kept up, electrons will continue to move and an electrical flow will exist.

Many people have been taught electrical theory using an analogy with water. In a comparison with water flowing through a pipe, the electrons would be the water and the wire is the pipe.

The flow of electricity can be measured much like the flow of water through a pipe. The unit of measurement used is amperes, frequently abbreviated as amps (a). You can compare amperage to the volume of water flowing through a pipe. When connected to a circuit, an ammeter will measure the actual amount of current flowing through the circuit. When relatively few electrons flow through a circuit, the amperage is low. When many electrons flow, the amperage is high.

Water pressure is measured in units such as pounds per square inch (psi); The electrical pressure is measured in units called volts (v). When a voltmeter is connected to a circuit, it is measuring the electrical pressure.

The actual flow of electricity depends not only on voltage and amperage, but also on the resistance of the circuit. The higher the resistance, the higher the force necessary to push the current through the circuit. The standard unit for measuring resistance is an ohm. Resistance in a circuit varies depending on the amount and type of components used in the circuit. The main factors which determine resistance are:

- Material—some materials have more resistance than others. Those with high resistance are said to be insulators. Rubber materials (or rubber-like plastics) are some of the most common insulators used in vehicles as they have a very high resistance to electricity. Very low resistance materials are said to be conductors. Copper wire is among the best conductors. Silver is actually a superior conductor to copper and is used in some relay contacts, but its high cost prohibits its use as common wiring. Most automotive wiring is made of copper.
- Size—the larger the wire size being used, the less resistance the wire will have. This is why components which use large amounts of electricity usually have large wires supplying current to them.
- Length—for a given thickness of wire, the longer the wire, the greater the resistance. The shorter the wire, the less the resistance. When determining the proper wire for a circuit, both size and length must be considered to design a circuit that can handle the current needs of the component.
- Temperature—with many materials, the higher the temperature, the greater the resistance (positive temperature coefficient). Some materials exhibit the opposite trait of lower resistance with higher temperatures (negative temperature coefficient). These principles are used in many of the sensors on the engine.

OHM'S LAW

There is a direct relationship between current, voltage and resistance. The relationship between current, voltage and resistance can be summed up by a statement known as Ohm's law.

Voltage (E) is equal to amperage (I) times resistance (R): $E = I \times R$

Other forms of the formula are $R = E/I$ and $I = E/R$

In each of these formulas, E is the voltage in volts, I is the current in amps and R is the resistance in ohms. The basic point to remember is that as the resistance of a circuit goes up, the amount of current that flows in the circuit will go down, if voltage remains the same.

The amount of work that the electricity can perform is expressed as power. The unit of power is the watt (w). The relationship between power, voltage and current is expressed as:

Power (w) is equal to amperage (I) times voltage (E): $W = I \times E$

This is only true for direct current (DC) circuits; The alternating current formula is a tad different, but since the electrical circuits in most vehicles are DC type, we need not get into AC circuit theory.

Electrical Components

POWER SOURCE

Power is supplied to the vehicle by two devices: The battery and the alternator. The battery supplies electrical power during starting or during periods when the current demand of the vehicle's electrical system exceeds the output capacity of the alternator. The alternator supplies electrical current when the engine is running. Just not does the alternator supply the current needs of the vehicle, but it recharges the battery.

The Battery

In most modern vehicles, the battery is a lead/acid electrochemical device consisting of six 2 volt subsections (cells) connected in series, so that the unit is capable of producing approximately 12 volts of electrical pressure. Each subsection consists of a series of positive and negative plates held a short distance apart in a solution of sulfuric acid and water.

The two types of plates are of dissimilar metals. This sets up a chemical reaction, and it is this reaction which produces current flow from the battery when its positive and negative terminals are connected to an electrical load. The

CHASSIS ELECTRICAL 6-3

power removed from the battery is replaced by the alternator, restoring the battery to its original chemical state.

The Alternator

On some vehicles there isn't an alternator, but a generator. The difference is that an alternator supplies alternating current which is then changed to direct current for use on the vehicle, while a generator produces direct current. Alternators tend to be more efficient and that is why they are used.

Alternators and generators are devices that consist of coils of wires wound together making big electromagnets. One group of coils spins within another set and the interaction of the magnetic fields causes a current to flow. This current is then drawn off the coils and fed into the vehicles electrical system.

GROUND

Two types of grounds are used in automotive electric circuits. Direct ground components are grounded to the frame through their mounting points. All other components use some sort of ground wire which is attached to the frame or chassis of the vehicle. The electrical current runs through the chassis of the vehicle and returns to the battery through the ground (-) cable; if you look, you'll see that the battery ground cable connects between the battery and the frame or chassis of the vehicle.

➡ It should be noted that a good percentage of electrical problems can be traced to bad grounds.

PROTECTIVE DEVICES

♦ See Figure 2

It is possible for large surges of current to pass through the electrical system of your vehicle. If this surge of current were to reach the load in the circuit, the surge could burn it out or severely damage it. It can also overload the wiring, causing the harness to get hot and melt the insulation. To prevent this, fuses, circuit breakers and/or fusible links are connected into the supply wires of the electrical system. These items are nothing more than a built-in weak spot in the system. When an abnormal amount of current flows through the system, these protective devices work as follows to protect the circuit:

- Fuse—when an excessive electrical current passes through a fuse, the fuse "blows" (the conductor melts) and opens the circuit, preventing the passage of current.
- Circuit Breaker—a circuit breaker is basically a self-repairing fuse. It will open the circuit in the same fashion as a fuse, but when the surge subsides, the circuit breaker can be reset and does not need replacement.
- Fusible Link—a fusible link (fuse link or main link) is a short length of special, high temperature insulated wire that acts as a fuse. When an excessive electrical current passes through a fusible link, the thin gauge wire inside the link melts, creating an intentional open to protect the circuit. To repair the circuit, the link must be replaced. Some newer type fusible links are housed in plug-in modules, which are simply replaced like a fuse, while older type fusible links must be cut and spliced if they melt. Since this link is very early in the electrical path, it's the first place to look if nothing on the vehicle works, yet the battery seems to be charged and is properly connected.

✱✱ CAUTION

Always replace fuses, circuit breakers and fusible links with identically rated components. Under no circumstances should a component of higher or lower amperage rating be substituted.

SWITCHES & RELAYS

♦ See Figures 3 and 4

Switches are used in electrical circuits to control the passage of current. The most common use is to open and close circuits between the battery and the various electric devices in the system. Switches are rated according to the amount of amperage they can handle. If a sufficient amperage rated switch is not used in a circuit, the switch could overload and cause damage.

Some electrical components which require a large amount of current to operate use a special switch called a relay. Since these circuits carry a large amount of current, the thickness of the wire in the circuit is also greater. If this large wire were connected from the load to the control switch, the switch would have to carry the high amperage load and the fairing or dash would be twice as large to accommodate the increased size of the wiring harness. To prevent these problems, a relay is used.

Relays are composed of a coil and a set of contacts. When the coil has a current passed though it, a magnetic field is formed and this field causes the contacts to move together, completing the circuit. Most relays are normally open, prevent-

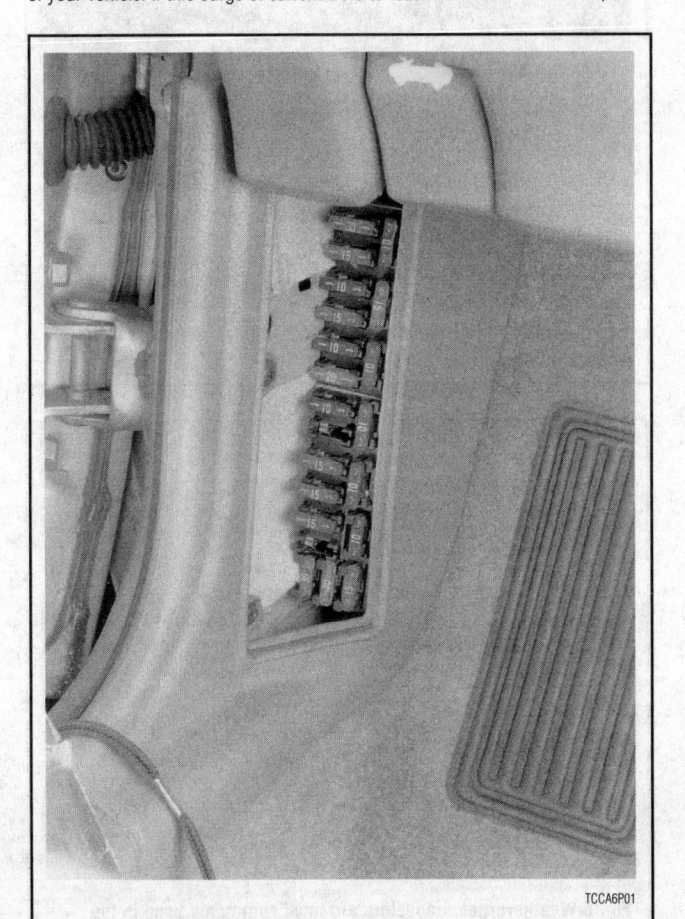

Fig. 2 Most vehicles use one or more fuse panels. This one is located on the driver's side kick panel

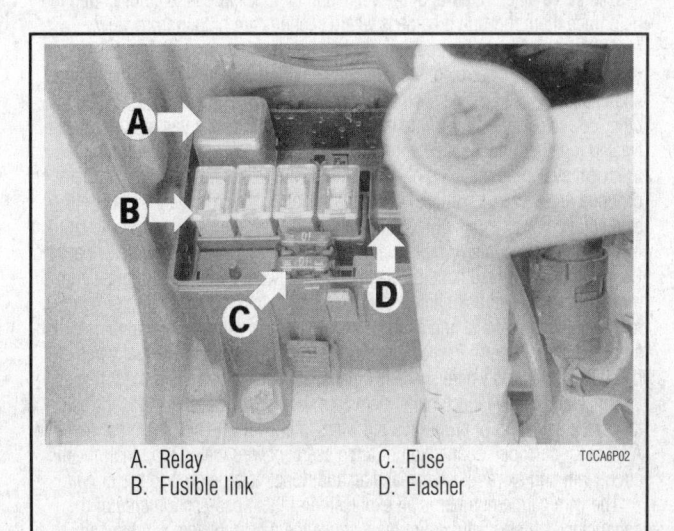

A. Relay   C. Fuse
B. Fusible link   D. Flasher

Fig. 3 The underhood fuse and relay panel usually contains fuses, relays, flashers and fusible links

6-4 CHASSIS ELECTRICAL

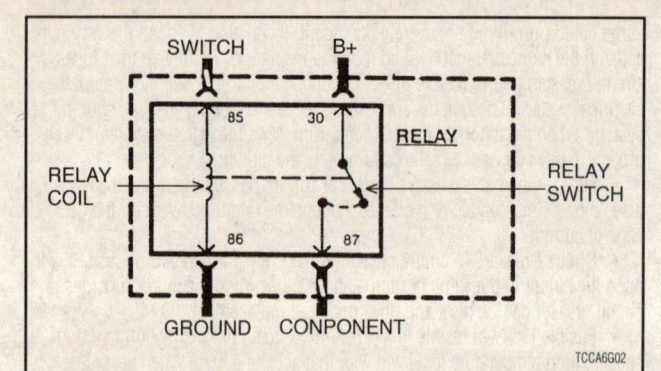

Fig. 4 Relays are composed of a coil and a switch. These two components are linked together so that when one operates, the other operates at the same time. The large wires in the circuit are connected from the battery to one side of the relay switch (B+) and from the opposite side of the relay switch to the load (component). Smaller wires are connected from the relay coil to the control switch for the circuit and from the opposite side of the relay coil to ground

ing current from passing through the circuit, but they can take any electrical form depending on the job they are intended to do. Relays can be considered "remote control switches." They allow a smaller current to operate devices that require higher amperages. When a small current operates the coil, a larger current is allowed to pass by the contacts. Some common circuits which may use relays are the horn, headlights, starter, electric fuel pump and other high draw circuits.

LOAD

Every electrical circuit must include a "load" (something to use the electricity coming from the source). Without this load, the battery would attempt to deliver its entire power supply from one pole to another. This is called a "short circuit." All this electricity would take a short cut to ground and cause a great amount of damage to other components in the circuit by developing a tremendous amount of heat. This condition could develop sufficient heat to melt the insulation on all the surrounding wires and reduce a multiple wire cable to a lump of plastic and copper.

WIRING & HARNESSES

The average vehicle contains meters and meters of wiring, with hundreds of individual connections. To protect the many wires from damage and to keep them from becoming a confusing tangle, they are organized into bundles, enclosed in plastic or taped together and called wiring harnesses. Different harnesses serve different parts of the vehicle. Individual wires are color coded to help trace them through a harness where sections are hidden from view.

Automotive wiring or circuit conductors can be either single strand wire, multi-strand wire or printed circuitry. Single strand wire has a solid metal core and is usually used inside such components as alternators, motors, relays and other devices. Multi-strand wire has a core made of many small strands of wire twisted together into a single conductor. Most of the wiring in an automotive electrical system is made up of multi-strand wire, either as a single conductor or grouped together in a harness. All wiring is color coded on the insulator, either as a solid color or as a colored wire with an identification stripe. A printed circuit is a thin film of copper or other conductor that is printed on an insulator backing. Occasionally, a printed circuit is sandwiched between two sheets of plastic for more protection and flexibility. A complete printed circuit, consisting of conductors, insulating material and connectors for lamps or other components is called a printed circuit board. Printed circuitry is used in place of individual wires or harnesses in places where space is limited, such as behind instrument panels.

Since automotive electrical systems are very sensitive to changes in resistance, the selection of properly sized wires is critical when systems are repaired. A loose or corroded connection or a replacement wire that is too small for the circuit will add extra resistance and an additional voltage drop to the circuit.

The wire gauge number is an expression of the cross-section area of the conductor. Vehicles from countries that use the metric system will typically describe the wire size as its cross-sectional area in square millimeters. In this method, the larger the wire, the greater the number. Another common system for expressing wire size is the American Wire Gauge (AWG) system. As gauge number increases, area decreases and the wire becomes smaller. An 18 gauge wire is smaller than a 4 gauge wire. A wire with a higher gauge number will carry less current than a wire with a lower gauge number. Gauge wire size refers to the size of the strands of the conductor, not the size of the complete wire with insulator. It is possible, therefore, to have two wires of the same gauge with different diameters because one may have thicker insulation than the other.

It is essential to understand how a circuit works before trying to figure out why it doesn't. An electrical schematic shows the electrical current paths when a circuit is operating properly. Schematics break the entire electrical system down into individual circuits. In a schematic, usually no attempt is made to represent wiring and components as they physically appear on the vehicle; switches and other components are shown as simply as possible. Face views of harness connectors show the cavity or terminal locations in all multi-pin connectors to help locate test points.

CONNECTORS

▶ See Figures 5 and 6

Three types of connectors are commonly used in automotive applications—weatherproof, molded and hard shell.

- Weatherproof—these connectors are most commonly used where the connector is exposed to the elements. Terminals are protected against moisture and dirt by sealing rings which provide a weathertight seal. All repairs require the use of a special terminal and the tool required to service it. Unlike standard blade type terminals, these weatherproof terminals cannot be straightened once they are bent. Make certain that the connectors are properly seated and all of the sealing rings are in place when connecting leads.

Fig. 5 Hard shell (left) and weatherproof (right) connectors have replaceable terminals

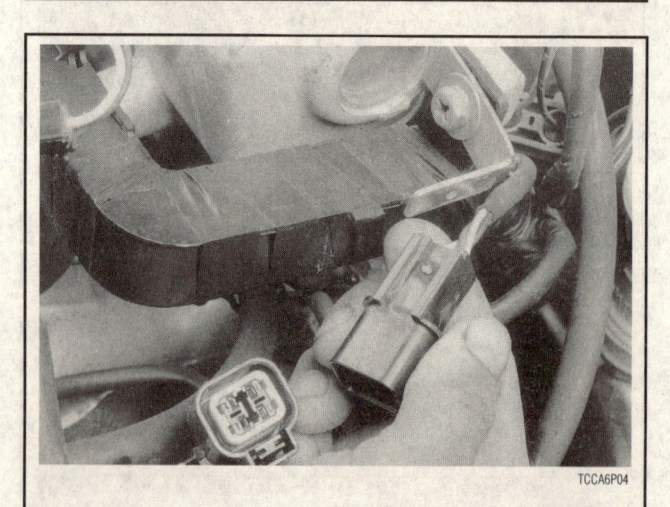

Fig. 6 Weatherproof connectors are most commonly used in the engine compartment or where the connector is exposed to the elements

CHASSIS ELECTRICAL 6-5

- Molded—these connectors require complete replacement of the connector if found to be defective. This means splicing a new connector assembly into the harness. All splices should be soldered to insure proper contact. Use care when probing the connections or replacing terminals in them, as it is possible to create a short circuit between opposite terminals. If this happens to the wrong terminal pair, it is possible to damage certain components. Always use jumper wires between connectors for circuit checking and NEVER probe through weatherproof seals.
- Hard Shell—unlike molded connectors, the terminal contacts in hard-shell connectors can be replaced. Replacement usually involves the use of a special terminal removal tool that depresses the locking tangs (barbs) on the connector terminal and allows the connector to be removed from the rear of the shell. The connector shell should be replaced if it shows any evidence of burning, melting, cracks, or breaks. Replace individual terminals that are burnt, corroded, distorted or loose.

Test Equipment

Pinpointing the exact cause of trouble in an electrical circuit is most times accomplished by the use of special test equipment. The following describes different types of commonly used test equipment and briefly explains how to use them in diagnosis. In addition to the information covered below, the tool manufacturer's instructions booklet (provided with the tester) should be read and clearly understood before attempting any test procedures.

JUMPER WIRES

✱✱ CAUTION

Never use jumper wires made from a thinner gauge wire than the circuit being tested. If the jumper wire is of too small a gauge, it may overheat and possibly melt. Never use jumpers to bypass high resistance loads in a circuit. Bypassing resistances, in effect, creates a short circuit. This may, in turn, cause damage and fire. Jumper wires should only be used to bypass lengths of wire or to simulate switches.

Jumper wires are simple, yet extremely valuable, pieces of test equipment. They are basically test wires which are used to bypass sections of a circuit. Although jumper wires can be purchased, they are usually fabricated from lengths of standard automotive wire and whatever type of connector (alligator clip, spade connector or pin connector) that is required for the particular application being tested. In cramped, hard-to-reach areas, it is advisable to have insulated boots over the jumper wire terminals in order to prevent accidental grounding. It is also advisable to include a standard automotive fuse in any jumper wire. This is commonly referred to as a "fused jumper". By inserting an in-line fuse holder between a set of test leads, a fused jumper wire can be used for bypassing open circuits. Use a 5 amp fuse to provide protection against voltage spikes.

Jumper wires are used primarily to locate open electrical circuits, on either the ground (-) side of the circuit or on the power (+) side. If an electrical component fails to operate, connect the jumper wire between the component and a good ground. If the component operates only with the jumper installed, the ground circuit is open. If the ground circuit is good, but the component does not operate, the circuit between the power feed and component may be open. By moving the jumper wire successively back from the component toward the power source, you can isolate the area of the circuit where the open is located. When the component stops functioning, or the power is cut off, the open is in the segment of wire between the jumper and the point previously tested.

You can sometimes connect the jumper wire directly from the battery to the "hot" terminal of the component, but first make sure the component uses 12 volts in operation. Some electrical components, such as fuel injectors or sensors, are designed to operate on about 4 to 5 volts, and running 12 volts directly to these components will cause damage.

TEST LIGHTS

▶ See Figure 7

The test light is used to check circuits and components while electrical current is flowing through them. It is used for voltage and ground tests. To use a 12 volt test light, connect the ground clip to a good ground and probe wherever

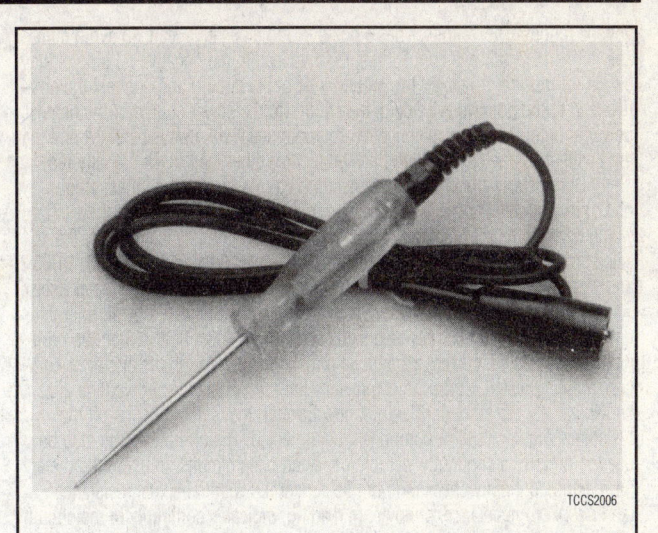

Fig. 7 A 12 volt test light is used to detect the presence of voltage in a circuit

necessary with the pick. The test light will illuminate when voltage is detected. This does not necessarily mean that 12 volts (or any particular amount of voltage) is present; it only means that some voltage is present. It is advisable before using the test light to touch its ground clip and probe across the battery posts or terminals to make sure the light is operating properly.

✱✱ WARNING

Do not use a test light to probe electronic ignition, spark plug or coil wires. Never use a pick-type test light to probe wiring on computer controlled systems unless specifically instructed to do so. Any wire insulation that is pierced by the test light probe should be taped and sealed with silicone after testing.

Like the jumper wire, the 12 volt test light is used to isolate opens in circuits. But, whereas the jumper wire is used to bypass the open to operate the load, the 12 volt test light is used to locate the presence of voltage in a circuit. If the test light illuminates, there is power up to that point in the circuit; if the test light does not illuminate, there is an open circuit (no power). Move the test light in successive steps back toward the power source until the light in the handle illuminates. The open is between the probe and a point which was previously probed.

The self-powered test light is similar in design to the 12 volt test light, but contains a 1.5 volt penlight battery in the handle. It is most often used in place of a multimeter to check for open or short circuits when power is isolated from the circuit (continuity test).

The battery in a self-powered test light does not provide much current. A weak battery may not provide enough power to illuminate the test light even when a complete circuit is made (especially if there is high resistance in the circuit). Always make sure that the test battery is strong. To check the battery, briefly touch the ground clip to the probe; if the light glows brightly, the battery is strong enough for testing.

➥**A self-powered test light should not be used on any computer controlled system or component. The small amount of electricity transmitted by the test light is enough to damage many electronic automotive components.**

MULTIMETERS

Multimeters are an extremely useful tool for troubleshooting electrical problems. They can be purchased in either analog or digital form and have a price range to suit any budget. A multimeter is a voltmeter, ammeter and ohmmeter (along with other features) combined into one instrument. It is often used when testing solid state circuits because of its high input impedance (usually 10 megaohms or more). A brief description of the multimeter main test functions follows:

- Voltmeter—the voltmeter is used to measure voltage at any point in a circuit, or to measure the voltage drop across any part of a circuit. Voltmeters usually have various scales and a selector switch to allow the reading of different

6-6 CHASSIS ELECTRICAL

voltage ranges. The voltmeter has a positive and a negative lead. To avoid damage to the meter, always connect the negative lead to the negative (-) side of the circuit (to ground or nearest the ground side of the circuit) and connect the positive lead to the positive (+) side of the circuit (to the power source or the nearest power source). Note that the negative voltmeter lead will always be black and that the positive voltmeter will always be some color other than black (usually red).

• Ohmmeter—the ohmmeter is designed to read resistance (measured in ohms) in a circuit or component. Most ohmmeters will have a selector switch which permits the measurement of different ranges of resistance (usually the selector switch allows the multiplication of the meter reading by 10, 100, 1,000 and 10,000). Some ohmmeters are "auto-ranging" which means the meter itself will determine which scale to use. Since the meters are powered by an internal battery, the ohmmeter can be used like a self-powered test light. When the ohmmeter is connected, current from the ohmmeter flows through the circuit or component being tested. Since the ohmmeter's internal resistance and voltage are known values, the amount of current flow through the meter depends on the resistance of the circuit or component being tested. The ohmmeter can also be used to perform a continuity test for suspected open circuits. In using the meter for making continuity checks, do not be concerned with the actual resistance readings. Zero resistance, or any ohm reading, indicates continuity in the circuit. Infinite resistance indicates an opening in the circuit. A high resistance reading where there should be none indicates a problem in the circuit. Checks for short circuits are made in the same manner as checks for open circuits, except that the circuit must be isolated from both power and normal ground. Infinite resistance indicates no continuity, while zero resistance indicates a dead short.

✱✱ WARNING

Never use an ohmmeter to check the resistance of a component or wire while there is voltage applied to the circuit.

• Ammeter—an ammeter measures the amount of current flowing through a circuit in units called amperes or amps. At normal operating voltage, most circuits have a characteristic amount of amperes, called "current draw" which can be measured using an ammeter. By referring to a specified current draw rating, then measuring the amperes and comparing the two values, one can determine what is happening within the circuit to aid in diagnosis. An open circuit, for example, will not allow any current to flow, so the ammeter reading will be zero. A damaged component or circuit will have an increased current draw, so the reading will be high. The ammeter is always connected in series with the circuit being tested. All of the current that normally flows through the circuit must also flow through the ammeter; if there is any other path for the current to follow, the ammeter reading will not be accurate. The ammeter itself has very little resistance to current flow and, therefore, will not affect the circuit, but it will measure current draw only when the circuit is closed and electricity is flowing. Excessive current draw can blow fuses and drain the battery, while a reduced current draw can cause motors to run slowly, lights to dim and other components to not operate properly.

Troubleshooting Electrical Systems

When diagnosing a specific problem, organized troubleshooting is a must. The complexity of a modern automotive vehicle demands that you approach any problem in a logical, organized manner. There are certain troubleshooting techniques, however, which are standard:

• Establish when the problem occurs. Does the problem appear only under certain conditions? Were there any noises, odors or other unusual symptoms? Isolate the problem area. To do this, make some simple tests and observations, then eliminate the systems that are working properly. Check for obvious problems, such as broken wires and loose or dirty connections. Always check the obvious before assuming something complicated is the cause.

• Test for problems systematically to determine the cause once the problem area is isolated. Are all the components functioning properly? Is there power going to electrical switches and motors. Performing careful, systematic checks will often turn up most causes on the first inspection, without wasting time checking components that have little or no relationship to the problem.

• Test all repairs after the work is done to make sure that the problem is fixed. Some causes can be traced to more than one component, so a careful verification of repair work is important in order to pick up additional malfunctions that may cause a problem to reappear or a different problem to arise. A blown fuse, for example, is a simple problem that may require more than another fuse to repair. If you don't look for a problem that caused a fuse to blow, a shorted wire (for example) may go undetected.

Experience has shown that most problems tend to be the result of a fairly simple and obvious cause, such as loose or corroded connectors, bad grounds or damaged wire insulation which causes a short. This makes careful visual inspection of components during testing essential to quick and accurate troubleshooting.

Testing

OPEN CIRCUITS

▶ See Figure 8

This test already assumes the existence of an open in the circuit and it is used to help locate the open portion.
1. Isolate the circuit from power and ground.
2. Connect the self-powered test light or ohmmeter ground clip to the ground side of the circuit and probe sections of the circuit sequentially.
3. If the light is out or there is infinite resistance, the open is between the probe and the circuit ground.
4. If the light is on or the meter shows continuity, the open is between the probe and the end of the circuit toward the power source.

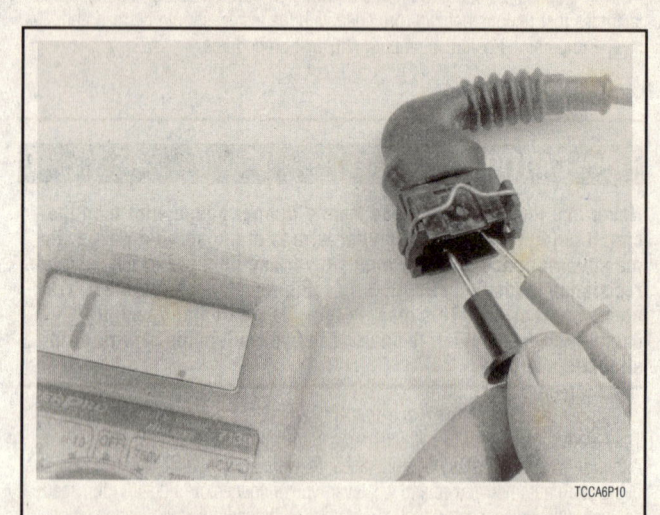

Fig. 8 The infinite reading on this multimeter indicates that the circuit is open

SHORT CIRCUITS

➡ **Never use a self-powered test light to perform checks for opens or shorts when power is applied to the circuit under test. The test light can be damaged by outside power.**

1. Isolate the circuit from power and ground.
2. Connect the self-powered test light or ohmmeter ground clip to a good ground and probe any easy-to-reach point in the circuit.
3. If the light comes on or there is continuity, there is a short somewhere in the circuit.
4. To isolate the short, probe a test point at either end of the isolated circuit (the light should be on or the meter should indicate continuity).
5. Leave the test light probe engaged and sequentially open connectors or switches, remove parts, etc. until the light goes out or continuity is broken.
6. When the light goes out, the short is between the last two circuit components which were opened.

VOLTAGE

This test determines voltage available from the battery and should be the first step in any electrical troubleshooting procedure after visual inspection. Many electrical problems, especially on computer controlled systems, can be caused by a low state of charge in the battery. Excessive corrosion at the battery cable

CHASSIS ELECTRICAL 6-7

terminals can cause poor contact that will prevent proper charging and full battery current flow.

1. Set the voltmeter selector switch to the 20V position.
2. Connect the multimeter negative lead to the battery's negative (-) post or terminal and the positive lead to the battery's positive (+) post or terminal.
3. Turn the ignition switch **ON** to provide a load.
4. A well charged battery should register over 12 volts. If the meter reads below 11.5 volts, the battery power may be insufficient to operate the electrical system properly.

VOLTAGE DROP

▶ See Figure 9

When current flows through a load, the voltage beyond the load drops. This voltage drop is due to the resistance created by the load and also by small resistances created by corrosion at the connectors and damaged insulation on the wires. The maximum allowable voltage drop under load is critical, especially if there is more than one load in the circuit, since all voltage drops are cumulative.

1. Set the voltmeter selector switch to the 20 volt position.
2. Connect the multimeter negative lead to a good ground.
3. Operate the circuit and check the voltage prior to the first component (load).
4. There should be little or no voltage drop in the circuit prior to the first component. If a voltage drop exists, the wire or connectors in the circuit are suspect.
5. While operating the first component in the circuit, probe the ground side of the component with the positive meter lead and observe the voltage readings. A small voltage drop should be noticed. This voltage drop is caused by the resistance of the component.
6. Repeat the test for each component (load) down the circuit.
7. If a large voltage drop is noticed, the preceding component, wire or connector is suspect.

3. Where necessary, also isolate at least one side of the circuit to be checked, in order to avoid reading parallel resistances. Parallel circuit resistances will always give a lower reading than the actual resistance of either of the branches.
4. Connect the meter leads to both sides of the circuit (wire or component) and read the actual measured ohms on the meter scale. Make sure the selector switch is set to the proper ohm scale for the circuit being tested, to avoid misreading the ohmmeter test value.

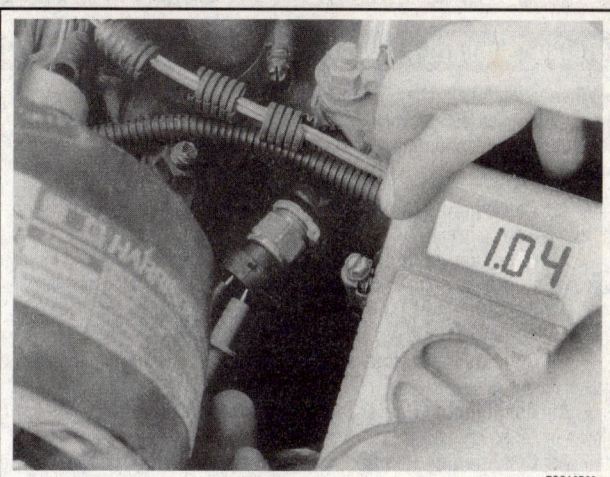

Fig. 10 Checking the resistance of a coolant temperature sensor with an ohmmeter. Reading is 1.04 kilohms

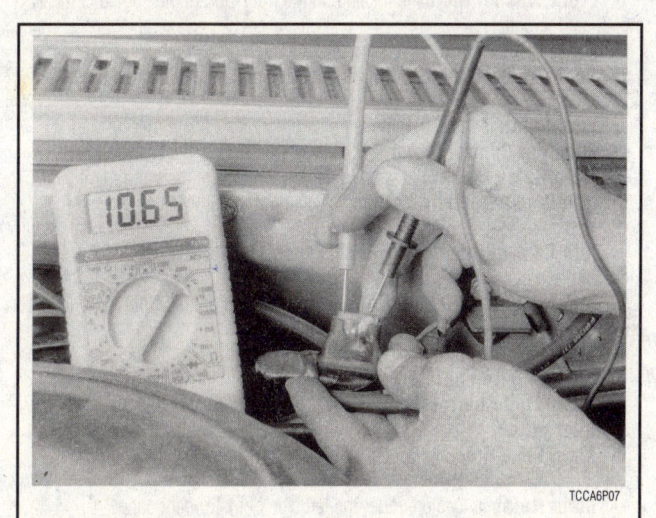

Fig. 9 This voltage drop test revealed high resistance (low voltage) in the circuit

RESISTANCE

▶ See Figures 10 and 11

✱✱ WARNING

Never use an ohmmeter with power applied to the circuit. The ohmmeter is designed to operate on its own power supply. The normal 12 volt electrical system voltage could damage the meter!

1. Isolate the circuit from the vehicle's power source.
2. Ensure that the ignition key is **OFF** when disconnecting any components or the battery.

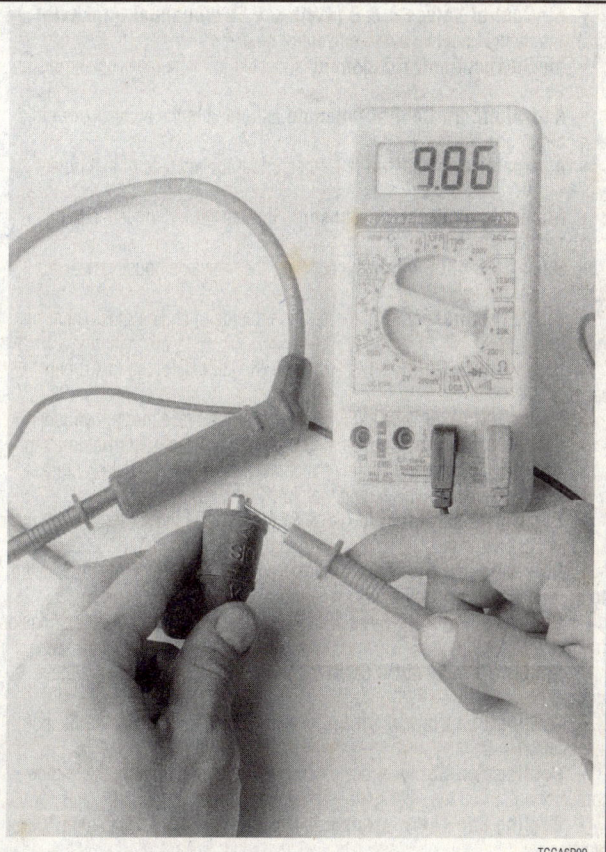

Fig. 11 Spark plug wires can be checked for excessive resistance using an ohmmeter

CHASSIS ELECTRICAL

Wire and Connector Repair

Almost anyone can replace damaged wires, as long as the proper tools and parts are available. Wire and terminals are available to fit almost any need. Even the specialized weatherproof, molded and hard shell connectors are now available from aftermarket suppliers.

Be sure the ends of all the wires are fitted with the proper terminal hardware and connectors. Wrapping a wire around a stud is never a permanent solution and will only cause trouble later. Replace wires one at a time to avoid confusion. Always route wires exactly the same as the factory.

➡ If connector repair is necessary, only attempt it if you have the proper tools. Weatherproof and hard shell connectors require special tools to release the pins inside the connector. Attempting to repair these connectors with conventional hand tools will damage them.

COMPONENT LOCATIONS

Relay, Sensors And Computer Locations

➡ When using this section, understand that some of the components may not be used on a particular vehicle. This is because the particular component in question may be used on an earlier model or a later model.

- **Air Conditioning Clutch Relay** is located on the right front corner of the engine compartment.
- **Air Conditioning Condenser Fan Relay** is located on the right front corner of the engine compartment.
- **Air Conditioning Diode** is located on the right front corner of the engine compartment.
- **Air Conditioning-Heater Blower Motor** is located under the right side of the instrument panel.
- **Air Conditioning Heater Blower Motor Resistor** is located on the blower motor housing.
- **Air Conditioning Idle Boost Solenoid (Si Models)** is located on the firewall near the hood latch.
- **Air Conditioning Thermostat** is located on the bottom side of the evaporator core housing.
- **Air Intake Temperature Sensor** is located in the air intake duct.
- **Air Control Solenoid Valve (Except 4WD & 1.5L)** is located on the in the number one emission control box.
- **Air Control Valves A & B (4WD & 1.5L w/manual transaxle)** are located in the number two emission control box.
- **Air Suction Control Solenoid** is located in the number one emission control box.
- **Anti-Afterburn Control Solenoid** is located in the number one emission control box.
- **Alternator Control Unit** is located under the left side of instrument panel.
- **Atmospheric Pressure Sensor** is located behind the left side of instrument cluster.
- **Back-Up Light Switch** is located on the transaxle housing near the speedometer holder.
- **Block Thermosensor** is located on the side of engine block, near oil the filter.
- **Brake Check Relay** is located under the left side of the instrument panel.
- **Brake Fluid Level Sensor** is located on the brake master cylinder.
- **Brake Fluid Level Switch** is located in the brake fluid reservoir cap.
- **Brake/Stoplight Switch** is located above the brake pedal on bracket.
- **Clutch Switch** is located on the clutch pedal bracket.
- **Cold Advance Solenoid (CRX Si Models)** is located on the left front inner fender panel.
- **Cold Advance Solenoid Valve (Federal 1.5L HF Only)** is located on the In emission control box No. 1.
- **Control Solenoid Valves A & B (4WD)** is located in the transmission control box.
- **Coolant Temperature Switch** is located on the left center of the engine.
- **Cooling Fan Control Unit** is located below the right side of the instrument cluster.
- **Cooling Fan Relay** is located on the right front corner of the engine compartment, or on the right front inner fender panel.
- **Cooling Fan Thermosensor** is located in the lower right side of the radiator.
- **Crank Angle Sensor** is located in the distributor assembly.
- **Cranking Leak Solenoid Valve** is located in the number one emission control box.
- **Dashlight Brightness Controller** is located in the left hand side of the instrument panel.
- **EGR Control Solenoid Valve A** is located in the number one emission control box.
- **EGR Control Solenoid Valve B** is located in the number one emission control box.
- **EGR Control Solenoid Valve C (Federal 1.5L HF Only)** is located in the number one emission control box.
- **Electric Load Detect Unit** is located in the main fuse block.
- **Electronic Control Unit** is located under the passenger seat.
- **Emission Control Box Number One** is located on the left side of the engine compartment.
- **Emission Control Box Number Two** is located on the right side of the engine compartment.
- **Emission Control Switch** is located in the number one emission control box.
- **Fast Idle Control Solenoid (CRX Si Models)** is located on the left front inner fender panel.
- **Feedback Control Solenoid Valve** is located in the number two emission control box.
- **Frequency Solenoid Valves A & B** are located in the number two emission control box.
- **Front Washer Motor** is located on the right side of the engine compartment near the battery.
- **Front Wiper Motor** is located under the left side of the air scoop.
- **Fuel Cut-Off Solenoid Valve** is located on the fuel inlet side of carburetor.
- **Fuel Pump** is located at the left side of fuel tank (right rear on wagon).
- **Function Control Motor** is located on the side of the blower housing.
- **Fuse/Relay Block** is located under the left side of the instrument panel.
- **Idle Boost Solenoid Valve** is located in the number one emission control box.
- **Idle Control Solenoid (CRX Si Models)** is located on the left front inner fender panel.
- **Idle Mixture Adjusting Sensor** is located on the left front inner fender panel.
- **Ignitor Unit** is located on the side of the ignition distributor.
- **Intake Air Temperature Sensor** is located in the air cleaner housing.
- **Integrated Control Unit** is located in the fuse/relay block.
- **Intermittent Wiper Relay** is located in the fuse/relay block.
- **Main Fuse Block** is located right rear corner of the engine compartment.
- **Main Relay** is located under the left side of instrument panel.
- **Manifold Air Pressure Sensor** is located at the rear of the engine compartment.
- **Neutral/Back-Up Light Switch** is located under the selector lever console.
- **Oil Pressure Sending Unit** is located on the side of the cylinder block, near the oil filter.
- **Opener Solenoid Valve (4WD)** is located in the transmission control box.
- **Oxygen Sensor** is located in the exhaust manifold.
- **Parking Brake Switch** is located under the rear of parking brake lever.
- **PGM-FI Electronic Control Unit** is located under the passenger seat, or behind the right hand side of the instrument panel, near the kick panel.
- **PGM-FI Relay** is located on the left hand side kick panel, near the hood opener.
- **Primary Main Air Cut Solenoid Valve (Federal 1.5L HF)** is located in the number two emission control box.
- **Primary Slow Mixture Solenoid Valve (Federal 1.5L HF)** is located in the number two emission control box.

CHASSIS ELECTRICAL 6-9

- **Purge Cut-Off Solenoid** is located on the left front inner fender panel.
- **Power Door Lock Control Unit** is located in the middle of the driver's side door.
- **Power Valve Control Solenoid** is located in the number one emission control box.
- **Power Window Relay** is located in the fuse/relay block.
- **Radiator Fan Relay** is located on the right front corner of the engine compartment.
- **Rear Defogger Relay** is located in the fuse/relay block, under the left side of the instrument panel.
- **Rear Wiper Motor** is located behind the trim panel in the liftgate.
- **Recirculation Control Motor** is located on the side of the blower housing.
- **RPM Sensor** is incorporated into the tachometer.
- **Seat Belt Buzzer/Timer** is located under the instrument panel, near the steering column.
- **Shift Indicator Control Unit** is located under the left side of the instrument panel.
- **Side Mark Control Unit** is located under the left side of the instrument panel.
- **Speed Sensor** is incorporated into the speedometer.

- **Sunroof Relay (CRX Models)** is located under the left side of the instrument panel.
- **System Resistor** is located on the left front inner fender panel.
- **Temperature Gauge Sending Unit** is located in the thermostat housing.
- **Throttle Angle Sensor** is located on the throttle body.
- **Throttle Control Solenoid Valve** is located in the number one emission control box.
- **Turn Signal/Hazard Flasher** is located in the fuse/relay block.
- **Vacuum Brake Switch A** is located in either the number one or number two emission control box depending on the model and year.
- **Vacuum Brake Switch B** is located in the number one emission control box.
- **Vacuum Brake Switch C** is located in the number one emission control box.
- **Vacuum Brake Switch D** is located in the number one emission control box.
- **Vacuum Brake Switch E** is located in the number one emission control box.
- **Vacuum Brake Switch F** is located in the number one emission control box.
- **Vacuum Holding Solenoid Valve** is located in the number one emission control box.

SUPPLEMENTAL RESTRAINT SYSTEM (SRS)

General Information

SYSTEM OPERATION

The SRS (air bag) is a safety device designed to be used in conjunction with the seatbelt. Its purpose is to help protect the driver and passenger in a frontal impact exceeding a certain set limit. The system consists of the SRS unit (including the safing sensor and impact sensor), the cable reel and driver's air bag, and, in most cases, a passenger's air bag. On models equipped with dual air bags, one air bag is stored in the steering wheel and the other in the passenger's side of the dashboard, above the glove box. An air bag system is ineffective if seat belts are not worn.

The air bag is a fabric bag or balloon with an explosive inflator unit attached. The system employs impact sensors and a safing sensor as well as an inflator circuit and control module. A back-up power system is connected in parallel with the battery. No single sensor can trigger SRS deployment. The safing sensor, and at least one of the impact sensors must engage together for at least 0.002 seconds. This "agreement" of the sensors is required to help prevent accidental deployment.

When the control unit receives these signals, power is supplied to the inflator circuit, either from the battery or back-up system. A small heater causes a chemical reaction in the igniter; the non-toxic gas from the chemical mixture expands very rapidly (in milliseconds), filling the bag and forcing it through the cover pad. Since all this is happening very rapidly, the expanding bag should reach the occupant before he/she reaches the steering wheel/dashboard during a frontal collision. The chemical reaction is complete by the time the air bag is fully inflated; as the occupant hits the bag, the gas is allowed to escape slowly through vents in the back of the bag.

The control unit contains a self-diagnostic circuit. When the ignition switch is turned **ON**, the SRS dash warning lamp should light for about 6 seconds, then go out. If the light does not come on, does not go out or comes on when driving, the system must be diagnosed and repaired by a Honda dealer or reputable shop. The system is NOT repairable at home.

1992–95 SUPPLEMENTAL RESTRAINT SYSTEM (SRS) PRECAUTIONS

♦ See Figures 12 and 13

This manual does not cover SRS repairs or replacement as such work should be left to a trained professional. The following precautions then, are only to inform the do-it-yourselfer and give him a greater appreciation for the system when required to work in proximity to SRS components.

➙All SRS wiring can easily be identified by its yellow outer protective covering.

- Replacement of the multi-function light and wiper/washer switches and cruise control switch can be done without removing the steering wheel. See the appropriate portion of this section for these procedures.
- To disconnect the SRS connector, do so at the under-dash fuse/relay box. First lift the connector lid with a thin pry tool, then press the connector tab down and pull the connector out. To re-install the connector, push it back until it positively clicks into position.

✳✳ WARNING

Be careful to not break the double-locked connector.

- Connect the short connectors before working below the dashboard near the SRS unit. To connect the driver's side short connector, first disconnect the battery negative cable, then the positive cable, then wait a minimum of three minutes. Next connect the (red) short connectors for the driver's side by removing the steering wheel access panel and removing the red short connectors from the panel. Then disconnect the 3-pin connector between the driver's air bag and cable reel and connect the short connector to the air bag side of the connector.
- To connect the short connector to passenger side, remove the glove box and access panel and disconnect the 3-pin connector between the front passenger's air bag and SRS main harness. Then connect the (red) short connector to the air bag side of the connector.
- If air bag has been deployed, do not repair the wiring harness, replace everything damaged with new parts. Inspect also the cable reel for heat damage. If any, replace.
- No SRS component should be used if it shows any sign of being dropped, dented or otherwise damaged.
- SRS components formerly installed in another car should never be used. Only new components should be installed.
- Whenever working on SRS components (except for electrical inspections), always disconnect both the positive and negative battery terminals and then wait at least three minutes before beginning (and taking other precautions as necessary).
- Whenever the air bag has been activated, replace the SRS unit.
- Whenever the ignition switch is ON or has been turned OFF for less than three minutes, be careful not to bump the SRS unit; the air bags could accidently deploy and do damage or cause injuries.
- Do not try to take apart the air bag assembly. Once deployed, it cannot be re-used or repaired.

6-10 CHASSIS ELECTRICAL

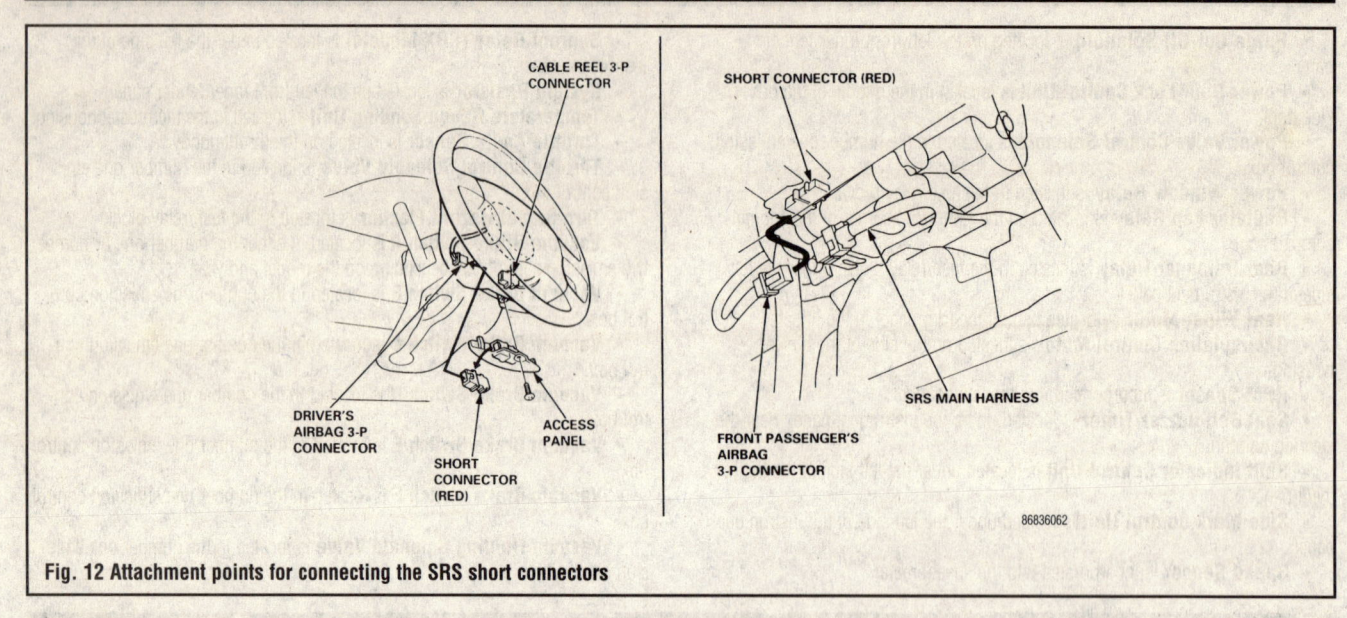

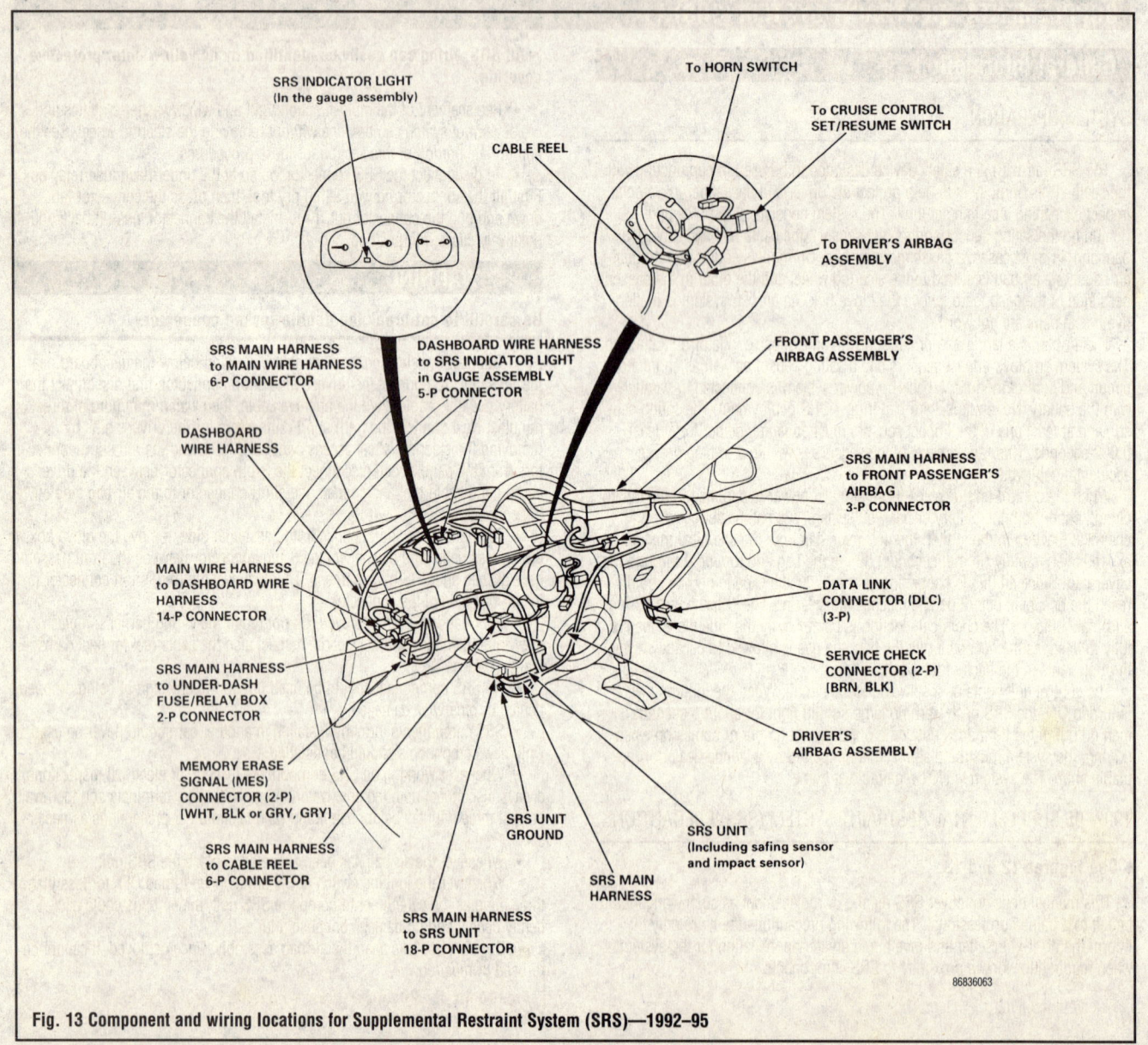

Fig. 12 Attachment points for connecting the SRS short connectors

Fig. 13 Component and wiring locations for Supplemental Restraint System (SRS)—1992–95

CHASSIS ELECTRICAL 6-11

• For temporary storage of the air bag assembly while servicing the vehicle, place it with the pad surface UP. Store the air bag assembly on a secure, clean, flat surface away from heat, oil, grease, water or detergent.

✲✲ CAUTION

If the air bag is stored face down, it could spontaneously deploy and cause serious injury and damage.

• Take extra care when doing paint or body work near the air bag assembly and keep away heat guns, welding and spray equipment from the air bag assembly.
• Make sure SRS wiring harnesses are not pinched, and all ground contacts are clean. Poor grounding can cause intermittent problems that are difficult to diagnose.

➡ If removing the steering wheel, make sure the road wheels are straight ahead to avoid misalignment of the steering wheel or air bag on reassembly.

✲✲ WARNING

Before removing the steering column, first disconnect the connector between the cable reel and the SRS main harness. If the steering column is going to be removed without dismounting the steering wheel, lock the steering by turning the ignition key to 0-LOCK position, or remove the key from the ignition so that the steering wheel will not turn.

• Upon re-installation of the steering wheel, rotate the SRS cable reel clockwise until it stops. Then rotate it counterclockwise (approximately two turns) until the yellow gear tooth lines up with the alignment mark on the cover and the arrow mark on the cable reel label points straight up.

DISARMING THE SYSTEM

✲✲ CAUTION

This procedure MUST be followed whenever working on or near SRS components and when otherwise noted. The shorting connector must be installed whenever the air bag is disconnected from the harness. If equipped with dual air bags, always disarm the passenger's air bag as well.

1. Disconnect the negative battery cable, then the positive cable from the battery. Wait at least three minutes before proceeding.
2. Remove the maintenance lid below the air bag on the steering wheel; it is secured by two screws. Remove the red shorting connector stored inside the lid.
3. Disengage the connector between the air bag and the cable reel in the steering wheel. The connector can be reached through the opening provided by removing the maintenance lid. Install the shorting connector into the air bag side of the wiring connector.
4. To disarm the passenger's side air bag, first remove the glove box. Now disengage the connector between the SRS main harness and the air bag. Install the red shorting connector into the air bag side of the connector.
5. DON'T forget to enable the system after the repairs are finished.

ENABLING THE SYSTEM

1. Remove the shorting connector from the air bag, then engage the air bag and cable reel (driver's side) or SRS main harness (passenger side) connectors.
2. Reinstall the maintenance lid and glove box (as applicable).
3. Reconnect the battery cables, positive cable first.
4. Turn the ignition switch **ON**, the SRS dash warning lamp should light for about 6 seconds, then go out. If the light does not come on, does not go out or comes on when driving, the system must be diagnosed and repaired by a Honda dealer or reputable shop.

HEATING AND AIR CONDITIONING

General Information

The heater unit is located in the center of the vehicle along the firewall. The heater system is a bi-level system designed to direct warm air through the vents to either the windshield or the floor and cool air through the panel outlet. The air conditioning system is designed to be activated in combination with a separate air conditioning switch installed in the control assembly and the fan speed switch. The system incorporates a compressor, condenser, evaporator, receiver/drier, pressure switch, expansion valve, thermo-switch, refrigerant lines and some models are equipped with an electronic control head assembly versus the standard cable operated control head.

On many of the Civic series models the original equipment Honda air conditioning system is installed at the car dealership. All of the vehicles come pre-wired to accept the installation of this system. This accessory kit contains all of the needed components to install the air conditioning system, this allows cars originally sold without air conditioning, in some cases, to be retrofitted with the system.

Air Conditioning Components

REMOVAL & INSTALLATION

▶ See Figure 14

Repair or service of air conditioning components is not covered by this manual, because of the risk of personal injury or death, and because of the legal ramifications of servicing these components without the proper EPA certification and experience. Cost, personal injury or death, environmental damage, and legal considerations (such as the fact that it is a federal crime to vent refrigerant into the atmosphere), dictate that the A/C components on your vehicle should be serviced only by a Motor Vehicle Air Conditioning (MVAC) trained, and EPA certified automotive technician.

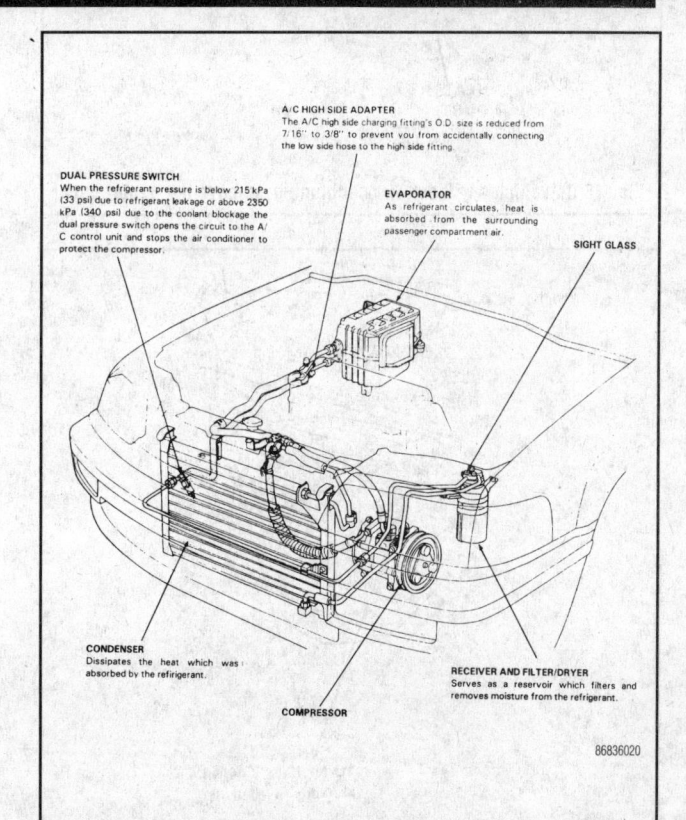

Fig. 14 Air conditioning system components—similar for all models

6-12 CHASSIS ELECTRICAL

→ If your vehicle's A/C system uses R-12 refrigerant and is in need of recharging, the A/C system can be converted over to R-134a refrigerant (less environmentally harmful and expensive). Refer to Section 1 for additional information on R-12 to R-134a conversions, and for additional considerations dealing with your vehicle's A/C system.

Blower Motor

REMOVAL & INSTALLATION

♦ See Figures 15 and 16

1. Disconnect the negative battery cable and disable the SRS system (if equipped.)

❄❄ WARNING

All SRS wire harnesses are covered with a yellow insulation. Read the SRS precautions found in this section and install the short connectors as described.

2. If equipped with air conditioning, properly discharge the air conditioning system into a refrigerant recovery/recycling machine.
3. Remove the glove box assembly and the glove box frame.
4. Remove the blower duct. This is the white plastic duct between the blower case and heater case on models not equipped with air conditioning.
5. On models equipped with air conditioning, remove the evaporator assembly.
6. Disconnect the wire connectors at the blower motor assembly.
7. Remove the blower motor housing attaching bolts.
8. Disassemble the housing and remove the blower motor. The blower motor can be removed from the housing and the blower cage separated if needed.

To install:

9. Install the blower assembly in position under the instrument panel. Install the retaining bolts and tighten to 7 ft. lbs. (10 Nm).
10. Connect the wires at the blower housing.
11. Install the air duct or evaporator housing, depending on how the vehicle is equipped. Make sure that the foam seals seat completely.
12. Install the glove box assembly and frame.
13. Properly evacuate and recharge the air conditioning system. Connect the battery cable.

Blower Motor Resistor

REMOVAL & INSTALLATION

The blower resistor is located on the front of the blower housing.
1. Disconnect the negative battery cable.
2. Remove the glove box assembly, if required.
3. Disconnect the resistor wire connector.
4. Remove the resistor attaching screws.
5. Remove the blower motor resistor.
6. Installation is the reverse of removal.
7. Install the glove box if it was removed and connect the negative battery cable.

Heater Core

REMOVAL & INSTALLATION

♦ See Figures 17, 18 and 19

1. Disconnect the negative battery cable and disable the SRS system (if equipped.)

❄❄ WARNING

All SRS wire harnesses are covered with a yellow insulation. Read the SRS precautions found in this section and install the short connectors as described.

2. Properly drain the cooling system.
3. Disconnect and plug the heater hoses at the firewall.
4. Disconnect the heater valve control cable.
5. Remove the heater unit attaching nut on the firewall.
6. Remove the dashboard as follows:
   a. Remove the center panel.
   b. Disconnect the wire connectors at the fuse box.
   c. Disconnect the sunroof switch connector, if equipped.
   d. Disconnect the ground cable at the right of the steering column and the power mirror switch wire connector.
   e. Remove the side air vent knob and face plate.
   f. Remove the screws attaching the air vent control lever.
   g. Remove the center panel, radio and heater control head.
   h. Remove the gauge upper panel (4-door) or instrument panel (3-door).
   i. Disconnect the speedometer cable.
   j. Remove the center upper lid.
   k. Remove the side defroster garnishes on both ends of the dash.
   l. Lower the steering column.
   m. Remove the dashboard mounting bolts.

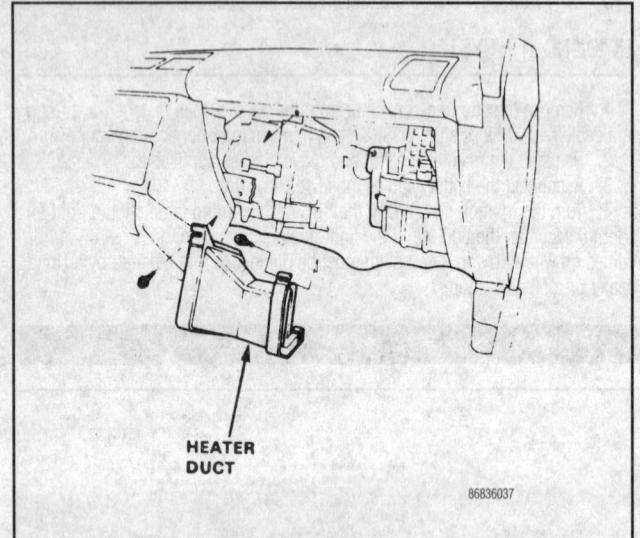

Fig. 15 On vehicles without air conditioning, remove the heater duct

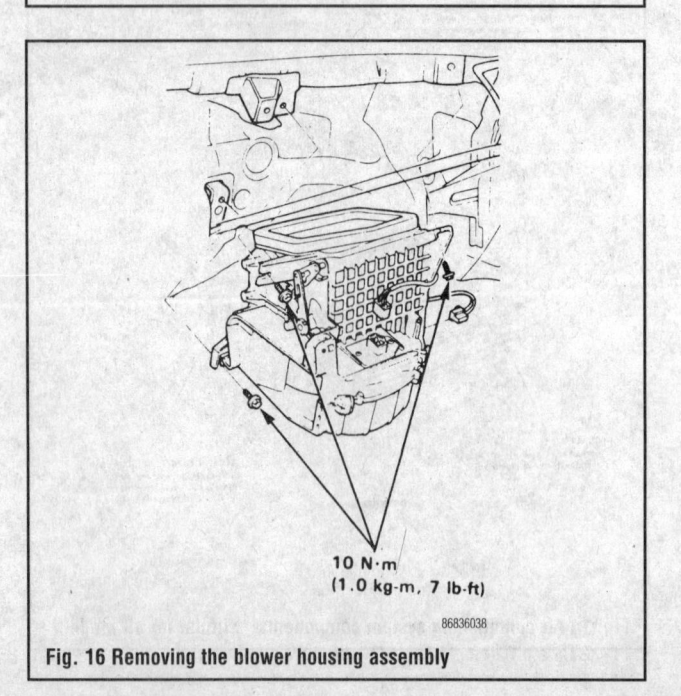

Fig. 16 Removing the blower housing assembly

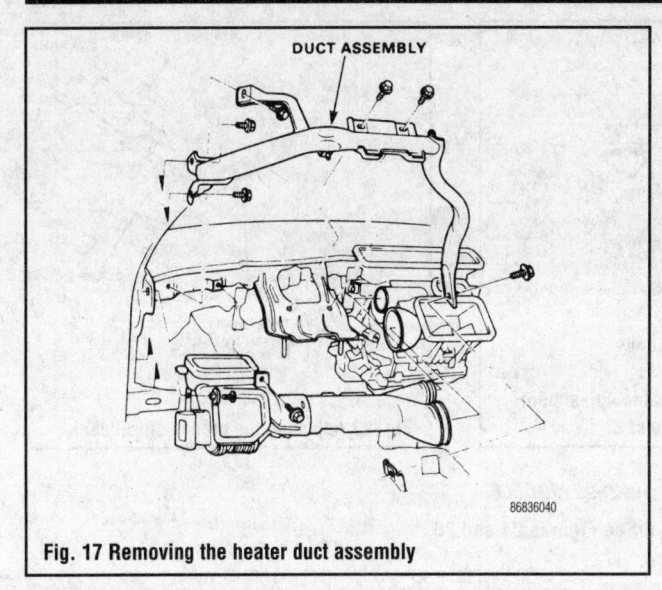

CHASSIS ELECTRICAL 6-13

n. Lift and remove the dashboard assembly.
7. Remove the heater duct.
8. Remove the steering column bracket and duct assembly.
9. Remove the heater assembly mounting bolts.
10. Remove the heater assembly and remove the heater core.

To install:
11. Install the heater core into the heater unit.
12. Install the heater assembly and the attaching bolts.
13. Install the heater duct and steering column bracket.
14. Install the dashboard assembly. Tighten the mounting bolts until just snug.
15. Raise and attach the steering column.
16. Install the center upper panel.
17. Connect the speedometer cable and install the instrument gauge assembly.
18. Install the control head, radio and center panel.
19. Install the side vent and face plate.
20. Connect the ground cable at the right of the steering column and the power mirror switch wire connector.
21. Connect the sunroof switch connector, if equipped.
22. Connect wire connectors at fuse panel.
23. Install the heater unit attaching nut on the firewall.
24. Connect the heater valve control cable.
25. Connect the heater hoses at the firewall.
26. Connect the negative battery cable.
27. Fill, bleed and leak check the cooling system.

Manual Control Head

REMOVAL & INSTALLATION

▶ See Figure 20

1. Disconnect the negative battery cable.
2. Remove the control head face panel attaching screws.
3. Disconnect the cigarette lighter connector and remove the face panel.
4. Remove the radio attaching screws and remove the assembly.

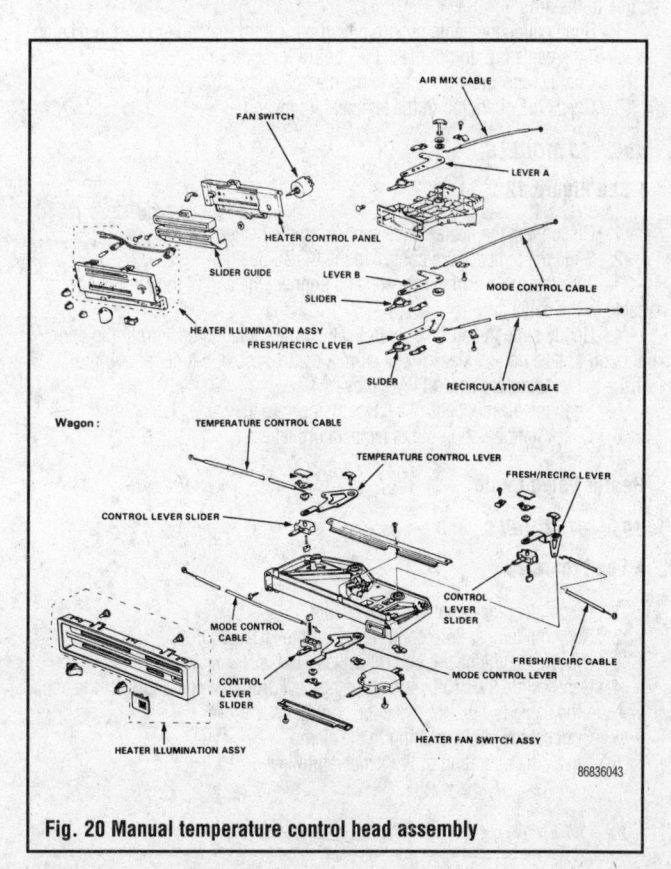

Fig. 17 Removing the heater duct assembly

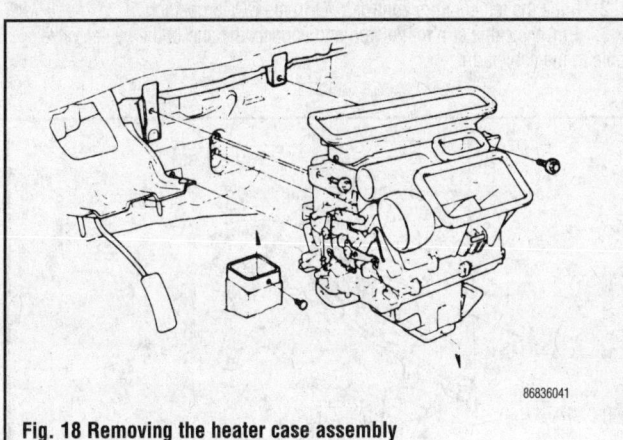

Fig. 18 Removing the heater case assembly

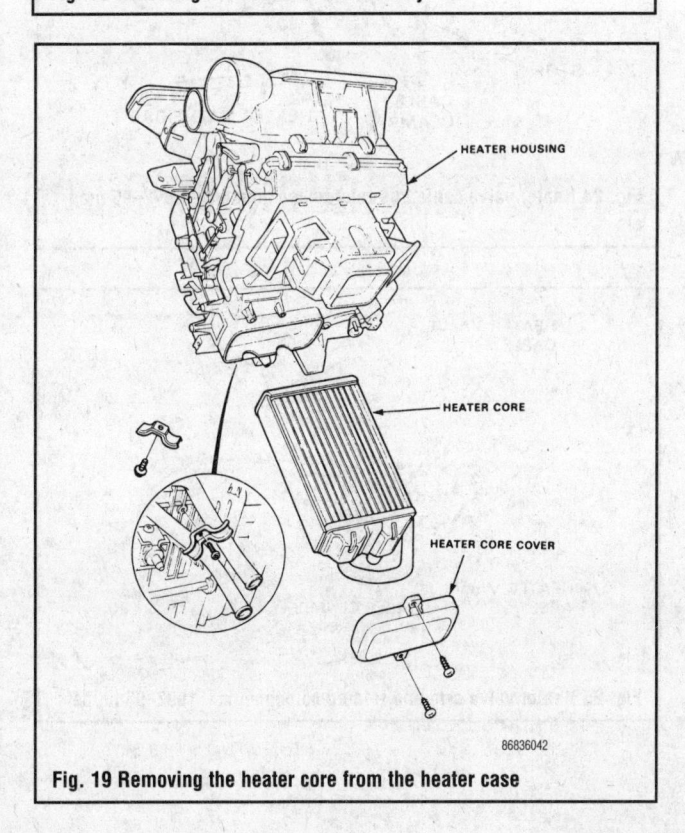

Fig. 19 Removing the heater core from the heater case

Fig. 20 Manual temperature control head assembly

6-14 CHASSIS ELECTRICAL

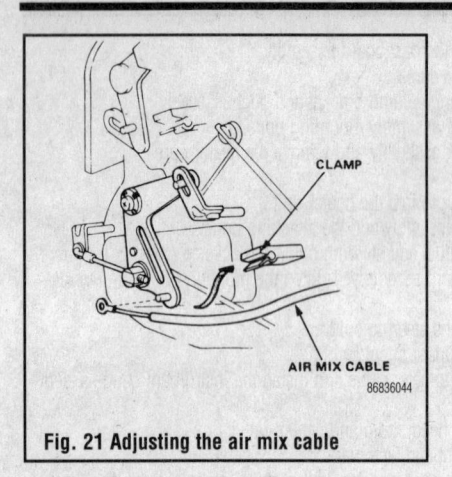

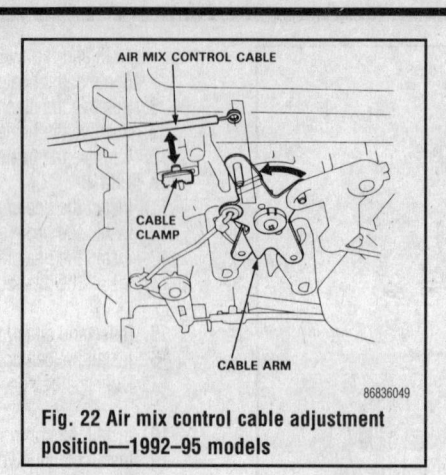

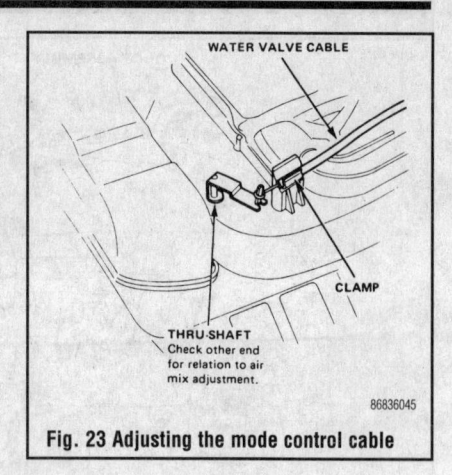

Fig. 21 Adjusting the air mix cable

Fig. 22 Air mix control cable adjustment position—1992-95 models

Fig. 23 Adjusting the mode control cable

5. Disconnect the control cables at the heater unit.
6. Remove the attaching screws at the control head and pull the assembly out.
7. Disconnect the control head wire connectors and remove the unit.
8. Installation is the reverse of removal.

Control Cables

ADJUSTMENT

Air Mix Cable

1984-91 MODELS

♦ See Figure 21

1. Disconnect the negative battery cable.
2. Slide the temperature control to the HOT position.
3. Turn the air mix door shaft arm to the left and connect the end of the cable to the arm.
4. Gently slide the cable outer housing back from the end enough to take up slack in the cable but do not make the control lever move.
5. Connect the cable housing to the clamp.
6. Check the operation of the temperature lever.

1992-95 MODELS

♦ See Figure 22

1. Disconnect the heater valve cable from the heater valve.
2. Slide the temperature control to the COOL position.
3. Turn the cable arm to the stop and connect the end of the air mix control cable to the cable arm.
4. Gently slide the air mix control cable outer housing back from the end enough to take up any slack in the air mix control cable, but not enough to make the temperature control lever move.
5. Snap the air mix control cable housing into the cable clamp.
6. Check the operation of the temperature lever.

Heater Valve Cable

1984-91 MODELS

♦ See Figure 23

1. Disconnect the negative battery cable.
2. Slide the temperature control lever to the HOT position.
3. Disconnect the heater control cable at the valve.
4. Gently slide the cable outer housing back from the end enough to take up slack in the cable but do not make the control lever move.
5. Connect the cable housing to the clamp.
6. Check the operation of the temperature lever.

1992-95 MODELS

♦ See Figures 24 and 25

1. Disconnect the heater valve cable from the heater valve.
2. Slide the temperature control lever to the COOL position.
3. Turn the cable arm to the stop and connect the end of the heater valve cable to the cable arm.

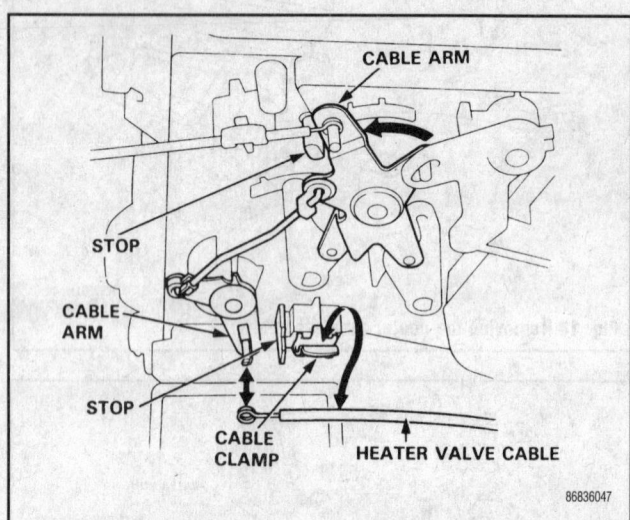

Fig. 24 Heater valve cable and related components—1992-95 models

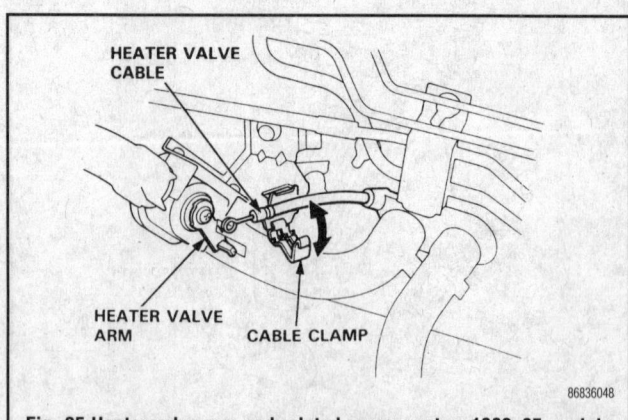

Fig. 25 Heater valve arm and related components—1992-95 models

CHASSIS ELECTRICAL 6-15

4. Gently slide the heater valve cable outer housing back from the end enough to take up slack in the heater valve cable but do not make the control lever move.

5. Hold the end of the heater valve cable housing against the stop, then snap the heater valve cable housing into the cable clamp.

6. Turn the heater valve cable to shut and connect the end of the heater valve cable to the cable arm.

7. Gently slide the heater valve cable outer housing back from the end enough to take up slack in the heater valve cable but do not make the control lever move, then snap the heater valve cable housing into the cable clamp.

➡ The air mix control cable should always be adjusted whenever the heater valve cable has been disconnected.

Function Control Cable

♦ See Figures 26 and 27

1. Disconnect the negative battery cable.
2. Slide the function control lever to the DEFROST position.
3. Turn the function control shaft to the front and connect the end of the cable to the arm.
4. Gently slide the cable outer housing back from the end enough to take up slack in the cable but do not make the control lever move.
5. Connect the cable housing to the clamp.
6. Check the operation of the temperature lever.

REMOVAL & INSTALLATION

♦ See Figure 28

1. Disconnect the negative battery cable.
2. Remove the control head assembly.
3. Disconnect the control cable being replaced, at the control head.
4. Remove the glove box, if required.

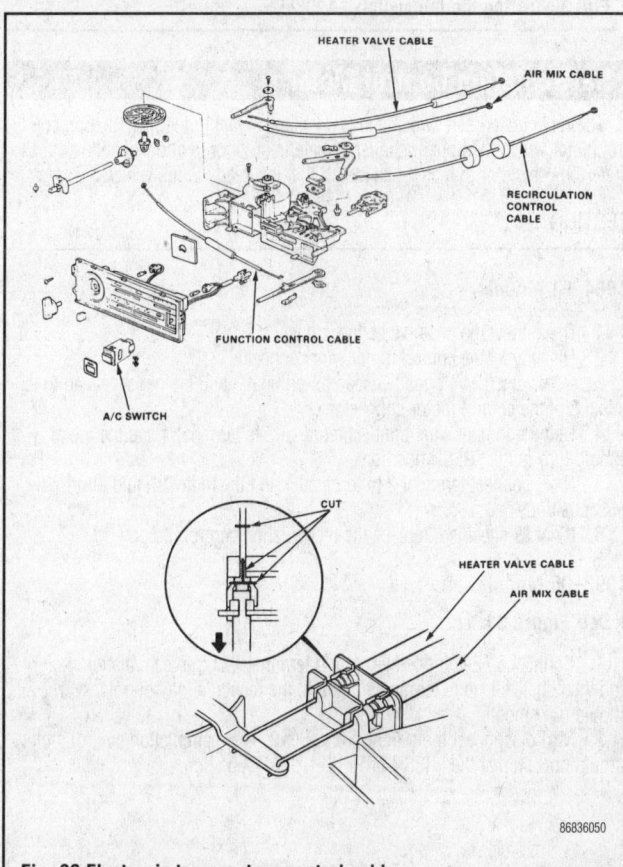

Fig. 26 Electronic temperature control cable

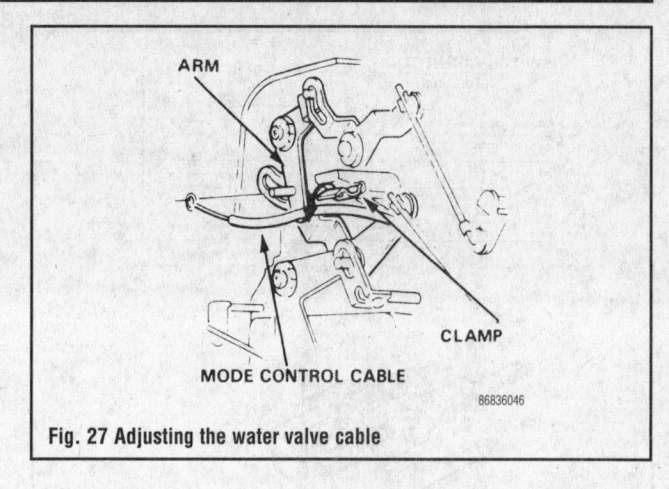

Fig. 27 Adjusting the water valve cable

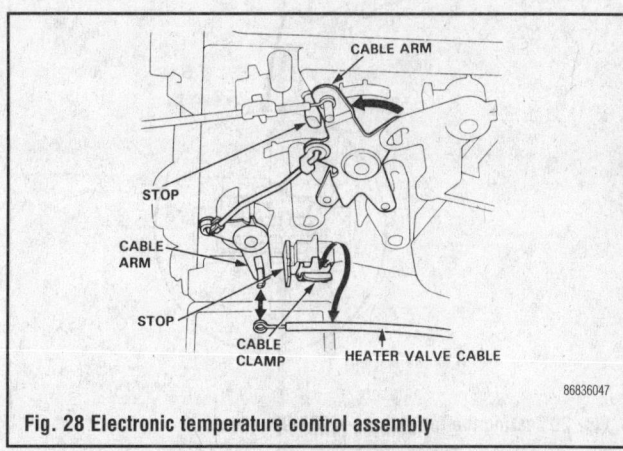

Fig. 28 Electronic temperature control assembly

5. Disconnect the control cables at the heater unit.
6. Disconnect the heater control valve cable located in the engine compartment.
7. Install the cable in position and adjust it.
8. Connect the battery cable.

Fan Switch

The fan switch is located with the control head and has 5 or 6 speeds including the **OFF** position. The blower motor always has approximately 12 volts and the fan switch then completes the circuit by supplying a ground through a resistor, thus varying the fan speed.

TESTING

1984–91 Models

♦ See Figure 29

1. Disconnect the negative battery cable.
2. Remove the fan switch and disconnect the wire connector.
3. Test for continuity as follows:
   a. With the switch in the OFF position, there should be no continuity between any of the terminals.
   b. With the switch in the LOW position, there should be continuity between terminals 1, 4 and 6.
   c. With the switch in the MED/LOW, there should be continuity between terminals 1, 2 and 6.
   d. With the switch in the MED/HIGH position, there should be continuity between terminals 1, 5 and 6.
   e. With the switch in the HIGH position, there should be continuity between terminals 1, 3 and 6.
4. If not as specified, replace the fan switch.

6-16 CHASSIS ELECTRICAL

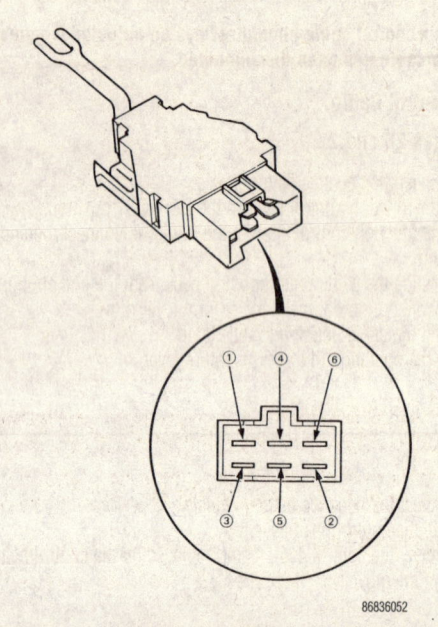

Fig. 29 Testing the fan switch—1984-92 models

Fig. 30 Testing the fan switch—1992-95

1992–95 Models

▶ See Figure 30

1. Disconnect the negative battery cable.
2. Remove the fan switch and disconnect the wire connector.
3. Test for continuity as follows:
    a. With the switch in the OFF position, there should be no continuity between any of the terminals.
    b. With the switch in the LOW position, there should be continuity between terminals 1, 2 and 3.
    c. With the switch in the MED LOW position, there should be continuity between terminals 1, 2 and 4.
    d. With the switch in the MED HIGH position, there should be continuity between terminals 1, 2 and 5.
    e. With the switch in the HIGH position, there should be continuity between terminals 1, 2 and 6.
4. If not as specified, replace the fan switch.

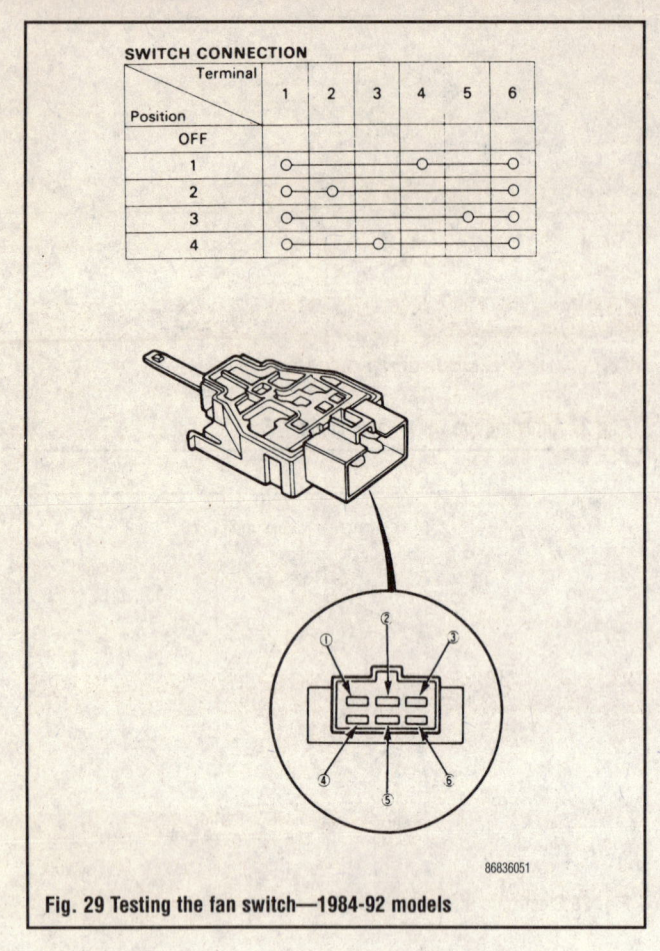

REMOVAL & INSTALLATION

1. Disconnect the negative battery cable.
2. Remove the control head assembly.
3. Remove the fan switch attaching screws.
4. Disconnect the fan switch wire connectors.
5. Remove the fan switch.
6. Install the new switch in position and connect the wires. Check the switch operation.

Recirculation Control Motor

The recirculate/fresh switch sends a voltage signal to the recirculation control motor which then opens the recirculate/fresh door to allow outside air to enter the vehicle or to recirculate the air that is already in the vehicle.

TESTING

1984–91 Models

1. Disconnect the negative battery cable.
2. Disconnect the control motor wire connector.
3. Connect a 12 volt positive lead to terminal 3 and the negative lead to terminal 2 of the control motor connector.
4. Using a jumper wire, connect terminal 2 to terminal 4 and the motor should turn to the FRESH position.
5. Next, connect terminal 2 to terminal 1 and the motor should turn to the REC position.
6. If not as specified, replace the recirculation control motor.

1992–95 Models

▶ See Figure 31

1. Connect a 12 volt positive lead to terminal 1 of the recirculation motor and the negative lead to terminals 2 and 3. the recirculation control motor should run smoothly.
2. Next disconnect the ground from 2 or 3. The recirculation control motor should now stop at the FRESH or RECIRCULATE position.

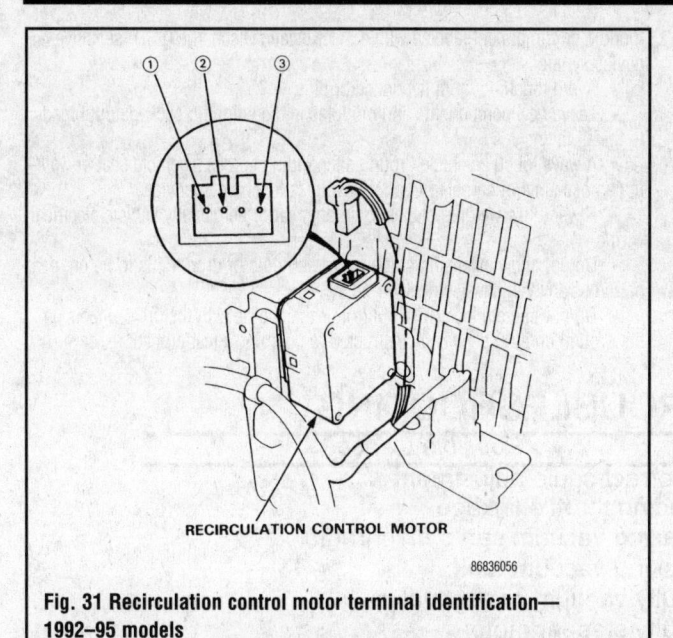

CHASSIS ELECTRICAL 6-17

➡ Do not cycle the re-circulation motor for any more time than is necessary for testing.

3. If the recirculation control motor does not run, remove it. Check the recirculation control linkage and door for smooth movement. If the linkage moves smoothly, replace the recirculation control motor.

REMOVAL & INSTALLATION

1. Disconnect the negative battery cable and disable the SRS system (if equipped.)

✴✴ WARNING

All SRS wire harnesses are covered with a yellow insulation. Read the SRS precautions found in this section and install the short connectors as described.

2. Remove the blower motor assembly.
3. Remove screws attaching recirculation control motor.
4. Remove wire connectors.
5. Disconnect all linkage.
6. Install the control motor in position and reinstall the blower housing assembly.
7. Connect the negative battery cable.

Fig. 31 Recirculation control motor terminal identification—1992–95 models

CRUISE CONTROL

General Description

▶ See Figures 32 and 33

The cruise control system found on some 1988 and later Civic series models maintains the vehicle speed at a setting selected by the driver by means of mechanical, electrical, and vacuum operated devices.

The cruise control unit receives command signals from the cruise control main switch and the cruise control set/resume switch. The control unit also receives information about operating conditions from the brake switch, the distributor, speed sensor, the clutch switch with manual transaxle, or the shift lever position switch with automatic transaxle. The cruise control unit, in turn, sends operational signals to the devices that regulate the throttle position. The throttle position maintains the selected vehicle speed. The cruise control compares the actual speed of the vehicle to the selected speed. Then, the control unit uses the result of that comparison to open or close the throttle.

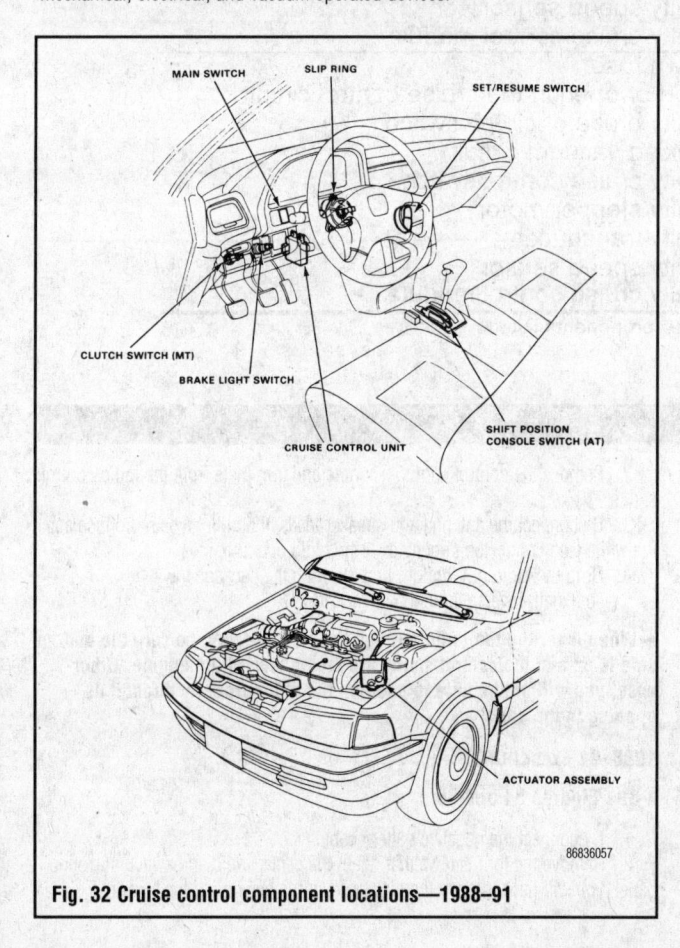

Fig. 32 Cruise control component locations—1988–91

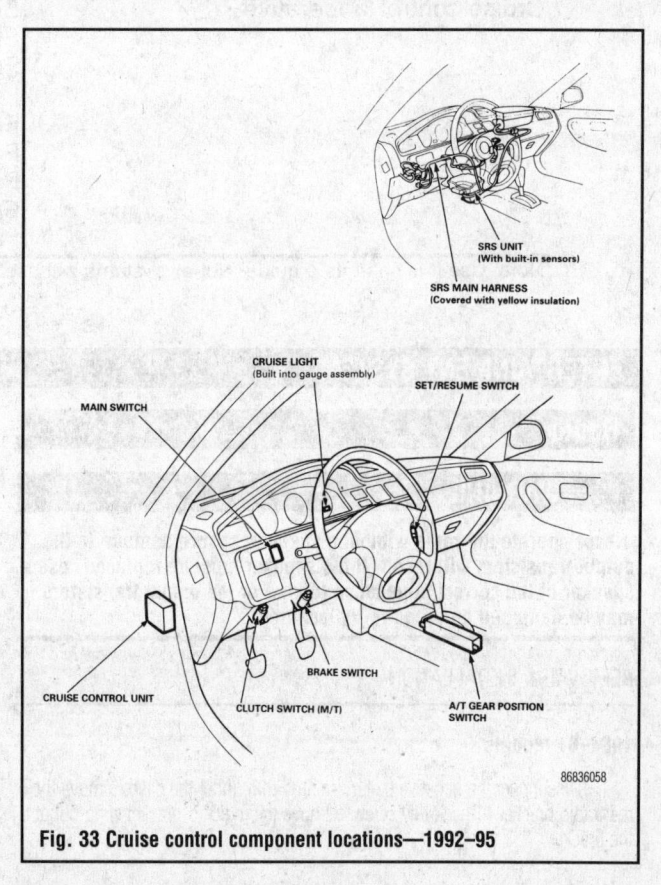

Fig. 33 Cruise control component locations—1992–95

6-18 CHASSIS ELECTRICAL

The control unit will disengage the instant the driver depresses the brake pedal. The brake switch sends an electronic signal to the control unit when the brake pedal is depressed; the control unit responds by allowing the throttle to close. The shift lever position switch (automatic transaxle) or the clutch switch (manual transaxle) sends a disengage signal input to the control unit that also allows the throttle to close.

SERVICE PRECAUTIONS

- Never disconnect any electrical connection with the ignition switch **ON** unless instructed to do so in a test.
- Always wear a grounded wrist static strap when servicing any control module or component labeled with a Electrostatic Discharge (ESD) sensitive device symbol.
- Avoid touching module connector pins.
- Leave new components and modules in the shipping package until ready to install them.
- Always touch a vehicle ground after sliding across a vehicle seat or walking across vinyl or carpeted floors to avoid static charge damage.
- Never allow welding cables to lie on, near or across any vehicle electrical wiring.
- Do not allow extension cords for power tools or drop lights to lie on, near or across any vehicle electrical wiring.
- Do not operate the cruise control or the engine with the drive wheels off the ground unless specifically instructed to do so by a test procedure.

CRUISE CONTROL TROUBLESHOOTING

Problem	Possible Cause
Will not hold proper speed	Incorrect cable adjustment
	Binding throttle linkage
	Leaking vacuum servo diaphragm
	Leaking vacuum tank
	Faulty vacuum or vent valve
	Faulty stepper motor
	Faulty transducer
	Faulty speed sensor
	Faulty cruise control module
Cruise intermittently cuts out	Clutch or brake switch adjustment too tight
	Short or open in the cruise control circuit
	Faulty transducer
	Faulty cruise control module
Vehicle surges	Kinked speedometer cable or casing
	Binding throttle linkage
	Faulty speed sensor
	Faulty cruise control module
Cruise control inoperative	Blown fuse
	Short or open in the cruise control circuit
	Faulty brake or clutch switch
	Leaking vacuum circuit
	Faulty cruise control switch
	Faulty stepper motor
	Faulty transducer
	Faulty speed sensor
	Faulty cruise control module

Note: Use this chart as a guide. Not all systems will use the components listed.

TCCA6C01

ENTERTAINMENT SYSTEM

Radio

✱✱ WARNING

Never operate the radio without a speaker; severe damage to the output transistors will result. If the speaker must be replaced, use a speaker of the correct impedance (ohms) or the output transistors may be damaged and require replacement.

REMOVAL & INSTALLATION

1984–87 Models

1. Disconnect the negative battery cable. From under the dash, remove the rear radio bracket-to-back tray screw. Remove the radio-to-bracket wing nut and the bracket.

2. Remove the control knobs, hex nuts and trim plate from the radio control shafts.
3. Disconnect the antenna and speaker leads, the bullet type radio fuse and the white lead connected directly over the radio opening.
4. Drop the radio out, bottom first, through the package tray.
5. To install, reverse the removal procedures.

➞When inserting the radio through the package tray, be sure the bottom side is up and the control shafts are facing toward the engine. Otherwise, you will not be able to position the radio properly through its opening in the dash.

1988–91 Except CRX

◆ See Figures 34 and 35

1. Disconnect the negative battery cable.
2. Remove the four screws that retain the center radio and switch surround panel, you will have to open the glove box to remove one of the screws.

CHASSIS ELECTRICAL 6-19

Fig. 34 Remove the center console to remove the radio—1988–91

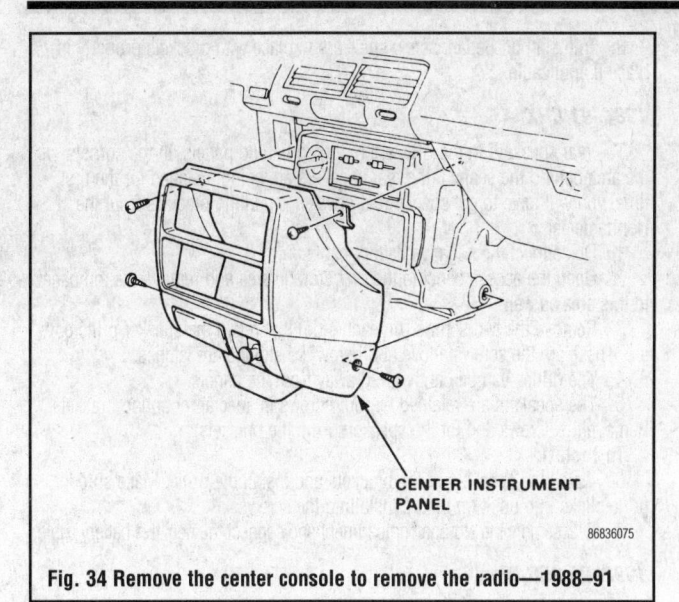

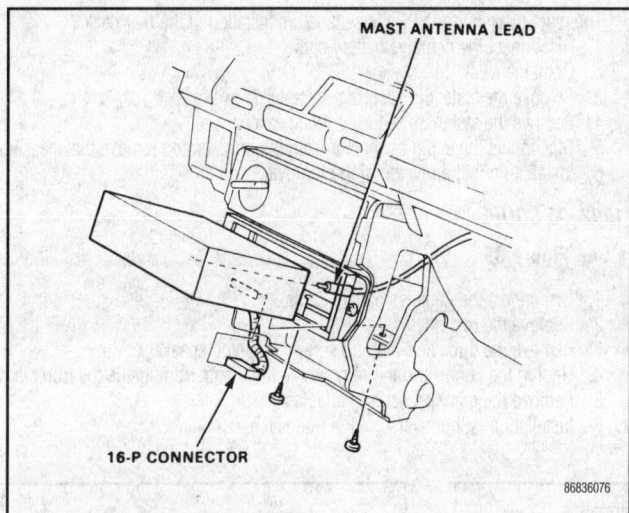

Fig. 35 Remove the mounting screws and carefully slide the radio out—1988–91

3. Remove the two screws that retain the radio unit (they are on the underneath of the radio), slide it forward from the dash.
4. Disconnect the connector and the antenna lead from the unit and remove it from the dash.
5. When installing the unit, make sure that the connector is firmly seated and that the antenna wire is connected properly.
6. Do not overtighten any of the screws or the dash panel may be cracked.

1988–91 CRX

Removing the radio will require removal of the front part of the center console.

1. Disconnect the negative battery cable.
2. On standard shift models, remove the shifter knob.
3. Remove the 6 screws that retain the center console assembly, one of the screws is accessible through the glove compartment.
4. Remove the console from the vehicle, you will have to tilt the rear of the console upward and slide the seats to their furthest back position to get it out.
5. Remove the screws that retain the radio, they are underneath of the unit.
6. Slide the radio forward and disconnect the harness connector and the antenna lead. remove the radio from the vehicle.
7. Installation is the reverse of removal.

1992–95 Models

Removing the radio will require removal of the front part of the center console.

The original equipment radio has a coded theft protection circuit. Before beginning removal, be sure to know the code for re-installation. The code is erased whenever the battery is disconnected or the No. 32 (7.5A) fuse from the under hood fuse/relay box is removed. When power is re-connected, the word "CODE" will be displayed and will require entry of the 5-digit code for that particular unit.

✱✱ WARNING

All SRS wire harnesses are covered with a yellow insulation. Read the SRS precautions found in this section and install the short connectors as described.

1. Disable the SRS system as described in this section.
2. Remove the center instrument panel by removing the four philips screws on the top and sides.
3. Disconnect the 4-pin connector from the cigarette lighter.
4. Remove the two screws, then disconnect the 16-pin connector and the antenna lead.
5. Carefully pull the radio/cassette player out.

To Install:

6. Plug in the 16 pin connector and install the two retaining screws.
7. Plug in the antenna lead, the cigarette lighter and carefully install the unit.
8. Install the center instrument panel. Re-connect power to the car, and enter the 5-digit theft protection code for the radio/cassette player.
9. Enable the SRS system.

Stereo Speakers

▶ See Figure 36

All Civic series models come pre-wired from the factory for four speakers. Many models are equipped with a stereo unit standard with two or four speakers already installed. The rear set of speakers, particularly on 1984–87 models are usually dealer installed and require some additional mounting components.

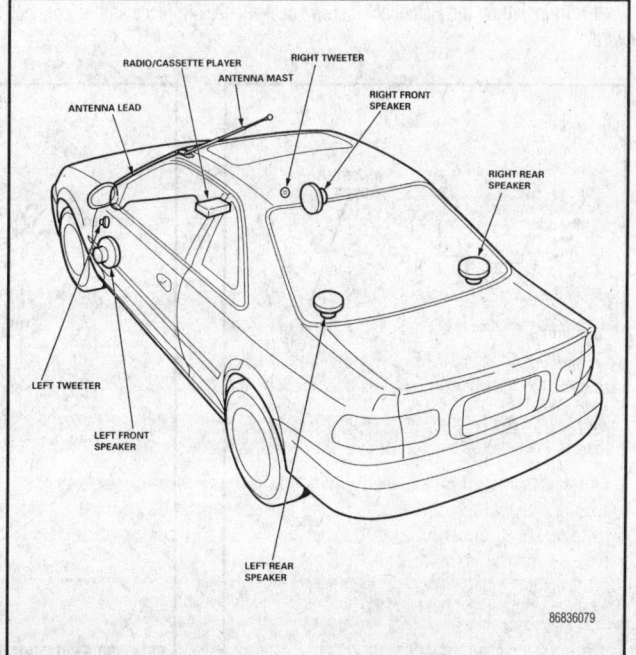

Fig. 36 Common speaker and stereo component location for 6-speaker stereo. A 4-speaker setup is similar, except it comes without the tweeters

6-20 CHASSIS ELECTRICAL

REMOVAL & INSTALLATION

Front

The front speakers on all models covered, are located in the lower part of the front doors.
1. Disconnect the negative battery cable.
2. Remove the two screws that retain the speaker grille covers.
3. Remove the grille by gently prying the grille from the door panel.
4. Remove the four screws that retain the speaker and tilt the speaker forward. Disconnect the plug from the speaker. Remove the speaker from the door.

To Install:
5. Install the speaker in position on the door, make sure the plug is securely connected.
6. When installing the grilles, make sure that they click into position before tightening the screws.

Front Tweeters

♦ See Figure 37

Some more fully equipped models of the 1992–95 Civic series come equipped with dedicated tweeter speakers that play only the higher frequency notes.
1. Disconnect the negative battery cable.
2. Remove the door panel and disconnect the tweeter connector
3. Remove the two self-tapping screws, then the tweeter and cover.
4. Remove the remaining screw to separate the tweeter from the bracket.
5. Install by securing the tweeter to the bracket, then engaging the connector. Mount the bracket back in place and install the door panel

Rear

1988–91—EXCEPT CRX

The rear speakers in the Civic Hatchback and 1984–87 CRX models are mounted on separate panels in the rear of the hatch area, they are removed by removing the mounting screws on these panels. The speakers can then be taken off of the panel.
The rear speakers on the Civic Sedans are installed in the rear deck. They can be removed by opening the trunk and removing the mounting screws.
1. To install on the Hatchback, screw speakers back in place and re-connect wire.
2. To install on Sedan, screw speakers in place with brackets properly in place if applicable.

1984–91 CRX

The rear speakers in the CRX are mounted behind panels, in the corners, of the area behind the seats. If the rear speakers are being installed for the first time, you will have to get special rear speaker mounting brackets from the Honda dealer.
1. Disconnect the negative battery cable.
2. Open the access panel to the rear storage area and remove the top panel to this area as well.
3. Remove the two screws (on each side) that retain the speaker grille covers. These are the screws above and below the storage are latches.
4. The grilles can be gently pried away from the panels.
5. The speakers are retained by four screws in special mounting brackets. Remove the screws and lift the speakers from the brackets.

To Install:
6. Install the speakers in the brackets and install the grille. Make sure the grille clicks into position before installing the screws.
7. Reassemble the storage compartment and connect the negative battery cable.

1992–95 SEDAN

♦ See Figure 38

The rear speakers on the Civic sedans are installed in the rear deck.
1. Disconnect the negative battery cable.
2. Open the trunk lid.
3. Remove the fasteners securing the speaker grilles
4. Remove the fasteners securing the speakers.
5. Disconnect the 2-pin connector from the speaker and remove the speaker.
6. Installation is the reverse of the removal.

1992–95 COUPE

♦ See Figure 39

1. Disconnect the negative battery cable.
2. Remove the rear shelf
3. Remove the three self-tapping screws, clip and speaker cover.
4. Unplug the connector and remove the three nuts from inside the trunk
5. Remove the rear speaker and adapter.
6. Installation is the reverse of the removal.

Fig. 37 Tweeter installation—1992–95

Fig. 38 Exploded view of rear speaker mounting—1992–95 4-door models

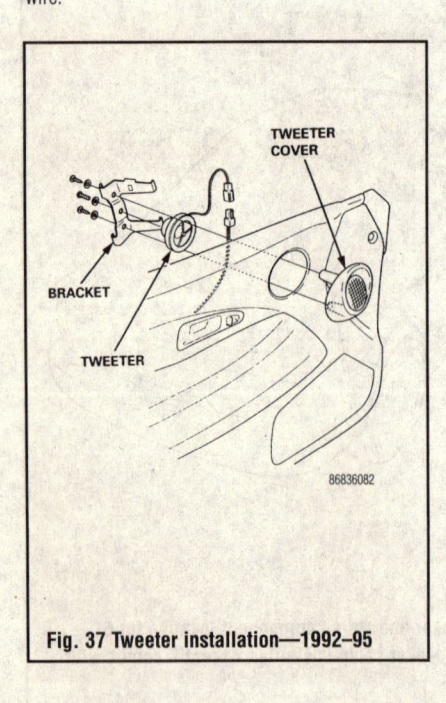

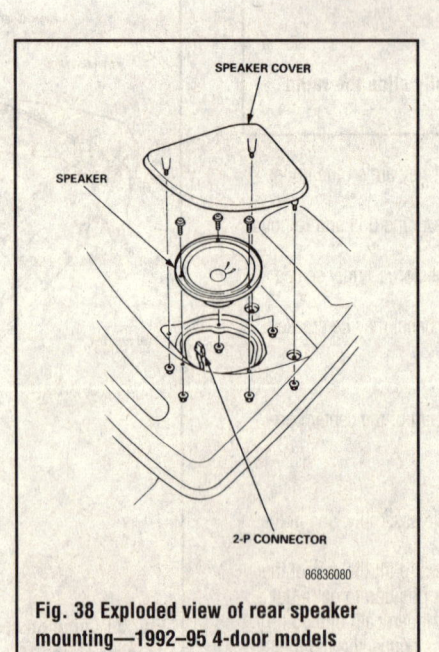

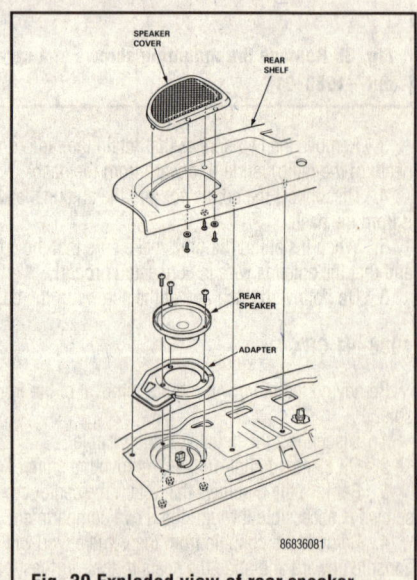

Fig. 39 Exploded view of rear speaker mounting—1992–95 coupe

CHASSIS ELECTRICAL 6-21

WINDSHIELD WIPERS AND WASHERS

Windshield Wiper Arms

REMOVAL & INSTALLATION

♦ See Figures 40, 41 and 42

➡On models equipped with a rear wiper system, the removal of the wiper arm design is the same as for the front wiper arms.

1. Lift the caps at the wiper arm pivots.
2. Remove the wiper arm retaining nuts.
3. Gently pry the arms from the pivots.
4. Install the arms in position, make sure that they are positioned correctly (not too low or too high). Tighten the retaining nuts to 10 ft. lbs. (14 Nm).

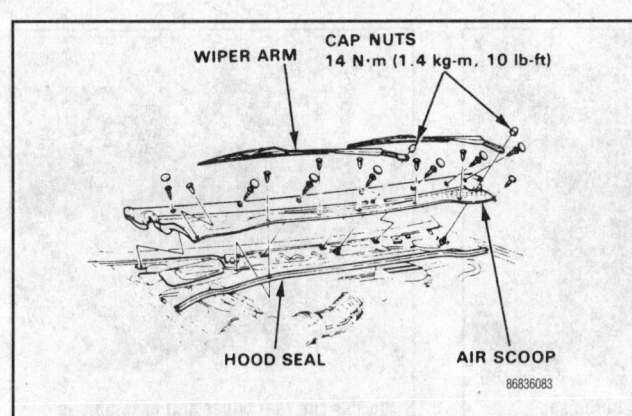

Fig. 40 Exploded view of the wiper arm assembly

Windshield Wiper Motor

REMOVAL & INSTALLATION

♦ See Figures 43 thru 49

Front

1. Note the installation angles of the wiper arms, then remove the cap nuts.
2. Remove the air inlet grille and hood seal.
3. Using a small pry tool, carefully pry the wiper linkage connection from the wiper motor. Remove the linkage by pushing the wiper pivots through the cowl. Pull off the clip and remove the motor cover.
4. Disconnect the electrical connector. Remove the mounting bolts from the cowl. Remove the motor, being careful not to lose the mounting bolts.
5. To install, grease the linkage joints and reverse the removal procedures. Make sure the linkage operates smoothly and that wipers are installed at the correct angle. Tighten the motor mounting bolts to 7 ft. lbs. (10 Nm) and the wiper arm retaining nuts to 10 ft. lbs. (14 Nm).

Rear

♦ See Figures 50 thru 59

1. Note the installation position of the wiper arm. Disconnect the negative battery cable.
2. Remove the rear hatch trim panel.
3. Remove the rear wiper arm.
4. Disconnect the electrical connector and remove the mounting bolts.
5. Remove the motor from the hatch by pulling it back.
6. Install the motor in position, making sure that the rubber seal is correctly installed.

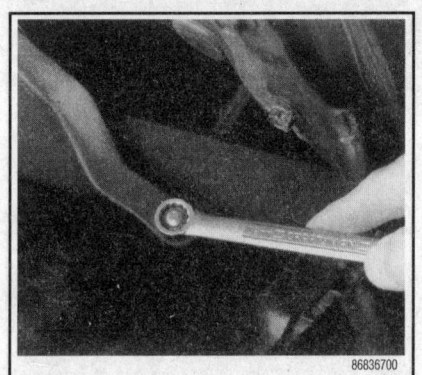

Fig. 41 Unbolt the wiper arm using a wrench

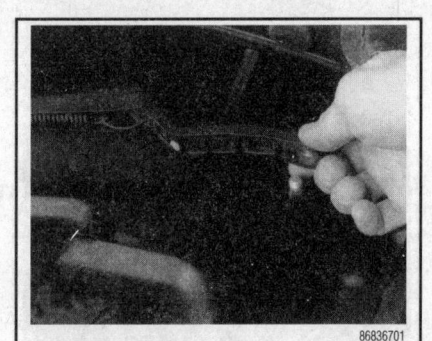

Fig. 42 The aluminum arm is held on the steel splines of the pivot point by the retaining nut

Fig. 43 Remove the shroud and clip to access wiper motor connector

Fig. 44 Detach connector from wiper motor

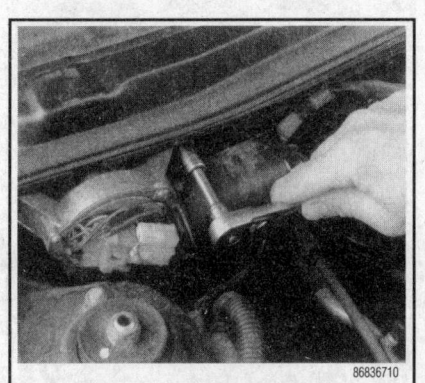

Fig. 45 Unbolt motor from the cowling

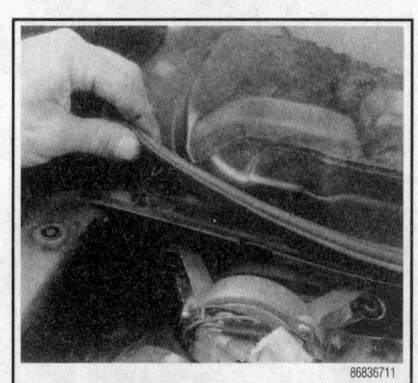

Fig. 46 Remove the rubber sealing gasket between cowl and engine bay

6-22 CHASSIS ELECTRICAL

Fig. 47 Remove cowl air scoop for access to the wiper motor linkage

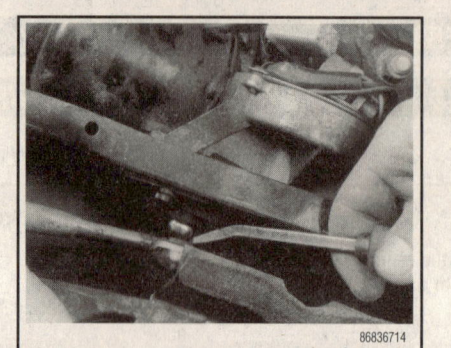

Fig. 48 After unbolting the linkage from the motor shaft, carefully pry off the linkage

Fig. 49 Close-up of the motor shaft flat

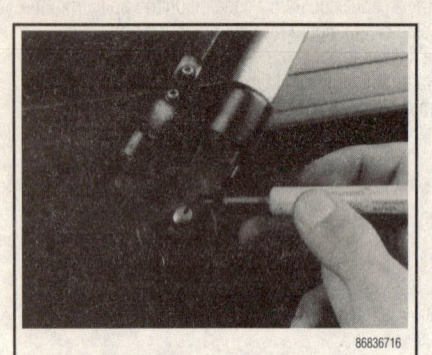

Fig. 50 Here the nut to the rear wiper arm has been removed and the pivot shaft is being marked for re-alignment later

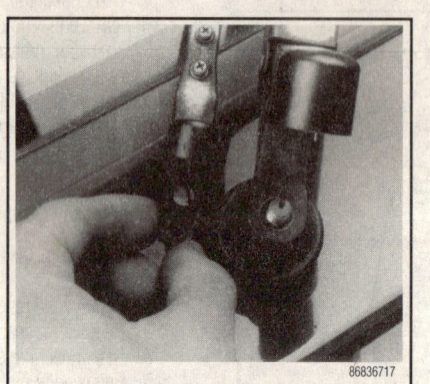

Fig. 51 Removing the clip to the secondary arm

Fig. 52 The rear wiper arm assembly is now free to remove to a safe place

Fig. 53 Check the dust/moisture cap for damage

Fig. 54 Unscrew the large nut from the motor shaft

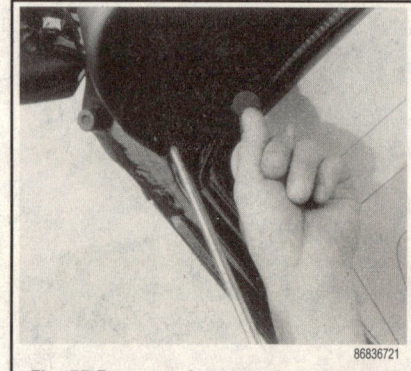

Fig. 55 Pry out and remove the trim caps that hide the fasteners holding the panel

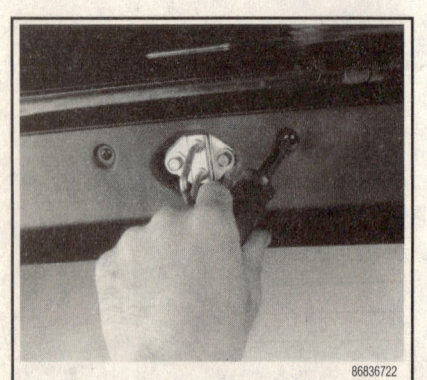

Fig. 56 Remove all screws and store all together in a safe place

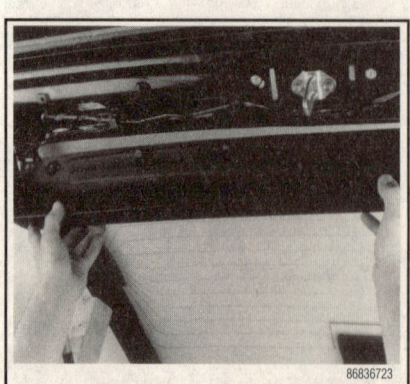

Fig. 57 Removing the panel to expose wiring and motor

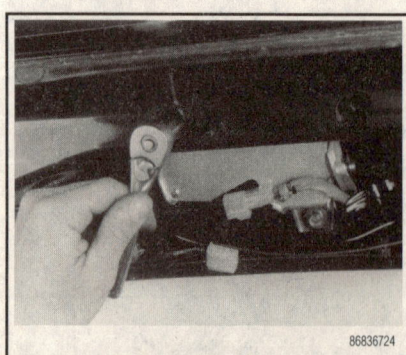

Fig. 58 A socket and ratchet will be needed to access the rear wiper motor retaining hardware inside the cutouts

CHASSIS ELECTRICAL   6-23

Fig. 59 The rear wiper motor electrical connector should be detached after unbolting it from the hatch

7. Tighten the mounting bolts to 7 ft. lbs. (10 Nm). Tighten the wiper arm nut to 10 ft. lbs. (14 Nm).

Washer Fluid Reservoir and Washer Pumps

REMOVAL & INSTALLATION

♦ See Figure 60

The washer fluid reservoir on 1984–87 models can be removed through the engine compartment. The washer fluid reservoir on 1988–95 models requires the removal of the front bumper cover. The reservoir for all models includes the front and rear washer motors.

1. Disconnect the negative battery cable.
2. On 1992–95 models, disconnect the receiver-dryer line, then remove the receiver-dryer.
3. On 1988–95 models, remove the front bumper cover.
4. Disconnect the electrical leads from the washer pump motors.
5. Disconnect one of the hoses and drain the washer fluid into a suitable container for reuse.

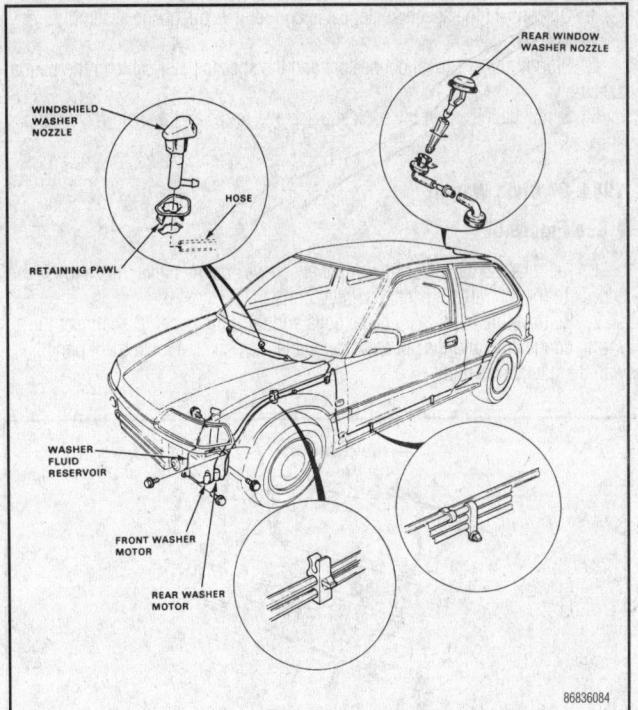

Fig. 60 Windshield washer system components. The front bumper needs to be removed for reservoir removal on 1988-95 models

6. Disconnect the other hose from the tank. Remove the tank retaining bolts (two on 1984–87 models and three on 1988–95 models).
7. On 1984–87 models, remove the tank from above, through the engine compartment. On 1988–95 models, remove the tank from beneath the car.
8. The washer pumps can now be removed if needed.

To install:
9. Install the pumps in the tank, if removed.
10. Install the tank into the vehicle and install the mounting bolts. Tighten the mounting bolts to 7 ft. lbs. (10 Nm).
11. Connect the electrical lead and the hoses.
12. Install the front bumper cover on late model vehicles and connect the negative battery cable.
13. Refill the reservoir and test the operation of the system.

INSTRUMENTS AND SWITCHES

Instrument Cluster

REMOVAL & INSTALLATION

1984–87 CRX

♦ See Figure 61

1. Remove the lower dash panel screws, clips and the panel.
2. Remove the lower heater control knob and the lower panel.
3. Remove the heater control mount screws and the upper screws from the instrument panel.
4. Pull the panel out and disconnect the wire connectors. Remove the instrument panel.
5. Remove the four screws, then lift out the gauge assembly and disconnect the wire connectors.
6. Disconnect the speedometer cable, then remove the gauge assembly.

To install:
7. Connect the speedometer cable, then connect the electrical connectors to the gauge assembly and screw in place.
8. Connect the panel, replace the panel, screw in place and reinsert the heater knob and clips.

1984–87 Civic Hatchback and Sedan

♦ See Figure 61

1. Remove the upper instrument panel caps, screws and the panel.
2. Remove the gauge assembly screws and lift out the assembly to disconnect the wire connectors.

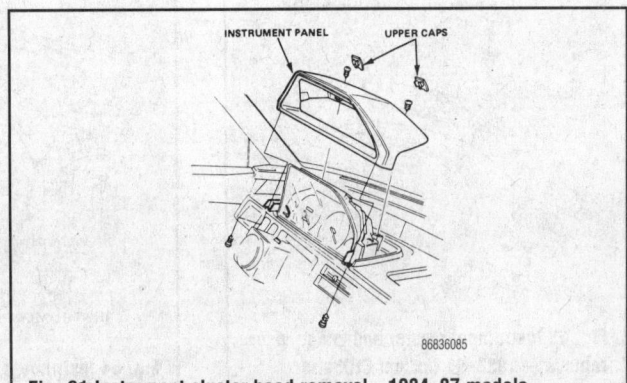

Fig. 61 Instrument cluster hood removal—1984–87 models

6-24 CHASSIS ELECTRICAL

3. Disconnect the speedometer cable and remove the gauge assembly.

To Install:

4. Plug in the electrical connectors and the speedometer cable to the gauge assembly.

5. Screw gauge assembly back in place and re-fasten the instrument panel above it.

1984–87 Civic Wagon

♦ See Figure 62

1. Remove the dashboard lower panel screws and the panel; this allows access to the instrument panel retaining bolts.

2. Remove the instrument panel bolts and raise the panel. Disconnect the wiring connectors and the speedometer cable. Remove the instrument panel with the gauge assembly.

3. Remove the gauge assembly screws and the assembly from the instrument panel.

To Install:

4. Plug in the electrical connectors and the speedometer cable to the gauge assembly.

5. Screw gauge assembly back in place and re-fasten the instrument panel above it.

1988–91 Civic Hatchback and Sedan

♦ See Figures 63, 64 and 65

1. Disconnect the negative battery cable.
2. Remove the two caps and screw from the top of the instrument housing.
3. Remove the four screws that retain the instrument cluster surround and remove the surround panel.
4. Remove the four screws that retain the instrument cluster and pull it forward. Disconnect the speedometer cable and the electrical connections.
5. Remove the instrument cluster from the dashboard.
6. Install the instrument cluster in position and connect the speedometer cable, and electrical leads. Make sure the speedometer cable seats fully.

1988–91 Civic Wagon and CRX

1. Disconnect the negative battery cable.
2. Remove the two covers and the screws from the top of the instrument cluster surround.
3. Remove the two screws from under the surround and pull it away from the dashboard. Disconnect the electrical leads from the switches.
4. Remove the instrument cluster retaining screws and pull the cluster forward. Disconnect the electrical leads and the speedometer cable.

To Install:

5. Put the instrument cluster in position and connect the speedometer cable, and electrical leads. Make sure the speedometer cable seats fully. Connect all switch leads before securing the surround panel.
6. Connect the negative battery cable.

1992–95 Models

※※ WARNING

All SRS wire harnesses are covered with a yellow insulation. Read the SRS precautions found in this section and install the short connectors as described.

1. Disconnect the negative battery cable and disable the SRS system.
2. Pry the hazard warning switch from the instrument panel taking care to not damage it.
3. Remove the three instrument panel screws and then the panel from the dashboard.
4. Disconnect the 4-pin connector from the instrument panel.

➡ Use a clean protective cloth to lay on steering column for the gauge assembly to rest upon.

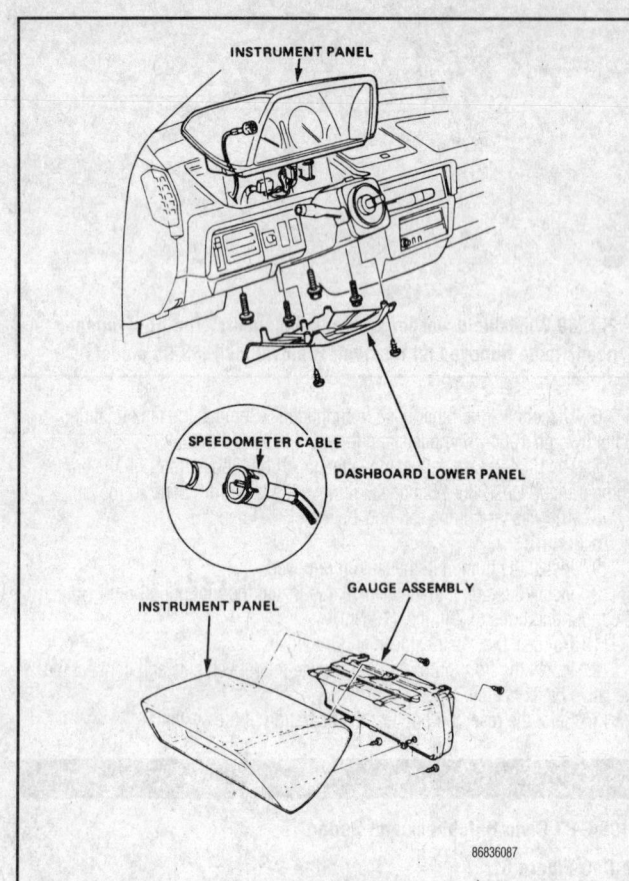

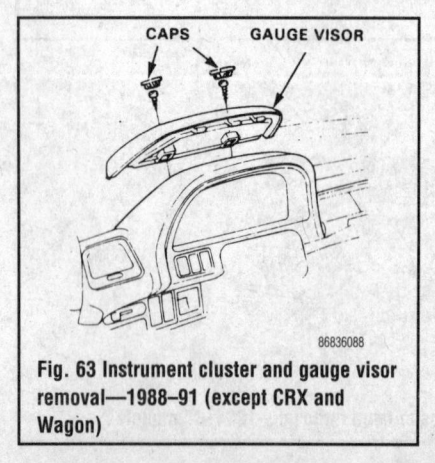

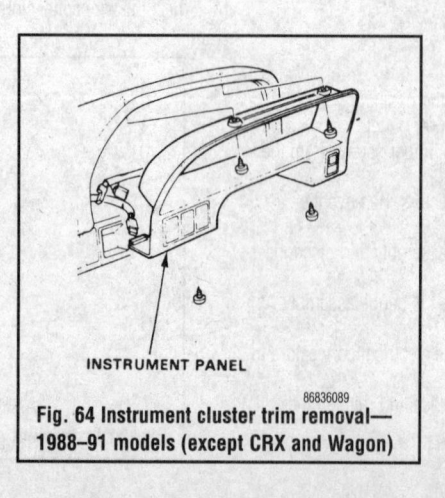

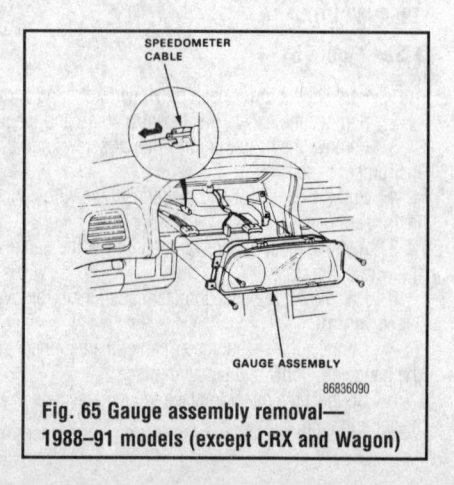

Fig. 62 Instrument cluster and gauge assembly removal—1984–87 Civic Wagon

Fig. 63 Instrument cluster and gauge visor removal—1988–91 (except CRX and Wagon)

Fig. 64 Instrument cluster trim removal—1988–91 models (except CRX and Wagon)

Fig. 65 Gauge assembly removal—1988–91 models (except CRX and Wagon)

CHASSIS ELECTRICAL 6-25

5. Remove the four gauge assembly screws and pull out the assembly.
6. Disconnect the gauge assembly and remove.

To install:

7. Place the gauge assembly in dashboard.
8. Connect the 4-pin connector to the dashboard.
9. Install the retaining screws.
10. Install the instrument panel.
11. Enable the SRS system.

Gauges

REMOVAL & INSTALLATION

♦ See Figure 66

The gauges are part of the instrument cluster assembly, each of the individual units is screwed to the cluster and connected by a printed circuit board and wires. The cluster housing can be separated to remove the light bar or replace the lens, it is clipped together by plastic tangs.

➡ If replacement is necessary for 1992–95 models (except del Sol) replace speedometer and tachometer as a unit. If replacement is necessary for 1993-95 del Sol models, replace speedometer and odometer/trip meter as a unit.

Avoid handling the printed circuit board, as it may be damaged by oils and acids from the skin.

❋❋ WARNING

All SRS wire harnesses are covered with a yellow insulation. Read the SRS precautions found in this section and install the short connectors as described.

1. Disconnect the negative battery cable.
2. Remove the instrument cluster as described earlier in this section.

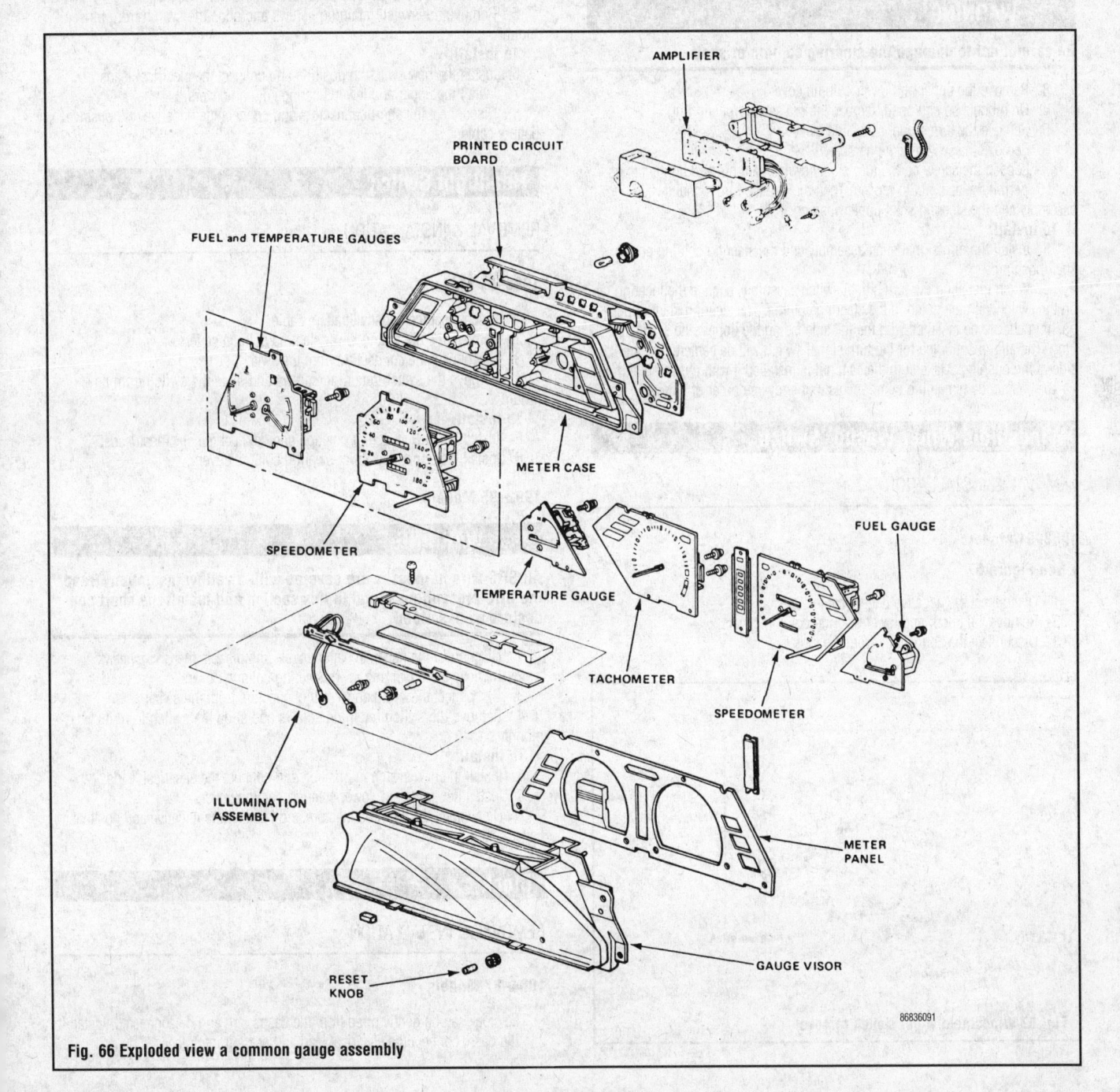

Fig. 66 Exploded view a common gauge assembly

6-26 CHASSIS ELECTRICAL

3. Locate the gauge to be removed and remove its mounting screws. Disconnect any wires leading to it.
4. Remove the gauge from the cluster assembly.
5. Install the gauge in position and install the retaining screws. Do not overtighten the mounting screws. Reconnect any disconnected leads.
6. Disconnect the short connectors and connect the negative and positive battery cable.

Combination Switch

REMOVAL & INSTALLATION

This switch housing contains the wiper, washer and headlight switches as well as the turn signal actuator. It is found only on 1984–87 models.
1. Remove the steering wheel.
2. Disconnect the column wiring harness and coupler.

※※ WARNING

Be careful not to damage the steering column or shaft.

3. Remove the upper and lower column covers.
4. On models so equipped, remove the cruise control slip ring.
5. Remove the turn signal canceling sleeve.
6. Remove the switch retaining screws.
7. Loosen the screw on the turn signal switch cam nut and lightly tap its head to permit the cam nut to loosen. Then remove the turn signal switch assembly and the steering shaft upper bushing.

To install:
8. Install the turn signal shaft assembly and the steering column upper shaft bushing.
9. When installing the turn signal switch assembly, engage the locating tab on the switch with the notch in the steering column. The steering shaft upper bushing should be installed with the flat side facing the upper side of the column. The alignment notch for the turn signal switch will be centered on the flat side of the bushing. Make sure the cam nut is installed it with the small end up.
10. Install the remaining components in the reverse order of removal.

Windshield Wiper/Washer Switch

REMOVAL & INSTALLATION

1988–91 Models

▶ See Figure 67

1. Disconnect the negative battery cable.
2. Remove the upper and lower steering column covers.
3. Disconnect the connectors from the switch.

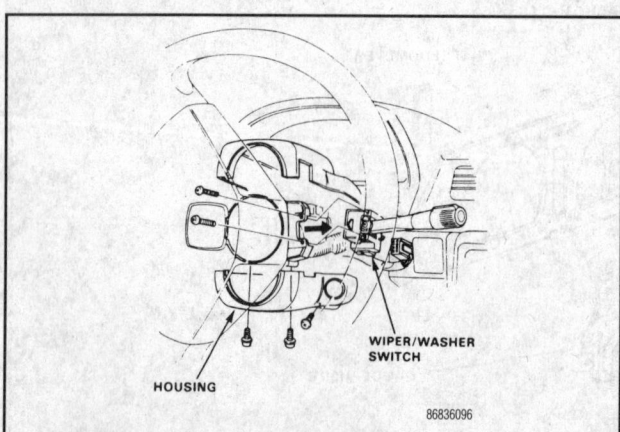

Fig. 67 Windshield wiper switch removal

4. Remove the switch retaining screws and slide the switch from its mount.

To install:
5. Install the new switch in position and connect the electrical leads.
6. Install the upper and lower steering column covers.

1992–95 Models

※※ WARNING

All SRS wire harnesses are covered with a yellow insulation. Read the SRS precautions found in this section and install the short connectors as described.

1. Disconnect the negative battery cable and disable the SRS system.
2. Remove the three screws that attach the lower dashboard cover.
3. Remove the upper and lower steering column covers.
4. Disconnect the 6-pin and 8-pin connectors from the switch.
5. Remove the switch retaining screws and slide the switch from its mount.

To install:
6. Install the new switch in position and connect the electrical leads.
7. Install the upper and lower steering column covers.
8. Disconnect the short connectors and connect the negative and positive battery cable.

Headlight Switch

REMOVAL & INSTALLATION

1988–91 Models

1. Disconnect the negative battery cable.
2. Remove the upper and lower steering column covers.
3. Disconnect the connectors from the switch.
4. Remove the switch retaining screws and slide the switch from its mount.

To install:
5. Install the new switch in position and connect the electrical leads.
6. Install the upper and lower steering column covers.

1992–95 Models

※※ WARNING

All SRS wire harnesses are covered with a yellow insulation. Read the SRS precautions found in this section and install the short connectors as described.

1. Disconnect the negative battery cable and disable the SRS system.
2. Remove the upper and lower steering column covers.
3. Disconnect the 4-pin and 7-pin connectors from the switch.
4. Remove the switch retaining screws and slide the switch from its mount.

To install:
5. Install the new switch in position and connect the electrical leads.
6. Install the upper and lower steering column covers.
7. Disconnect the short connectors and connect the negative and positive battery cable.

Clock

REMOVAL & INSTALLATION

1984–87 Models

The clock can be gently pried from the dashboard and disconnected. Install it into the dash, connect the electrical lead and press it into place.

CHASSIS ELECTRICAL 6-27

1988–91 Models

♦ See Figure 68

1. Disconnect the negative battery cable.
2. Remove the center duct assembly, by gently prying it out.
3. Disconnect the electrical lead from the clock.
4. Remove the two screws that retain the clock to the center duct outlets.

To install:

5. Install the clock to the center duct outlets and connect the electrical lead.
6. Insert the center duct into the dashboard and press it in firmly. Connect the negative battery cable.

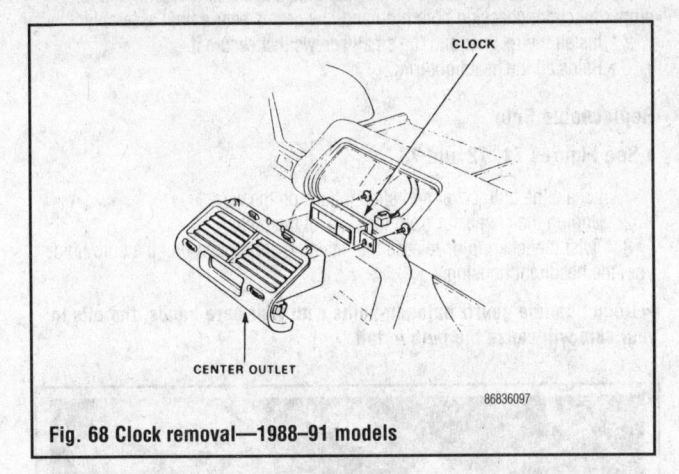

Fig. 68 Clock removal—1988–91 models

1992–95 Models

⚠ WARNING

All SRS wire harnesses are covered with a yellow insulation. Read the SRS precautions found in this section and install the short connectors as described.

1. Disconnect the negative battery cable and disable the SRS system.
2. Pry out the hazard warning switch from the instrument panel taking care to not damage it.
3. Remove the screws from the instrument panel and then remove it.
4. Disconnect the 4-pin electrical connector from the panel.
5. Remove the screws that retain the clock to the instrument panel and remove.

To install:

6. Install the clock to the instrument panel and connect the electrical lead.
7. Install the instrument panel and the hazard warning switch.
8. Disconnect the short connectors and connect the negative and positive battery cable.

Rear Window Defogger System

The rear window defogger system consists of a rear window with two vertical bus bars and a series of electrically connected grid lines baked on the inside surface. A control switch and a relay are also used in this system.

➡ Since the grid lines can be damaged or scraped off with sharp instruments, caution should be used when cleaning the glass or removing foreign materials, decals or stickers. Normal glass cleaning solvents or hot water used with rags or toweling is recommended.

REMOVAL & INSTALLATION

Rear Window Defogger Switch

⚠ WARNING

All SRS wire harnesses are covered with a yellow insulation. Read the SRS precautions found in this section and install the short connectors as described.

1. Disconnect the negative battery cable and remove the lower dashboard panel (except for the wagon models).
2. Disconnect the wire harness from behind the console. Depress the switch locking pawls and remove the switch from the front of the dash panel.
3. Installation is the reverse order of the removal procedure.

COMPONENT TESTING

Rear Defroster Relay

1984–91 MODELS

♦ See Figure 69

1. Check for continuity between top left A terminal and the bottom left B terminal, when applying battery voltage to upper right C terminal and grounding the lower right D terminals.
2. Once the voltage is removed from the C and D terminals, there should be no continuity in the relay. If the relay fails this test, replace it. The relays are located under the dash on the left side.

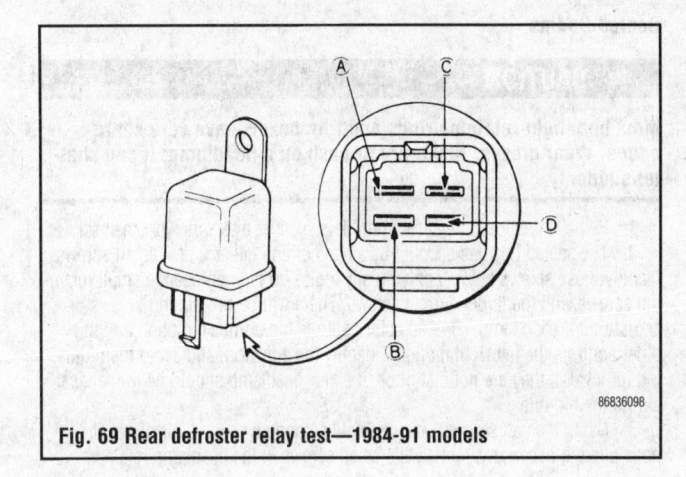

Fig. 69 Rear defroster relay test—1984-91 models

1992–95 MODELS

♦ See Figure 70

1. Check for continuity between top right C terminal and the bottom right D terminal. There should be continuity between the upper left A and B terminals when power and ground are connected to the C and D terminals.
2. Once the voltage is removed from the C and D terminals, there should be no continuity in the relay. If the relay fails this test, replace it. The relay is located in the under-hood relay box.

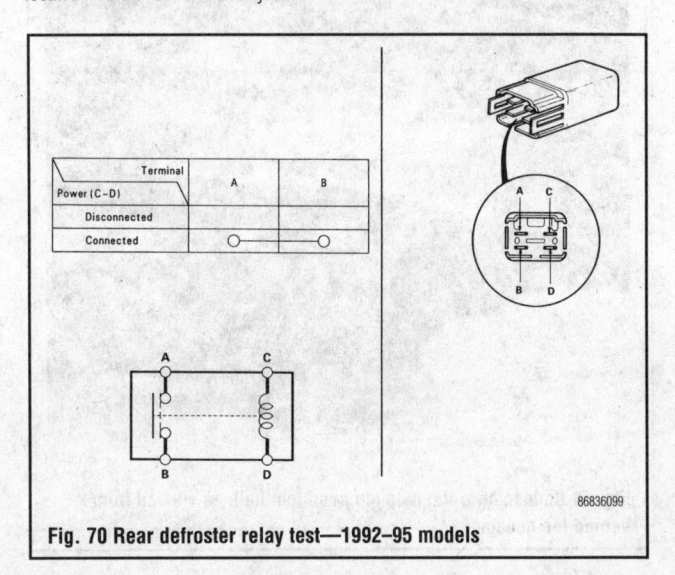

Fig. 70 Rear defroster relay test—1992–95 models

6-28 CHASSIS ELECTRICAL

Rear Window Defogger Grid

1. Turn **ON** the rear window defroster switch. Connect the positive lead of a voltmeter to the center of each filament and connect the negative lead to the body of the vehicle.
2. The standard voltage at the center of the filament is 6 volts. If the meter is higher than 6 volts, the problem exists in the ground side of the filament.
3. If the meter indication is low or 0, the problem is between the center and the power side. Isolate the problem grid line and mark the break in the grid wire.
4. If necessary, repair the grid wire. Kits are usually available from parts stores or your Honda dealer.

LIGHTING

Headlights

There are two general styles of headlamps, the sealed beam and replaceable bulb type. The sealed beam is by far the most common and includes almost all of the circular and rectangular lamps found on vehicles built through the 1980s. The sealed beam is so named because it includes the lamp (filament), the reflector and the lens in one sealed unit. Sealed beams are available in several sizes and shapes.

REMOVAL & INSTALLATION

Sealed Beams

※※ CAUTION

Most headlight retaining rings and trim bezels have very sharp edges. Wear gloves. Never pry or push on a headlamp; it can shatter suddenly.

1. Remove the headlight bezel (trim.) It is secured by a series of small screws.
2. The sealed beam is secured by a retainer and either 2 or 4 small screws. Identify these screws before applying any tools. Do not confuse the small retaining screws with the larger aiming screws. There will be two aiming screws or adjustors for each lamp. (One adjustor controls the up/down motion and the other controls the left/right motion.) Identify the adjustors and avoid them during removal. If they are not disturbed, the new headlamp should be in identical aim to the old one.
3. Using a small screwdriver (preferably magnetic) and a pair of needle-nose pliers if necessary, remove the small screws in the headlamp retainer.

➡ **A good kitchen or household magnet placed on the shank of the screwdriver will provide enough grip to hold the screw during removal.**

4. Remove the retainer and the headlamp may be gently pulled free from its mounts. Detach the connector from the back of the sealed beam unit and remove the unit from the car.

To install:

5. Place the new headlamp in position and connect the wiring harness.

Remember to install the rubber boot on the back of the new lamp. Also, make sure the headlight is right-side up.

6. Have an assistant turn on the headlights and check the new lamp for proper function, checking both high and low beams before final assembly.
7. Install the retainer and the small screws that secure it.
8. Reinstall the headlight trim.

Replaceable Bulb

▶ See Figures 71, 72 and 73

1. Locate the bulb cover and slide it back on the wire.
2. Unplug the connector from the back of the bulb.
3. Twist the lockring or release the retaining spring and pull the bulb holder from the headlight housing.

➡ **Do not handle quartz halogen bulbs with your bare hands, the oils in your skin will cause the bulb to fail.**

Fig. 72 Removing the halogen headlight bulb from the back of the headlight housing

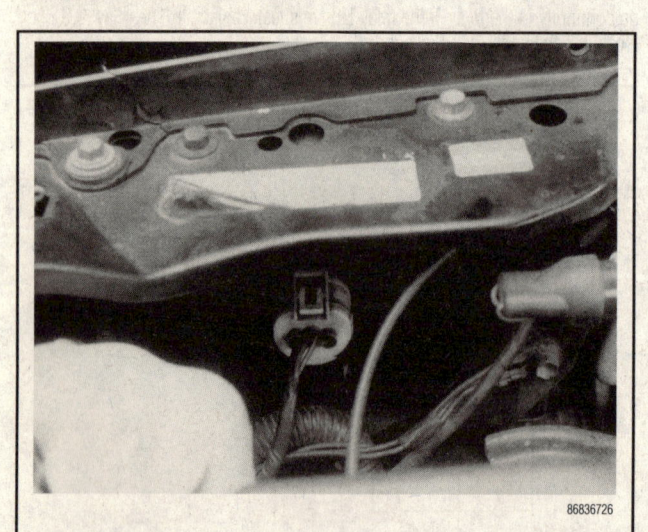

Fig. 71 Connector to the halogen headlight bulb as viewed from behind the housing

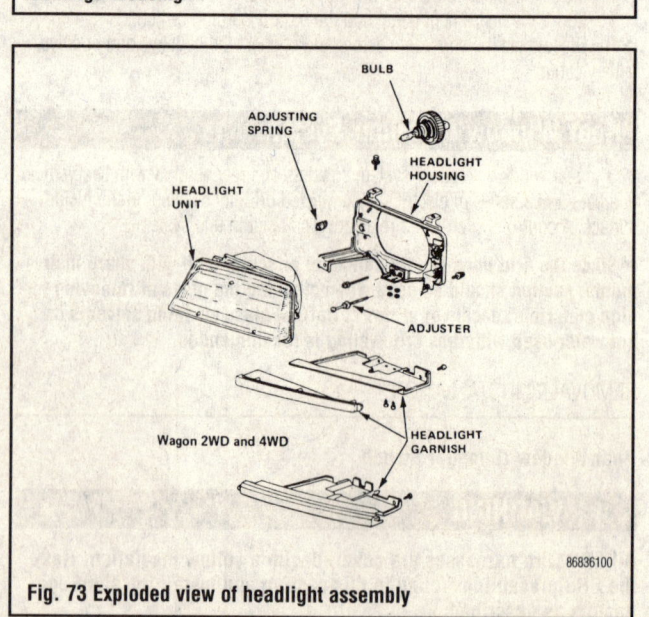

Fig. 73 Exploded view of headlight assembly

CHASSIS ELECTRICAL 6-29

4. Remove the bulb from the holder.
To Install:
5. Insert the replacement bulb.
6. Reinsert the holder into the housing and turn the lockring. Reconnect the electrical lead. Slide the rubber boot back into position.

Tail Lights, Reverse Lights, Rear Marker and Signal Lights

REMOVAL & INSTALLATION

▶ See Figures 74 thru 81

All of these lights can be replaced by removing the access covers in the trunk/hatch area of the vehicle and twisting the bulb from its socket. The housing assemblies are retained by nuts that can also be removed through the access panels.

Front Turn Signal, Side Marker and License Plate Lights

REMOVAL & INSTALLATION

▶ See Figures 82 and 83

These can be removed by removing the screws that retain the housing and then twisting the bulbs from the sockets. When replacing theses housings, make sure that the sealing gaskets are properly seated.

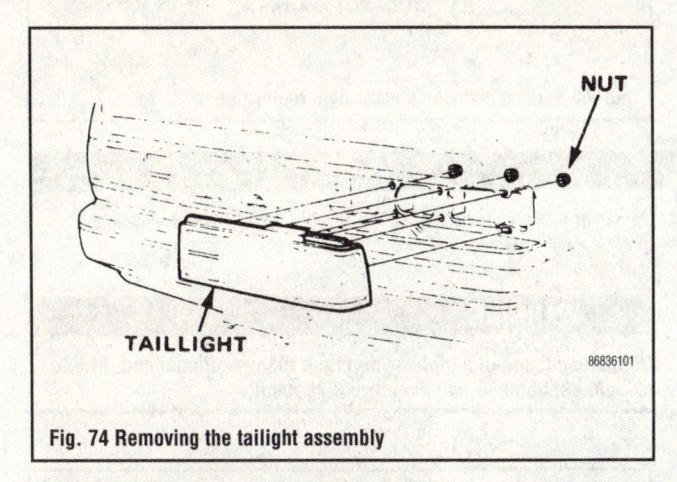

Fig. 74 Removing the tailight assembly

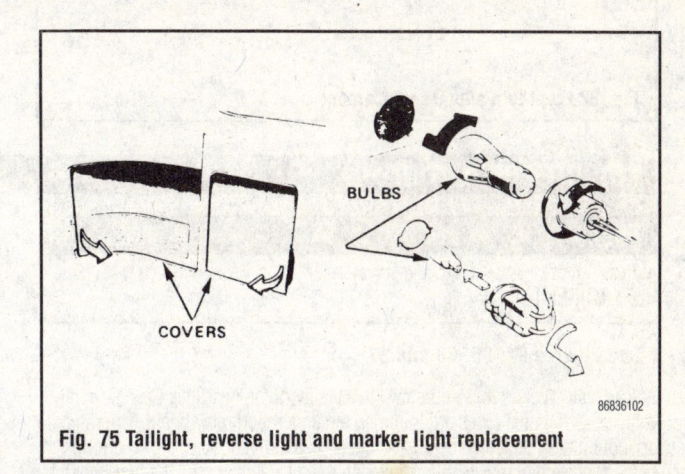

Fig. 75 Tailight, reverse light and marker light replacement

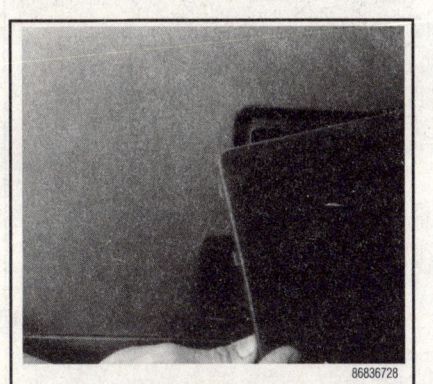

Fig. 76 The side marker lights are being accessed through the panel

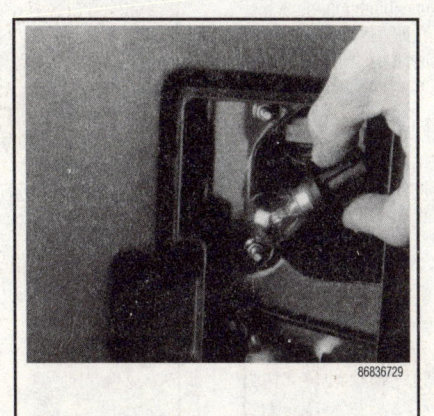

Fig. 77 Removing the bulb holder

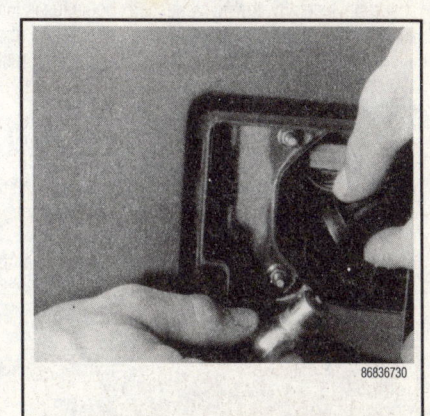

Fig. 78 Removing the bulb from the holder

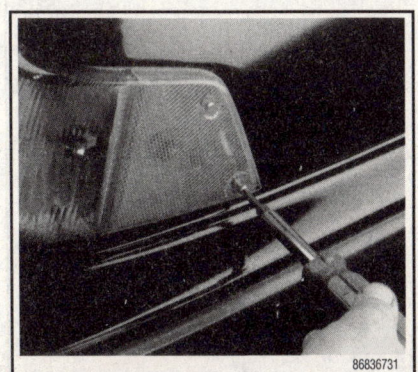

Fig. 79 Unscrewing the front turn signal lens

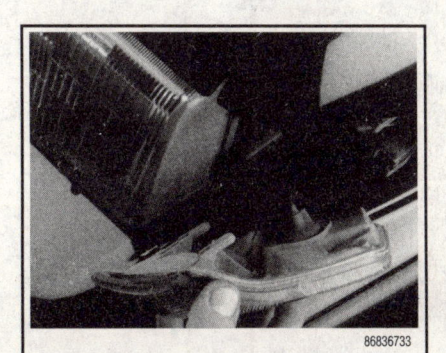

Fig. 80 Pull away the lens assembly to expose the bulb protective cover and wiring

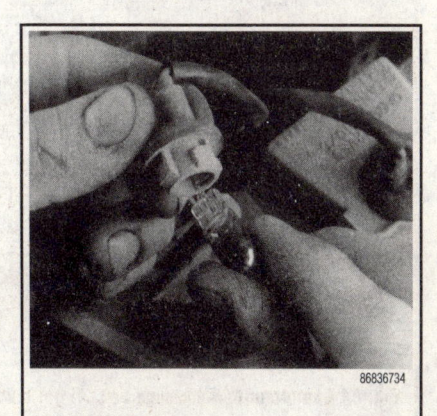

Fig. 81 Pull the bulb out for replacement

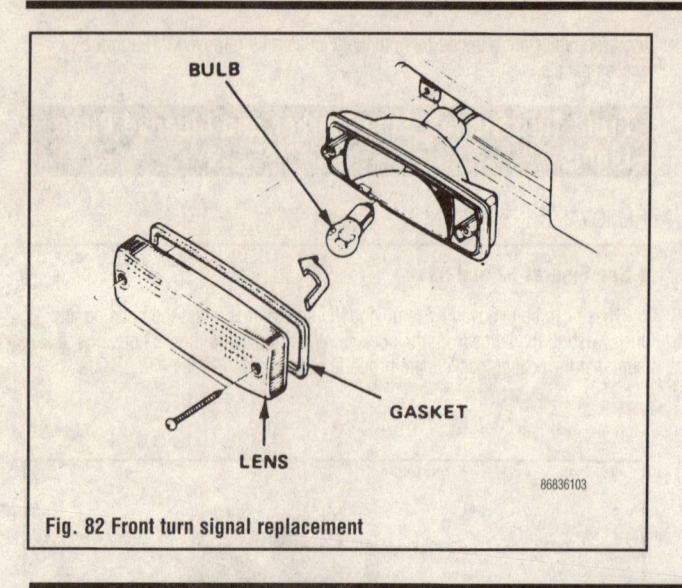

6-30 CHASSIS ELECTRICAL

Fig. 82 Front turn signal replacement

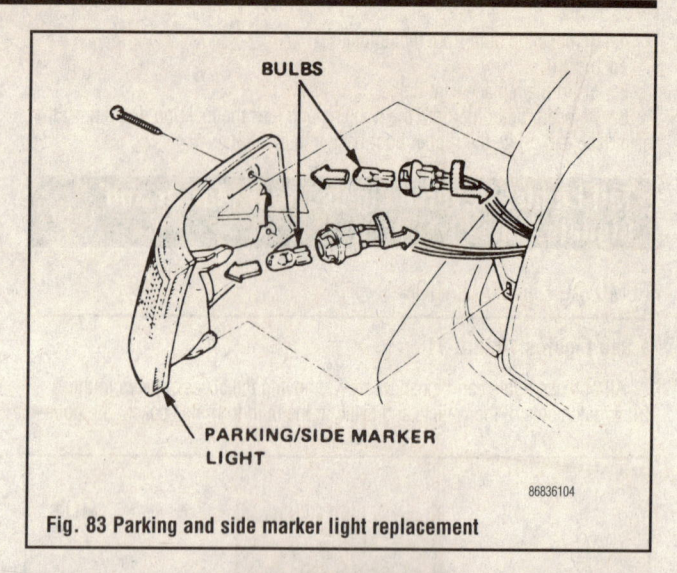

Fig. 83 Parking and side marker light replacement

CIRCUIT PROTECTION

Fuses

REPLACMENT

♦ See Figures 84, 85, 86 and 87

The fuse block is located below the left side of the instrument panel on all vehicles. Additional fuses are found on the underhood relay board. The radio or audio unit is protected by an additional fuse in the body of the unit. In the event that anything electrical isn't working, the fuse should be the first item checked. Generally, the in-car fusebox holds fuses for the cabin circuits such as wipers, cigarette lighter, rear defogger, etc. The underhood fusebox generally contains fuses and relays for engine and major electrical systems such as air conditioning, headlights (left and right separately), fuel injection components, etc.

Each fuse location is labeled on the fuseblock, identifying its primary circuit, but designations such as BACKUP, CCON or IG 1 may not tell you what you need to know. A fuse can control more than one circuit, so check related circuits as well. This sharing of fuses is necessary to conserve space and wiring.

The underdash or underhood fusebox contains a fuse puller which can be used to grip and remove the fuse. The fuse cannot be checked while in the fuseblock; it must be removed. View the fuse from the side, looking for a broken element in the center.

✱✱ WARNING

Do not use fuses of a higher amperage than recommended. Major component damage and even fire may result.

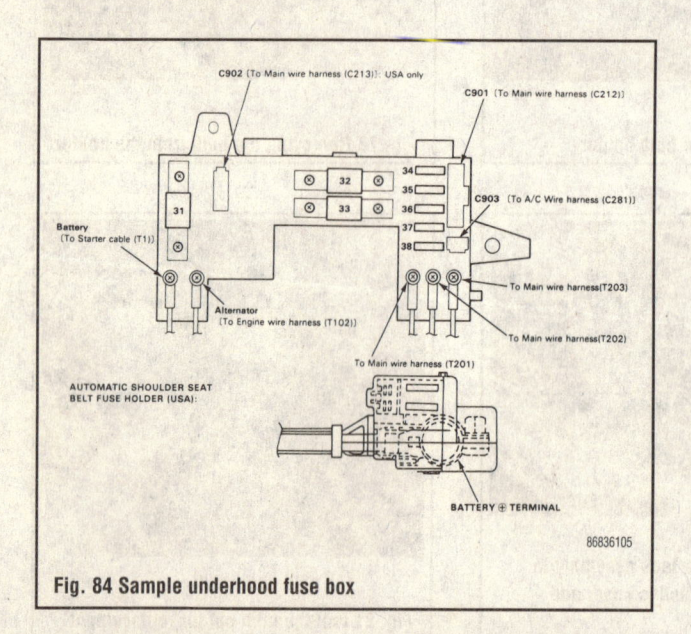

Fig. 84 Sample underhood fuse box

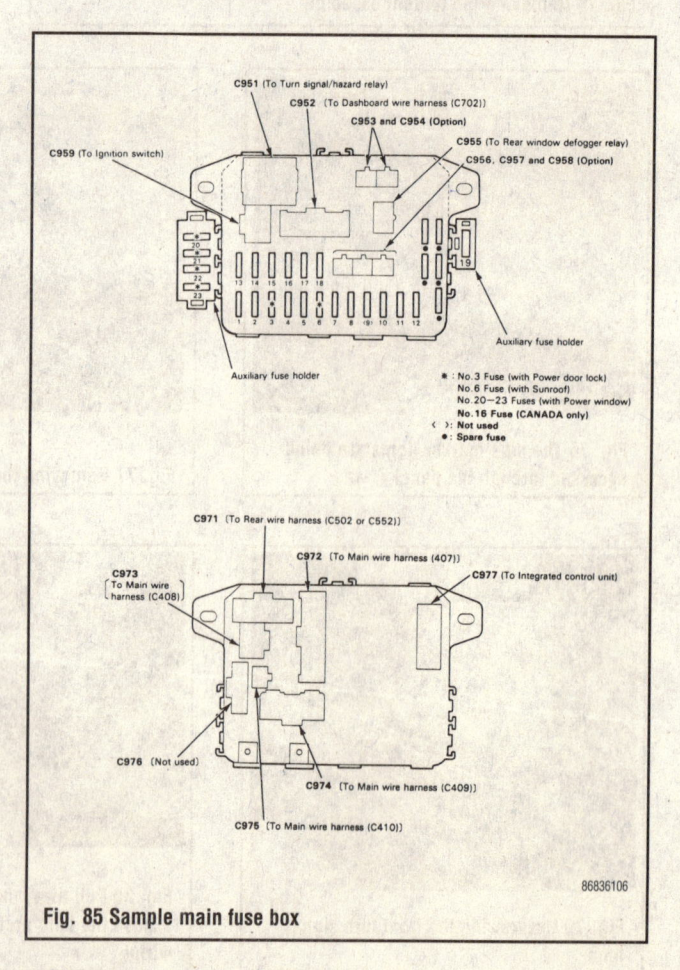

Fig. 85 Sample main fuse box

CHASSIS ELECTRICAL 6-31

Fusible Links

REPLACEMENT

The fusible links serve as main circuit protection for certain high amperage circuits. The Honda fusible or main links are located at the in the under-hood fuse and relay box. On earlier models, these are large strips of conductive metal held by bolts between two terminals. The thickness and length of the material determines the amperage rating of the link. To replace one of these, disconnect the negative battery cable, unscrew the failed one and install the new one.

On later cars, the main links appear about the same size as the fuses; they are square cases with a small window on top. Close examination will reveal the amperage rating, usually 50 or 80 amps. Remove these links as you would a fuse after loosening their retaining screws.

If the link failed, it was due to a high amperage load passing through the circuit; find the cause or the new link will fail again.

Relays

REPLACEMENT

As vehicles rely more and more on electronic systems and electrically operated options, the number of relays grows steadily. Many relays are located in logical positions on the relay and fuse board under the dash or in the engine compartment. However, some relays may be located near the component they control.

Relays can be removed by pulling them from their socket.

Fig. 86 The fuse box in the car protects many components and includes a fuse puller tool and spare fuse from the factory

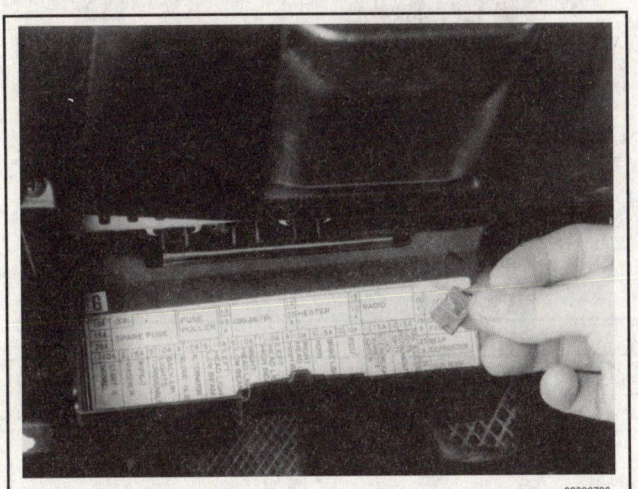

Fig. 87 Fuses are color coded according to amperage rating. Be sure to replace with the same type

6-32 CHASSIS ELECTRICAL

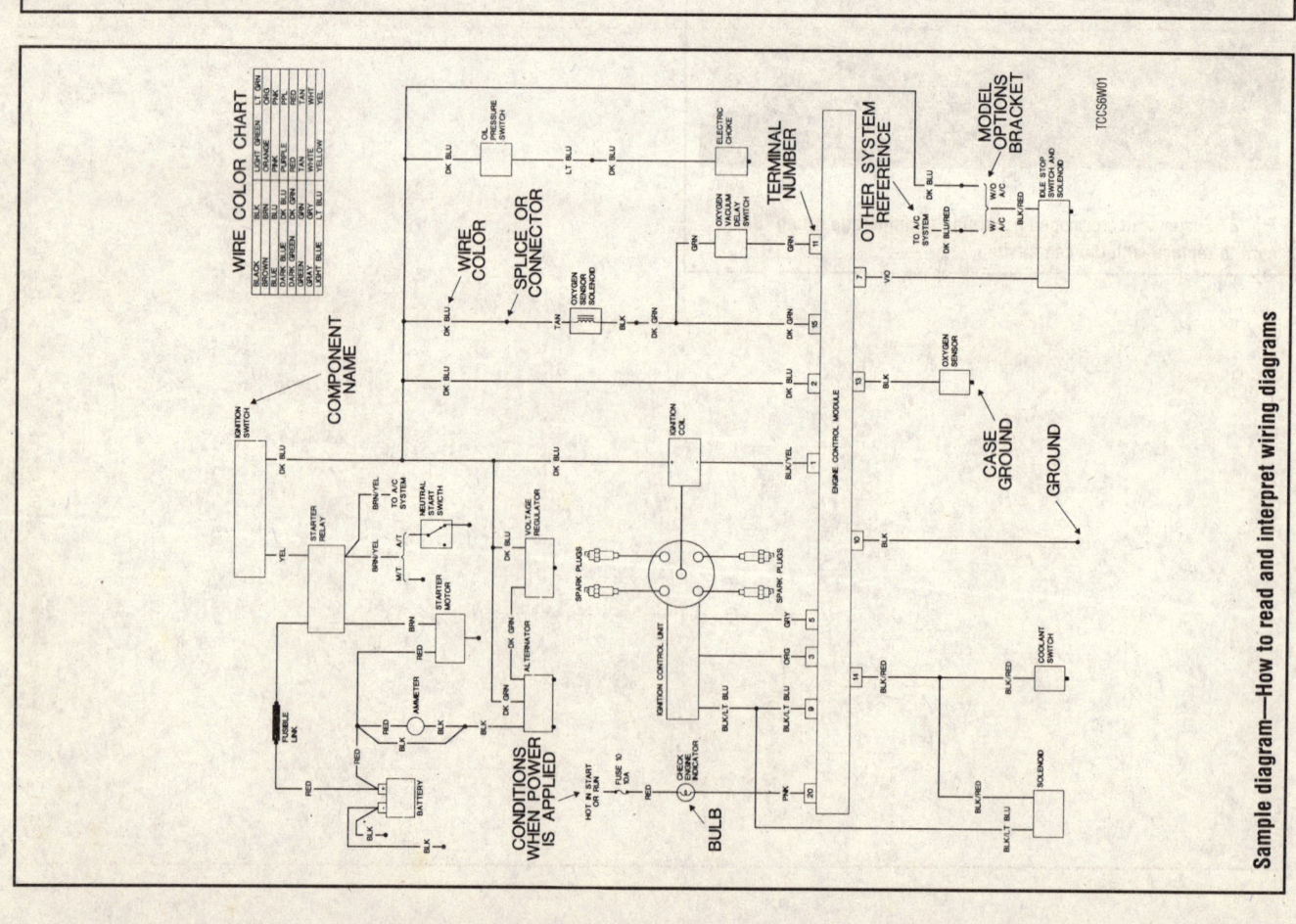

WIRING DIAGRAMS

Wiring diagram symbols

Sample diagram—How to read and interpret wiring diagrams

CHASSIS ELECTRICAL 6-33

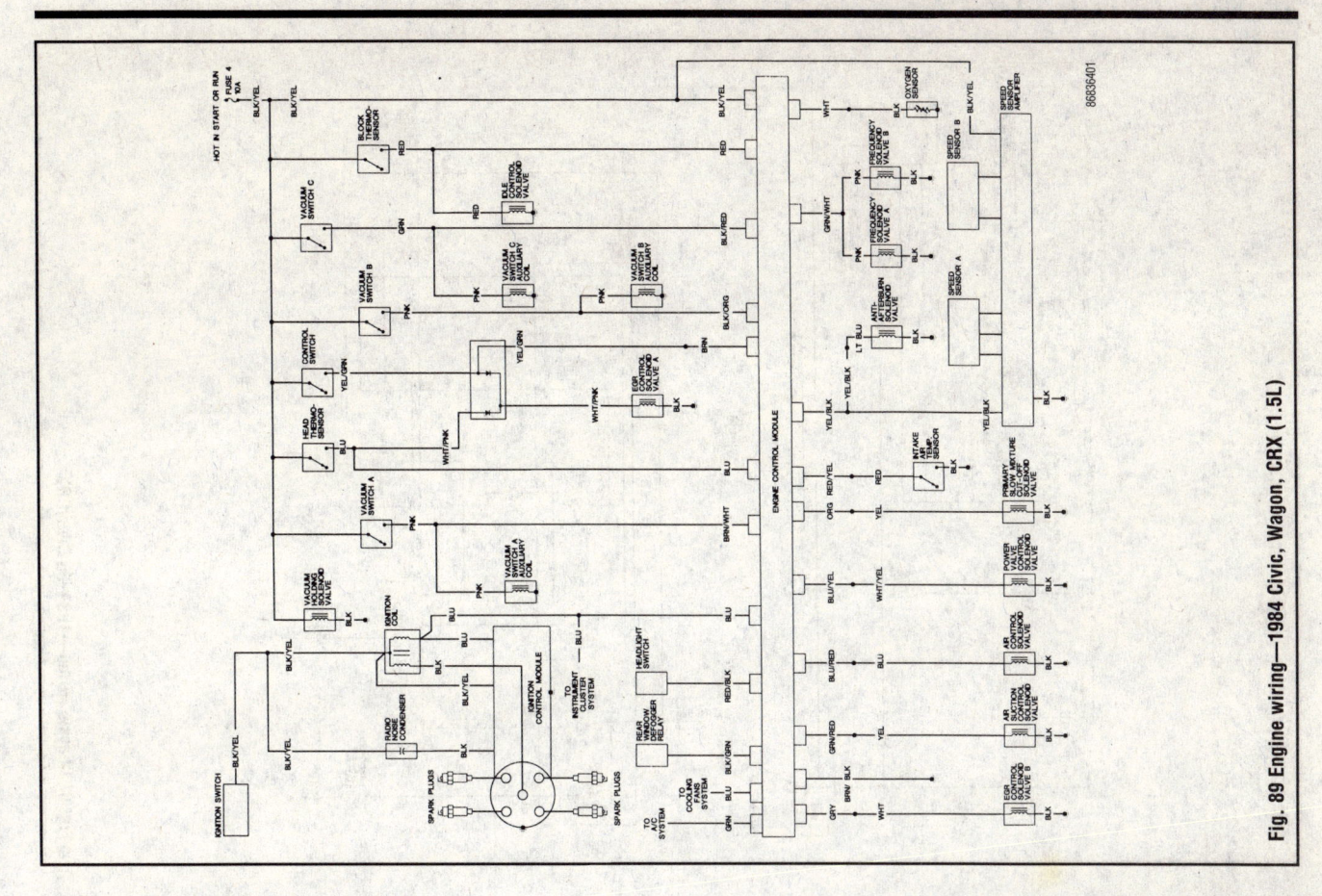

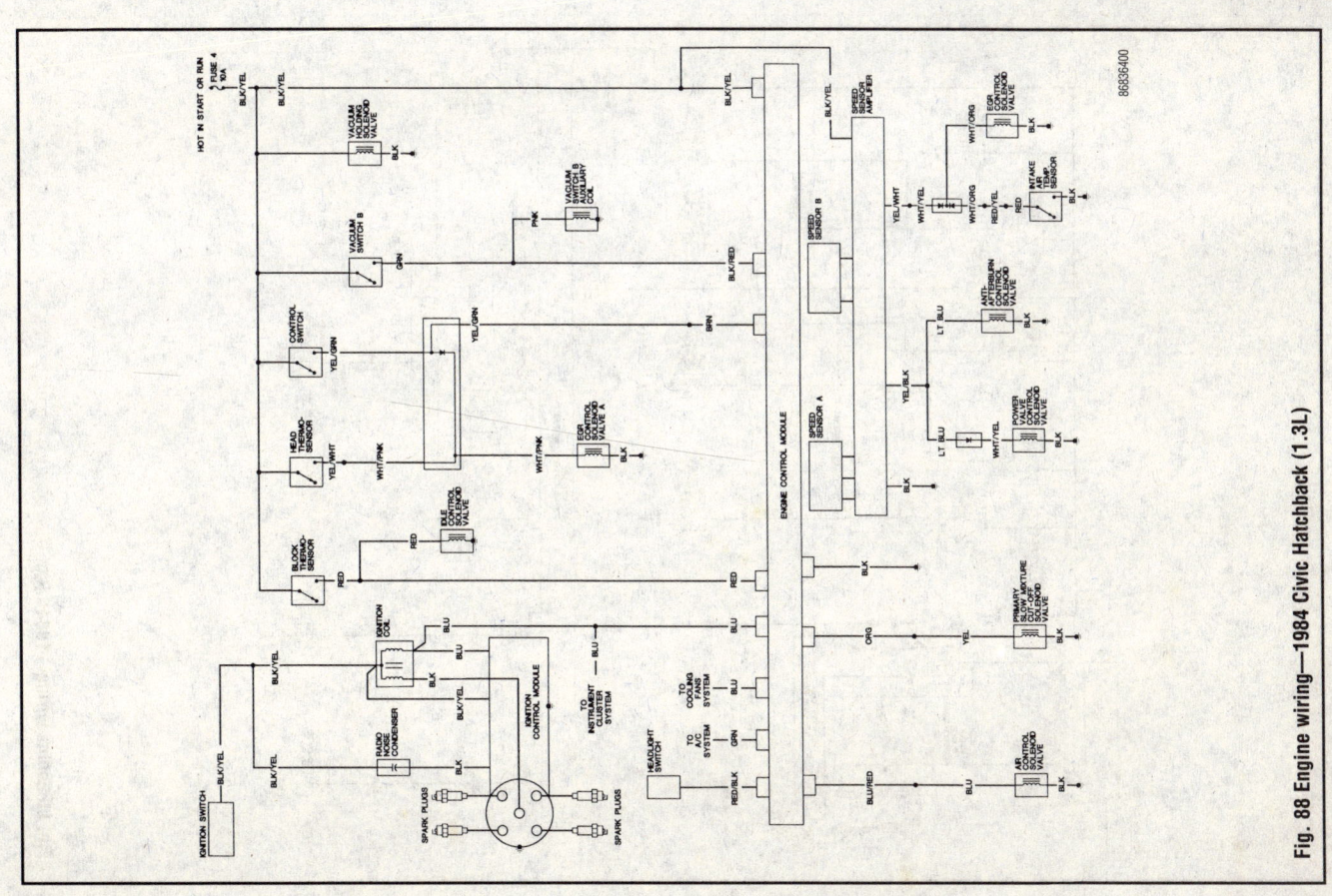

Fig. 89 Engine wiring—1984 Civic, Wagon, CRX (1.5L)

Fig. 88 Engine wiring—1984 Civic Hatchback (1.3L)

6-34 CHASSIS ELECTRICAL

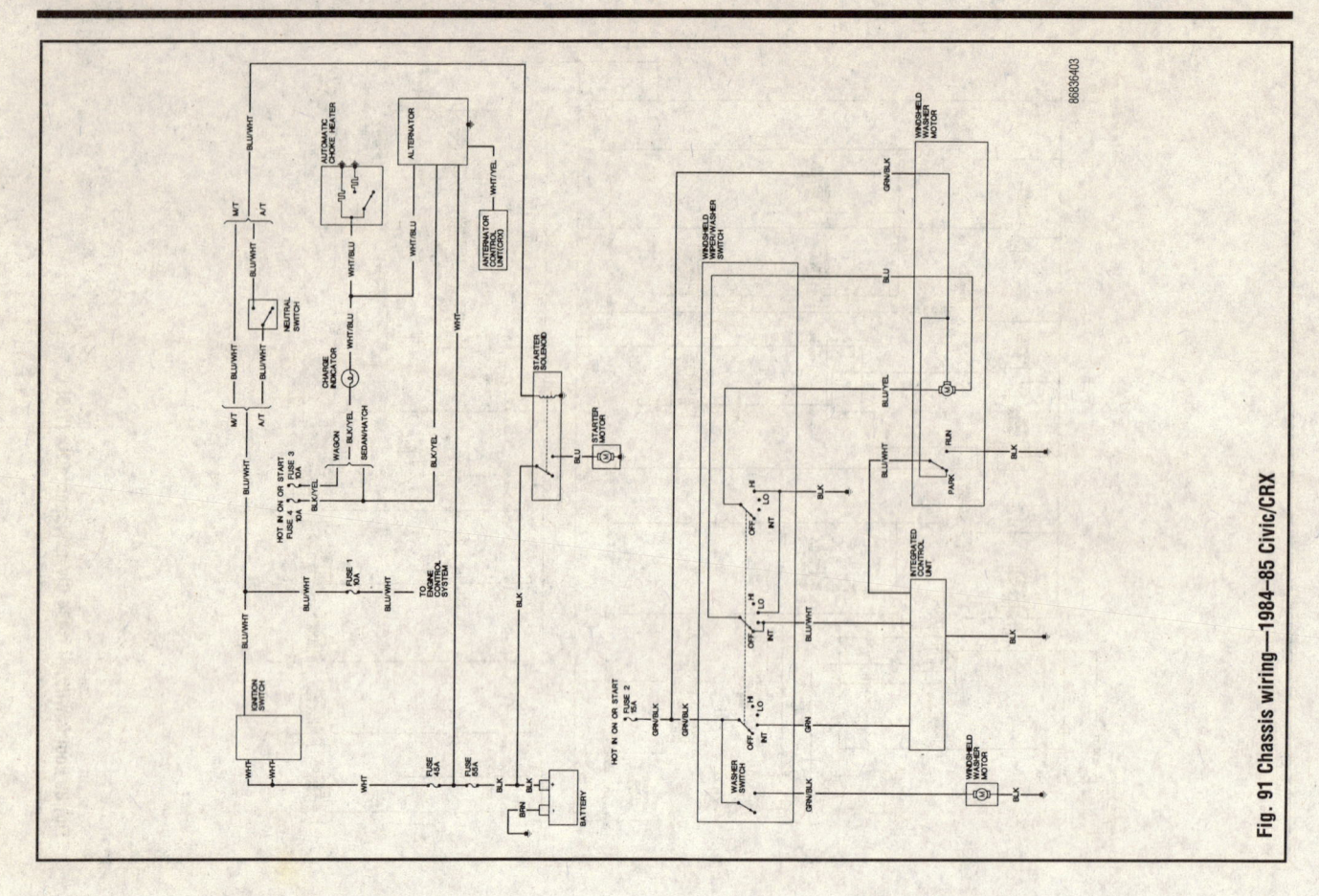

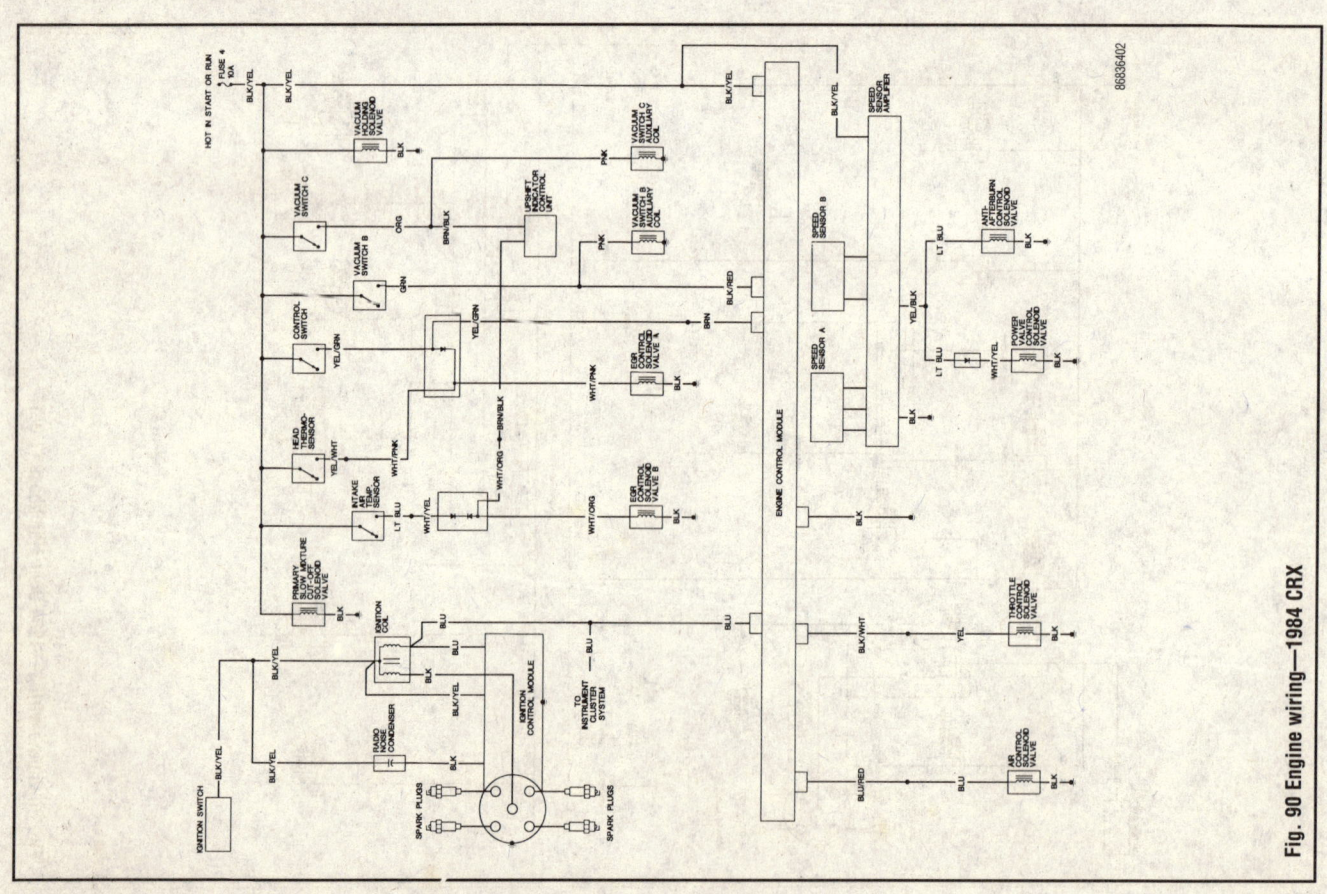

Fig. 91 Chassis wiring—1984-85 Civic/CRX

Fig. 90 Engine wiring—1984 CRX

CHASSIS ELECTRICAL 6-35

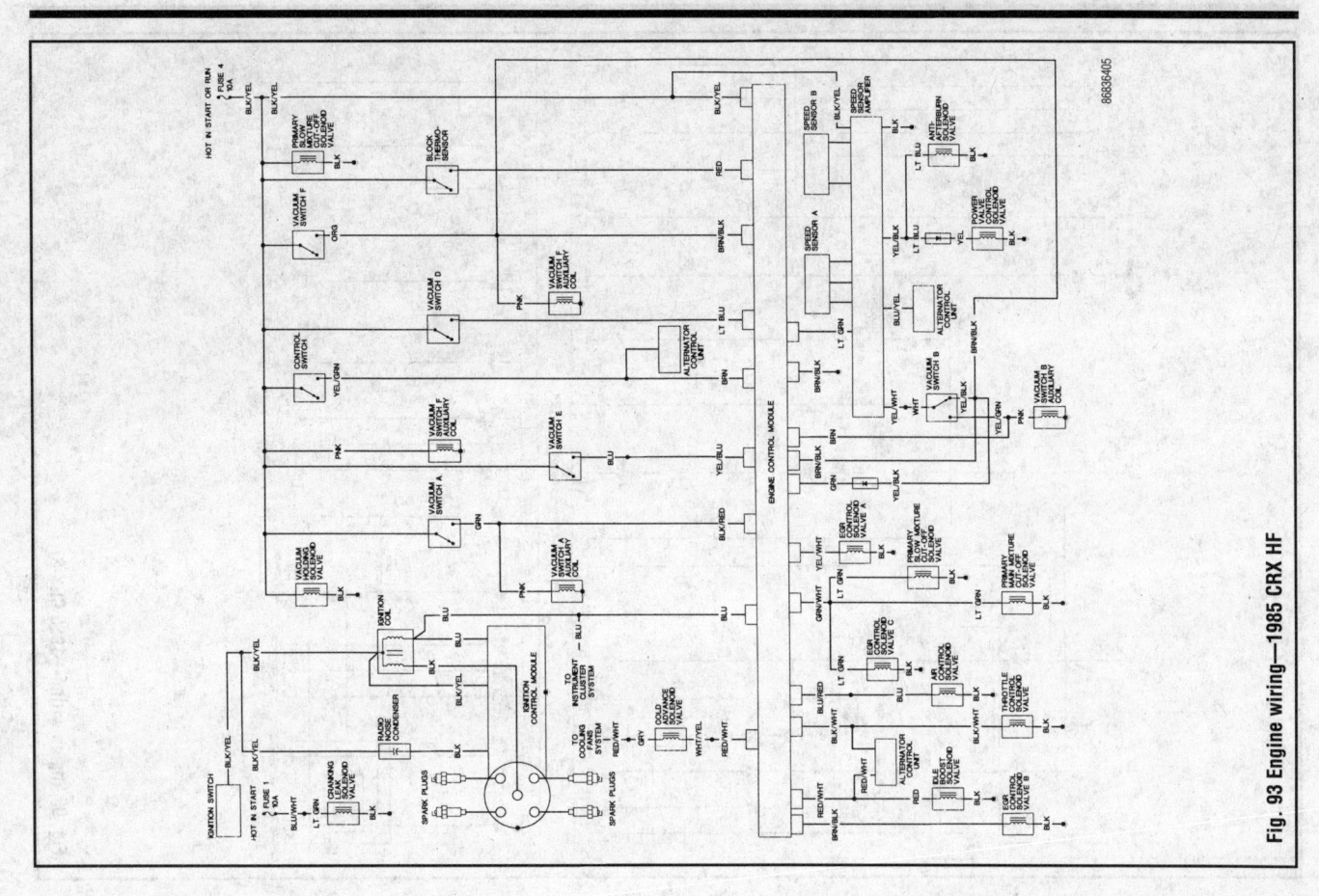

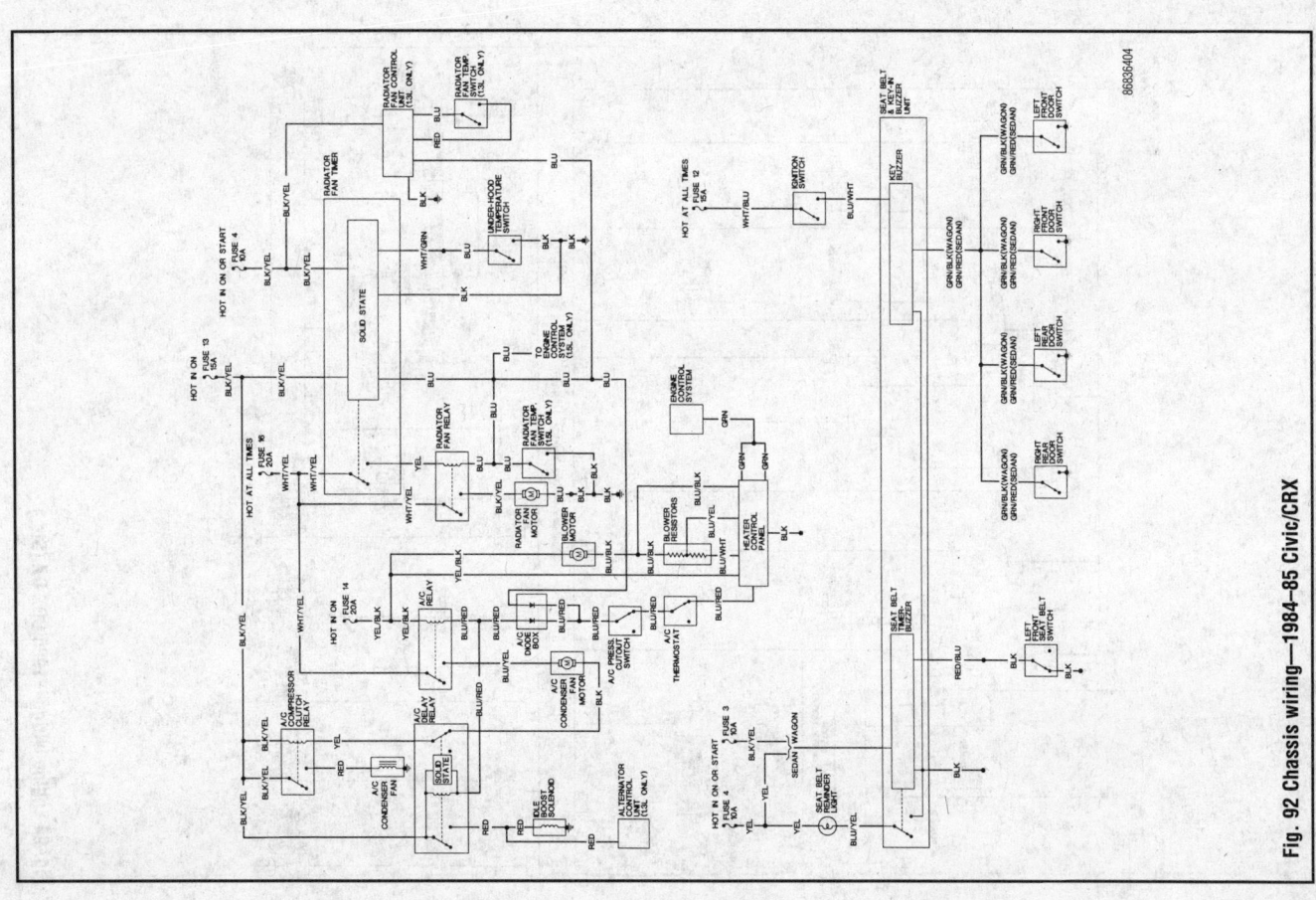

Fig. 93 Engine wiring—1985 CRX HF

Fig. 92 Chassis wiring—1984-85 Civic/CRX

6-36 CHASSIS ELECTRICAL

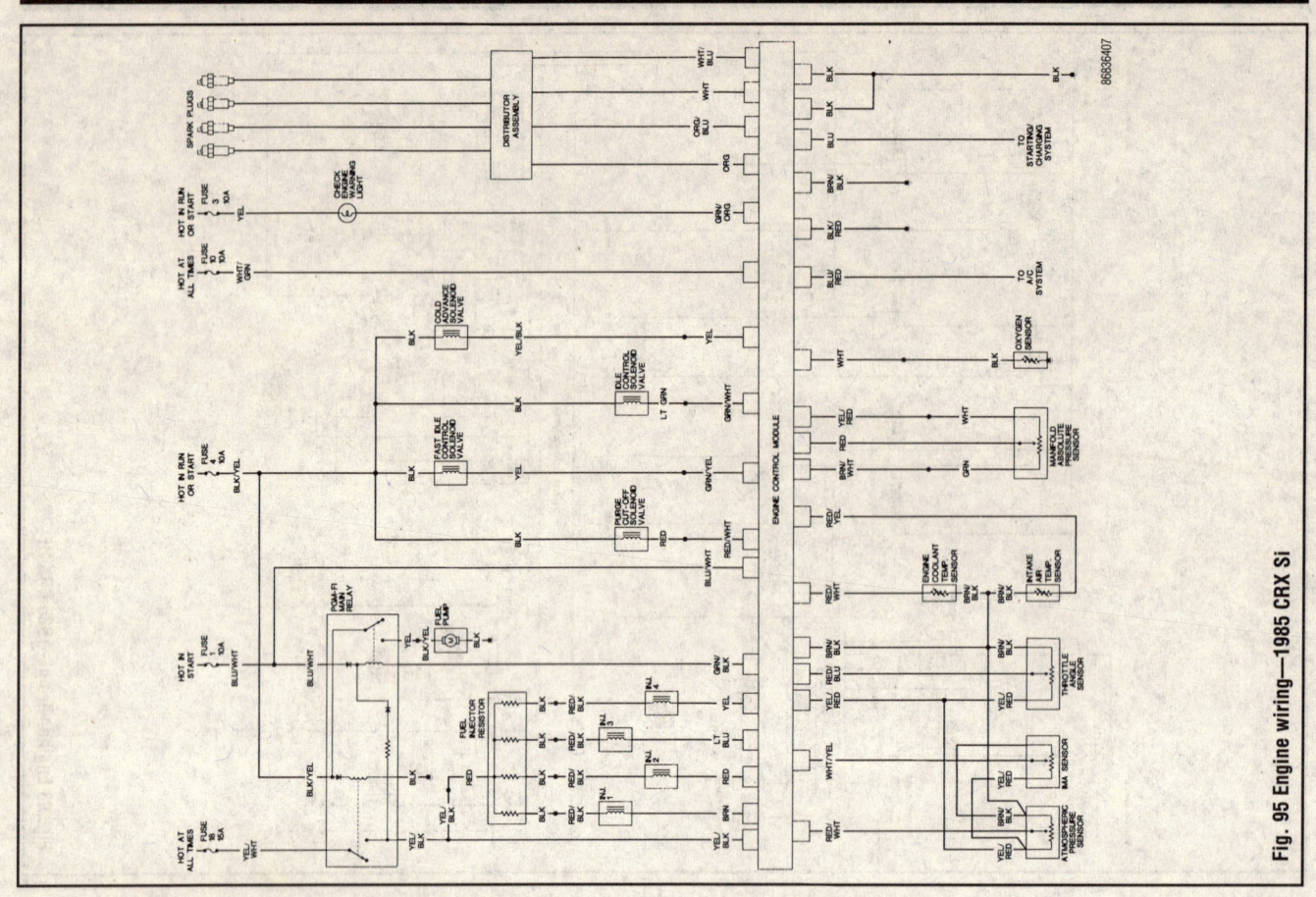

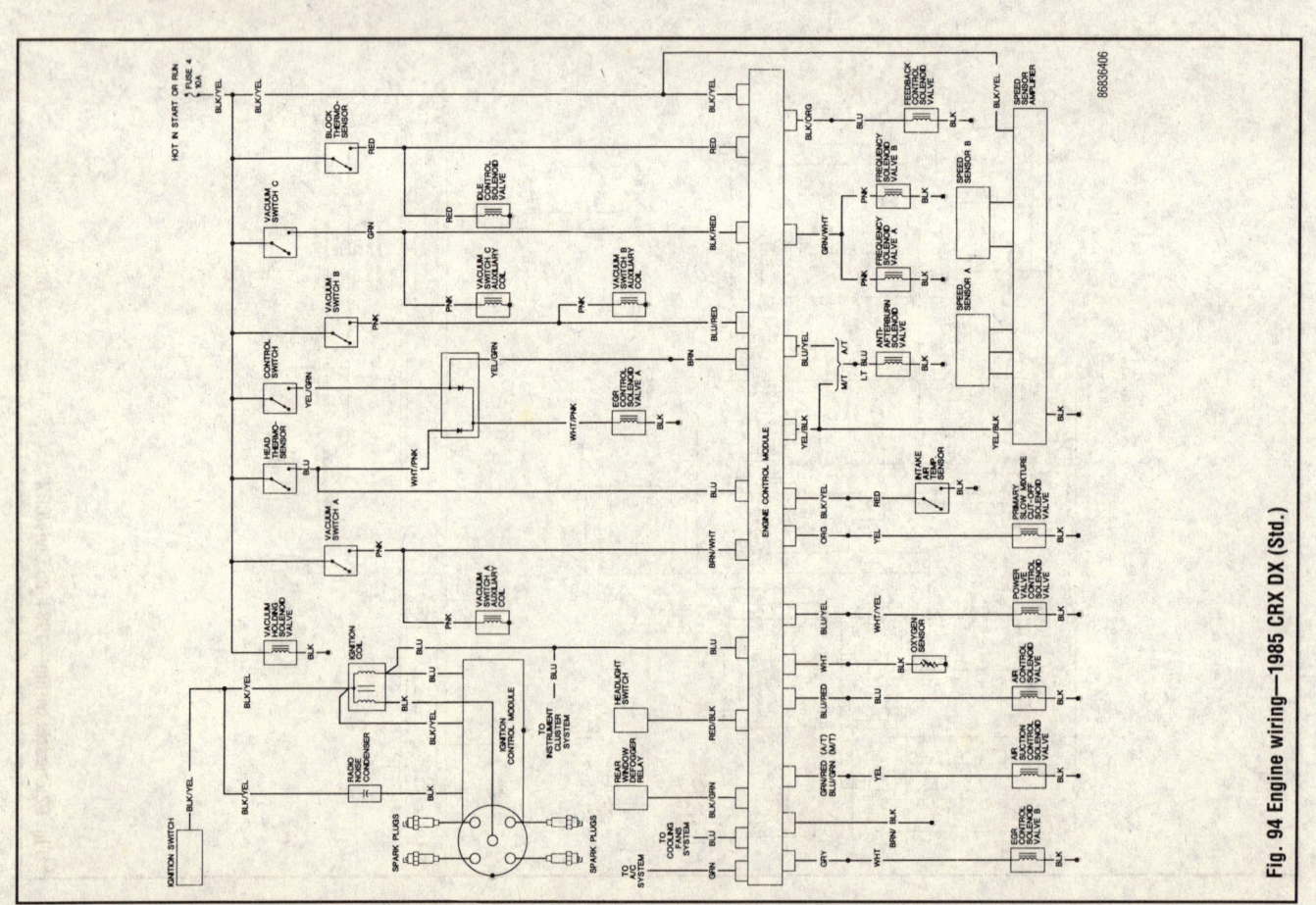

Fig. 95 Engine wiring—1985 CRX Si

Fig. 94 Engine wiring—1985 CRX DX (Std.)

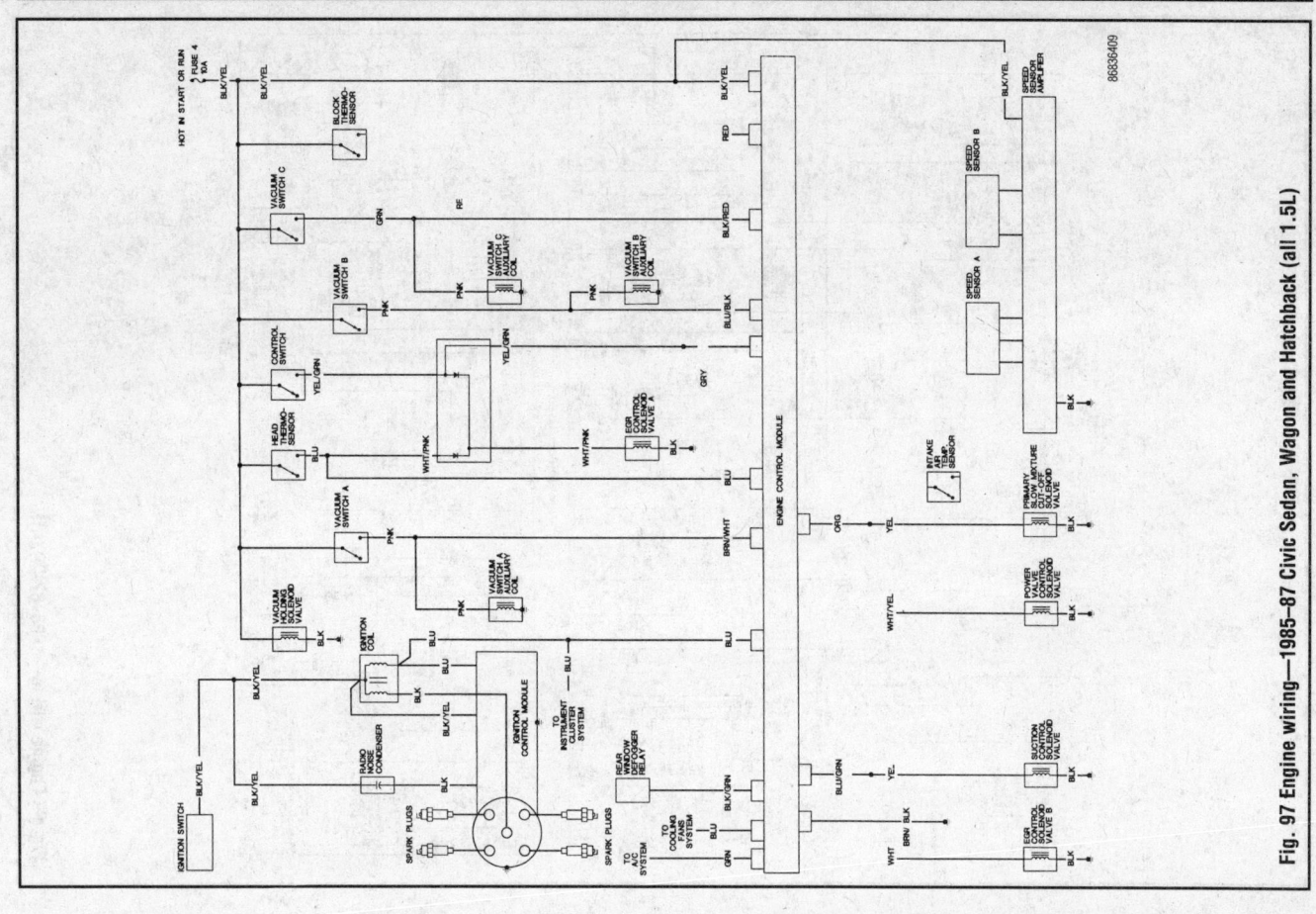

CHASSIS ELECTRICAL 6-37

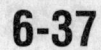

Fig. 97 Engine wiring—1985-87 Civic Sedan, Wagon and Hatchback (all 1.5L)

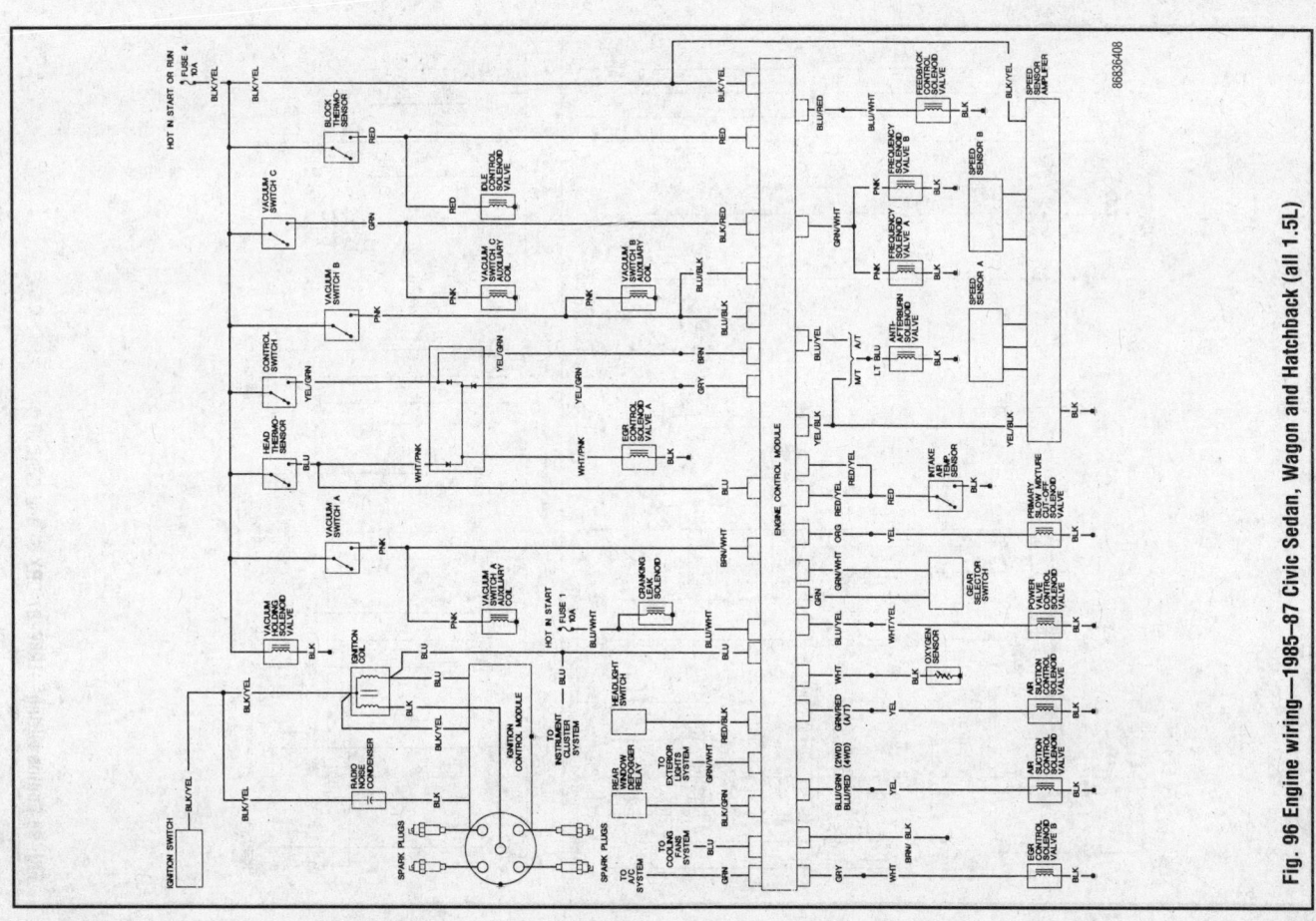

Fig. 96 Engine wiring—1985-87 Civic Sedan, Wagon and Hatchback (all 1.5L)

6-38 CHASSIS ELECTRICAL

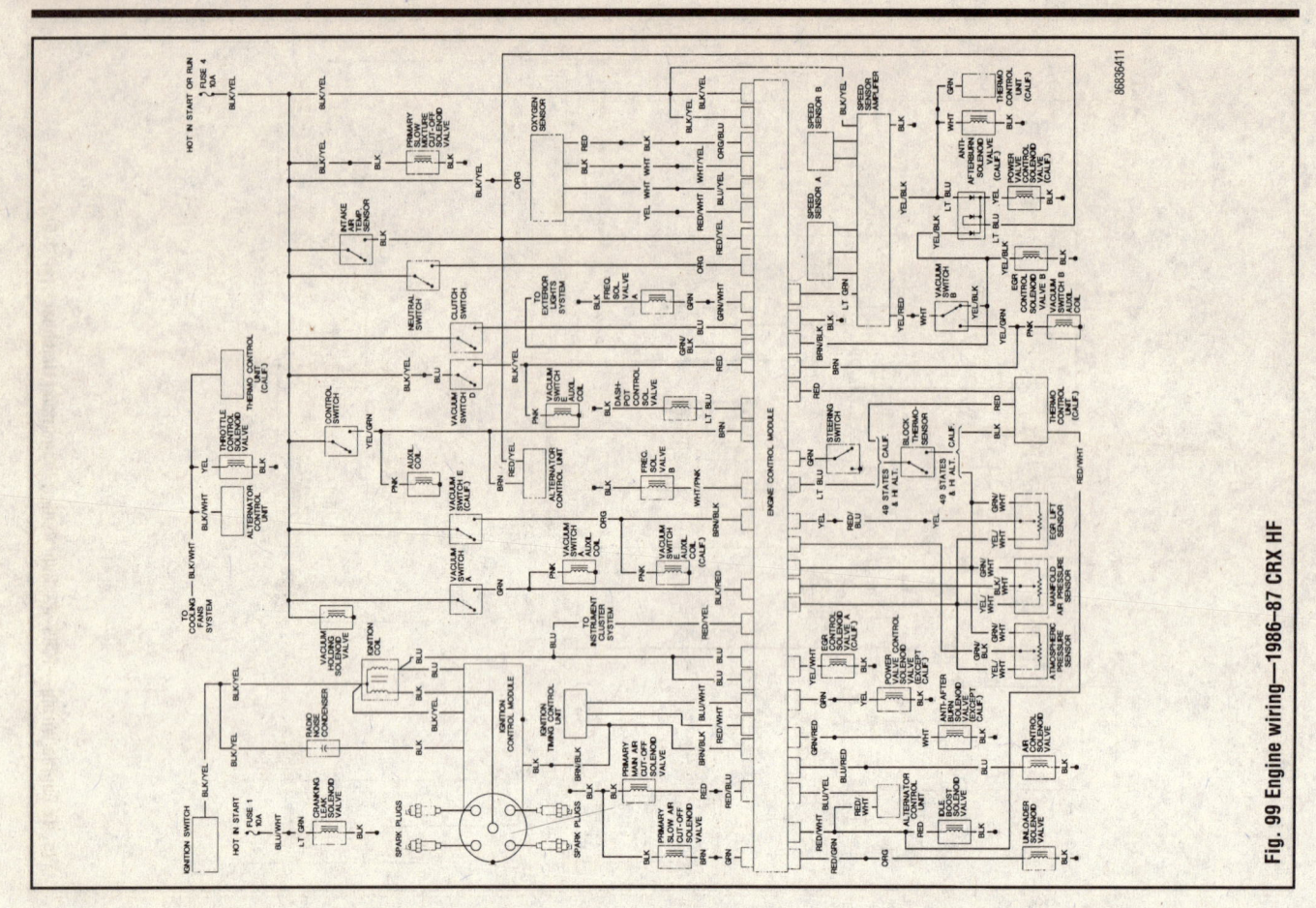

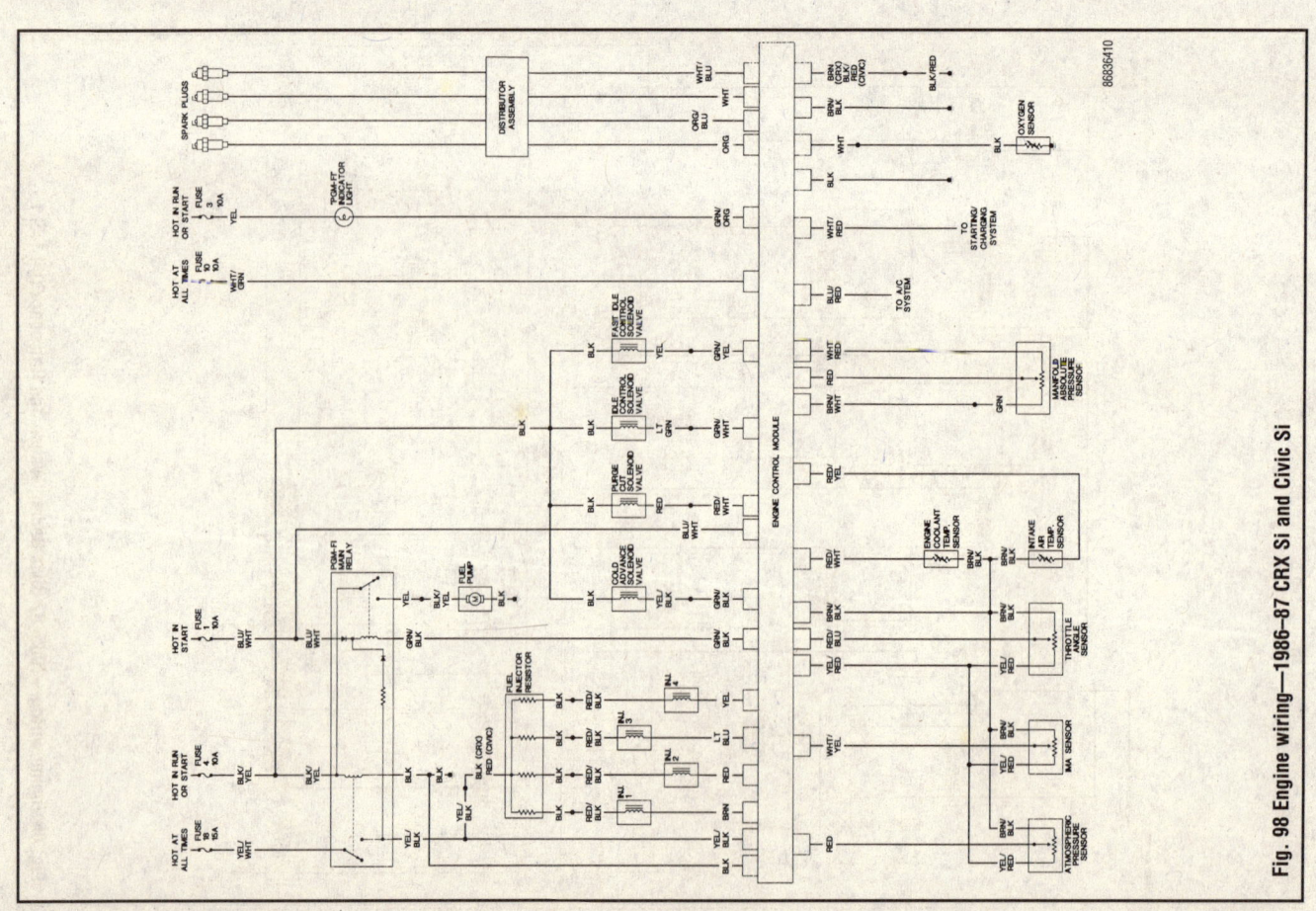

Fig. 99 Engine wiring—1986–87 CRX HF

Fig. 98 Engine wiring—1986–87 CRX Si and Civic Si

CHASSIS ELECTRICAL 6-39

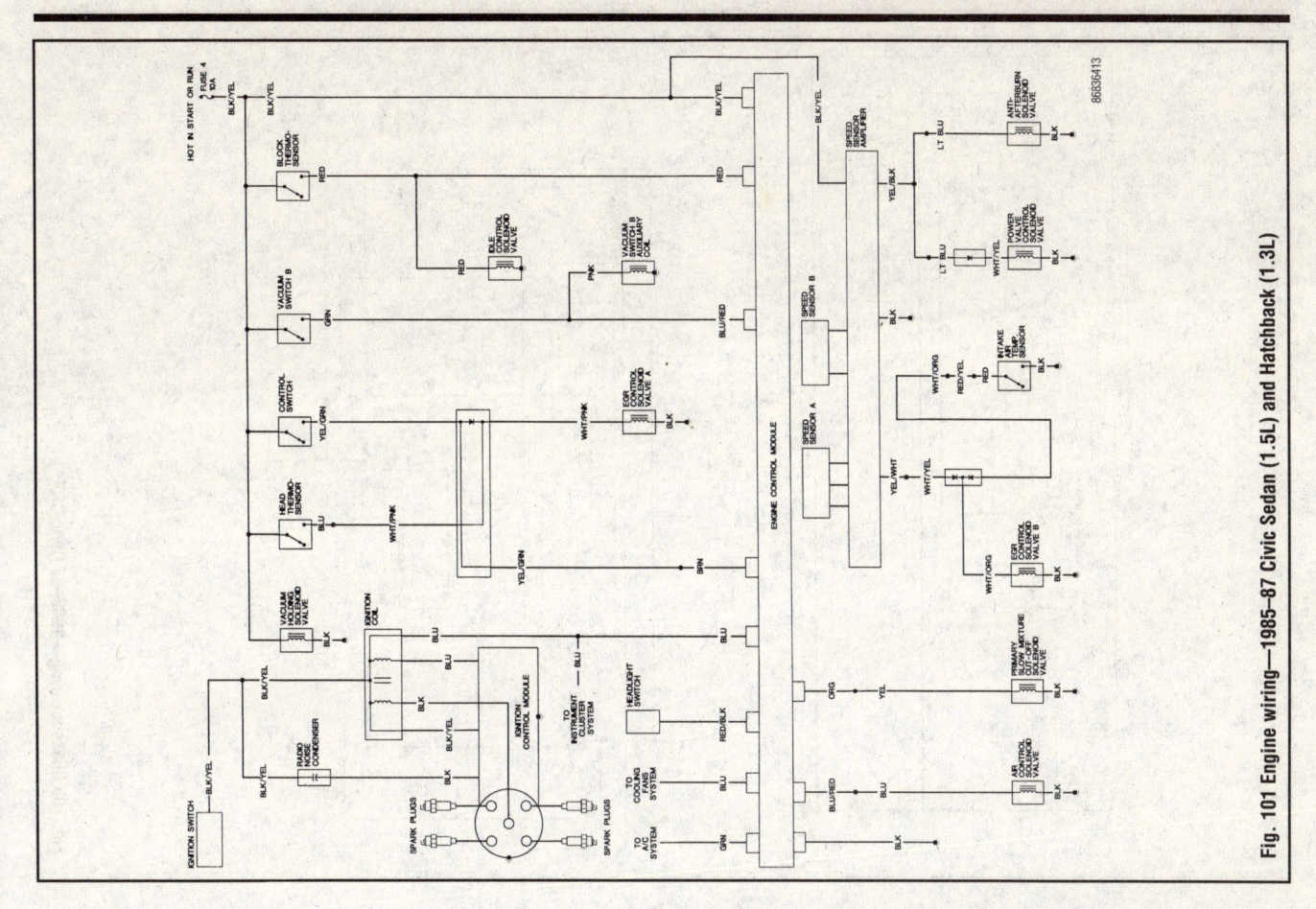

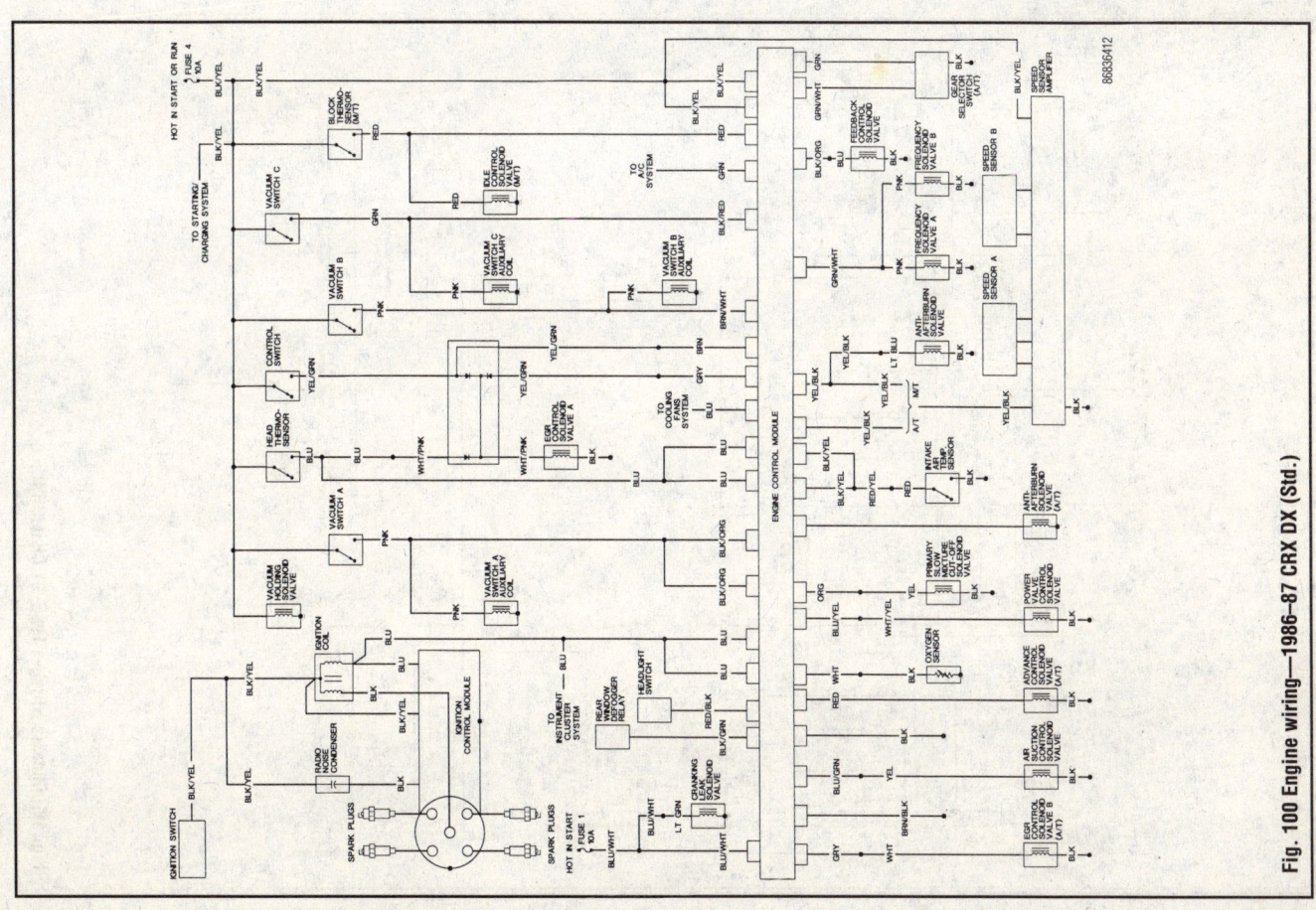

Fig. 101 Engine wiring—1985–87 Civic Sedan (1.5L) and Hatchback (1.3L)

Fig. 100 Engine wiring—1986–87 CRX DX (Std.)

6-40 CHASSIS ELECTRICAL

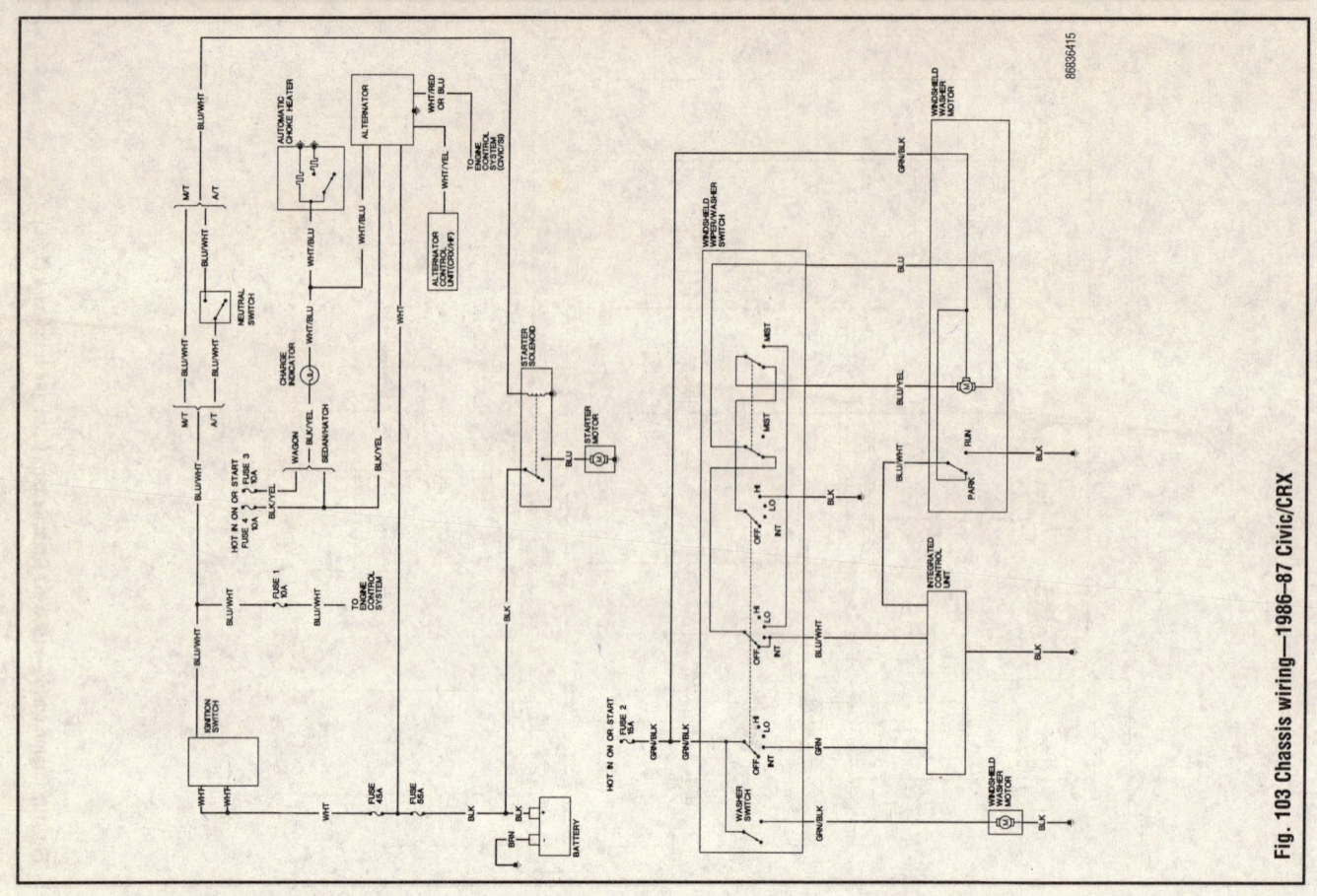

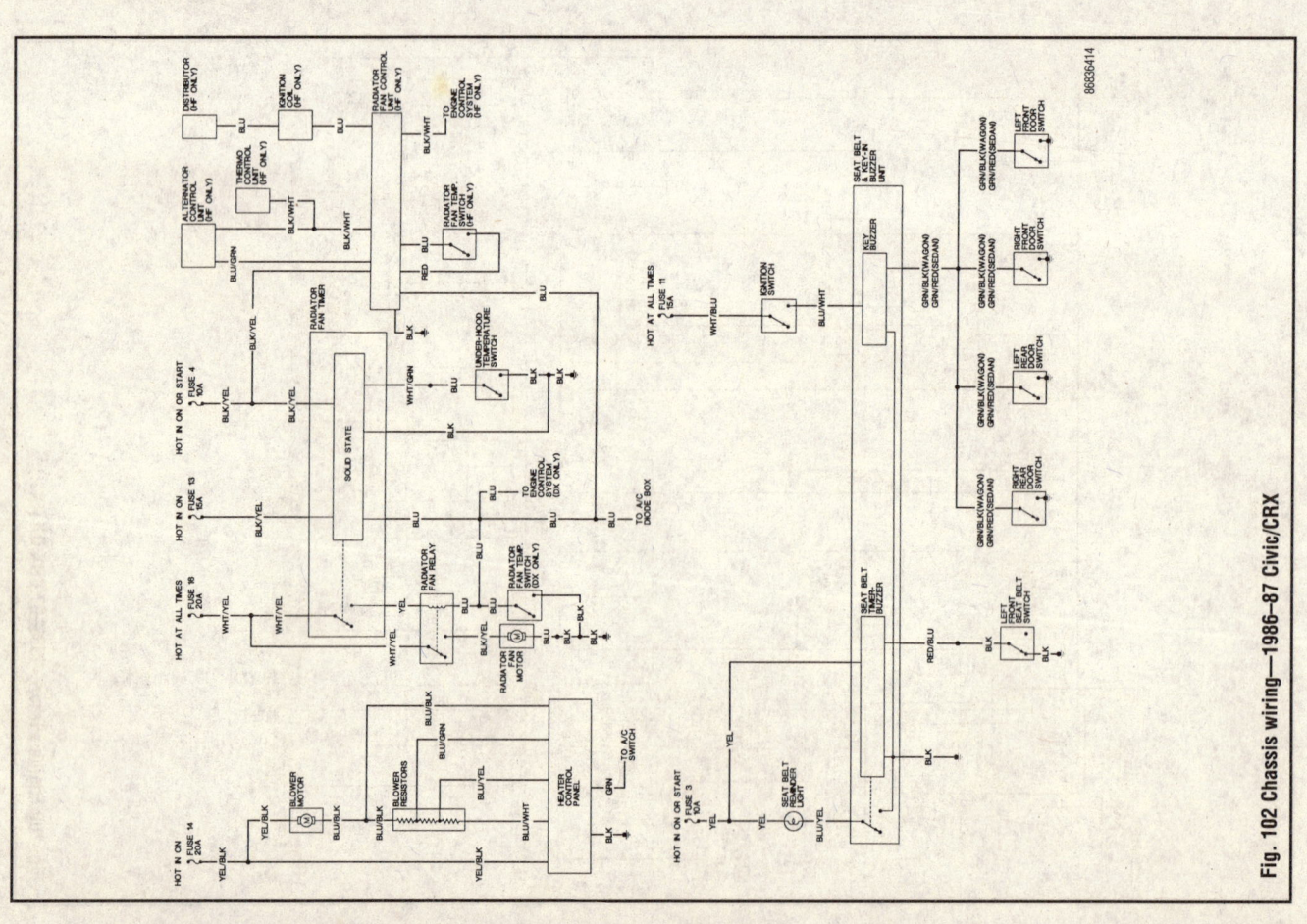

Fig. 103 Chassis wiring—1986–87 Civic/CRX

Fig. 102 Chassis wiring—1986–87 Civic/CRX

CHASSIS ELECTRICAL 6-41

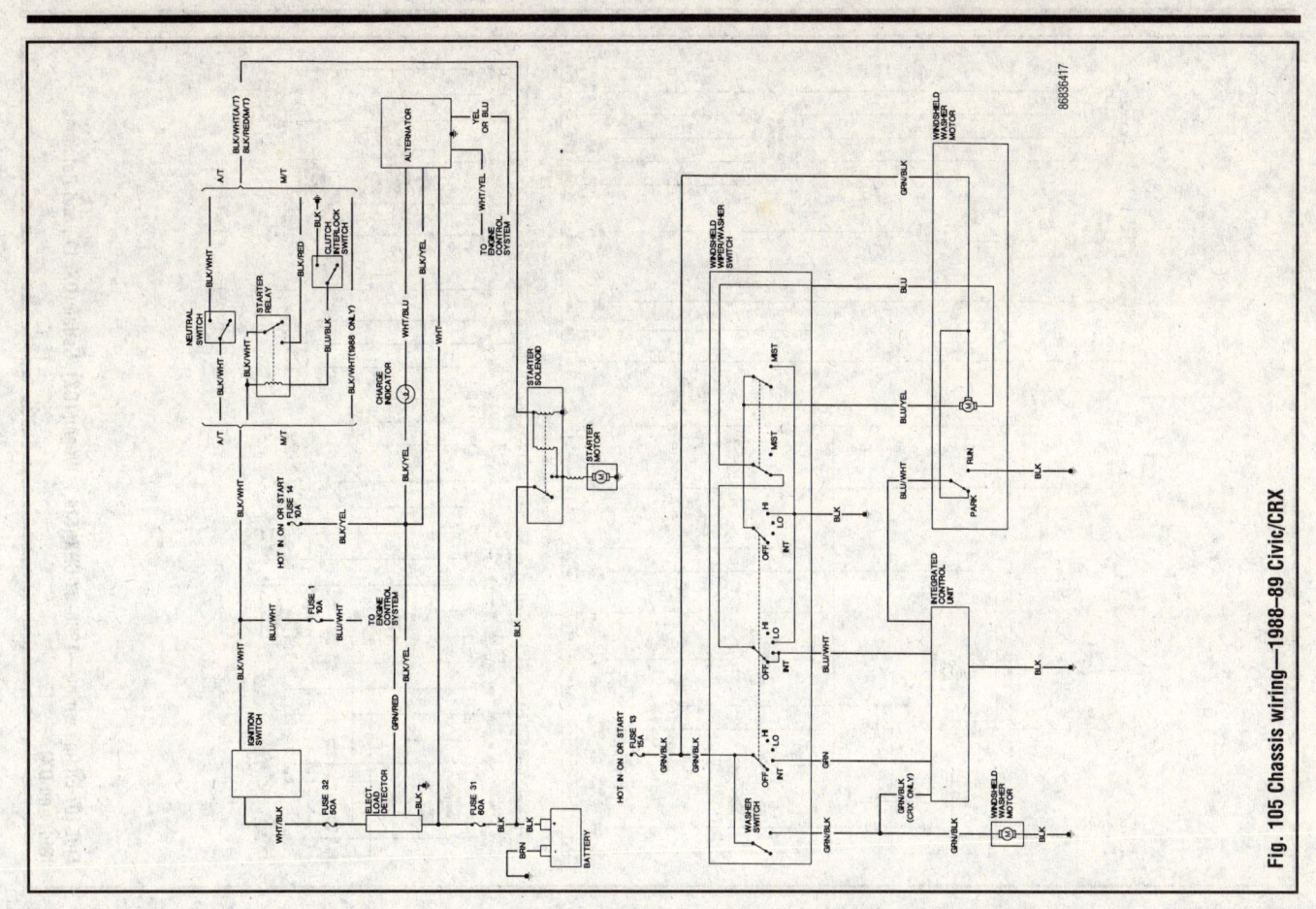

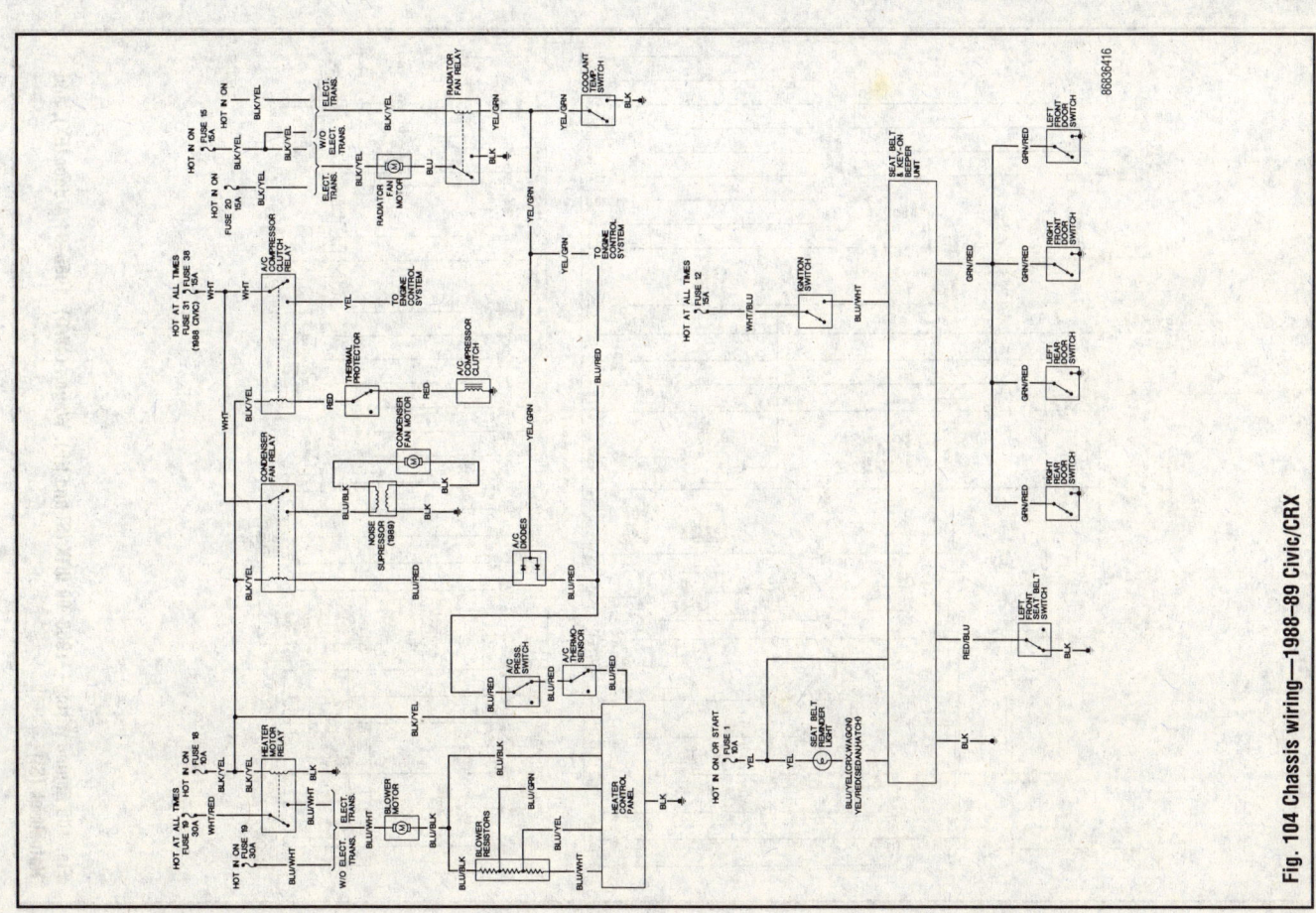

Fig. 105 Chassis wiring—1988-89 Civic/CRX

Fig. 104 Chassis wiring—1988-89 Civic/CRX

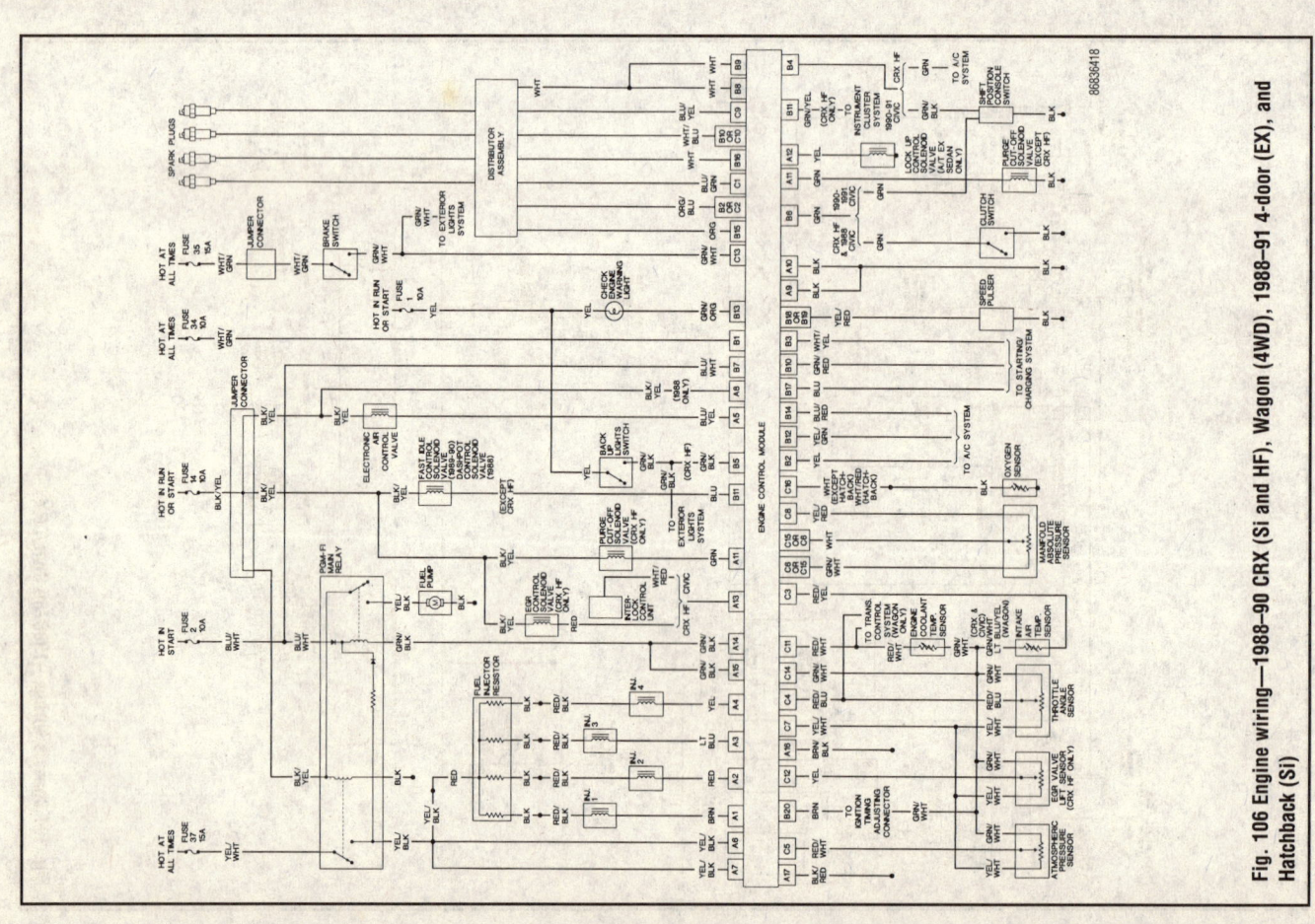

6-42 CHASSIS ELECTRICAL

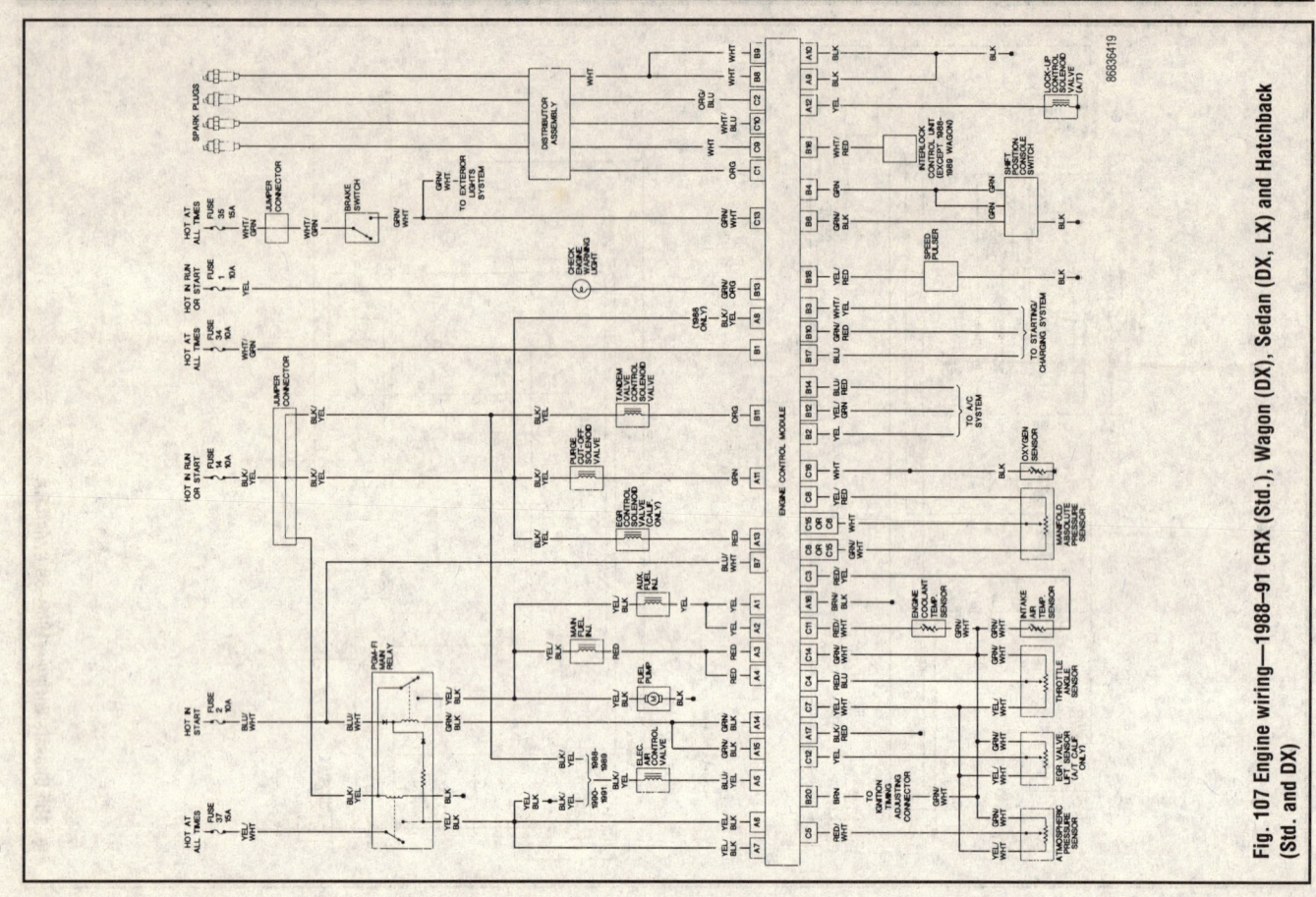

Fig. 107 Engine wiring—1988-91 CRX (Std.), Wagon (DX), Sedan (DX, LX) and Hatchback (Std. and DX)

Fig. 106 Engine wiring—1988-90 CRX (Si and HF), Wagon (4WD), 1988-91 4-door (EX), and Hatchback (Si)

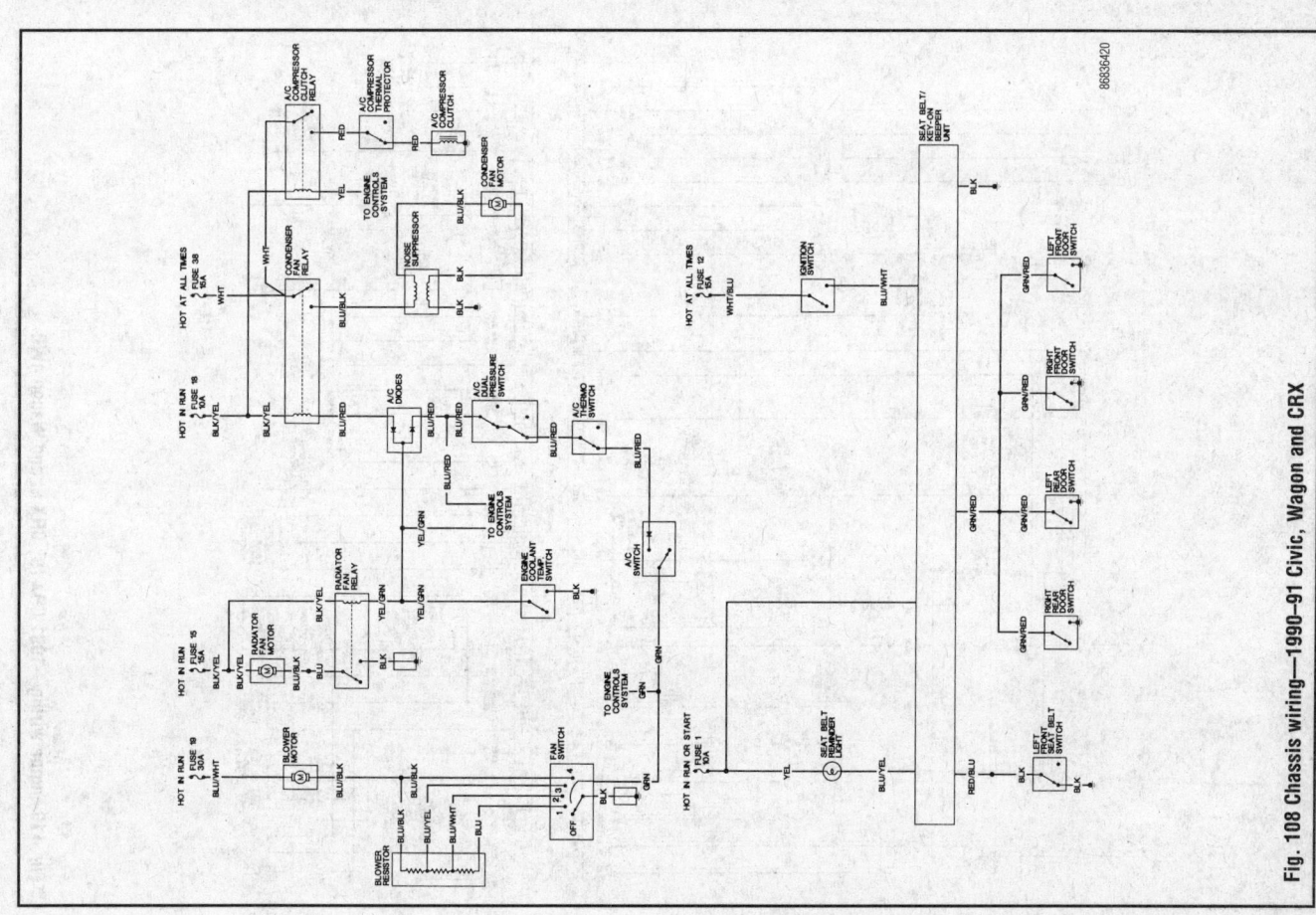

CHASSIS ELECTRICAL 6-43

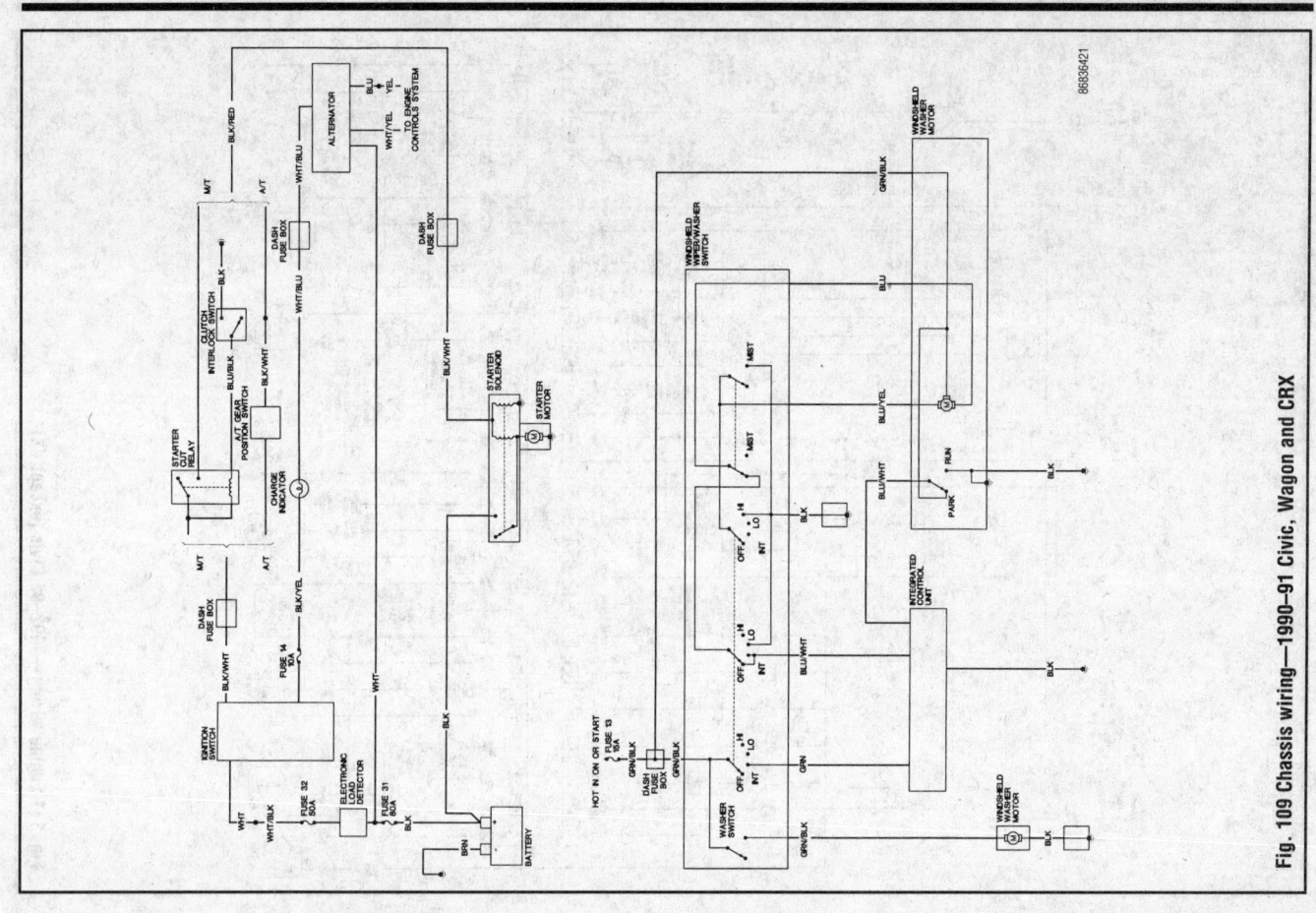

Fig. 109 Chassis wiring—1990–91 Civic, Wagon and CRX

Fig. 108 Chassis wiring—1990–91 Civic, Wagon and CRX

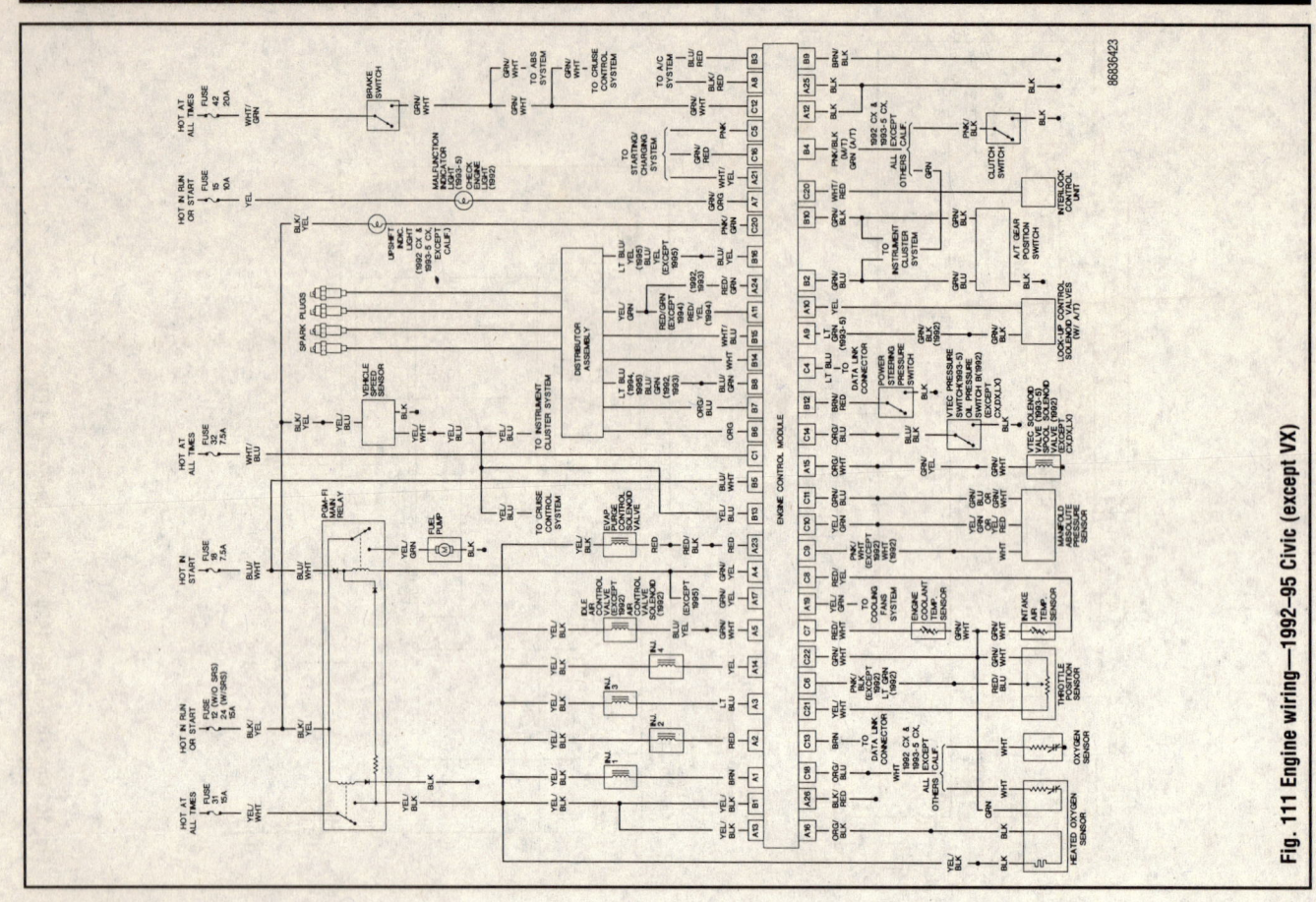

6-44 CHASSIS ELECTRICAL

Fig. 111 Engine wiring—1992-95 Civic (except VX)

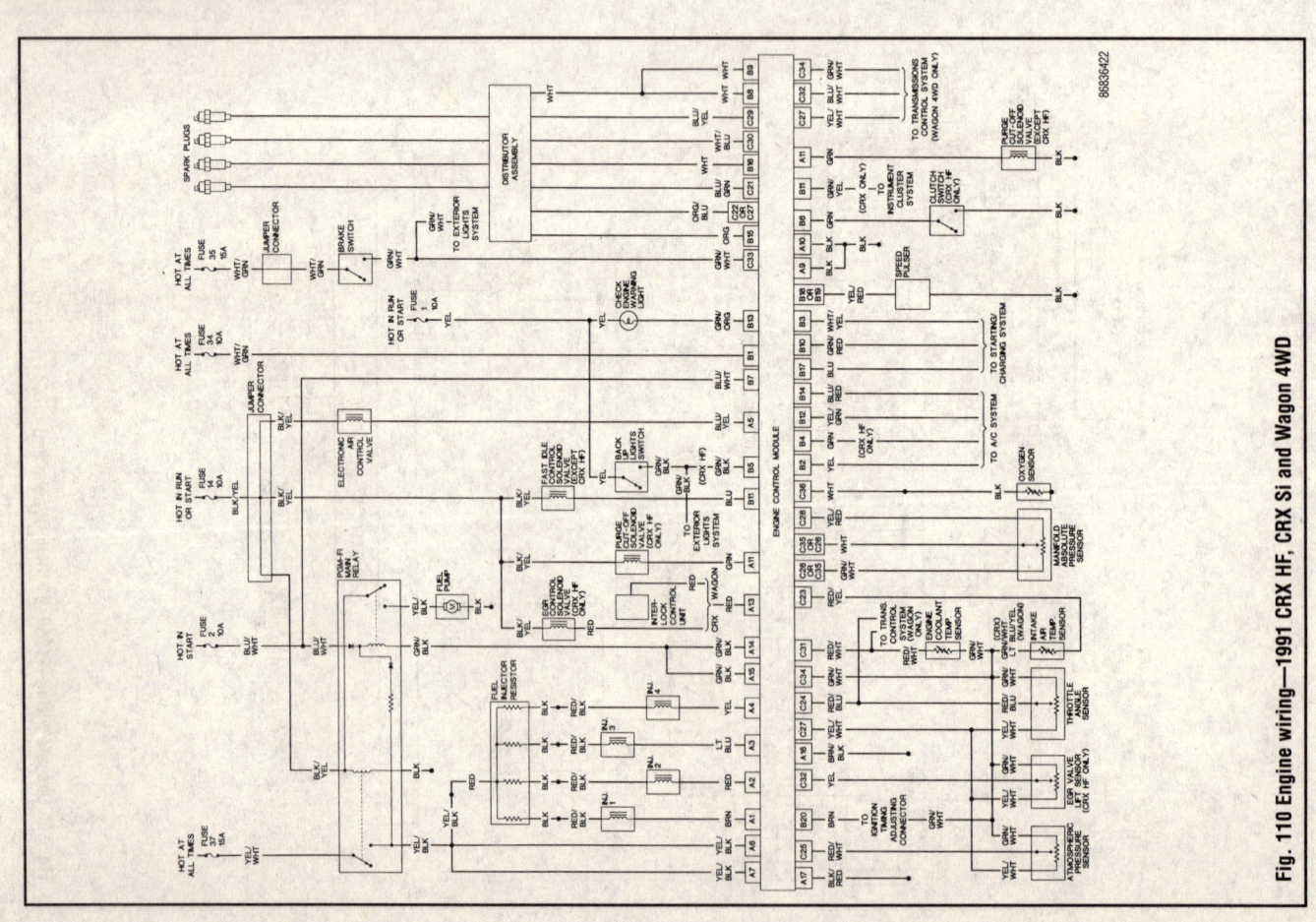

Fig. 110 Engine wiring—1991 CRX HF, CRX Si and Wagon 4WD

CHASSIS ELECTRICAL 6-45

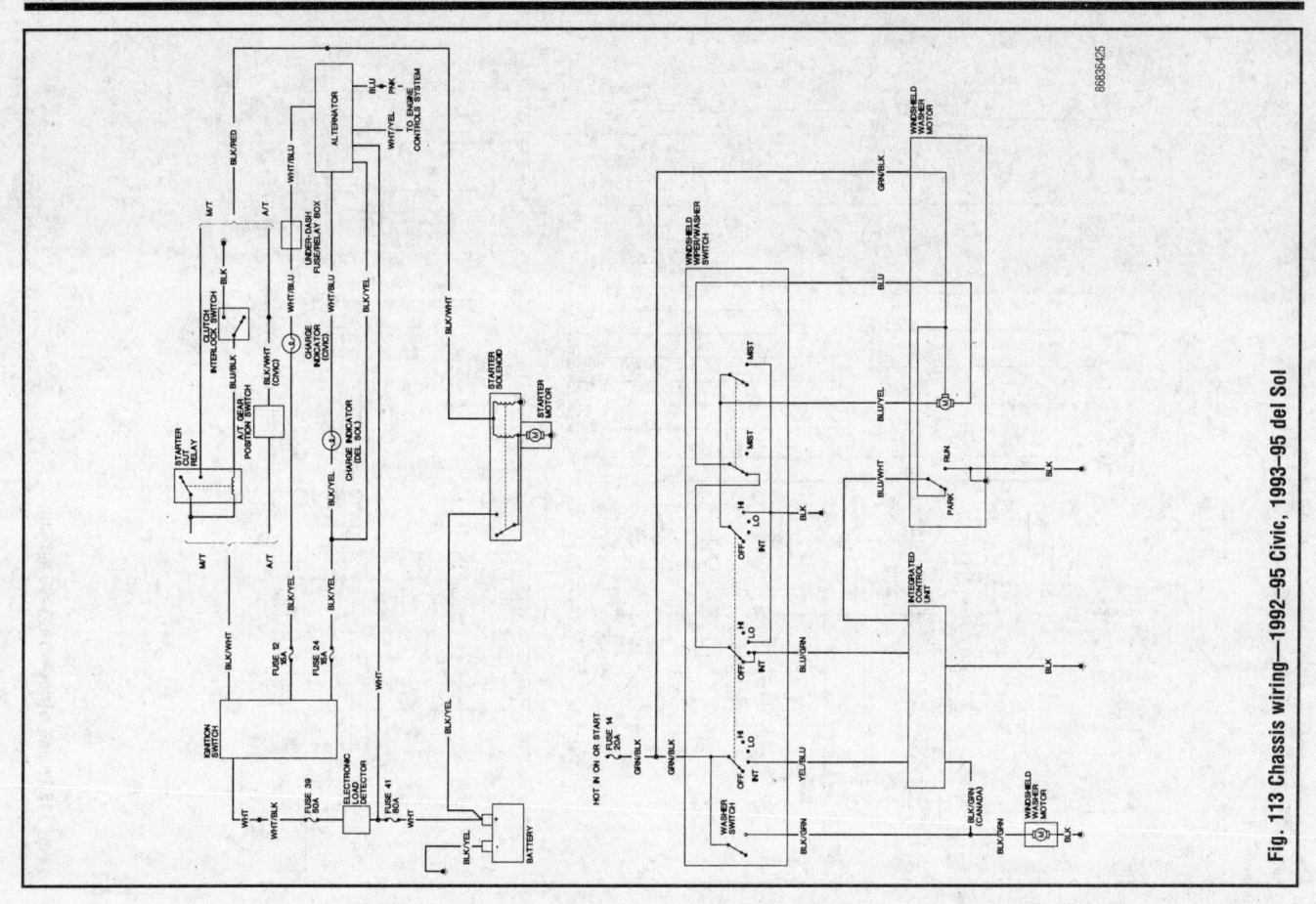

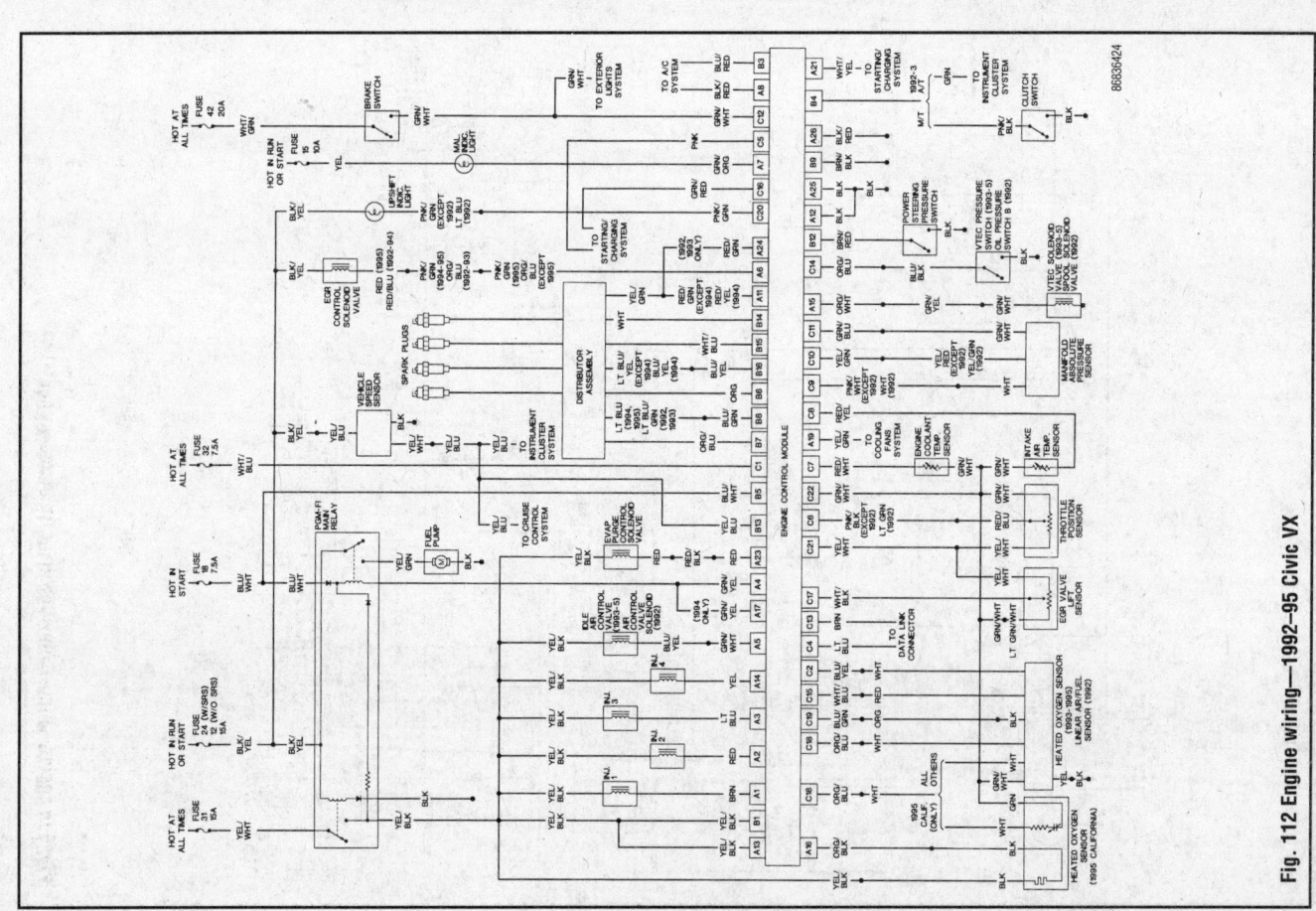

Fig. 113 Chassis wiring—1992–95 Civic, 1993–95 del Sol

Fig. 112 Engine wiring—1992–95 Civic VX

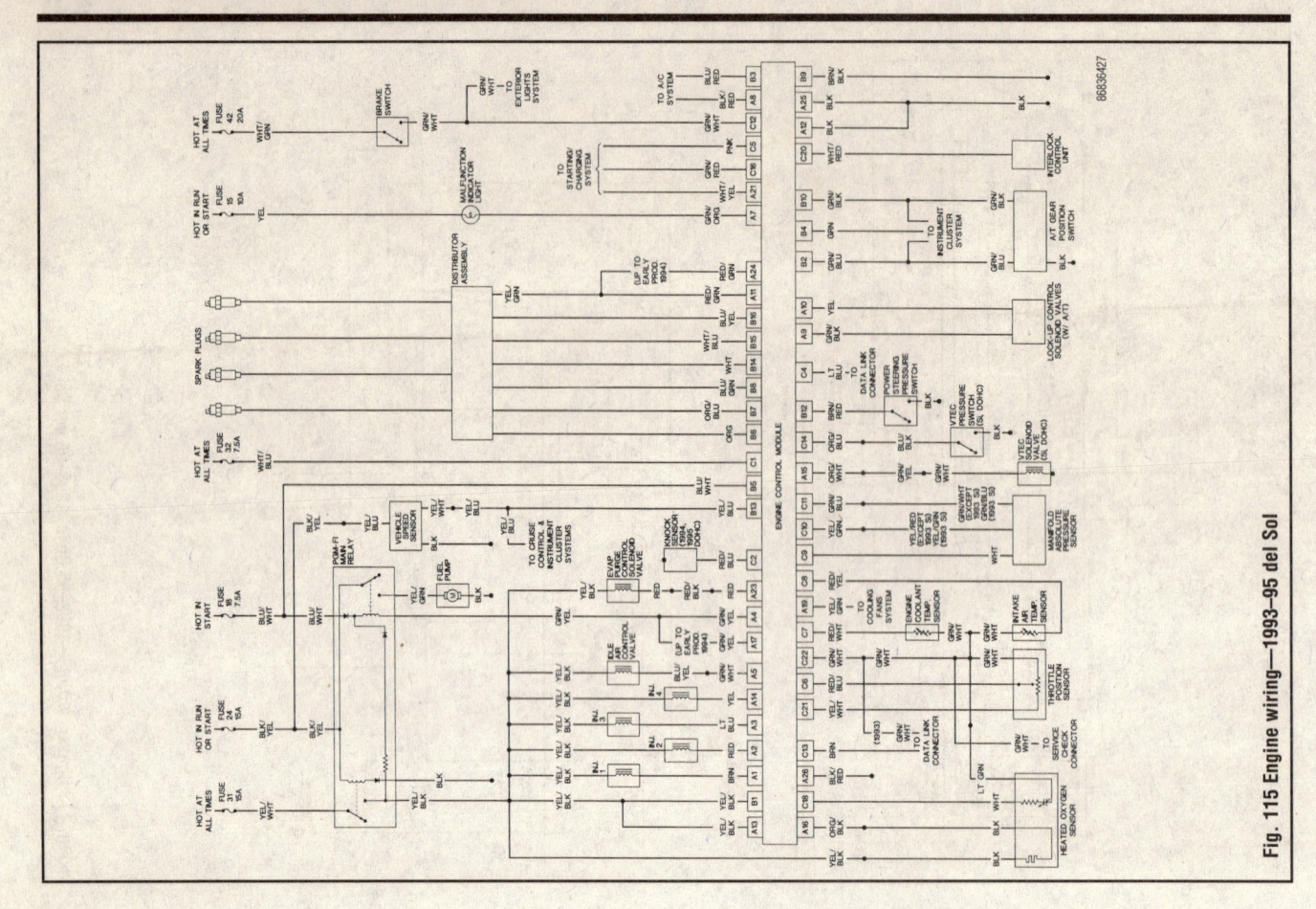

6-46 CHASSIS ELECTRICAL

Fig. 115 Engine wiring—1993-95 del Sol

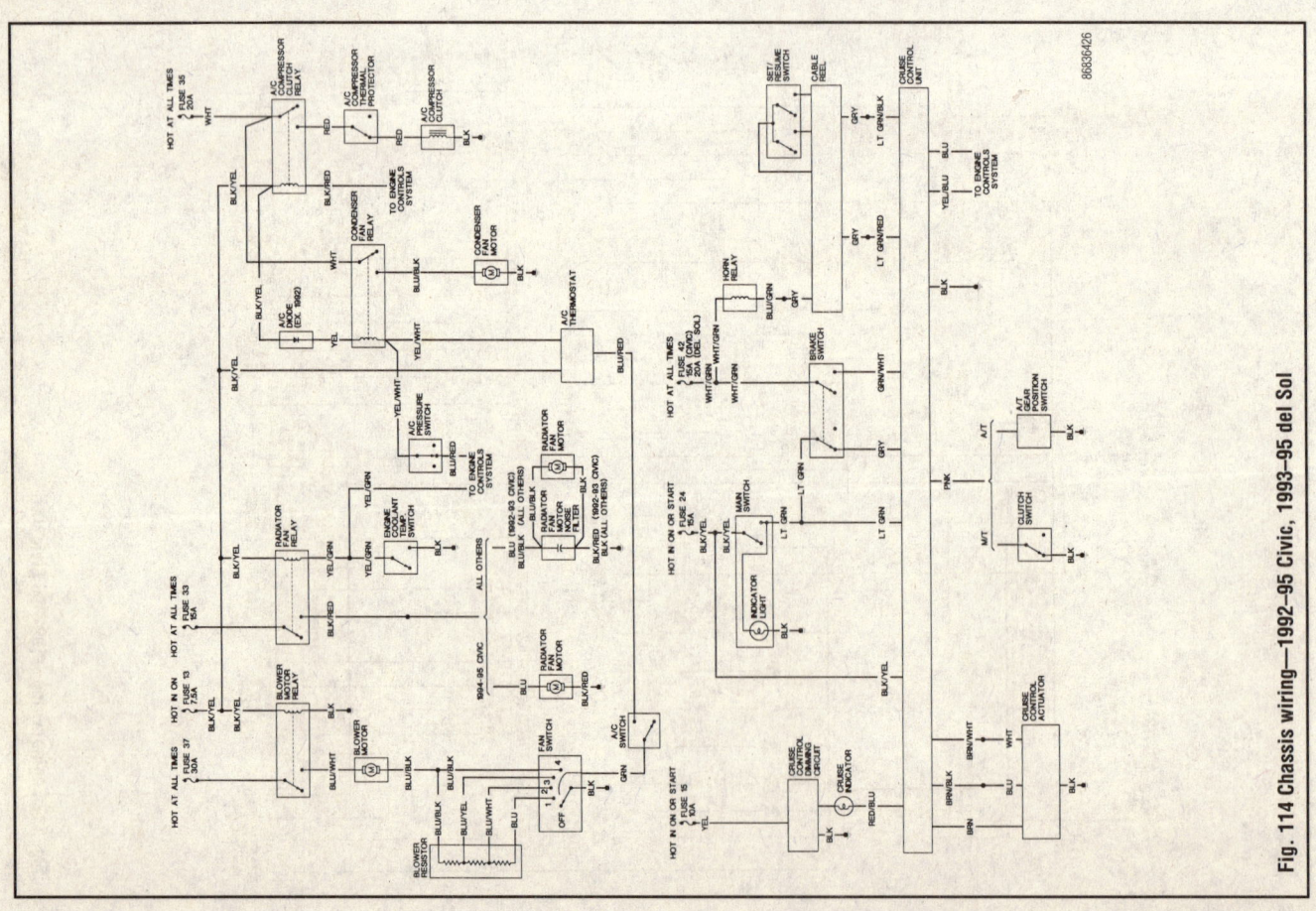

Fig. 114 Chassis wiring—1992-95 Civic, 1993-95 del Sol

MANUAL TRANSAXLE 7-2
UNDERSTANDING THE MANUAL
  TRANSAXLE 7-2
HONDA TRANSAXLE DESCRIPTION 7-2
IDENTIFICATION 7-2
ADJUSTMENTS 7-2
  SHIFT LINKAGE 7-2
GEARSHIFT ASSEMBLY 7-2
  REMOVAL & INSTALLATION 7-2
BACK-UP LIGHT SWITCH 7-2
  REMOVAL & INSTALLATION 7-4
TRANSAXLE 7-4
  REMOVAL & INSTALLATION 7-4
SPEEDOMETER DRIVEN GEAR 7-12
  REMOVAL & INSTALLATION 7-12
CLUTCH 7-12
UNDERSTANDING THE CLUTCH 7-12
ADJUSTMENTS 7-13
  PEDAL HEIGHT/FREE-PLAY
    ADJUSTMENT 7-13
CLUTCH CABLE 7-13
  REMOVAL & INSTALLATION 7-13
DRIVEN DISC AND PRESSURE
  PLATE 7-13
  REMOVAL & INSTALLATION 7-13
CLUTCH MASTER CYLINDER 7-15
  REMOVAL & INSTALLATION 7-15
  OVERHAUL 7-16
SLAVE CYLINDER 7-16
  REMOVAL & INSTALLATION 7-16
  OVERHAUL 7-16
  SYSTEM BLEEDING 7-16
AUTOMATIC TRANSAXLE 7-16
UNDERSTANDING AUTOMATIC
  TRANSAXLES 7-16
    THE TORQUE CONVERTER 7-16
    THE PLANETARY GEARBOX 7-17
    THE SERVOS AND
      ACCUMULATORS 7-17
    THE HYDRAULIC CONTROL
      SYSTEM 7-17
IDENTIFICATION 7-17
ADJUSTMENTS 7-18
  SHIFT CABLE 7-18
  THROTTLE LINKAGE 7-19
TRANSAXLE 7-20
  REMOVAL & INSTALLATION 7-20
DRIVELINE 7-22
HALFSHAFTS 7-22
  REMOVAL & INSTALLATION 7-22
CV-JOINT AND BOOT 7-23
  OVERHAUL 7-23
DRIVESHAFT AND U-JOINTS 7-24
  REMOVAL & INSTALLATION 7-24
REAR AXLE 7-25
APPLICATION 7-25
UNDERSTANDING DRIVE AXLES 7-25
DETERMINING AXLE RATIO 7-25
REAR HALFSHAFTS 7-26

REAR AXLE SHAFTS AND
  BEARINGS 7-26
  REMOVAL & INSTALLATION 7-26
PINION SEAL 7-27
  REMOVAL & INSTALLATION 7-27
DIFFERENTIAL CARRIER 7-27
  REMOVAL & INSTALLATION 7-27

7

DRIVE TRAIN

MANUAL TRANSAXLE 7-2
CLUTCH 7-12
AUTOMATIC TRANSAXLE 7-16
DRIVELINE 7-22
REAR AXLE 7-25

7-2 DRIVE TRAIN

MANUAL TRANSAXLE

Understanding the Manual Transaxle

Because of the way an internal combustion engine breathes, it can produce torque, or twisting force, only within a narrow speed range. Most modern, overhead valve pushrod engines must turn at about 2500 rpm to produce their peak torque. By 4500 rpm they are producing so little torque that continued increases in engine speed produce no power increases. The torque peak on overhead camshaft engines is generally much higher, but much narrower.

The manual transaxle and clutch are employed to vary the relationship between engine speed and the speed of the wheels so that adequate engine power can be produced under all circumstances. The clutch allows engine torque to be applied to the transaxle input shaft gradually, due to mechanical slippage. Consequently, the vehicle may be started smoothly from a full stop. The transaxle changes the ratio between the rotating speeds of the engine and the wheels by the use of gears. The gear ratios allow full engine power to be applied to the wheels during acceleration at low speeds and at highway/passing speeds.

In a front wheel drive transaxle, power is usually transmitted from the input shaft to a mainshaft or output shaft located slightly beneath and to the side of the input shaft. The gears of the mainshaft mesh with gears on the input shaft, allowing power to be carried from one to the other. All forward gears are in constant mesh and are free from rotating with the shaft unless the synchronizer and clutch is engaged. Shifting from one gear to the next causes one of the gears to be freed from rotating with the shaft and locks another to it. Gears are locked and unlocked by internal dog clutches which slide between the center of the gear and the shaft. The forward gears employ synchronizers; friction members which smoothly bring gear and shaft to the same speed before the toothed dog clutches are engaged.

Honda Transaxle Description

The Honda Civic series utilizes a transaxle arrangement where the transaxle and the differential are contained within the same housing. Power is transmitted from the engine to the transaxle and in turn, to the differential. The front drive axle halfshafts transfer the power from the differential to the front wheels.

The GW transaxle is a five-speed version of the GV four-speed transaxle. They consist of mainshaft and countershaft assemblies, synchronizer assemblies, shift forks and a differential. A reverse light switch is incorporated into the unit to complete the reverse light circuit when backing up.

The L3 Series four-speed and five-speed of 1988-91 and the S20 series five-speed of 1992-95 are later versions of the GW transaxle.

The 1985-87 Civic Wagon 4WD employs a modified version of the conventional five-speed. Under normal usage, the vehicle operates in the same manner as other Civics in 2WD, transmitting the engine's power through halfshafts to the front wheels. When the 4WD mode is engaged, power is transmitted to the rear wheels through the use of a transfer case.

The 1988-91 Civic Wagon 4WD employs a viscous coupling that engages the rear wheels as needed, depending on front wheel traction. This "Real Time 4WD" system is intended to make power transfer between 4WD and 2WD essentially automatic. The viscous coupling unit is mated to the front of the driveshaft which connects to the rear differential. Inside the unit's housing are 79 plates, which have 0.0078 in. (0.2mm) of clearance between one another, surrounded by silicone oil. Forty of these oil-bathed plates are splined to the housing, and 39 are splined to the shaft in alternate succession. Whenever the front wheels lose traction and spin faster than the rear, the friction (and pressure) in the viscous coupling forces the plates to drag, engaging the rear wheels as needed. Once engine torque is delivered to the rear wheels, front and rear wheel traction stabilizes and the wheels begin to turn at the same speed. As this happens, the temperature of the silicon oil and the pressure in the viscous coupling unit decreases and the power to the rear wheels decreases proportionally.

Identification

The transaxle can be identified by a number located on the clutch housing. The 1984-87 models use a 4 or five-speed manual transaxle. The four-speed transaxle is identified as the GV type and the five-speed as the GW type. The 1985-87 4WD Wagons use a modified GW five-speed and the 1988-91 4WD Wagons use a modified L3 manual transaxle. Both of these 4WD models also employ a Super Low (SL) gear, making it essentially a six-speed. The 1988-91 models use a 4 or five-speed manual transaxle which is identified as the L3 type. The 1992-95 models use a five-speed transaxle identified as the S20 type.

Adjustments

SHIFT LINKAGE

Except 4WD Civic Wagon

♦ See Figures 1 and 2

The Honda shift linkage on these models is non-adjustable. However, if the linkage is binding, or if there is excessive play, check the linkage bushings and pivot points. Lubricate with light oil, or replace worn bushings as necessary.

4WD Civic Wagon

♦ See Figure 3

SELECTOR CABLE

1. Remove the console.
2. With the transaxle in neutral, check that the groove in the lever bracket is aligned with the index mark on the selector cable.
3. If the index mark is not aligned with the groove in the cable, loosen the locknuts and turn the adjuster as necessary.

➡After adjustment, check the operation of the gearshift lever. Also check that the threads of the cables do not extend out of the cable adjuster by more than 0.39 in. (10mm.)

GEARSHIFT CABLE

♦ See Figure 4

1. Remove the console.
2. Place the transaxle in 4th gear.
3. Measure the clearance between the gearshift lever bracket and the stopper while pushing the lever forward. It should be between 0.02–0.26 in. (0.5–6.5mm).
4. If the clearance is outside of the specification, loosen the locknuts and turn the adjuster in or out until the correct clearance is obtained.

➡After adjustment, check the operation of the gearshift lever. Also check that the threads of the cables do not extend out of the cable adjuster by more than 0.39 in. (10mm.)

Gearshift Assembly

REMOVAL & INSTALLATION

The gearshift assembly uses a torque rod and shift rod assembly. The torque rod bolts to the chassis, through a rubber mount and to the back of the transaxle, it absorbs the twisting motion of the transaxle and serves to support the gearshift assembly.

The gearshift assembly pivots on a ball joint attached to the torque rod. The shift rod attaches the gearshift assembly and to the transaxle.

The shift change lever and its mounting can be reached by removing the center console assembly and removing the dust boot. The rest of the mechanism can be reached from underneath the vehicle.

1. Remove the console assembly.
2. Remove the shift boot attaching screws and remove the boot. Remove the torque rod front extension mount bolts.
3. From beneath the vehicle remove the torque rod rear mount by removing the rear extension collar bolt and removing the extension collar. The rod will now drop from the mount.
4. Remove the torque rod from the shift lever by removing the retaining bolts. Pull off the change ball dust seal A and remove the change lever ball seat, and seal B.

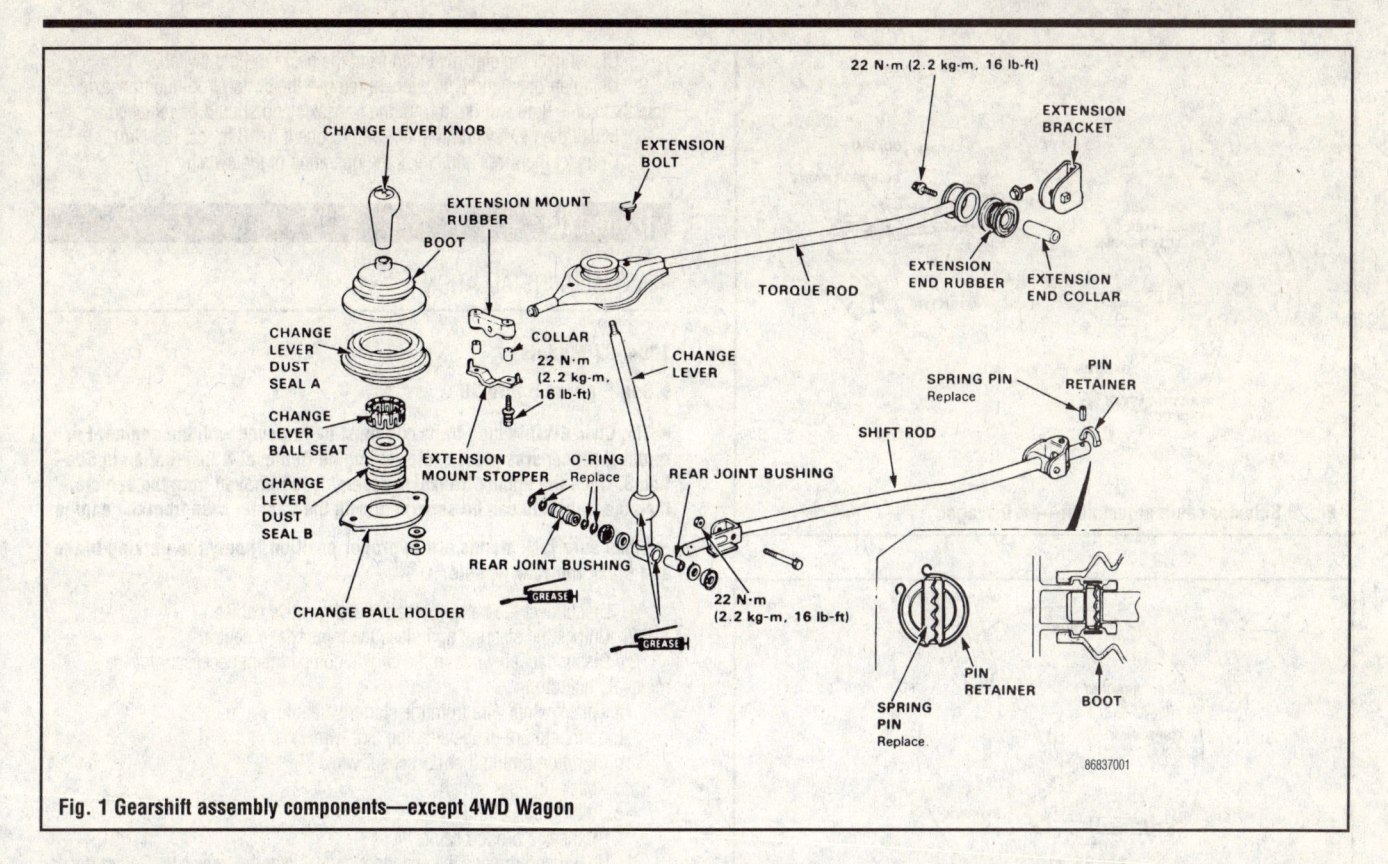

DRIVE TRAIN   7-3

Fig. 1 Gearshift assembly components—except 4WD Wagon

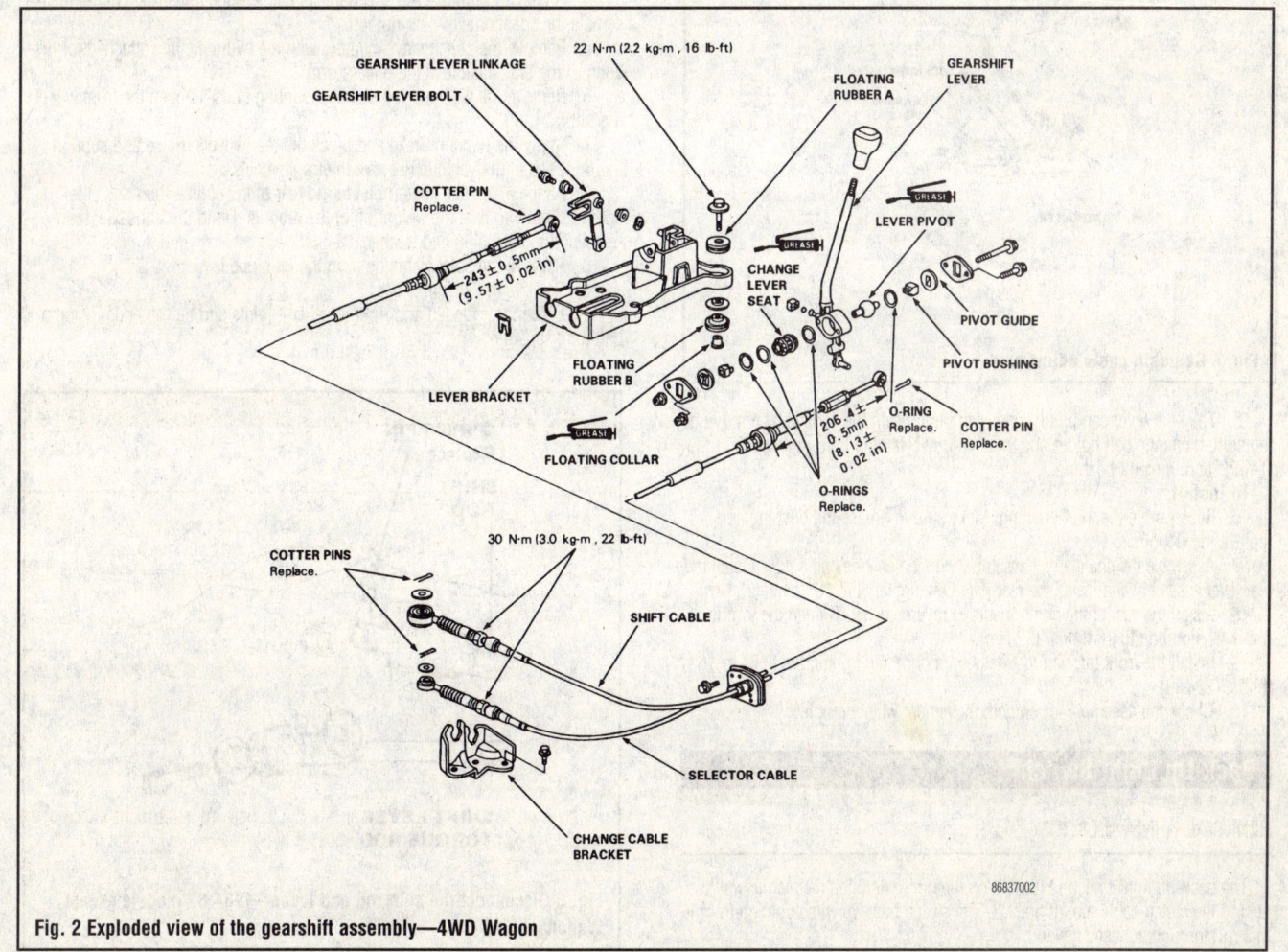

Fig. 2 Exploded view of the gearshift assembly—4WD Wagon

7-4 DRIVE TRAIN

Fig. 3 Selector cable adjustment—4WD Wagon

Fig. 4 Gearshift cable adjustment

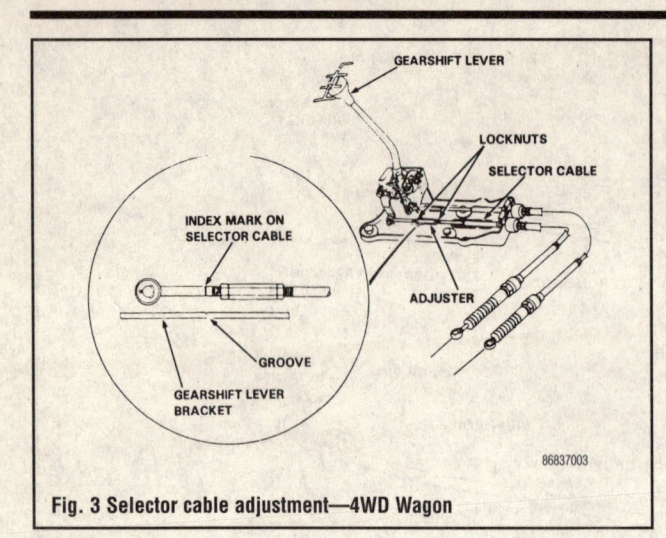

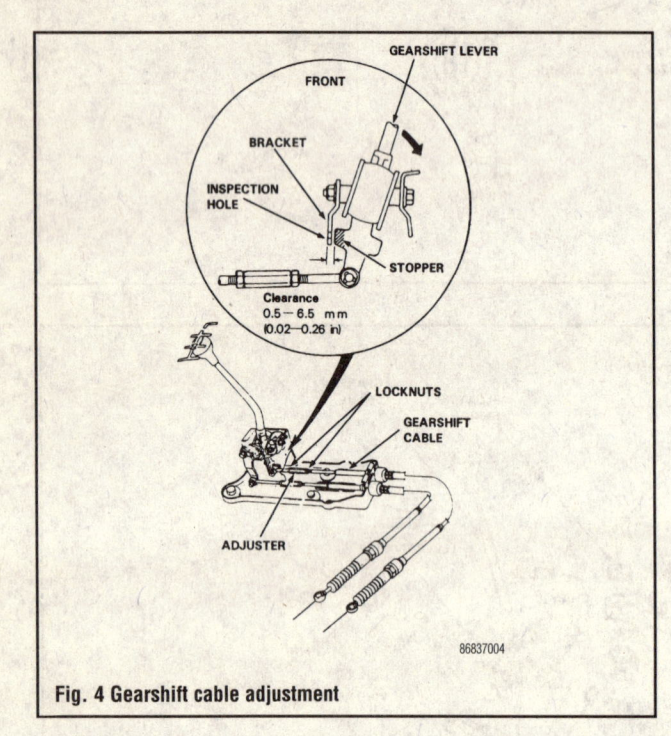

5. The shift lever can be disconnected underneath the vehicle by removing the bottom change ball holder and the rear joint retaining bolt. Keep track of the O-rings and the joint bushing.

To install:

6. Reassemble the joint bushing. Make sure to grease the bushing and replace the O-rings.
7. Assemble the change ball assembly with the change ball holder. Install the rear lever assembly and install the rear joint bolt, tighten it to 16 ft. lbs. (22 Nm).
8. Install the rest of the components in reverse order. Tighten the torque rod extension bolt to 16 ft. lbs. (22 Nm).
9. Install the front torque rod extension mount and tighten the bolts to 16 ft. lbs. (22 Nm).
10. Check the assembly for smooth movement when complete.

Back-Up Light Switch

REMOVAL & INSTALLATION

The back-up light switch is located on the transaxle (side nearest firewall) next to the clutch cable mount bracket. The switch has two wires coming from it that run through a plastic cover.

1. Disconnect the electrical lead from the back-up light switch.
2. Using an open end type wrench, remove the switch from the transaxle, note that some fluid will drain from the transaxle and should be replaced.
3. Install the new switch in place and torque it to 18 ft. lbs. (25 Nm).
4. Connect the wires and check the operation of the switch.

Transaxle

REMOVAL & INSTALLATION

1984–87 Models

▶ See Figures 5 thru 16

➡ The Civic 4WD Wagon transaxle must be removed with the engine. For removal procedures please refer to Engine Removal & Installation in Section 3. Once the engine/transaxle assembly is removed from the vehicle, then the transaxle can be separated with the transfer case from the engine.

➡ Make sure jack stands are in proper position. Apply the parking brake and block the rear wheels.

1. Disconnect the battery negative and positive cables.
2. Unlock the steering and place the transaxle in neutral.
3. Disconnect all wires in the engine compartment necessary for the removal, including:
   a. black/white wire from the starter solenoid.
   b. temperature gauge sending unit wire.
   c. ignition timing thermo-sensor wire.
   d. back-up light switch.
   e. distributor wiring.
   f. transaxle ground cable.
4. Unclip and remove the speedometer cable at the transaxle. Do not disassemble the speedometer gear holder!
5. Remove the clutch slave cylinder with the hydraulic line attached, or disconnect the clutch cable at the release arm.
6. Remove the side and top starter mounting bolts. Loosen the front wheel lug nuts.
7. Apply the parking brake and block the rear wheels. Raise and support the front end of the vehicle. Remove the front wheels.
8. Attach a suitable chain hoist to the rear of the engine then raise the engine slightly to take the weight off of the mounts. Drain the transaxle, then reinstall the drain plug and washer.
9. Remove the splash shields from the underside.
10. Remove the stabilizer bar.
11. Disconnect the left and right lower ball joints and tie rod ends, using a ball joint separator.

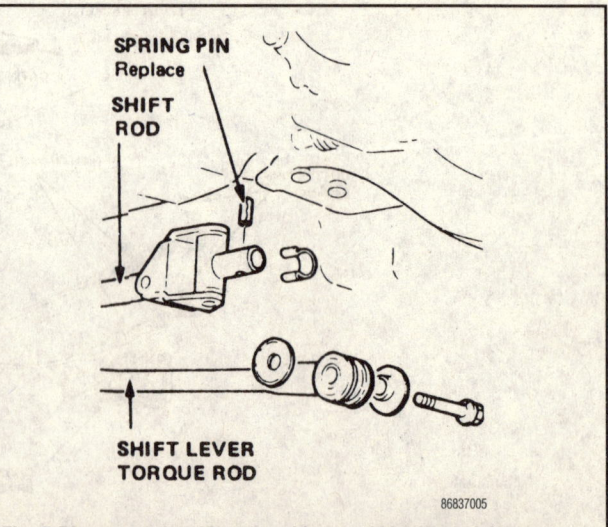

Fig. 5 Disconnect the shift rod and lever—1984-87 models except Wagon

DRIVE TRAIN 7-5

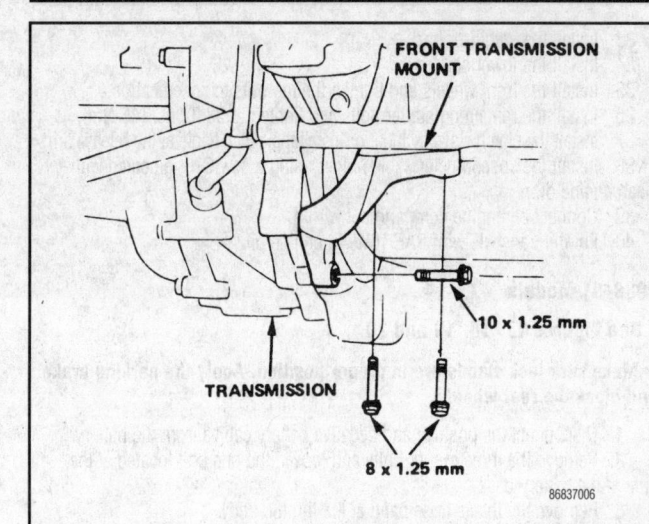

Fig. 6 Removing the front transaxle mount—1984-87 models except Wagon

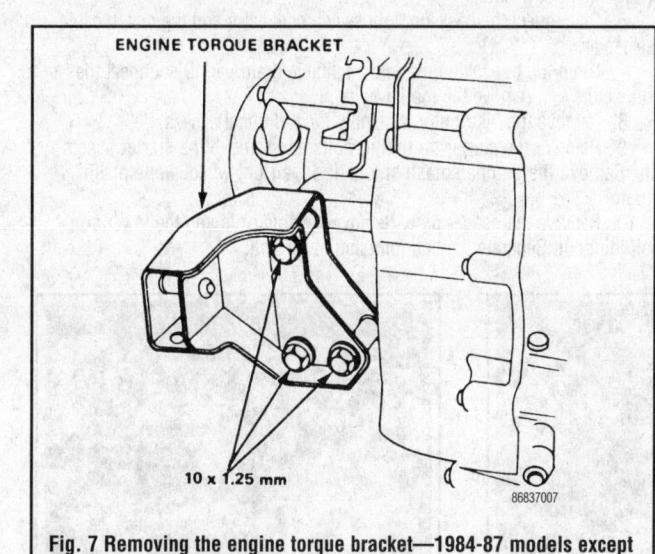

Fig. 7 Removing the engine torque bracket—1984-87 models except Wagon

Fig. 8 Removing the rear transaxle mount—1984-87 models except Wagon

❋❋ CAUTION

Use caution when separating the ball joints. Place a floor jack under the lower control arm securely at the ball joint. Otherwise, the lower control arm may jump suddenly away from the steering knuckle as the ball joint is removed!

12. Turn the right steering knuckle out as far as it will go. Place a prybar against the inboard CV-joint. Pry the right axle out of the transaxle about 1/2 inch (12mm). This will force the spring clip out of the groove inside the differential gear splines. Pull it out the rest of the way. Repeat this procedure on the other side.

13. Disconnect the shift lever torque rod from the clutch housing.

14. Slide the pin retainer back and drive out the spring pin using a pin punch, then disconnect the shift rod. Remove the bolt from the shift rod clevis, if so equipped.

15. Place a transaxle jack under the transaxle and raise the jack securely against the transaxle to take up the weight.

16. Remove the engine torque rods and brackets. Remove the bolts from the front transaxle mount. Remove the transaxle housing bolts from the engine torque bracket.

17. Remove the remaining starter mounting bolts and take out the starter.

18. Remove the remaining transaxle mounting bolts and the upper bolt from the engine damper bracket. Remove the clutch housing bolts from the

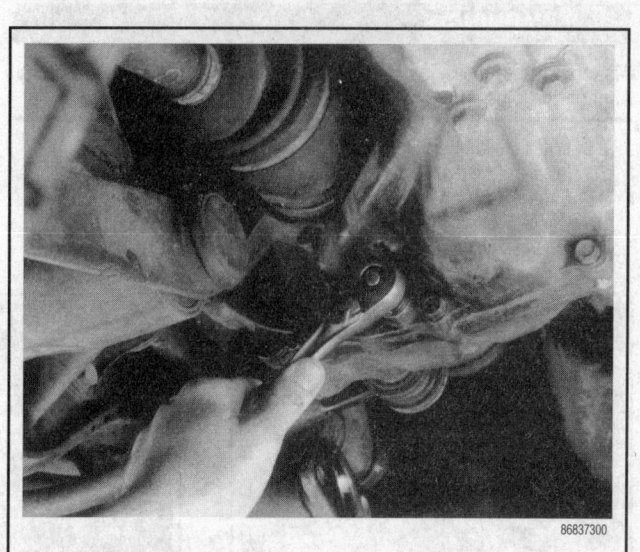

Fig. 9 Unbolting the transaxle drain plug—1984-87 style CRX shown

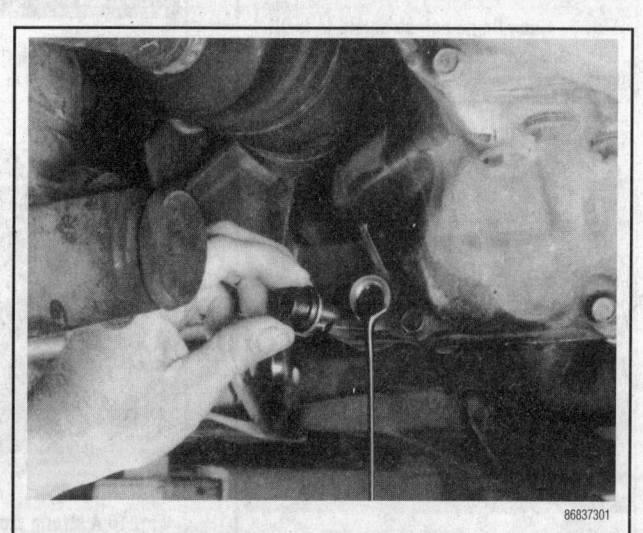

Fig. 10 Drain the transaxle into a suitable container—warm transaxle oil will drain much faster

7-6 DRIVE TRAIN

rear transaxle mounting bracket. Remove the one remaining bolt from the engine.

19. Start backing the transaxle away from the engine and remove the two lower damper bolts.

20. Pull the transaxle clear of the engine and lower the jack.

To install:

21. To ease installation, fabricate two 14mm diameter dowel pins and install them in the clutch housing.

22. Raise the transaxle and slide it onto the dowels. Slide the transaxle onto position aligning the mainshaft splines with the clutch plate.

23. Attach the damper lower bolts when the positioning allows. Tighten both bolts until the clutch housing is seated against the block.

24. Install two lower mounting bolts and tighten them to 33 ft. lbs. (45 Nm).

25. Install the front and rear torque rod brackets. Tighten the front torque rod bolts to 54 ft. lbs., the front bracket bolts to 33 ft. lbs. (45 Nm), the rear torque rod bolts to 54 ft. lbs., and the rear bracket bolts to 47 ft. lbs. (65 Nm).

26. Remove the transaxle jack.

27. Install the starter and tighten the mounting bolts to 33 ft. lbs. (45 Nm).

28. Turn the right steering knuckle out far enough to fit the CV-joint end into the transaxle. Use new 26mm spring clips on both axles. Repeat procedure for the other side.

✱✱ WARNING

Make sure that the axles bottom fully so that you feel the spring clip engage the differential.

29. Install the lower ball joints. Tighten the nuts to 32 ft. lbs. (43.4 Nm).
30. Install the tie rods. Tighten the nuts to 32 ft. lbs. (43.4 Nm).
31. Connect the shift linkage.
32. Connect the shift lever torque rod to the clutch housing and tighten the bolt to 7 ft. lbs. (9.5 Nm).
33. Install the stabilizer bar.
34. Install the lower shields.
35. Install the front wheels and tighten the lug nuts to specification.
36. Install the remaining starter bolts and tighten to 33 ft. lbs. (45 Nm).
37. Install the clutch slave cylinder or install the clutch cable at the release arm.
38. Install the speedometer cable holder using a new O-ring coated with clean engine oil.
39. Connect all engine compartment wiring.
40. Fill the transaxle with SAE 10W-40 engine oil.

1988–91 Models

♦ See Figures 17, 18, 19 and 20

➥**Make sure jack stands are in proper position. Apply the parking brake and block the rear wheels.**

1. Disconnect the positive and negative battery cables from the battery.
2. Remove the three mount bolts and loosen the one bolt located at the side of the battery base.
3. Remove the intake hose band at the throttle body.
4. Remove the air cleaner case complete with the intake hose. Disconnect the starter and transaxle ground cables.
5. Disconnect the speedometer, but be sure not to disassemble the speedometer gear holder.
6. Disconnect the back-up light switch connector and the clutch cable release arm.
7. Drain the transaxle fluid into a suitable drain pan. Disconnect the connectors and remove the mount bolts.
8. Remove the distributor assembly for additional access.
9. Remove the starter mounting bolts and remove the starter assembly. Remove the engine splash shield and the right wheel well splash shield.
10. Remove the header pipe. Remove the cotter pin and the lower arm ball joint nut. Separate the ball joint and lower arm.

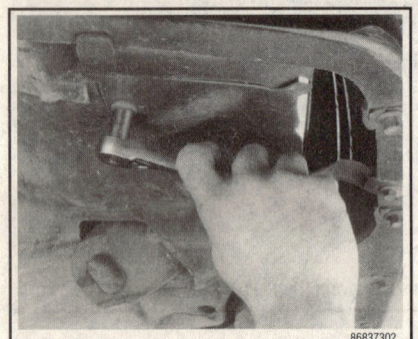

Fig. 11 Unbolting a splash shield to facilitate transaxle removal—1984-87 style CRX shown

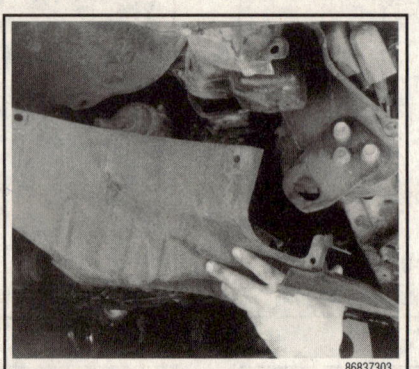

Fig. 12 Removing the right-side splash shield—1984-87 style CRX shown

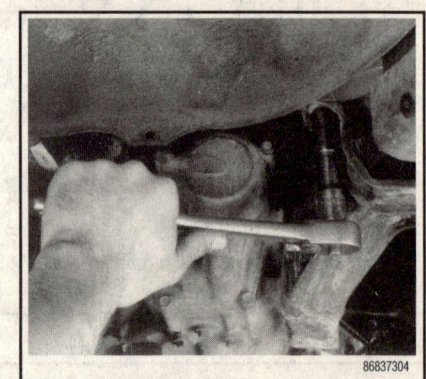

Fig. 13 Unbolt the front transaxle mount . . .

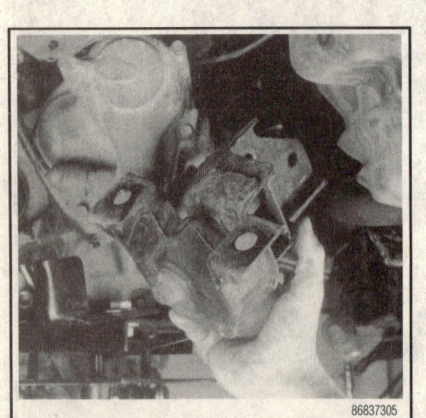

Fig. 14 . . . then remove it from the vehicle

Fig. 15 A strong cross bar and tie-strap suspended on pads between the front fenders can be used in lieu of a hoist

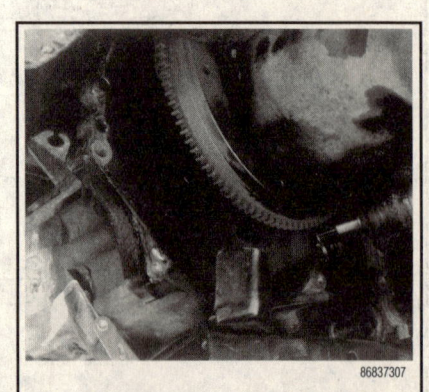

Fig. 16 The transaxle is here being pulled away from the engine—note the support strap securing the transaxle

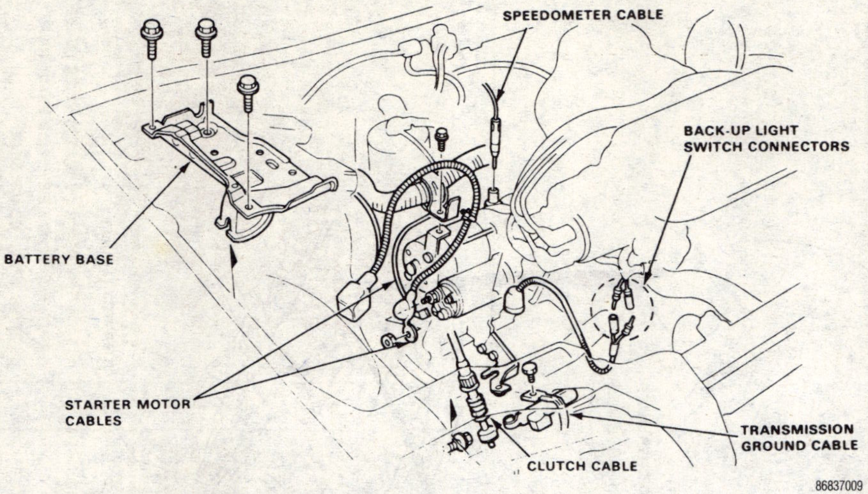

DRIVE TRAIN 7-7

Fig. 17 Disconnect the components before removing the transaxle

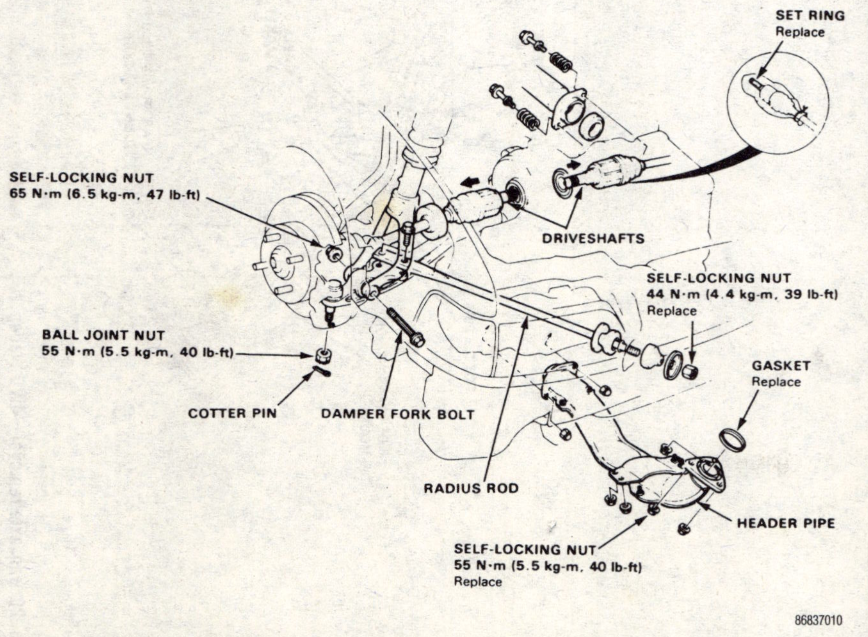

Fig. 18 Disconnect the halfshafts and header pipe before removing the transaxle

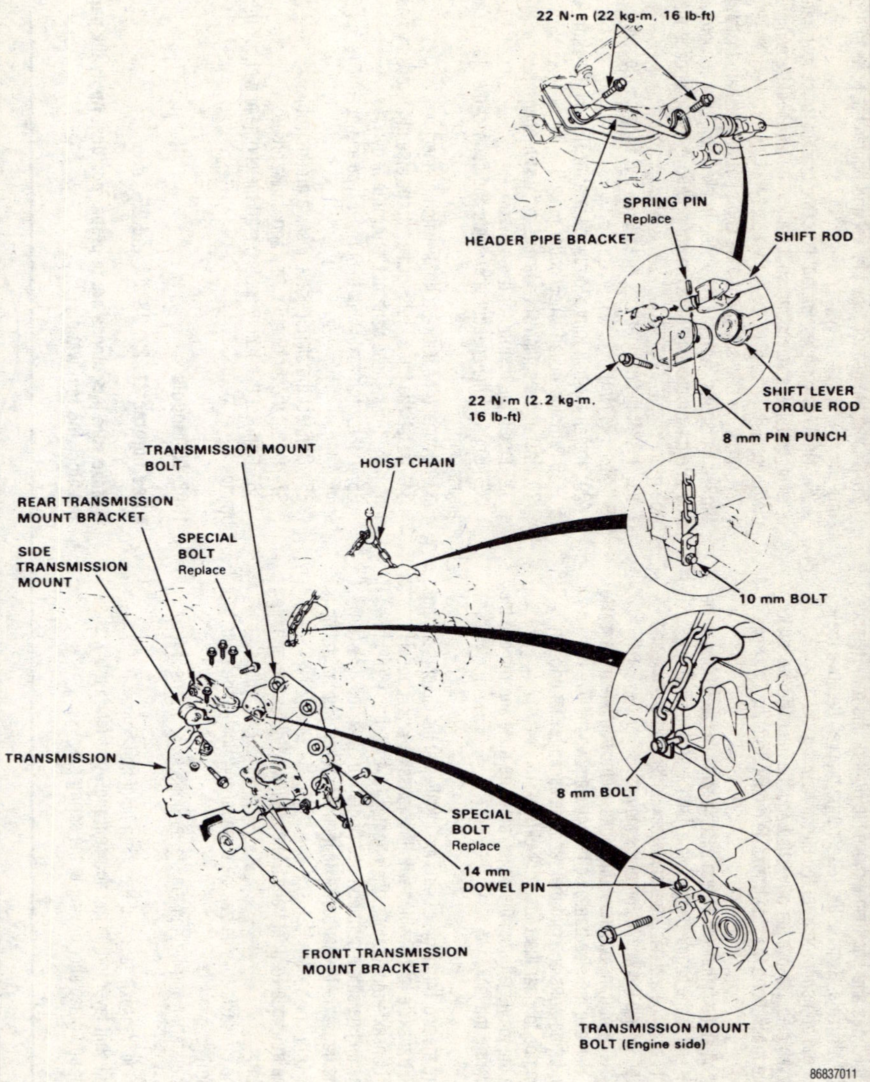

Fig. 19 Removing the transaxle from the vehicle

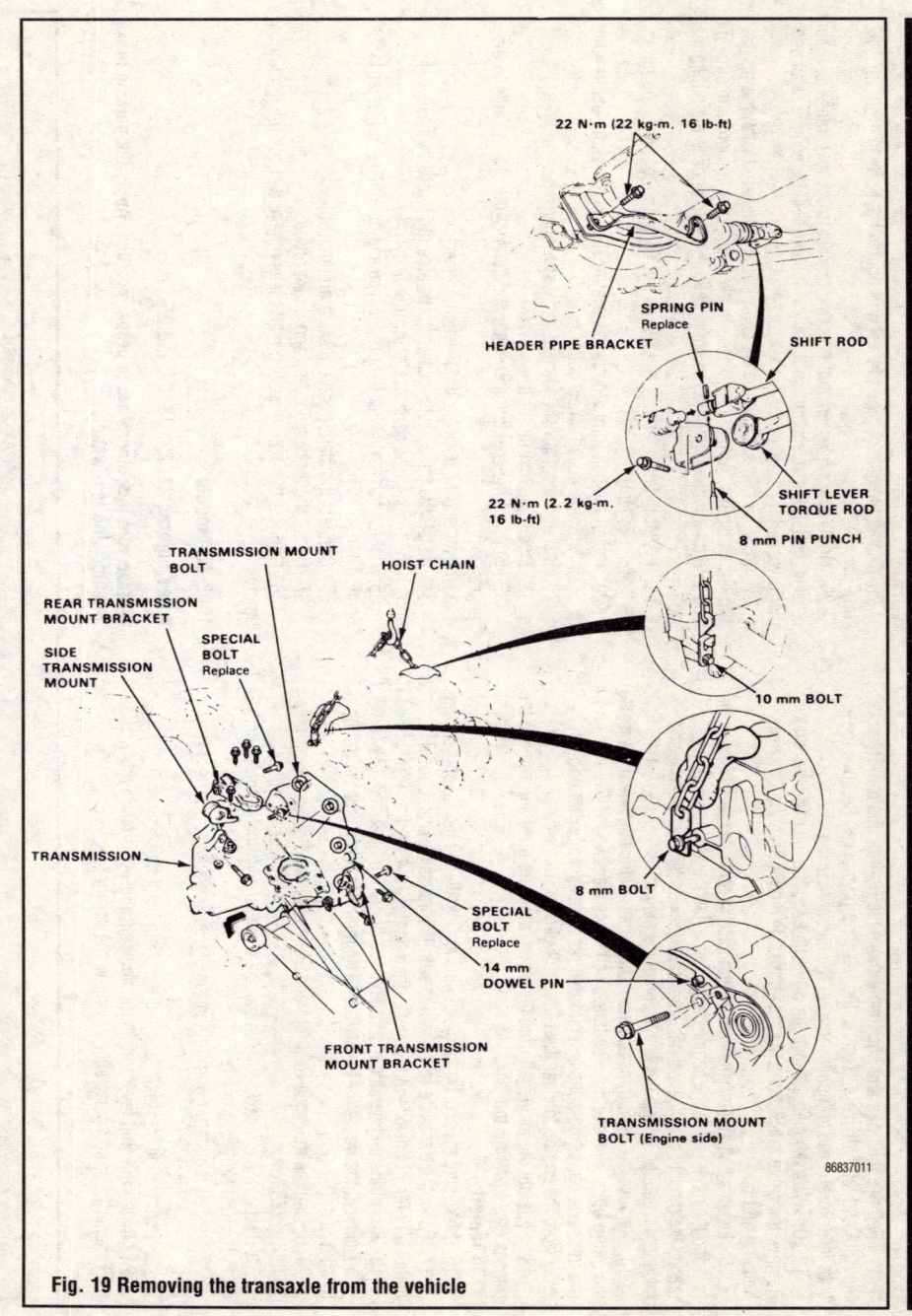

7-8 DRIVE TRAIN

11. Remove the bolts and nut, then remove the right radius rod. Remove the right and left halfshafts. Remove the header pipe bracket. Remove the shift lever torque rod and shift rod from the clutch housing.
12. On 4WD vehicles, remove the driveshaft and the intermediate shaft. Remove the cable bracket and the side transaxle mount from the transaxle housing and body.
13. Install a bolt at the cylinder head and attach a suitable chain hoist to the bolt and the other end to the engine hanger plate. Lift the engine slightly to unload the mounts or place a suitable transaxle jack under the transaxle and raise it just enough to take the weights off of the mounts.
14. Remove the front transaxle mounting bolts. Remove the rear transaxle mounting bolts. Remove the side transaxle mount. Remove the five remaining transaxle mounting bolts and pull the transaxle assembly far enough away from the engine to clear the 14mm dowel pins.
15. Separate the mainshaft from the clutch pressure plate and remove the transaxle by lowering the jack.

To install:

16. Make sure the two 14mm dowel pins are installed in the clutch housing.
17. Raise the transaxle into position with the transaxle jack. Loosely install all of the transaxle mount bolts, then tighten to 43 ft. lbs. (60 Nm).
18. Install the engine side mount bolt and tighten to 50 ft. lbs. (68 Nm).
19. Install the transaxle-to-rear transaxle mount bracket. Tighten the bolts to 40 ft. lbs. (55 Nm).
20. Install the transaxle-to-front transaxle mount and tighten the bolts to 29 ft. lbs. (40 Nm).
21. Install the transaxle-to-side transaxle mount and tighten the bolts to 43 ft. lbs. (60 Nm).
22. Install the starter and tighten the starter bolts to 33 ft. lbs. (45 Nm).
23. Remove the transaxle jack and the chain hoist and bolts.
24. Install the shift lever torque rod and shift rod. After reassembly, slide the retainer back into place after driving in the spring pin.
25. Install the header pipe bracket and tighten the bolts to 16 ft. lbs. (22 Nm).
26. On 4WD vehicles, install the cable bracket to the rear transaxle mount bracket. Install the select and shift cables. Install the intermediate shaft and the driveshaft.
27. Install a new set ring on the end of each halfshaft. Install the right and left halfshafts. Turn the right steering knuckle fully outward and slide the axle into the differential, until the spring clip is felt to engage the differential side gear.
28. Install the damper fork and radius rod. Tighten the radius rod nut to 39 ft. lbs. (44 Nm).
29. Install the ball joints to the lower arm. Tighten the stud nuts to 40 ft. lbs. (55 Nm) and install a new cotter pin.
30. Install the splash shields and exhaust header pipe.
31. Install the distributor if removed.
32. Connect the speedometer cable and connect the clutch cable to the release arm.
33. Connect the back-up light switch connector.
34. Install the three bolts located at the side of the battery base and tighten the intake hose band of the throttle body.
35. Fill the transaxle with the proper type and quantity of oil (refer to Section 1).
36. Connect the starter and transaxle ground cable.
37. Install the air cleaner case and intake hose.
38. Connect the battery cables, start the engine and check the ignition timing.

1992–95 Models

▶ See Figures 21, 22, 23, 24 and 25

➡ Make sure jack stands are in proper position. Apply the parking brake and block the rear wheels.

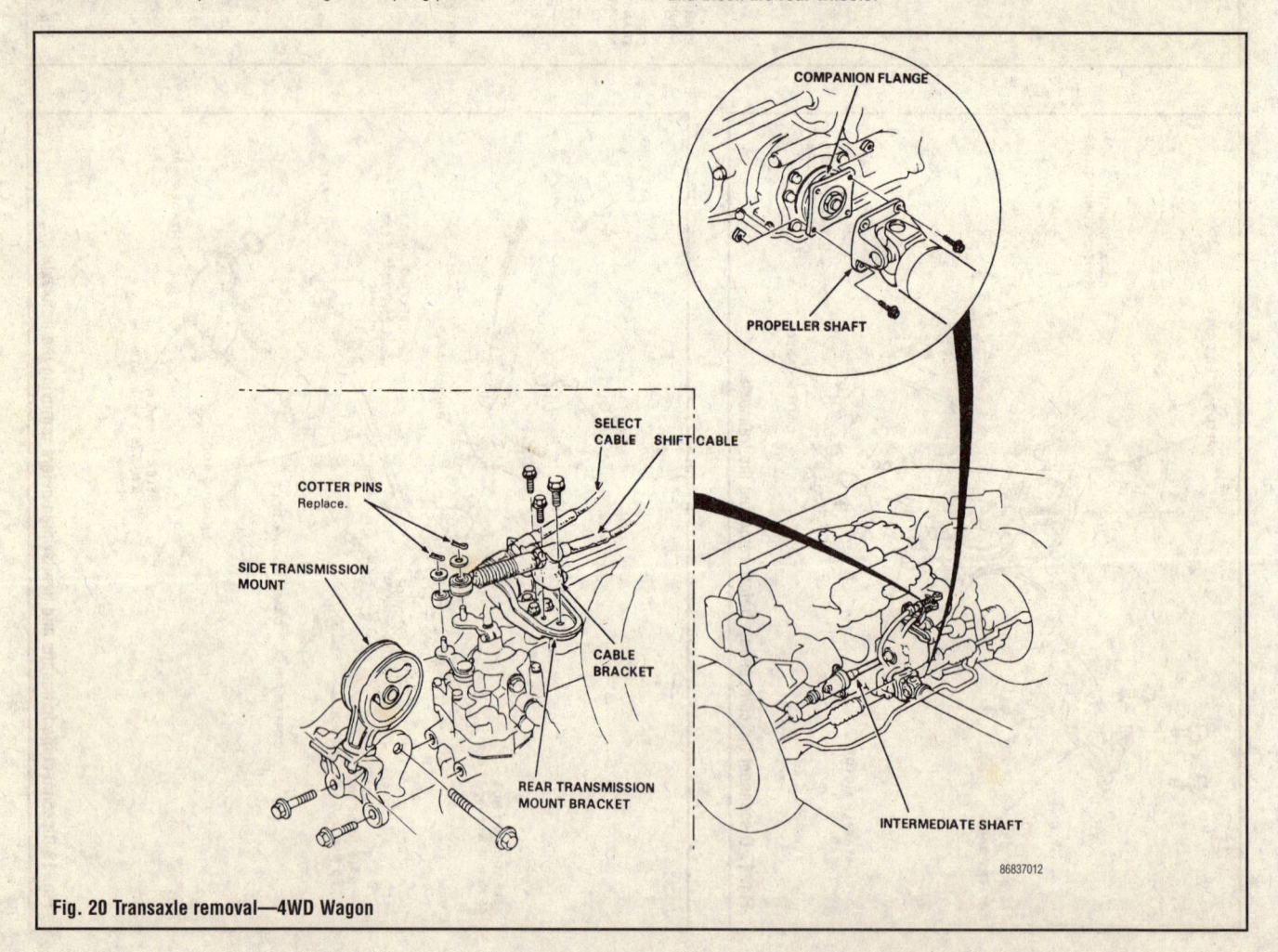

Fig. 20 Transaxle removal—4WD Wagon

DRIVE TRAIN   7-9

> ✲✲ **WARNING**
>
> Use fender covers to avoid damaging painted surfaces.

1. Disconnect the positive and negative battery cables from the battery.
2. Remove the resonator, intake air duct, and air cleaner assembly.
3. Drain the transaxle oil.
4. Disconnect the starter motor cables, transaxle ground wire, and remove the wire harness clamp.
5. Disconnect the Vehicle Speed Sensor (VSS) and back up light switch connector.
6. Remove the clutch pipe bracket and slave cylinder.

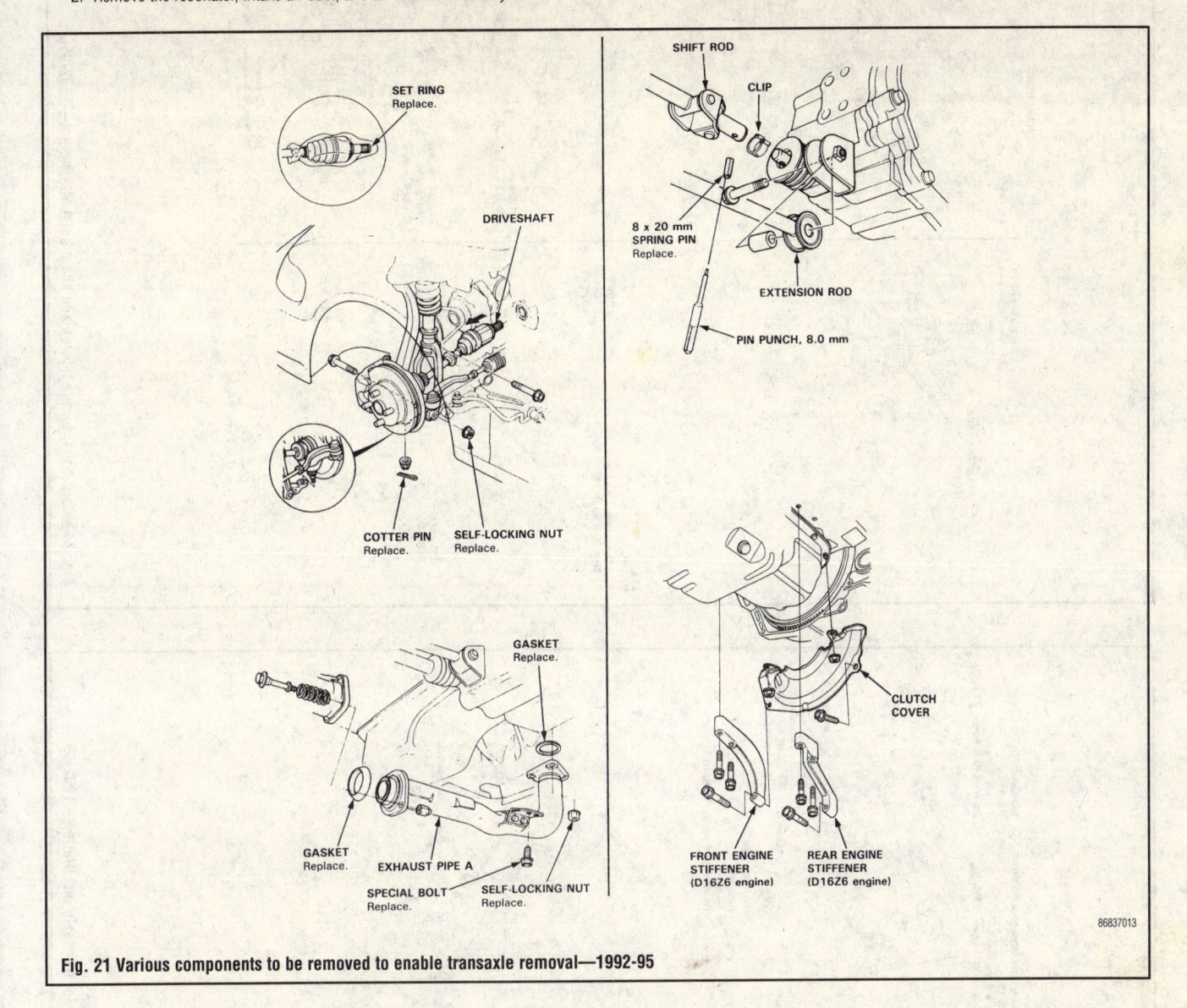

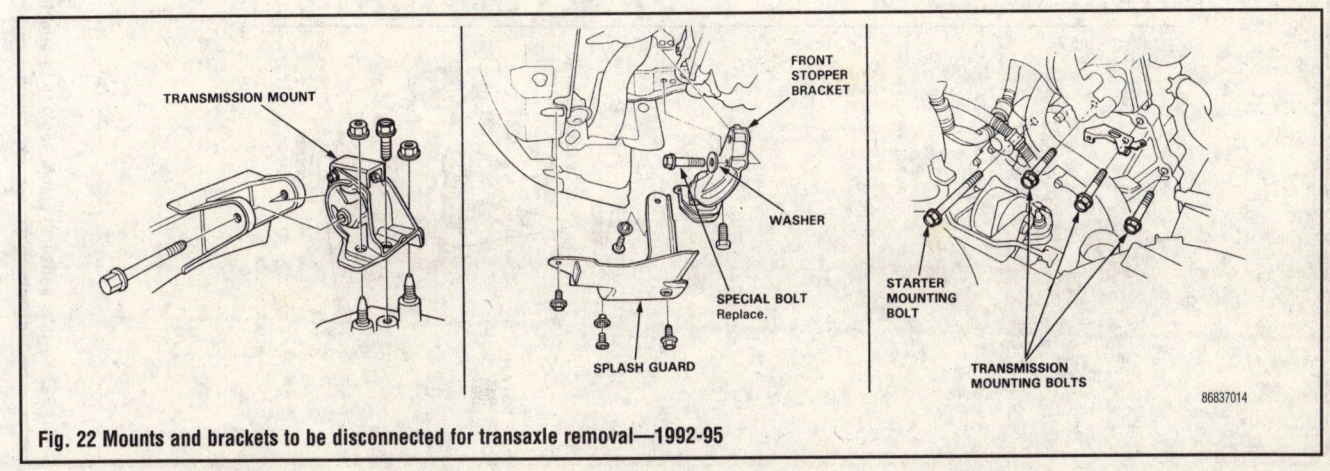

Fig. 21 Various components to be removed to enable transaxle removal—1992-95

Fig. 22 Mounts and brackets to be disconnected for transaxle removal—1992-95

7-10 DRIVE TRAIN

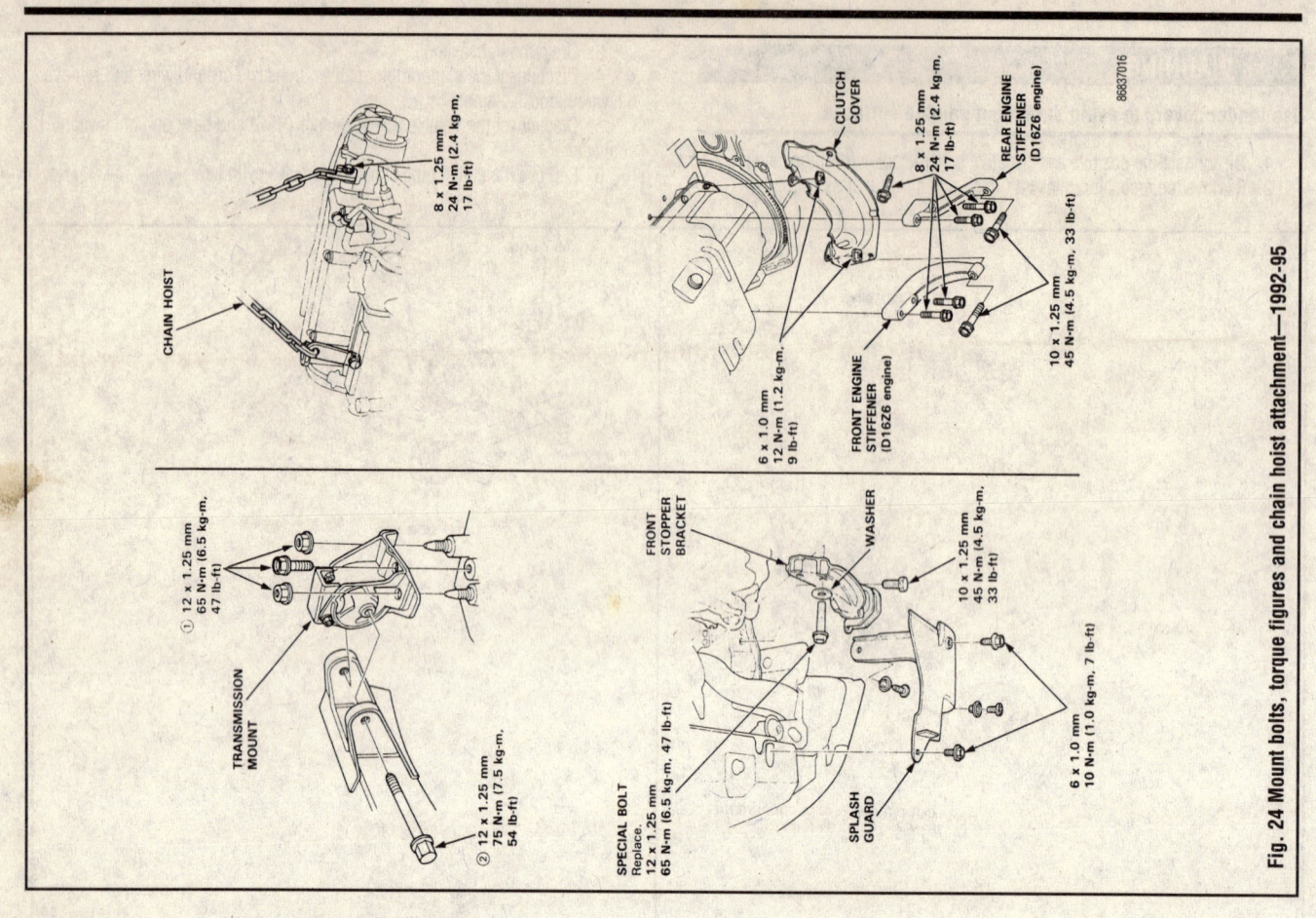

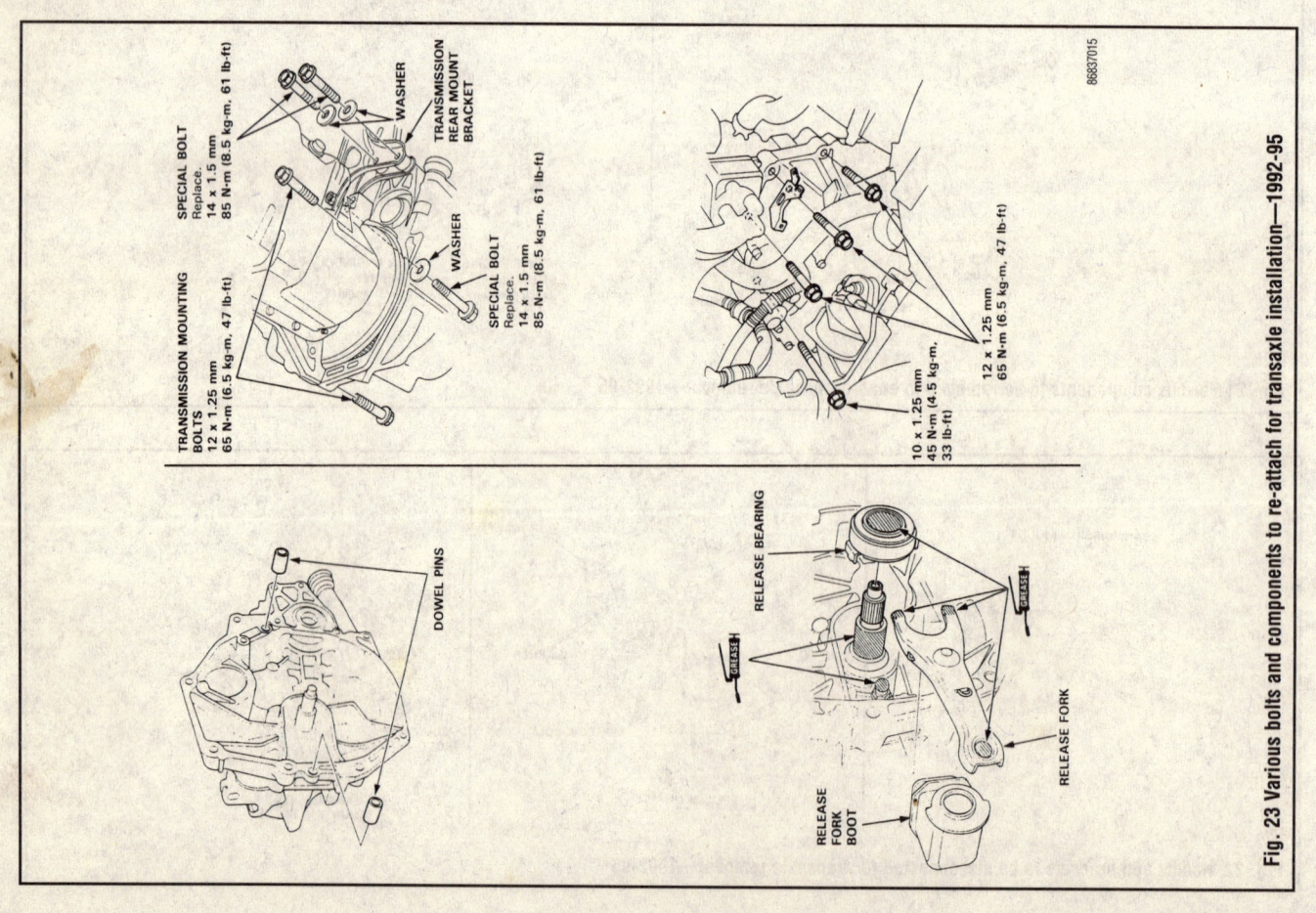

Fig. 24 Mount bolts, torque figures and chain hoist attachment—1992-95

Fig. 23 Various bolts and components to re-attach for transaxle installation—1992-95

DRIVE TRAIN 7-11

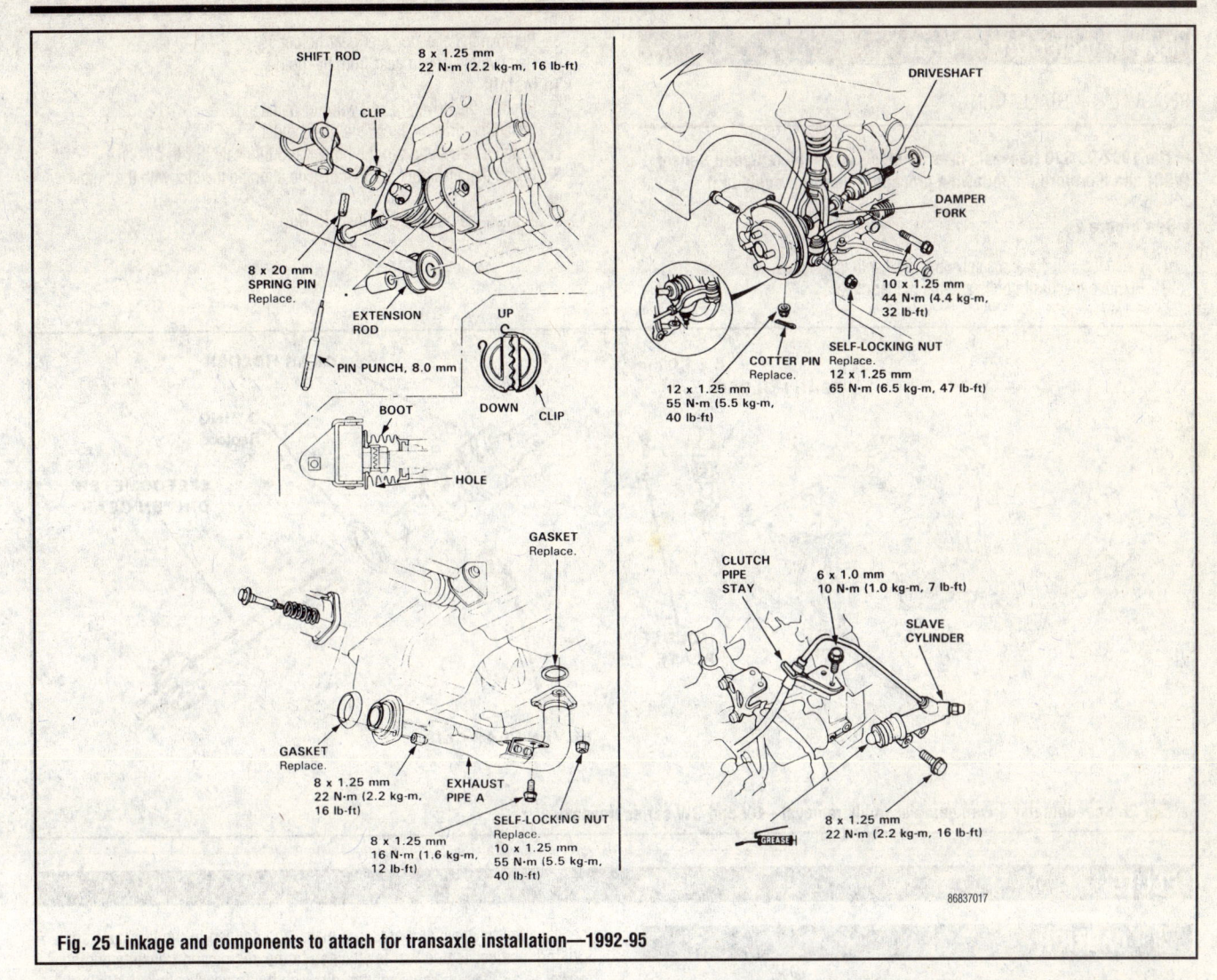

Fig. 25 Linkage and components to attach for transaxle installation—1992-95

7. Remove the transaxle halfshafts and coat the precision finished surfaces with clean grease. Tie sturdy plastic bags over the halfshaft ends.
8. Remove exhaust pipe A (header pipe).
9. Remove the shift rod and extension rod.
10. For D16Z6 engines, remove the front and rear engine stiffeners.
11. Remove the clutch cover.
12. Install a bolt at the cylinder head and attach a suitable chain hoist to the bolt and the other end to the engine hanger plate. Lift the engine slightly to unload the mounts or place a suitable transaxle jack under the transaxle and raise it just enough to take the weights off of the mounts.
13. Remove the splash gaurds and front stopper bracket.
14. Place a jack under the transaxle.
15. Remove the transaxle mount.
16. Remove the three upper transaxle mounting bolts and the lower starter mounting bolt.
17. Remove the transaxle rear mount bracket bolts and the transaxle mounting bolts. Pull away the transaxle from the engine until it clears the mainshaft, then lower the transaxle using the transmission jack.

To install:

18. Make sure the two dowel pins are installed in the clutch housing.
19. Apply Super High Temp Urea Grease (Honda part No. 080798-9002 or suitable aftermarket equivalent) as shown in the accompanying drawing, then install the release fork, release bearing and release fork boot.
20. Raise the transaxle into position with the transaxle jack. Loosely install the transaxle mount bolts and rear mount bracket bolts. Then tighten the mount bolts to 47 ft. lbs. (65 Nm) and tighten new rear mount bolts to 61 ft. lbs. (85 Nm).
21. Place the three upper transaxle mounting bolts in position, then tighten to 47 ft. lbs. (65 Nm). Install the lower starter mounting bolts in position and tighten to 33 ft. lbs. (45 Nm).
22. Install the engine side mount bolt and tighten to 54 ft. lbs. (75 Nm).
23. Install the front stopper bracket and install the splash gaurd.
24. Install the distributor mounting bolt.
25. Install the clutch cover.
26. For D16Z6 engines, install the front and rear engine stiffeners as shown in the drawing.
27. Install the shift rod, spring pin and clip making sure the boot is installed on the shift rod and the hole is facing down.
28. Install the extension rod.
29. Install exhaust pipe A (header pipe), and tighten the bolts to 16 ft. lbs. (22 Nm).
30. Install the halfshafts using new set rings.
31. Install the ball joint onto the lower arm.
32. Install the clutch slave cylinder and then the clutch pipe stay.
33. Plug in the Vehicle Speed Sensor connector, transaxle ground, back-up light switch connector, transaxle ground wire and starter motor cables. Make sure that the crimped side of the ring is facing out.
34. Install the resonator, air cleaner asembly and air intake duct.
35. Turn the breather cap so that the "F" mark points at an angle of 30° towards the front of the car.
36. Refill the transaxle with oil, connect the battery cables and check for proper clutch and shifting function.
37. Check ignition timing and adjust.
38. Check front wheel alignment.

7-12 DRIVE TRAIN

Speedometer Driven Gear

REMOVAL & INSTALLATION

➡ The 1992-95 S20 transaxle uses an electronic Vehicle Speed Sensor (VSS) and therefore the following procedure does not apply to it.

◆ See Figure 26

1. Raise the vehicle and safely support it with jackstands.
2. Remove the 10mm bolts and the holder plate.
3. Remove the speedometer gear holder.
4. Remove the driven gear from the holder.

To install:

5. Replace the driven gear if worn or damaged.
6. Install the driven gear to the gear holder.
7. Install a new O-ring to the holder and lubricate it with oil.
8. Install the assembly in the transaxle aligning the slot with the holder plate.
9. Tighten the bolts to 9 ft. lbs. (12 Nm).

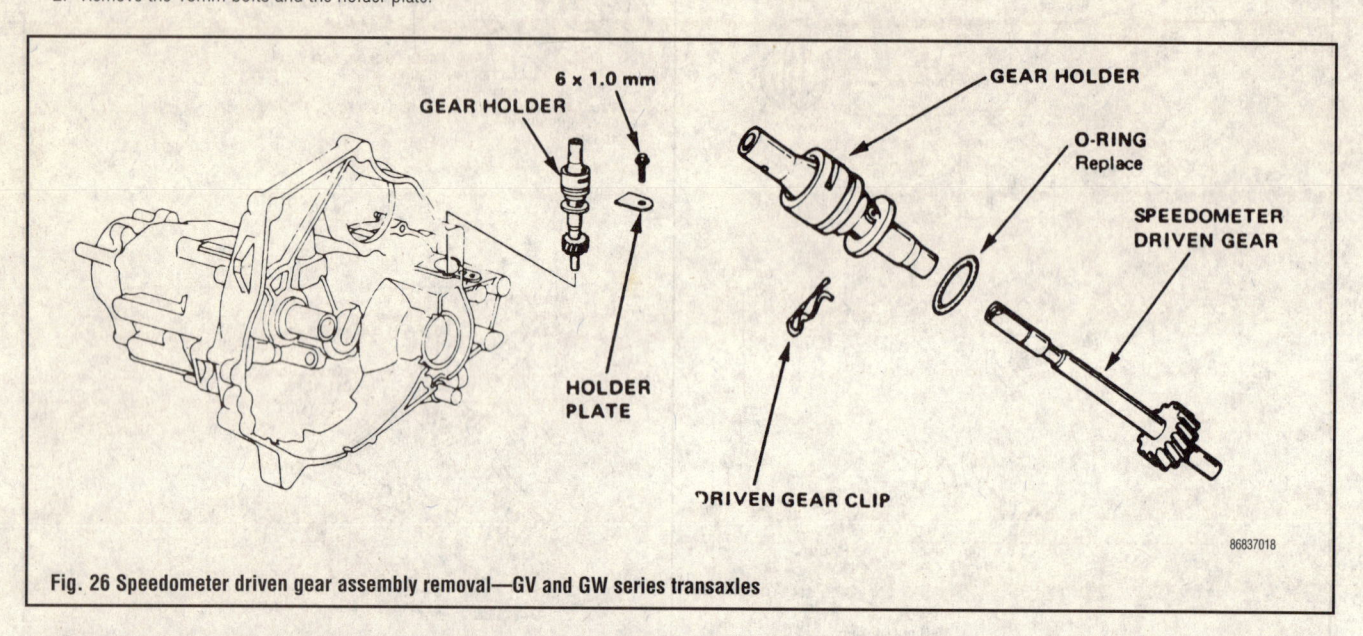

Fig. 26 Speedometer driven gear assembly removal—GV and GW series transaxles

CLUTCH

✶✶ CAUTION

The clutch driven disc may contain asbestos, which has been determined to be a cancer causing agent. Never clean clutch surfaces with compressed air! Avoid inhaling any dust from any clutch surface! When cleaning clutch surfaces, use a commercially available brake cleaning fluid.

Understanding the Clutch

The purpose of the clutch is to disconnect and connect engine power from the transaxle. A car at rest requires a lot of engine torque to get its weight moving. An internal combustion engine does not develop a high starting torque (unlike steam engines), so it must be allowed to operate without any load until it builds up enough torque to move the car. Torque increases with engine rpm. The clutch allows the engine to build up torque by physically disconnecting the engine from the transaxle and relieving the engine of any load or resistance. The transfer of engine power to the transaxle (the load) must be smooth and gradual. If it were not, drive line components would wear out or break quickly. This gradual power transfer is made possible by gradually releasing the clutch pedal. The clutch disc and pressure plate are the connecting link between the engine and transaxle. When the clutch pedal is released, the disc and plate contact each other (clutch engagement), physically joining the engine and transaxle. When the pedal is pushed in, the disc and plate separate (the clutch is disengaged), disconnecting the engine from the transaxle.

The clutch assembly consists of the flywheel, the clutch disc, the clutch pressure plate, the throwout bearing and fork, the actuating linkage and the pedal. The flywheel and clutch pressure plate (driving members) are connected to the engine crankshaft and rotate with it. The clutch disc is located between the flywheel and pressure plate, and splined to the transaxle shaft. A driving member is one that is attached to the engine and transfers engine power to a driven member (clutch disc) on the transaxle shaft. A driving member (pressure plate) rotates (drives) a driven member (clutch disc) on contact and, in so doing, turns the transaxle shaft. There is a circular diaphragm spring within the pressure plate cover (transaxle side). In a relaxed state (when the clutch pedal is fully released), this spring is convex; that is, it is dished outward toward the transaxle. Pushing in the clutch pedal actuates an attached linkage rod. Connected to the other end of this rod is the throwout bearing fork. The throwout bearing is attached to the fork. When the clutch pedal is depressed, the clutch linkage pushes the fork and bearing forward to contact the diaphragm spring of the pressure plate. The outer edges of the spring are secured to the pressure plate and are pivoted on rings so that when the center of the spring is compressed by the throwout bearing, the outer edges bow outward and, by so doing, pull the pressure plate in the same direction — away from the clutch disc. This action separates the disc from the plate, disengaging the clutch and allowing the transaxle to be shifted into another gear. A coil type clutch return spring attached to the clutch pedal arm permits full release of the pedal. Releasing the pedal pulls the throwout bearing away from the diaphragm spring resulting in a reversal of spring position. As bearing pressure is gradually released from the spring center, the outer edges of the spring bow outward, pushing the pressure plate into closer contact with the clutch disc. As the disc and plate move closer together, friction between the two increases and slippage is reduced until, when full spring pressure is applied (by fully releasing the pedal), the speed of the disc and plate are the same. This stops all slipping, creating a direct connection between the plate and disc which results in the transfer of power from the engine to the transaxle. The clutch disc is now rotating with the pressure plate at engine speed and, because it is splined to the transaxle shaft, the shaft now turns at the same engine speed. Understanding clutch operation can be rather difficult at first; if you

DRIVE TRAIN 7-13

are still do not clearly understand after reading this, consider the following analogy. The action of the diaphragm spring can be compared to that of an oil can bottom. The bottom of an oil can is shaped very much like the clutch diaphragm spring and pushing in on the can bottom and then releasing it produces a similar effect. As mentioned earlier, the clutch pedal return spring permits full release of the pedal and reduces linkage slack due to wear. As the linkage wears, clutch free-pedal travel will increase and free-travel will decrease as the clutch wears. Free-travel is actually throwout bearing lash.

The diaphragm spring type clutches used are available in two different designs: flat diaphragm springs or bent spring. The bent fingers are bent back to create a centrifugal boost ensuring quick re-engagement at higher engine speeds. This design enables pressure plate load to increase as the clutch disc wears and makes low pedal effort possible even with a heavy duty clutch. The throwout bearing used with the bent finger design is 32mm long and is shorter than the bearing used with the flat finger design. These bearings are not interchangeable. If the longer bearing is used with the bent finger clutch, free-pedal travel will not exist. This results in clutch slippage and rapid wear.

The transaxle varies the gear ratio between the engine and drive wheels. It can be shifted to change engine speed as driving conditions and loads change. The transaxle allows disengaging and reversing power from the engine to the wheels.

Adjustments

PEDAL HEIGHT/FREE-PLAY ADJUSTMENT

▶ See Figure 27

1. Measure the clutch pedal disengagement height.
2. Measure the clutch pedal free-play.
3. Adjust the clutch pedal free-play by turning the clutch cable adjusting nut, found at the end of the clutch cable housing near the release shaft.
4. Turn the adjusting nut until the clutch pedal free-play is 0.6–0.8 in. (15–20mm).
5. After adjustment, make sure the free-play at the tip of the release arm is 0.12–0.16 in. (3.0–4.0mm).

Clutch Cable

▶ See Figure 28

REMOVAL & INSTALLATION

1. Disconnect the negative battery cable.
2. Disconnect the cable end from the brake pedal.
3. Remove the adjuster nut assembly from its mounting.
4. Raise and support the vehicle safely.
5. Disconnect the cable end from the release arm. Remove the cable from the vehicle.
6. Installation is the reverse of the removal procedure. Adjust the cable to specification.

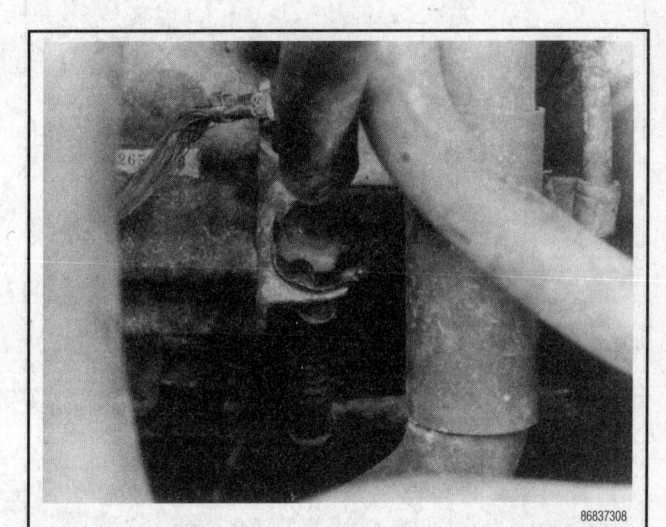

Fig. 28 Clutch cable and adjuster nut asembly at the release arm—1984-87 style CRX shown

Driven Disc and Pressure Plate

REMOVAL & INSTALLATION

▶ See Figures 29 thru 40

✶✶ CAUTION

The clutch driven disc may contain asbestos, which has been determined to be a cancer causing agent. Never clean clutch surfaces with compressed air! Avoid inhaling any dust from any clutch surface! When cleaning clutch surfaces, use a commercially available brake cleaning fluid.

1. Disconnect the negative battery cable. Raise and safely support the vehicle, refer to the procedures found earlier in this section. Remove the transaxle from the vehicle. Matchmark the flywheel and clutch for reassembly.

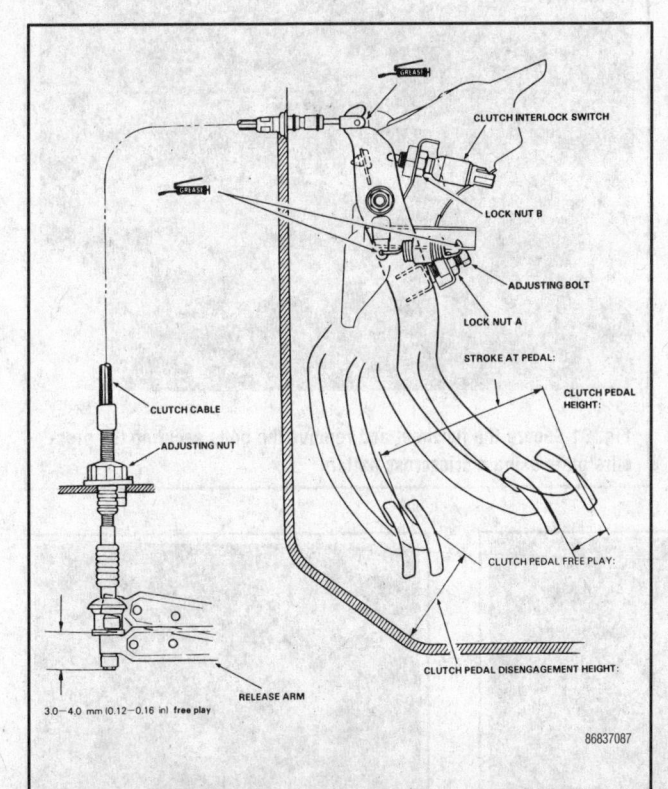

Fig. 27 Clutch pedal height and free-play adjustment

7-14 DRIVE TRAIN

2. Hold the flywheel ring gear with a tool made for this purpose or an equivalent method of holding the flywheel. Remove the retaining bolts and remove the pressure plate and clutch disc. Remove the bolts two turns at a time working in a crisscross pattern, to prevent warping the pressure plate.

3. At this time, inspect the flywheel for wear, cracks or scoring and resurface or replace, as necessary.

4. If the clutch release bearing is to be replaced, perform the following procedure:
   a. Remove the 8mm special bolt.
   b. Remove the release shaft and the release bearing assembly.
   c. Separate the release fork from the bearing by removing the release fork spring from the holes in the release bearing.

5. Check the release bearing for excessive play by spinning it by hand. Do not wash with solvent as it is packed with grease. Replace if there is excessive play.

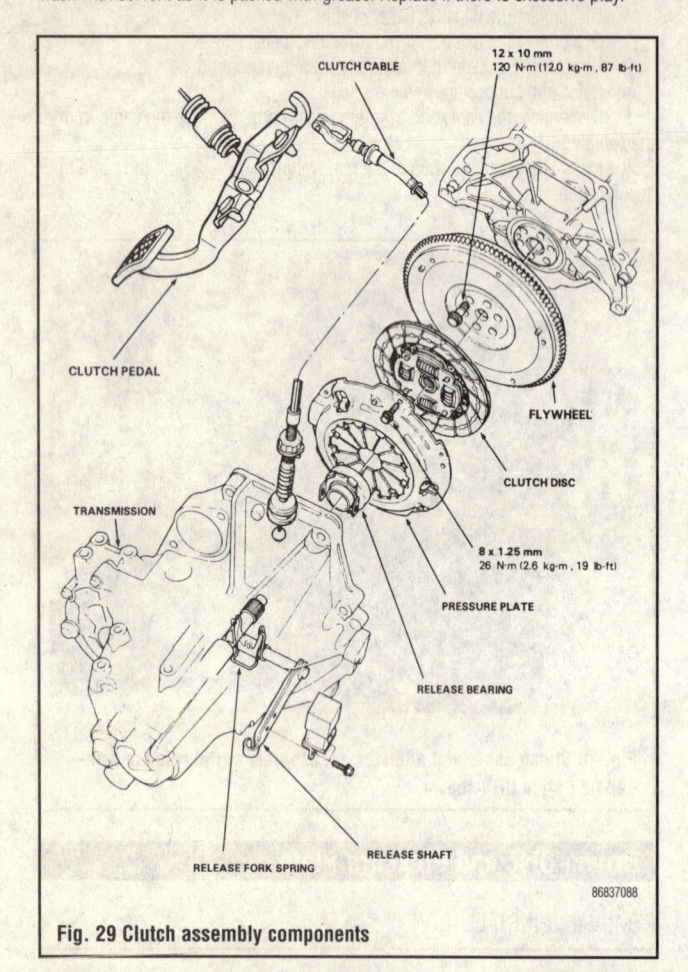

To install:

6. If the flywheel was removed, make sure the flywheel and crankshaft mating surfaces are clean. Align the hole in the flywheel with the crankshaft dowel pin and install the flywheel bolts finger tight. Install the ring gear holder and tighten the flywheel bolts in a crisscross pattern. Tighten the flywheel bolts to 87 ft. lbs. (120 Nm).

Fig. 29 Clutch assembly components

Fig. 30 View of the clutch assembly mounted on the end of the crankshaft as it appears upon removal of the transaxle housing

Fig. 31 Secure the flywheel and remove the bolts securing the pressure plate using a crisscross pattern

Fig. 32 During removal, make a a note as to which side of the clutch disc faces outward

Fig. 33 Inspect the flywheel for cracks, scoring or excessive wear—the mounting bolts shown are removable with a 12-point socket

Fig. 34 Inspect the bolt threads. Replace any damaged bolts

DRIVE TRAIN 7-15

Fig. 35 After its attaching bolts are removed, the flywheel can be removed from the crankshaft

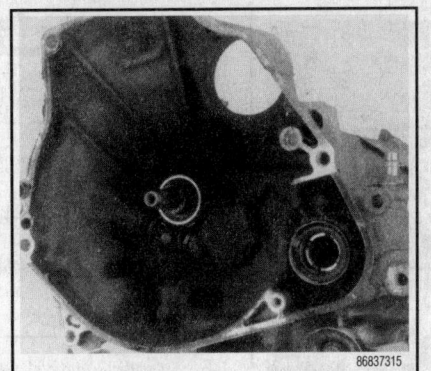

Fig. 36 The transaxle housing with release bearing and release fork intact

Fig. 37 Removing the release bearing

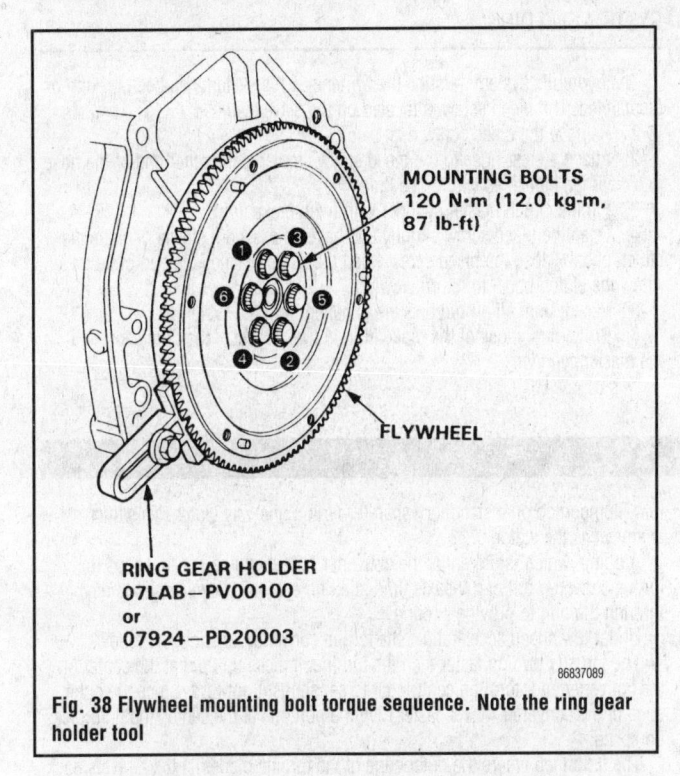

Fig. 38 Flywheel mounting bolt torque sequence. Note the ring gear holder tool

MOUNTING BOLTS
120 N·m (12.0 kg-m, 87 lb-ft)
FLYWHEEL
RING GEAR HOLDER
07LAB—PV00100
or
07924—PD20003

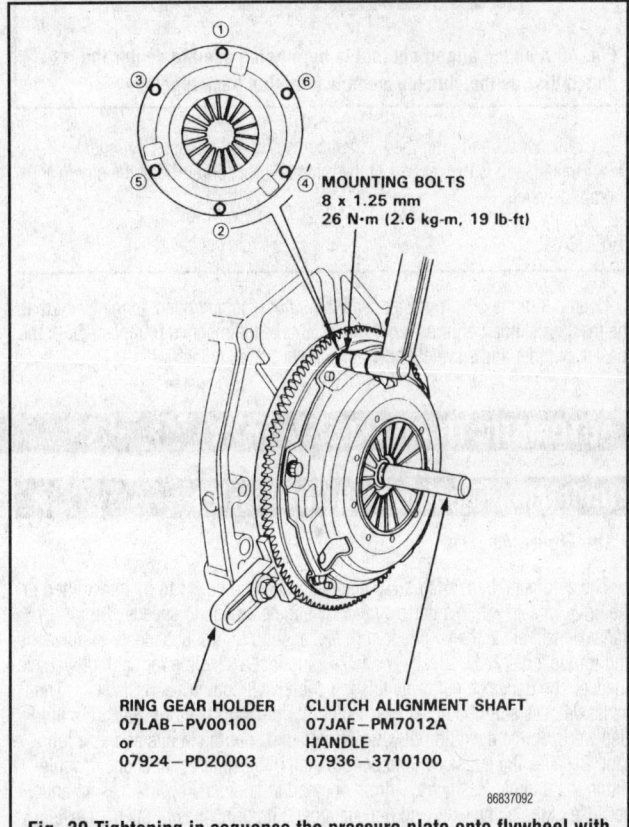

Fig. 39 Tightening in sequence the pressure plate onto flywheel with clutch alignment shaft in place

MOUNTING BOLTS
8 x 1.25 mm
26 N·m (2.6 kg-m, 19 lb-ft)
RING GEAR HOLDER   CLUTCH ALIGNMENT SHAFT
07LAB—PV00100   07JAF—PM7012A
or   HANDLE
07924—PD20003   07936—3710100

7. Install the clutch disc and pressure plate by aligning the dowels on the flywheel with the dowel holes in the pressure plate. If the same pressure plate is being installed that was removed, align the marks that were made during the removal procedure. Install the pressure plate bolts finger tight.

8. Insert a suitable clutch disc alignment tool into the splined hole in the clutch disc. Tighten the pressure plate bolts in a crisscross pattern two turns at a time to prevent warping the pressure plate. The final torque should be 20 ft. lbs. (26 Nm).

9. Remove the alignment tool and ring gear holder.

10. If the release bearing was removed, replace it in the reverse order of the removal procedure. Place a light coat of molybdenum disulfide grease on the inside diameter of the bearing prior to installation.

11. Install the transaxle, making sure the mainshaft is properly aligned with the clutch disc splines and the transaxle case is properly aligned with the cylinder block, before tightening the transaxle case bolts.

12. Adjust the clutch pedal free-play and connect the negative battery cable.

Clutch Master Cylinder

REMOVAL & INSTALLATION

1. Disconnect the negative battery cable and disable the SRS system (if equipped.) At the top of the clutch pedal, remove the cotter pin and pull the pedal pin out of the yoke.

2. Remove the nuts and bolts holding the clutch master cylinder and remove the cylinder from the engine compartment.

7-16 DRIVE TRAIN

Fig. 40 A clutch alignment tool is here being used to center the clutch disc as the clutch assembly is bolted back together

3. Disconnect and plug the hydraulic lines from the master cylinder.
4. Installation is the reverse of the removal procedure. Bleed the clutch hydraulic system.

OVERHAUL

Overhaul of the clutch master cylinder is not recommended. In the event that the master cylinder requires replacement, it is recommended to replace both the master cylinder and slave cylinder as a set for added reliability.

Slave Cylinder

REMOVAL & INSTALLATION

1. Disconnect the negative battery cable and disable the SRS system, (if equipped.)
2. Disconnect and plug the clutch hose from the slave cylinder.
3. Remove the two retaining bolts and remove the slave cylinder.
4. Install in reverse order. Bleed the clutch hydraulic system.

OVERHAUL

Overhaul of the slave cylinder is not recommended. In the event that the cylinder requires replacement, it is recommended to replace both the master cylinder and slave cylinder as a set for added reliability.

SYSTEM BLEEDING

The hydraulic system must be bled whenever the system has been leaking or dismantled. The bleed screw is located on the slave cylinder.
1. Remove the bleed screw dust cap.
2. Attach a clear hose to the bleed screw. Immerse the other end of the hose in a clear jar, half-filled with brake fluid.
3. Fill the clutch master cylinder with fresh brake fluid.
4. Open the bleed screw slightly and have an assistant slowly depress the clutch pedal. Close the bleed screw when the pedal reaches the end of its travel. Allow the clutch pedal to return slowly.
5. Repeat until all air bubbles are expelled from the system.
6. Properly dispose of the brake fluid in the jar. Install the dust cap. Refill the master cylinder.

AUTOMATIC TRANSAXLE

Understanding Automatic Transaxles

♦ See Figure 41

The automatic transaxle allows engine torque and power to be transmitted to the drive wheels within a narrow range of engine operating speeds. The transaxle will allow the engine to turn fast enough to produce plenty of power and torque at very low speeds, while keeping it at a sensible rpm at high vehicle speeds. The transaxle performs this job entirely without driver assistance. The transaxle uses a light fluid as the medium for the transaxle of power. This fluid also works in the operation of various hydraulic control circuits and as a lubricant. Because the transaxle fluid performs all of these three functions, trouble within the unit can easily travel from one part to another. For this reason, and because of the complexity and unusual operating principles of the transaxle, a very sound understanding of the basic principles of operation will simplify troubleshooting.

THE TORQUE CONVERTER

The torque converter replaces the conventional clutch. It has three functions:
1. It allows the engine to idle with the vehicle at a standstill, even with the transaxle in gear.
2. It allows the transaxle to shift from range to range smoothly, without requiring the driver to close the throttle during the shift.
3. It multiplies engine torque to an increasing extent as vehicle speed drops and throttle opening is increased. This has the effect of making the transaxle more responsive and reduces the amount of shifting required.

The torque converter is a metal case which is shaped like a sphere that has been flattened on opposite sides. It is bolted to the rear end of the engine's crankshaft. Generally, the entire metal case rotates at engine speed and serves as the engine's flywheel.

The case contains three sets of blades. One set is attached directly to the case. This set forms the torus or pump. Another set is directly connected to the output shaft, and forms the turbine. The third set is mounted on a hub which, in turn, is mounted on a stationary shaft through a one-way clutch. This third set is known as the stator.

A pump, which is driven by the converter hub at engine speed, keeps the torque converter full of transaxle fluid at all times. Fluid flows continuously through the unit to provide cooling.

Under low speed acceleration, the torque converter functions as follows:
The torus is turning faster than the turbine. It picks up fluid at the center of the converter and, through centrifugal force, slings it outward. Since the outer edge of the converter moves faster than the portions at the center, the fluid picks up speed.

The fluid then enters the outer edge of the turbine blades. It then travels back toward the center of the converter case along the turbine blades. Upon contact with the turbine blades, the fluid loses the energy picked up in the torus.

If the fluid was now to immediately be returned directly into the torus, both halves of the converter would have to turn at approximately the same speed at all times, and torque input and output would both be the same.

In flowing through the torus and turbine, the fluid picks up two types of flow, or flow in two separate directions. It flows through the turbine blades, and it spins with the engine. The stator, whose blades are stationary when the vehicle is being accelerated at low speeds, converts one type of flow into another. Instead of allowing the fluid to flow straight back into the torus, the stator's curved blades turn the fluid almost 90° toward the direction of rotation of the engine. Thus the fluid does not flow as fast toward the torus, but is already spinning when the torus picks it up. This has the effect of allowing the torus to turn much faster than the turbine. This difference in speed may be compared to the difference in speed between the smaller and larger gears in any gear train. The result is that engine power output is higher, and engine torque is multiplied.

As the speed of the turbine increases, the fluid spins faster and faster in the direction of engine rotation. As a result, the ability of the stator to redirect the fluid flow is reduced. Under cruising conditions, the stator is eventually forced to rotate on its one-way clutch in the direction of engine rotation. Under these conditions, the torque converter begins to behave almost like a solid shaft, with the torus and turbine speeds being almost equal.

DRIVE TRAIN 7-17

Fig. 41 Cutaway view of the automatic transaxle

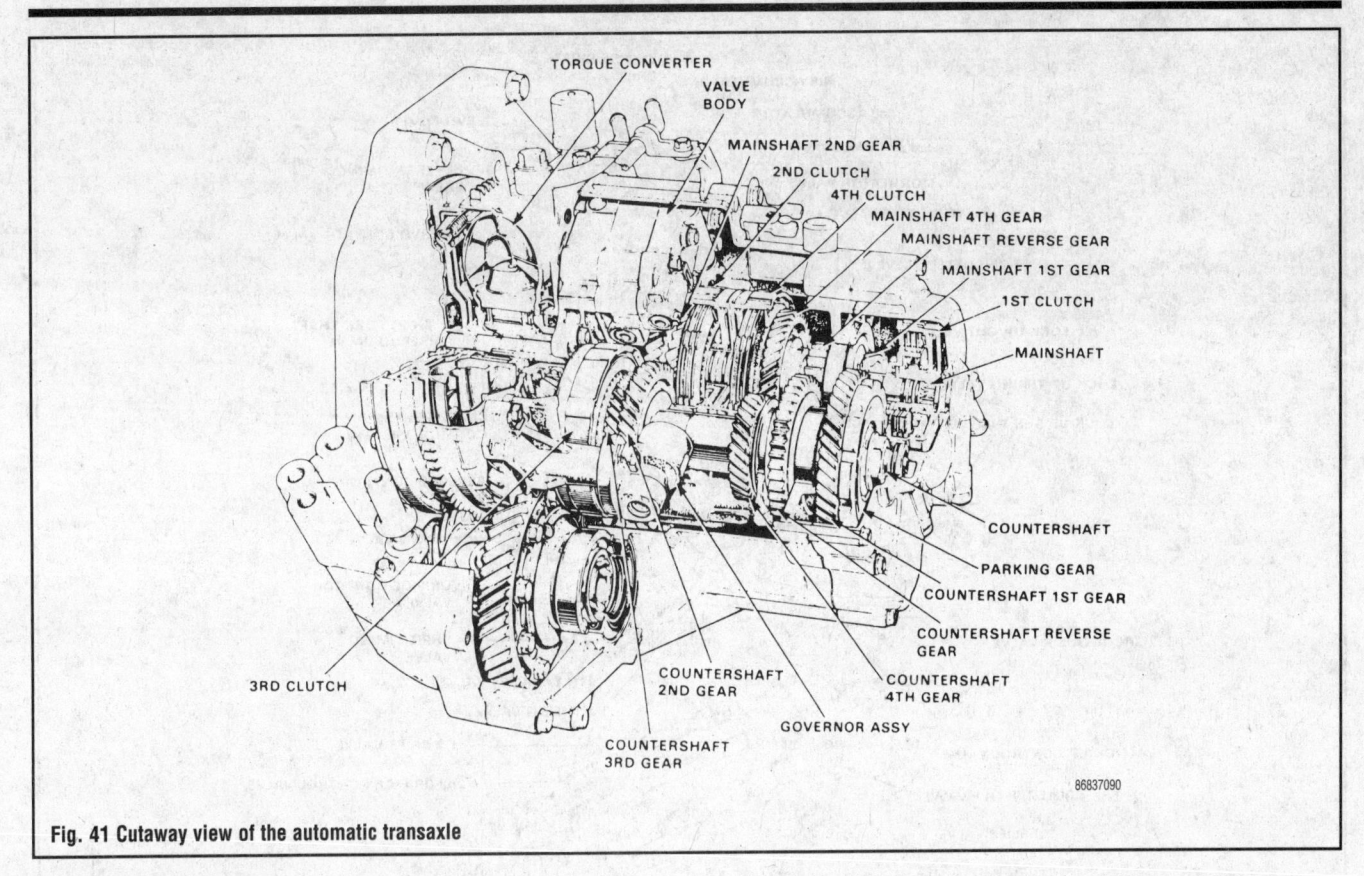

THE PLANETARY GEARBOX

The ability of the torque converter to multiply engine torque is limited. Also, the unit tends to be more efficient when the turbine is rotating at relatively high speeds. Therefore, a planetary gearbox is used to carry the power output of the turbine to the halfshafts.

Planetary gears function very similarly to conventional transaxle gears. However, their construction is different in that three elements make up one gear system, and, in that all three elements are different from one another. The three elements are: an outer gear that is shaped like a hoop, with teeth cut into the inner surface; a sun gear, mounted on a shaft and located at the very center of the outer gear; and a set of three planet gears, held by pins in a ring-like planet carrier, meshing with both the sun gear and the outer gear. Either the outer gear or the sun gear may be held stationary, providing more than one possible torque multiplication factor for each set of gears. Also, if all three gears are forced to rotate at the same speed, the gearset forms, in effect, a solid shaft.

Most modern automatics use the planetary gears to provide either a single reduction ratio of about 1.8:1, or two reduction gears: a low of about 2.5:1, and an intermediate of about 1.5:1. Bands and clutches are used to hold various portions of the gearsets to the transaxle case or to the shaft on which they are mounted. Shifting is accomplished, then, by changing the portion of each planetary gearset which is held to the transaxle case or to the shaft.

THE SERVOS AND ACCUMULATORS

The servos are hydraulic pistons and cylinders. They resemble the hydraulic actuators on many familiar machines, such as bulldozers. Hydraulic fluid enters the cylinder, under pressure, and forces the piston to move to engage the band or clutches.

The accumulators are used to cushion the engagement of the servos. The transaxle fluid must pass through the accumulator on the way to the servo. The accumulator housing contains a thin piston which is sprung away from the discharge passage of the accumulator. When fluid passes through the accumulator on the way to the servo, it must move the piston against spring pressure, and this action smooths out the action of the servo.

THE HYDRAULIC CONTROL SYSTEM

▶ See Figure 42

The hydraulic pressure used to operate the servos comes from the main transaxle oil pump. This fluid is channeled to the various servos through the shift valves. There is generally a manual shift valve which is operated by the transaxle selector lever and an automatic shift valve for each automatic upshift the transaxle provides: i.e., 2-speed automatics have a low/high shift valve, while 3-speeds have a 1-2 valve, and a 2-3 valve.

There are two pressures which effect the operation of these valves. One is the governor pressure which is affected by vehicle speed. The other is the modulator pressure which is affected by intake manifold vacuum or throttle position. Governor pressure rises with an increase in vehicle speed, and modulator pressure rises as the throttle is opened wider. By responding to these two pressures, the shift valves cause the upshift points to be delayed with increased throttle opening to make the best use of the engine's power output.

Most transaxles also make use of an auxiliary circuit for downshifting. This circuit may be actuated by the throttle linkage or the vacuum line which actuates the modulator, or by a cable or solenoid. It applies pressure to a special downshift surface on the shift valve or valves.

The transaxle modulator also governs the line pressure, used to actuate the servos. In this way, the clutches and bands will be actuated with a force matching the torque output of the engine.

Identification

The transaxle can be identified by a number plate located on the top of the transaxle housing. The transaxle used in the 1984–87 models is the CA type. For the 1988–91 models the L4 type is used. And for 1992–95, the M24A type is used. In the 4WD Wagon, a version of the 2WD transaxle is used, it uses extra gearing for the Super-low modes of driving.

7-18 DRIVE TRAIN

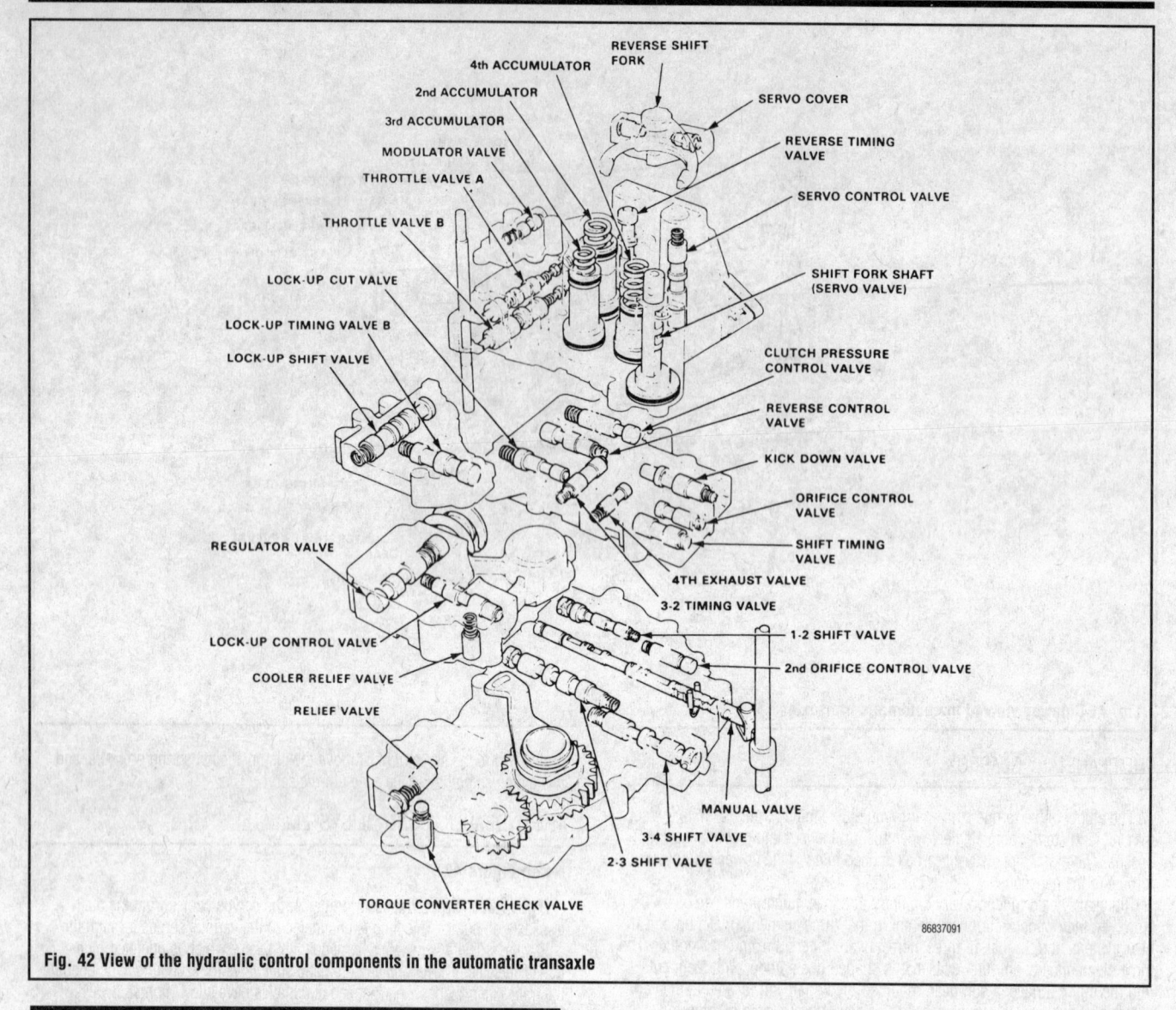

Fig. 42 View of the hydraulic control components in the automatic transaxle

Adjustments

SHIFT CABLE

▶ See Figures 43 and 44

1. Start the engine. Shift the transaxle to R, to see if the reverse gear engages.
2. Shut the engine off and disconnect the negative battery cable.
3. Remove the console.
4. On 1988 models, place the selector lever in D. On 1989–91 models, place the selector lever in N or R. On 1992–95 models, shift to N. Remove the lock pin from the cable adjuster.
5. Check that the hole in the adjuster is perfectly aligned with the hole in the shift cable.

➡ There are two holes in the end of the shift cable. They are positioned 90 degrees apart to allow cable adjustments in ¼ turn increments.

6. If not perfectly aligned, loosen the locknut on the shift cable and adjust as required.
7. Tighten the locknut and install the lock pin on the adjuster.

➡ If the lock pin feels like it is binding when being installed, the cable is still out of adjustment and must be adjusted again.

8. Connect the negative battery cable, start the engine and check the shift lever in all gears. Install the console.

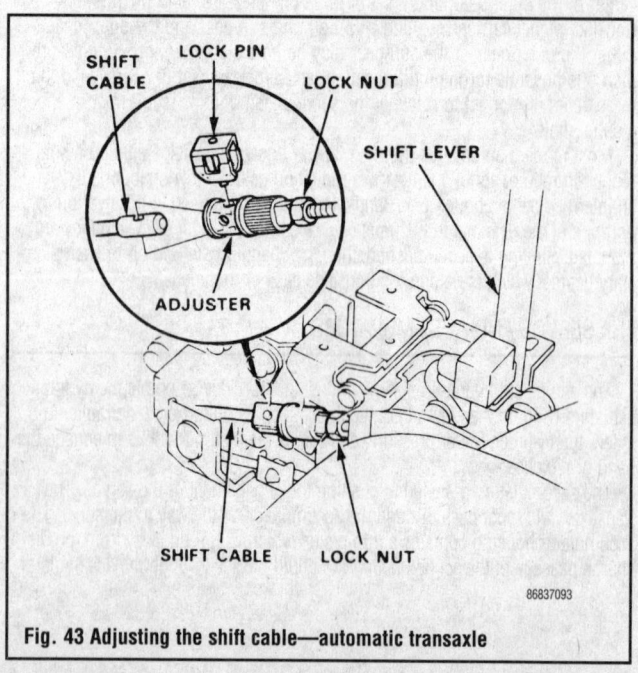

Fig. 43 Adjusting the shift cable—automatic transaxle

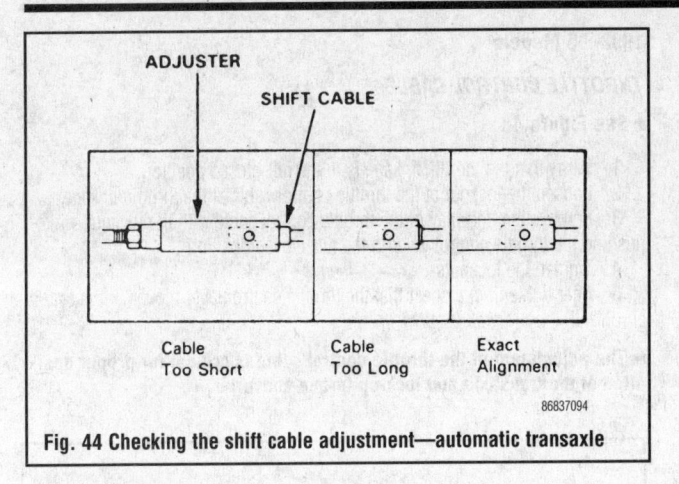

DRIVE TRAIN 7-19

Fig. 44 Checking the shift cable adjustment—automatic transaxle

THROTTLE LINKAGE

Carbureted Engines

THROTTLE CONTROL CABLE BRACKET

1. Disconnect the negative battery cable.
2. Disconnect the throttle control cable from the throttle control lever.
3. Bend down the lock tabs of the lock plate and remove the two 6mm bolts to free the bracket.
4. Loosely install a new lock plate.
5. Adjust the position of the bracket by measuring the distance between the cable housing side of the bracket and the bracket side edge of the throttle control lever. Measure between the same points that the cable would pass through the bracket and lever.
6. Tighten the two 6mm bolts when the measurement is 3.287 in. (83.5mm). The bolts should be tightened to 9 ft. lbs. (12 Nm).

➡ Make sure the control lever does not get pulled toward the bracket side as the bolts are tightened.

7. Bend up the lock plate tabs against the bolt heads, connect the throttle control cable and connect the negative battery cable.

THROTTLE CONTROL CABLE

♦ See Figures 45 and 46

1. Start the engine and warm it up to normal operating temperature. The cooling fan must come on at least once.
2. Make sure the throttle cable play, idle speed and automatic choke operation are correct.
3. Check the distance between the throttle control lever and the throttle control bracket and adjust as necessary.
4. Turn the engine off and disconnect the negative battery cable.
5. Disconnect the throttle control cable from the control lever.
6. If the vehicle is equipped with a dash pot, disconnect the vacuum hose from the dash pot, connect a vacuum pump and apply vacuum. This simulates a normal operating amount of pull by the dash pot, as if the engine were running.
7. Attach a weight of about 3 lbs. (1.35 kg) to the accelerator pedal. Raise the pedal, then release it. This will allow the weight to remove the normal free-play from the throttle cable.
8. Secure the throttle cable with clamps.
9. Lay the end of the throttle cable aside on Civic and CRX.
10. Adjust the distance between the throttle control cable end and the locknut closest to the cable housing to 3.366 in. (85.5mm).
11. Insert the end of the throttle control cable in the groove of the throttle control lever. Insert the throttle control cable in the bracket and secure with the other locknut.

➡ Make sure the cable is not kinked or twisted.

12. Check that the cable moves freely by depressing the accelerator.
13. Remove the weight on the accelerator pedal and push the pedal to make sure there is at least 0.08 in. (2.0mm) play at the throttle control lever.
14. Connect the negative battery cable and the vacuum hose to the dash pot.

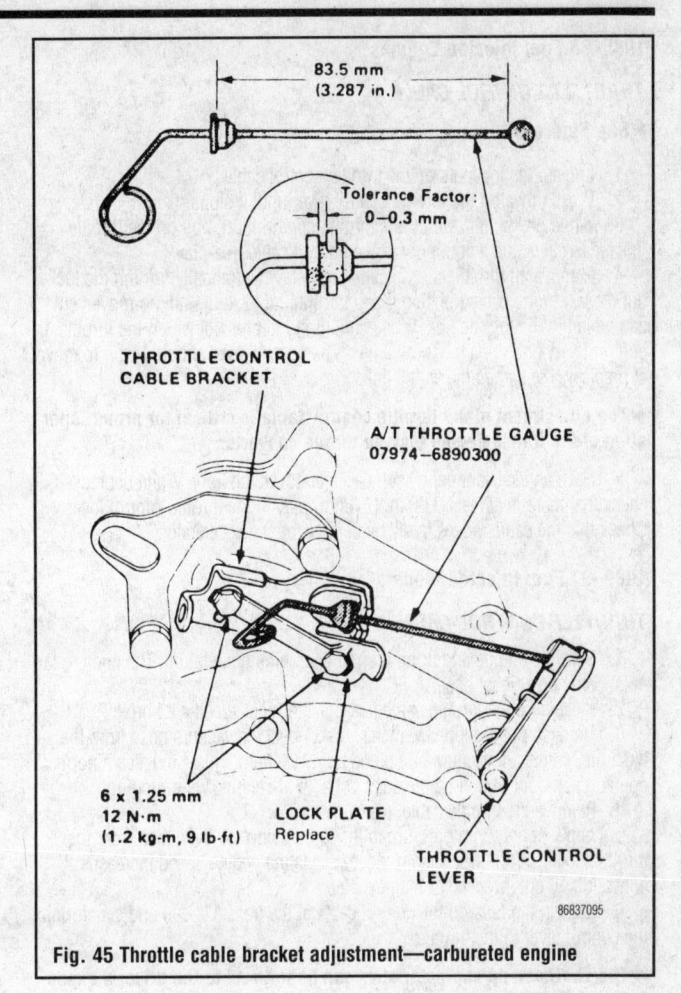

Fig. 45 Throttle cable bracket adjustment—carbureted engine

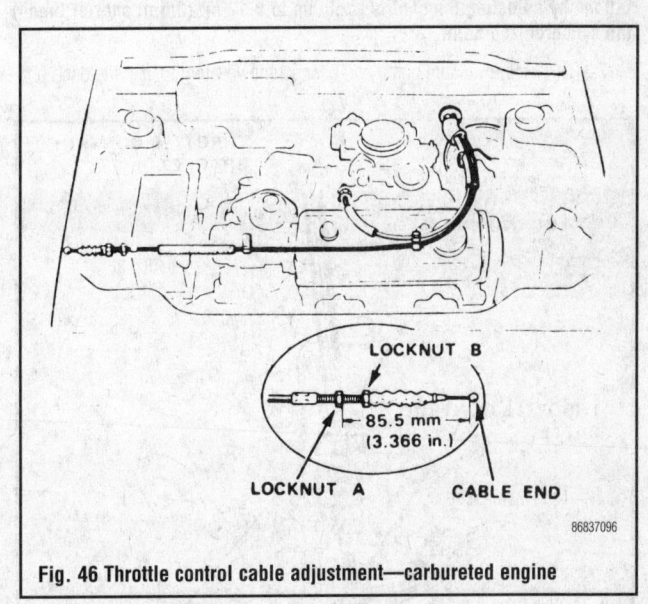

Fig. 46 Throttle control cable adjustment—carbureted engine

15. Start the engine and check the synchronization between the carburetor and the throttle control cable. The throttle control lever should start to move as the engine speed increases.
16. If the throttle control lever starts to move before the engine speed increases, turn the cable locknut closest to the cable housing counterclockwise and retighten the locknut closest to the cable end.
17. If the throttle control lever moves after the engine speed increases, turn the locknut closest to the cable housing clockwise and retighten the locknut closest to the cable end.

7-20 DRIVE TRAIN

1985–88 Fuel Injected Engines

THROTTLE CONTROL CABLE

▶ See Figure 47

1. Loosen the locknuts on the throttle control cable.
2. Press down on the throttle control lever until it stops.
3. While pressing down on the throttle control lever, pull on the throttle linkage to check the amount of throttle control cable free-play.
4. Remove all throttle control cable free-play by gradually turning the locknut closest to the cable housing. Keep turning the locknut until no movement can be felt in the throttle link. While continuing to press down on the throttle control lever, pull open the throttle link. The control lever should begin to move at precisely the same time as the link.

➡ The adjustment of the throttle control cable is critical for proper operation of the transaxle and lock-up torque converter.

5. Have an assistant depress the accelerator to the floor. While depressed, check that there is at least 0.08 in. (2.0mm) play in the throttle control lever. Check that the cable moves freely by depressing the accelerator.

1989–91 Fuel Injected Models

THROTTLE CONTROL CABLE

1. Start the engine and bring it up to operating temperature. The cooling fan must come on at least once.
2. Make sure the throttle cable free-play and idle speed are correct.
3. On dash pot equipped vehicles, disconnect the vacuum hose from the dash pot, connect a vacuum pump and apply vacuum. This simulates a normal operating amount of pull by the dash pot as if the engine were running.
4. Remove the throttle cable free-play.
5. Apply light thumb pressure to the throttle control lever, then work the throttle linkage. The lever should move just as the engine speed increases above idle. If not, proceed to the next step.
6. Loosen the nuts on the control cable at the transaxle end and synchronize the control lever to the throttle.

➡ The shift/lock-up characteristics can be tailored to the driver's expectations by adjusting the control cable up to 0.12 in. (3mm) shorter than the synchronized point.

7. Remove the vacuum pump and connect the vacuum hose to the dash pot.

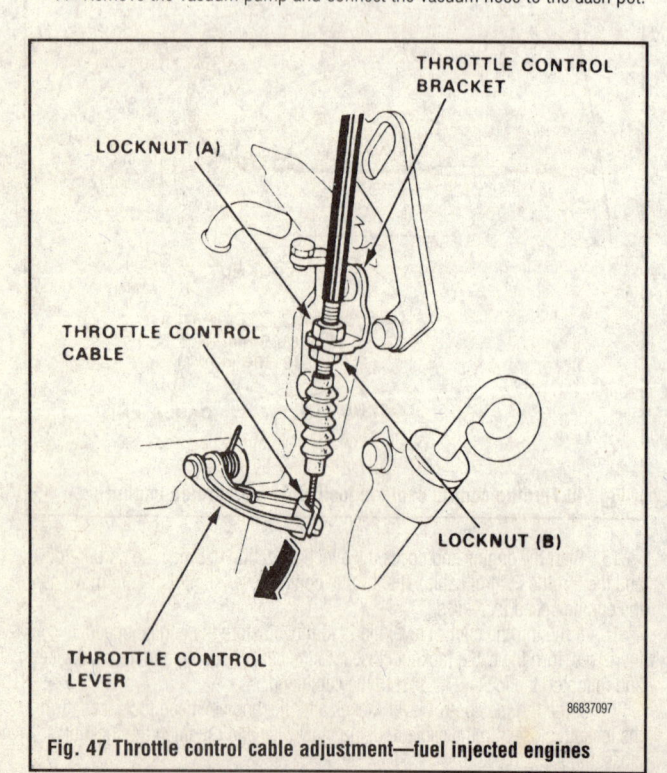

Fig. 47 Throttle control cable adjustment—fuel injected engines

1992–95 Models

THROTTLE CONTROL CABLE

▶ See Figure 48

1. Verify that the throttle linkage is in the full-closed position.
2. Loosen the locknut of the throttle control cable at the throttle linkage.
3. Remove the free-play of the throttle control cable with the locknut, while pushing the throttle control lever to the full-closed position.
4. Tighten the locknuts.
5. After tightening, inspect that the throttle control lever movement is synchronized.

➡ The adjustment of the throttle control cable is critical for proper operation of the transaxle and lock-up torque converter.

Fig. 48 Throttle control cable—1992-95 models

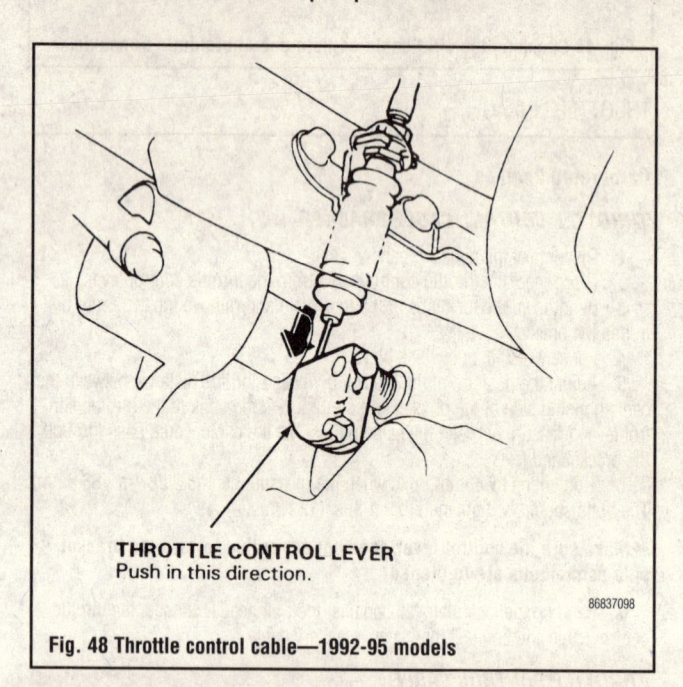

Transaxle

REMOVAL & INSTALLATION

▶ See Figure 49

➡ Use fender covers to avoid damaging painted surfaces. Be sure to know the 5-digit radio "CODE" to input for re-connecting the original radio on 1992-95 models.

1. Disconnect the battery cables and remove the battery.
2. Raise and safely support the vehicle.
3. Remove the three mount bolts and loosen the one bolt located at the side of the battery base and intake hose band of the throttle body.
4. Remove the air cleaner case, complete with the intake hose and resonator, if equipped.
5. Remove the starter and the transaxle ground cable.
6. Disconnect the lock-up control solenoid valve wire connector. On 4WD Civic Wagon, disconnect the lock-up control solenoid valve and shift control solenoid valve wire connectors and disconnect the automatic transaxle speed pulser connector.
7. Disconnect the control cable at the control lever.
8. Drain the transaxle fluid. Remove the filler plug to speed draining. After draining, reinstall the drain plug with a new washer.
9. Disconnect and plug the cooler hoses at the joint pipes.
10. Mark the position of the distributor housing in relation to the cylinder head. Disconnect the connectors and remove the mount bolts, then remove the distributor from the cylinder head.

DRIVE TRAIN 7-21

Fig. 49 Removing the automatic transaxle

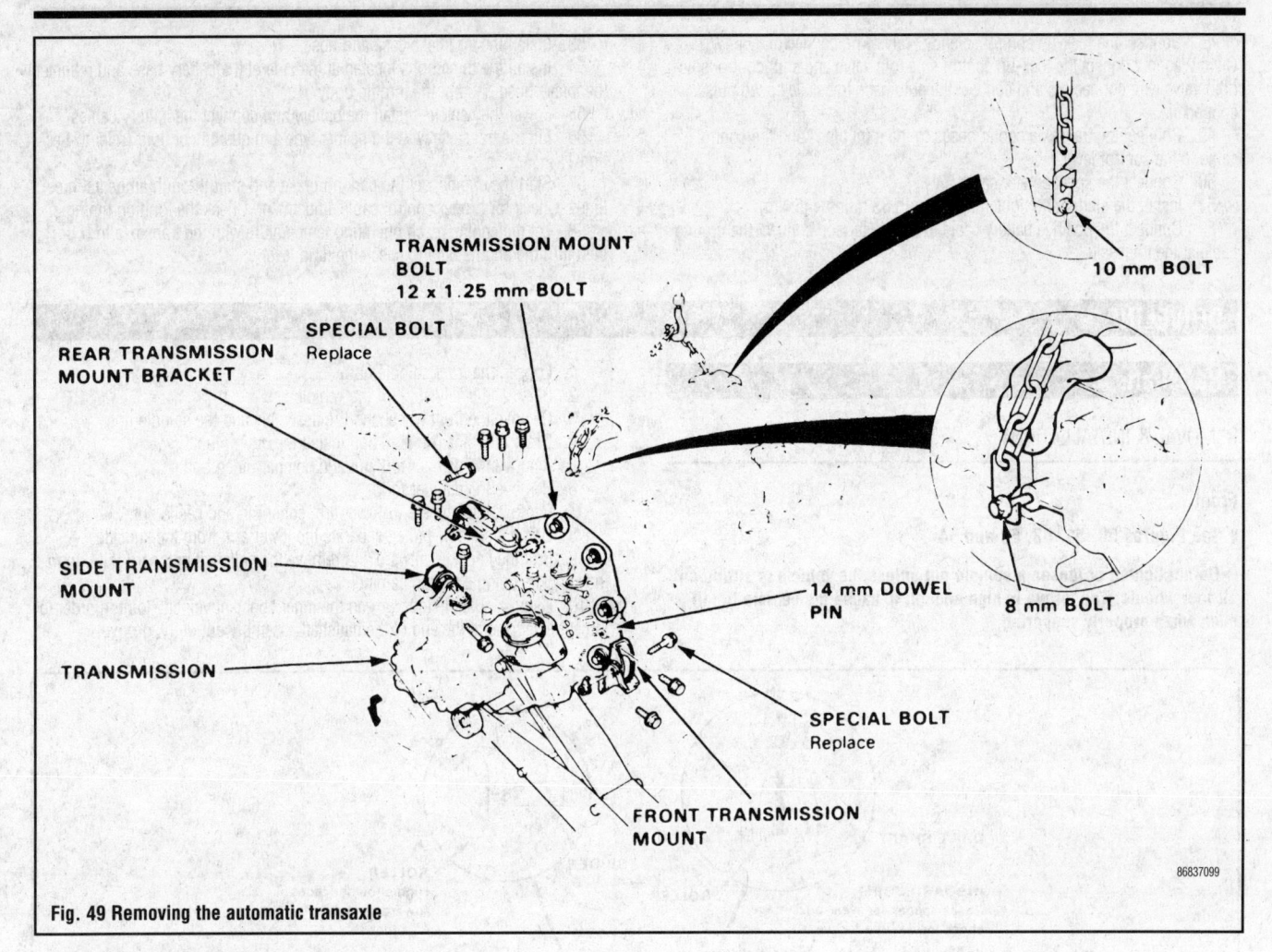

11. Remove the shift cable by removing the cotter pin, control pin, control lever roller and loosening the locknut.
12. Disconnect the speedometer cable. Do not disassemble the speedometer gear holder.
13. On 4WD Civic Wagon, remove the driveshaft.
14. Remove the torque converter cover. Remove the plug, then remove the driveplate bolts one at a time while rotating the crankshaft pulley.
15. Remove the engine splash shield and the right wheel well splash shield.
16. Remove the header pipe.
17. Remove the cotter pins and lower arm ball joint nuts and then separate the ball joints using a suitable tool.
18. Remove the bolts and nut, then remove the right radius rod.
19. Remove the right and left halfshafts. On 4WD Civic Wagon, remove the intermediate shaft.
20. On 4WD Civic Wagon, remove the nine mounting bolts, then remove the transaxle cover.
21. Install bolts in each end of the cylinder head and attach a suitable hoist to the bolts. Lift the engine slightly to unload the mounts.
22. Place a suitable jack under the transaxle and raise the transaxle just enough to take weight off of the mounts.
23. On 4WD Civic Wagon, remove the transaxle mounting bolt on the engine side.
24. Remove the bolts from the front transaxle mount.
25. Remove the rear transaxle mount bracket by removing the four mounting bolts.
26. Remove the four mounting bolts. Remove the side transaxle mount.
27. Remove the transaxle-to-cylinder block mounting bolts.
28. On 4WD Civic Wagon, loosen the side engine mounting bolt and tilt the engine.
29. Pull the transaxle away from the engine until it clears the 14mm dowel pins, then lower the transaxle jack.

To install:
30. Make sure the two 14mm dowel pins are installed in the torque converter housing.
31. Raise the transaxle into position and loosely install the transaxle-to-cylinder block bolts. Tighten them to 43 ft. lbs. (60 Nm).
32. Install the engine side mounting bolt and tighten to 50 ft. lbs. (68 Nm).
33. Install the transaxle-to-rear transaxle mount bracket and tighten the bolts to 40 ft. lbs. (55 Nm).
34. Install the transaxle to the front and side transaxle mounts. Tighten the mount retaining bolts to 29 ft. lbs. (40 Nm) and the mount through bolts to 40 ft. lbs. (55 Nm).
35. Remove the transaxle jack. On 4WD Civic Wagon, tighten the side engine mount bolt to 40 ft. lbs. (55 Nm).
36. Remove the hoist chain and bolts from the cylinder head.
37. Attach the torque converter to the driveplate with the eight 12mm bolts. Tighten the bolts in two steps, first to 4.5 ft. lbs. (6 Nm) in a crisscross pattern and finally to 9 ft. lbs. (12 Nm) in the same pattern.
38. Install the shift cable and cable holder.
39. Install the torque converter cover and header pipe bracket.
40. On 4WD Civic Wagon, install the transaxle cover and the intermediate shaft.
41. Install a new set ring on the end of each halfshaft.
42. Turn the right steering knuckle fully outward and slide the axle into the differential, until the spring clip can be felt engaging the differential side gear. Repeat the procedure on the left side.
43. Install the damper fork bolt and radius rod.
44. Install the ball joints to the lower arms and tighten the nuts to 40 ft. lbs. (55 Nm). Install new cotter pins.
45. On 4WD Civic Wagon, install the driveshaft.
46. Install the splash shields and header pipe.
47. Install the distributor.

7-22 DRIVE TRAIN

48. Connect the lock-up control solenoid valve wire connector. On 4WD Civic Wagon, connect the lock-up control solenoid valve and shift control solenoid valve wire connectors and connect the automatic transaxle speed pulser connector.
49. Connect the transaxle cooler hoses to the joint pipes and the control cable to the control lever.
50. Connect the speedometer cable.
51. Install the starter and tighten the bolts to 33 ft. lbs. (45 Nm).
52. Connect the positive battery cable to the starter and connect the ground cable at the transaxle.
53. Install the air intake case and hose.
54. Install the three bolts located at the side of the battery base and retighten the intake hose band of the throttle body.
55. Lower the vehicle, install the battery and connect the battery cables.
56. Fill the transaxle with the proper type and quantity of fluid (refer to Section 1).
57. Start the engine, set the parking brake and shift through all gears three times. Check for proper control cable adjustment. Check the ignition timing.
58. Let the engine reach operating temperature with the transaxle in N or P, then turn the engine off and check the fluid level.

DRIVELINE

Halfshafts

REMOVAL & INSTALLATION

Front

▶ See Figures 50, 51, 52, 53 and 54

➥ Do not tighten or loosen a spindle nut unless the vehicle is sitting on all four wheels. The torque is high enough to cause the vehicle to fall even when properly supported.

1. Loosen the front spindle nut.
2. Raise and safely support the vehicle.
3. Remove the front wheel and tire assembly and the spindle nut.
4. Drain the transaxle fluid and install the drain plug.
5. Remove the damper fork nut and damper pinch bolt.
6. Remove the damper fork.
7. Remove the knuckle-to-lower arm cotter pin and castle nut.
8. Using a suitable puller, separate the lower arm from the knuckle.
9. Pull the knuckle outward and remove the halfshaft outboard joint, from the knuckle, using a plastic hammer.
10. Using a suitable tool, pry on the inner constant velocity joint in order to force the set ring at the end of the halfshaft assembly out of the groove.

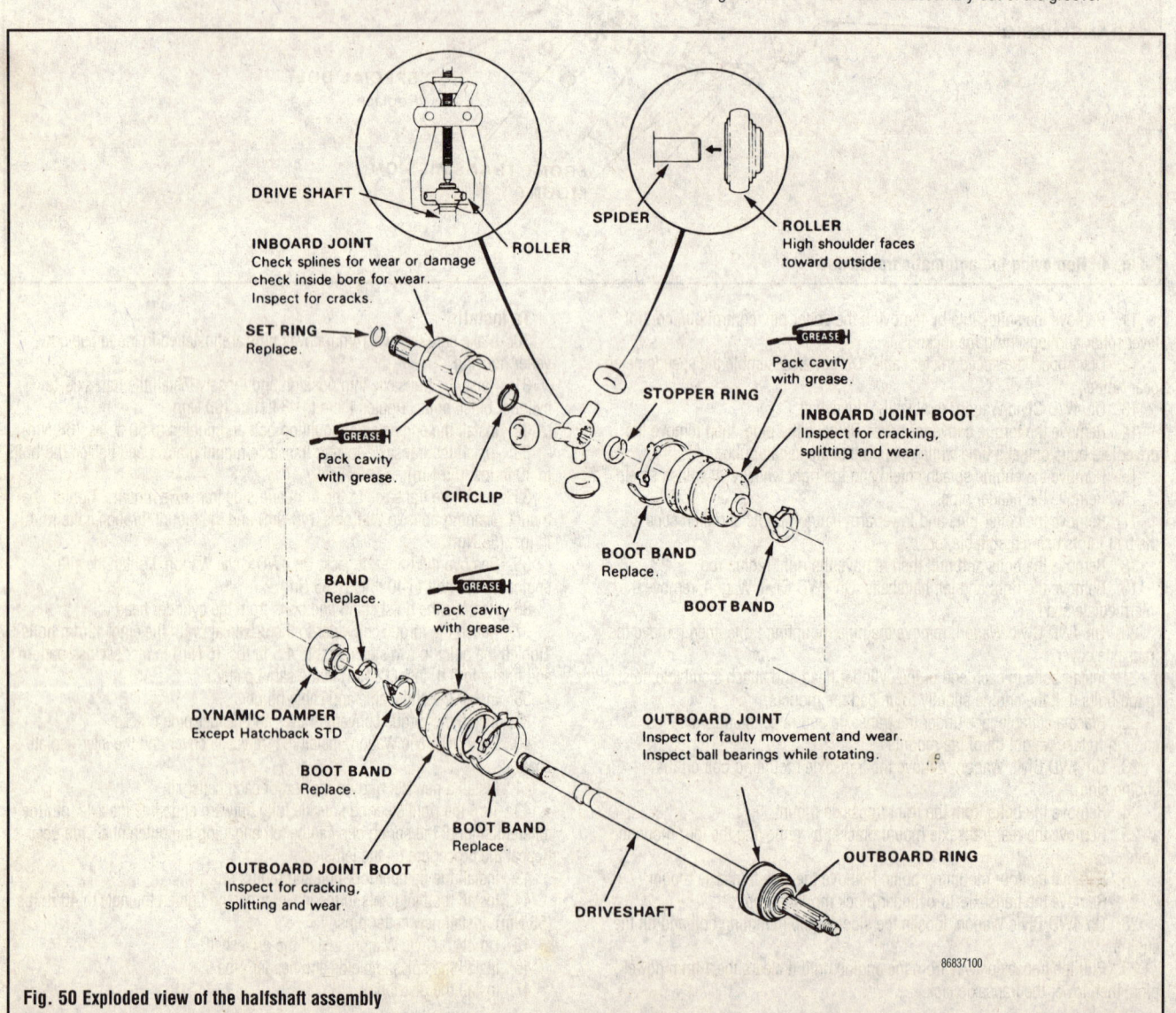

Fig. 50 Exploded view of the halfshaft assembly

DRIVE TRAIN 7-23

Fig. 51 Outboard constant velocity boot—this one is cracked from age

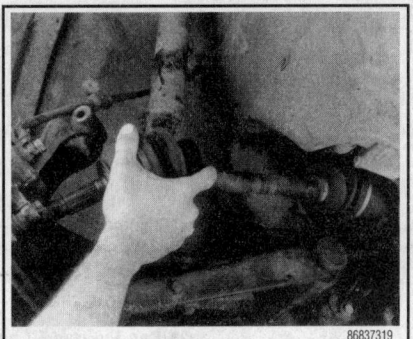

Fig. 52 The outer constant velocity joint is here being removed from the splined wheel hub

Fig. 53 Prying the inner constant velocity joint from its grooved mounting in the differential case

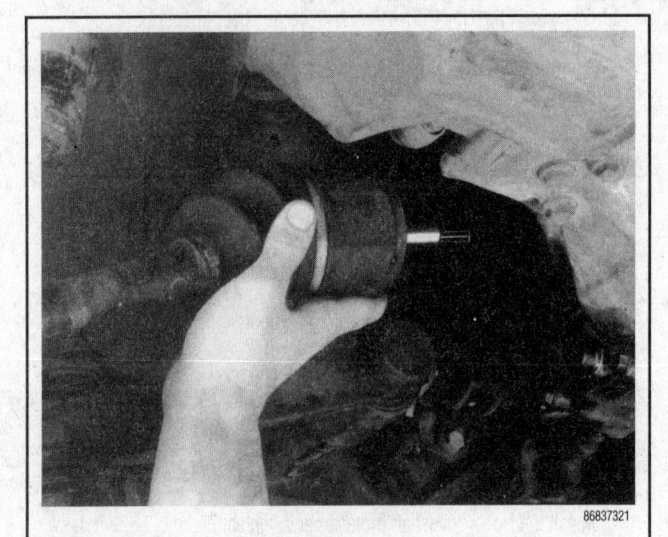

Fig. 54 Removing the inner CV-joint from the differential case

11. Pull on the inboard CV-joint and remove the halfshaft and joint from the differential case or intermediate shaft.

➡ Do not pull on the halfshaft as the CV-joint may come apart. Use care when prying out the assembly and pull it straight to avoid damaging the differential oil seal or intermediate shaft dust seal.

To install:

12. Always install new set rings on the ends of the halfshafts.
13. Make sure the set ring locks in the differential side gear groove and the axle bottoms in the differential or intermediate shaft.
14. Install the spindle nut but do not torque it yet.
15. Assemble the suspension in the reverse order. Tighten the upper damper pinch bolt to 32 ft. lbs. (44 Nm) and the fork nut to 47 ft. lbs. (65 Nm). Install the axle.
16. Tighten the ball joint nut to 40 ft. lbs. (55 Nm), then tighten as required to install a new cotter pin.
17. With the vehicle resting on all 4 wheels, tighten the spindle nut to the proper specification and stake the nut in place to 134 ft. lbs. (185 Nm).
18. Fill the transaxle with the proper type and quantity of fluid.

Rear

1988–91 4WD WAGON

➡ Do not tighten or loosen a spindle nut unless the vehicle is sitting on all four wheels. The torque is high enough to cause the vehicle to fall even when properly supported.

1. Pry the spindle nut stake away from the spindle. Loosen the nut. Loosen the wheel nuts.
2. Raise and support the vehicle safely. Remove the tire and wheel assemblies.
3. Disconnect the brake hose from the brake pipe.
4. Using a floor jack raise the rear suspension until the weight of the lower arm is relieved.
5. Remove the trailing arm bushing bolts. Disconnect the upper arm and the lower arm from the trailing arm.
6. Pull the trailing arm outward. Remove the rear halfshaft outboard joint from the trailing arm, using the proper tool.
7. Using a suitable tool, pry the halfshaft assembly to force the set ring at the halfshaft end past the groove.
8. Pull the inboard joint and remove the halfshaft and the CV-joint from the differential case as an assembly.

To install:

9. Always install a new set ring on the end of the axle. Make sure the set ring locks in the differential side gear groove and the CV-joint subaxle bottoms in the differential.
10. Reassemble the suspension and tighten the upper and lower arm bushing bolts to 40 ft. lbs. (55 Nm) and the trailing arm bushing bolts to 47 ft. lbs. (65 Nm).
11. When the vehicle is resting on all four wheels, tighten the spindle nut to 134 ft. lbs. (185 Nm).

CV-Joint and Boot

OVERHAUL

▸ See Figures 55 and 56

1. Raise and safely support the vehicle.
2. Remove the halfshaft.
3. If replacing the inboard CV-joint boot, perform the following procedure:
   a. Place the halfshaft in a suitable holding fixture where it will remain in position during disassembly.

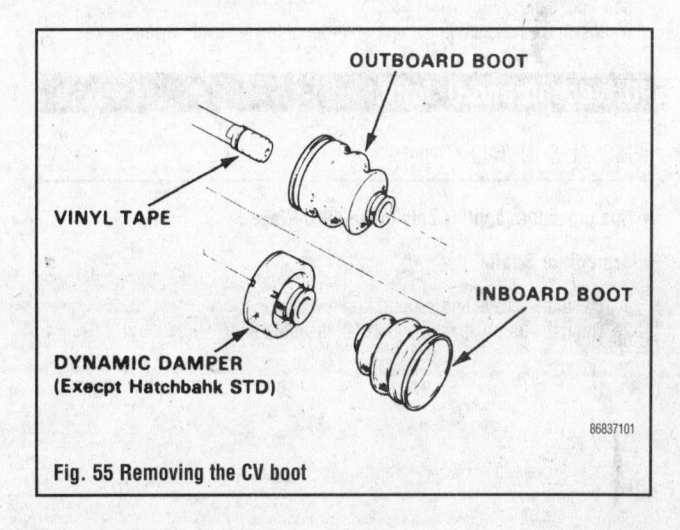

Fig. 55 Removing the CV boot
(OUTBOARD BOOT, VINYL TAPE, DYNAMIC DAMPER (Execpt Hatchbahk STD), INBOARD BOOT)

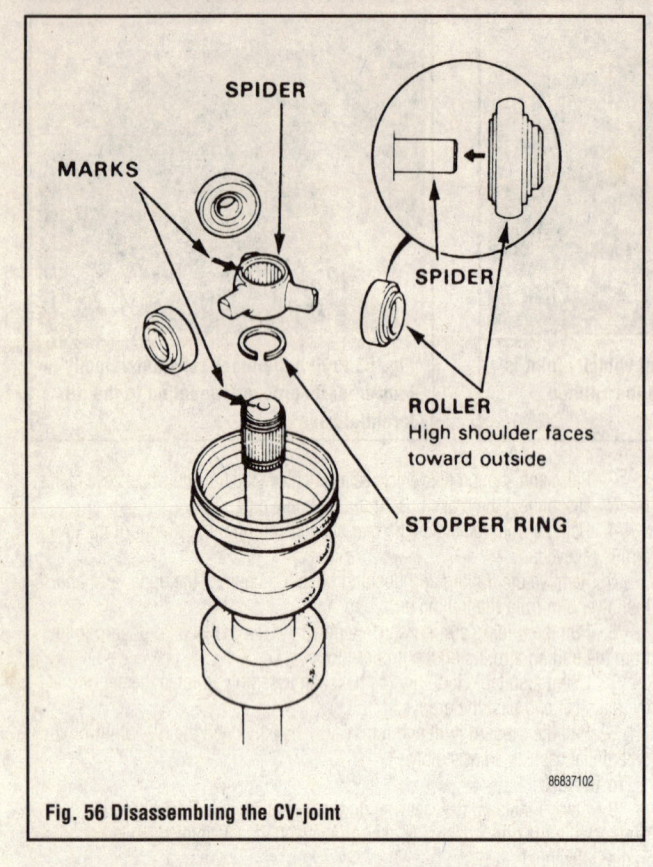

7-24 DRIVE TRAIN

Fig. 56 Disassembling the CV-joint

b. Remove the boot bands. If the boot bands are the welded type, they must be cut to be removed. After removing the bands, push the CV-joint boot away from the end of the halfshaft to gain access to the CV-joint.

c. Remove the inboard CV-joint. Mark the components during disassembly to ensure proper positioning during reassembly.

d. Remove the CV-joint boot.

e. Installation is the reverse of the removal procedure. Check the CV-joint components for wear prior to installation and replace as necessary.

f. Thoroughly pack the inboard CV-joint and boot with molybdenum disulfide grease. Always install new boot bands.

4. If replacing the outboard CV-joint boot, perform the following procedure:

a. Place the halfshaft in a suitable holding fixture where it will remain in position during disassembly.

b. Remove the inboard CV-joint and boot. Do not try to remove or disassemble the outboard CV-joint.

c. Remove the boot bands and the outboard CV-joint boot.

d. Installation is the reverse of the removal procedure. Thoroughly pack the outboard CV-joint boot with molybdenum disulfide grease. Always install new boot bands.

5. Install the halfshaft in the reverse order of the removal procedure.

Driveshaft and U-Joints

REMOVAL & INSTALLATION

➡ This procedure applies only to the 4WD Wagon.

Intermediate Shaft

1. Raise the vehicle and support it safely.
2. Drain the transaxle fluid and install the drain plug.
3. Remove the driver's side halfshaft.
4. Remove the three 10mm bolts.
5. Lower the bearing support close to the steering gearbox and remove the intermediate shaft from the differential.

➡ To avoid damage to the differential oil seal, hold the intermediate shaft in a horizontal position until it is clear of the differential.

6. Installation is the reverse of the removal procedure. Tighten the three 10mm bolts to 29 ft. lbs. (40 Nm). Fill the transaxle with the proper type and quantity of fluid (refer to Section 1).

Rear Driveshafts

▸ See Figures 57 and 58

1. Raise the vehicle and support it safely.
2. Mark the position of the driveshafts in relation to the flanges for reassembly.
3. Remove the No. 1 driveshaft protector.
4. Disconnect the No. 1 driveshaft and viscous coupling.
5. Remove the front center bearing support from the body.
6. Remove the No. 1 driveshaft by disconnecting the U-joint.
7. Remove the No. 3 driveshaft protector.
8. Disconnect the No. 3 driveshaft and rear differential.
9. Remove the rear center bearing support from the body, then remove the viscous coupling and No. 3 driveshaft.

To install:

10. Install the rear center bearing support on the frame and tighten the bolts to 29 ft. lbs. (40 Nm).
11. Temporarily connect the No. 3 driveshaft and rear differential using the 12-point bolts and yoke nuts. Tighten the bolts to 24 ft. lbs. (33 Nm).
12. Install the No. 3 driveshaft protector and tighten the bolts to 29 ft. lbs. (40 Nm).
13. Temporarily connect the No. 1 driveshaft and front differential using the 12-point bolts and yoke nuts. Tighten the bolts to 24 ft. lbs. (33 Nm).
14. Install the front center bearing support on the frame and tighten the bolts to 29 ft. lbs. (40 Nm). Temporarily connect the No. 1 driveshaft and viscous coupling using 12-point bolts. Tighten the bolts to 24 ft. lbs. (33 Nm).
15. Install the No. 1 driveshaft protector and tighten the bolts to 29 ft. lbs. (40 Nm).

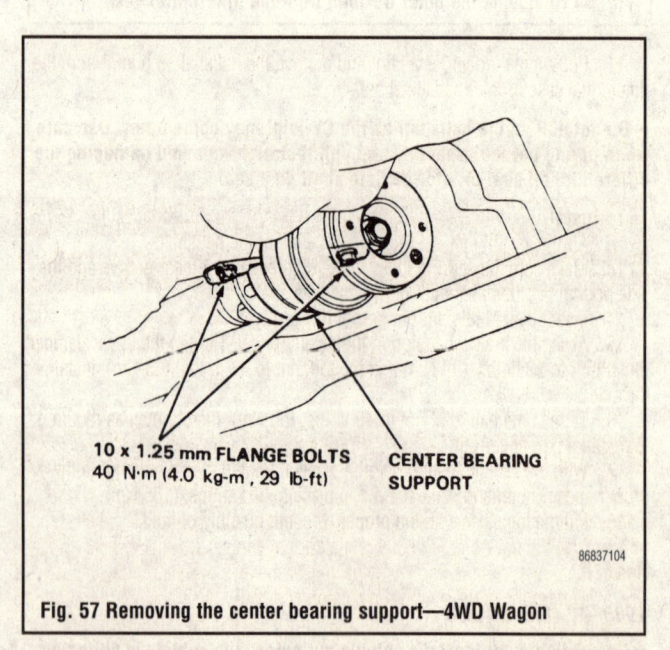

Fig. 57 Removing the center bearing support—4WD Wagon

DRIVE TRAIN 7-25

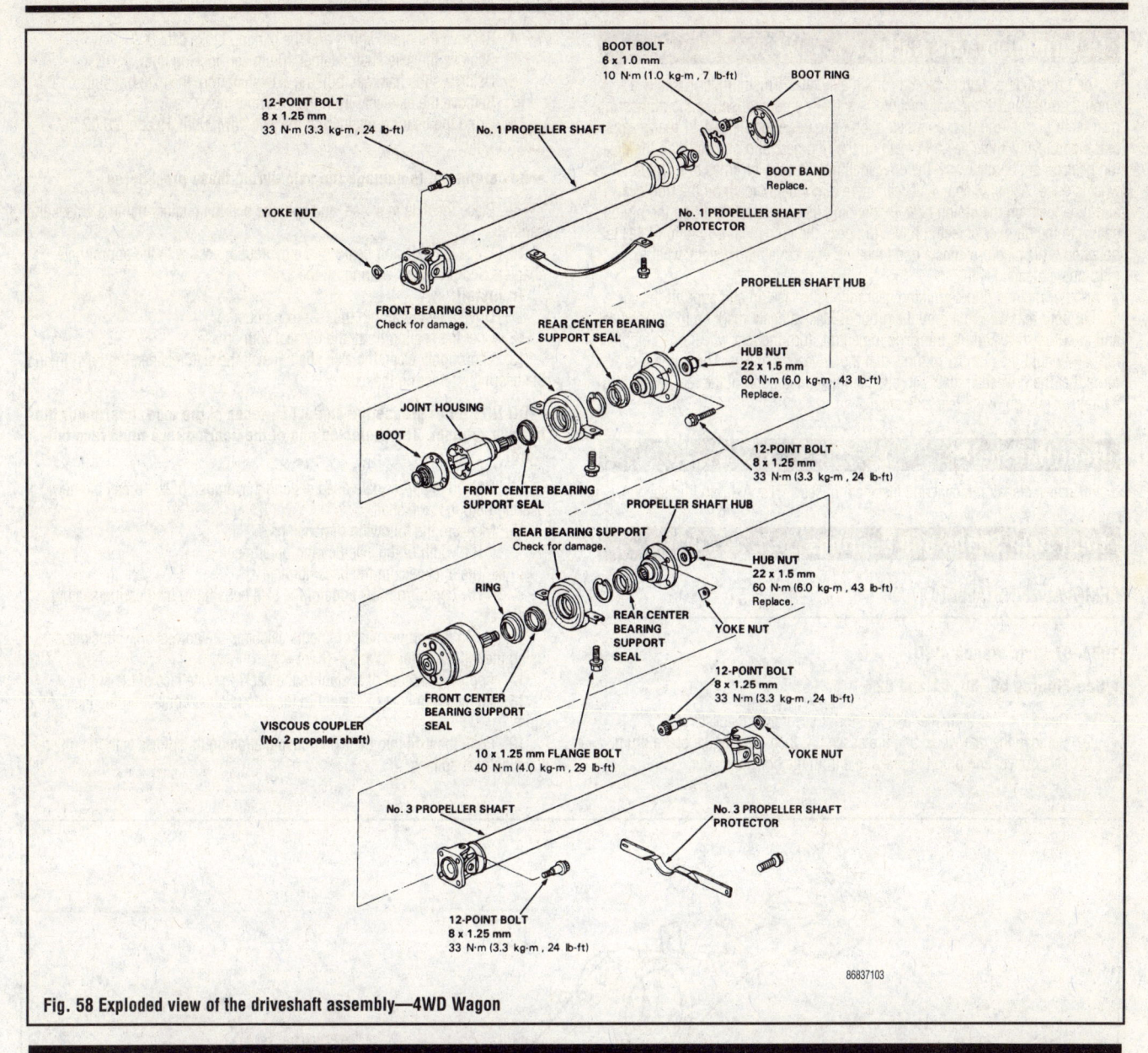

Fig. 58 Exploded view of the driveshaft assembly—4WD Wagon

REAR AXLE

Application

The 4WD Civic Wagon is the only Honda Civic model which uses a rear axle. The 1984–87 models use a solid rear axle and the 1988–91 models use an axle housing and halfshafts.

The gear reduction of the rear axle is 2.529.

Understanding Drive Axles

Power enters the axle from the driveshaft via the companion flange. The flange is mounted on the drive pinion shaft. The drive pinion shaft and gear which carry the power into the differential turn at engine speed. The gear on the end of the pinion shaft drives a large ring gear the axis of rotation of which is 90° away from the of the pinion. The pinion and gear reduce the gear ratio of the axle, and change the direction of rotation to turn the axle shafts which drive both wheels. The axle gear ratio is found by dividing the number of pinion gear teeth into the number of ring gear teeth.

The ring gear drives the differential case. The case provides the two mounting points for the ends of a pinion shaft on which are mounted two pinion gears. The pinion gears drive the two side gears, one of which is located on the inner end of each axle shaft.

By driving the axle shafts through the arrangement, the differential allows the outer drive wheel to turn faster than the inner drive wheel in a turn.

The main drive pinion and the side bearings, which bear the weight of the differential case, are shimmed to provide proper bearing preload, and to position the pinion and ring gears properly.

✸✸ WARNING

The proper adjustment of the relationship of the ring and pinion gears is critical. It should be attempted only by those with extensive equipment and/or experience.

Limited slip differentials include clutches which tend to link each axle shaft to the differential case. Clutches may be engaged either by spring action or by pressure produced by the torque on the axles during a turn. During turning on a dry pavement, the effects of the clutches are overcome, and each wheel turns at the required speed. When slippage occurs at either wheel, however, the clutches will transmit some of the power to the wheel which has the greater amount of traction. Because of the presence of clutches, limited slip units require a special lubricant.

7-26 DRIVE TRAIN

Determining Axle Ratio

The drive axle is said to have a certain axle ratio. This number (usually a whole number and a decimal fraction) is actually a comparison of the number of gear teeth on the ring gear and the pinion gear. For example, a 4.11 rear means that theoretically, there are 4.11 teeth on the ring gear and one tooth on the pinion gear or, put another way, the driveshaft must turn 4.11 times to turn the wheels once. Actually, on a 4.11 rear, there might be 37 teeth on the ring gear and nine teeth on the pinion gear. By dividing the number of teeth on the pinion gear into the number of teeth on the ring gear, the numerical axle ratio (4.11) is obtained. This also provides a good method of ascertaining exactly what axle ratio one is dealing with.

Another method of determining gear ratio is to jack up and support the car so that both rear wheels are off the ground. Make a chalk mark on the rear wheel and the driveshaft. Put the transaxle in neutral. Turn the rear wheel one complete turn and count the number of turns that the driveshaft makes. The number of turns that the driveshaft makes in one complete revolution of the rear wheel is an approximation of the rear axle ratio.

Rear Halfshafts

For rear halfshaft removal and installation, see "Drive Axle" in this section.

Rear Axle Shafts and Bearings

REMOVAL & INSTALLATION

1984–87 Civic Wagon 4WD

▶ See Figures 59, 60, 61 and 62

1. Raise the rear of the car and support it securely on jackstands.
2. Remove the rear wheel and tire assembly, then remove the brake drum.
3. Disconnect and plug the brake line from the brake cylinder.
4. Remove the brake shoes and the parking brake cable.
5. Remove the axle shaft retainer 10mm self-locking nuts.
6. Using a slide hammer, pull the axle shaft from the axle housing.
7. Remove the axle seal from the axle housing.
8. Grind the bearing retainer on the axle down until it's about 0.02 in. (0.5mm) thick.

➡**Be careful not to damage the axle during these procedures.**

9. Place the axle in a vise, and split the bearing retainer using a chisel and hammer.
10. Place the axle and bearing in a hydraulic press with the appropriate adapters and press the bearing off the axle.

To install:

11. Drive a new seal into the rear axle housing.
12. Coat the sealing lip of the oil seal with grease.
13. Thoroughly clean the axle shaft; install the axle retainer, the bearing and the bearing retainer on the axle.

➡**DO NOT oil or grease the contact surfaces of the axle, bearing or the bearing retainer. The projected end of the bearing race must face outward.**

14. Using the appropriate adapters and a hydraulic press, press the new bearing onto the axle shaft.
15. Measure the following dimensions:
    a. The width of the axle bearing outer race.
    b. The thickness of the backing plate.
    c. The depth from the edge of the axle housing to the bearing seating surface.
16. Using the above measurements calculate the correct shim thickness using the following formula: a − (b+c) = X.
17. Apply a thin coat of sealant to the backing plate face of the shim.
18. Apply a thin coat of sealant to the axle retainer contacting the face of the shim.
19. Push the axle into the axle housing aligning its splines with the differential side gear splines.

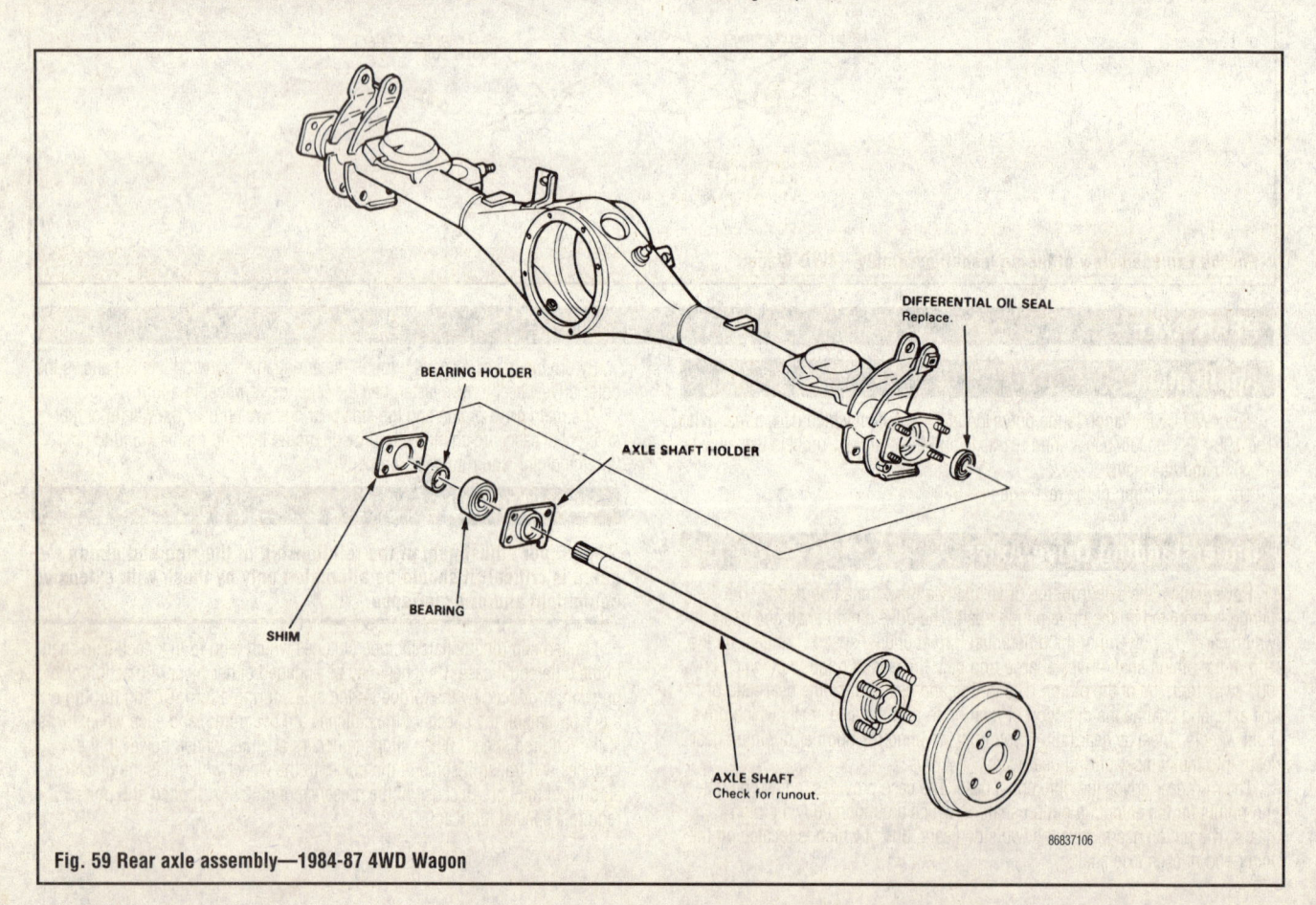

Fig. 59 Rear axle assembly—1984-87 4WD Wagon

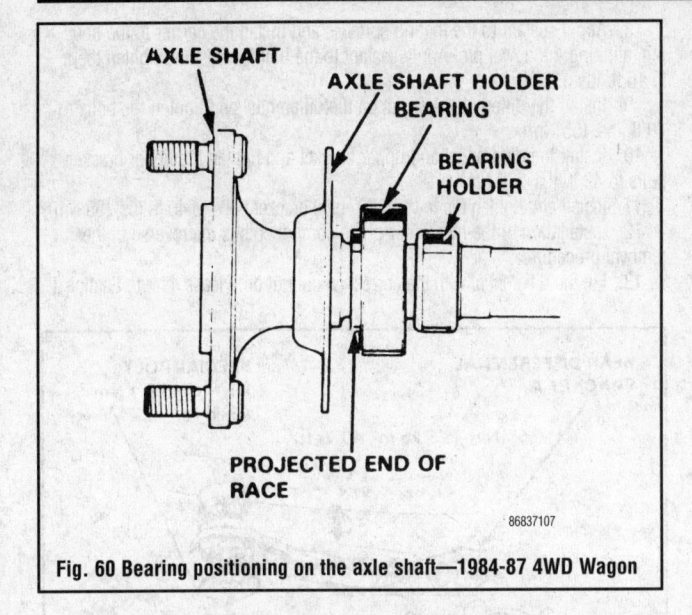

DRIVE TRAIN 7-27

Fig. 60 Bearing positioning on the axle shaft—1984-87 4WD Wagon

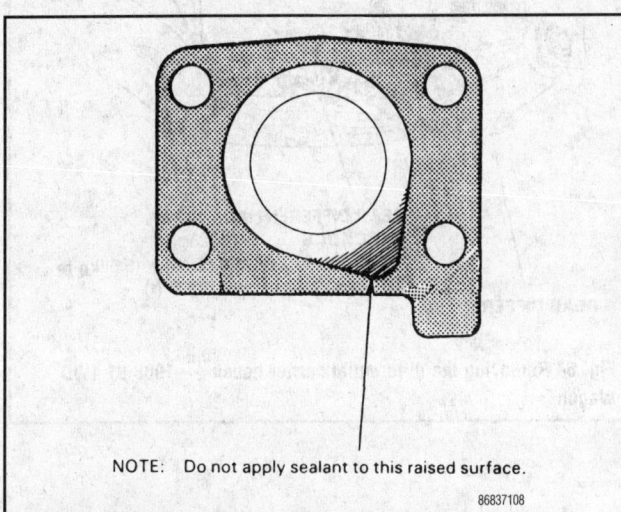

Fig. 61 Apply sealant to the shaded area of the axle holder, but do not apply sealant to the raised area—1984-87 4WD Wagon

NOTE: Do not apply sealant to this raised surface.

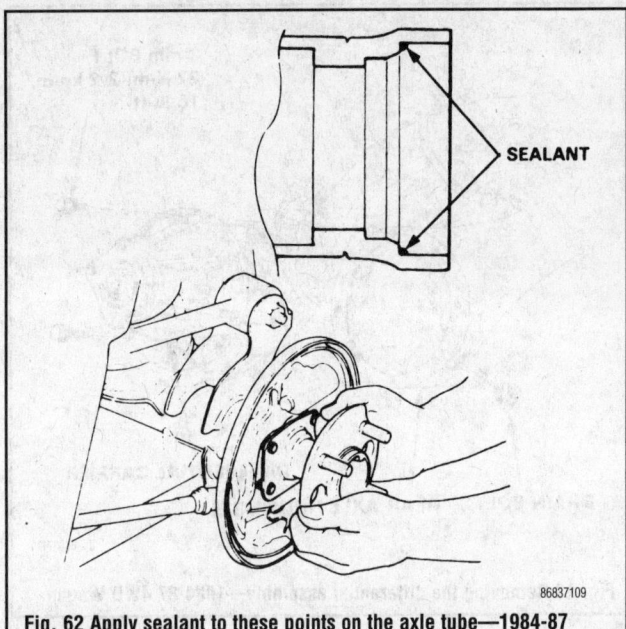

Fig. 62 Apply sealant to these points on the axle tube—1984-87 4WD Wagon

➡ Before installing the axle, coat the inner corner of the bearing housing seat with sealant.

20. Using a slide hammer with the appropriate adapters install the axle into the axle housing.
21. Install and tighten the bolts on the axle retainer. Tighten to 30 ft. lbs. (41 Nm).
22. Install the brake shoes and the parking brake cable.
23. Reconnect the brake line to the wheel cylinder.
24. Install the brake drum and bleed the brakes.
25. Install the wheel and lower the car to the ground.

Pinion Seal

REMOVAL & INSTALLATION

1. Raise and safely support the vehicle.
2. Remove the rear wheel and tire assemblies.
3. Keep the rear halfshafts in their normal horizontal position by raising the lower arms to their normal road level position.
4. Mark the position of the driveshaft in relation to the pinion flange, then disconnect the driveshaft from the pinion flange.
5. Using a suitable inch lb. torque wrench, turn the pinion flange by the pinion nut and record the reading. This is the total bearing preload.
6. Hold the pinion flange using a suitable holding tool and remove the pinion nut.
7. Remove the pinion flange and use a suitable removal tool to remove the pinion seal.

To install:
8. Lubricate the lip of a new pinion seal.
9. Use a suitable installation tool to install the pinion seal.
10. Install the pinion flange and pinion nut. Tighten the pinion nut until the reading on an inch lb. torque wrench is the same as that recorded prior to removal. The actual torque specification for the pinion flange nut is 127 ft. lbs. (175 Nm).
11. Connect the driveshaft to the pinion flange, aligning the marks that were made during the removal procedure. Tighten the bolts to 24 ft. lbs. (33 Nm).
12. Install the wheel and tire assemblies.
13. Lower the vehicle.

Differential Carrier

REMOVAL & INSTALLATION

◆ See Figure 63

1984-87 4WD Civic Wagon

1. Raise and support the rear of the vehicle.
2. Drain the differential oil into a suitable container by removing the drain plug.
3. Disconnect the driveshaft from the companion flange.
4. Remove the axle shafts, the procedure can be found eralier in this section.
5. Remove the differential carrier retaining bolts from the axle housing.
6. Remove the carrier from the axle housing, you may have to loosen it with a soft mallet.

To install:
7. Install the carrier in position and install the mounting bolts. Tighten the bolts to 16 ft. lbs. (22 Nm).
8. Install the axle shafts and connect the driveshaft to the companion flange.
9. Fill the differential with the correct amount of oil, that is until it just starts to come out of the filler plug hole (refer to Section 1).
10. Lower the vehicle.

7-28 DRIVE TRAIN

Fig. 63 Removing the differential assembly—1984-87 4WD Wagon

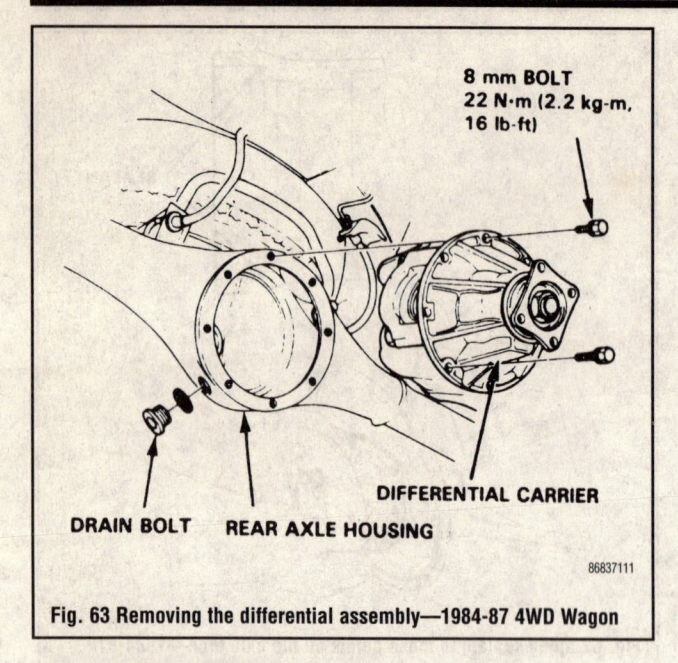

1988–91 4WD Civic Wagon

◆ See Figure 64

1. Raise and safely support the vehicle.
2. Drain the oil from the differential and replace the drain bolt.
3. Remove the driveshaft and the right and left rear halfshafts.
4. Remove the differential mounting bolts from the lower differential mounting bracket.
5. Remove the two bolts from the upper differential bracket and remove the differential and bracket as an assembly.
6. Remove the upper differential bracket from the differential and remove the differential carrier from the differential housing.

To install:

7. Clean the mating surfaces of the differential carrier and differential housing.
8. Apply sealant to the mating surfaces and install the carrier to the housing, aligning the dowel pin. Apply sealant to the bolt threads and tighten them to 16 ft. lbs. (22 Nm).
9. Install the differential bracket on the differential and tighten the bolts to 40 ft. lbs. (55 Nm).
10. Install the differential and upper bracket and tighten the upper bracket bolts to 43 ft. lbs. (59 Nm).
11. Install and tighten the lower differential bracket bolts to 40 ft. lbs. (55 Nm).
12. Installation of the remainder of the components is the reverse of the removal procedure.
13. Fill the differential with the correct amount of oil. Please refer to Section 1.

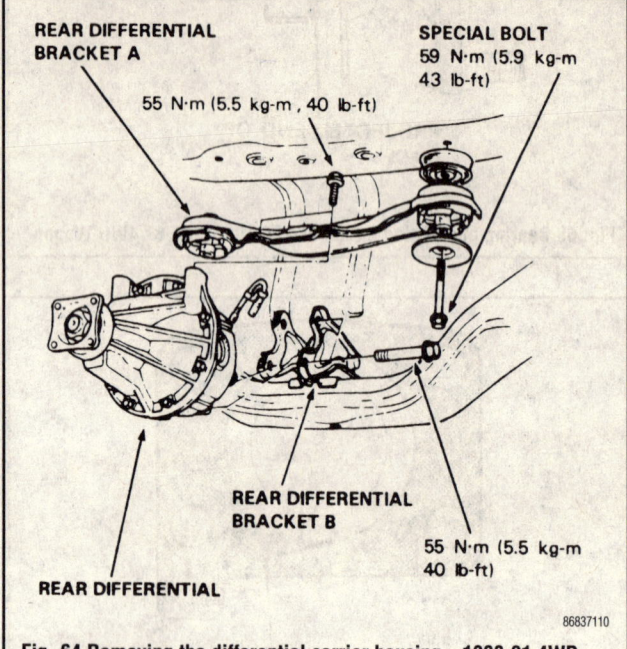

Fig. 64 Removing the differential carrier housing—1988-91 4WD Wagon

WHEELS 8-2
WHEELS AND TIRES 8-2
  REMOVAL & INSTALLATION 8-2
FRONT SUSPENSION 8-2
TORSION BARS 8-2
  REMOVAL & INSTALLATION 8-2
  ADJUSTMENT 8-3
SHOCK ABSORBERS 8-3
  REMOVAL & INSTALLATION 8-3
  INSPECTION 8-5
FRONT DAMPER 8-5
  INSPECTION 8-5
  REMOVAL & INSTALLATION 8-6
  DISASSEMBLY AND INSPECTION 8-7
LOWER BALL JOINT 8-7
  INSPECTION 8-7
  LUBRICATION 8-7
  REMOVAL & INSTALLATION 8-7
UPPER BALL JOINTS 8-8
  INSPECTION 8-8
  REMOVAL & INSTALLATION 8-8
UPPER CONTROL ARM AND BUSHING 8-8
  REMOVAL & INSTALLATION 8-8
RADIUS ARM 8-8
  REMOVAL & INSTALLATION 8-8
LOWER CONTROL ARM 8-8
  REMOVAL & INSTALLATION 8-8
FRONT WHEEL HUB, KNUCKLE AND SPINDLE 8-10
  REMOVAL & INSTALLATION 8-10
WHEEL ALIGNMENT 8-11
  CASTER 8-12
  CAMBER 8-12
  TOE 8-12
REAR SUSPENSION 8-13
GENERAL INFORMATION 8-13
  1984–87 MODELS 8-13
  1988–95 MODELS 8-13
COIL SPRINGS AND SHOCK ABSORBER 8-16
  REMOVAL & INSTALLATION 8-16
DAMPER STRUTS 8-16
  REMOVAL & INSTALLATION 8-16
  OVERHAUL 8-17
CONTROL ARMS 8-17
  REMOVAL & INSTALLATION 8-17
REAR WHEEL BEARINGS 8-19
  REPLACEMENT 8-19
STEERING 8-20
STEERING WHEEL 8-20
  REMOVAL & INSTALLATION 8-20
COMBINATION SWITCH 8-21
  REMOVAL & INSTALLATION 8-21
IGNITION SWITCH 8-21
  REMOVAL & INSTALLATION 8-21
  REMOVAL & INSTALLATION 8-21
IGNITION LOCK 8-22
  REMOVAL & INSTALLATION 8-22
STEERING COLUMN 8-22
  REMOVAL & INSTALLATION 8-22
STEERING LINKAGE 8-23
  REMOVAL & INSTALLATION 8-23
MANUAL RACK AND PINION 8-24
  ADJUSTMENT 8-24
  REMOVAL & INSTALLATION 8-25
  OVERHAUL 8-25
POWER STEERING GEAR 8-26
  ADJUSTMENT 8-26
  REMOVAL & INSTALLATION 8-26
  OVERHAUL 8-28
POWER STEERING PUMP 8-30
  REMOVAL & INSTALLATION 8-30
  BELT ADJUSTMENT 8-30
  SYSTEM BLEEDING 8-30
SPECIFICATIONS CHART
WHEEL ALIGNMENT 8-13

8

SUSPENSION AND STEERING

WHEELS 8-2
FRONT SUSPENSION 8-2
REAR SUSPENSION 8-13
STEERING 8-20

8-2 SUSPENSION AND STEERING

WHEELS

Wheels and Tires

REMOVAL & INSTALLATION

➡ Before removing the wheel and tire assembly, make sure that the vehicle is properly supported. Use the recommended jacking points and always block the wheel opposite the point you are jacking. When possible, use jackstands as an added safety precaution.

1. If using a scissors jack or similar floor jack, block the wheel diagonally opposite the side being lifted.
2. Remove the hub cap or center cap, note that on some models this cannot be done until the lug nuts are removed.
3. Loosen the lug nuts slightly, using the proper size wrench.
4. Lift the vehicle with the jack and remove the lug nuts. Remove the tire.
5. Install the tire in position and install the lug nuts, do not completely tighten the lug nuts until the vehicle is lowered.
6. Lower the vehicle and tighten the lug nuts. The nuts should be tightened in an "X" pattern to 80 ft. lbs. (110 Nm).

FRONT SUSPENSION

♦ See Figures 1, 2 and 3

The 1984–87 models use a Torsion bar front suspension. This change was made to lower the hood line, thus making the vehicle more aerodynamic and improving the driver's view of the road. The suspension consists of two independent torsion bars and front shock absorbers similar to a front strut assembly but without a spring. Both lower forged radius arms are connected with a stabilizer bar.

The 1988–91 models use an upper and lower control arm configuration which uses a coil-over shock absorber assembly. This suspension configuration allows the wheel to remain more parallel with the road surface thus allowing the vehicle to handle better.

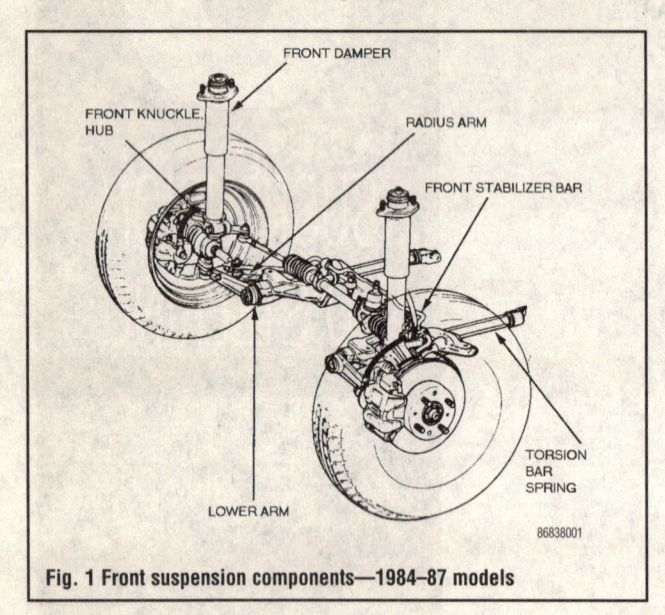

Fig. 1 Front suspension components—1984–87 models

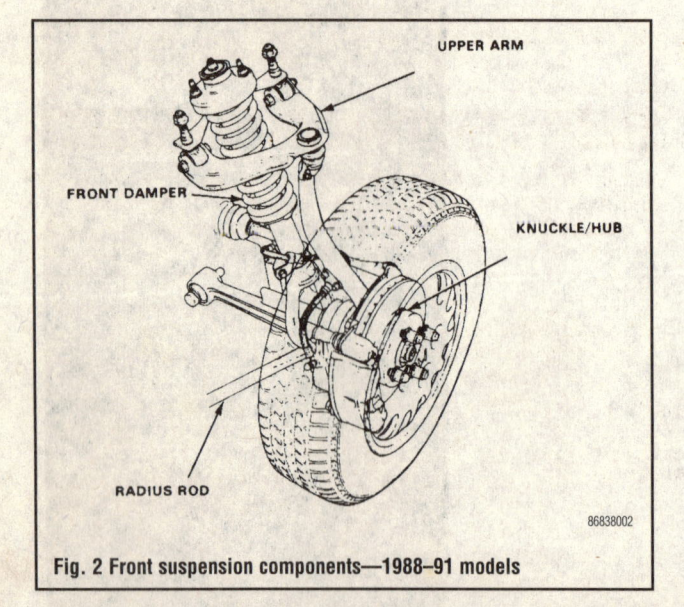

Fig. 2 Front suspension components—1988–91 models

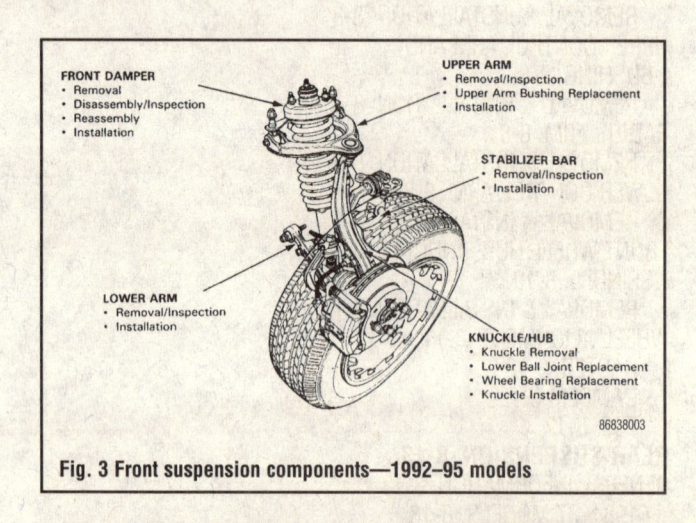

Fig. 3 Front suspension components—1992–95 models

Torsion Bars

REMOVAL & INSTALLATION

1984–87 Models

♦ See Figures 4, 5 and 6

1. Raise and support the front of the vehicle on jackstands.
2. Remove the height adjusting nut and the torque tube holder.
3. Remove the 33mm circlip.
4. Remove the torsion bar cap, then the torsion bar clip by tapping the bar out of the torque tube.

➡ The torsion bar will slide easier if you move the lower arm up and down.

5. Tap the torsion bar backward, out of the torque tube and remove the torque tube.

To install:

6. Install a new seal onto the torque tube. Coat the torque tube seal and torque with grease, then install them on the rear beams.

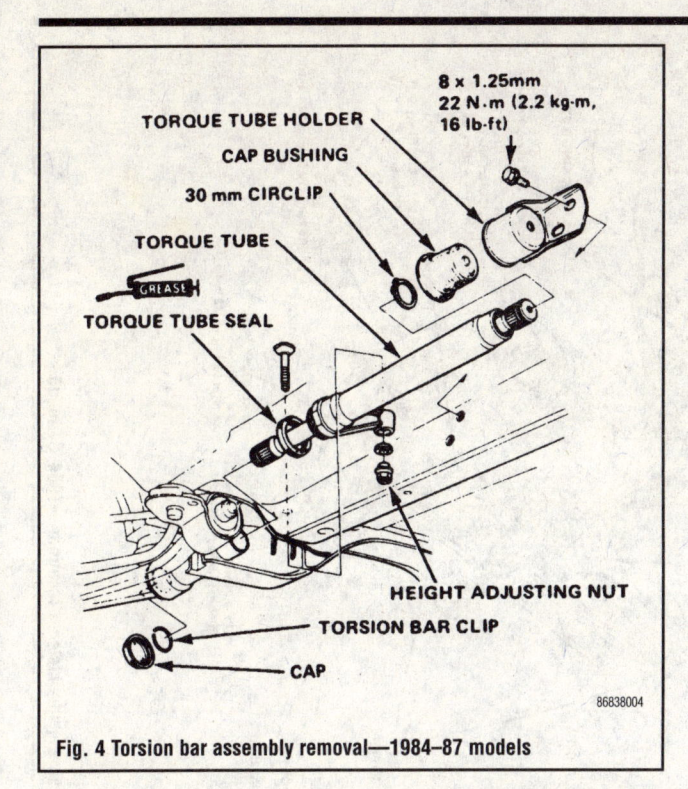

SUSPENSION AND STEERING 8-3

9. Install the torsion bar clip and cap, then install the 30mm circlip and the torque tube cap.

➡ Push the torsion bar to the front so there is no clearance between the torque tube and the 30mm circlip.

10. Coat the cap bushing with grease and install it on the torque tube. Install the torque tube holder and tighten the retaining bolts to 16 ft. lbs. (22 Nm).
11. Temporarily tighten the height adjusting nut.
12. Remove the jackstands and lower the vehicle to the ground. Adjust the torsion bar spring height.

ADJUSTMENT

1. Measure the torsion bar spring height between the ground and the highest point of the wheel arch.
   - CRX 25.15–25.55 in. (639–649mm).
   - HATCHBACK 25.24–25.63 in. (641–651mm).
   - SEDAN 25.43–25.8 in. (646–656mm).
   - WAGON 25.35–25.75 in. (644–654mm).
2. If the spring height for the models being worked on does not meet the specifications above, make the following adjustment.
   a. Raise and support the front of the vehicle on jackstands with the wheels off the ground.
   b. Adjust the spring height by turning the height adjusting nut. Tightening the nut raises the height and loosening the nut lowers the height.

➡ The height varies 0.2 in. (5mm) per revolution of the adjusting nut.

3. Lower the front wheels to the ground, then bounce the vehicle up and down several times and recheck the spring height to see if it is within specifications.

Shock Absorbers

REMOVAL & INSTALLATION

1984–87 Models

◆ See Figures 7 thru 13

1. Raise and support the front of the vehicle on jackstands. Remove the front wheels.
2. Remove the brake hose clamp bolt.
3. Using a floor jack, place it beneath the lower control arm to support it.
4. Remove the lower shock absorber retaining bolt from the steering knuckle, then slowly lower the jack.

✳✳ WARNING

Be sure the jack is positioned securely beneath the lower control arm at the ball joint. Otherwise, the tension from the torsion bar may cause the lower control arm to suddenly "jump" away from the shock absorber as the pinch bolt is removed.

5. Compress the shock absorber by hand, then remove the two upper lock nuts and shock absorber.
6. Perform the inspection procedure following or replace the damper.
7. To install, take note of the following:
   a. Use new self-locking nuts on the top of the shock absorber assembly mount and tighten to 28 ft. lbs. (39 Nm). Install a new self-locking nut on the shock absorber shaft, tighten it to 33 ft. lbs. (45 Nm).
   b. With the floor jack under the steering knuckle, carefully raise it to allow proper re-insertion of the damper tube into the steering knuckle and tighten the lower pinch bolt to 47 ft. lbs. (65 Nm).
   c. Install and tighten the brake hose clamp to 16 ft. lbs. (22 Nm).
8. Install the wheel/tire assembly, then perform the following inspection procedure, if satisfied, road test for proper function.

Fig. 4 Torsion bar assembly removal—1984–87 models

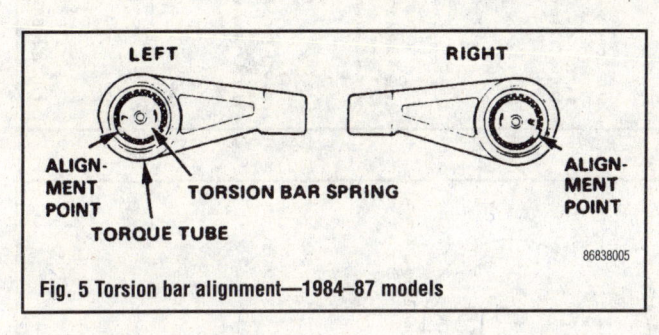

Fig. 5 Torsion bar alignment—1984–87 models

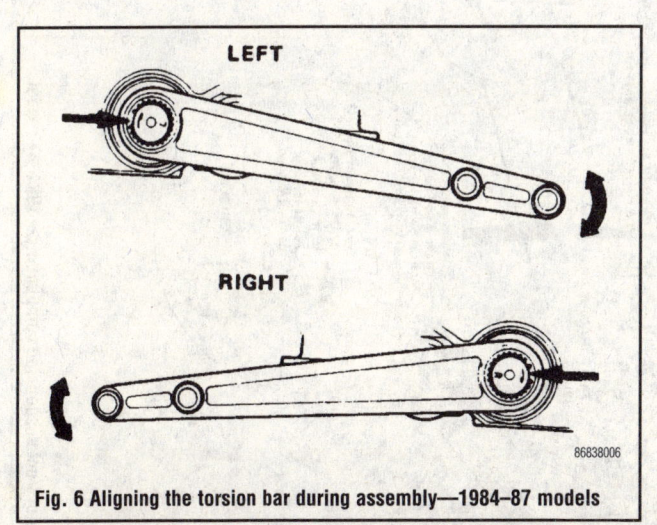

Fig. 6 Aligning the torsion bar during assembly—1984–87 models

7. Grease the ends of the torsion bar and insert into the torque tube from the back.
8. Align the projection on the torque tube splines with the cutout in the torsion bar splines and insert the torsion bar approximately 10mm.

➡ The torsion bar will slide easier if the lower arm is moved up and down.

8-4 SUSPENSION AND STEERING

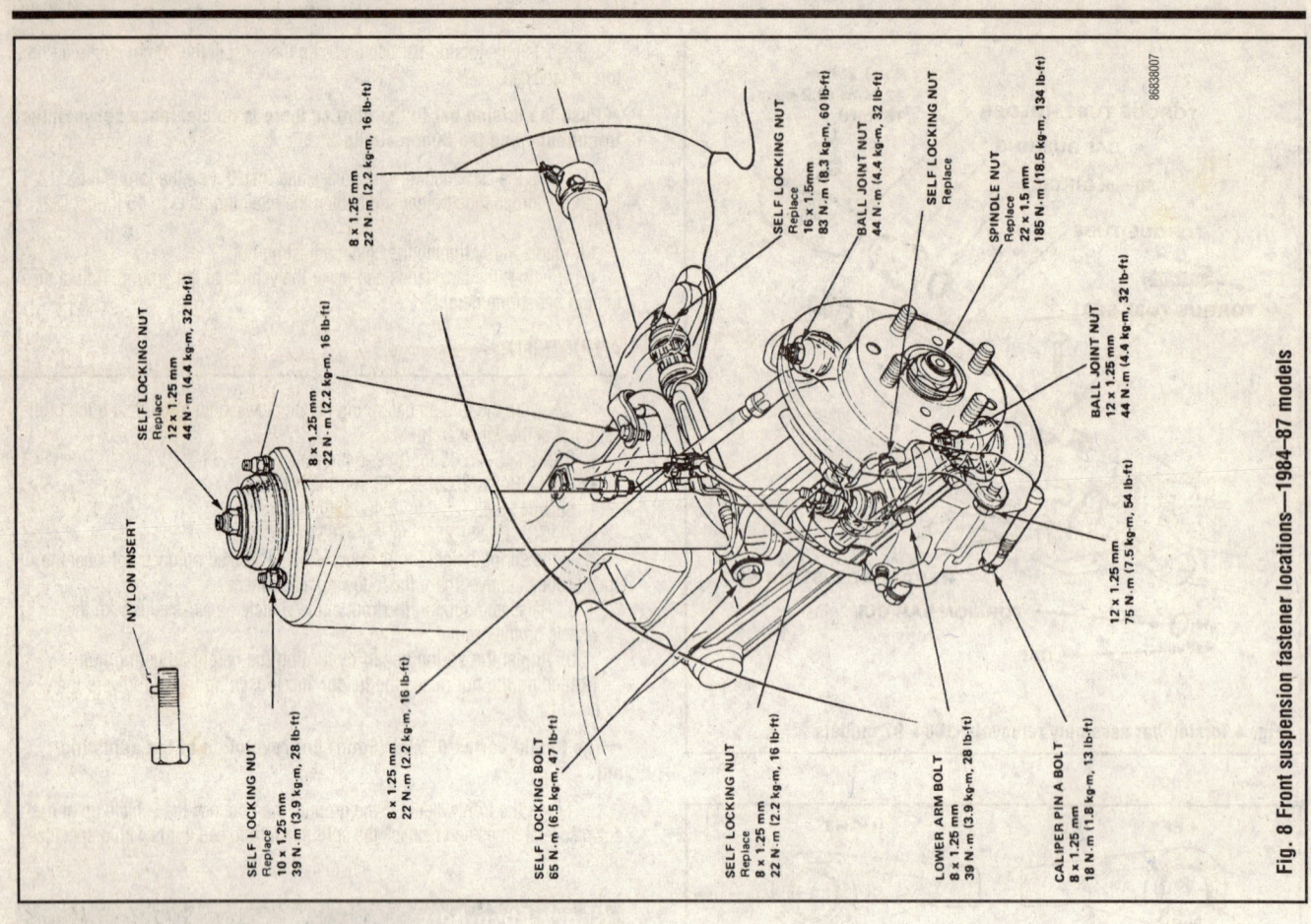

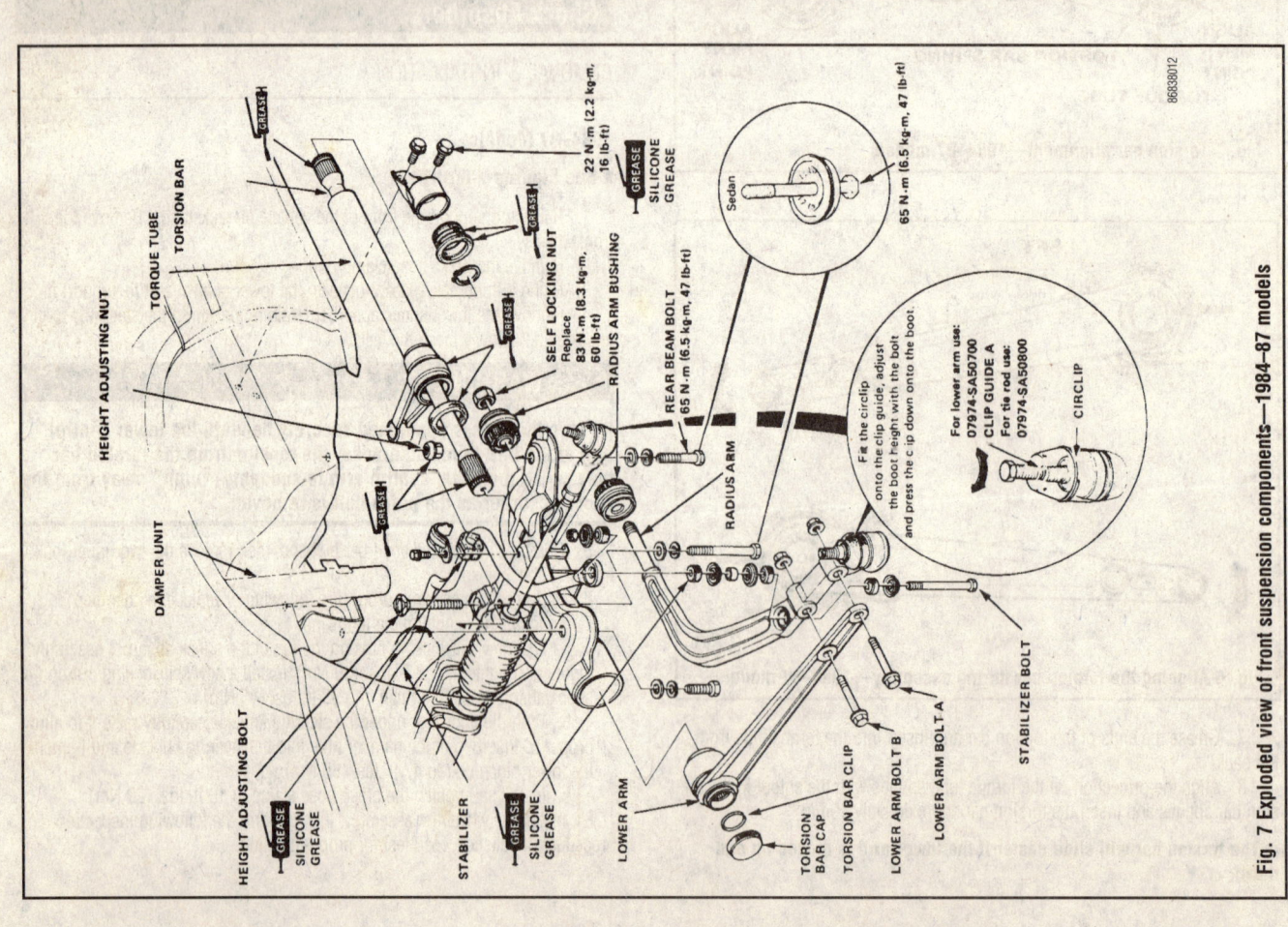

Fig. 8 Front suspension fastener locations—1984–87 models

Fig. 7 Exploded view of front suspension components—1984–87 models

SUSPENSION AND STEERING 8-5

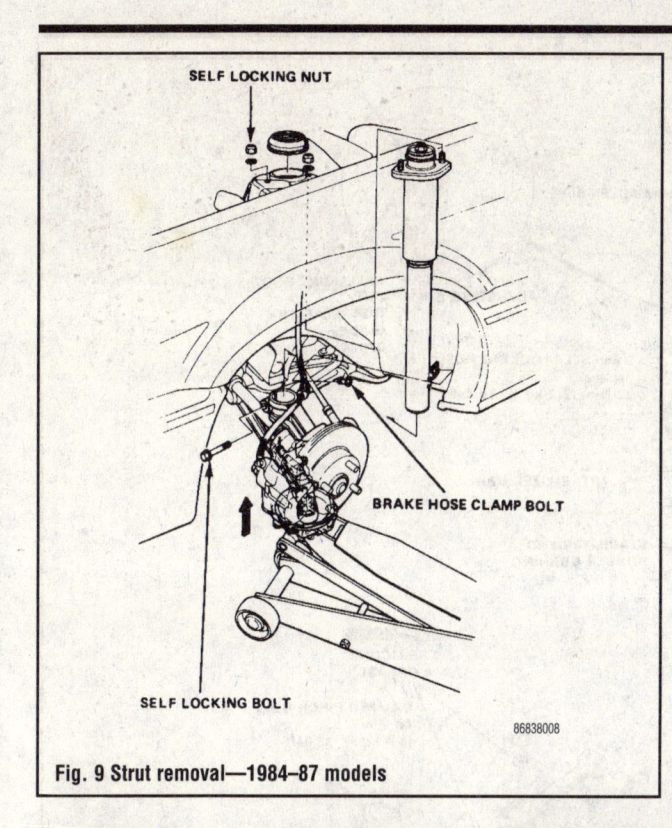

Fig. 9 Strut removal—1984–87 models

INSPECTION

▶ See Figure 14

1. Check for wear or damage to bushings.
2. Check for oil leaks from the shock absorbers.
3. Check all rubber parts for wear or damage.
4. Bounce the vehicle to check shock absorbing effectiveness. The vehicle should continue to bounce for no more than two cycles.

Front Damper

Honda began using their four wheel independent Double Wishbone suspension on the Civic series in 1988. This system is an adaptation of the suspension which was first used by Honda on the larger Accords of 1986. Unlike previous Civics, the hydraulic damper tube is assisted by a coil-over spring much like a MacPherson Strut. The Double Wishbone system distinguishes itself by the use of two parallel "wishbones" (or A-arms), one above and one below the strut. Honda claims the system is more compact, allowing a lower hoodline, and that it is designed to maximize grip and control by keeping the plane of the tire at a 90° angle to the road at various cornering speeds.

INSPECTION

1988–95 Models

▶ See Figure 15

➡ It may help to raise and safely support the vehicle to look more closely at the condition of the dampers. If the damper assembly shows

Fig. 10 With the vehicle supported, begin shock absorber removal by detaching the brake hose clamp

Fig. 11 The lower control arm (directly under the brake rotor) is supported by a floor jack

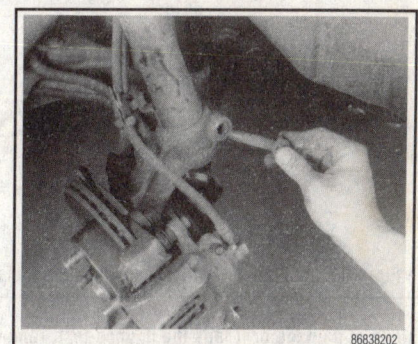

Fig. 12 Remove the lower shock retaining bolt to free the lower portion of the damper

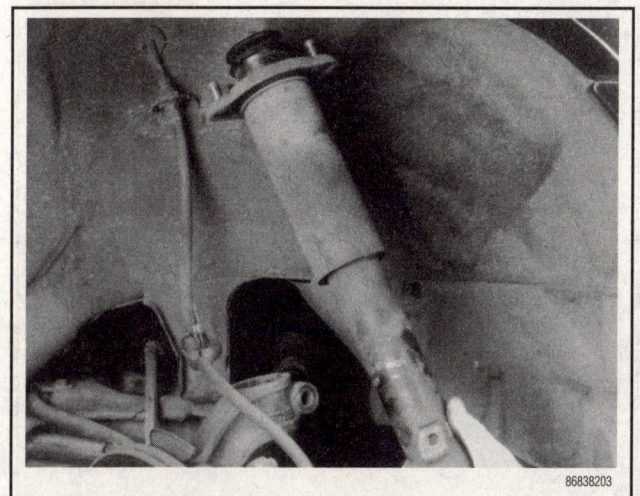

Fig. 13 Once detached, the shock absorber can be removed for inspection or replacement

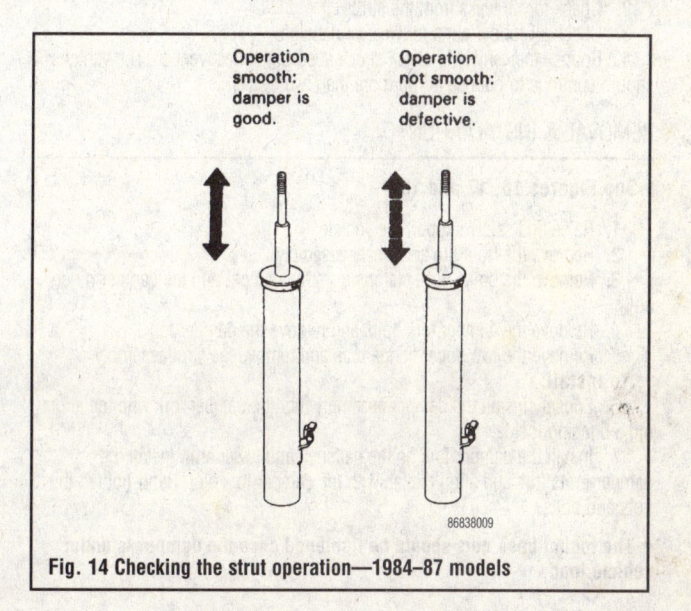

Fig. 14 Checking the strut operation—1984–87 models

8-6 SUSPENSION AND STEERING

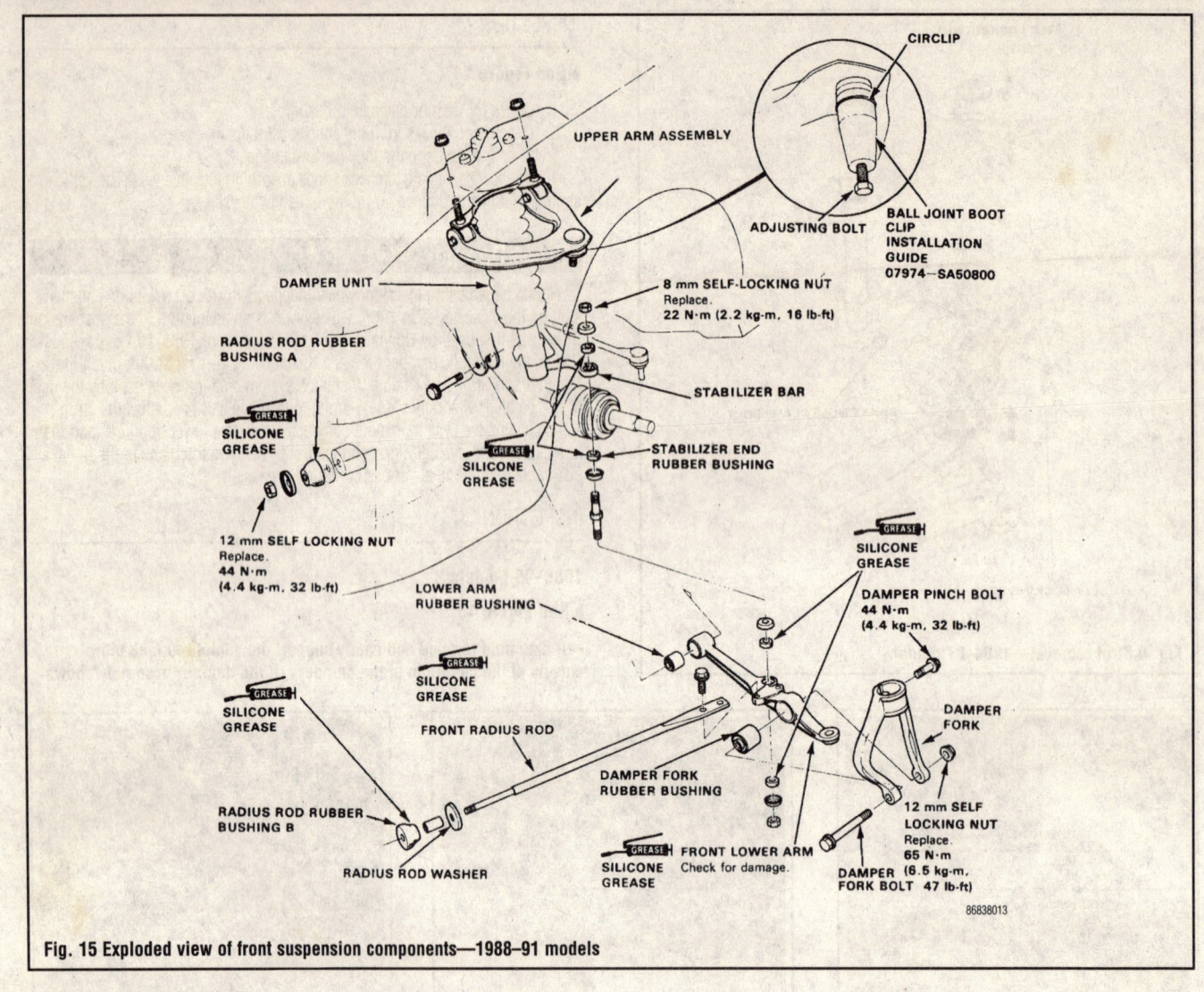

Fig. 15 Exploded view of front suspension components—1988–91 models

signs of damage or it fails the bounce-test, removal and replacement may be in order.

1. Looking under the vehicle, check for wear or damage to bushings and needle bearings.
2. Check for oil leaks from the struts.
3. Check all rubber parts for wear or damage.
4. Bounce the vehicle to check shock absorbing effectiveness. The vehicle should continue to bounce for no more than two cycles.

REMOVAL & INSTALLATION

▶ See Figures 16, 17 and 18

1. Raise and safely support the vehicle.
2. Remove the front tire and wheel assembly.
3. Remove the brake hose clamps from the damper and the damper pinch bolt.
4. Remove the damper fork bolts and remove the damper fork.
5. Remove the two upper flange nuts and remove the strut assembly.

To install:

6. Loosely install the damper assembly into the damper fork with the aligning tab facing inside.
7. Install the damper fork on the halfshaft and lower arm. Install the damper in its fork and align the tab with the damper fork slot. Hand-tighten the nuts and bolts.

➡ The mount base nuts should be tightened once the damper is under vehicle load.

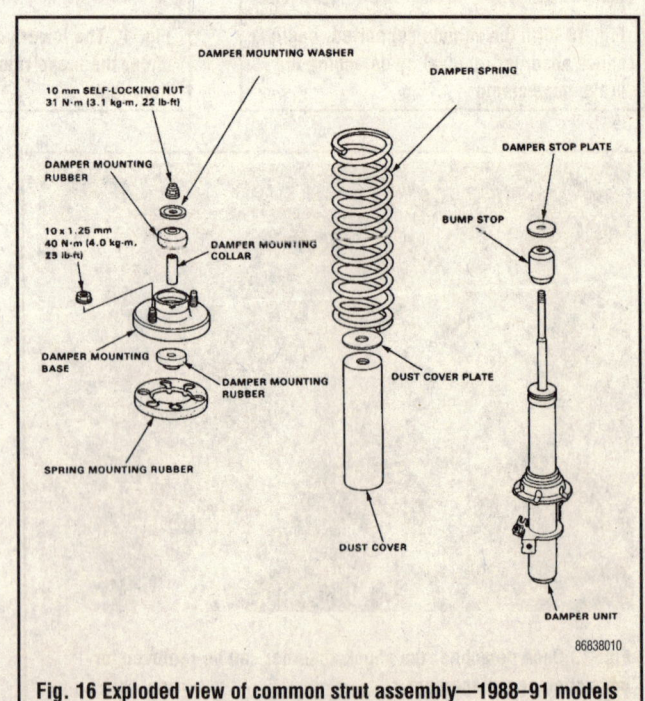

Fig. 16 Exploded view of common strut assembly—1988–91 models

SUSPENSION AND STEERING 8-7

8. Tighten the self-locking damper pinch bolt to 32 ft. lbs. (44 Nm).
9. Tighten a new self-locking damper fork nut to 47 ft. lbs. (65 Nm) while holding the damper fork bolt.
10. Install the brake hose clamps with the two bolts.
11. Install the two upper flange nuts for the damper assembly to secure it to the frame, but do not fully tighten them.

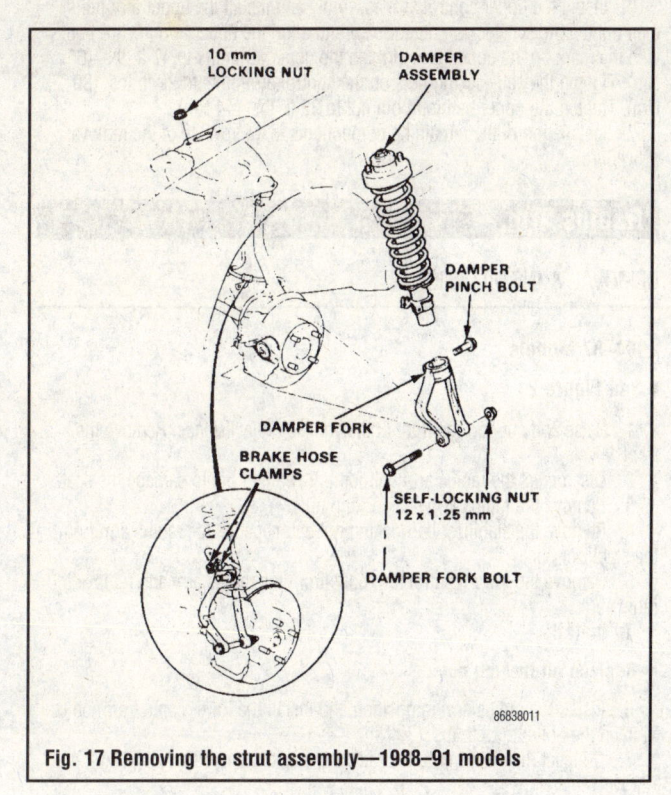

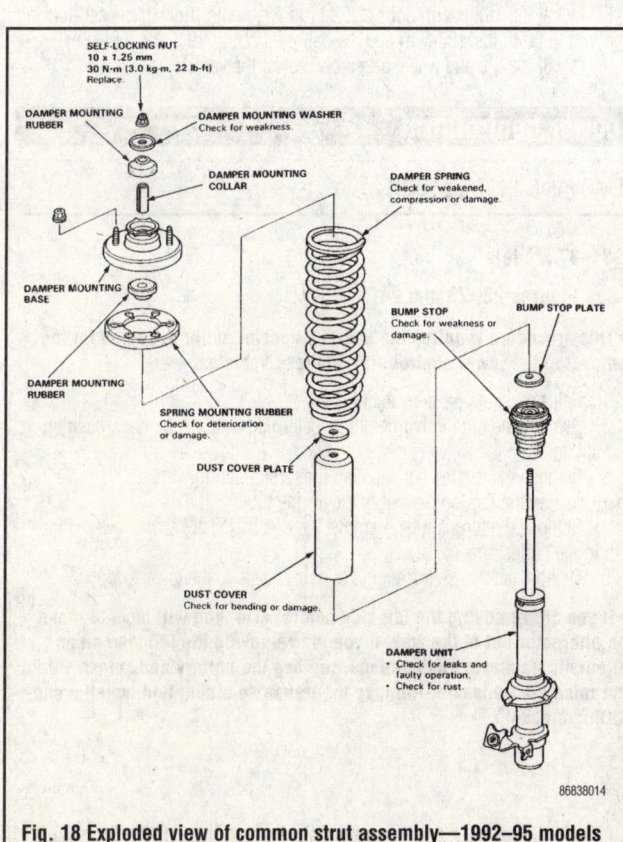

12. Lower the vehicle and tighten the flange nuts to the strut under vehicle load. Tighten the flange nuts for 1988–91 models to 29 ft. lbs. (40 Nm), and for 1992–95 models to 36 ft. lbs. (50 Nm).

DISASSEMBLY AND INSPECTION

1988–95 Models

This procedure will allow you to more carefully examine the damper assembly if unsure whether replacement is in order. The damper works to control the ride in the downward (compression) stroke, as well as the upward (rebound) stroke.

1. Remove the damper from the vehicle. Install a spring compressor safely on the spring according to the manufacturer's instructions and remove the self-locking nut from the damper. Take care to not compress the spring any more than is needed to remove the nut.
2. Remove the rubber cover and the center retaining nut.
3. Slowly release the compressor and remove the spring.
4. Remove the upper mounting cap, washers, thrust plates, bushing and dust cover.
5. Check for smooth operation throughout the range of motion. Next pull the strut shaft all the way out, hold it in this position and slide the rubber bumper down the shaft to the strut body. This should hold the shaft in the extended position. If the damper fails, replace it with a new one.

➡ Before discarding any parts, check a parts list to determine which parts are available as replacements.

To Install:

6. Position the spring and its top plate. Make sure the spring seats properly.
7. Install the partially assembled strut in the compressor. Compress the strut until the shaft protrudes through the top plate about 1 min. (25 mm).
8. Now install the bushings, thrust plates, top mounting cap washers and retaining nuts. Using an open wrench to turn the nut, and an Allen wrench to prevent the damper shaft from spinning, tighten the retaining nut to 22 ft. lbs. (31 Nm).
9. Once the retaining nut is installed, release the tension on the compressor and loosen the thumbscrew on the bottom plate. Separate the bottom plates and remove the compressor.

Lower Ball Joint

INSPECTION

1. Check ball joint play as follows:
   a. Raise and support the front of the vehicle on jackstands.
   b. Using a dial indicator clamp it onto the lower control arm and place the indicator tip on the steering knuckle, near the ball joint.
   c. Using a prybar, place it between the lower control arm and the steering knuckle.
   d. Work the ball joint to check for looseness; if the play exceeds 0.2 in. (0.5 mm), replace the ball joint.

LUBRICATION

1. Remove the screw plug from the bottom of the ball joint and install a grease nipple.
2. Lubricate the ball joint with NLGI No. 2 multipurpose type grease.
3. Remove the nipple and reinstall the screw plug.
4. Repeat for the other ball joint.

REMOVAL & INSTALLATION

1984–87 Models

1. Raise and safely support the vehicle.
2. Remove the wheel and tire assembly.
3. Remove the steering knuckle.
4. Pry off the snapring and remove the boot.
5. Remove the circlip.

Fig. 17 Removing the strut assembly—1988–91 models

Fig. 18 Exploded view of common strut assembly—1992–95 models

8-8 SUSPENSION AND STEERING

6. Install a suitable ball joint remover/installer 07965–SB00100, or equivalent, on the ball joint and tighten the ball joint nut.
7. Position ball joint remover base 07JAF–SH20200 or equivalent, over the ball joint and then set the assembly in a suitable vise. Press the ball joint out of the knuckle.

To install:

8. Place the ball joint in position by hand.
9. Install a ball joint remover/installer 07965–SB00100 or suitable equivalent, and ball joint installer base 07965–SB00200 or equivalent, over the ball joint and position in a suitable vise. Press the ball joint in.
10. Install the steering knuckle.

Upper Ball Joints

INSPECTION

1988–91 Models

1. Raise and safely support the vehicle.
2. Remove the front wheel and tire assembly.
3. Grasp the steering knuckle and move it back and forth.
4. Replace the upper control arm if any play is detected.

REMOVAL & INSTALLATION

The upper ball joint is an integral component of the upper control arm. If the ball joint is defective the entire upper control arm must be replaced.

Upper Control Arm and Bushing

REMOVAL & INSTALLATION

1988–91 Models

♦ See Figures 19 and 20

1. Raise and support the vehicle safely.
2. Remove the front wheels. Properly support the lower control arm assemblies.
3. Remove the self-locking nuts, upper control arm bolts and upper control anchor bolts. Separate the upper ball joint using a suitable ball joint separator tool.
4. Place the upper control arm assembly into a suitable holding fixture and drive out the upper arm bushing.

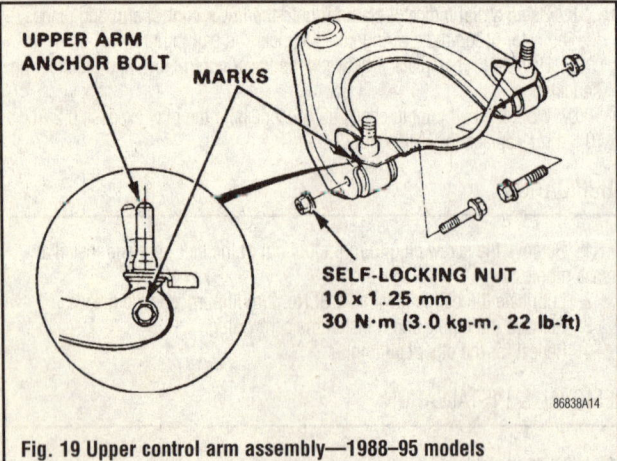

Fig. 19 Upper control arm assembly—1988–95 models

To install:

5. Drive the new upper arm bushing into the upper arm anchor bolts. On 1988–91 Civic and CRX, center the bushing so 0.3543 in. (9mm) protrudes from each side of the anchor bolt. On 1992–95 Civic and del Sol models, center the bushing so that 0.4 in. (11 mm) protrudes from each side of the anchor bolt as shown.
6. Install the upper control arm assembly and install the upper arm bolts, then tighten the self-locking nuts. Be sure to align the upper arm anchor bolt with the mark on the upper arm. Tighten the upper arm nuts to 47 ft. lbs. (65 Nm). Tighten the self-locking nuts on the control arm bolts to 22 ft. lbs. (30 Nm). Tighten the upper ball joint pin nut to 32 ft. lbs. (44 Nm).
7. Installation of the remaining components is the reverse of the removal procedure.

Radius Arm

REMOVAL & INSTALLATION

1984–87 Models

♦ See Figure 21

1. Raise and support the front of the vehicle on jackstands. Remove the front wheels.
2. Disconnect the radius arm ball joint. Be careful not to damage the seal.
3. Remove the radius arm self-locking nuts.
4. Remove the stabilizer bar locking nut and separate the radius arm from the stabilizer bar.
5. Remove the lower control arm bolts from the radius arm and remove the radius arm.

To install:

➡ Replace all locking nuts.

6. Install the radius arm in position and install the lower control arm bolts. Tighten the bolts to 28 ft. lbs. (39 Nm).
7. Connect the stabilizer to the radius arm and tighten the bolt to 16 ft. lbs. (22 Nm).
8. Install the radius arm bolts and self-locking nuts. Tighten the self-locking pivot nut to 60 ft. lbs. (83 Nm) and the ball joint nut to 32 ft. lbs. (44 Nm).
9. Install the tire and wheel assembly. Lower the vehicle.

Lower Control Arm

REMOVAL & INSTALLATION

1984–87 Models

♦ See Figures 22, 23 and 24

➡ This procedure requires the use of a special puller designed for the removal of the lower control arm on these vehicles.

1. Raise and support the vehicle.
2. Remove the torsion bar height adjuster nut, torque tube hold, bushing and circlip.
3. Remove the torsion bar cap. Tap the torsion bar forward and remove the snapring. Tap the torsion bar out of the torque tube.
4. Remove the torque tube. Remove the bolt that attaches the lower control arm to the radius arm.
5. Using a jack, raise the engine to gain access to the lower arm.

➡ If you are removing the left side control arm, you will have to move the alternator out of the way. If you are removing the left side on an automatic transaxle equipped car, remove the battery and splash shield, and raise the transaxle. Remove the transaxle mount and install a supporting bracket.

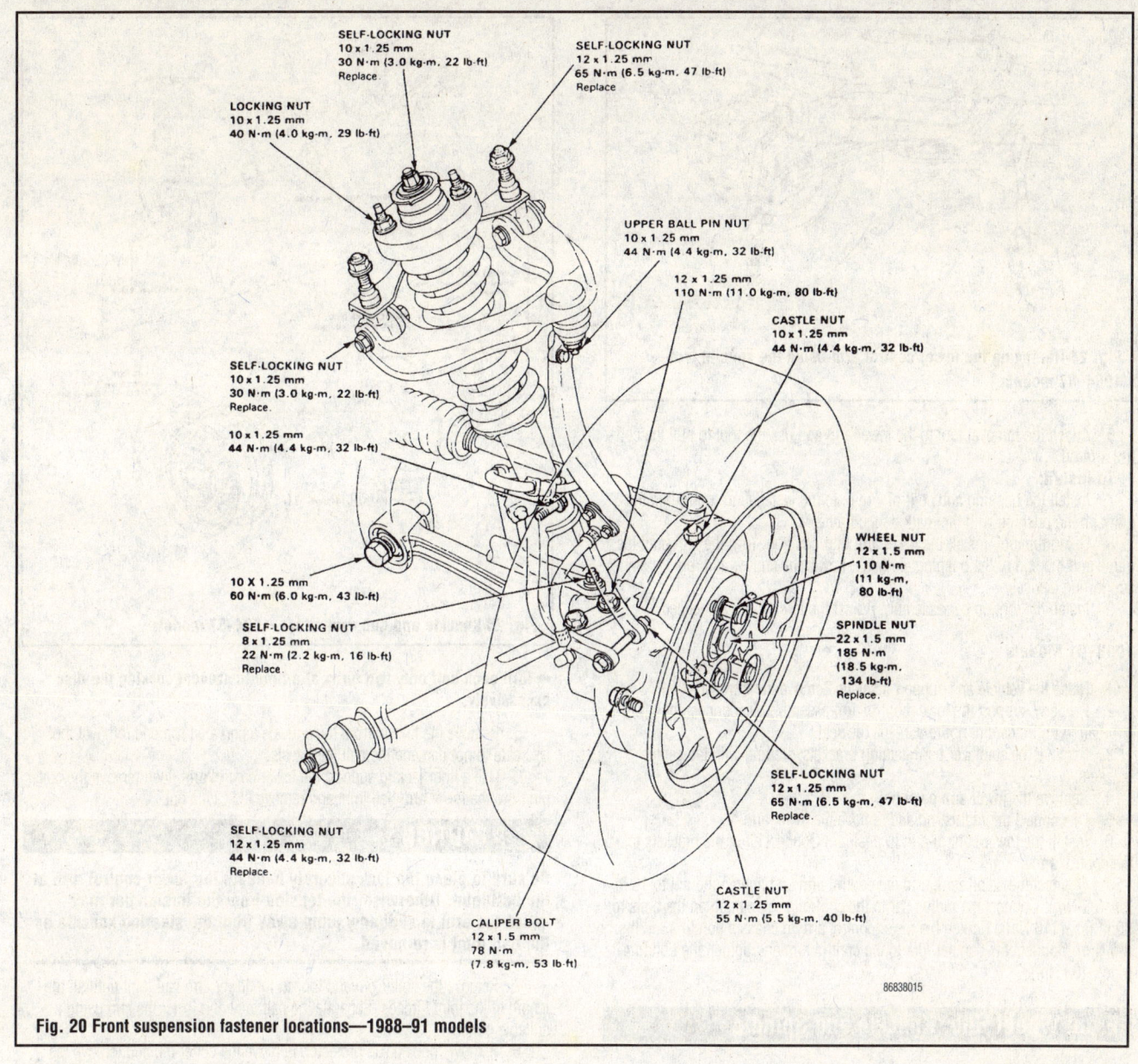

SUSPENSION AND STEERING 8-9

Fig. 20 Front suspension fastener locations—1988–91 models

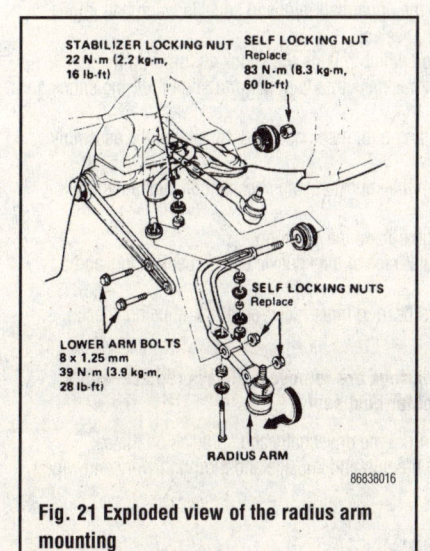

Fig. 21 Exploded view of the radius arm mounting

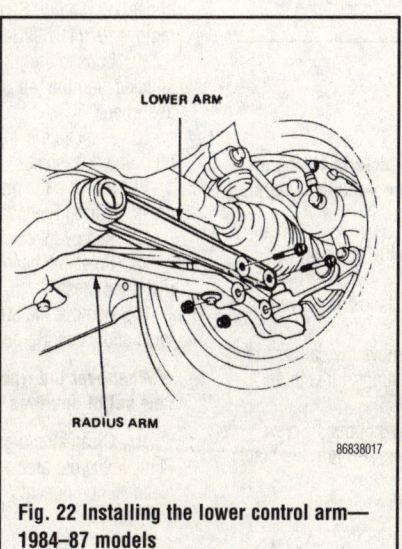

Fig. 22 Installing the lower control arm—1984–87 models

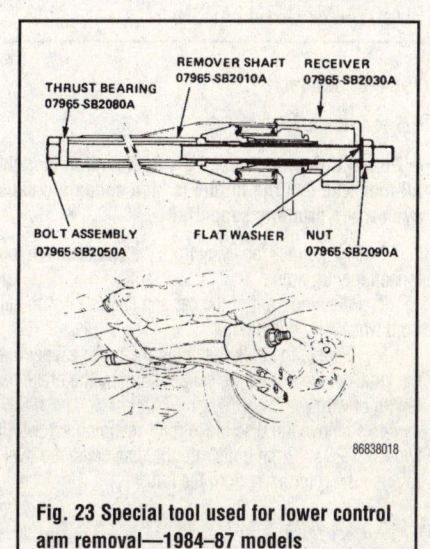

Fig. 23 Special tool used for lower control arm removal—1984–87 models

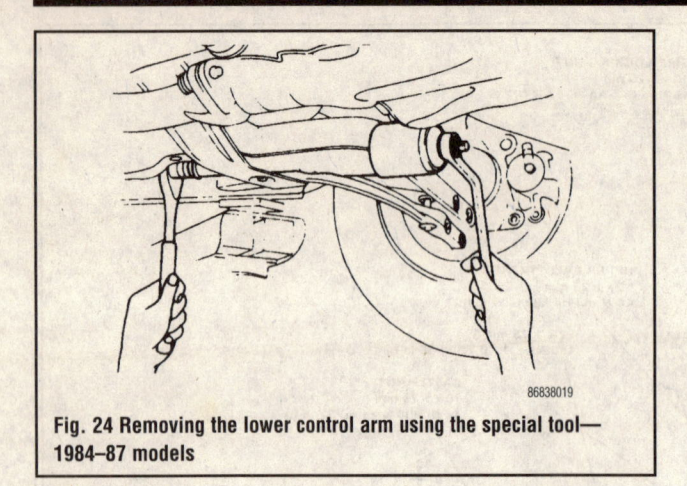

8-10 SUSPENSION AND STEERING

Fig. 24 Removing the lower control arm using the special tool—1984–87 models

6. Attach the removal tool to the lower arm and use the tool to pull the control arm off.

To install:
7. Install the bushing and control arm onto the installation tool. Align the mark on the bushing with the mark on the frame.
8. Using the tool, install the lower arm into position. Turn the tool until the frame and the end of the bushing are flush. Lower the engine and remove any supporting devices.
9. Install the torsion bar assembly. Adjust the torsion bar as needed.

1988–91 Models

1. Raise the vehicle and support it safely. Remove the front wheels.
2. Properly support the lower control arm assembly. Disconnect the lower arm ball joint. Be careful not to damage the seal.
3. Remove the stabilizer bar retaining brackets, starting with the center brackets.
4. Remove the lower arm pivot bolt.
5. Disconnect the radius rod and remove the lower arm.
6. Install the lower control arm in position. Connect all of the brackets to the control arm.
7. Connect the stabilizer bar to the control arm and tighten the nut to 16 ft. lbs. (22 Nm). Connect the radius bar to the control arm and tighten the bolts to 80 ft. lbs. (110 Nm). Tighten the lower control arm to chassis bolt to 43 ft. lbs. (60 Nm). Connect the damper fork to the control arm and tighten the bolt to 47 ft. lbs. (65 Nm).

Front Wheel Hub, Knuckle and Spindle

REMOVAL & INSTALLATION

1984–87 Models

▶ See Figure 25

➡ Do not tighten or loosen a spindle nut unless the vehicle is sitting on all four wheels. The torque is high enough to cause the vehicle to topple even when properly supported.

1. Pry the lock tab away from the spindle, then loosen the nut. Slightly loosen the lug nuts.
2. Raise the front of the car and support it with safety stands. Remove the front wheel and spindle nut.
3. Remove the bolts retaining the brake caliper and remove the caliper from the knuckle. Do not let the caliper hang by the brake hose, support it with a length of wire.
4. Remove the disc brake rotor retaining screws (if so equipped). Screw two 8 x 1.25 x 12mm bolts into the disc brake removal holes, and turn the bolts to push the rotor away from the hub.

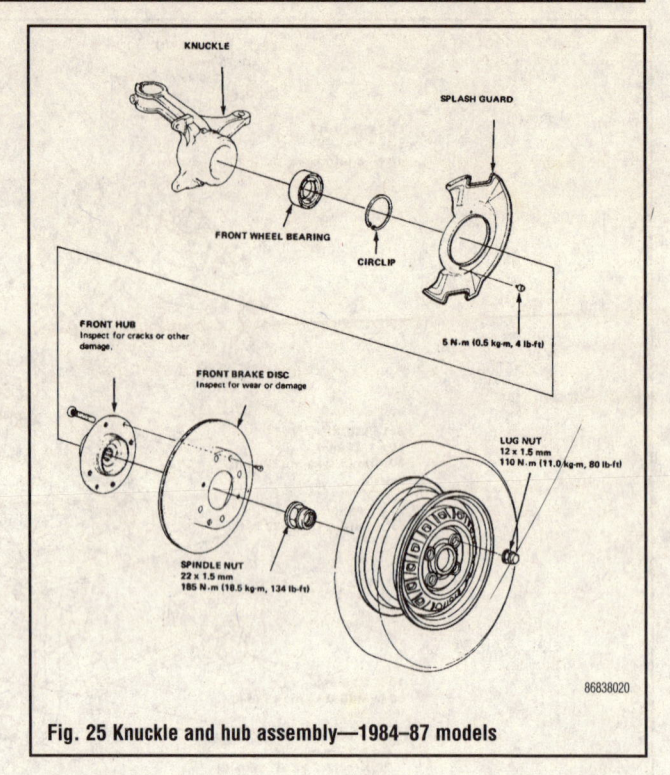

Fig. 25 Knuckle and hub assembly—1984–87 models

➡ Turn each bolt only two turns at a time to prevent cocking the disc excessively.

5. Remove the tie rod from the knuckle using a tie rod end removal tool. Use care to not damage the ball joint seals.
6. Use a floor jack to support the lower control arm, then remove the cotter pin from the lower arm ball joint and remove the castle nut.

※※ CAUTION

Be sure to place the jack securely beneath the lower control arm at the ball joint. Otherwise, the tension from the torsion bar may cause the arm to suddenly jump away from the steering knuckle as the ball joint is removed.

7. Remove the cotter pin and loosen the lower arm ball joint nut half the length of the joint threads. Separate the ball joint and lower the arm using a suitable puller.
8. Remove the knuckle protector. Remove the cotter pin and remove the upper ball pin nut. Separate the upper ball joint and knuckle using ball joint remover 07941-6920-2 or equivalent.
9. Loosen the pinch bolt which retains the shock absorber in the knuckle. Tap the top of the knuckle with a hammer and slide it off the shock absorber.
10. Remove the knuckle and hub, If still attached, by sliding the assembly off of the driveshaft.
11. Remove the hub from the knuckle using special tools and a hydraulic press. Bearing Removal:
12. Remove the splash guard and the snapring.
13. Press the bearing outer race of the knuckle using special tools and a hydraulic press.
14. Remove the outboard bearing inner race from the hub using special tools and a bearing puller.

➡ Whenever the wheel bearings are removed, always replace with a new set of bearings and outer dust seal.

15. Clean all old grease from the driveshafts and spindles on the car. Remove the old grease from the hub and knuckle and thoroughly dry and wipe clean all components.

SUSPENSION AND STEERING 8-11

16. To install the bearings, press the bearing outer race into the knuckle using the special tools as used above, plus the installing base tool.
17. Install the snapring, then install the splash guard.
18. Place the hub in the special tool fixture, then set the knuckle in position on the press and apply downward pressure.
19. The remaining steps are the reverse of the removal procedure. Use a new spindle nut, and stake after torquing. Tighten the shock absorber pinch bolt to 47 ft. lbs. (65 Nm), and the ball joint nut to 32 ft. lbs. (44 Nm).

1988–95 Models

▶ See Figures 26 and 27

➡ Do not tighten or loosen a spindle nut unless the vehicle is sitting on all four wheels. The torque is high enough to cause the vehicle to topple even when properly supported.

1. Pry the spindle nut stake away from the spindle, then loosen the nut.
2. Raise and safely support the vehicle.
3. Remove the wheel and tire assembly and the spindle nut.
4. Remove the caliper mounting bolts and the caliper. Support the caliper out of the way with a length of wire. Do not let the caliper hang from the brake hose.
5. Remove the 6mm brake disc retaining screws. Screw two 8 x 1.25 x 12mm bolts into the disc to push it away from the hub.

➡ Turn each bolt two turns at a time to prevent cocking the brake disc.

6. Remove the cotter pin from the tie rod castle nut, then remove the nut. Break the tie rod ball joint using a suitable ball joint remover, then lift the tie rod out of the knuckle.
7. Remove the cotter pin and loosen the lower arm ball joint nut half the length of the joint threads.
8. Separate the ball joint and lower arm using a suitable puller with the pawls applied to the lower arm.

➡ Avoid damaging the ball joint boot. If necessary, apply penetrating type lubricant to loosen the ball joint.

9. Remove the knuckle protector.

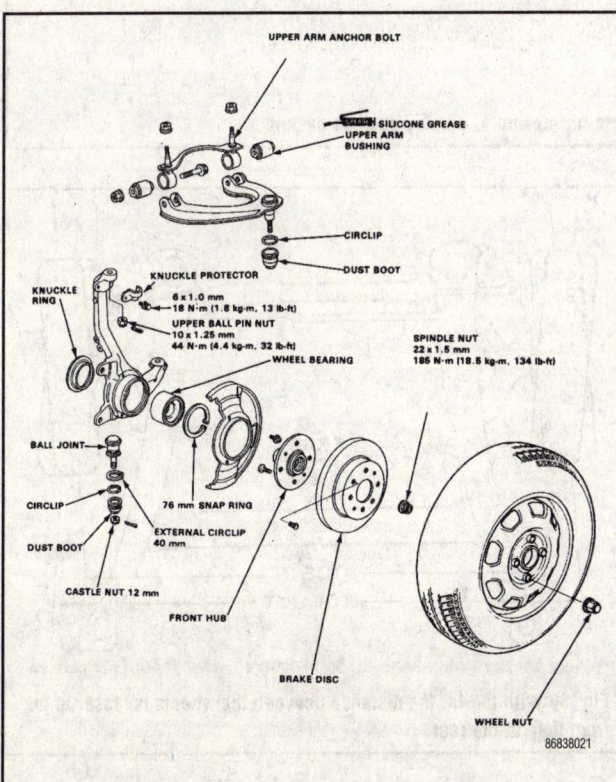

Fig. 26 Knuckle and hub assembly—1988–91 models

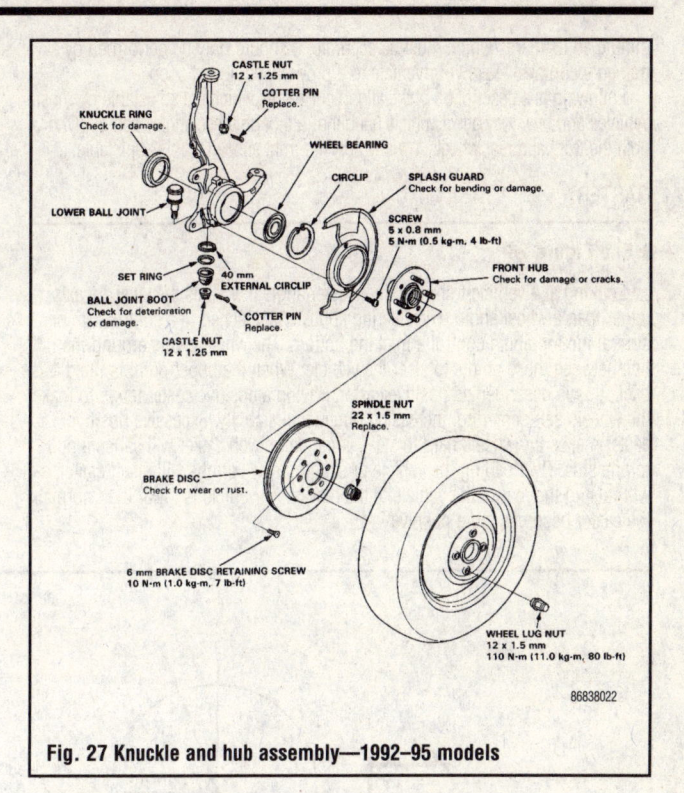

Fig. 27 Knuckle and hub assembly—1992–95 models

10. Remove the cotter pin and remove the upper ball pin nut.
11. Separate the upper ball joint and knuckle using a suitable tool.
12. Remove the knuckle and hub by sliding them off the halfshaft.
13. Remove the splash guard screws from the knuckle.
14. Position the knuckle/hub assembly on a hydraulic press. Press the hub from the knuckle using a suitable driver while supporting the knuckle with a suitable base.

➡ The bearing must be replaced with a new one after removal.

15. Remove the 76mm snapring and knuckle ring from the knuckle.
16. Press the wheel bearing out of the knuckle using a suitable driver while supporting the knuckle with a suitable base.
17. Remove the outboard bearing inner race from the hub using a suitable bearing puller.

To install:

18. Clean the knuckle and hub thoroughly before reassembly.
19. Press a new wheel bearing into the hub using a suitable driver while supporting the knuckle with a suitable base.
20. Install the 76mm snapring securely in the knuckle groove.
21. Install the splash guard and tighten the screws to 7 ft. lbs. (10 Nm).
22. Place the knuckle into position on the hydraulic press and press onto the hub using a suitable driver. The maximum press load should be two tons.
23. Install the front knuckle ring on the knuckle.
24. Install the knuckle/hub assembly onto the vehicle in the reverse order of the removal procedure. Tighten the upper ball pin nut and tie rod nut to 32 ft. lbs. (44 Nm) and the lower ball joint castle nut to 40 ft. lbs. (55 Nm).
25. With all four wheels resting on the ground, tighten the spindle nut to 134 ft. lbs. (185 Nm).

Wheel Alignment

If the tires are worn unevenly, if the vehicle is not stable on the highway or if the handling seems uneven in spirited driving, the wheel alignment should be checked. If an alignment problem is suspected, first check for improper tire inflation and other possible causes. These can be worn suspension or steering components, accident damage or even unmatched tires. If any worn or damaged components are found, they must be replaced before the wheels can be properly aligned. Wheel alignment requires very expensive equipment and involves

8-12 SUSPENSION AND STEERING

minute adjustments which must be accurate; it should only be performed by a trained technician. Take your vehicle to a properly equipped shop.

Following is a description of the alignment angles which are adjustable on most vehicles and how they affect vehicle handling. Although these angles can apply to both the front and rear wheels, usually only the front suspension is adjustable.

CASTER

▶ **See Figure 28**

Looking at a vehicle from the side, caster angle describes the steering axis rather than a wheel angle. The steering knuckle is attached to a control arm or strut at the top and a control arm at the bottom. The wheel pivots around the line between these points to steer the vehicle. When the upper point is tilted back, this is described as positive caster. Having a positive caster tends to make the wheels self-centering, increasing directional stability. Excessive positive caster makes the wheels hard to steer, while an uneven caster will cause a pull to one side. Overloading the vehicle or sagging rear springs will affect caster, as will raising the rear of the vehicle. If the rear of the vehicle is lower than normal, the caster becomes more positive.

CAMBER

▶ **See Figure 29**

Looking from the front of the vehicle, camber is the inward or outward tilt of the top of wheels. When the tops of the wheels are tilted in, this is negative camber; if they are tilted out, it is positive. In a turn, a slight amount of negative camber helps maximize contact of the tire with the road. However, too much negative camber compromises straight-line stability, increases bump steer and torque steer.

TOE

▶ **See Figure 30**

Looking down at the wheels from above the vehicle, toe angle is the distance between the front of the wheels, relative to the distance between the back of the wheels. If the wheels are closer at the front, they are said to be toed-in or to have negative toe. A small amount of negative toe enhances directional stability and provides a smoother ride on the highway.

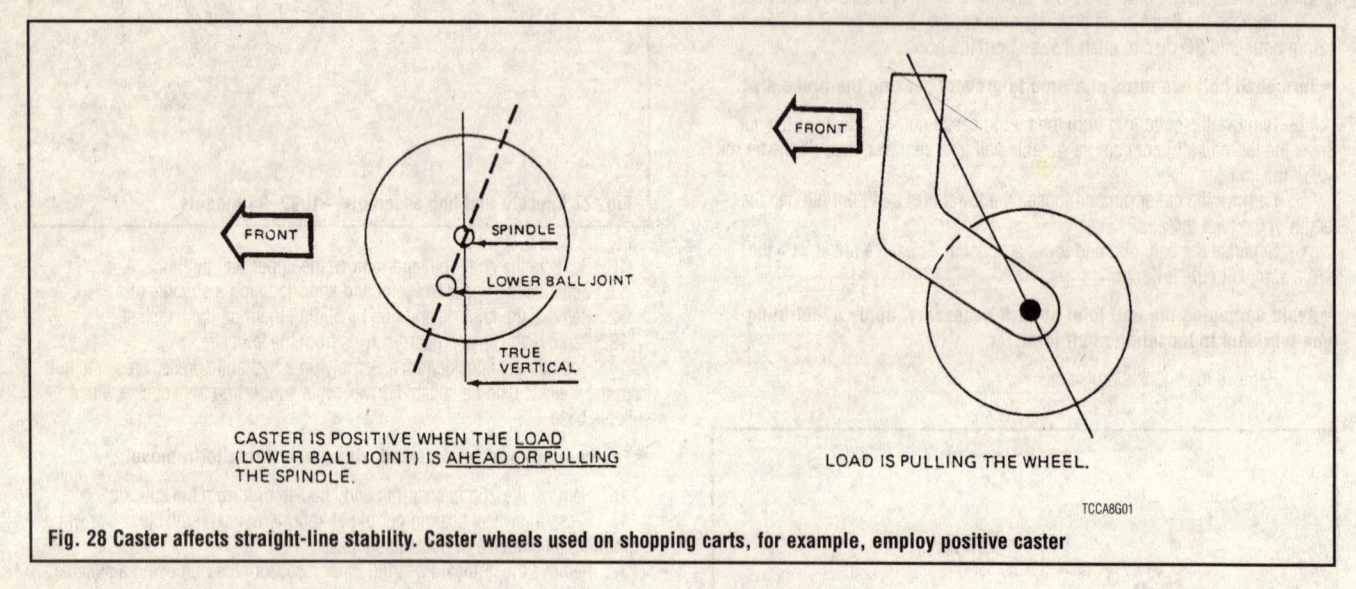

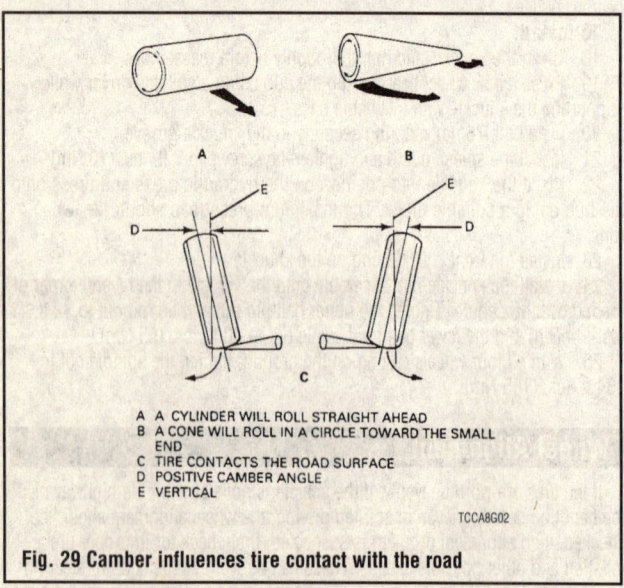

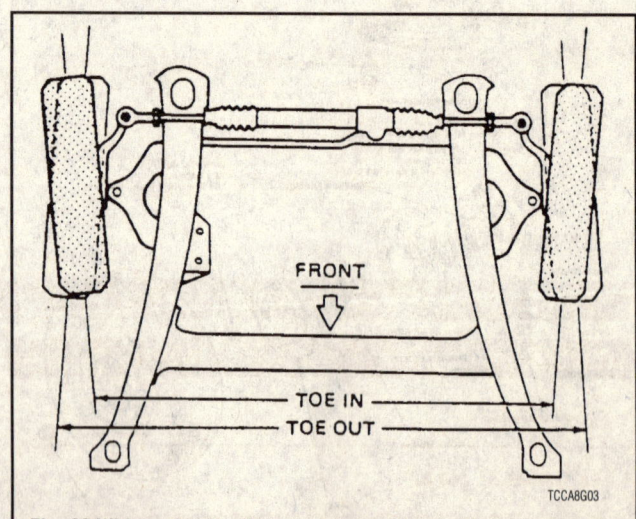

Fig. 28 Caster affects straight-line stability. Caster wheels used on shopping carts, for example, employ positive caster

Fig. 29 Camber influences tire contact with the road

Fig. 30 With toe-in, the distance between the wheels is closer at the front than at the rear

SUSPENSION AND STEERING   8-13

WHEEL ALIGNMENT

Year	Model		Caster Range (deg.)	Caster Preferred Setting (deg.)	Camber Range (deg.)	Camber Preferred Setting (deg.)	Toe-in (in.)	Steering Axis Inclination (deg.)
1984	Civic		1 1/2P-3 1/2P [1]	2 1/2P [2]	1N-1P	0	0	13
	Civic SW		1P-3P	2P	1N-1P	0	0	12
1985	Civic		1 1/2P-3 1/2P [1]	2 1/2P [2]	1N-1P	0	0	13
	Civic SW		1P-3P	2P	1N-1P	0	0	12
1986	Civic		1 1/2P-3 1/2P [1]	2 1/2P [2]	1N-1P	0	0	13
	Civic SW		1P-3P	2P	1N-1P	0	0	12
1987	Civic		1 1/2P-3 1/2P [1]	2 1/2P [2]	1N-1P	0	0	13
	Civic SW		1P-3P	2P	1N-1P	0	0	12
1988	Civic		2P-4P	3P	1N-1P	0	0	7 5/16
	Civic SW		1 15/16P-3 15/16P	2 15/16P	11/16N-1 5/16P	5/16P	0	7 1/4
	Civic 4WD		1 15/16P-3 15/16P	2 15/16P	7/16N-1 9/16P	9/16P	0	6 15/16
1989	Civic		2P-4P	3P	1N-1P	0	0	-
	Civic SW		1 15/16P-3 15/16P	2 15/16P	11/16N-1 5/16P	5/16P	0	-
	Civic 4WD		1 15/16P-3 15/16P	2 15/16P	7/16N-1 9/16P	9/16P	0	-
1990	Civic		2P-4P	3P	1N-1P	0	0	-
	Civic SW		1 15/16P-3 15/16P	2 15/16P	11/16N-1 5/16P	5/16P	0	-
	Civic 4WD		1 15/16P-3 15/16P	2 15/16P	7/16N-1 9/16P	9/16P	0	-
1991	Civic		2P-4P	3P	1N-1P	0	0	-
	Civic SW		1 15/16P-3 15/16P	2 15/16P	11/16N-1 5/16P	5/16P	0	-
	Civic 4WD		1 15/16P-3 15/16P	2 15/16P	7/16N-1 9/16P	9/16P	0	-
1992	Civic		1/10P-2 1/10P	1 1/10P	1N-1P	0	0	-
1993	Civic	F	0.17P-2.17P	1.17P	1N-1P	0	0 [3]	-
		R	-	-	1.33N-0.67P	0.33N	0.08 [4]	-
	del Sol	F	0.17P-2.17P	1.17P	1.25N-0.75P	0.25N	0 [3]	-
		R	-	-	1.50N-0.50P	0.50N	0.08 [4]	-
1994	Civic	F	0.17P-2.17P	1.17P	1N-1P	0	0 [3]	-
		R	-	-	1.33N-0.66P	0.33N	0.08 [4]	-
	del Sol	F	0.17P-2.17P	1.17P	1.25N-0.75P [5]	0.25N [5]	0 [3]	-
		R	-	-	1.50N-0.50P	0.50N	0.08 [4]	-
1995	Civic	F	0.17P-2.17P	1.17P	1N-1P	0	0 [3]	-
		R	-	-	1.33N-0.67P	0.33N	0.08 [4]	-
	del Sol	F	0.17P-2.17P	1.17P	1.25N-0.75P [5]	0.25N [5]	0 [3]	-
		R	-	-	1.50N-0.50P	0.50N	0.08 [4]	-

P: Positive
N: Negative
F - Front
R - Rear

1. With power steering: 2P-4P
2. With power steering: 3P
3. Range: 0.08N-0.08P
4. Range: 0.04P-0.16P
5. With DOHC engine:
   Camber preferred setting: 0.33N
   Camber range: 1.33N-0.67P

86838C01

REAR SUSPENSION

General Information

1984–87 MODELS

▶ See Figures 31 and 32

All 1984–87 sedan, hatchback and CRX models utilize an independent strut arrangement for the rear wheels. Each suspension unit consists of a combined coil spring/shock absorber strut, a lower control arm and a radius rod.

The 2WD and 4WD Civic Wagon uses a live (or solid) rear axle located by upper and lower fore and aft control arms and a panhard rod. The fore and aft arms keep the axle located fore and aft and vertically. The panhard rod runs from the right side of the axle up to a bushed mounting point on the wagon's body and stabilizes the axle in terms of side-to-side motion. Coil springs which are separate from conventional shock absorbers complete the rear suspension system on the 4WD models and on the 2WD models a coil over damper setup is used.

1988–95 MODELS

▶ See Figures 33 and 34

The 1988–95 models use an independent trailing arm suspension system. Upper and lower controls arms are used to locate a trailing arm which uses a coil-over damper unit to control suspension travel. A unique feature of this sus-

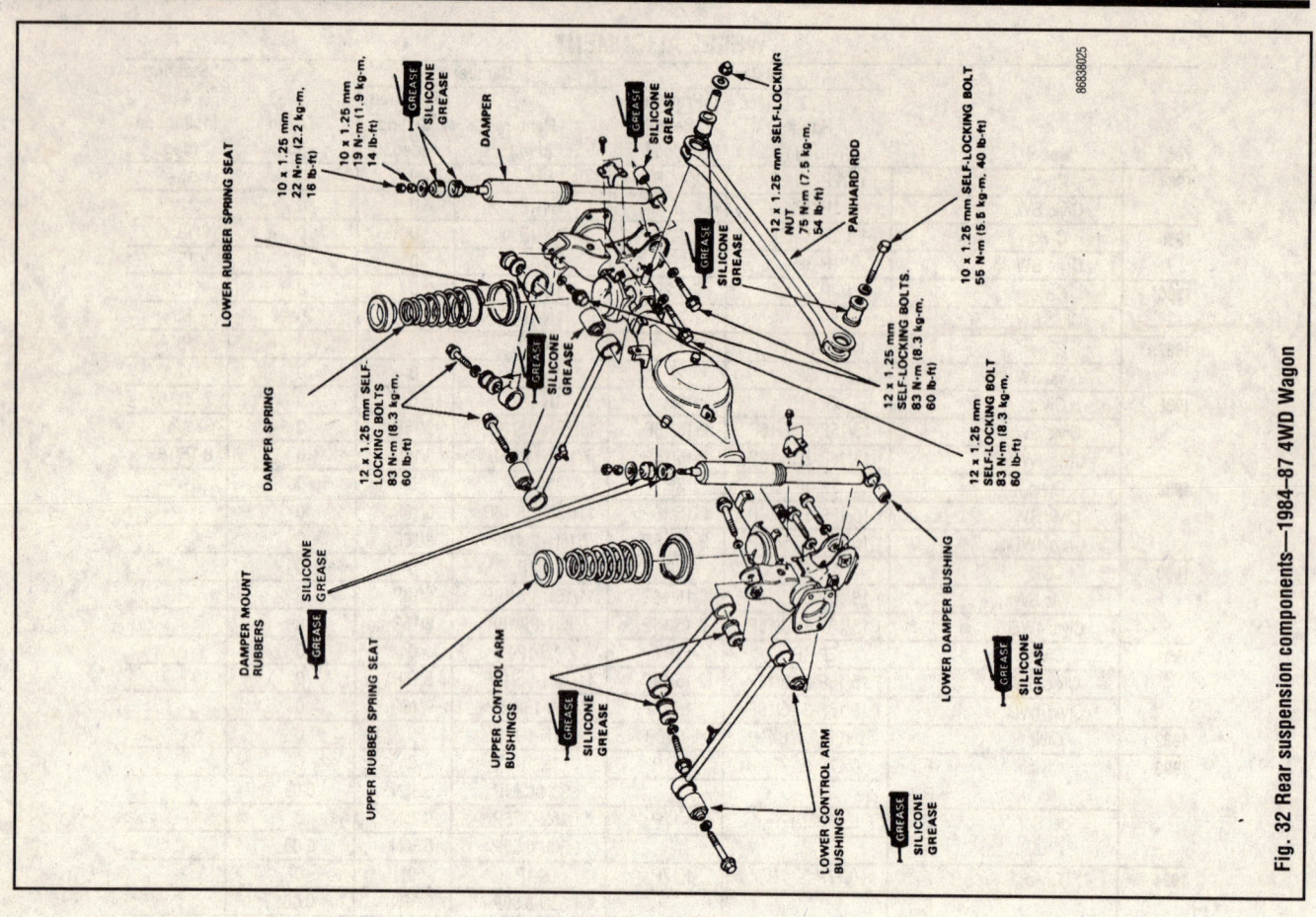

8-14 SUSPENSION AND STEERING

Fig. 32 Rear suspension components—1984-87 4WD Wagon

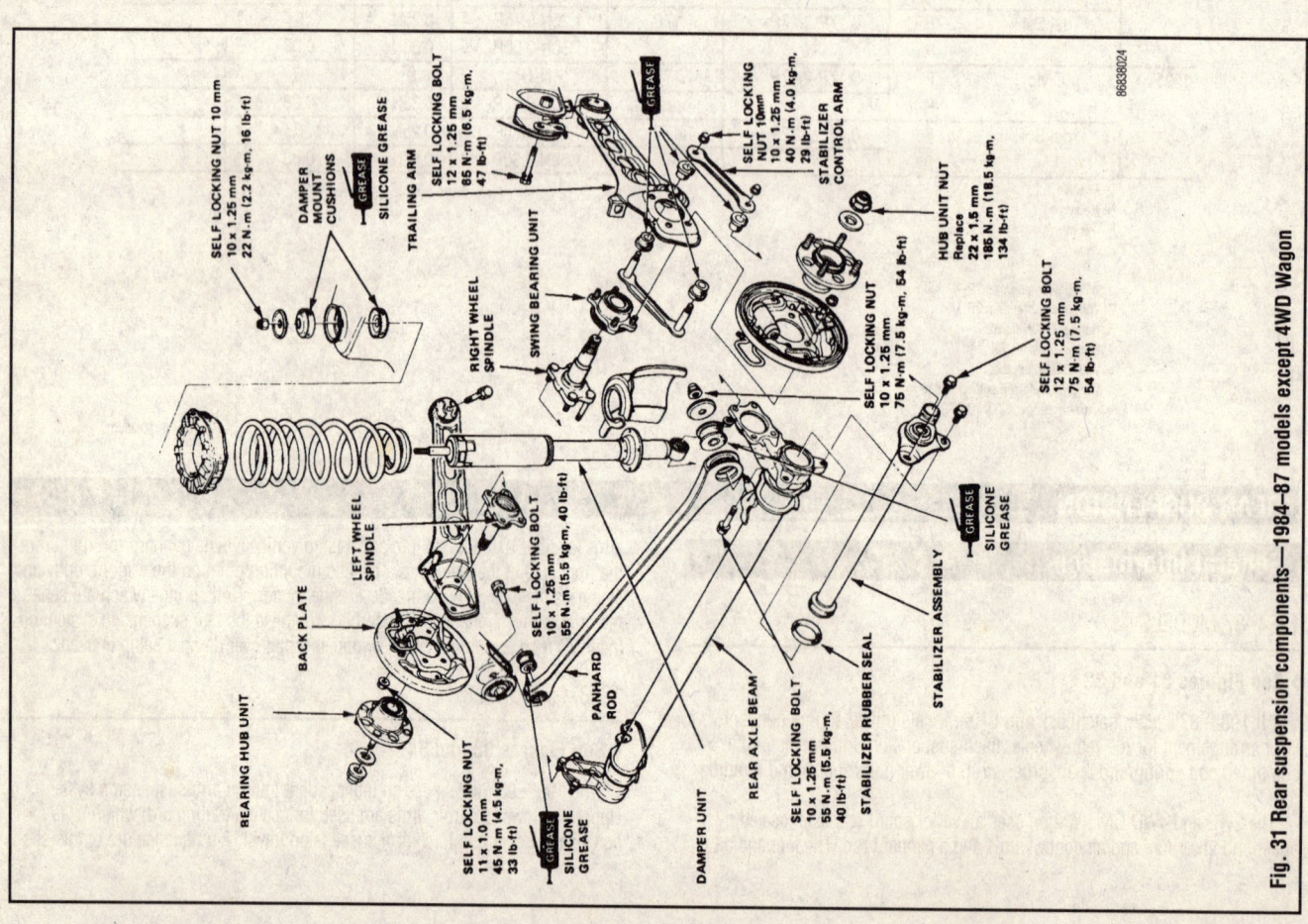

Fig. 31 Rear suspension components—1984-87 models except 4WD Wagon

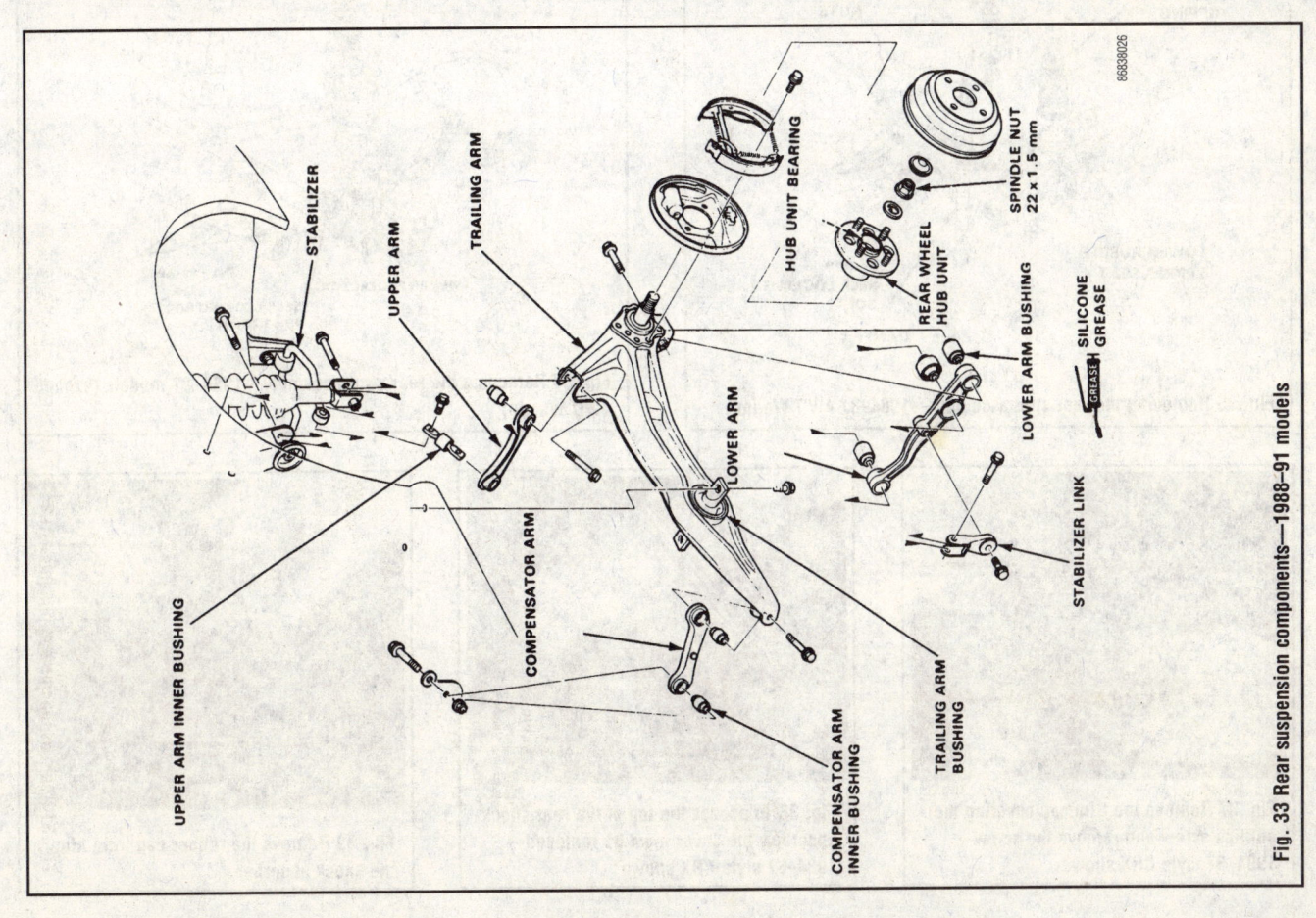

SUSPENSION AND STEERING 8-15

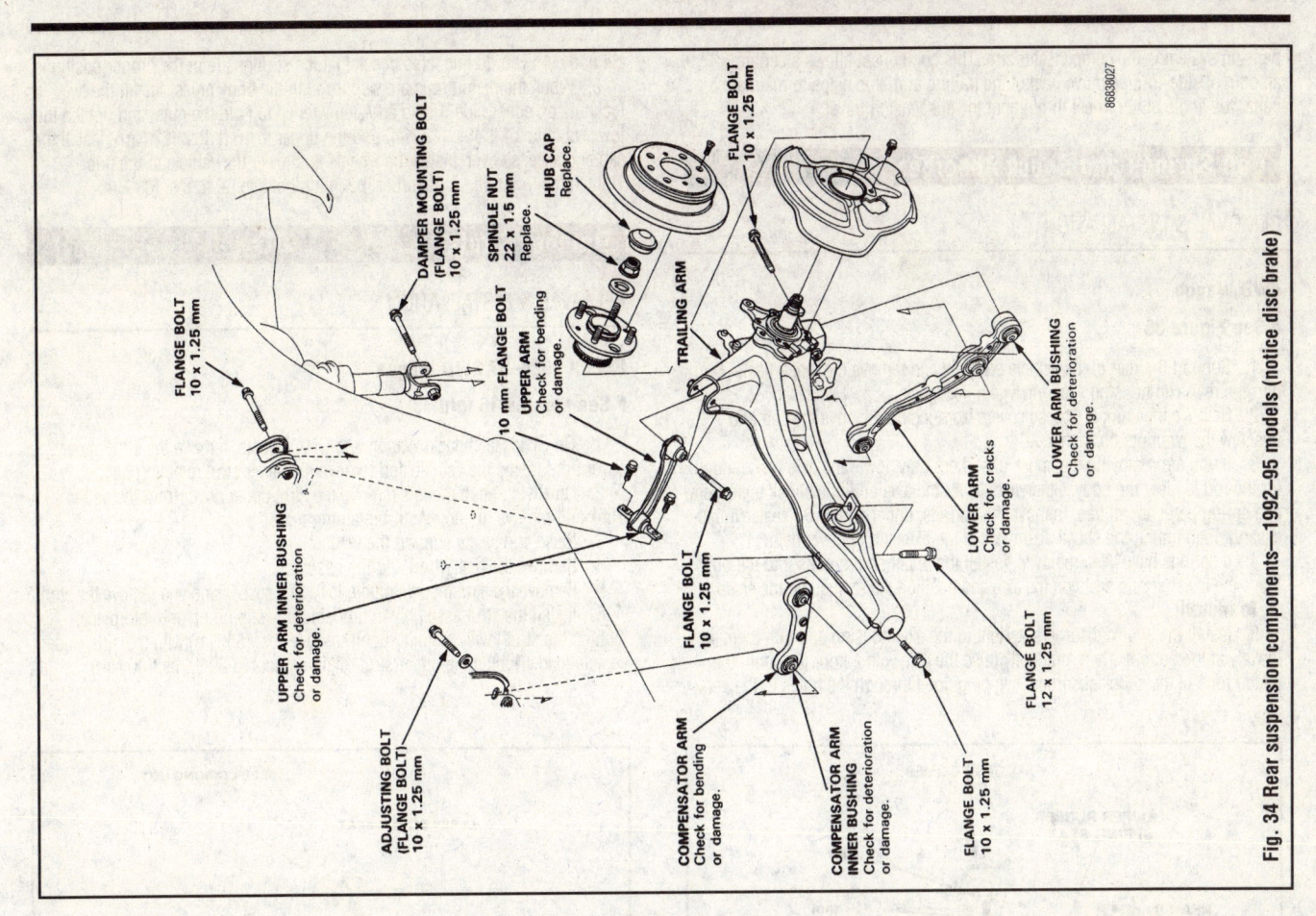

Fig. 34 Rear suspension components—1992–95 models (notice disc brake)

Fig. 33 Rear suspension components—1988–91 models

8-16 SUSPENSION AND STEERING

pension system is the compensator arm. This component allows a certain amount of side-to-side movement of the trailing arm, this helps to maintain a better toe angle of the wheel throughout the suspension travel.

Coil Springs and Shock Absorber

REMOVAL & INSTALLATION

4WD Wagon

♦ See Figure 35

1. Support the rear of the vehicle securely and remove the rear wheels. Support the rear axle housing securely with a floor jack.
2. Remove the damper access covers from inside. Remove the cap and unscrew the two nuts on each side.
3. If you are removing the shock absorbers only, lower the axle just enough for the rod to clear the body. Otherwise, lower the axle until the shock absorbers and spring parts are all free. Remove the springs, upper and lower seats (if necessary), then unbolt the shock absorbers at the axle and remove them.
4. If springs have been removed, install the upper spring seats into the body. Place the lower spring seats on the axle and position the springs on the seats.

To install:

5. Install the shock absorbers upright in the axle housing and bolt them loosely at the bottom. Then, gradually raise the axle with a floor jack until the piston rods of the shock absorbers can be guided through the holes in the vehicle body. As you do this, also check that the springs are in the proper positions.
6. When the piston rods are well through the body holes, install the two locknuts on either side. Using an Allen wrench to hold the nuts, and tighten the lower nuts to 14 ft. lbs. (19 Nm) and the upper to 16 ft. lbs. (22 Nm). Install the cap and access cover. Install the wheels and lower the vehicle to the floor. Tighten the lower shock absorber mounting bolts to 60 ft. lbs. (83 Nm).

Damper Struts

REMOVAL & INSTALLATION

Except 1984–87 2WD Wagon

♦ See Figures 36 thru 45

1. On Civic Hatchback, Wagon and CRX models, remove the strut upper cover from inside the vehicle and remove the upper strut retaining nuts.
2. On Civic Sedan models remove the strut upper cover from inside the trunk and remove the upper strut retaining nuts.
3. Raise and safely support the vehicle.
4. Remove the rear wheel and tire assembly.
5. Remove the strut mounting bolt, lower the suspension and remove the strut.
6. Install the strut assembly into position by sliding up and into place. Tighten the strut lower mounting bolt to 40 ft. lbs. (55 Nm) with the strut under vehicle load. Tighten the upper strut retaining nuts to 29 ft. lbs. (40 Nm)

Fig. 35 Removing the rear strut assembly—1984–87 4WD Wagon

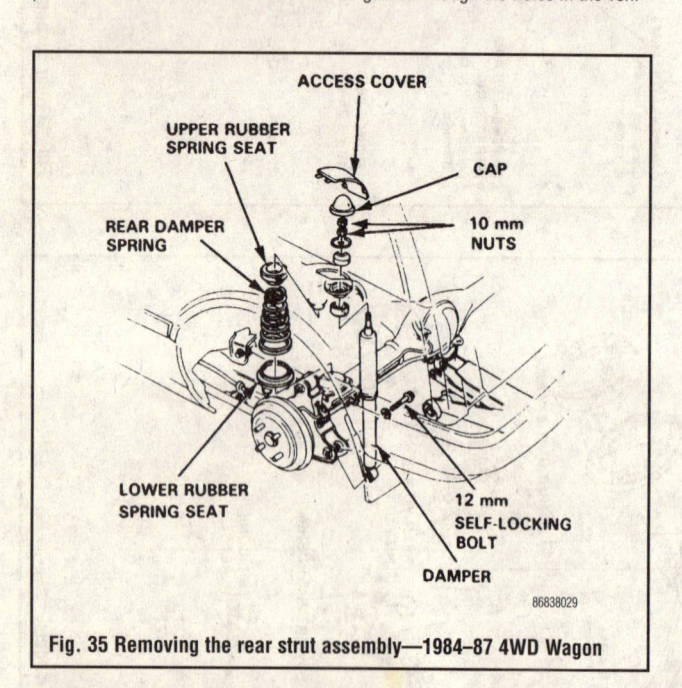

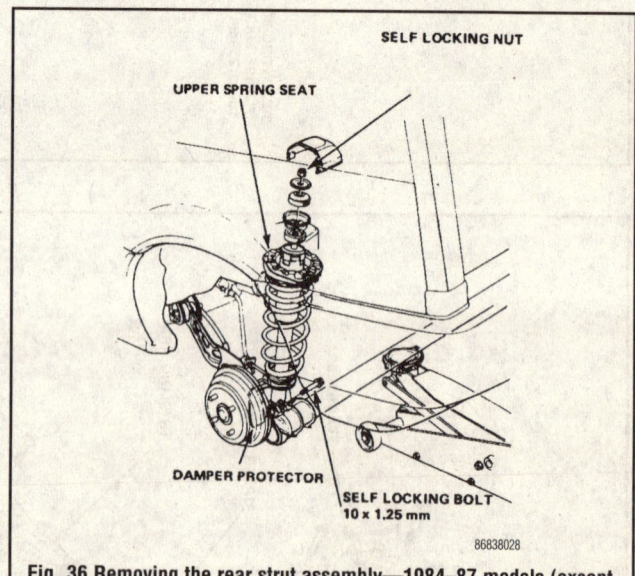

Fig. 36 Removing the rear strut assembly—1984–87 models (except 4WD Wagon)

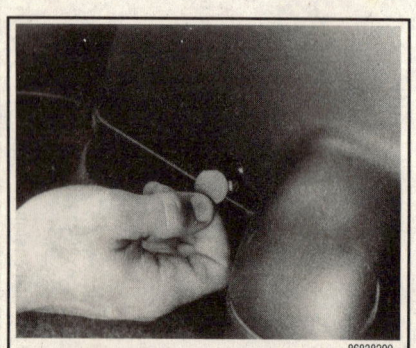

Fig. 37 Remove the trim cap covering the phillips screw and remove the screw—1984–87 style CRX shown

Fig. 38 To access the top of the rear shock absorber, the cover must be removed—1984–87 style CRX shown

Fig. 39 Remove the rubber cap from atop the shock absorber

SUSPENSION AND STEERING   8-17

Fig. 40 Using an Allen wrench to keep the shaft from spinning and a box-end wrench, loosen the nut

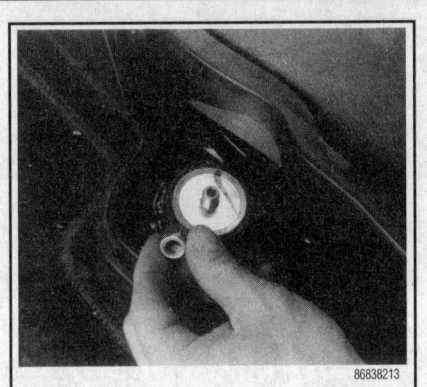

Fig. 41 Remove the nut and the washer beneath it

Fig. 42 Remove the bushing and the shock absorber is free to be removed, provided the lower mount is detached as well

Fig. 43 The lower through-bolt that secured the original shock absorber to the suspension on this 8-year old car was rusted solid and had to be cut off

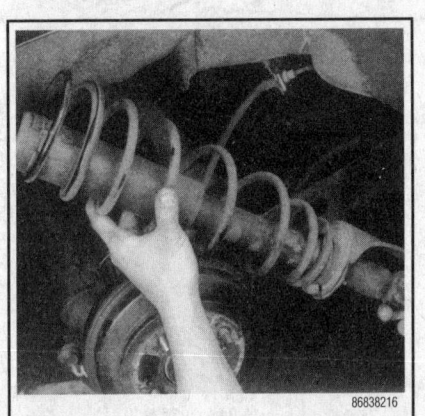

Fig. 44 Once completely detached remove the shock absorber from the car

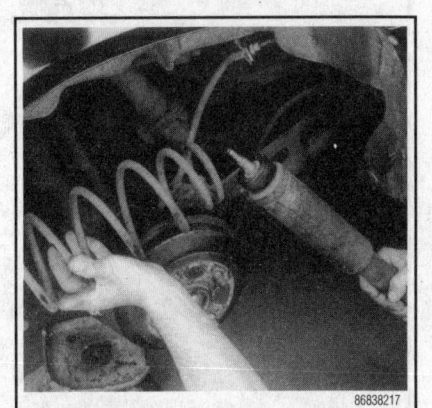

Fig. 45 Once free, the spring is easily removed from the damper body

1984–87 2WD Wagon

1. Raise the rear of the vehicle and support is securely on stands. Remove the rear wheels.
2. Support the rear axle securely on a floor jack. On each side, remove the damper cover from inside the vehicle; then remove the self-locking nut. To remove the self-locking nut, use an Allen wrench to hold the shock absorber rod.
3. Lower the axle carefully until all spring pressure is released. Remove the self-locking bolt from the lower end of the damper. Remove the damper assembly and the spring and upper spring seat above it.
4. To install the struts, first fit the upper spring seat into the frame. Install the spring and then the bolt fastening the lower strut to the axle. Just tighten the bolt slightly. Perform this step on both sides.
5. Fit the damper upper rubber mount into the body (on both sides). Then, carefully jack up the axle, making sure the damper shaft fits into the hole in the body far enough that you can install the self-locking retaining nut just far enough so that all threads are engaged. Carefully lower the vehicle.
6. Final tighten the self-locking bolts at the lower ends of the struts to 40 ft. lbs. (55 Nm) and the self-locking nuts on top to 16 ft. lbs. (22 Nm). Install the damper covers.

OVERHAUL

1. Use a coil spring compressor to disassemble the strut. When assembling the compressor onto the strut, the long studs should be installed so that they are flush with the bottom plate and also flush with the retaining nut on the top end. The adjustable plate in the center cup should be screwed all the way in.
2. Insert the strut in the compressor and compress the strut about 2 in. (51mm). Then remove the center retaining nut.
3. Loosen the compressor and remove the strut.
4. Remove the top plate, rubber protector, spring and rubber bumper.
5. Check the shock absorber for oil leaks and check all rubber parts for damage, wear and deterioration. If significant damage is found or the shock absorber leaks, replace it.
6. To assemble, compress the spring and rubber parts with the spring compressor according to the manufacturer's directions, and replace the hardware, tightening the self-locking nut to 22 ft. lbs. (30 Nm.).

Control Arms

REMOVAL & INSTALLATION

1984–87 4WD Wagon

1. Remove the control arm outboard and inboard pivot bolts.
2. Pull the inboard side of the arm down until it clears the body.
3. Slide the arm towards the center of the vehicle until it is free of the hub carrier.
4. Check the Bushing at each end of the control arm and the control arm for damage and wear. To install, position the control arm and install the retaining bolts.
5. Tighten the control arm retaining bolts to 60 ft. lbs. (83 Nm).

1988–95 Models

▶ See Figures 46 and 47

TRAILING ARM

> ※※ CAUTION

When the trailing arms are to be removed with the vehicle raised at both the front and rear, place additional weight in the trunk. The vehicle could otherwise become unbalanced and tip forward.

8-18 SUSPENSION AND STEERING

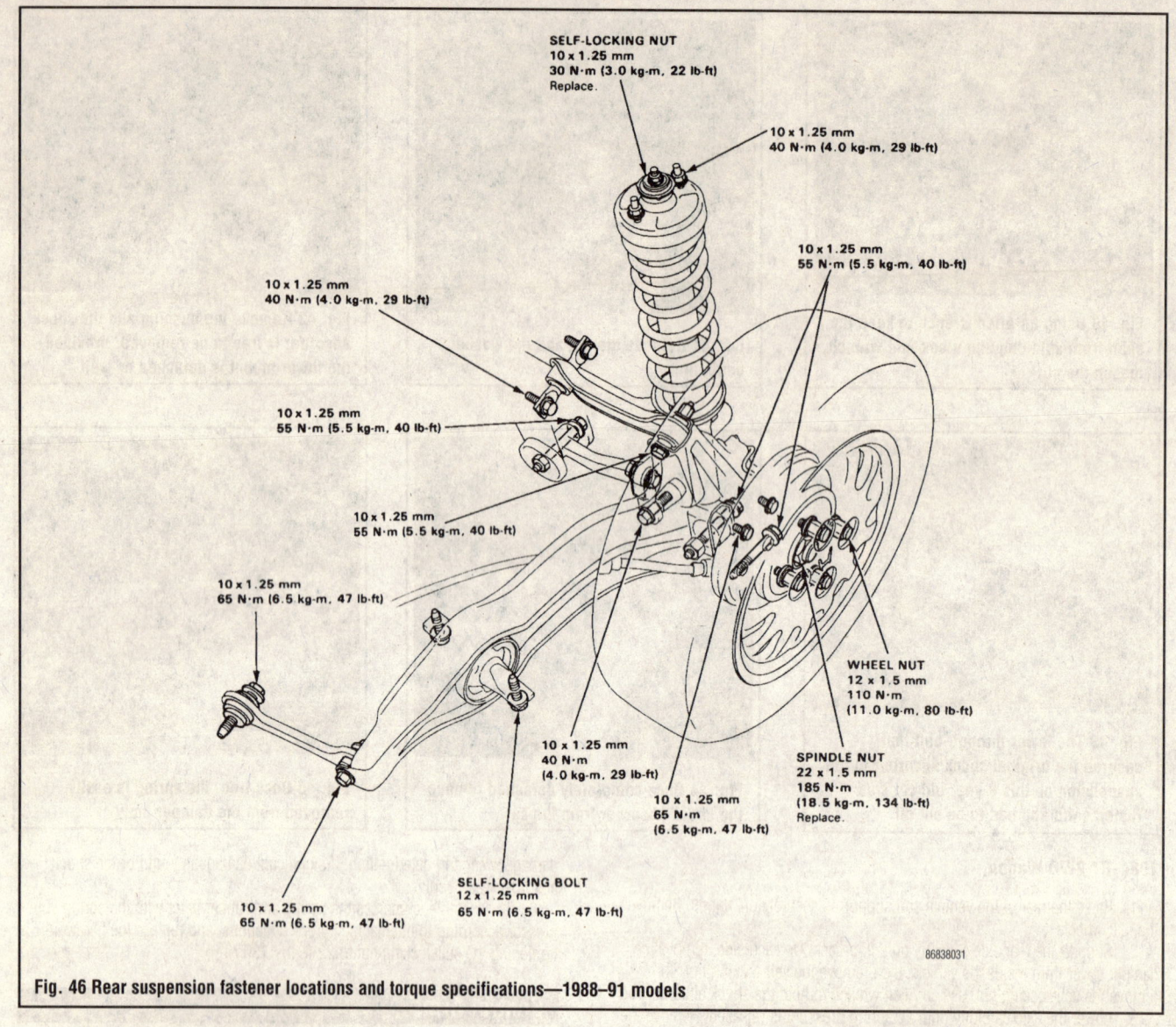

Fig. 46 Rear suspension fastener locations and torque specifications—1988–91 models

1. Raise and safely support the vehicle.
2. Remove the rear wheel and tire assembly.
3. On all except Civic Wagon with 4WD, perform the following procedure:
   a. Remove the brake drum or rotor.
   b. Remove the spindle nut and hub unit.
   c. Disconnect the parking brake cable and the brake line from the wheel cylinder. Plug the line.
   d. Remove the brake backing plate.
4. On 4WD Civic Wagon, remove the brake drum and spindle nut and disconnect the parking brake cable.
5. Disconnect the brake hose from the brake line. Plug the line.
6. Support the lower control arm or beam axle with a suitable jack.
7. Remove the trailing arm bushing mounting bolts.
8. Disconnect the upper arm and compensator arm from the trailing arm, if equipped.
9. On Civic Wagon with 4WD, remove the rear halfshaft outboard CV-joint from the trailing arm using a suitable puller.
10. Remove the trailing arm from the vehicle.

To install:

11. Install the trailing arm in position.

➡ Tighten all bolts and nuts with the vehicle on the ground. This will ensure proper tightening.

12. Install the halfshaft outboard CV-joint, if removed on Civic Wagon.
13. Reconnect the upper control arm and compensator arm, if equipped.
14. Install the trailing arm bushing bolts. Reconnect all hoses and cables to the trailing arm. Reconnect the brake line.
15. Install the backing plate and any brake components.
16. Tighten all bolts to the appropriate specifications.
17. Bleed the brake system.

UPPER CONTROL ARM

1. Raise and safely support the vehicle.
2. Remove the rear wheel and tire assembly.
3. Support the lower control arm or rear axle, as necessary.
4. Unbolt the 10mm flange bolt that attaches the arm to the body and remove the self-locking bolt that attached the upper arm to the trailing arm.
5. Remove the upper control arm mounting bolts and the upper control arm.
6. To install, place the bushing side of the control arm to the body and the trailing arm to the end of the control arm.
7. Install the wheel and tire assembly and tighten the bolts with the wheels on the ground. Tighten the control arm bolts-to-body to 29 ft. lbs. (40 Nm), and the control arm-to-trailing arm bolt to 40 ft. lbs. (55 Nm).

LOWER CONTROL ARM

1. Raise and safely support the vehicle.
2. Remove the rear wheel and tire assembly.

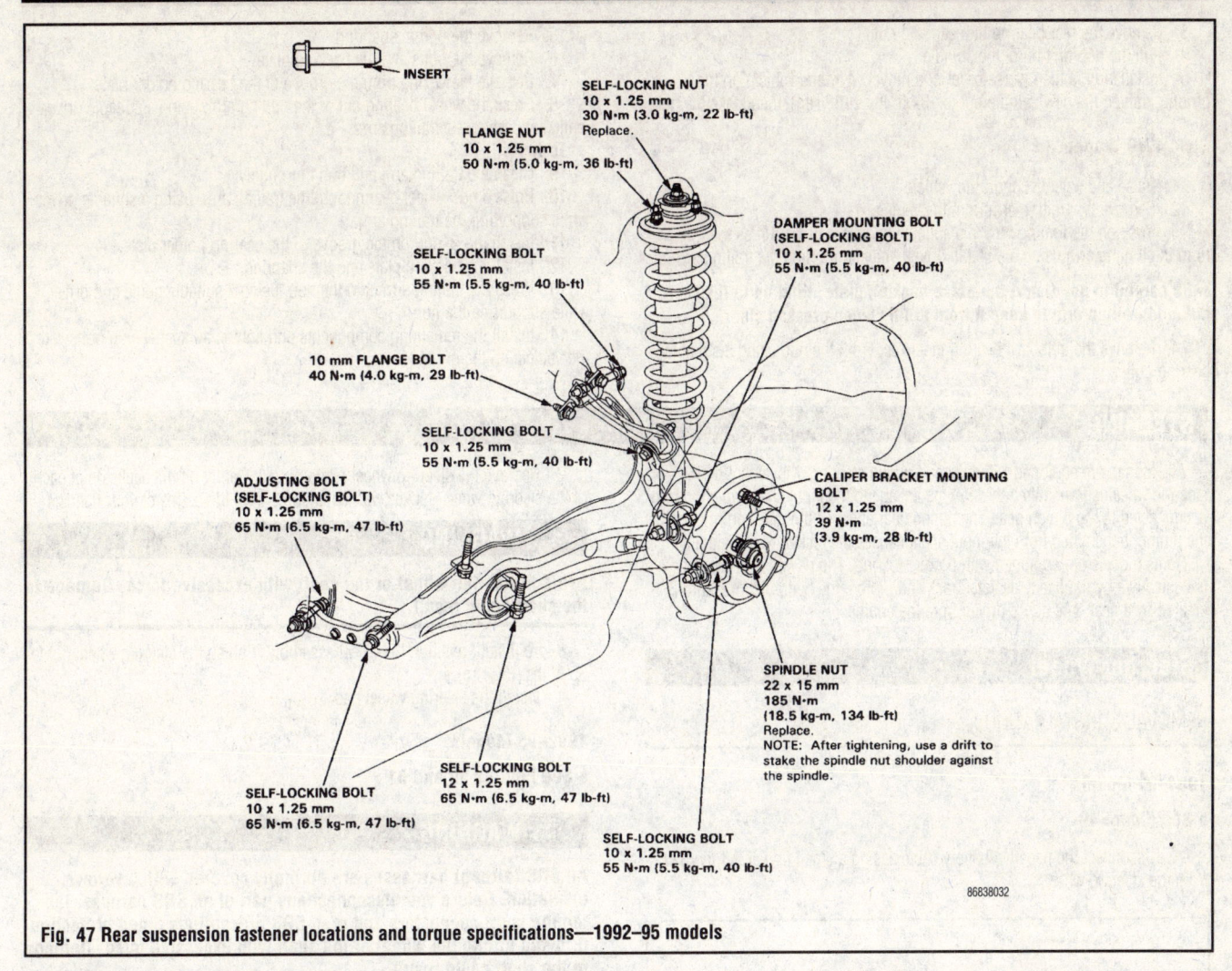

SUSPENSION AND STEERING   8-19

Fig. 47 Rear suspension fastener locations and torque specifications—1992-95 models

3. Remove the strut and/or radius rod mounting bolts from the lower control arm, if necessary.
4. Remove the lower arm mounting bolts and remove the lower control arm.
5. To install, install the lower arm and replace bolts. Replace the wheel and tire. Tighten the bolts with the vehicle on the ground.
6. Tighten the lower control arm-to-body bolt to 40 ft. lbs. (55 Nm) and the control arm-to-trailing arm bolt to 29 ft. lbs. (40 Nm).

Rear Wheel Bearings

REPLACEMENT

Except Civic 4WD Wagon

♦ See Figure 48

➥Do not tighten or loosen a spindle nut unless the vehicle is sitting on all four wheels. The torque is high enough to cause the vehicle to fall even when properly supported.

1. Loosen the rear lug nuts and the spindle nut. Raise the vehicle and support it safely.
2. Release the parking brake. Remove the rear wheel and the brake drum.

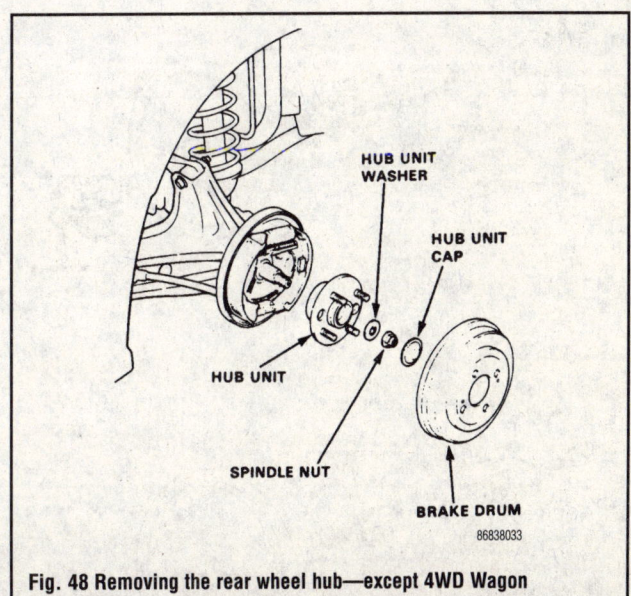

Fig. 48 Removing the rear wheel hub—except 4WD Wagon

8-20 SUSPENSION AND STEERING

3. Remove the rear bearing hub cap and nut.
4. Pull the hub unit off of the spindle.
5. Installation is the reverse order of removal. With the vehicle on the ground, tighten the new spindle nut to 134 ft. lbs. (185 Nm), then stake the nut.

Civic 4WD Wagon

1. Raise and safely support the vehicle.
2. Remove the trailing arm from the vehicle.
3. Position the trailing arm in a suitable hydraulic press. Press the hub from the trailing arm using a suitable driver while supporting the trailing arm.

➡ Be careful to not distort the brake backing plate. Hold on to the rear hub and trailing arm to keep it from falling when pressed clear.

4. Remove the outboard bearing inner race from the hub using a suitable bearing puller.

STEERING

All Honda's are equipped with rack and pinion steering. Movement of the steering wheel is transmitted through the linkage to the input shaft, which in turn is connected to the pinion gear. The pinion gear engages the rack and rotation of the pinion, transmitted from the input shaft, causes the rack to move laterally.

Some models are equipped with power steering. The power steering system is basically a hydraulically assisted rack and pinion, the hydraulic pressure is created by a gear type pump driven from the engine.

Steering Wheel

REMOVAL & INSTALLATION

1984–91 Models

▶ See Figure 49

1. Disconnect the negative battery cable. Using a small prybar, lift the pad from the steering wheel.

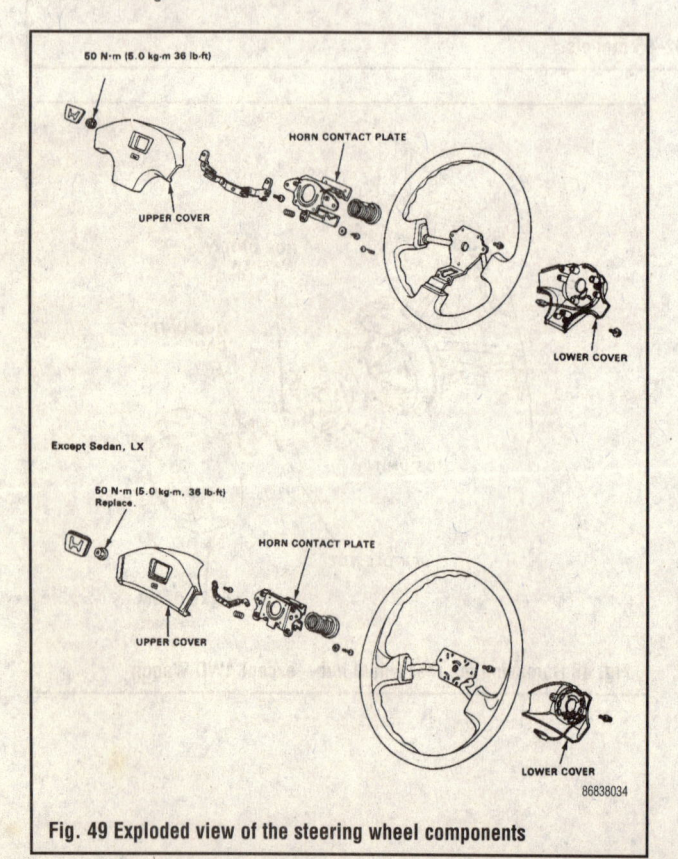

5. Remove the 64mm snapring.
6. Remove the bolts and the backing plate.
7. Remove the O-ring from the groove of the bearing holder plate.
8. Press the wheel bearing out of the trailing arm, using a suitable driver, while supporting the trailing arm.

To install:

9. Clean the trailing arm and hub thoroughly.
10. Press a new wheel bearing into the trailing arm, using a suitable driver, while supporting the trailing arm.
11. Install the O-ring on the groove of the bearing holder plate.
12. Install the backing plate and the snapring.
13. Press the trailing arm onto the hub, using a suitable guide and driver, while supporting the hub.
14. Install the remaining components and tighten all suspension bolts with the vehicle on the ground.

2. Remove the steering wheel retaining nut. Gently hit the backside of each of the steering wheel spokes with equal force from the palms of your hands.

✸✸ WARNING

Avoid hitting the wheel or the shaft with excessive force. Damage to the shaft could result.

3. To install, position the wheel assembly. Tighten the steering wheel nut to 36 ft. lbs. (50 Nm).
4. Install the steering wheel pad.

1992–95 Models

▶ See Figures 50 and 51

✸✸ WARNING

All SRS (airbag) harnesses are normally covered with a yellow insulation. Before you disconnect any part of an SRS harness, fasten the short connectors and read SRS precautions listed in Section 6. Avoid hitting the wheel or the shaft with excessive force. Damage to the shaft could result.

✸✸ CAUTION

Severe injury could result if the airbag is mishandled or improperly connected and an accidental deployment occurs.

➡ Make sure you have the 5-digit radio security "CODE" before disconnecting the battery.

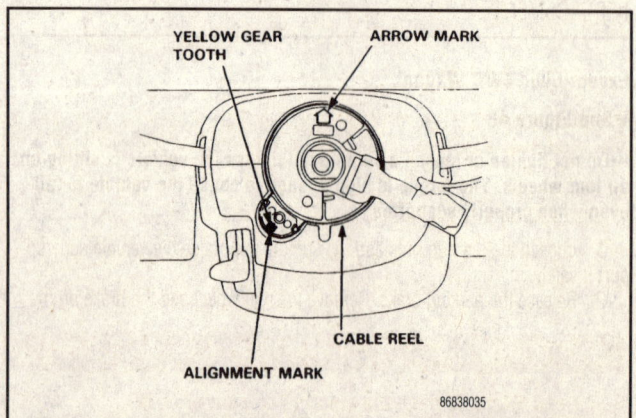

Fig. 49 Exploded view of the steering wheel components

Fig. 50 The yellow gear tooth and the cable reel MUST be in the proper position for the cable reel to be centered—1992–95 models

SUSPENSION AND STEERING 8-21

Fig. 51 Proper steering wheel shaft-to-cable reel and canceling sleeve engagement—1992–95 models

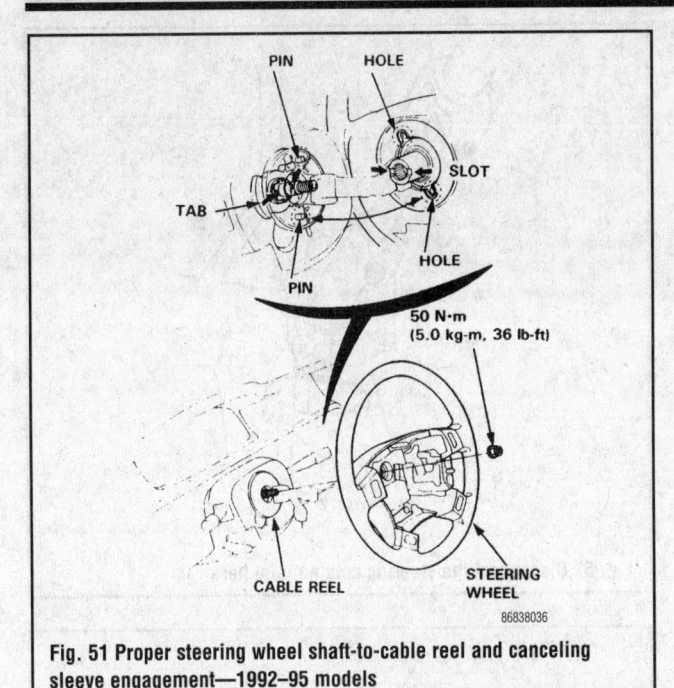

1. To connect the driver's side short connector, first disconnect the battery negative cable, then the positive cable, then wait a minimum of three minutes.
2. Fasten together the (red) short connectors for the driver's side by removing the steering wheel access panel and removing the red short connectors from the panel. Then disconnect the 3-pin connector between the driver's airbag and cable reel and attach the short connector to the airbag side of the connector.
3. Once the short connectors are in place, begin to remove the steering wheel by removing lid **B** and the cruise control set/resume switch cover.
4. Unfasten the Torx® T30 bit bolts and remove the driver's airbag assembly.
5. Detach the electrical connectors from the horn and cruise control switch, remove the steering wheel nut and remove the wheel by gently rocking it from side-to-side as you pull the wheel towards you with with both hands.

To install:
6. Make sure that the driver's airbag assembly is properly attached and the cable reel is centered.
7. To do so, first rotate the cable reel clockwise until it stops, then rotate it counterclockwise (about two turns) until
   a. The yellow gear tooth lines up with the alignment mark on the cover.
   b. The arrow mark on the cable reel label points straight up.
8. Install the steering wheel and attach the nut, tightening it to 36 ft. lbs. (50 Nm.).

➡ Make sure the cable reel and canceling sleeve are properly engaged on the steering wheel shaft.

9. Reconnect the horn and cruise control switch.
10. Using new Torx® bolts, install the airbag assembly.
11. Unfasten the short connector, attach it to the access panel and install the access panel on the steering lower cover.

Combination Switch

REMOVAL & INSTALLATION

1. Refer to the "Steering Wheel, Removal and Installation" procedures in this section and remove the steering wheel.
2. Disconnect the column wiring harness and coupler.

※※ WARNING

Be careful to not damage the steering column or shaft.

3. Remove the upper and lower column covers.
4. For 1992–95 models, disconnect the 4-pin and 7-pin connectors, remove the two retaining screws and lift out the switch.
5. For pre-1992 models, if equipped with cruise control, remove the slip ring.
6. Remove the turn signal canceling sleeve.
7. Remove the switch retaining screws, then remove the switch.
8. Loosen the turn signal switch cam nut screw and lightly tap its head to permit the cam nut to loosen. Remove the turn signal switch assembly and the steering shaft upper bushing.

To install:
9. Replace the turn signal switch cam nut (small end up) if equipped, and the screws that secure the combination switch. When installing the turn signal switch assembly on pre-1992 models, engage the locating tab on the switch with the notch in the steering column. The steering shaft upper bushing should be installed with the flat side facing the upper side of the column. The alignment notch for the turn signal switch will be centered on the flat side of the bushing.

➡ If the cam nut has been removed on the earlier models, be sure to install it with the small end up.

Ignition Switch

REMOVAL & INSTALLATION

1984–91 Models

1. Disconnect the negative battery cable.
2. Remove the steering wheel and steering column covers as necessary, to gain access to the switch.
3. Disconnect the electrical connector.
4. Insert the ignition key and turn it to the **0** position.
5. Remove the two screws and remove the base of the switch.

To install:
6. Install the new switch in position and install the two screws.
7. Reconnect the electrical connectors.
8. Install the column cover and lower dashboard panels.
9. Connect the negative battery cable. Check the operation of the switch in all positions.

REMOVAL & INSTALLATION

1992–95 Models

※※ WARNING

SRS (airbag) harnesses are normally covered with a yellow insulation. See Section 6 for safety precautions. Avoid hitting the wheel or the shaft with excessive force, as damage to the shaft could result.

※※ CAUTION

Severe injury could result if the airbag is mishandled or improperly connected and an accidental deployment occurs.

1. Disconnect the negative and positive battery cables and wait at least three minutes.

➡ Make sure you have the 5-digit security "CODE" for the original radio before disconnecting the battery.

2. Remove the dashboard lower cover and knee bolster.
3. Disconnect the 5-pin electrical connector from under the dash in the fuse/relay box, and the 7-pin connector from the main wire harness.
4. steering wheel and steering column covers as necessary, to gain access to the switch.
5. Insert the ignition key and turn it to the **0** position.
6. Remove the two screws and remove the switch.

To install:
7. Install the new switch in position and install the two screws.
8. Fasten the electrical connectors.
9. Install the column cover and lower dashboard panels.

8-22 SUSPENSION AND STEERING

10. Connect the negative battery cable. Check the operation of the switch in all positions.

Ignition Lock

REMOVAL & INSTALLATION

1984–91 Models

1. Disconnect the negative battery cable.
2. Remove the steering wheel and the steering column covers.
3. Disconnect the ignition switch connector.
4. Center punch each of the two shear bolts and drill their heads off with a suitable drill bit.

➡ Do not damage the switch body when removing the shear bolt heads.

5. Remove the shear bolts from the switch body and remove the switch.

To install:

6. Install the new ignition switch without the key inserted.
7. Loosely tighten the new shear bolts.

➡ Make sure the projection on the ignition switch is aligned with the hole in the steering column.

8. Insert the ignition key and check for proper operation of the steering wheel lock and that the ignition key turns freely.
9. Tighten the shear bolts until the hex heads twist off.

1992–95 Models

※ WARNING

SRS (airbag) harnesses are normally covered with a yellow insulation. See Section 6 for safety precautions. Avoid hitting the wheel or the shaft with excessive force. Damage to the shaft could result.

1. Disconnect the negative and positive battery cables and wait at least three minutes.
2. Remove the dashboard lower cover and knee bolster.
3. Disconnect the 5-pin and 7-pin connectors from under the dash.
4. Remove the steering column covers, remove the column holder mounting bolts and nuts and lower the steering column.
5. Center punch each of the two shear bolts and drill their heads off with a 3/16 in. (5mm) drill bit.

➡ Do not damage the switch body when removing the shear bolt heads.

6. Remove the shear bolts from the switch body and remove the switch.

To install:

7. Install the new ignition switch without the key inserted.
8. Loosely tighten the new shear bolts.

➡ Make sure the projection on the ignition switch is aligned with the hole in the steering column.

9. Insert the ignition key and check for proper operation of the steering wheel lock and that the ignition key turns freely.
10. Tighten the shear bolts until the hex heads twist off.

Steering Column

REMOVAL & INSTALLATION

◆ See Figures 52, 53 and 54

※ WARNING

SRS (airbag) harnesses are normally covered with a yellow insulation. See Section 6 for safety precautions. Avoid hitting the wheel or the shaft with excessive force, as damage to the shaft could result.

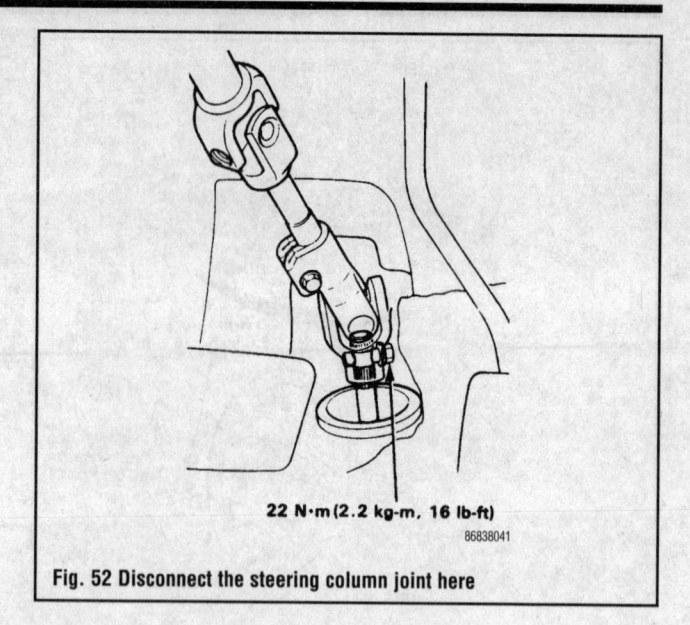

Fig. 52 Disconnect the steering column joint here

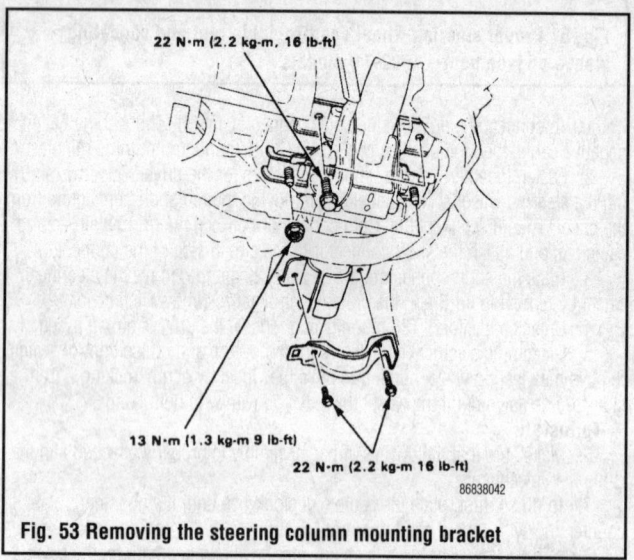

Fig. 53 Removing the steering column mounting bracket

1. Disconnect the negative battery cable. If working on a 1992–95 model, also disconnect the positive cable and wait at least three minutes
2. Remove the steering wheel according to the procedure found earlier in this section.
3. Remove the lower cover panel. Remove the driver's knee bolster, if equipped.
4. Remove the upper and lower column covers.
5. If equipped with SRS, remove the connector holder and disconnect the SRS main harness from the cable reel wire harness and remove the cable.
6. Disconnect the wire couplers from the combination switch. Remove the turn signal canceling sleeve and the combination switch.
7. Remove the steering joint cover and remove the steering joint bolt(s).
8. Disconnect each wire coupler from the fuse box under the left side of the dashboard.
9. Remove the steering column retaining brackets.
10. Remove the nuts attaching the bending plate guide and bending plate and remove the steering column assembly.

To install:

11. Fit the column into place and secure the bracket, bending plate and guide. Tighten the nuts to 16 ft. lbs. (22 Nm).
12. Install the steering joint bolts and tighten to 16 ft. lbs. (22 Nm).
13. Connect the wiring at the fuse box.
14. Install the switches and connect the wiring.
15. Install the knee bolster and steering wheel following all procedures, especially for SRS equipped models.

SUSPENSION AND STEERING   8-23

Fig. 54 Steering column assembly—exploded view

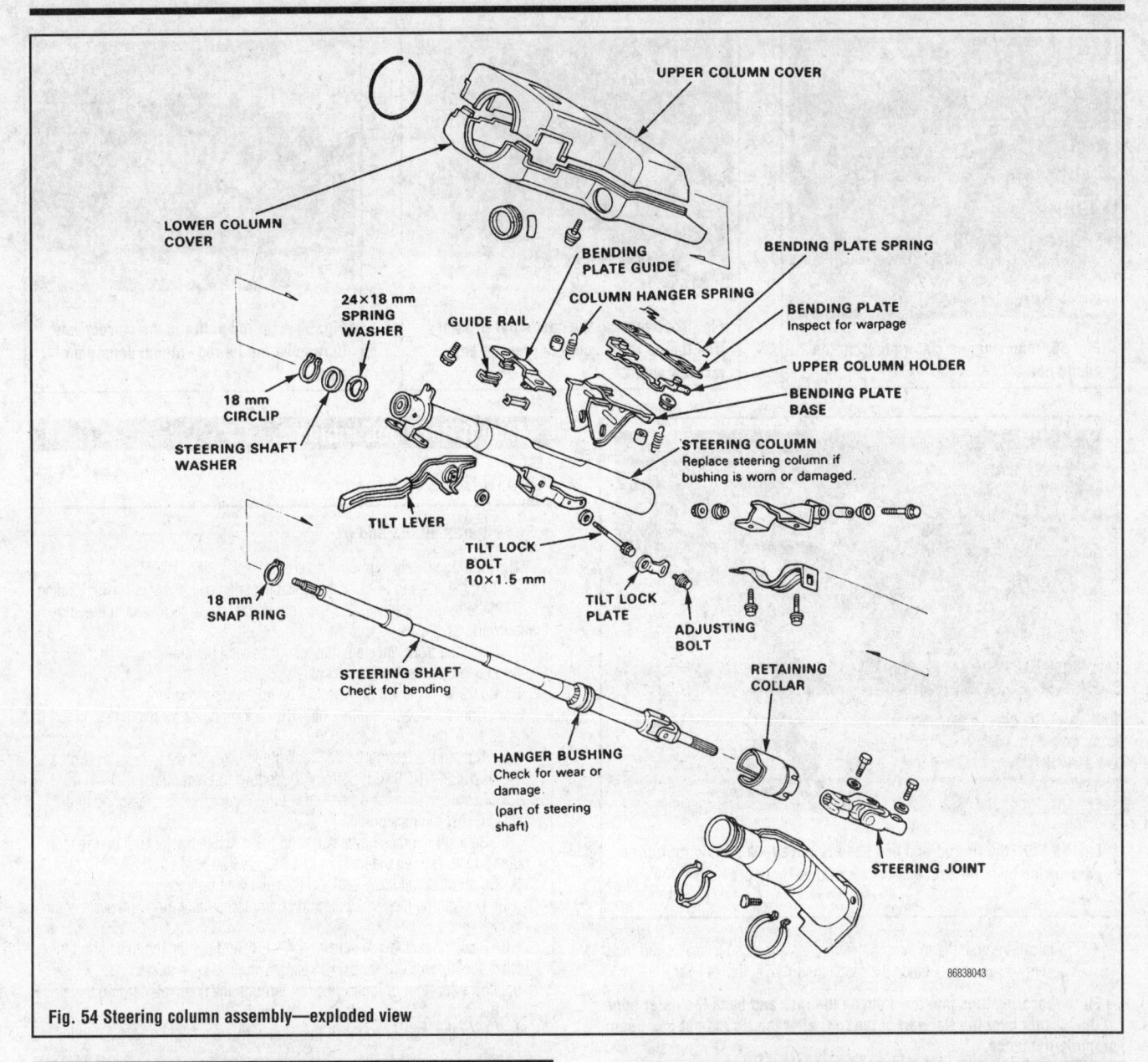

Steering Linkage

REMOVAL & INSTALLATION

Tie Rod Ends

▶ See Figures 55, 56, 57, 58 and 59

1. Raise and support the front of the vehicle on jackstands. Remove the front wheels.
2. Using a Ball Joint Remover tool, press the tie rod end from the steering knuckle.

➡If you have to remove the tie rod end as well, loosen the locking nut before removing the tie rod. Do not do so unless necessary, however, because tie-rod adjustment is one of the factors affecting front-end alignment.

3. Remove the tie rod dust seal bellows clamps and move the rubber bellows on the tie rod and rack joints. Disconnect the air tube, if equipped, at the dust seal joint.
4. Straighten the tie rod lockwasher tabs at the tie rod-to-rack joint and remove the tie rod by turning it with a wrench.

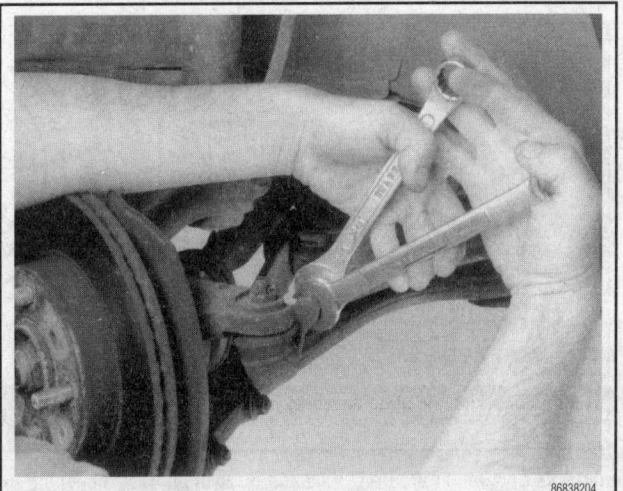

Fig. 55 Removing the tie rod end from the tie rod is easier if the locking nut is cracked loose first

8-24 SUSPENSION AND STEERING

Fig. 56 Removing the cotter pin from the castle nut

Fig. 57 Removing the castle nut will not free the tie rod since it is pressed onto a tapered shaft

Fig. 58 A tie rod puller is the correct way to remove the tie rod without damaging parts

Fig. 59 Here the tie rod end has been removed—a simple matter of unscrewing by hand because the locking nut was previously loosened

5. To install, reverse the removal procedures. Always use a new tie rod lockwasher during reassembly. Tighten the castle nut to 32 ft. lbs. (44 Nm).

➡ Fit the locating lugs into the slots on the rack and bend the outer edge of the washer over the flat part of the rod, after the tie rod nut has been properly tightened.

Manual Rack and Pinion

ADJUSTMENT

◆ See Figures 60, 61 and 62

1. Raise and safely support the vehicle.
2. Turn the steering wheel with a suitable spring gauge and check the reading.
3. If the reading exceeds 3.3 lbs., adjust the steering gear as outlined in the following Steps.
4. Place the front wheels in the straight ahead position.
5. Loosen the rack screw locknut.
6. On 1988–91 Civic Si, perform the following procedure:
   a. Tighten, loosen and retighten the rack guide screw two times, to 3.6 ft. lbs. (5 Nm), then back it off 15°.
   b. Tighten the locknut on the rack guide screw to 49 ft. lbs. (68 Nm).
7. On 1988–91 CRX Si, perform the following procedure:
   a. Retighten the rack guide screw until it compresses the spring and seats against the rack guide.
   b. Back off the rack guide screw 15° and install the locknut on the rack guide screw. Tighten the locknut to 18 ft. lbs. (25 Nm).
8. On all other vehicles, perform the following procedure:
   a. Tighten the rack screw until it compresses the spring and seats against the rack guide.
   b. Back the rack guide screw off 40–60° and tighten the locknut on the rack guide screw to 18 ft. lbs. (25 Nm).
9. Check for tight or loose steering through the complete turning travel.
10. Recheck the steering effort.
11. Lower the vehicle and with the wheels in the straight ahead position,

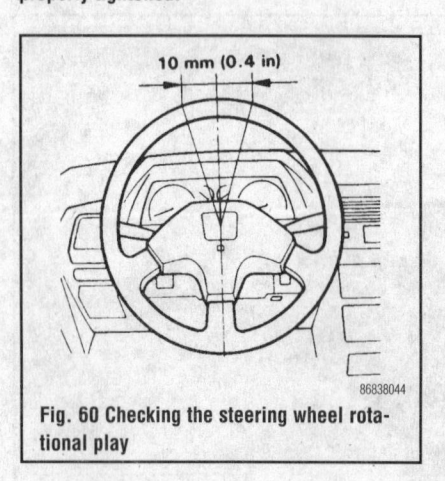

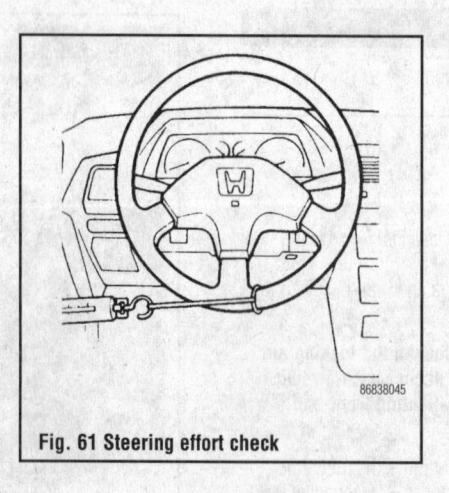

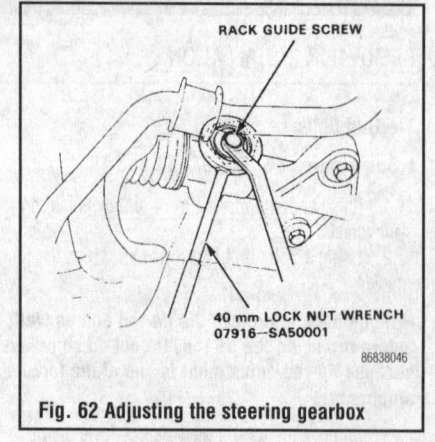

Fig. 60 Checking the steering wheel rotational play

Fig. 61 Steering effort check

Fig. 62 Adjusting the steering gearbox

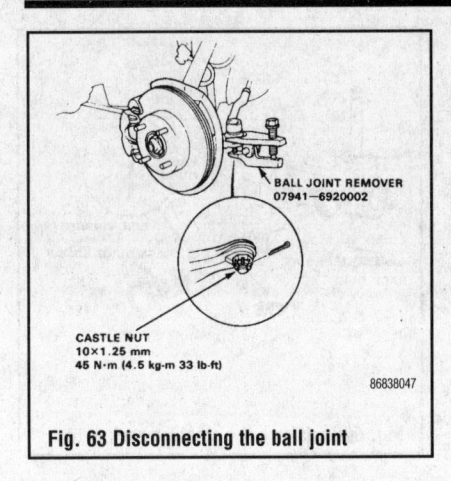

SUSPENSION AND STEERING 8-25

Fig. 63 Disconnecting the ball joint

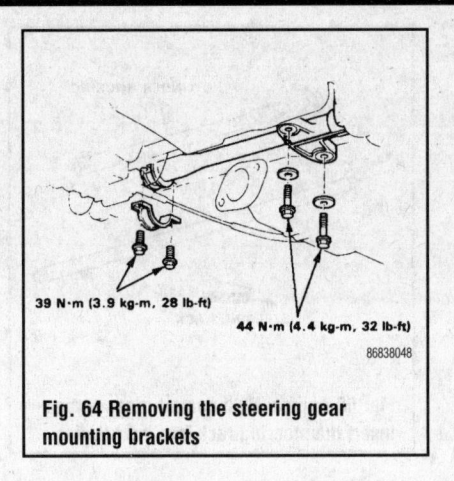

Fig. 64 Removing the steering gear mounting brackets

Fig. 65 Remove the steering gear to the left

measure the distance the steering wheel can be turned without moving the front wheels. The play should not exceed 0.4 in. (10mm). If it does, check all steering components.

REMOVAL & INSTALLATION

▶ See Figures 63, 64 and 65

1. Raise the vehicle and support it safely.
2. Remove the cover panel and steering joint cover. Unbolt and separate the steering shaft at the coupling.
3. Remove the front wheels.
4. Remove the cotter pins and unscrew the castle nuts on the tie rod ends. Using a ball joint tool disconnect the tie rod ends. Lift the tie rod ends out of the steering knuckles.
5. If equipped with manual transaxle, disconnect the shift lever torque rod from the clutch housing. Slide the pin retainer out of the way, drive out the spring pin and disconnect the shift rod.
6. If equipped with an automatic transaxle, remove the shift cable guide from the floor and pull the shift cable down by hand.
7. Remove the two nuts connecting the exhaust header pipe to the exhaust pipe and move the exhaust pipe out of the way.
8. Push the rack all the way to the right and remove the brackets or mounting bolts. Slide the tie rod ends all the way to the right.
9. Drop the rack far enough to permit the end of the pinion shaft to come out of the hole in the frame channel, then rotate it forward until the shaft is pointing rearward.
10. Slide the rack to the right until the left tie rod clears the exhaust pipe, then drop it down and out of the vehicle to the left.

To install:

11. Position the rack into the vehicle and tighten the mounting bolts to 29 ft. lbs. (40 Nm).
12. Connect the exhaust pipe using a new gasket.
13. Reconnect the shift linkage and adjust as required.
14. Install the tie rod ends into the steering knuckle. Tighten the nuts to 32 ft. lbs. (44 Nm), then tighten as required to install a new cotter pin.
15. Connect the steering shaft coupling and tighten the bolt to 22 ft. lbs. (30 Nm).

OVERHAUL

▶ See Figures 66, 67, 68, 69 and 70

Disassembly, Inspection and Assembly

1. Remove the rack and pinion from the vehicle and support it in a bench vise.
2. Loosen the band that retain the bellows on the ends of the rack and pull the bellows back.
3. Unbend the lockwashers on the end of the tie rods.
4. Using a 22mm wrench to hold the rack, remove the tie rods using a 17mm wrench.
5. Remove the lock nut that retains the rack guide components and remove the pieces from the rack body. Be careful not to lose the spring.
6. Remove the pinion assembly dust seal, then remove the 35mm snapring that retains the pinion assembly. Pull the pinion from the body.
7. Slide the rack out of the gearbox body.

Inspection:

8. Inspect the pinion assembly for wear, scoring or damaged teeth. If any of these are apparent the pinion must be replaced.
9. Inspect the pinion bearing for freedom of movement and for signs of damaged balls in the bearing assembly. If the bearing is damaged, the entire pinion assembly must be replaced.
10. Inspect the rack for signs of excessive wear or for damaged teeth. If the rack shows signs of damage, it must be replaced.
11. Inspect the rack guide for scoring.
12. Inspect the steering gear housing for signs of cracking, or damage to the pinion bore.

Assembly:

13. Apply a thin coat of grease to the rack end bushing and install it in the housing by aligning the projections on the bushing with the hole in the housing.
14. Apply a coating of grease to the rack and carefully slide it into the housing. Do this carefully to avoid damaging the surface of the rack.

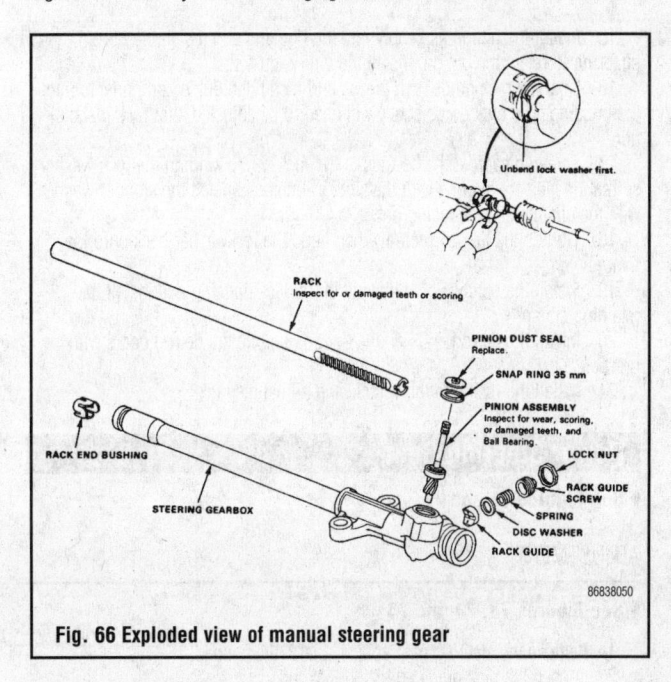

Fig. 66 Exploded view of manual steering gear

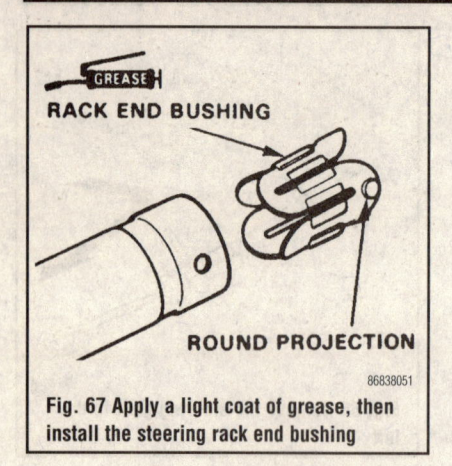

8-26 SUSPENSION AND STEERING

Fig. 67 Apply a light coat of grease, then install the steering rack end bushing

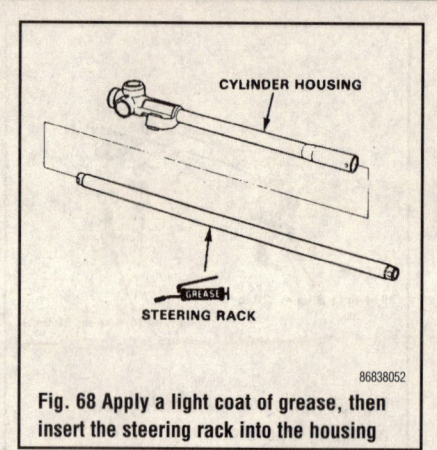

Fig. 68 Apply a light coat of grease, then insert the steering rack into the housing

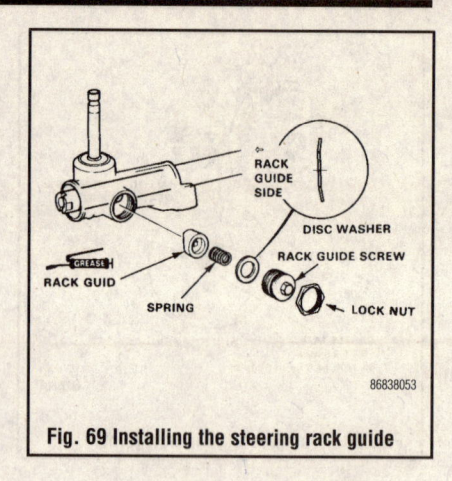

Fig. 69 Installing the steering rack guide

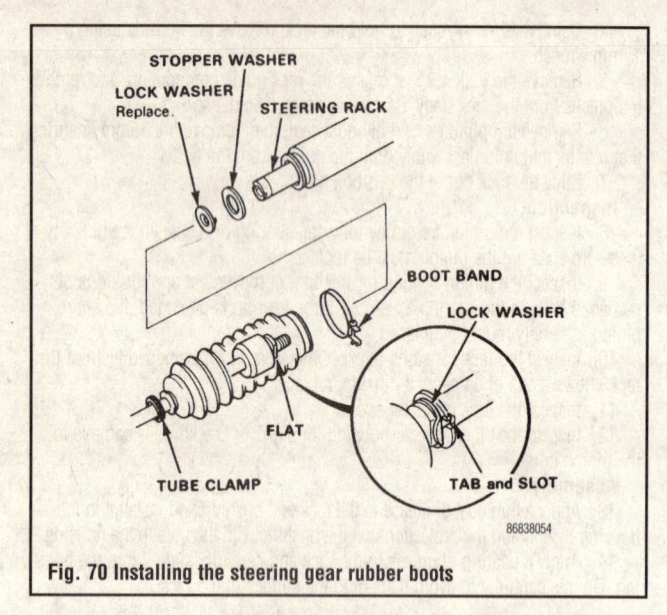

Fig. 70 Installing the steering gear rubber boots

15. Install the pinion assembly into the housing and install the 35mm snapring to retain the pinion. Install the pinion dust seal.
16. Coat the rack guide with grease and install the guide components into the housing. The rack guide screw will need to be adjusted after rack installation.
17. Screw in each of the tie rods to the rack, while holding the lock washer so that its tabs are aligned with the slots in the rack. Install the stopper washer with the chamfered side facing out.
18. Tighten the tie rod securely then bend the lockwasher back onto the flange.
19. Install the bellows over the end of the rack housing and install the retaining clamp.
20. Install the tie rod ends on the tie rods and fill the tie rod boots with grease.
21. Install the rack into the vehicle and adjust the rack.

Power Steering Gear

♦ See Figures 71 and 72

ADJUSTMENT

♦ See Figures 73, 74 and 75

1. Remove the steering gear splash shield, if equipped.
2. Loosen the rack guide adjusting locknut.
3. Tighten the adjusting screw until it compresses the spring and seats against the guide, then loosen it. Torque it to 35 inch lbs., then for 1988–91 models back it off 20–20 1/18 degrees, or for 1992–95 models, back it off to 40–60 degrees.
4. Hold it in that position while adjusting the 40mm locknut with a special locknut wrench, part No. 07916-SA50001, or equivalent, to 18 ft. lbs. (25 Nm).
5. Recheck the play, and then move the wheels lock-to-lock, to make sure the rack moves freely without undue effort.

➥Turning the steering wheel from lock-to-lock several times will warm up the fluid. Effort should be no more than 4 lbs. (1.8 kg) for 1988–91 models, or 3.3 lbs. (1.5 kg) for 1992–95 models when measured with a spring scale attached to the steering wheel.

REMOVAL & INSTALLATION

1. Disconnect the negative battery cable. Raise the vehicle and support it safely.
2. Remove the cover panel and steering joint cover. Unbolt and separate the steering shaft at the coupling.
3. Drain the power steering fluid by disconnecting the return hose at the box and running the engine while turning the steering wheel lock-to-lock until fluid stops draining. Remove the gearbox shield, if equipped. Remove the front wheels.
4. Remove the cotter pins and unscrew the castle nuts on the tie rod ends. Using a ball joint tool, disconnect the tie rod ends. Lift the tie rod ends out of the steering knuckles.
5. If equipped with manual transaxle, disconnect the shift lever torque rod from the clutch housing. Slide the pin retainer out of the way, drive out the spring pin and disconnect the shift rod.
6. If equipped with automatic transaxle, remove the shift cable guide from the floor and pull the shift cable down by hand.
7. Remove the two nuts connecting the exhaust header pipe to the exhaust pipe and remove the exhaust header pipe (called exhaust pipe A for 1992–95 models). Disconnect the three hydraulic lines from the control unit.
8. Slide the tie rod ends all the way to the right and remove the steering rack mounting bolts.
9. Drop the gearbox far enough to permit the end of the pinion shaft to come out of the hole in the frame channel, then rotate it forward until the shaft is pointing rearward.
10. Slide the gearbox to the right until the left tie rod clears the beam, then drop it down and out of the vehicle to the left.

To install:

11. Install the gearbox and tighten the bracket bolts to 29 ft. lbs. (40 Nm). If equipped with a manual transaxle, reinstall the pin retainer after driving in the pin and be sure the projection on the pin retainer is in the hole.
12. Reconnect the fluid lines and tighten the line from the pump to 28 ft. lbs. (38 Nm), the line to the reservoir to 9 ft. lbs. (13 Nm) and the line to the oil cooler to 20 ft. lbs. (29 Nm).
13. Reconnect the tie rod ends and tighten to the proper specification.
14. Fill the system with fluid and bleed the air from the system.

SUSPENSION AND STEERING 8-27

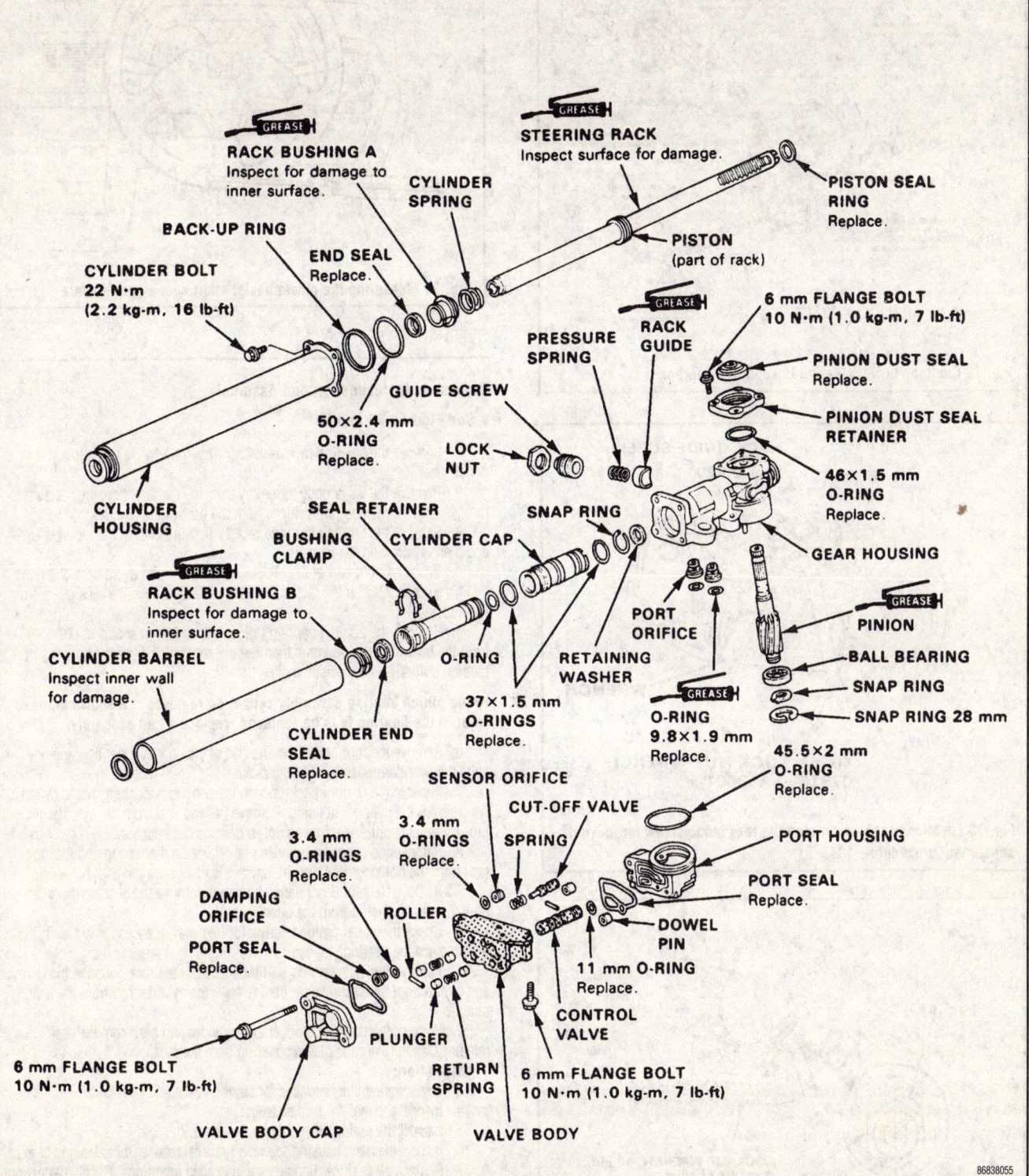

Fig. 71 Exploded view of common power steering rack

8-28 SUSPENSION AND STEERING

Fig. 72 Common power steering linkage and gearbox

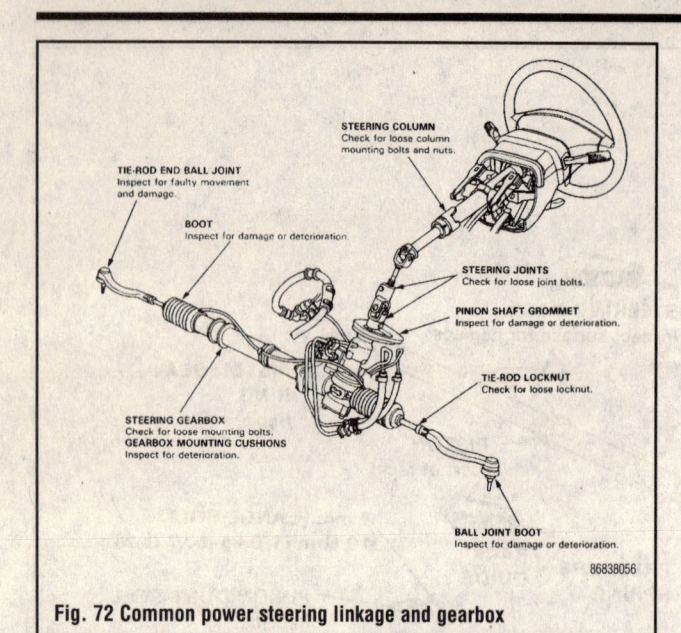

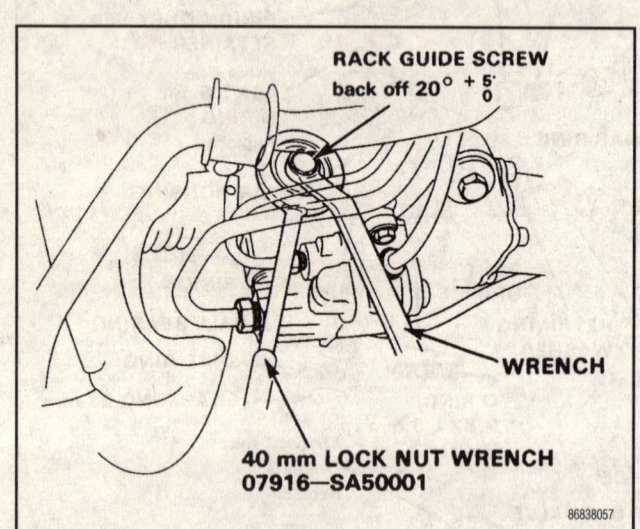

Fig. 73 Location of the power steering rack guide screw for locknut adjustment procedure—1988–91

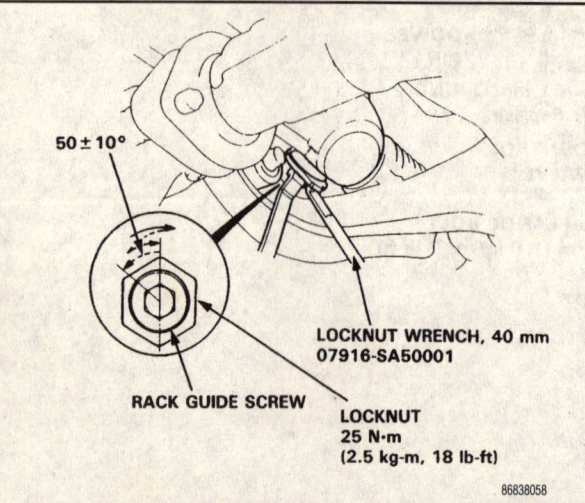

Fig. 74 Location of the power steering rack guide screw for locknut adjustment procedure—1992–95

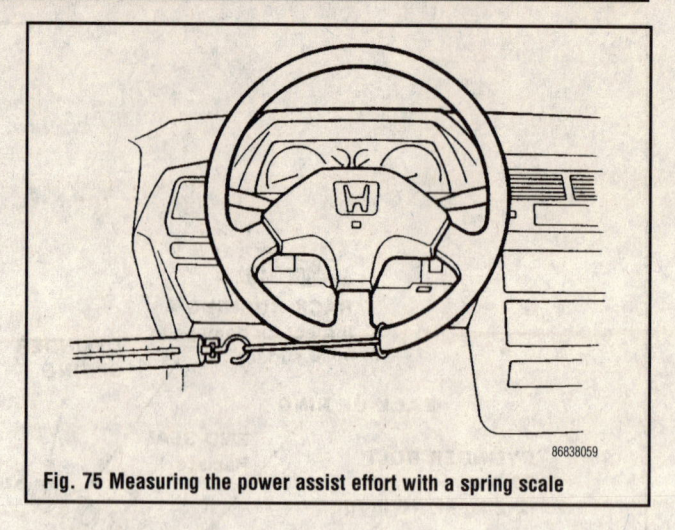

Fig. 75 Measuring the power assist effort with a spring scale

OVERHAUL

Disassembly, Inspection and Assembly

◆ See Figures 76 and 77

1. Remove the steering rack assembly from the vehicle and support it in a bench vise.
2. Remove the valve body assembly from the gearbox housing, be sure to remove the O-rings and port orifices from the housing.
3. Remove the tie rods from the rack by first unbending the lockwashers and then unscrewing the tie rods.
4. Push the right end of the rack into the housing to avoid scratching the sliding surface. Loosen the rack screw locknut and remove the rack guide from the housing.
5. Remove the 28mm snapring (1988–91) or 35mm snapring (1992–95) from the bottom of the housing, then remove the pinion from the housing by tapping it lightly with a plastic mallet.

➡The pinion bearing assembly cannot be removed if it shows signs of wear. If the bearing is to be replaced, replace as an assembly.

6. Remove the four bolts from the end of the rack cylinder housing and slide the cylinder housing off of the rack.
7. Remove the O-ring, back-up ring, steering rack bushing A and cylinder spring from the cylinder housing. Remove the end seal from the cylinder housing. Try to use your fingers so as not to damage the housing.
8. Remove the cylinder, cylinder seal retainer, cylinder cap and steering rack from the housing. Remove the retainer washer from the gear housing.
9. Remove the four 6mm flange bolts and remove the gear housing cap, pinion seal and O-ring from the gear housing.
10. Check the upper pinion bearing for free-play, if excess free-play is evident replace the bearing.
11. Remove the cylinder and seal retainer from the rack. Remove the O-rings and circlip from the seal retainer then remove the cylinder cap from the seal retainer.
12. Remove the bushing stopper and cylinder end seal from the seal retainer. Carefully pry the piston seal ring from the rack.

Inspection:

13. Inspect all sliding surfaces for signs of wear.
14. Inspect the rack for broken teeth.
15. Inspect the pinion assembly for broken teeth.
16. Inspect the main housing, cylinder and seal retainer for any signs of wear.
17. Replace all O-rings. Replace any damaged snaprings. Check all bearings for excessive free-play, replace any that are damaged.
18. Lubricate all sliding pieces with power steering fluid before assembly, unless otherwise noted.

Assembly:

19. Install a new O-ring on the rack, with its narrow edge facing out.
20. Install a piston seal guide tool, Honda 07GAG–SD40100, or equivalent onto the rack (big end first). Position the new piston seal ring on the tool and slide it onto the big end of the tool and onto the O-ring on the rack. Remove the tool.

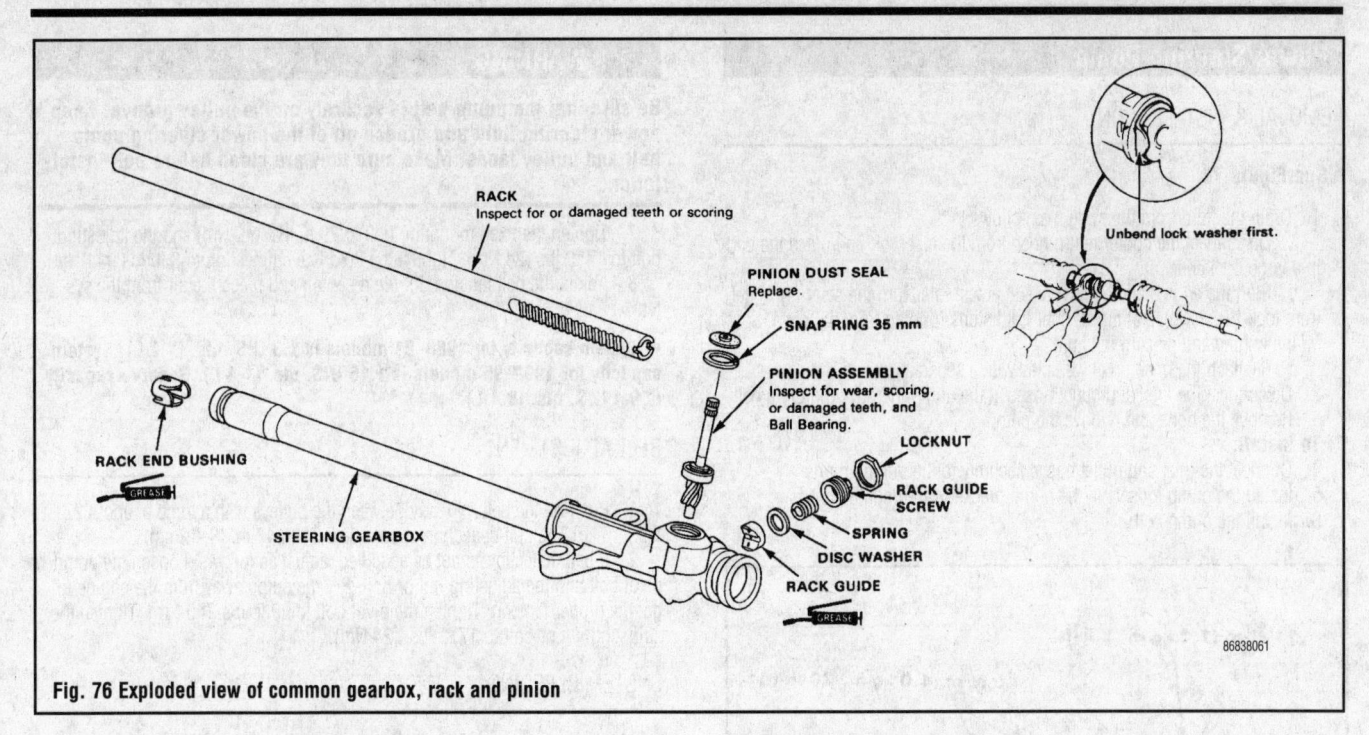

SUSPENSION AND STEERING 8-29

Fig. 76 Exploded view of common gearbox, rack and pinion

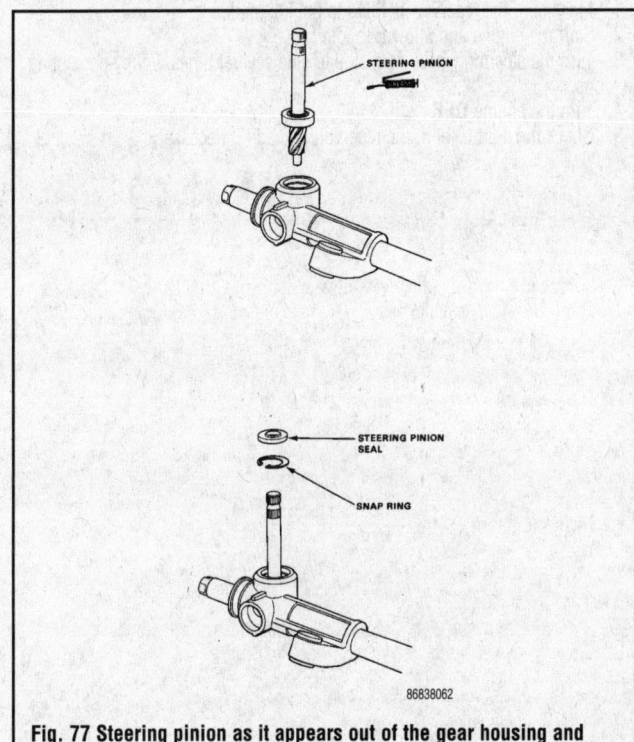

Fig. 77 Steering pinion as it appears out of the gear housing and installed

21. Coat new O-rings with grease and install them on the cylinder cap and slide the cylinder cap onto the seal retainer. Install the circlip onto the seal retainer.
22. Grease the sliding surface of the steering rack bushing and install the bushing on the steering rack with the groove facing the steering rack piston.

✴✴ WARNING

Do not fill the slots with grease. They serve as air passages and must remain clear of obstruction.

23. Grease the sliding surface of the cylinder end seal and slide it onto the rack with the grooves facing opposite the seal slider. Attach the seal retainer to the rack.
24. Push the rack bushing to the seal retainer until the cylinder end seal is seated in the retainer. Install the stopper ring in the groove of the seal retainer.
25. Install the retainer washer on the gear housing. Place the gear housing on a work bench and insert the rack and seal retainer assembly into the housing. Coat the inside of the cylinder with power steering fluid and slide it over the rack assembly, press it into the housing until it seats.
26. Install the cylinder spring over the rack, then coat the rack bushing with power steering fluid and install it on the spring.
27. Coat the inside surface of the cylinder with power steering fluid and install the cylinder end seal with the groove facing out.
28. Install the O-ring and back-up ring on the gear housing. Carefully position the cylinder on the gear housing and loosely install the four bolts.
29. Install the steering rack into the cylinder housing, be careful not to damage the rack.
30. Tighten the cylinder housing bolts to 16 ft. lbs. (22 Nm).
31. Press the lower bearing onto the pinion with its shielded side facing down. Apply a light coating of grease and install the correct snapring. Grease the pinion seal and install it on the gear housing using a seal driver.
32. Apply grease to the O-ring and install it in the groove in the housing cap. Install the gear housing cap over the pinion and tighten the flange bolts to 8 ft. lbs. (11 Nm).
33. Install the valve body on the assembly. Coat the rack guide with grease.
34. Install the rack guide assembly into the housing.
35. Screw in each of the tie rods to the rack, while holding the lockwasher so that its tabs are aligned with the slots in the rack. Install the stopper washer with the chamfered side facing out.
36. Tighten the tie rod securely, then bend the lockwasher back onto the flange.
37. Install the bellows over the end of the rack housing and install the retaining clamp.
38. Install the tie rod ends on the tie rods and fill the tie rod boots with grease.
39. Install the rack into the vehicle and connect the power steering lines. Adjust the rack as needed.

8-30 SUSPENSION AND STEERING

Power Steering Pump

REMOVAL & INSTALLATION

♦ See Figure 78

1. Drain the fluid from the system as follows:
    a. Disconnect the cooler return hose from the reservoir and place the end in a large container.
    b. Start the engine and allow it to run at fast idle. Turn the steering wheel from lock-to-lock several times, until fluid stops running from the hose. Shut off the engine and discard the fluid.
    c. Reattach the hose on all vehicles with a separate reservoir.
2. Disconnect the inlet and outlet hoses at the pump. Remove the drive belt.
3. Remove the bolts and remove the pump.

To install:

4. Connect the inlet and outlet hoses securely to the pump fittings.
5. Install the pump loosely in the pump bracket with mounting bolts.
6. Install the pump belt.

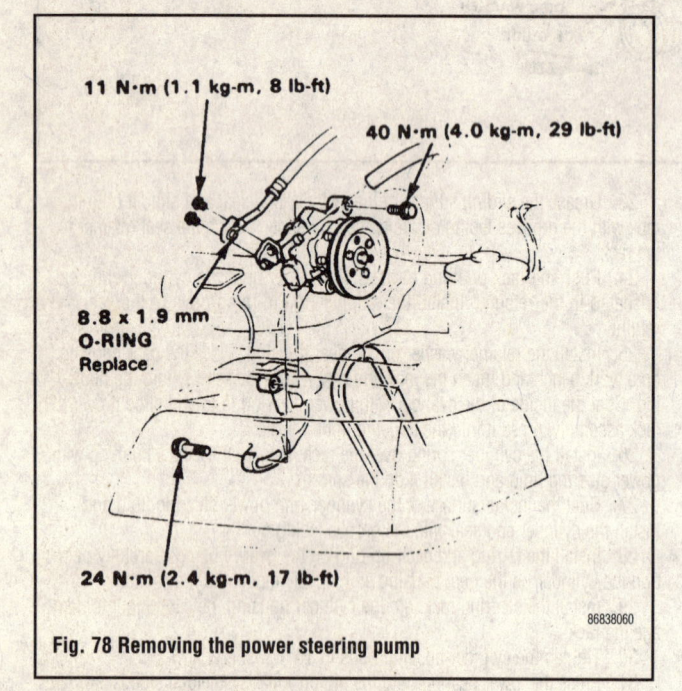

Fig. 78 Removing the power steering pump

✳✳✳ WARNING

Be sure that the pump belt is securely on the pulley groove. Keep power steering fluid and grease off of the power steering pump belt and pulley faces. Make sure they are clean before belt installation.

7. Tighten the fixed mounting bolt to 29 ft. lbs. (40 Nm) and the adjusting bolt to 17 ft. lbs (24 Nm). Tighten the fluid line connection to 8 ft. lbs. (11 Nm).
8. Adjust the belt tension, fill the reservoir and bleed the air from the system.

➡ System capacity for 1988–91 models is 1.3 U.S. qts. (1.2 L). System capacity for 1992–95 models is 1.16 U.S. qts. (1.1 L). Reserve capacity is 0.4 U.S. qts. (0.4 L).

BELT ADJUSTMENT

1. Push on the belt mid-way between the pulleys with a force of about 22 lbs. (10 kg) The belt deflection should be 0.31–0.47 in. (8–12mm).
2. If belt deflection is not as specified, adjust as follows: Loosen the pump pivot bolt and the adjusting nut or bolt. Pry the pump away from the engine to get the proper tension. Tighten the pivot bolt to 29 ft. lbs. (40 Nm). Tighten the adjusting nut or bolt to 17 ft. lbs. (24 Nm).

SYSTEM BLEEDING

1. Make sure the reservoir is filled to the full mark.
2. Start the engine and allow it to idle.
3. Turn the steering wheel from side to side several times, lightly contacting the stops.
4. Turn the engine **OFF**.
5. Check the fluid level in the reservoir and add if necessary.

BRAKE OPERATING SYSTEM 9-2
GENERAL DESCRIPTION 9-2
   HYDRAULIC SYSTEM 9-2
   DRUM BRAKES 9-2
   DISC BRAKES 9-3
   POWER BRAKE BOOSTERS 9-3
   HYDRAULIC SYSTEM 9-3
ADJUSTMENTS 9-3
   REAR DRUM BRAKE
     ADJUSTMENT 9-3
   BRAKE PEDAL FREE-PLAY 9-3
BRAKE LIGHT SWITCH 9-4
   REMOVAL & INSTALLATION 9-4
   TESTING 9-4
MASTER CYLINDER 9-4
   REMOVAL & INSTALLATION 9-4
   OVERHAUL 9-5
POWER BRAKE BOOSTER 9-7
   INSPECTION 9-7
   REMOVAL & INSTALLATION 9-7
DUAL PROPORTIONING VALVE 9-7
   REMOVAL & INSTALLATION 9-7
BRAKE HOSES AND PIPES 9-8
   REMOVAL & INSTALLATION 9-8
BLEEDING THE BRAKE SYSTEM 9-8
FRONT DISC BRAKES 9-9
BRAKE PADS 9-9
   REMOVAL & INSTALLATION 9-9
BRAKE CALIPER 9-11
   REMOVAL & INSTALLATION 9-11
   OVERHAUL 9-12
BRAKE DISC (ROTOR) 9-13
   REMOVAL & INSTALLATION 9-13
   INSPECTION 9-14
REAR DRUM BRAKES 9-14
BRAKE DRUMS 9-15
   REMOVAL & INSTALLATION 9-15
   INSPECTION 9-15
BRAKE SHOES 9-15
   REMOVAL & INSTALLATION 9-15
WHEEL CYLINDERS 9-16
   REMOVAL & INSTALLATION 9-16
   OVERHAUL 9-16
REAR DISC BRAKES 9-17
BRAKE CALIPER 9-17
   REMOVAL & INSTALLATION 9-17
BRAKE PADS 9-17
   REMOVAL & INSTALLATION 9-17
BRAKE DISC 9-17
   REMOVAL & INSTALLATION 9-17
   INSPECTION 9-18
PARKING BRAKE 9-19
CABLE 9-19
   REMOVAL & INSTALLATION 9-19
   ADJUSTMENT 9-20
ANTI-LOCK BRAKES
  (ALB/ABS) 9-21
DESCRIPTION AND OPERATION 9-21
TROUBLESHOOTING 9-21

GENERAL INFORMATION 9-21
SYSTEM TESTING 9-22
DIAGNOSTIC FLOW CHARTS 9-23
CONTROL UNIT 9-33
   REMOVAL & INSTALLATION 9-33
WHEEL SENSOR 9-33
   REMOVAL & INSTALLATION 9-33
GEAR PULSERS 9-33
   INSPECTION 9-33
MODULATOR UNIT/ALB/ABS PUMP
  ASSEMBLY 9-34
   RELIEVING LINE PRESSURE 9-34
   REMOVAL & INSTALLATION 9-34
ACCUMULATOR 9-36
   REMOVAL AND INSTALLATION 9-36
   DISPOSAL 9-36
FILLING AND BLEEDING 9-36
   FILLING THE SYSTEM 9-36
   MODULATOR UNIT BLEEDING WITH
     THE ALB/ABS CHECKER 9-36
   BLEEDING THE SYSTEM 9-37
SPECIFICATIONS CHART
BRAKE SPECIFICATIONS 9-38

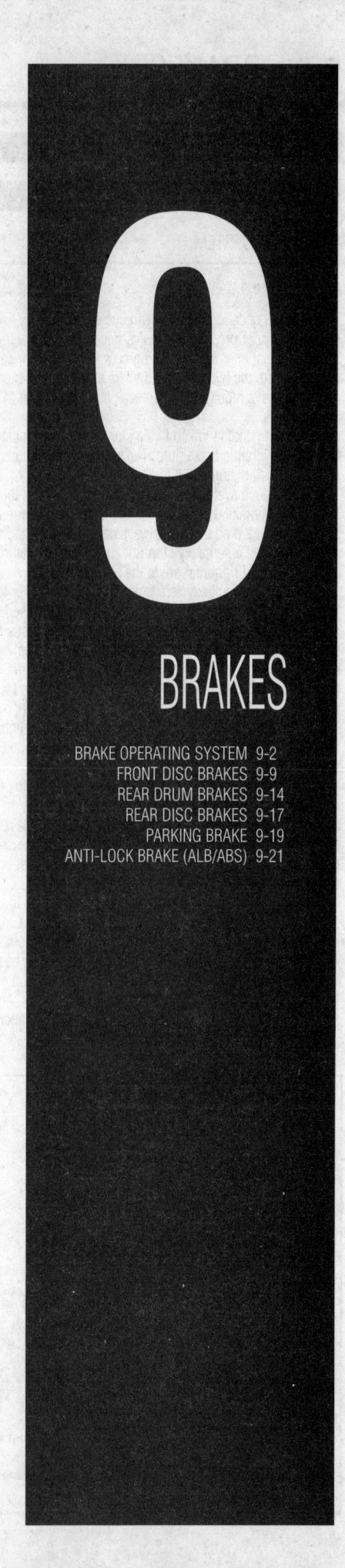

9
BRAKES

BRAKE OPERATING SYSTEM 9-2
FRONT DISC BRAKES 9-9
REAR DRUM BRAKES 9-14
REAR DISC BRAKES 9-17
PARKING BRAKE 9-19
ANTI-LOCK BRAKE (ALB/ABS) 9-21

9-2 BRAKES

BRAKE OPERATING SYSTEM

General Description

HYDRAULIC SYSTEM

♦ See Figure 1

The brake pedal operates a hydraulic system that is used for two reasons. First, fluid under pressure can be carried to all parts of the vehicle by small hoses or metal lines without taking up a lot of room or causing routing problems. Second, the hydraulic fluid offers a great mechanical advantage—little foot pressure is required on the pedal, but a great deal of pressure is generated at the wheels.

The brake pedal is linked to a piston in the brake master cylinder, which is filled with hydraulic brake fluid. The master cylinder consists of a cylinder, containing a small piston and a fluid reservoir.

The hydraulic system is a dual circuit type that has the advantage of retaining 50 percent of braking effectiveness in the event of failure in one system. The circuits are arranged so that you always have one front and one rear brake for a more controlled emergency stop (the right front and left rear are in one circuit; the left front and right rear are in the second circuit).

The entire hydraulic system from the master cylinder to the wheels is full of hydraulic brake fluid. When the brake pedal is depressed, the pistons in the master cylinder are forced to move, exerting tremendous force on the fluid in the lines. The fluid has nowhere to go and forces the wheel cylinder piston (drum brakes) or caliper pistons (disc brakes) to exert pressure on the brake shoes or pads. The resulting friction between the brake shoe and wheel drum or the brake pad and disc slows the vehicle and eventually stops it.

Also attached to the brake pedal is a switch which lights the brake lights as the pedal is depressed. The lights stay **ON** until the brake pedal is released and returns to its normal position.

Each wheel cylinder in a drum brake system contains two pistons, one at either end, which push outward in opposite directions. In disc brake systems, the wheel cylinders are part of the caliper; there can be as many as four or as few as one. Whether disc or drum type, all pistons use some type of rubber seal to prevent leakage around the piston and a rubber dust boot seals the outer ends of the wheel cylinders against dirt and moisture.

When the brake pedal is released, a spring pushes the master cylinder pistons back to their normal position. Check valves in the master cylinder piston allow fluid to flow toward the wheel cylinders or calipers as the piston returns. As the brake shoe return springs pull the brake shoes back to the released position, excess fluid returns to the master cylinder through compensating ports, which have been uncovered as the pistons move back. Any fluid that has leaked from the system will also be replaced through the compensating ports.

All dual circuit brake systems use a switch to activate a light, warning of brake failure. The switch is located in a valve mounted near the master cylinder. A piston in the valve receives pressure on each end from the front and rear brake circuits. When the pressures are balanced, the piston remains stationary but when one circuit has a leak, greater pressure during the application of the brakes will force the piston to one side or the other, closing the switch and activating the warning light.

The Anti-Lock Brake (ALB or ABS) system was made available on some Civics beginning in the 1992 models year. The system includes a hydraulic modulator, a pressure accumulator, speed sensors at each wheel and an electronic control unit. The control unit is equipped with a self-diagnostic program and is capable of storing "problem codes" (Honda's terminology). The code memory can only be accessed with the dealer's diagnostic equipment. The system operates at very high pressure and is capable of generating these pressures even when the engine is not running. Because of the need for special equipment and the dangerous pressures involved, owner service to this system is essentially limited to replacement of brake pads and adding and bleeding brake fluid.

Disc brake systems also have a metering valve to prevent the front disc brakes from engaging before the rear brakes have contacted the drums. This ensures that the front brakes will not normally be used alone to stop the vehicle. A proportioning valve is also used to limit pressure to the rear brakes to prevent rear wheel lock-up during hard braking.

DRUM BRAKES

Drum brakes use two brake shoes mounted on a stationary backing plate. These shoes are positioned inside a circular cast iron drum which rotates with the wheel assembly. The shoes are held in place by springs; this allows them to slide toward the drums (when they are applied) while keeping the linings and drums in alignment. The shoes are actuated by a wheel cylinder which is usually mounted at the top of the backing plate. When the brakes are applied, hydraulic pressure forces the wheel cylinder's two actuating links outward. Since these links bear directly against the top of the brake shoes, the tops of the shoes are forced outward against the inner side of the drum. This action forces the bottoms of the two shoes to contact the brake drum by rotating the entire assembly slightly (known as servo action). When pressure within the wheel cylinder is relieved, return springs pull the shoes back away from the drum.

Most modern drum brakes are designed to self-adjust during application when the vehicle is moving in reverse. This motion causes both shoes to rotate very slightly with the drum, rocking an adjusting lever. The self-adjusters are only intended to compensate for normal wear. Although the adjustment is "automatic," there is a definite method to actuate the self-adjuster, which is done during normal driving. Driving the vehicle in reverse and applying the brakes usually activates the automatic adjusters. If the brake pedal was low, you should be able to feel an increase in the height of the brake pedal.

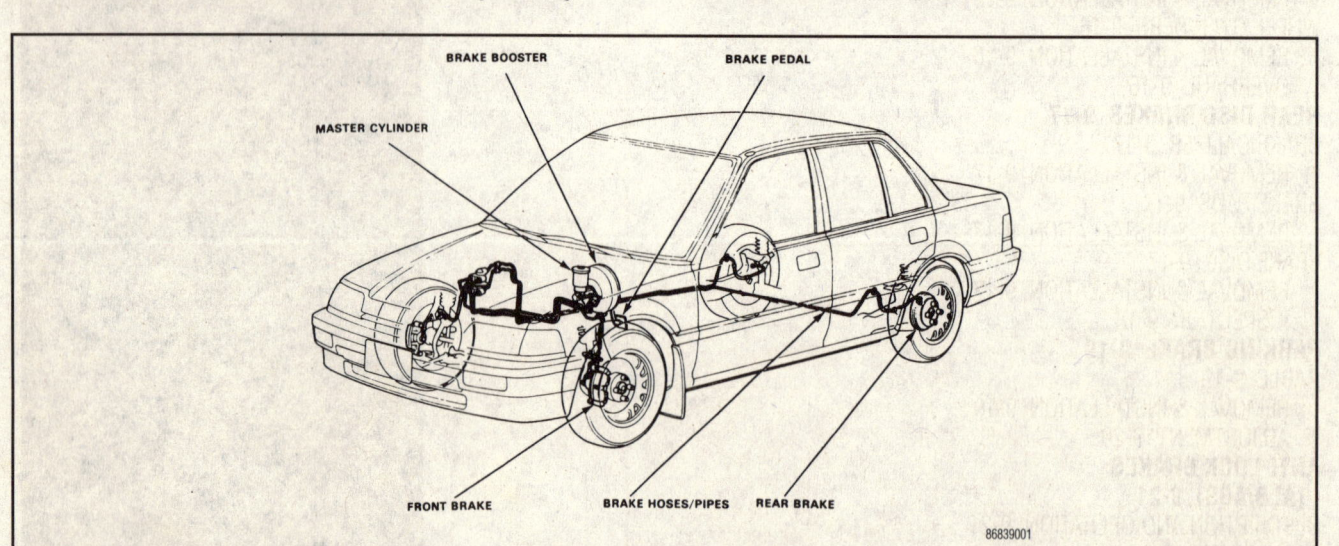

Fig. 1 Location of brake system components

BRAKES 9-3

DISC BRAKES

Instead of the traditional expanding brakes that press outward against a circular drum, disc brake systems utilize a cast iron disc with brake pads positioned on either side of it. Braking effect is achieved in a manner similar to the way you would squeeze a spinning disc between your fingers. The disc (rotor) is a one-piece casting with cooling fins between the two braking surfaces. This enables air to circulate between the braking surfaces making them less sensitive to heat buildup and more resistant to fade. Dirt and water do not affect braking action since contaminants are thrown off by the centrifugal action of the rotor or scraped off by the pads. Also, the equal clamping action of the two brake pads tends to ensure uniform, straight-line stops. All disc brakes are inherently self-adjusting. There are three general types of disc brake:
1. A fixed caliper, 4-piston type.
2. A floating caliper, single piston type.
3. A sliding caliper, single piston type.

The fixed caliper design uses two pistons mounted on either side of the rotor (in each side of the caliper). The caliper is mounted rigidly and does not move.

The sliding and floating designs are quite similar and often considered as one. The pad on the inside of the rotor is moved into contact with the rotor by hydraulic force. The caliper, which is not held in a fixed position, moves slightly, bringing the outside pad into contact with the rotor. There are various methods of attaching floating calipers; some pivot at the bottom or top and some slide on mounting bolts.

POWER BRAKE BOOSTERS

Power brakes operate just as standard brake systems except in the actuation of the master cylinder pistons. A vacuum diaphragm is located behind the master cylinder and assists the driver in applying the brakes, reducing both the effort and travel he must put into moving the brake pedal.

The vacuum diaphragm housing is connected to the intake manifold by a vacuum hose. A check valve at the point where the hose enters the diaphragm housing ensures that during periods of low manifold vacuum brake assist vacuum will not be lost.

Depressing the brake pedal closes off the vacuum source and allows atmospheric pressure to enter on one side of the diaphragm. This causes the master cylinder pistons to move and apply the brakes. When the brake pedal is released, vacuum is applied to both sides of the diaphragm and return springs return the diaphragm and master cylinder pistons to the released position. If the vacuum fails, the brake pedal rod will butt against the end of the master cylinder actuating rod and direct mechanical application will occur as the pedal is depressed.

The hydraulic and mechanical problems that apply to conventional brake systems also apply to power brakes.

Honda uses a dual hydraulic system, with the brakes connected diagonally. In other words, the right front and left rear brakes are on the same hydraulic line and the left front and right rear are on the other line. This has the added advantage of front disc emergency braking, should either of the hydraulic systems fail. The diagonal rear brake serves to counteract the sway from single front disc braking.

A leading/trailing drum brake is used for the rear brakes, with disc brakes for the front. All Hondas are equipped with a brake warning light, which is activated when a defect in the brake system occurs.

HYDRAULIC SYSTEM

The hydraulic system is composed of the master cylinder and brake booster, the brake lines, the brake pressure differential valve(s), the wheel cylinders (drum brakes) and calipers (disc brakes).

The master cylinder serves as a brake fluid reservoir and (along with the booster) as a hydraulic pump. Brake fluid is stored in the two sections of the master cylinder. Each section corresponds to each part of the dual braking system. This tandem master cylinder is required by Federal law as a safety device.

When the brake pedal is depressed, it moves a piston mounted in the bottom of the master cylinder. The movement of this piston creates hydraulic pressure in the master cylinder. This pressure is carried to the wheel cylinders or the calipers by brake lines, passing through the pressure differential or proportioning valve.

When the hydraulic pressure reaches the wheels, after the pedal has been depressed, it enters the wheel cylinders or calipers. Here it comes into contact with a piston(s). The hydraulic pressure causes the piston(s) to move, which moves the brake shoes or pads (disc brakes), causing them to contact the drums or rotors (disc brakes). Friction between the brake shoes and the drums causes the vehicle to slow. There is a relationship between the amount of pressure that is applied to the brake pedal and the amount of force which moves the brake shoes against the drums. Therefore, the harder the brake pedal is depressed, the quicker the vehicle will stop.

Since the hydraulic system is one which operates on fluids, air is a natural enemy of the brake system. Air in the hydraulic system retards the passage of hydraulic pressure from the master cylinder to the wheels. Anytime a hydraulic component below the master cylinder is opened or removed, the system must be bled of air to ensure proper operation. Air trapped in the hydraulic system can also cause the brake warning light to turn **ON**, even though the system has not failed. This is especially true after repairs have been performed on the system.

Adjustments

REAR DRUM BRAKE ADJUSTMENT

1. Block the front wheels and release the parking brake. Raise and support the rear of the vehicle on jackstands.
2. Depress the brake pedal two or three times and release.
3. The adjuster is located on the inboard side, underneath the control arm. Turn the adjuster clockwise until the wheel no longer turns.
4. Back off the adjuster two (2) clicks and turn the wheel to see if the brake shoes are dragging. If they are dragging, back off the adjuster one more click.

BRAKE PEDAL FREE-PLAY

♦ See Figure 2

Free-play is the distance the pedal travels from the stop (brake light switch) until the pushrod actuates the master cylinder.

To check free-play, first measure the distance (with the carpet removed) from the floor to the brake pedal. Then disconnect the return spring and again mea-

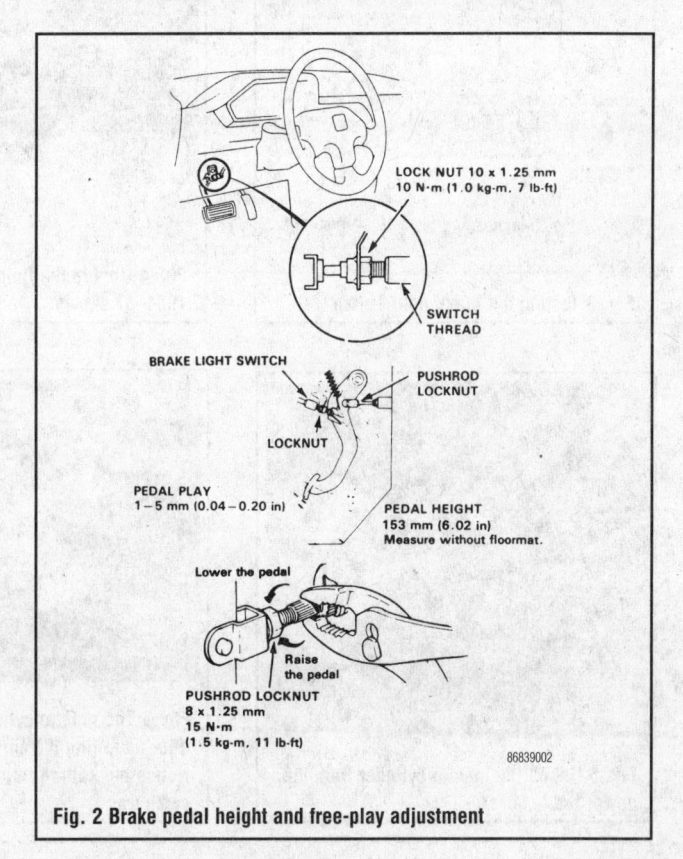

Fig. 2 Brake pedal height and free-play adjustment

9-4 BRAKES

sure the distance from the floor to the brake pedal. The difference between the two measurements is the pedal free-play. The specified free-play is 0.04–0.2 in (1.0–5.0mm). Free-play adjustment is made by loosening the locknut on the brake light switch and rotating the switch body until the specified clearance is obtained.

※ WARNING

If there is no free-play, the master cylinder pistons will not return to their stops. This can block the compensating ports, which will prevent the brake pads and linings from returning fully when the pedal is released. This will result in rapid brake burn-up. Free-play provides a safety factor against normal rubber swell and expansion or deflection of body parts and pedal linkage.

Brake Light Switch

REMOVAL & INSTALLATION

1. Disconnect the negative battery cable.
2. Locate the switch under the dash at the top of the brake pedal mount.
3. Disconnect the electrical lead from the switch.
4. Remove the nut from the switch and unscrew the switch from its mount.
5. Install the new switch in position and connect the electrical lead.
6. Connect the negative battery cable. Adjust the pedal height.

TESTING

▶ See Figure 3

1. Remove the switch from the vehicle.
2. Using an ohmmeter, check for continuity between both terminals. The readings should indicate no continuity with the plunger pushed in, and continuity with the plunger released.
3. If the switch fails to meet these conditions, replace it.

Master Cylinder

REMOVAL & INSTALLATION

▶ See Figures 4 thru 10

Before removing the master cylinder, cover the body surfaces with fender covers and shop towels or plastic sheeting to prevent damage to painted surfaces by brake fluid. If any fluid does contact painted or anodized surfaces, wipe the fluid off immediately.

➡ The brake fluid in the reservoir may be sucked out with a clean turkey baster or other similar device prior to removing the brake lines.

1. Disconnect the fluid sensor lead.
2. Disconnect and plug the brake lines from the master cylinder.
3. Remove the master cylinder-to-vacuum booster attaching bolts and the master cylinder from the vehicle.
4. Install the master cylinder to the booster and tighten the mounting nuts to 11 ft. lbs. (15 Nm).
5. Connect the brake lines to the master cylinder. Tighten them to 14 ft. lbs. (19 Nm).

➡ Be sure to use a line wrench when connecting the brake lines. Standard open end wrenches may slip and round over the fittings. Never use any other lubricant for brake line fittings than clean brake fluid.

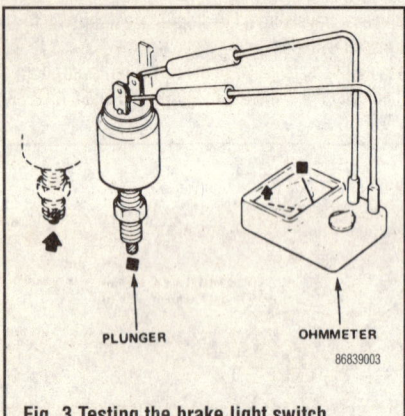

Fig. 3 Testing the brake light switch

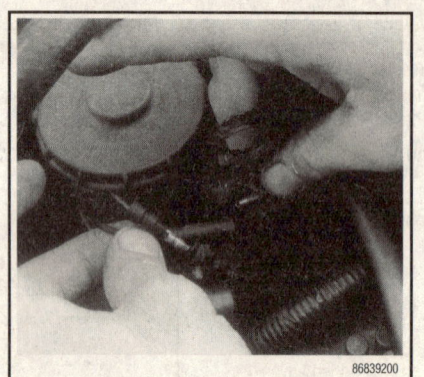

Fig. 4 Unplug the fluid sensor lead—1984–87 shown

Fig. 5 Disconnect the brake lines using the correct wrench. Plug the brake lines as soon as you remove them

Fig. 6 Unbolt the master cylinder from the power brake booster

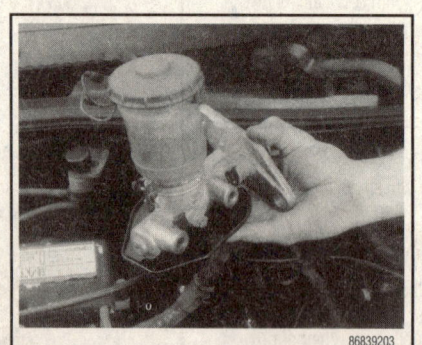

Fig. 7 The master cylinder comes free after unbolting it from the brake booster. Notice the spilled fluid on the edge of the catch tray

Fig. 8 This catch tray prevents fluid from spilling and comes in handy when refilling the reservoir. It simply unclips

BRAKES 9-5

Fig. 9 Master cylinder, brake booster and dual proportioning valve mounting

Fig. 10 Exploded view of the master cylinder—1984-87 model shown

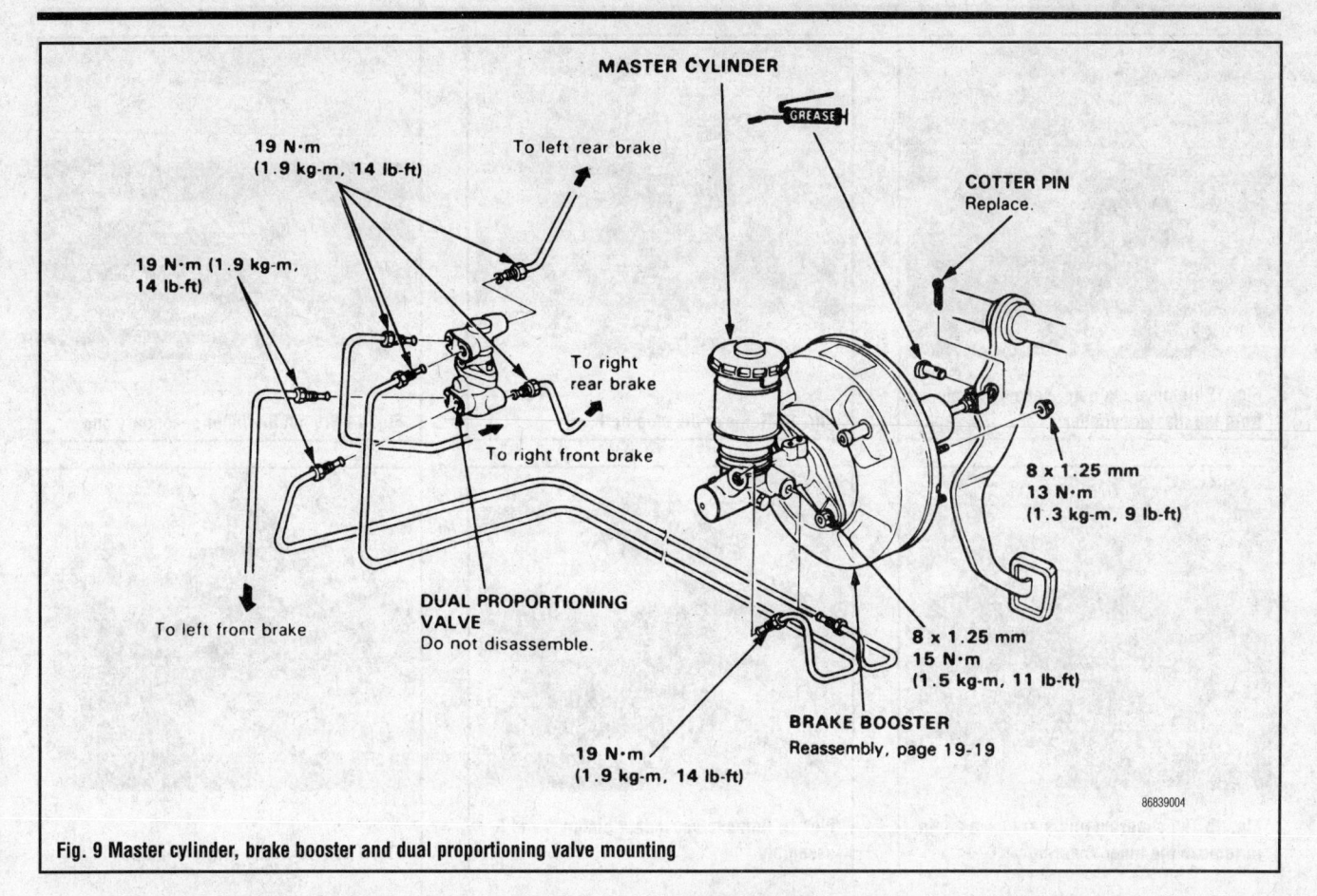

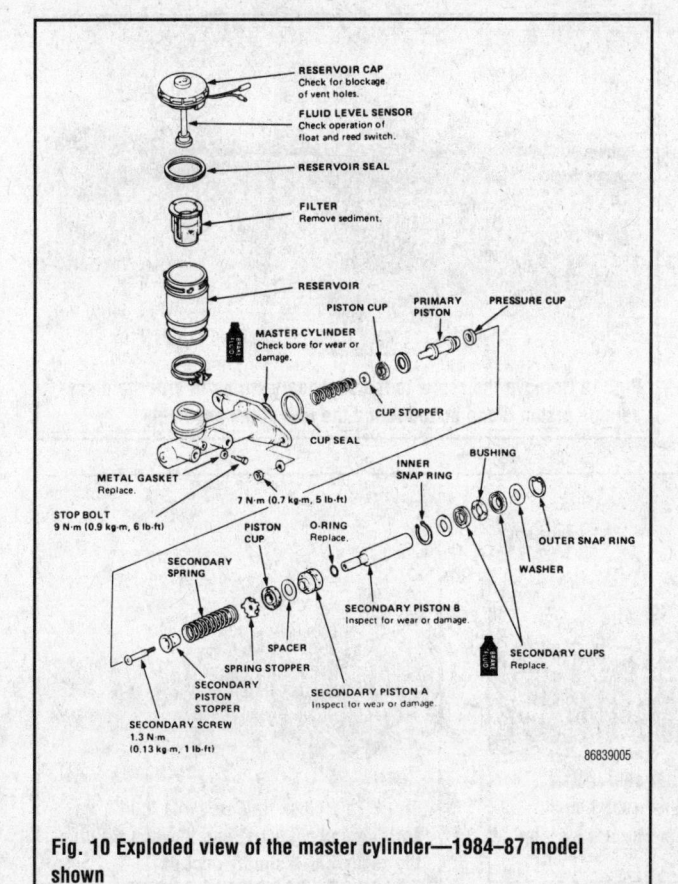

OVERHAUL

1984–87 Models

▶ See Figures 11 thru 21

Before beginning this procedure, make sure you have available a master cylinder rebuild kit, or the necessary internal replacement parts to repair what you take apart. If you are unable to obtain the parts needed, you may have to forego the following procedure and replace the faulty master cylinder with a new one.

1. Remove the master cylinder from the vehicle and mount it securely on a bench vise being cautious not to clamp it too hard.

Fig. 11 Remove the snapring from the secondary piston assembly—1984-87 models only

9-6 BRAKES

Fig. 12 Remove the outer secondary cup from the master cylinder

Fig. 13 Remove the stop bolt

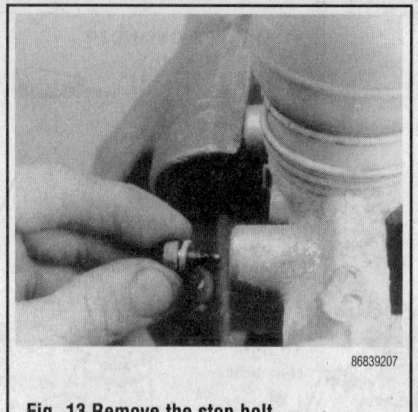

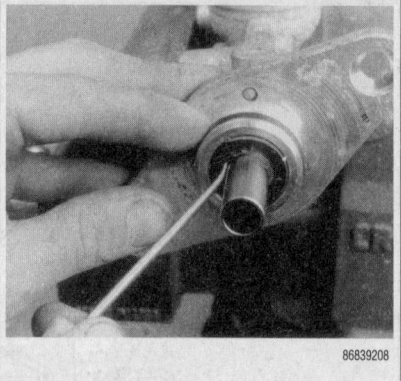

Fig. 14 Pry out the inner secondary cup

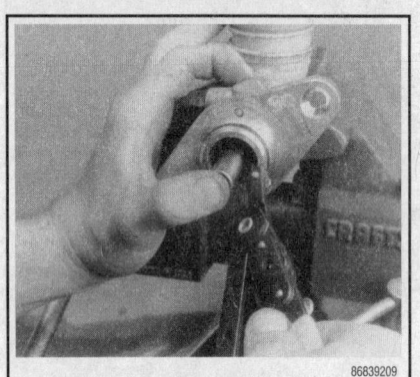

Fig. 15 The snapring pliers are used again to remove the inner snapring

Fig. 16 Remove secondary piston A and B assembly

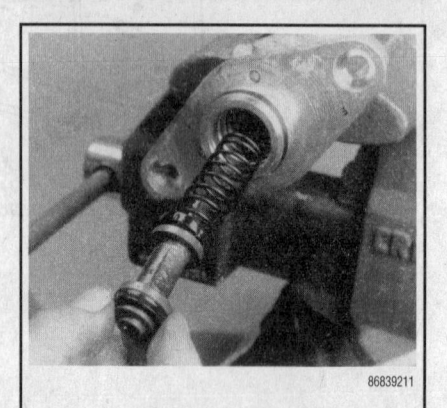

Fig. 17 Remove the primary piston

Fig. 18 The secondary and primary piston assemblies and related clips, washers and cups—1984–87 models

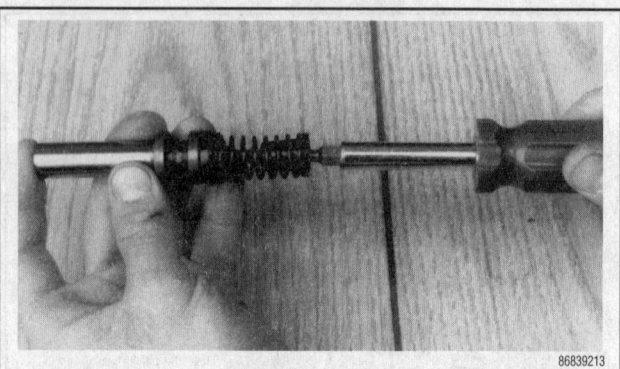

Fig. 19 Remove the screw to the secondary spring in order to disassemble piston B and piston A and the rest of the assembly

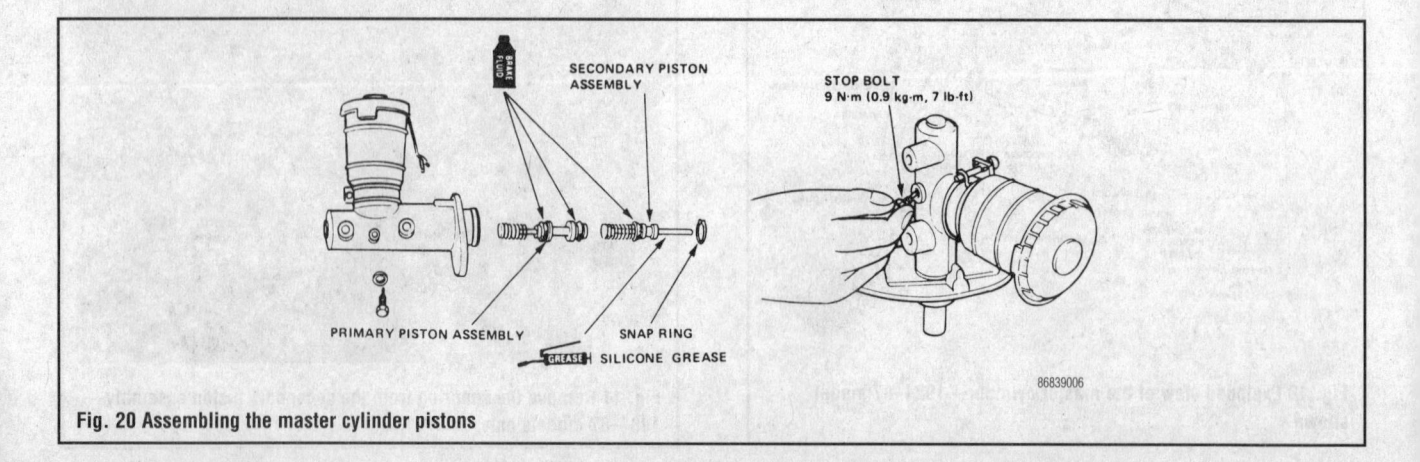

Fig. 20 Assembling the master cylinder pistons

BRAKES 9-7

Fig. 21 Install the master cylinder cups and seals during reassembly

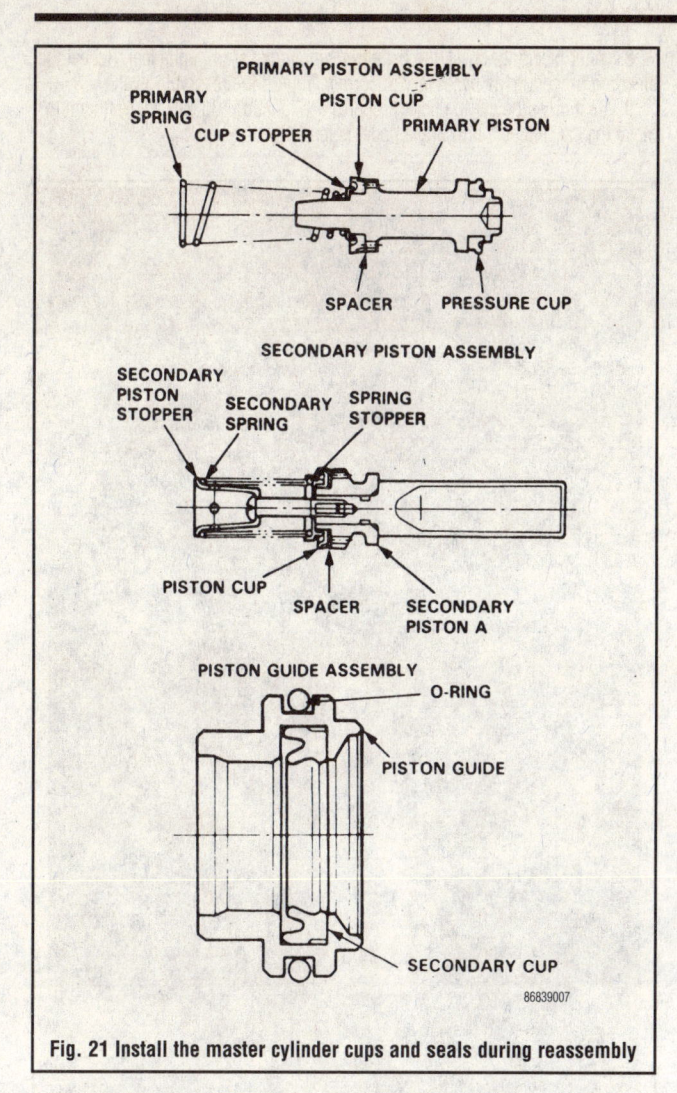

2. Remove the reservoir cap, fluid level sensor, filter and seal.
3. Remove the outer snapring from the secondary piston assembly. Remove the washer secondary cup and secondary piston bushing.
4. Remove the stop bolt, and remove the inner snapring while pushing on the secondary piston assembly.
5. Remove the primary piston assembly from the bore.
6. Remove the screw from the secondary piston and remove the spring.

To Install:

7. Clean all parts thoroughly, with clean brake fluid.

➡ Use only clean brake fluid to clean master cylinder components. Other solvents may damage the components.

8. Lubricate new piston assemblies with brake fluid then install them in the master cylinder body.
9. Press down on the piston assembly and install the stop bolt in the master cylinder body. Tighten the stop bolt to 7 ft. lbs. (10 Nm).
10. Again press down on the piston assembly and install the inner snapring.
11. Install the secondary cups, bushing and outer snapring.
12. Insert a new seal on the master cylinder mounting flange and install the master cylinder in the vehicle. Bleed the brake system when finished.

1988–95 Models

The Honda Motor Co. does not recommend disassembly of the master cylinder assembly for 1988–95 models. If a master cylinder for a 1988–95 Civic is found to be faulty, replace the unit with a new one.

Power Brake Booster

INSPECTION

▶ See Figure 22

1. A preliminary check of the vacuum booster can be made as follows:
   a. Depress the brake pedal several times using normal pressure. Make sure that the pedal height does not vary.
   b. Hold the pedal in the depressed position and start the engine. The pedal should drop slightly.
   c. Hold the pedal in the above position and stop the engine. The pedal should stay in the depressed position for approximately 30 seconds.
   d. If the pedal does not drop when the engine is started or rises after the engine is stopped, the booster is not functioning properly.
   e. To test the check valve, disconnect the vacuum hose at the booster. Start the engine and let it idle, There should be vacuum available at the end of the hose. If there is no vacuum available, the check valve is faulty.

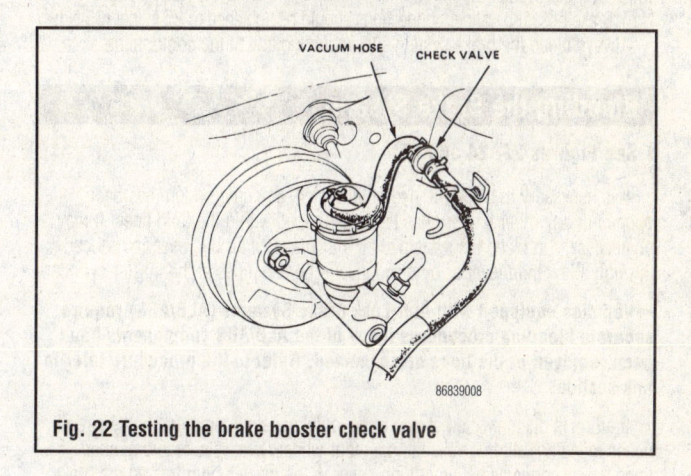

Fig. 22 Testing the brake booster check valve

REMOVAL & INSTALLATION

1. Disconnect the vacuum hose from the booster.
2. Disconnect the brake lines from the master cylinder.
3. Remove the brake pedal-to-booster link pin and the booster-to-fire wall nuts. The pushrod and nuts are located inside the vehicle on the driver's side, under the dash.
4. Remove the booster with the master cylinder attached.
5. Install the booster in position and tighten the mounting nuts to 9 ft. lbs. (13 Nm).
6. Connect the vacuum hose to the booster and the fluid lines to the master cylinder. Tighten the fluid lines to 14 ft. lbs. (19 Nm).
7. Bleed the brake system.

Dual Proportioning Valve

The brake proportioning valve is located on the passenger side of the vehicle, bolted to the fire wall or fender.

REMOVAL & INSTALLATION

1. Locate the valve and disconnect the fluid lines from it. Plug the lines. There are six lines coming from the valve.
2. Remove the mounting bolts and remove the valve.

➡ The valve is replaced as an assembly, it can not be overhauled.

3. Mount the valve in position and tighten the mounting bolts to 9 ft. lbs. (13 Nm).

9-8 BRAKES

4. Reconnect the six brake lines to the valve. Tighten the line fittings to 14 ft. lbs. (19 Nm).

Brake Hoses and Pipes

REMOVAL & INSTALLATION

All brake lines and hoses should be checked for rust, holes, kinks, tears or any sign of deterioration. Any hose showing these signs should be replaced. Always replace brake hose or line with parts specifically designed for braking systems. Parts that are not designed for brake systems can fail under the extreme pressure in the system.

The brake system lines can be replaced by disconnecting them at their mounting points along the underside of the vehicle and at the components that they attach to. Tighten brake lines-to-brake hoses to 10 ft. lbs. (15 Nm).

The brake hoses, which are rubber, are located at each wheel. They can be replaced by disconnecting them from the brake line and then from the caliper or wheel cylinder. Tighten all brake hose-to-wheel cylinder fittings to 14 ft. lbs. (19 Nm). Brake hoses-to-caliper connections should be tightened to 25 ft. lbs. (35 Nm).

Always bleed the brake system after replacing any of the components.

Bleeding the Brake System

♦ See Figures 23, 24 and 25

It is necessary to bleed the hydraulic system any time system has been opened or any other time air has become trapped within the fluid lines. It may be necessary to bleed the system at all four brakes if air has been introduced through a low fluid level or by disconnecting brake pipes at the master cylinder.

➡ **Vehicles equipped with Anti-Lock Brake Systems (ALB/ABS) require separate bleeding procedures if any of the ALB/ABS components have been removed or the lines disconnected. Refer to the procedure later in this section.**

If a line is disconnected at one wheel only, generally only that brake circuit needs bleeding, although bleeding all four wheels is always recommended. If lines are disconnected at any fitting between the master cylinder and the brake, the components fed by the disconnected pipe must be bled.

➡ **On vehicles with rear drum brakes, mis-adjusted brake shoes can cause a "long" pedal and reduced braking, giving the illusion of air in the lines. If you're trying to track down such a problem, adjust the rear brakes and test drive the car before bleeding the system.**

1. Start bleeding by filling the master cylinder reservoir to the MAX or FULL line with fresh brake fluid from a sealed container and keep it at least half full throughout the bleeding procedure.
2. It is strongly recommended that any time bleeding is to be performed all wheels be bled rather than just one or two. The correct order of bleeding is:
   - 1984–87 Models: left front, right rear, right front, left rear.
   - 1988–95 Models: right rear, left front, left rear, right front.
3. At the wheel to be bled, Place the correct size box-end or line wrench over the hex on the bleeder valve, then attach a tight-fitting hose over the bleeder opening. Allow the tube to hang submerged in a transparent container half full of clean brake fluid. The hose end must remain submerged in the fluid at all times.
4. Have an assistant pump the brake pedal several times slowly and hold it down.
5. Slowly unscrew the bleeder valve (¼–½ turn is usually enough). After the initial rush of air and fluid, tighten the bleeder and have the assistant slowly release the pedal.

➡ **DON'T allow your assistant to release the pedal while the bleeder valve is still loose or air may be drawn back into the system.**

6. Repeat until no air bubbles are seen flowing from the bleeder to the hose or container. If air is constantly appearing after repeated bleeding, the system must be examined for the source of the leak or loose fitting. When finished with a wheel, the bleeder screw should be tightened to 5–7 ft. lbs. (7–9 Nm).
7. Periodically check the reservoir on the master cylinder, topping it as needed to maintain the proper level. If the master cylinder level is allowed to drop significantly during this procedure, it is possible to draw air back into the system, thus defeating your bleeding efforts up to that point.
8. After bleeding, check the pedal for "sponginess" or vague feel. Repeat the bleeding procedure as necessary to correct. Top off the reservoir level.
9. Test drive the car to check proper brake function. But DO NOT attempt to move the car unless a firm brake pedal is felt.

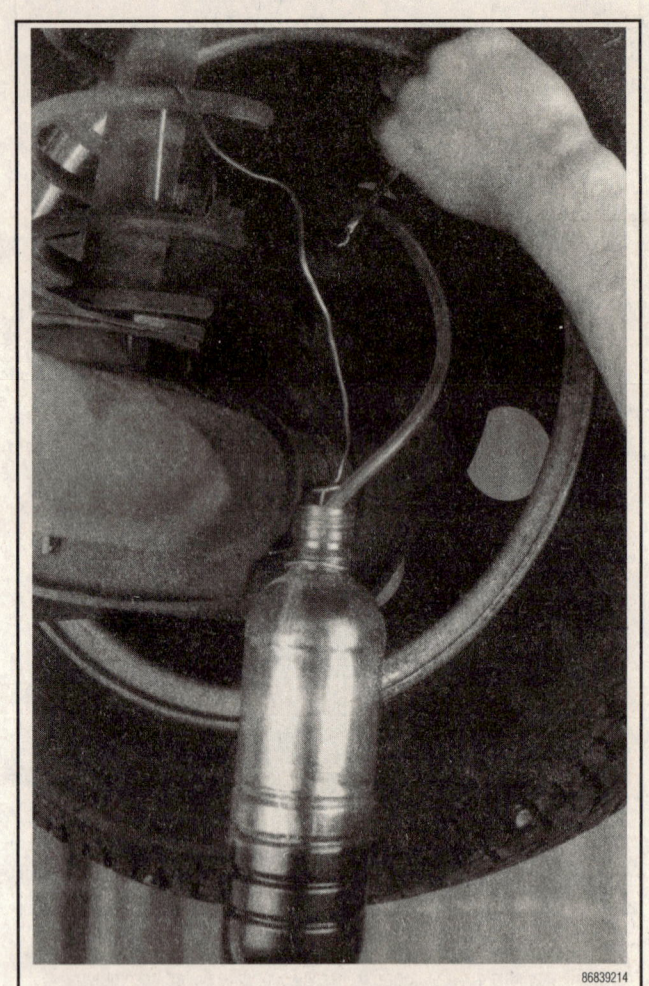

Fig. 23 Here is the quintessential homemade brake bleeder setup: a used plastic bottle, a clean hose submerged in the fluid and a stiff wire to securely fasten the arrangement to the underside of the car near the brakes

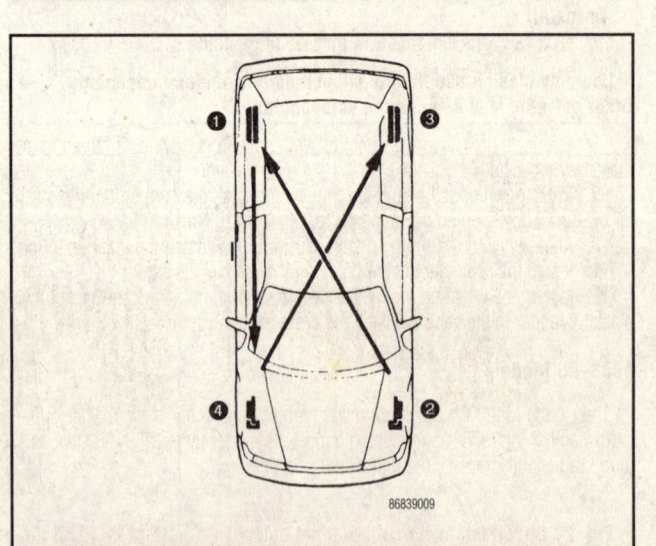

Fig. 24 Brake system bleeding sequence—1988–95 models

BRAKES 9-9

※※ WARNING

Brake fluid is adversely affected by contamination from dirt, automotive petroleum products and water. Contaminants can plug parts of the hydraulic system, causing rapid wear or swelling of rubber parts and lower the boiling point of the fluid. KEEP THE BRAKE FLUID CLEAN. NEVER reuse brake fluid which has been bled from a system.

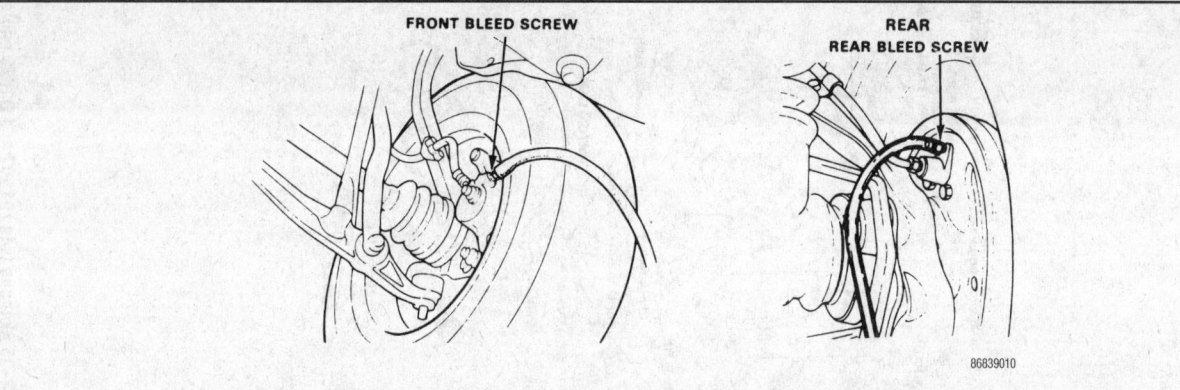

Fig. 25 Hose connected to bleeder—used in bleeding the brake system, the end of the hose is submerged in a jar of brake fluid

FRONT DISC BRAKES

The major components of the disc brake system are the brake pads, the caliper and the rotor (brake disc). The caliper is similar in function to the wheel cylinder used with drum brakes and the rotor is similar to the brake drum used in drum brakes.

The major difference between drum brakes and disc brakes is that with drum brakes, the wheel cylinder forces the brake shoes out against the brake drum to stop the vehicle, while with disc brakes, the caliper forces the brake pads inward to squeeze the rotor and stop the vehicle. The biggest advantage of disc brakes over drum brakes is that the caliper and brake pads enclose only a small portion of the rotor, leaving the rest of it exposed to outside air. This allows rapid heat dissipation, reduces brake fade and throws off water fast.

※※ CAUTION

Brake shoes may contain asbestos, which has been determined to cause cancer and lung disease. Never clean the brake surfaces with compressed air! Avoid inhaling any dust from any brake surface! When cleaning brake surfaces, use a commercially available brake cleaning fluid.

Brake Pads

REMOVAL & INSTALLATION

♦ See Figures 26 thru 32

1. Remove the master cylinder cover and, assuming the reservoir is full, remove half the quantity of brake fluid in the master cylinder.

➥Forcing the pistons back into the calipers will push the displaced brake fluid back into the master cylinder reservoir. Make sure the level is low enough to accommodate the added fluid.

Fig. 26 A large C-clamp is helpful tool to push the pistons back into the calipers—be careful as fluid will flow back to the master cylinder; make sure it's not full

Fig. 27 Unbolt the lower caliper pivot bolt to access the brake pads

Fig. 28 Pivot the caliper out of the way . . .

Fig. 29 . . . and remove the pads

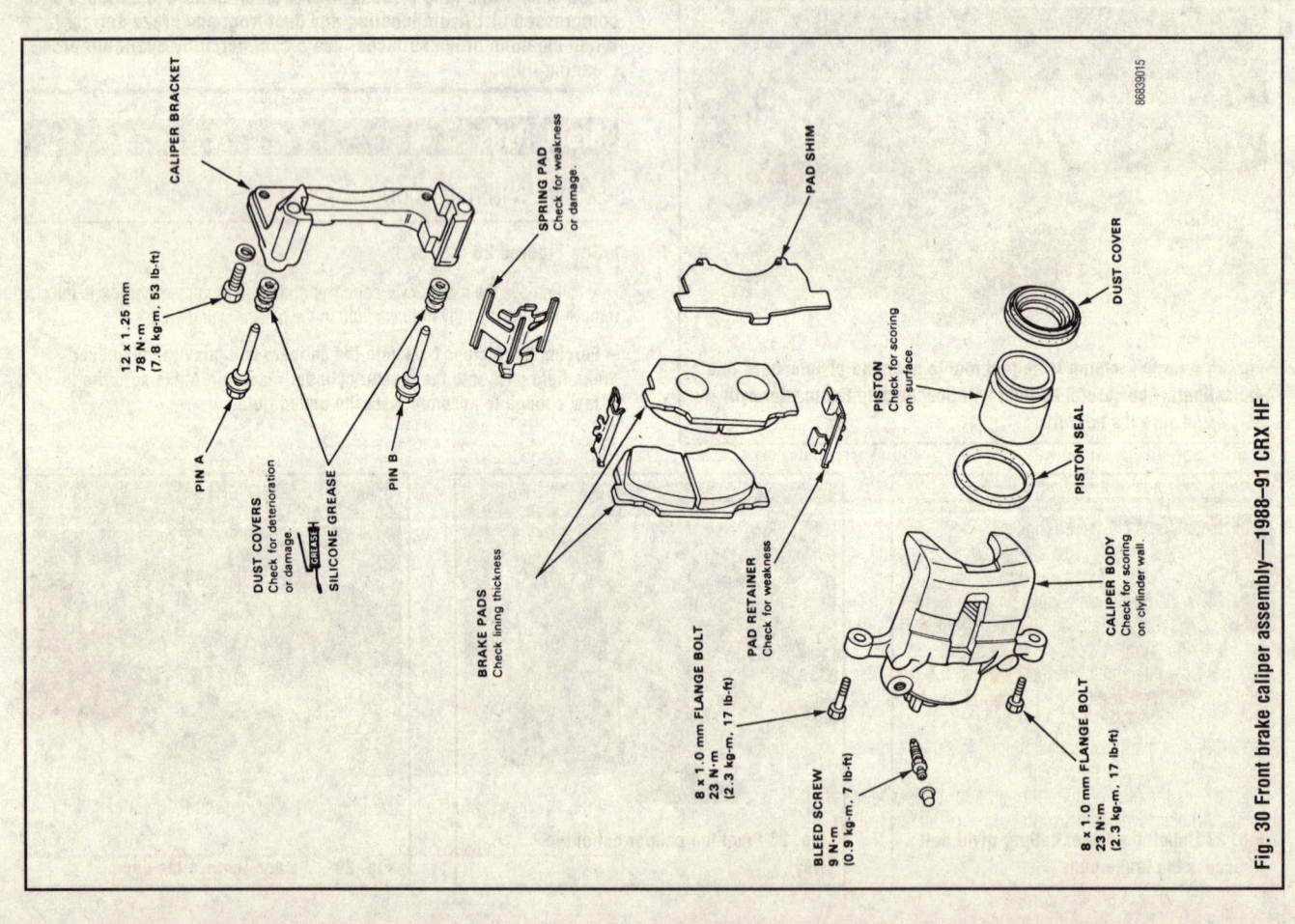

9-10 BRAKES

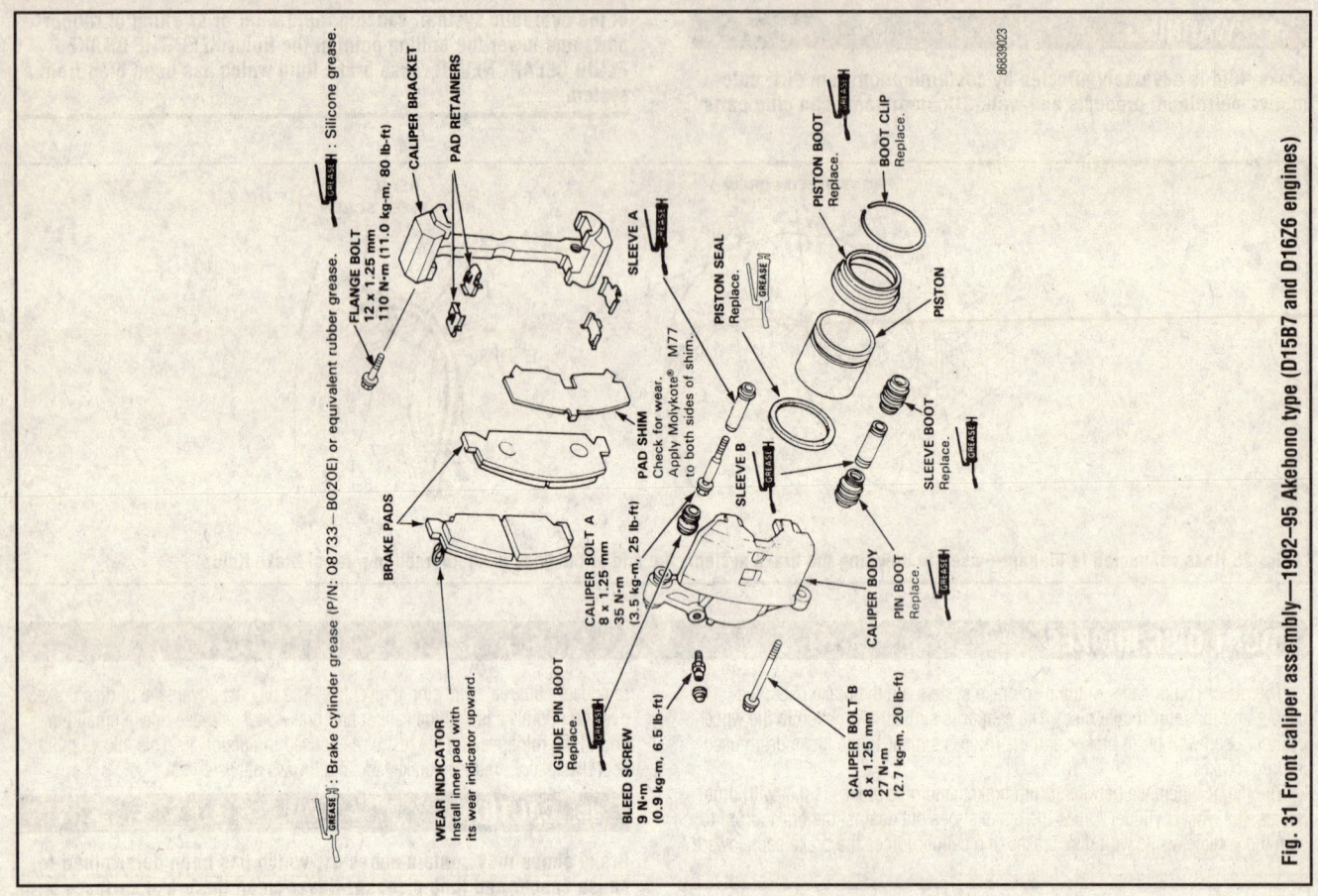

Fig. 31 Front caliper assembly—1992–95 Akebono type (D15B7 and D16Z6 engines)

Fig. 30 Front brake caliper assembly—1988–91 CRX HF

BRAKES 9-11

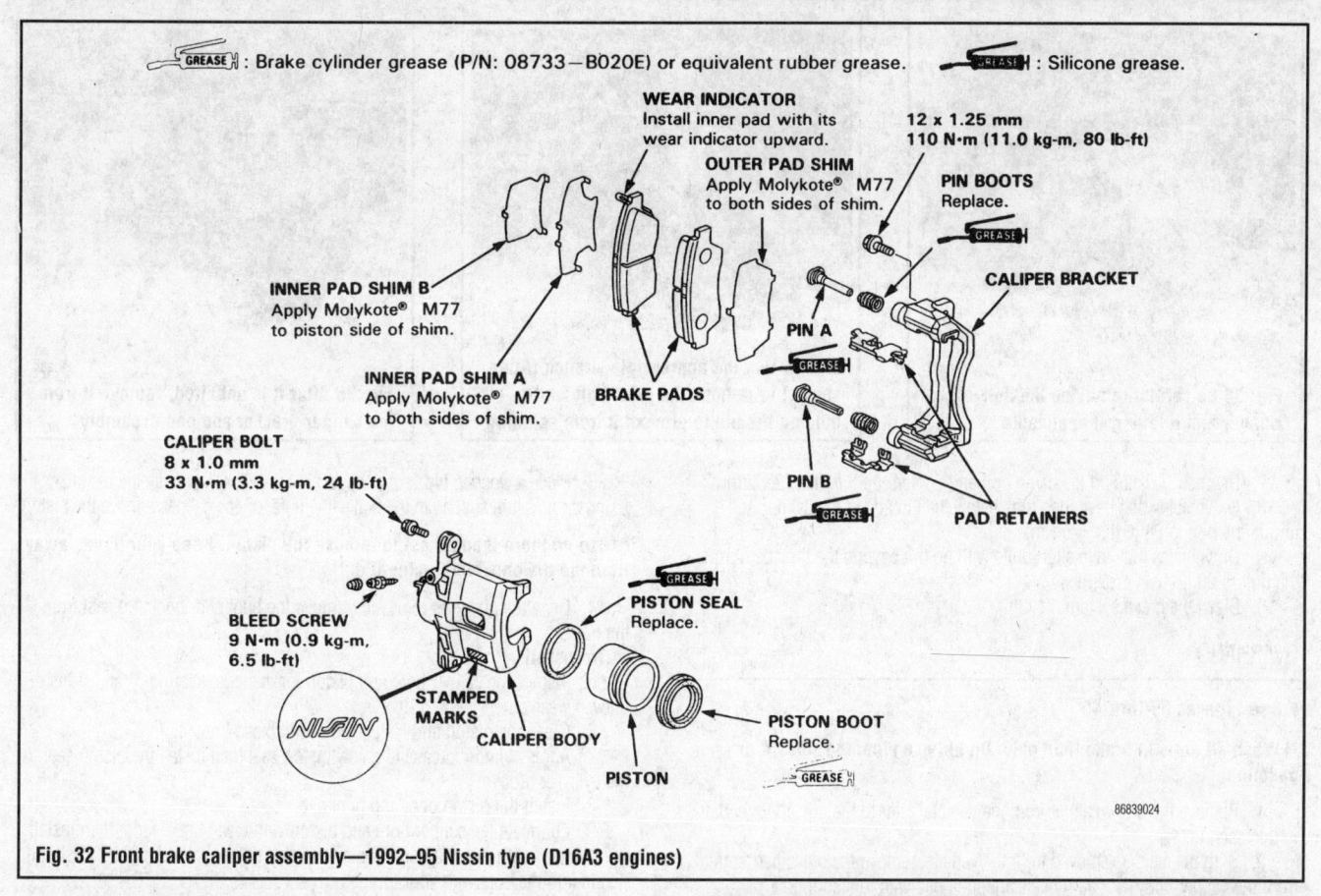

Fig. 32 Front brake caliper assembly—1992–95 Nissin type (D16A3 engines)

2. Raise and support the vehicle safely. Remove the tire and wheel assemblies.
3. As required, separate the brake hose clamp from the knuckle by removing the retaining bolts.
4. Remove the lower caliper retaining bolt and pivot the caliper out of the way.
5. Remove the pad shim and pad retainers. Remove the disc brake pads from the caliper.

To install:
6. Clean the caliper thoroughly; remove any rust. Check the brake rotor for grooves, cracks and proper thickness. Machine or replace, as necessary.
7. Install the pad retainers. Apply a thin coat of Molykote® or a suitable anti-seize disc brake pad lubricant to both sides of the shims and the back of the disc brake pads.

✹✹ CAUTION

Do not get any lubricant on the braking surface of the pad or rotor or the braking will be impaired. If necessary, use a fast-drying brake cleaning spray to dissolve all traces of lubricant.

8. Install the pads and shims.
9. Use a suitable tool to push in the caliper piston so the caliper will fit over the pads.
10. Pivot the caliper down into position and tighten the mounting bolts. For 1984–87 Civic models (all) tighten to: 13 ft. lbs (18 Nm) torque. For 1988–91 Civic EX 4-door, tighten the bolts to 24 ft. lbs. (33 Nm). For 1988–91 CRX DX (Std.) and Si, tighten the bolts to: 36 ft. lbs. (50 Nm). For 1988–95 Civic models (except CRX and and Civic 4D EX models), tighten the bolts to 20 ft. lbs. (27 Nm).
11. Connect the brake hose to the knuckle, if removed. Install the wheel and tire assembly and lower the vehicle.
12. Check the master cylinder and add fluid as required, then replace the master cylinder cover. Depress the brake pedal several times to seat the pads.

Brake Caliper

▶ See Figures 33, 34, 35 and 36

REMOVAL & INSTALLATION

1. Raise and safely support the vehicle.
2. Remove the front wheel and tire assembly.
3. Use a brake hose clamp or equivalent tool to gently pinch closed the rubber brake hose to prevent fluid loss when the caliper is unbolted.
4. Remove the banjo bolt and disconnect the brake hose from the caliper. Plug the hose.
5. Remove the mounting bolt(s) and the caliper.

To install:
6. Place the caliper over the installed pads and bolt to the pins in the caliper retainer.

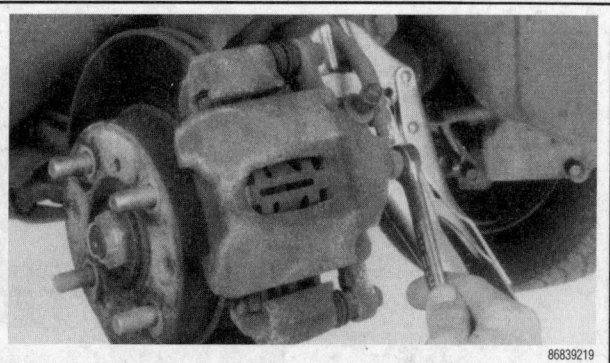

Fig. 33 Here the hose is clamped and the banjo bolt is being removed

9-12 BRAKES

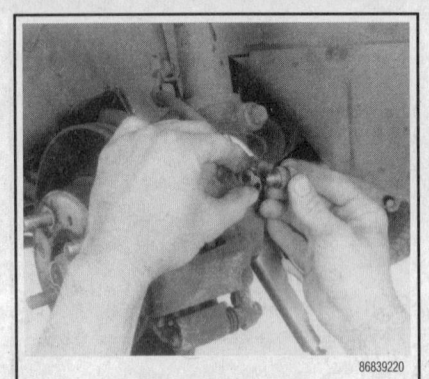

Fig. 34 Be careful removing the bolt, washer and retainer, if applicable

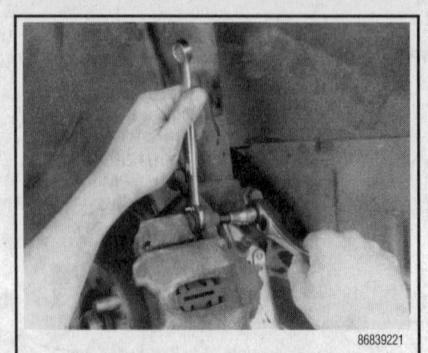

Fig. 35 Use the appropriate wrench (Allen shown) to remove the pivot bolt while holding the pin to prevent it from spinning

Fig. 36 After it is unbolted, remove it from the caliper bracket and pad assembly

  7. Use new gaskets on the banjo bolt and tighten to 25 ft. lbs. (35 Nm).
  8. On vehicles that have long pins below the threads of the caliper bolt, torque the bolt to 54 ft. lbs. (75 Nm).
  9. On vehicles that have short bolts with no pin beyond the threads, torque the bolt to 20 ft. lbs. (27 Nm).
  10. Bleed the brake system.

OVERHAUL

♦ See Figures 37 thru 45

➡ Wash all parts in brake fluid only. Do not use cleaning solvent or gasoline.

  1. Remove the brake caliper from the vehicle, refer to the procedure earlier in this section.
  2. With the caliper removed from the vehicle, remove the brake pad spring from the caliper body. Not all models are equipped with this spring.
  3. Place a wooden block or shop rag in the caliper, opposite the piston. Using air pressure applied to the brake line hole, on the caliper, remove the piston.

➡ Use no more than 30 psi to remove the piston. Keep your hands away from the piston when forcing it out.

  4. Once the piston is removed, remove the boot clip, boot and seal from the bore.

To install:
  5. Inspect the caliper bore and piston for damage, scoring or any signs of uneven wear.
  6. Clean the piston and caliper bore with clean brake fluid.
  7. Apply cylinder grease to a new piston seal then install the piston seal in the cylinder groove.
  8. Install the piston boot and boot clip.
  9. Lubricate the caliper bore and piston with clean brake fluid, then install the caliper piston into the bore. The dished end of the piston faces in. Push the piston into the bore with your thumbs.

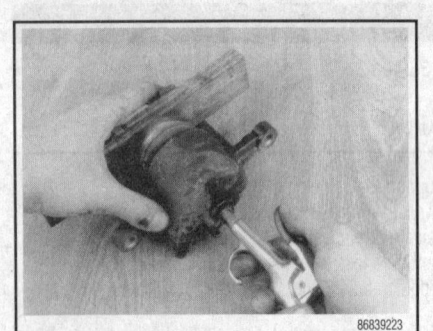

Fig. 37 Use low-pressure air to remove the piston from the caliper body and a wood block to safely control the piston as it is forced out of the bore

Fig. 38 With the compressed air nozzle just put aside, the piston is here being removed from the caliper

Fig. 39 An appropriate prytool helps in the removal of the boot clip

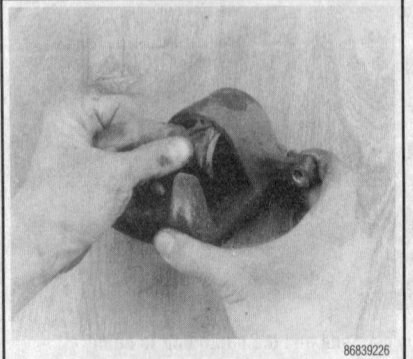

Fig. 40 Use caution as you pull the protective boot out of the groove

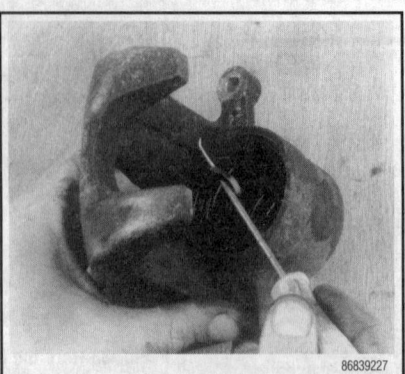

Fig. 41 Carefully manipulate the piston seal from the caliper using a small prytool

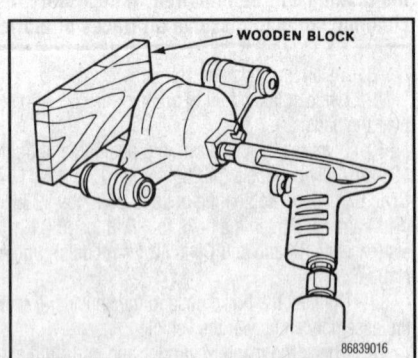

Fig. 42 Place a wood block in front of the piston and use 30 psi (207 kpa) of air pressure to remove the piston

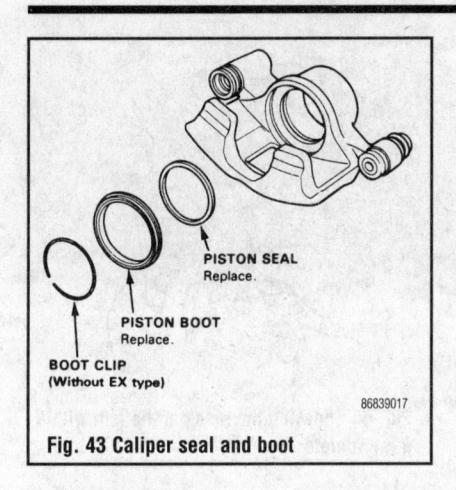

Brakes 9-13

Fig. 43 Caliper seal and boot

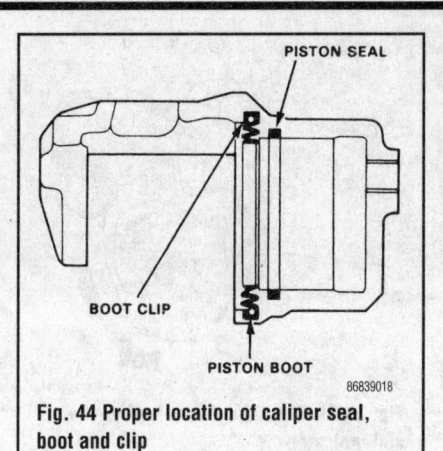

Fig. 44 Proper location of caliper seal, boot and clip

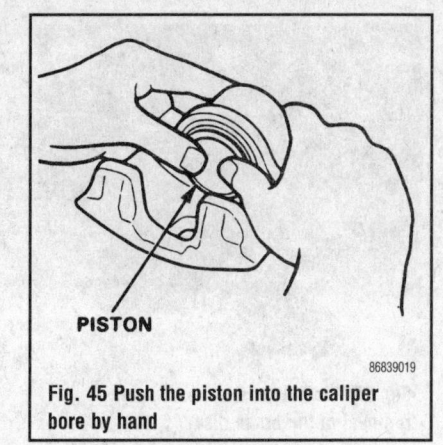

Fig. 45 Push the piston into the caliper bore by hand

10. Reinstall the caliper on the vehicle. Fill and bleed the brake system with fresh brake fluid.

Brake Disc (Rotor)

REMOVAL & INSTALLATION

♦ See Figures 46 thru 53

1. Raise and support the front of the vehicle on jackstands. Remove the front wheels.
2. Remove the caliper assembly; do not let the caliper assembly hang by the brake hose, instead support it with a piece of wire looped around the front spring or control arm.
3. Remove the retaining screw from the brake rotor and remove the rotor from the hub.

Fig. 46 The (front) rotor as it appears upon removal of the wheel

Fig. 47 Tie the caliper up out of the way so its weight is not supported by the rubber brake hose

Fig. 48 A penetrating lubricant may be used to loosen the screw

Fig. 49 An impact driver with a No. 3 Phillips bit is the best way to remove the set screw

Fig. 50 Spray-type penetrating lubricant may help free the disc—use it minimally and be sure to not spray the brake pads

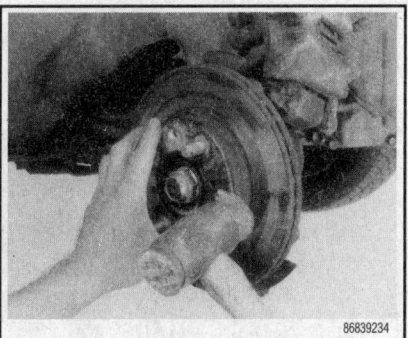

Fig. 51 A soft mallet may help free the disc, but if the disc does not come free in your hands at this point . . .

Fig. 52 . . . proper sized bolts can be alternately threaded into the disc holes in order to press the rotor off

9-14 BRAKES

Fig. 53 The hub as it appears upon removal of the brake disc

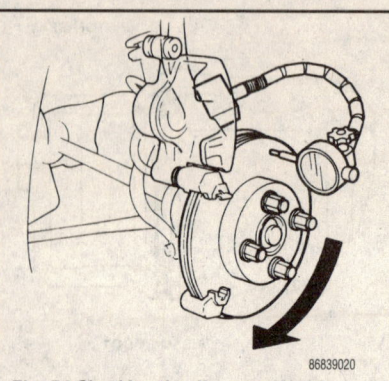

Fig. 54 Checking the disc runout with a dial indicator

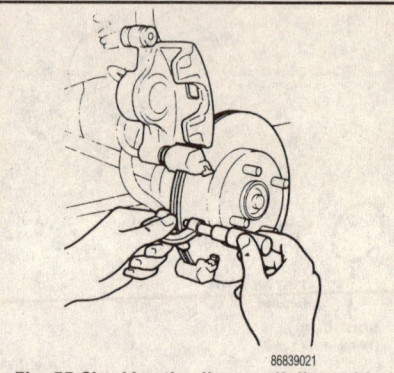

Fig. 55 Checking the disc parallelism with a micrometer

➥An impact driver, No. 3 phillips screwdriver bit, and a hammer may be needed to remove the retaining screw. It is fairly easy to destroy the screw slots using an ordinary phillips screwdriver to remove the tightly driven screw (see photo).

4. To install the disc, replace the disc on the hub, screw in the retaining screw and replace the caliper assembly.

INSPECTION

▶ See Figures 54 and 55

1. The brake disc develops circular scores after long or even short usage when there is frequent braking. Excessive scoring not only causes a squealing brake, but also shortens the service life of the brake pads. However, light scoring of the disc surface, not exceeding 0.0152 in. (0.38mm) in depth, will result from normal use and is not detrimental to brake operation.

➥Differences in the left and right disc surfaces can result in uneven braking.

2. Disc runout is the movement of the disc from side-to-side. Position a dial indicator in the middle of the pad wear area and turn the disc, while checking the indicator. If disc runout exceeds 0.004 in. (0.1mm), replace the disc.

3. Disc parallelism is the measurement of variations in disc thickness at several locations on the disc circumference. To measure parallelism, place a mark on the disc and measure the disc thickness with a micrometer. Repeat this measurement at eight (8) equal increments on the circumference of the disc. If the measurements vary more than .0028 in. (0.07mm), replace the disc.

➥Only the outer portion of the disc can be checked while installed on the vehicle. If the installed parallelism check is within specifications, but you have reason to suspect that parallelism is the problem, remove the disc and repeat the check using the center of pad wear for a checking point.

REAR DRUM BRAKES

▶ See Figures 56, 57 and 58

✳ CAUTION

Brake shoes may contain asbestos, which has been determined to cause cancer and lung disease. Never clean the brake surfaces with

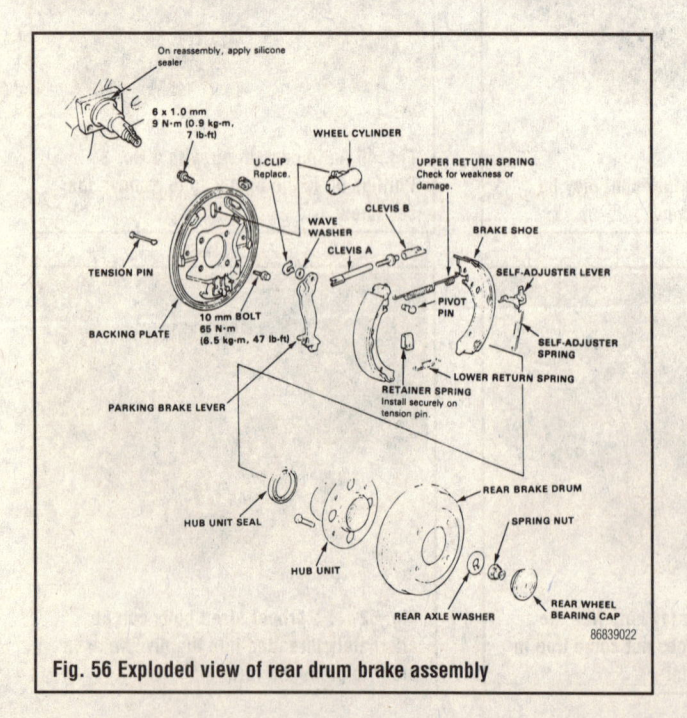

Fig. 56 Exploded view of rear drum brake assembly

Fig. 57 Screwing in two bolts and tightening alternately will press off the brake drum

Fig. 58 Removing the rear brake drum

BRAKES 9-15

compressed air! Avoid inhaling any dust from any brake surface! When cleaning brake surfaces, use a commercially available brake cleaning fluid.

All Honda Civic models not equipped with rear disc brakes employ a leading/trailing type of drum brake, in which there are two curved brake shoes supported by an anchor plate and activated by a wheel cylinder. When the brake pedal is depressed and hydraulic pressure is delivered to the wheel cylinder, the wheel cylinder expands to force the shoes against the drum.

Friction between the brake shoes and the drum causes the vehicle to slow and stop. When the brake pedal is released, the brake shoe return springs move the brakes away from the drum. If the lining on the brakes becomes contaminated or the lining/drum becomes grooved, the engagement of the brakes and drum will become very harsh, causing the brakes to lock up and/or squeal. If the brake shoes on one wheel contact the drum before the same action occurs in the other wheels, the brakes will pull to one side when applied.

Brake Drums

REMOVAL & INSTALLATION

1. Raise and safely support the vehicle.
2. Remove the rear wheel and tire assembly.
3. Remove the brake drum. You may need to screw two bolts into the threaded holes and tighten alternately to press the rear drum off.

※※ WARNING

Do not use a hammer to remove the rear brake drum.

4. To install, replace the drum on the hub assembly. Adjust the brakes if necessary.
5. Install the wheel and tire, and lower the vehicle.

INSPECTION

Check the drum for cracks and the inner surface of the shoe for excessive wear and damage. The inner diameter (I.D.) of the drum should be no more than specifications, nor should the drum be more than .0004 in. (0.10mm) out-of-round.

Brake Shoes

REMOVAL & INSTALLATION

▶ See Figures 59 thru 66

1. Raise and safely support the vehicle.
2. Remove the wheel and tire assemblies and the brake drums.

➡ A spray-type brake cleaning fluid allows a cleaner job and may be necessary to remove brake dust and leaking fluid that has accumulated on the brake assembly.

3. Remove the tension pins by pushing the retainer spring and turning them.
4. Lower the brake shoe assembly and remove the lower return spring.
5. Remove the brake shoe assembly.
6. Disconnect the parking brake cable from the lever.
7. Remove the upper return spring, self-adjuster lever and self-adjuster spring. Separate the brake shoes.
8. Remove the wave washer, parking brake lever and pivot pin from the brake shoe by removing the U-clip.

To install:

9. Apply brake cylinder grease to the sliding surface of the pivot pin and insert the pin into the brake shoe.
10. Install the parking brake lever and wave washer on the pivot pin and secure with the U-clip.

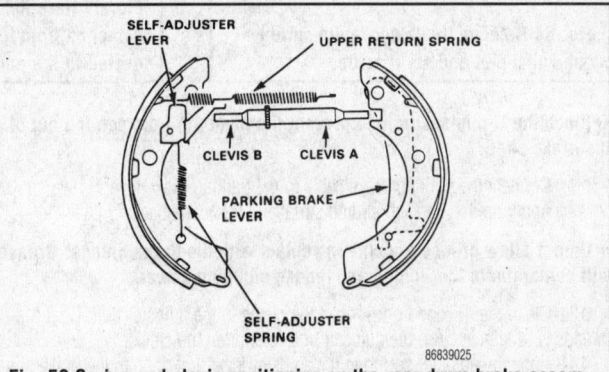

Fig. 59 Spring and clevis positioning on the rear drum brake assembly

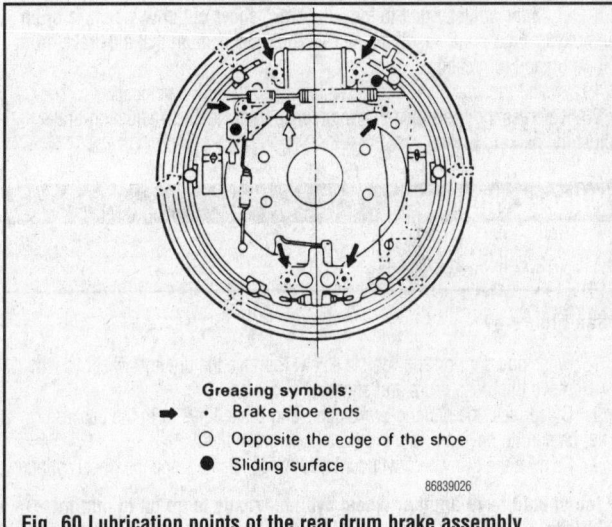

Fig. 60 Lubrication points of the rear drum brake assembly

Fig. 61 A can of fast-evaporating brake cleaner is useful for cutting down the mess before disassembly

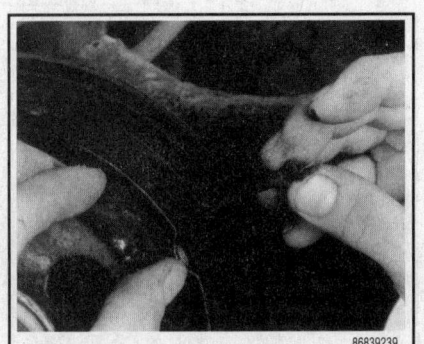

Fig. 62 Removing the tension pins is accomplished by depressing the retainer spring (in left hand) and turning

Fig. 63 The lower return spring must be detached to remove the brake shoes

9-16 BRAKES

Fig. 64 Remove the lower return spring once uncliped and set it aside

Fig. 65 Here the brake shoe assembly detached from its mounting but still connected to the emergency brake cable

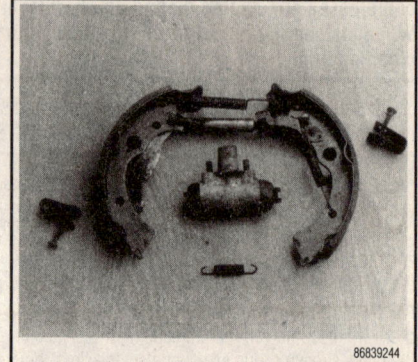

Fig. 66 The complete brake shoe assembly and the wheel cylinder which actuates it

➡ Pinch the U-clip securely to prevent the pivot pin from coming out of the brake shoe.

11. Connect the parking brake cable to the parking brake lever.
12. Apply grease on each sliding surface of the brake backing plate.

➡ Do not allow grease to come in contact with the brake linings. Grease will contaminate the linings and reduce stopping power.

13. Clean the threaded portions of the clevis of the adjuster bolt. Coat the threads with grease. Turn the adjuster bolt to shorten the clevis.
14. Hook the adjuster spring to the adjuster lever first, then to the brake shoe.
15. Install the adjuster bolt/clevis assembly and the upper return spring.
16. Install the brake shoes to the backing plate.
17. Install the lower return spring, the tension pins and retaining springs.
18. Turn the adjuster bolt to force the brake shoes out until the brake drum will not easily go on. Back off the adjuster bolt just enough that the brake drum will go on and turn easily.
19. Install the wheel and tire assemblies and lower the vehicle.
20. Depress the brake pedal several times to set the self-adjusting brake. Adjust the parking brake.

Wheel Cylinders

REMOVAL & INSTALLATION

◆ See Figure 67

1. Raise and support the vehicle safely. Remove the tire and wheel assembly. Remove the brake drum and shoes.
2. Disconnect the parking brake cable and brake lines from the backing plate; be sure to have a drip pan to catch the brake fluid.
3. Remove the two wheel cylinder-to-backing plate nuts and the wheel cylinder.

➡ You should have another wheel cylinder ready to go on to minimize fluid loss.

To install:
4. Apply a thin coat of grease to the grooves of the wheel cylinder piston and the sliding surfaces of the backing plate.
5. Replace the wheel cylinder and the brake assembly parking brake cable.

OVERHAUL

◆ See Figures 68 thru 73

Remove the wheel cylinder dust seals from the grooves to permit the removal of the cylinder pistons.

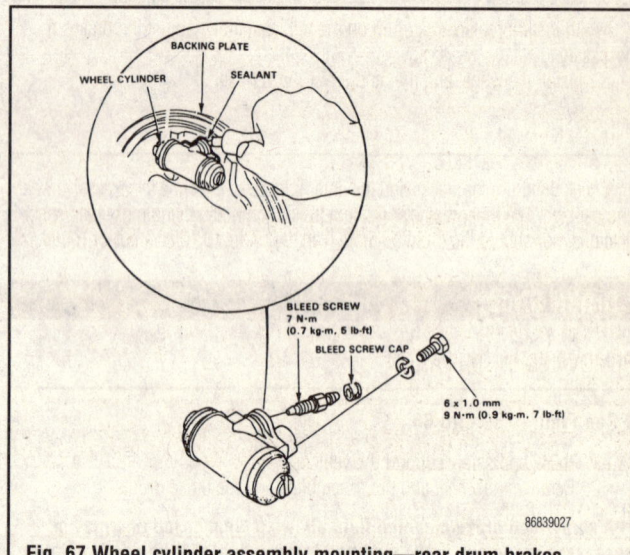

Fig. 67 Wheel cylinder assembly mounting—rear drum brakes

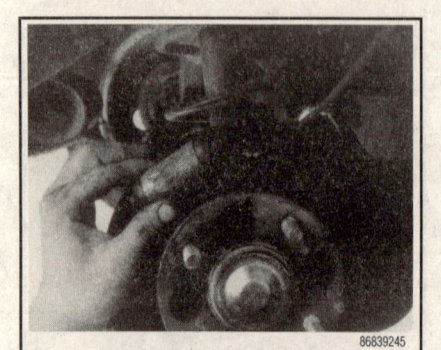

Fig. 68 Removing the disconnected and unbolted wheel cylinder from the wheel assembly

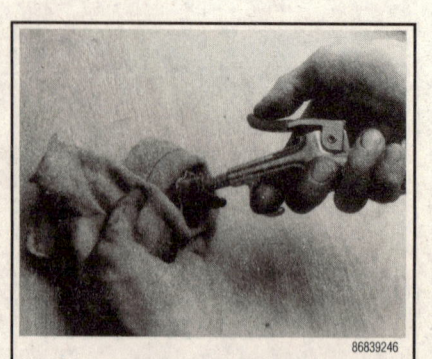

Fig. 69 Remove the wheel cylinder pistons with compressed air while holding them to prevent blowing them out

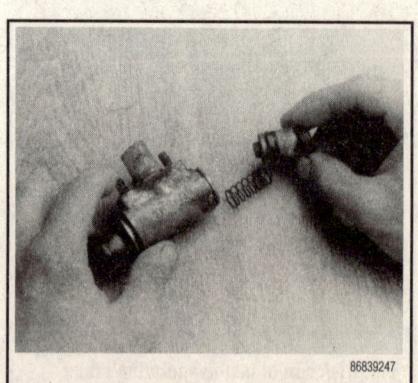

Fig. 70 Removing one of the wheel cylinder pistons

BRAKES 9-17

Wash all parts in fresh brake fluid and check the cylinder bore and pistons for scratches and other damage, replacing where necessary. Check the clearance between the piston and the cylinder bore by taking the difference between the piston diameter and the bore diameter. The specified clearance is 0.0008–0.004 in. (0.020–0.100mm).

When assembling the wheel cylinder, apply a coat of brake fluid to the pistons, piston cups and cylinder walls.

Fig. 71 Removing the boot from the piston

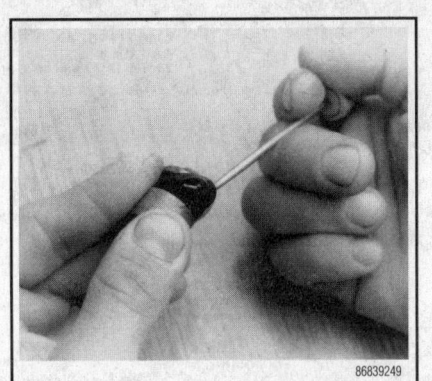

Fig. 72 Removing the fluid seal from the piston with an appropriate prytool

Fig. 73 The complete wheel cylinder assembly with both pistons removed and one disassembled

REAR DISC BRAKES

The major components of the disc brake system are the brake pads, the caliper and the rotor (disc). The caliper is similar in function to the wheel cylinder used with drum brakes and the rotor is similar to the brake drum used in drum brakes.

The major difference between drum brakes and disc brakes is that with drum brakes, the wheel cylinder forces the brake shoes out against the brake drum to stop the vehicle, while with disc brakes, the caliper forces the brake pads inward to squeeze the rotor and stop the vehicle. The biggest advantage of disc brakes over drum brakes is that the caliper and brake pads enclose only a small portion of the rotor, leaving the rest of it exposed to outside air. This aids in rapid heat dissipation, reducing brake fade and it throws off water fast as well.

✺✺ CAUTION

Brake shoes contain asbestos, which has been determined to cause cancer and lung disease. Never clean the brake surfaces with compressed air! Avoid inhaling any dust from any brake surface! When cleaning brake surfaces, use a commercially available brake cleaning fluid.

Brake Caliper

REMOVAL & INSTALLATION

♦ See Figure 74

1. Raise and safely support the vehicle.
2. Remove the rear wheel and tire assembly.
3. Remove the caliper shield.
4. Disconnect the parking brake cable from the lever on the caliper by removing the lock pin.
5. Remove the banjo bolt and disconnect the brake hose from the caliper. Plug the hose.
6. Remove the two caliper mounting bolts and the caliper from the bracket.

To install:

7. Replace the caliper and torque the caliper bracket bolts to 28 ft. lbs. (39 Nm).
8. Use new gaskets on the banjo bolt and tighten to 25 ft. lbs. (35 Nm).
9. Replace the parking brake cable and bleed the brakes.

Brake Pads

REMOVAL & INSTALLATION

1. Remove the master cylinder cover and remove half the quantity of brake fluid so the reservoir is half-full.
2. Raise and safely support the vehicle. Remove the rear wheel and tire assemblies.
3. Remove the two caliper mounting bolts and the caliper from the bracket.
4. Remove the pads, shims and pad retainers.

To install:

5. Clean the caliper thoroughly; remove any rust. Check the brake rotor for grooves or cracks and machine or replace, as necessary.
6. Install the pad retainers. Apply Molykote® or a suitable disc brake pad lubricant to both surfaces of the shims and the back of the disc brake pads. Do not get any lubricant on the braking surface of the pad.
7. Install the pads and shims.
8. Use a suitable tool to rotate the caliper piston clockwise into the caliper bore, enough to fit over the brake pads. Lubricate the piston boot with silicone grease to avoid twisting the piston boot.

➡ **The piston to the rear disc brake caliper must be screwed back into the bore. The correctly fitting tool is commercially available and should be sought to ensure you will be able to complete this procedure.**

9. Install the brake caliper, aligning the cutout in the piston with the tab on the inner pad. Tighten the mounting bolts.
10. Install the wheel and tire assemblies and lower the vehicle.
11. Check the fluid in the master cylinder and add, as required, then replace the master cylinder cover. Depress the brake pedal several times to seat the pads.

Brake Disc

REMOVAL & INSTALLATION

1. Raise and safely support the vehicle.
2. Remove the wheel and tire assembly.

9-18 BRAKES

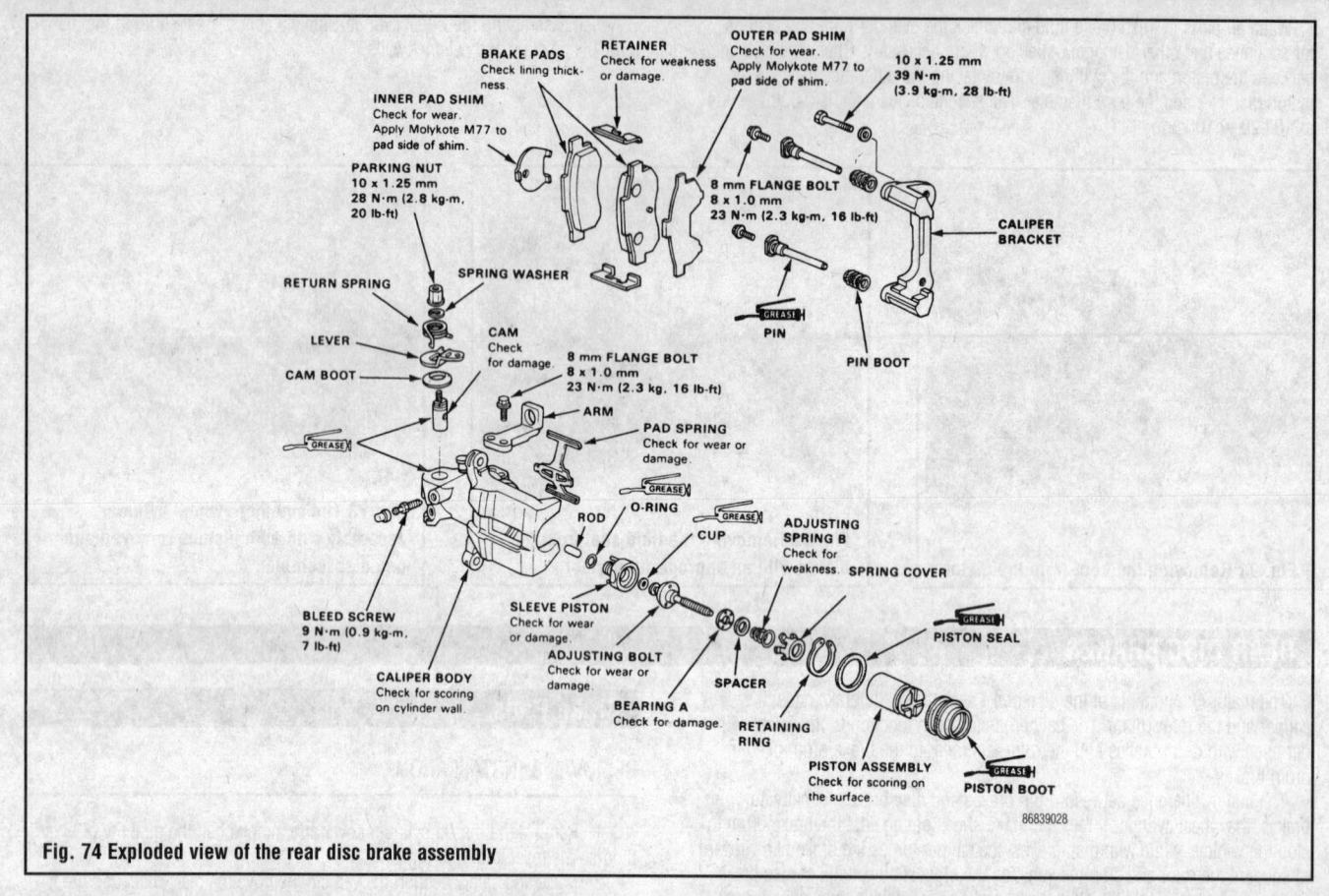

Fig. 74 Exploded view of the rear disc brake assembly

3. Disconnect the caliper from the caliper bracket. Support the caliper out of the way with a length of wire. Do not allow the caliper to hang from the brake hose.
4. Remove the caliper bracket.
5. Remove the two 6mm screws and the brake disc. If the brake disc is difficult to remove, install two 8mm bolts into the threaded holes and tighten them evenly to prevent cocking the rotor.
6. Install the rotor in position. Install the two 6mm retaining screws.
7. Install the brake caliper. Install the tire and wheel assembly.

INSPECTION

▶ See Figures 75 and 76

1. The brake disc develops circular scores after long or even short usage when there is frequent braking. Excessive scoring not only causes a squealing brake, but also shortens the service life of the brake pads. However, light scoring of the disc surface, not exceeding 0.0152 in. (0.38mm) in depth, will result from normal use and is not detrimental to brake operation.

➥Differences in the left and right disc surfaces can result in uneven braking.

2. Disc runout is the movement of the disc from side-to-side. Position a dial indicator in the middle of the pad wear area and turn the disc, while checking the indicator. If disc runout exceeds 0.006 in. (0.15mm), replace the disc.
3. Disc parallelism is the measurement of variations in disc thickness at several locations on the disc circumference. To measure parallelism, place a mark on the disc and measure the disc thickness with a micrometer. Repeat this measurement at eight (8) equal increments on the circumference of the disc. If the measurements vary more than 0.0006 in. (0.015mm), replace the disc.

➥Only the outer portion of the disc can be checked while installed on the vehicle. If the installed parallelism check is within specifications but you have reason to suspect that imperfect parallelism is the problem, remove the disc and repeat the check using the center of pad wear for a checking point.

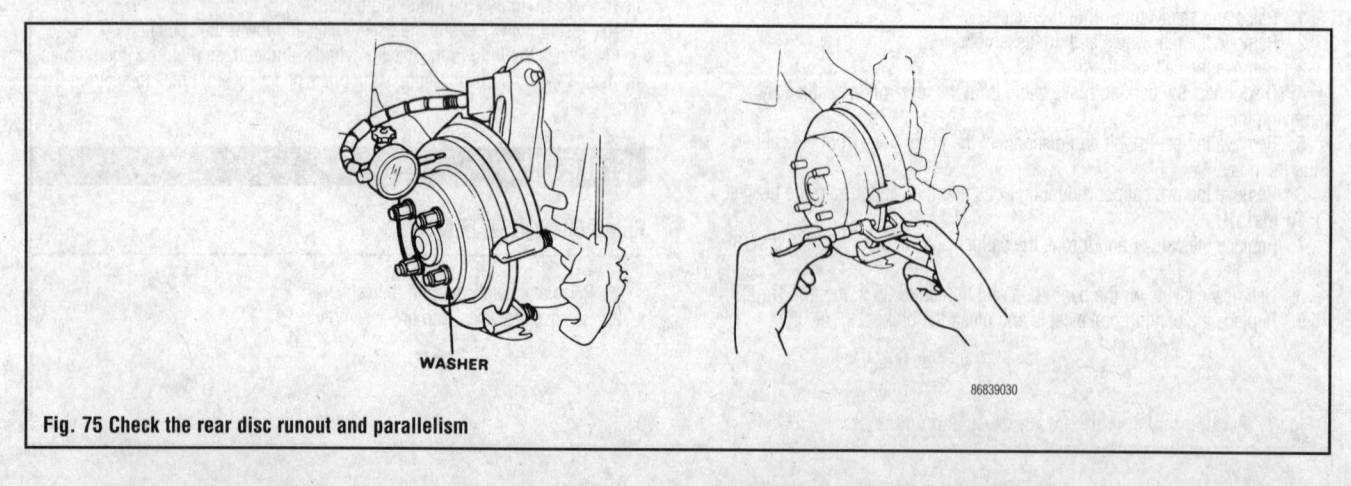

Fig. 75 Check the rear disc runout and parallelism

BRAKES 9-19

Fig. 76 Removing the rear disc brake rotor—use two 8mm bolts to assist the removal by screwing them in alternately to press off the rotor

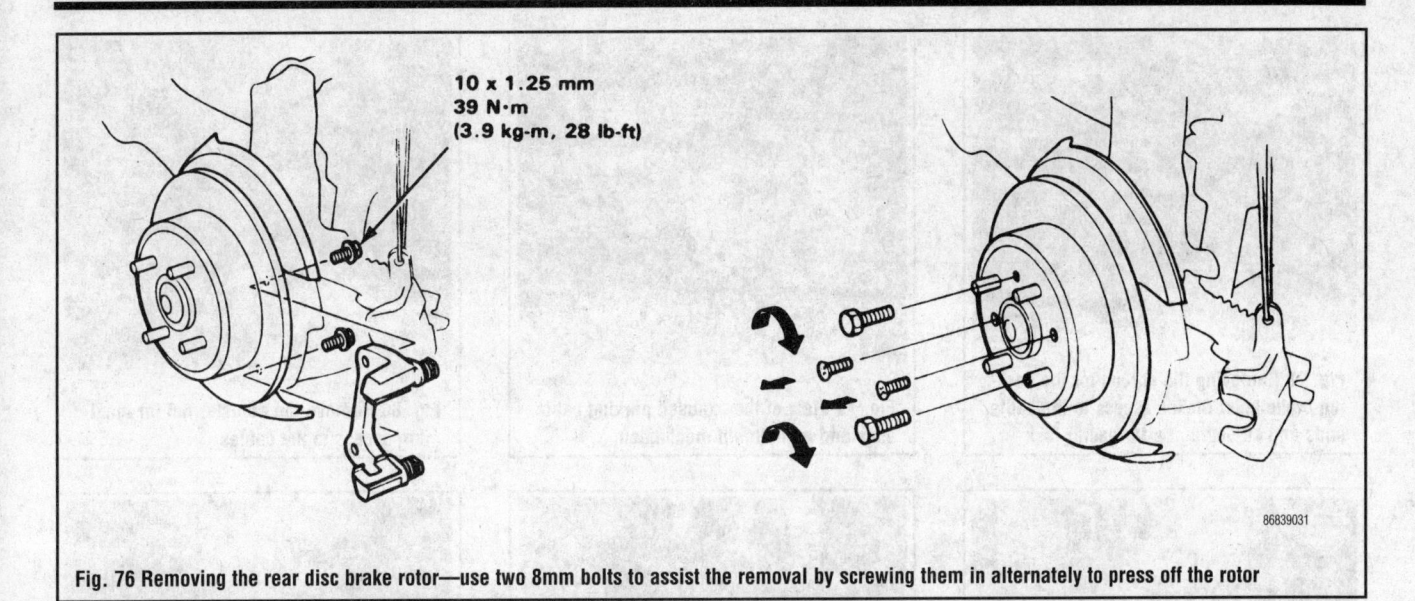

PARKING BRAKE

Cable

♦ See Figure 77

The parking brake is a mechanical type which applies braking force to the rear wheels, through the rear brake shoes (disc brake pads on rear disc equipped models). The cable, which is attached to the tail end of the parking brake lever, extends to the equalizer and to both rear brakes. When the lever is pulled, the attached cable pulls both the both parking brake arms fitted to the brake shoes. On rear disc equipped models, the cables are connected to a lever on the caliper.

REMOVAL & INSTALLATION

♦ See Figures 78 thru 83

1. Remove the access cover at the rear of the console. Loosen the adjusting nut until the cable ends can be disconnected from the equalizer.

➥On some models removing the shroud surrounding the parking brake lever may give greater access to the parking brake cable ends.

2. Raise and safely support the vehicle.
3. Remove the rear wheel and tire assemblies.
4. On disc brake equipped vehicles, pull out the lock pin, remove the clevis pin and remove the clip.
5. On drum brake equipped vehicles, remove the brake drum and brake shoes. Disconnect the cable from the backing plate.
6. Detach the cables from the cable guides or remove the cable guides if necessary and slide out the cables from the vehicle.

To install:

7. Connect the cables to the cable guides. Connect the cables at the brake drum and install the brake shoes. On rear disc equipped models, connect the cable to the caliper lever and install the lock pin and clevis pin.
8. Install the brake drum.
9. Install the wheel and tire assembly. Connect the cable at the equalizer assembly.
10. Adjust the parking brake.

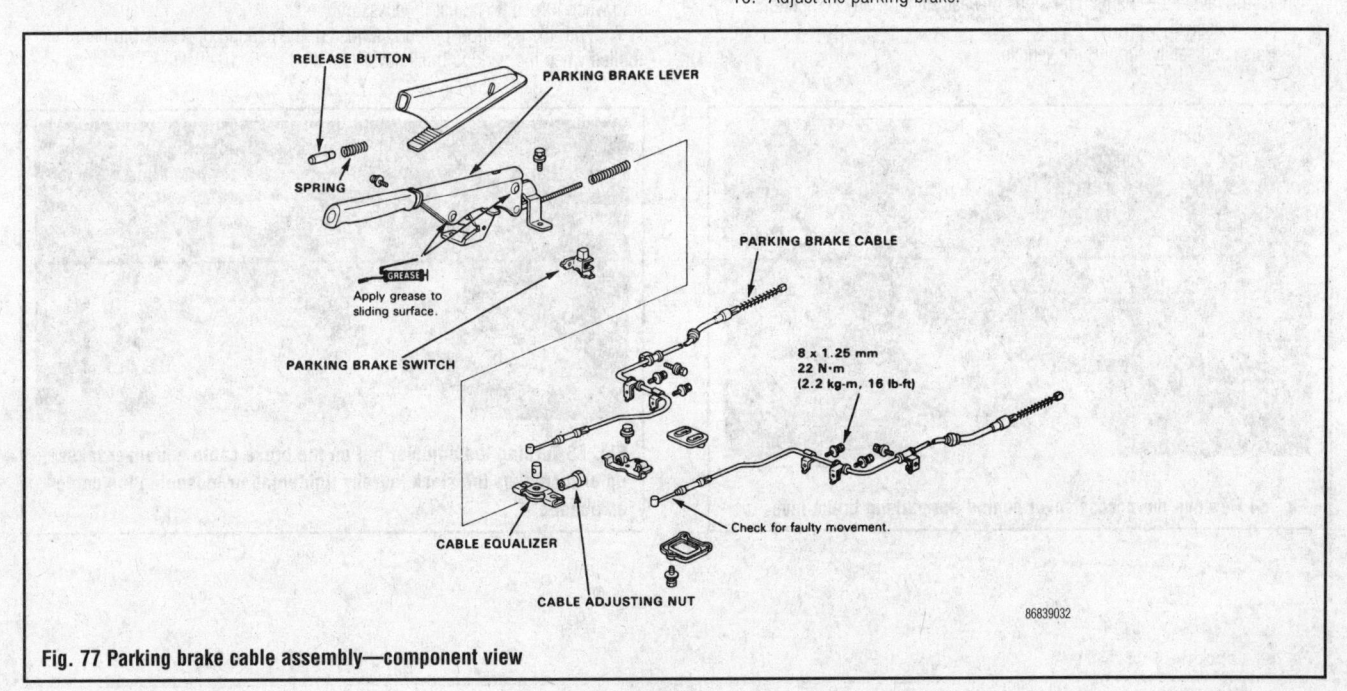

Fig. 77 Parking brake cable assembly—component view

9-20 BRAKES

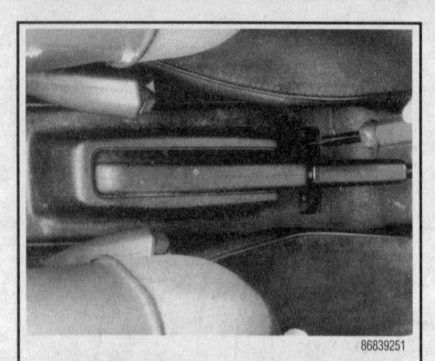

Fig. 78 Removing the shroud for the parking brake lever makes access to the cable ends and adjustment a far-easier task

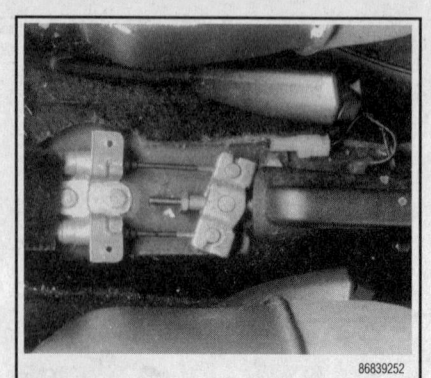

Fig. 79 View of the exposed parking cable ends and adjustment mechanism

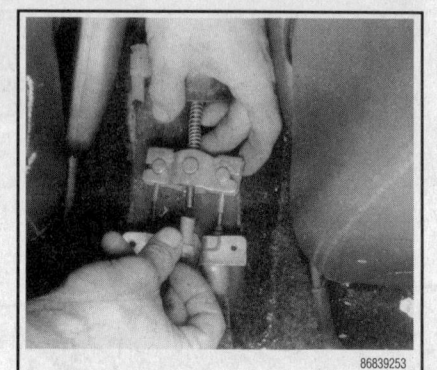

Fig. 80 Remove the adjuster nut for maximum access to the cables

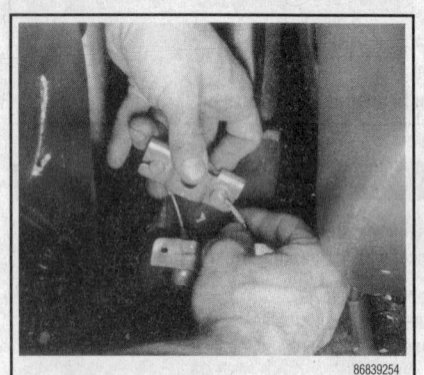

Fig. 81 Remove the barrel ends from the parking brake cables by turning and lifting

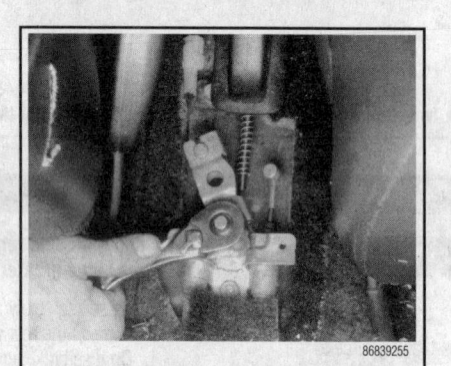

Fig. 82 Unbolting the cable guides in this case makes removal (and replacement) easier

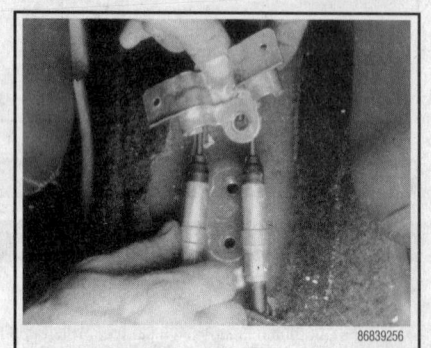

Fig. 83 Free at last: the cables are ready for removal—assuming they are detached at the rear brakes as well

ADJUSTMENT

▶ See Figures 84 and 85

Inspect the cables for excess wear, signs of fraying or any damage which may hamper proper operation. The cables should not bind through their travel. If a cable is worn to the point of fraying, its breakage is just a matter of time. Replace any cable which show signs of damage

Check the ratchet mechanism for signs of wear. Replace any components which are excessively worn.

1. Raise and safely support the vehicle.
2. On rear disc brake equipped vehicles, make sure the lever of the rear brake caliper contacts the brake caliper pin.
3. On drum brake equipped vehicles, make sure the rear brakes are properly adjusted.
4. Pull the parking brake lever up one notch.
5. Remove the access cover at the rear of the console and tighten the adjusting nut until the rear wheels drag slightly when turned.
6. Release the parking brake lever and check that the rear wheels do not drag when turned. Readjust if necessary.
7. With the equalizer properly adjusted, the parking brake should be fully applied when the parking brake lever is pulled up 6–10 clicks.

Fig. 84 Remove the access cover behind the parking brake lever

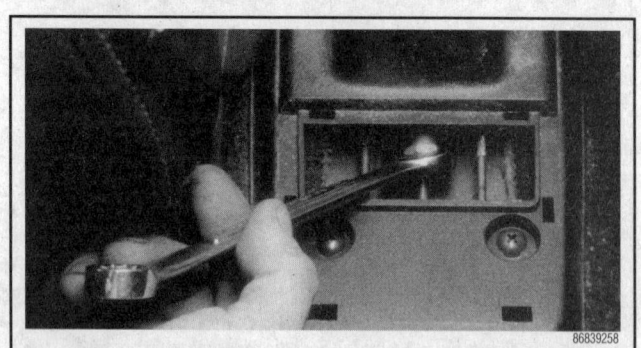

Fig. 85 Turning the adjuster nut on the brake cable equalizer takes up or increases the slack thereby tightening or loosening the cable as needed

BRAKES 9-21

ANTI-LOCK BRAKES (ALB/ABS)

Description and Operation

♦ See Figure 86

The Anti-Lock Brake (ALB/ABS) system is an answer to the real-world problem of panic stops and braking on slippery or uneven pavement. The need for brakes that can use what traction is available to stop the car without locking-up becomes apparent during hazardous braking situations. These situations can cause the front, rear or all four wheels to skid or lock on a conventional brake system. If the rear wheels skid, stability is lost or compromised and if the front wheels skid, it is maneuverability that is affected. Beyond theory, what often happens when braking under such hazardous conditions is simply an out-of-control four-wheel lock-up.

Honda's ALB/ABS is a computer-controlled system that augments otherwise conventional hydraulic brakes by monitoring each wheel's rotational speed as compared to vehicle road speed. When during braking it detects one or more wheels are rotating slower than the rate of travel, it compensates by modulating the brakes, allowing the car to slow but preventing the impending lock-up. In many cases, the goal to maximize stability and maneuverability under adverse braking conditions is met. The system became available on certain models in 1992.

There are conditions, however, for which the ALB/ABS system provides no benefit. Hydroplaning is possible when the tires ride on a film of water, losing contact with the paved surface. And the same is possible on snow, ice and loose terrain. This renders the vehicle totally uncontrollable until road contact is regained. Extreme steering maneuvers at high speed or cornering beyond the limits of tire adhesion can result in skidding which is independent of vehicle braking. For this reason, the system is named "anti-lock" rather than anti-skid.

Under normal conditions, the ALB/ABS system functions in the same manner as a conventional brake system and is not able to be detected by the operator. The system is a combination of electrical, mechanical and hydraulic components, working together to control the flow of brake fluid to the wheels when necessary.

The ALB/ABS control unit is the electronic brain of the system, receiving and interpreting signals from the wheel sensors which read and send a magnetic signal generated by the gear pulsers, toothed rings that rotate with the wheels. The unit will enter anti-lock mode when the wheel sensors detect impending wheel lock at any wheel, and will immediately control the brake line pressures to the affected wheel(s) by issuing output signals to the Modulator Unit.

The modulator unit contains solenoid valves which react to the signals from the control unit. The solenoids controls brake fluid pressure to the wheels. They work by allowing brake line pressure to build according to brake pedal pressure, hold (by isolating the system from the pedal and maintaining current pressure) or decrease (by isolating the pedal circuit and bleeding some fluid from the line).

The decisions regarding these functions are made very rapidly and each solenoid valve has a response time of five milliseconds or less. Honda provides three solenoid valves; the front wheels are controlled separately, while the rear wheels are controlled together.

Additionally the ALB/ABS system utilizes a pressure switch which sends the pressure reading to the control unit; an ALB/ABS pump assembly; a fail-safe relay which operates when a malfunction is detected by electronically disconnecting the solenoid valve ground circuit and illuminating the indicator light. The indicator light has a dual purpose of alerting the driver of system malfunction, and when switched to a different mode, it is used in system diagnosis by relaying the problem "codes" the control unit has detected with any of the components of the ALB/ABS system.

Although the ALB/ABS system prevents wheel lock-up under hard braking, as brake pressure increases, wheel slip is allowed to increase as well. This slip will result in some tire chirp during ALB/ABS operation. The sound should not be interpreted as lock-up but rather as an indication of the system holding the wheel(s) just outside the locking point. The final few feet of an ALB/ABS-engaged stop may be completed with the wheels locked, as the system is inoperative below 6 mph (10 kph).

When the ignition is **ON** and vehicle speed is over 6 mph (10 kph), the control unit monitors the function of the system. Should a fault be noted, such as loss of signal from a sensor, the ALB/ABS system is immediately disabled by the control unit and the ALB/ABS warning lamp illuminates to inform the operator. When the ALB/ABS system is disabled, the vehicle retains normal braking capacity without the benefits of anti-lock.

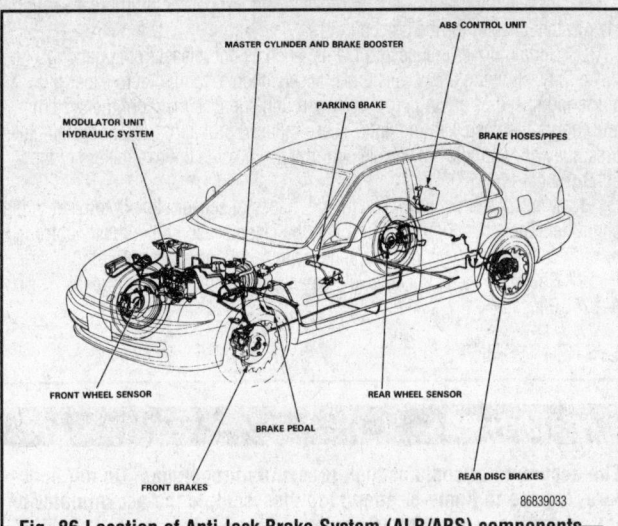

Fig. 86 Location of Anti-lock Brake System (ALB/ABS) components—1992–95 models

Troubleshooting

GENERAL INFORMATION

➡ **The key to successful troubleshooting is doing everything systematically and in order until all possibilities are eliminated and the real problem is found. Honda uses the term "problem" for a malfunction, and system diagnosis is done by tracking down problem "codes" which describe alleged malfunctions (problems) in the system. Special Honda factory tools may be needed for certain procedures involving the anti-lock brake system. Read the procedures thoroughly to determine whether you are prepared to undertake the repairs involved.**

Vehicles with ALB/ABS have an electronic "problem" memory and an indicator light on the instrument panel. When the engine is first started, the light will go on to indicate the system is pressurizing and performing a self-diagnostic check. After the system is at full pressure, the light will go out. If it remains lit, there is a problem detected in the system.

The fault(s) in the system must be verified by simulated operations on the various components of the ALB/ABS system which are performed by the Honda ALB/ABS Checker, or equivalent. If this diagnostic equipment is not available, most of the system can still be tested with a digital multimeter (volt/ohmmeter in one). Service for the system is quite limited. Most components cannot be repaired, only replaced.

Before diagnosing an apparent ALB/ABS problem, make absolutely certain that the normal braking system is in correct working order. Many common brake problems (dragging parking brake, seepage, etc.) will affect the ALB/ABS system. A visual check of specific system components may reveal problems that simulate an apparent ALB/ABS malfunction. Performing this inspection may reveal a simple failure, thus eliminating extended diagnostic time.

1. Inspect the tire pressures; they must be approximately equal for the system to operate correctly.
2. Inspect the wheels and tires on the vehicle. They must be of the same size and type to generate accurate speed signals.
3. Inspect the brake fluid level in the reservoir.
4. Inspect brake lines, hoses, master cylinder assembly and brake calipers for leakage.
5. Visually check brake lines and hoses for excessive wear, heat damage, punctures, contact with other parts, missing clips or holders, blockage or crimping.
6. Check the calipers for rust or corrosion. Check for proper sliding action if applicable.
7. Check the calipers for freedom of motion during application and release.

9-22 BRAKES

8. Inspect the speed sensors for proper mounting and connections (see section later in this chapter for the procedure).

9. Inspect the gear pulsers for broken teeth or poor mounting (see section later in this chapter for the procedure).

10. Certain driver induced problems, such as not releasing the parking brake fully, spinning the wheels under acceleration, sliding due to excessive cornering speed or driving on extremely rough surfaces may fool the system and trigger the dash warning light. These induced problems are not system failures; they are examples of vehicle performance outside the parameters of the control unit.

11. Many system shutdowns are due to loss of sensor signals to or from the controller. The most common cause is not a failed sensor, but a loose, corroded or dirty connector. Check harness and component connectors carefully.

12. Check for correct battery voltage and inspect the condition of all ALB/ABS fuses.

SYSTEM TESTING

✲✲ CAUTION

The accumulator contains high-pressure nitrogen gas. Do not puncture, expose to flame or attempt to disassemble the accumulator or it may explode, resulting in severe personal injury.

Reading Problem Codes

▶ See Figures 87 and 88

If the ALB/ABS indicator light comes on and remains on while the engine is running, there may or may not be a (real) problem with the system. Before assuming there is, go through the following steps to determine if there was a false alarm, or to determine exactly which problem code should be checked.

1. Turn off the engine and then turn the ignition key back to the **ON** position to see whether the indicator light comes on as it should. Then restart the engine to confirm whether the light will remain on or go off as normal.

2. If the light comes on with the ignition on, and goes off with the engine running, there is no problem.

➡**If the indicator light on the dashboard does not come on at all, stop here and refer to the following section "ABS Indicator Light."**

3. If the ALB/ABS indicator light remains on with the engine running, thus confirming that a problem has been detected, stop the engine. Disconnect the service check connector from the connector cover under the glove box and connect the two terminals with a jumper wire.

4. Turn the ignition **ON** without starting the engine to determine the problem. The ALB/ABS control unit will communicate by a series of blinks through the ALB/ABS indicator light which problem code(s) must be checked.

5. To understand the blinks, record the blinking frequency of the ALB/ABS indicator light and compare it to the Problem Code Chart following. The blinking frequency will indicate the problem code. If you miscount the blinking frequency, or want to make sure, turn **OFF** the ignition and then turn it **ON** again to repeat the cycle.

6. Once the problem code(s) have been determined by use of the Symptom-to-System Chart following, use a digital multimeter and the appropriate flowcharts (provided after the code chart) to test the system. Start at the beginning and work all the way towards the end before removing any components.

✲✲ WARNING

Before starting the engine, disconnect the jumper wire from the service check connector, or else the Check Engine light will remain on while the engine is running.

7. After repairs, make sure the warning light on the instrument panel operates properly. It should light when the ignition is first turned **ON**, then go out after the vehicle starts moving. If not, the system is still not repaired.

ALB/ABS Indicator Light

▶ See Figure 89

The ALB/ABS indicator light should always come on when the ignition key is turned **ON**, and then go out when the car is started. If the light fails to operate

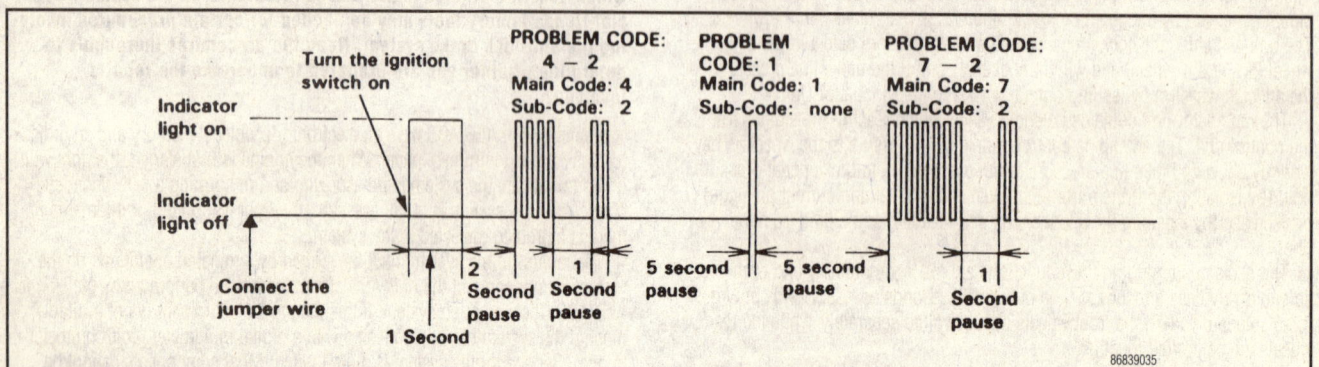

Fig. 87 Problem Code Chart: Use this chart to read the problem codes emitted by the blinking ALB/ABS light. The control unit can indicate three problem codes (one, two or three problems). If you miscount the blinks, turn the key off and then on again and the process will start again

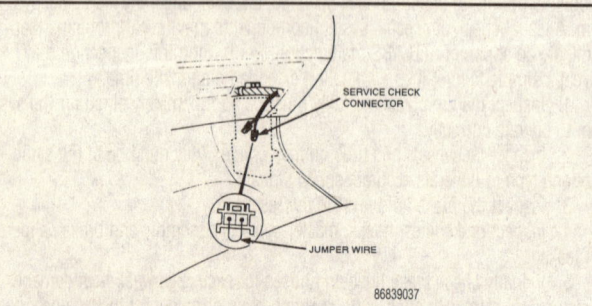

Fig. 88 To read ALB/ABS problem codes, connect a jumper wire between the service check connectors and the indicator light will communicate the problem by blinking the code as indicated by the ALB/ABS control unit

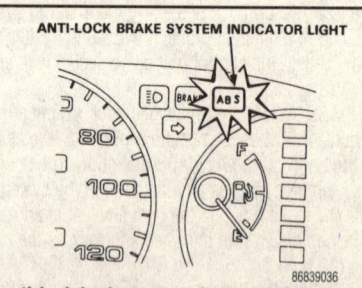

Fig. 89 The anti-lock brake system indicator light is an important part of the ALB/ABS system. In daily operation it tells the driver the system is working properly, but in the event of a malfunction, it can be used to indicate the specific problem code(s) for system troubleshooting

BRAKES 9-23

and the car's electrical system is energized, check the following items. If they are OK, check the control unit connectors. If not loose or disconnected, substitute a known-good control unit (if possible) and recheck:
- Blown anti-lock brake system indicator light bulb.
- Open circuit in YEL wire between the No. 15 backup light (10A) fuse and the gauge assembly.
- Open circuit in BLU/RED wire between the gauge assembly and control unit.
- Poor ground connection between the control unit and the body.

➡ The testing of the ALB/ABS system may require the Bleeder T-Wrench, Honda part No. 07HAA-SG00101 and a digital multimeter. If you intend to service and maintain your ALB/ABS system, investment in these tools may be necessary.

➡ After a repair is completed, the ALB/ABS B2 (15A) fuse must be disconnected for a minimum of three seconds to erase the control unit's memory. Then turn the ignition key ON and recheck. The memory is erased if the connector is disconnected from the control unit or the control unit is removed from the body.

DIAGNOSTIC FLOW CHARTS

▸ **See Figures 90 thru 108**

The accompanying charts should be used along with the information in this section to help troubleshoot system problem codes.

PROBLEM CODE MAIN	PROBLEM CODE SUB	PROBLEMATIC COMPONENT/ SYSTEM	AFFECTED FRONT RIGHT	AFFECTED FRONT LEFT	AFFECTED REAR RIGHT	AFFECTED REAR LEFT	OTHER COMPONENT
1	—	Pump motor over-run	—	—	—	—	Pressure switch
1	2	Pump motor circuit problem	—	—	—	—	Motor relay, Unit fuse, Motor fuse
1	3	High pressure leakage	—	—	—	—	Solenoid
1	4	Pressure switch	—	—	—	—	
1	8	Accumulator gas leakage	—	—	—	—	
2	1	Parking brake switch-related problem	—	—	—	—	Brake fluid level switch, BRAKE light
3	1	Pulser(s)	O				
3	2	Pulser(s)		O			
3	4	Pulser(s)			O	O	
4	1	Speed sensor	O				
4	2	Speed sensor		O			
4	4	Speed sensor			O		
4	8	Speed sensor				O	
5	—	Speed sensor(s)			O	O	Modulator
5	4	Speed sensor(s)			O		Modulator
5	8	Speed sensor(s)				O	Modulator
6	—	Fail-safe relay (Open, short)	—	—	—	—	Front or rear fail-safe relay
6	1	Fail-safe relay (Open, short)	—	—	—	—	Front fail-safe relay
6	4	Fail-safe relay (Open, short)	—	—	—	—	Rear fail-safe relay
7	1	Solenoid related problem (Open)	O				ABS B1 fuse
7	2	Solenoid related problem (Open)		O			Front fail-safe relay
7	4	Solenoid related problem (Open)			O	O	Rear fail-safe relay

Fig. 90 Troubleshooting symptom-to-system chart

9-24 BRAKES

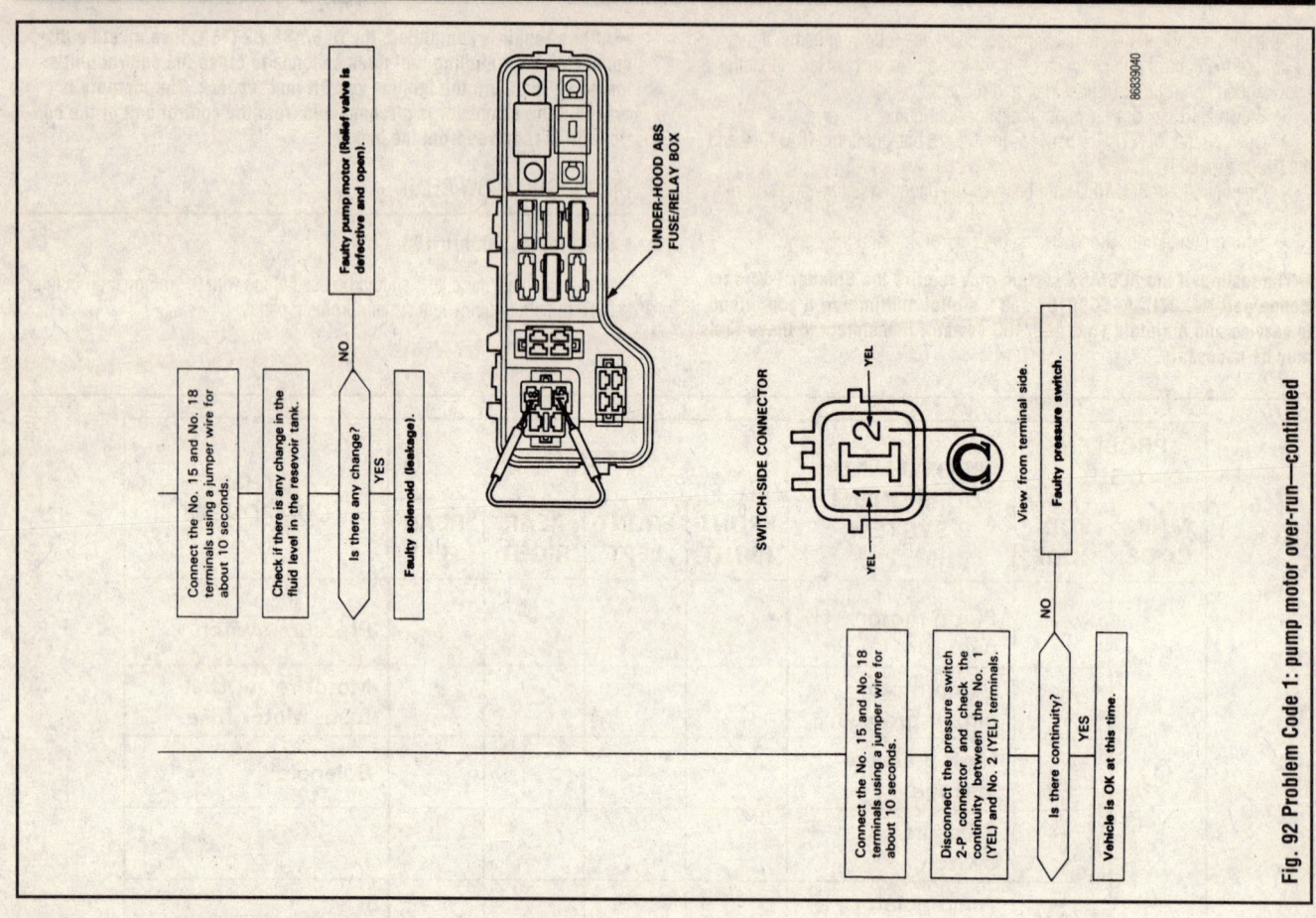

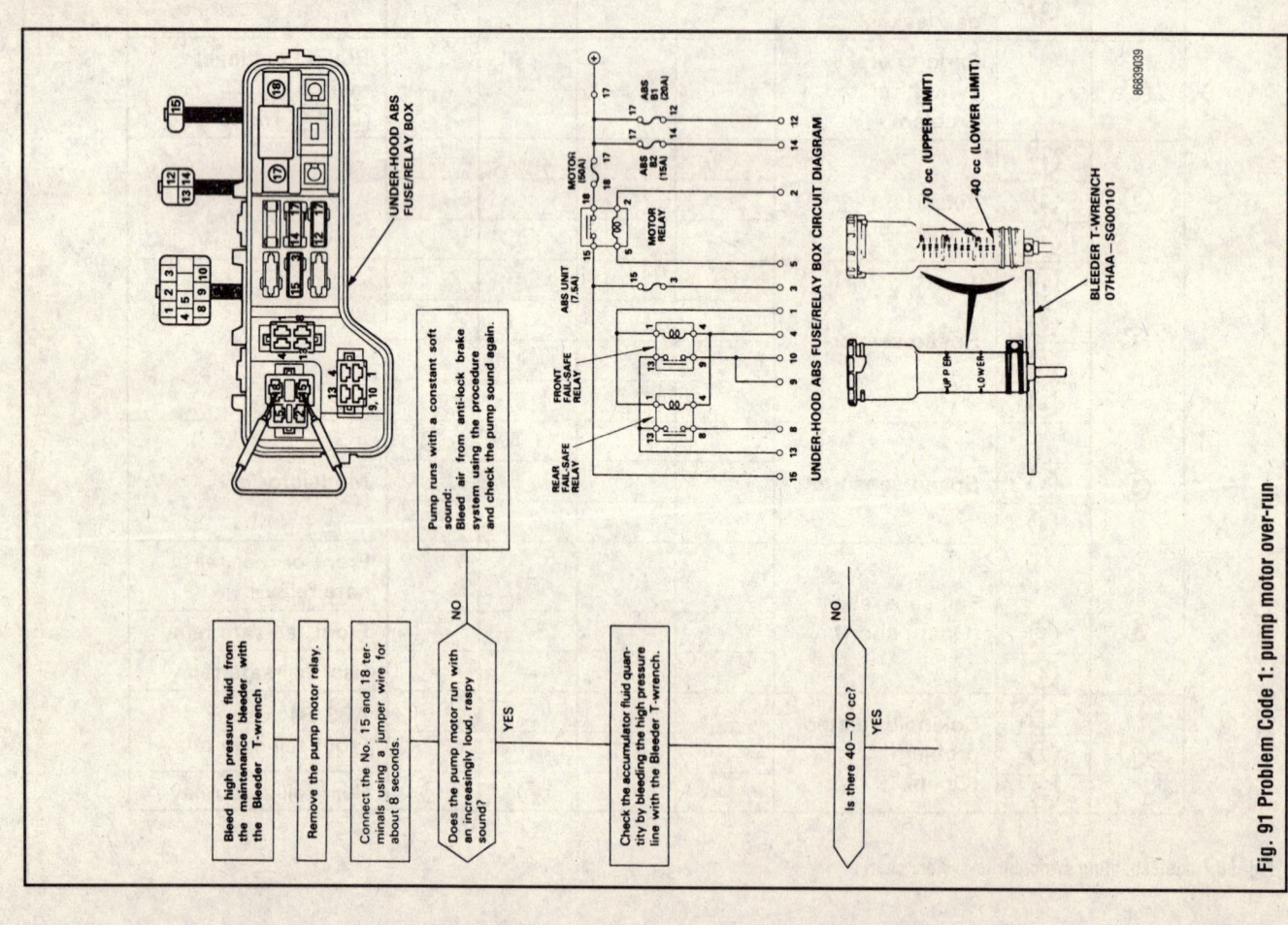

Fig. 92 Problem Code 1: pump motor over-run—continued

Fig. 91 Problem Code 1: pump motor over-run

BRAKES 9-25

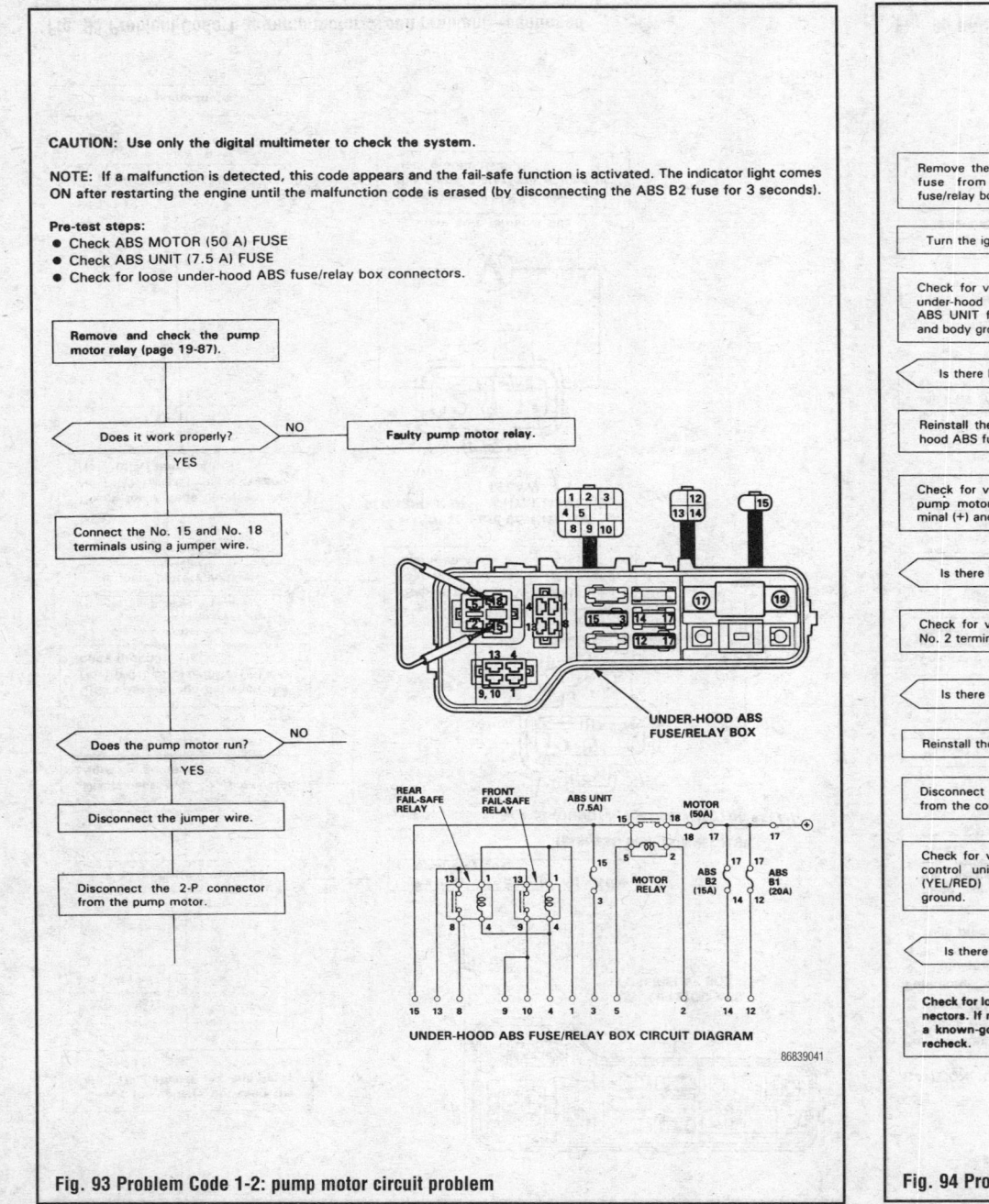

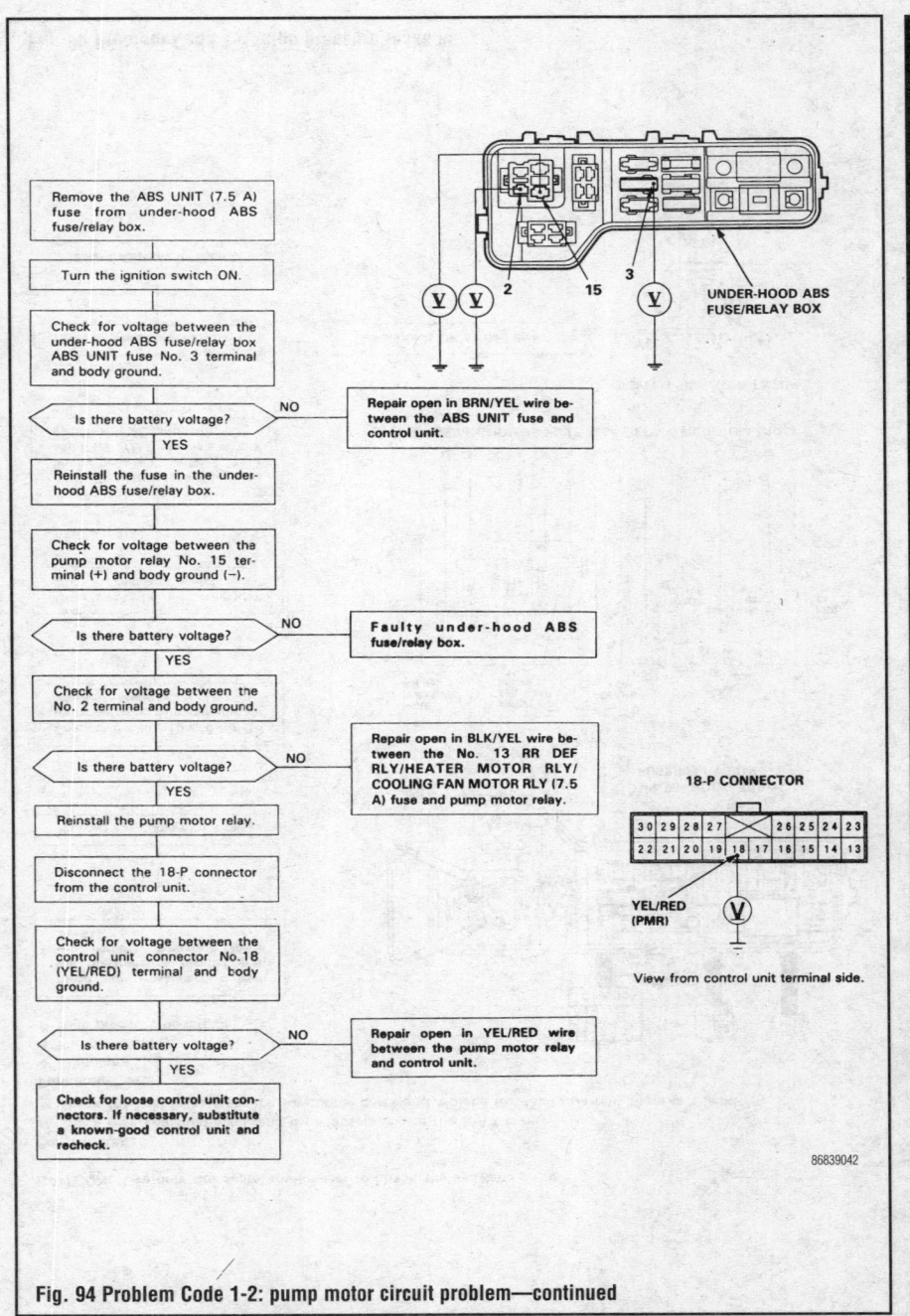

Fig. 94 Problem Code 1-2: pump motor circuit problem—continued

Fig. 93 Problem Code 1-2: pump motor circuit problem

9-26 BRAKES

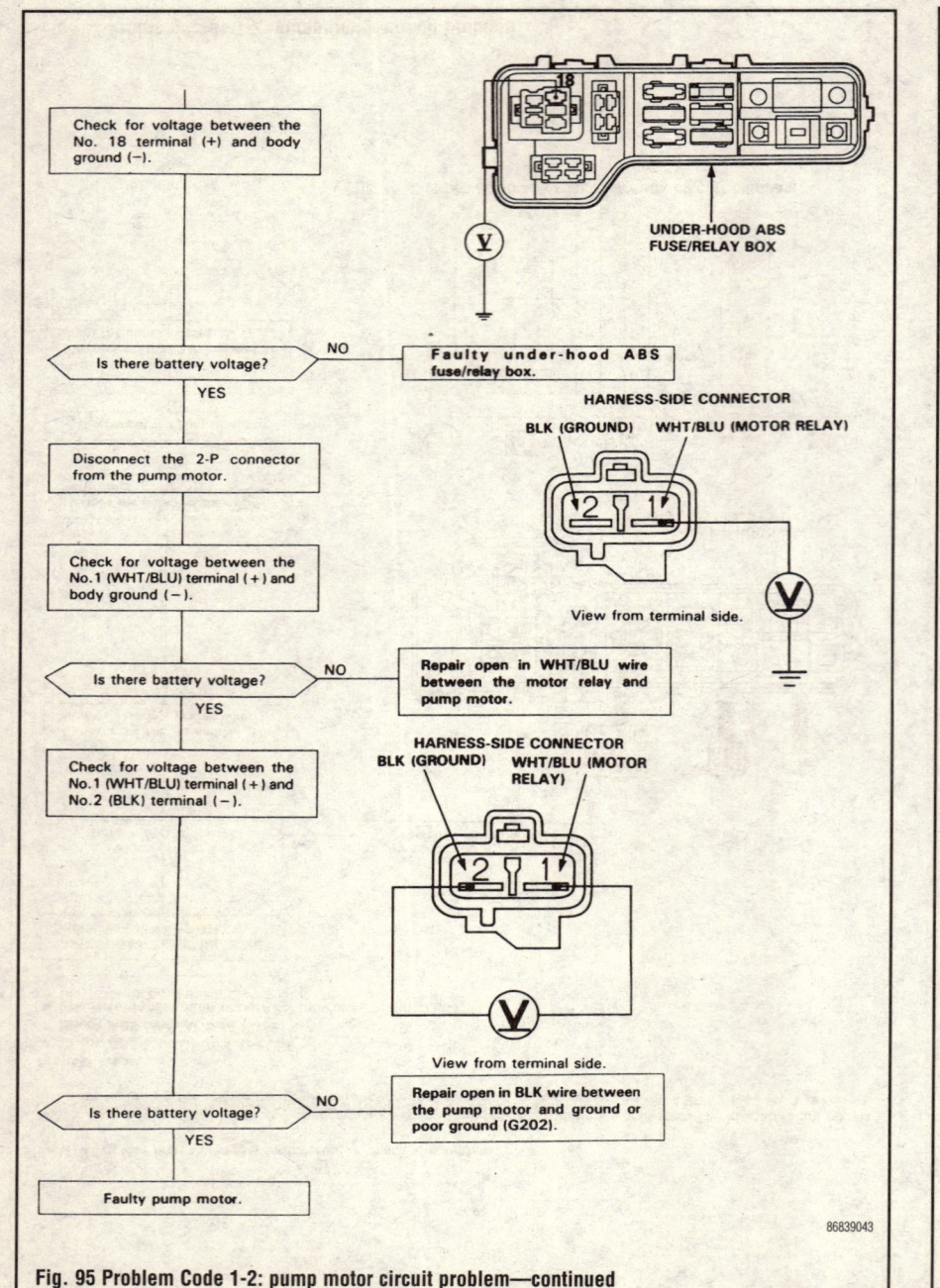

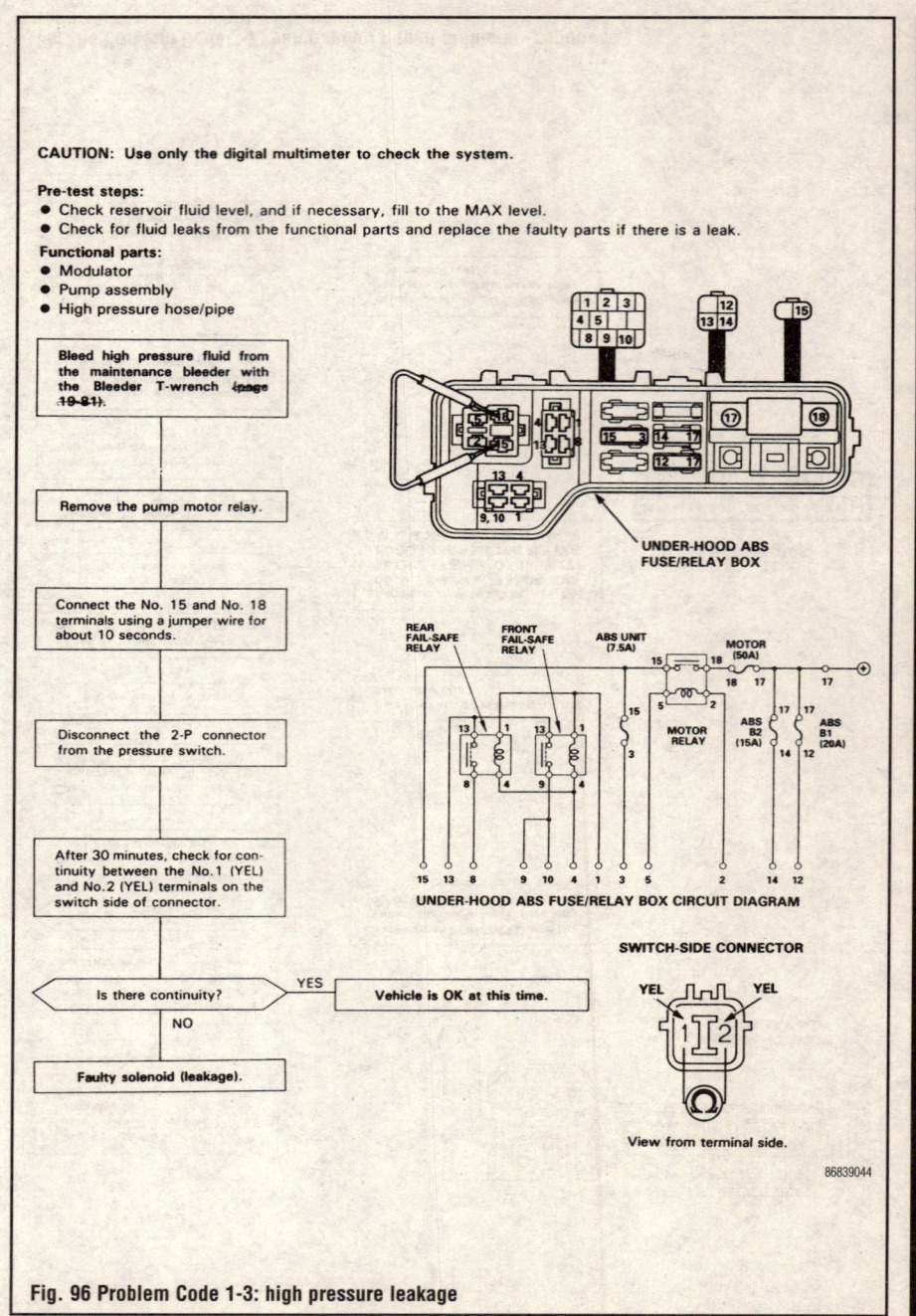

Fig. 96 Problem Code 1-3: high pressure leakage

Fig. 95 Problem Code 1-2: pump motor circuit problem—continued

BRAKES 9-27

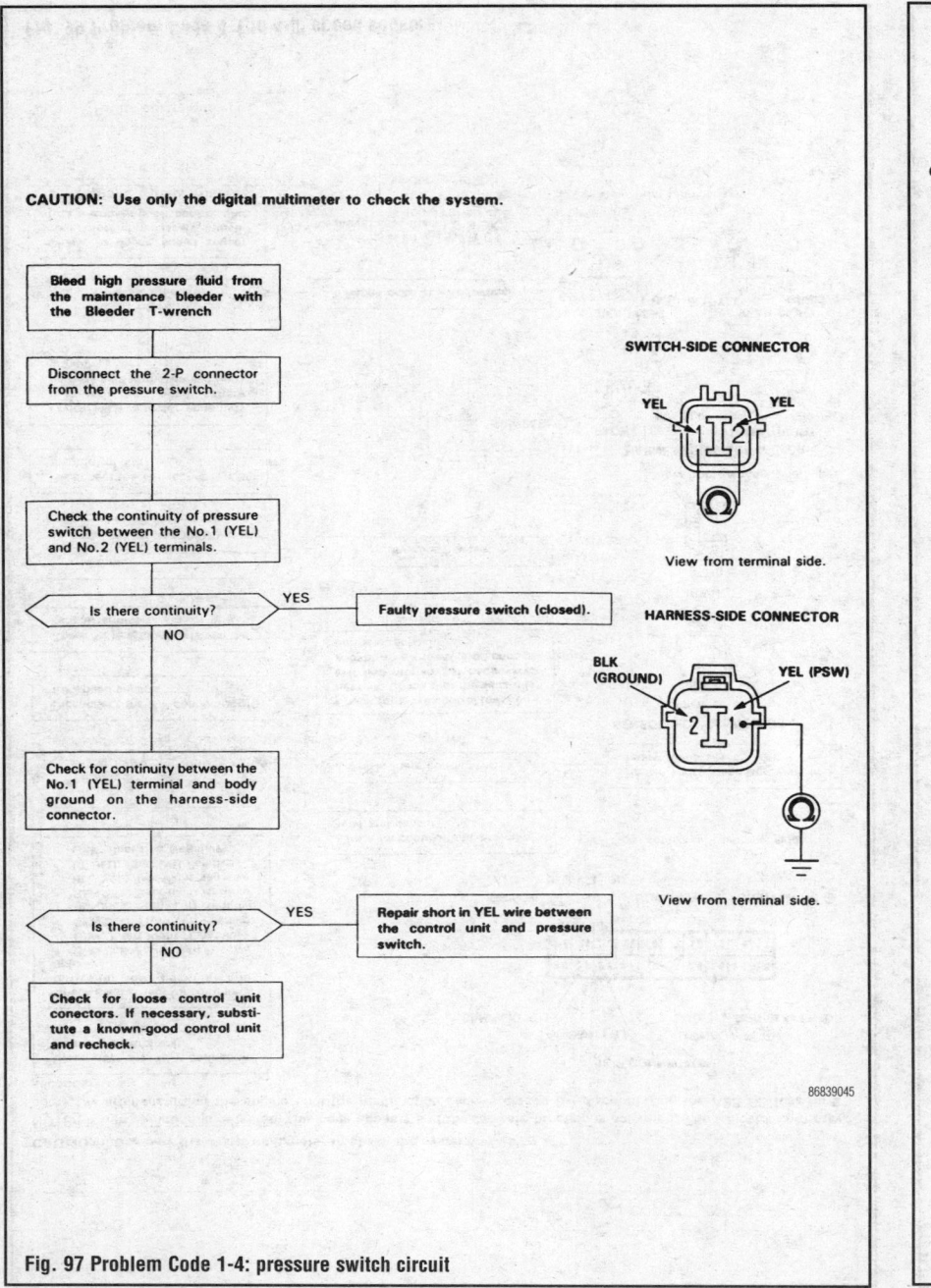

Check the following items:
- The relief plug is loose.
- The relief plug O-ring is out of place.
- Bleed the high pressure line with the Bleeder T-wrench. Operate the pump motor for 10 seconds and bleed the high pressure line again with the Bleeder T-wrench. If no fluid or more than 70 cc of fluid come out, it is likely that the gas has leaked out.

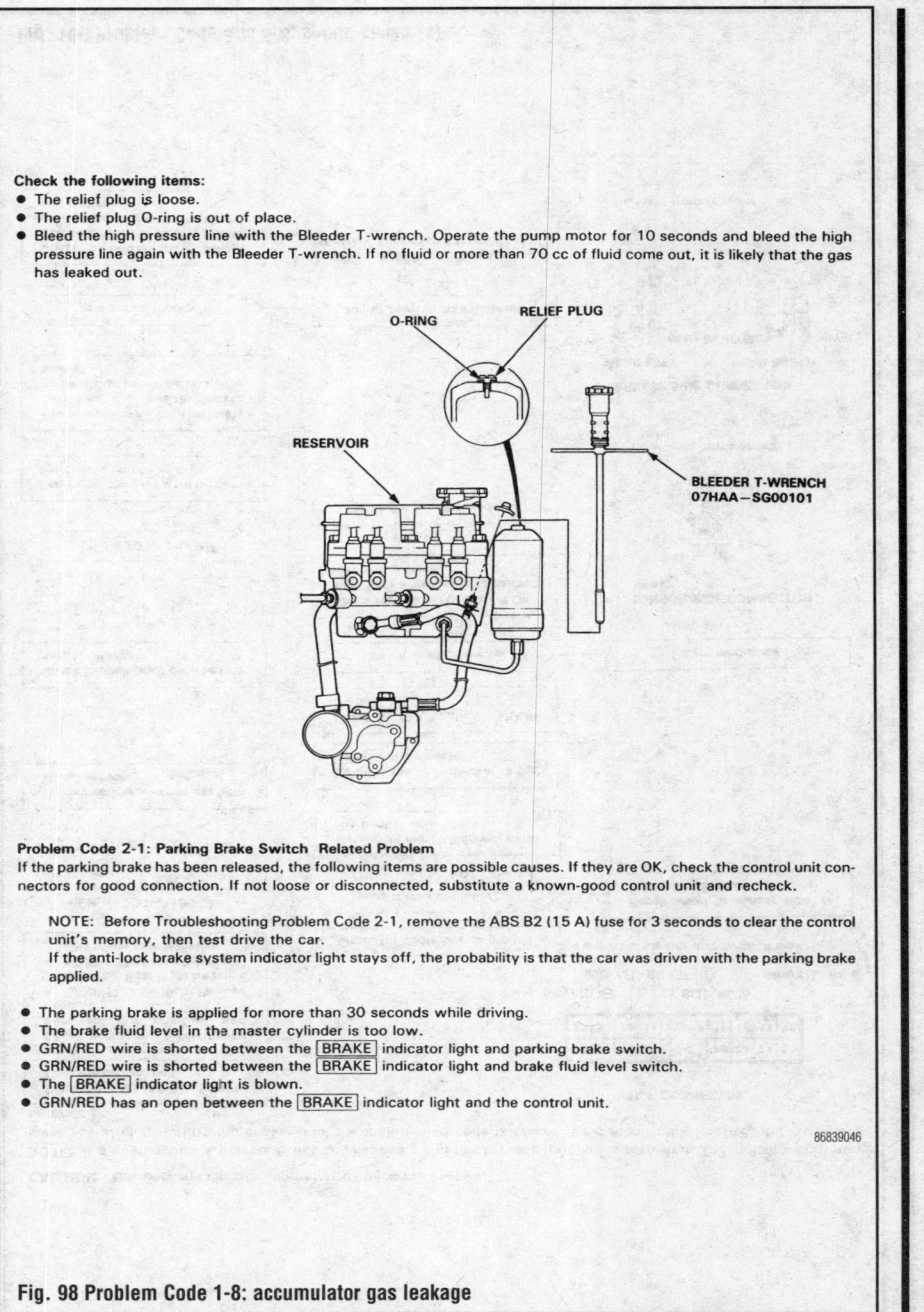

Problem Code 2-1: Parking Brake Switch Related Problem

If the parking brake has been released, the following items are possible causes. If they are OK, check the control unit connectors for good connection. If not loose or disconnected, substitute a known-good control unit and recheck.

NOTE: Before Troubleshooting Problem Code 2-1, remove the ABS B2 (15 A) fuse for 3 seconds to clear the control unit's memory, then test drive the car.
If the anti-lock brake system indicator light stays off, the probability is that the car was driven with the parking brake applied.

- The parking brake is applied for more than 30 seconds while driving.
- The brake fluid level in the master cylinder is too low.
- GRN/RED wire is shorted between the [BRAKE] indicator light and parking brake switch.
- GRN/RED wire is shorted between the [BRAKE] indicator light and brake fluid level switch.
- The [BRAKE] indicator light is blown.
- GRN/RED has an open between the [BRAKE] indicator light and the control unit.

Fig. 98 Problem Code 1-8: accumulator gas leakage

CAUTION: Use only the digital multimeter to check the system.

Fig. 97 Problem Code 1-4: pressure switch circuit

9-28 BRAKES

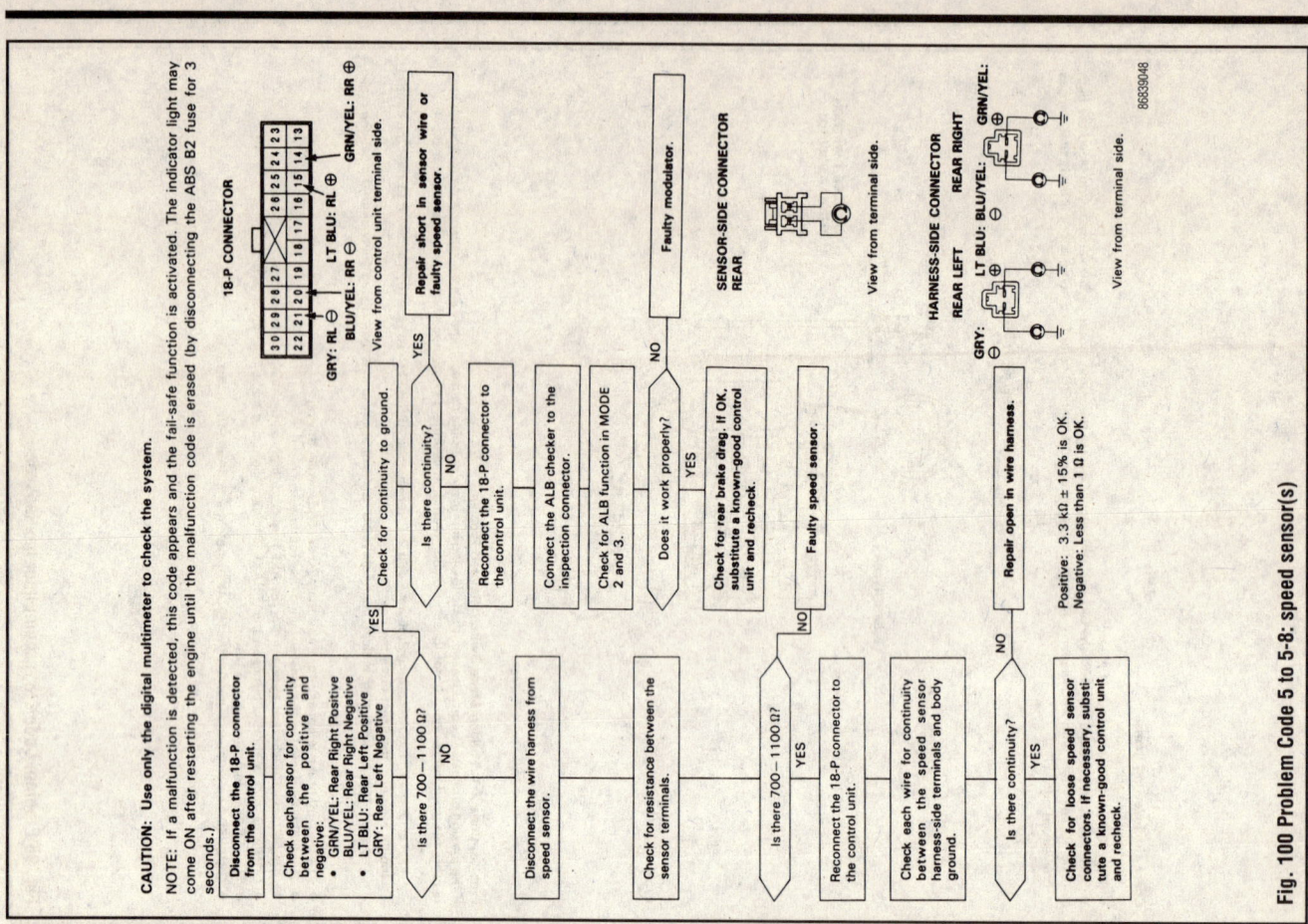

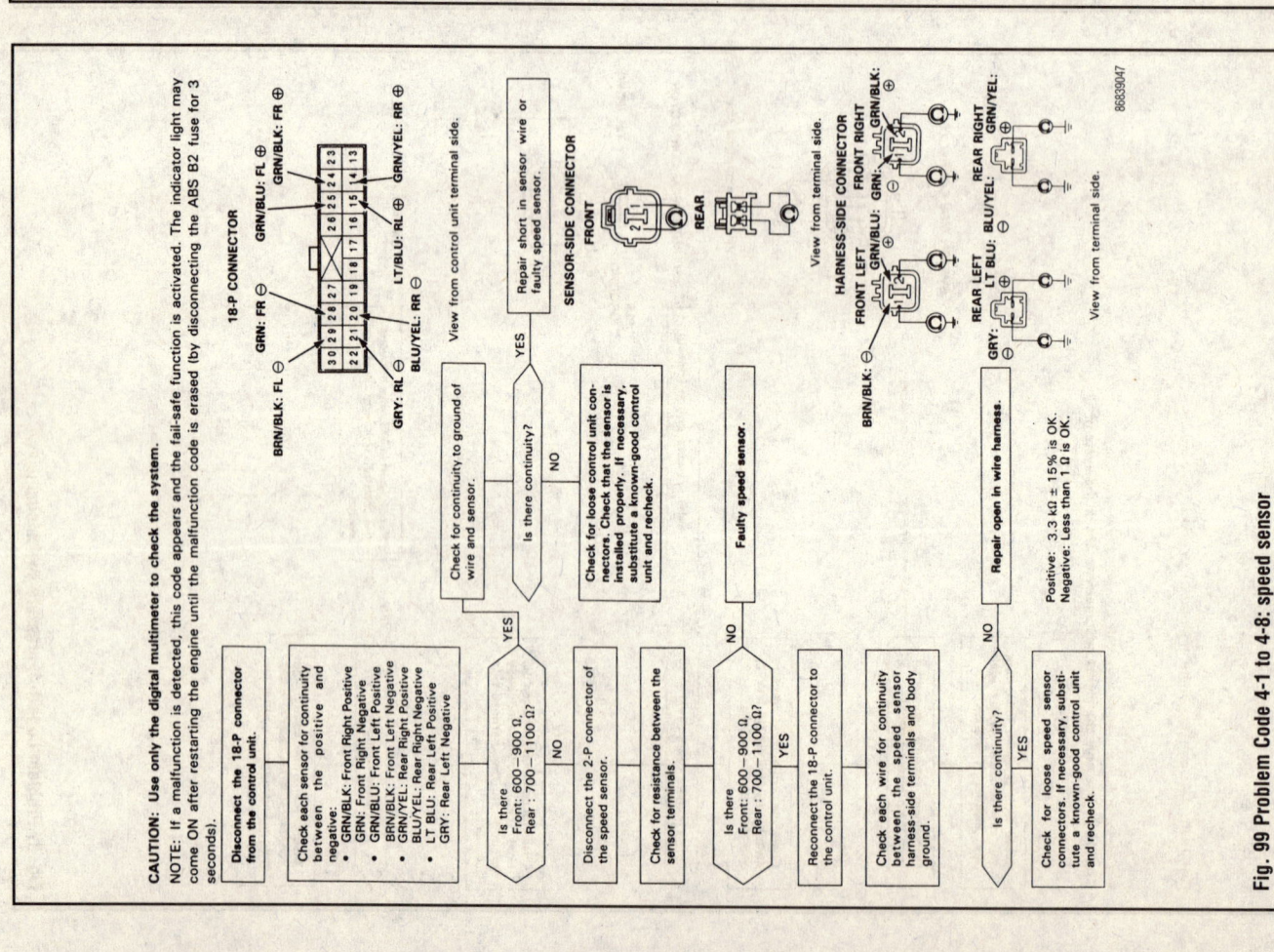

Fig. 99 Problem Code 4-1 to 4-8: speed sensor

Fig. 100 Problem Code 5 to 5-8: speed sensor(s)

Brakes 9-29

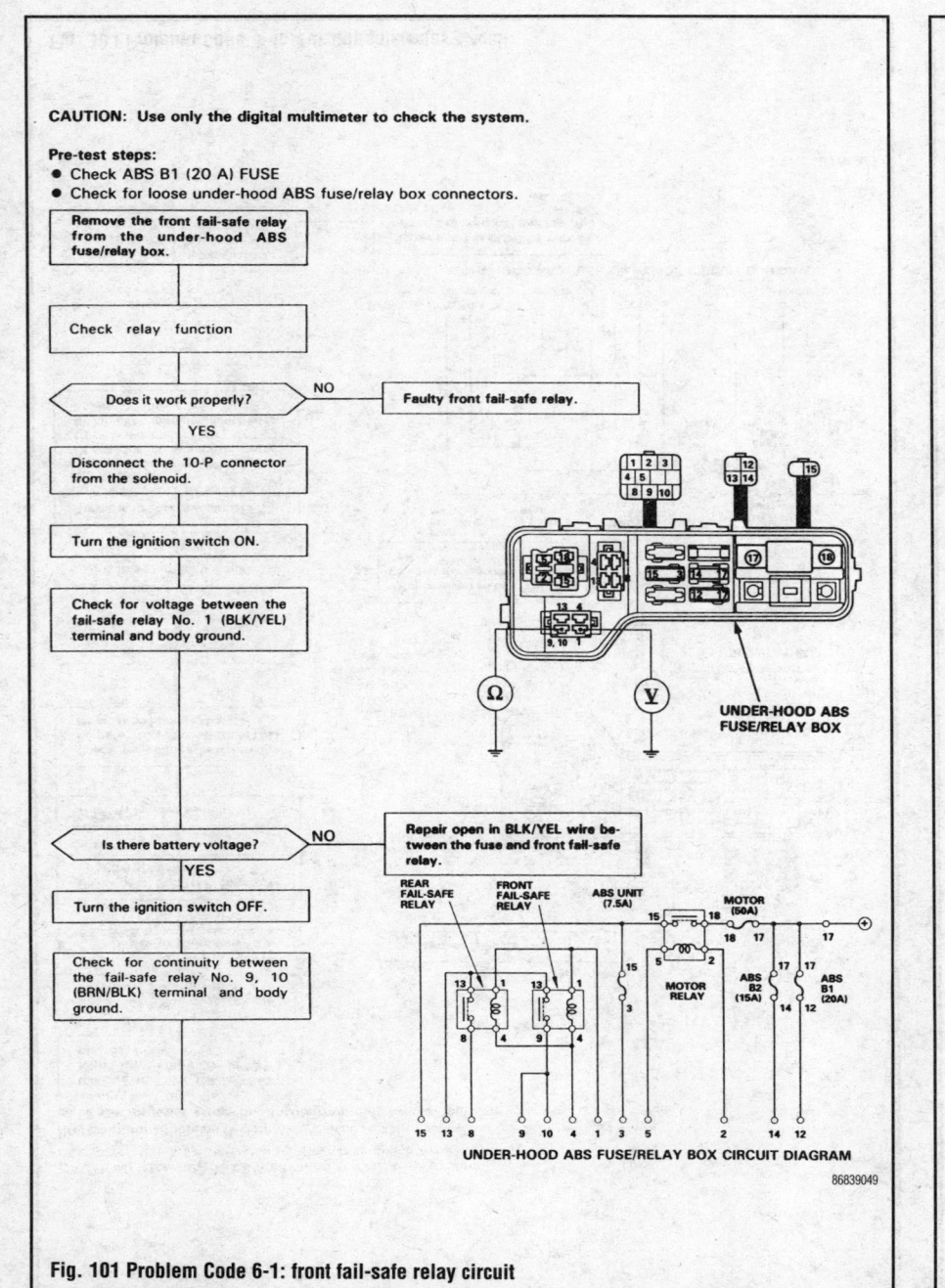

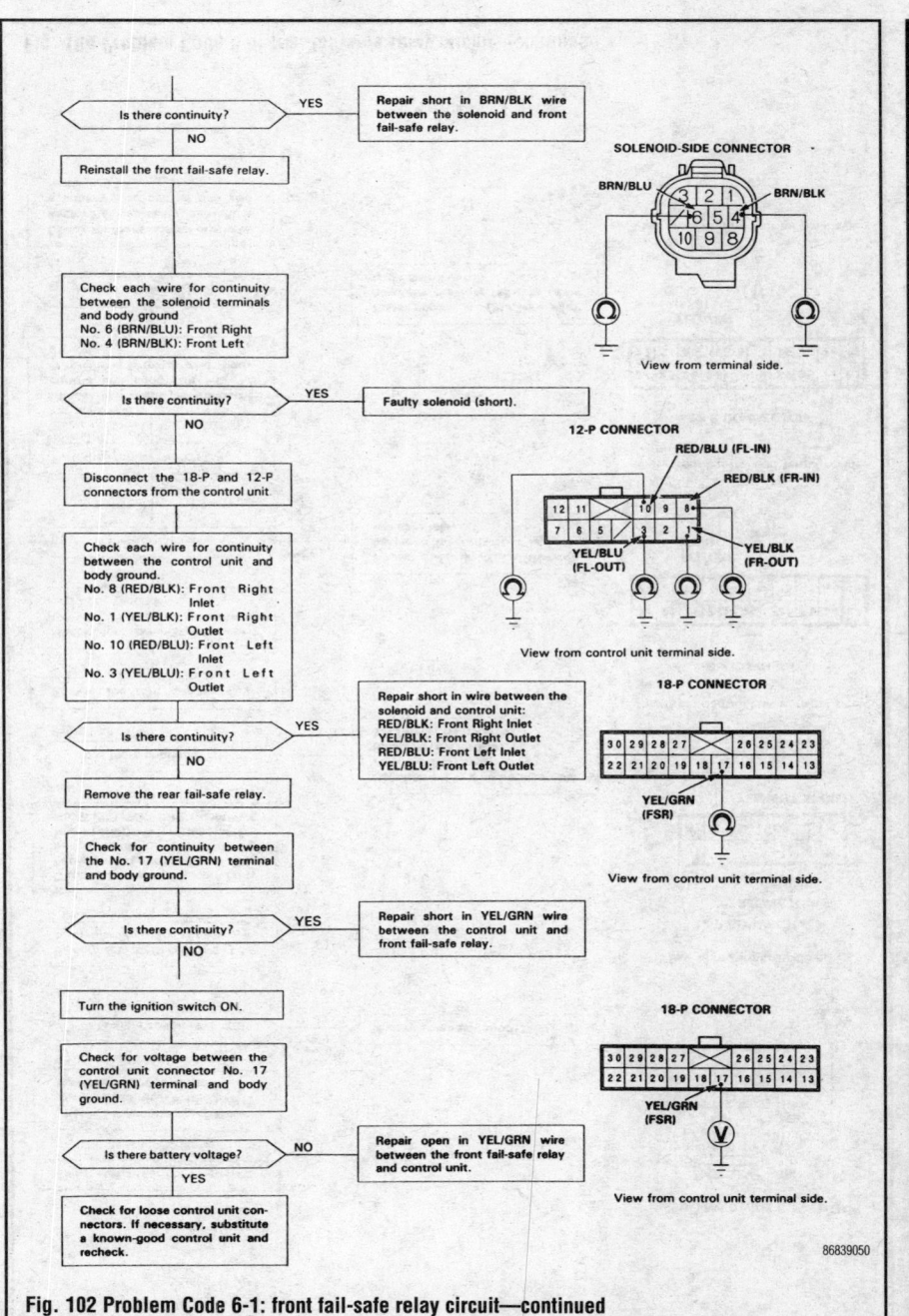

Fig. 102 Problem Code 6-1: front fail-safe relay circuit—continued

Fig. 101 Problem Code 6-1: front fail-safe relay circuit

9-30 BRAKES

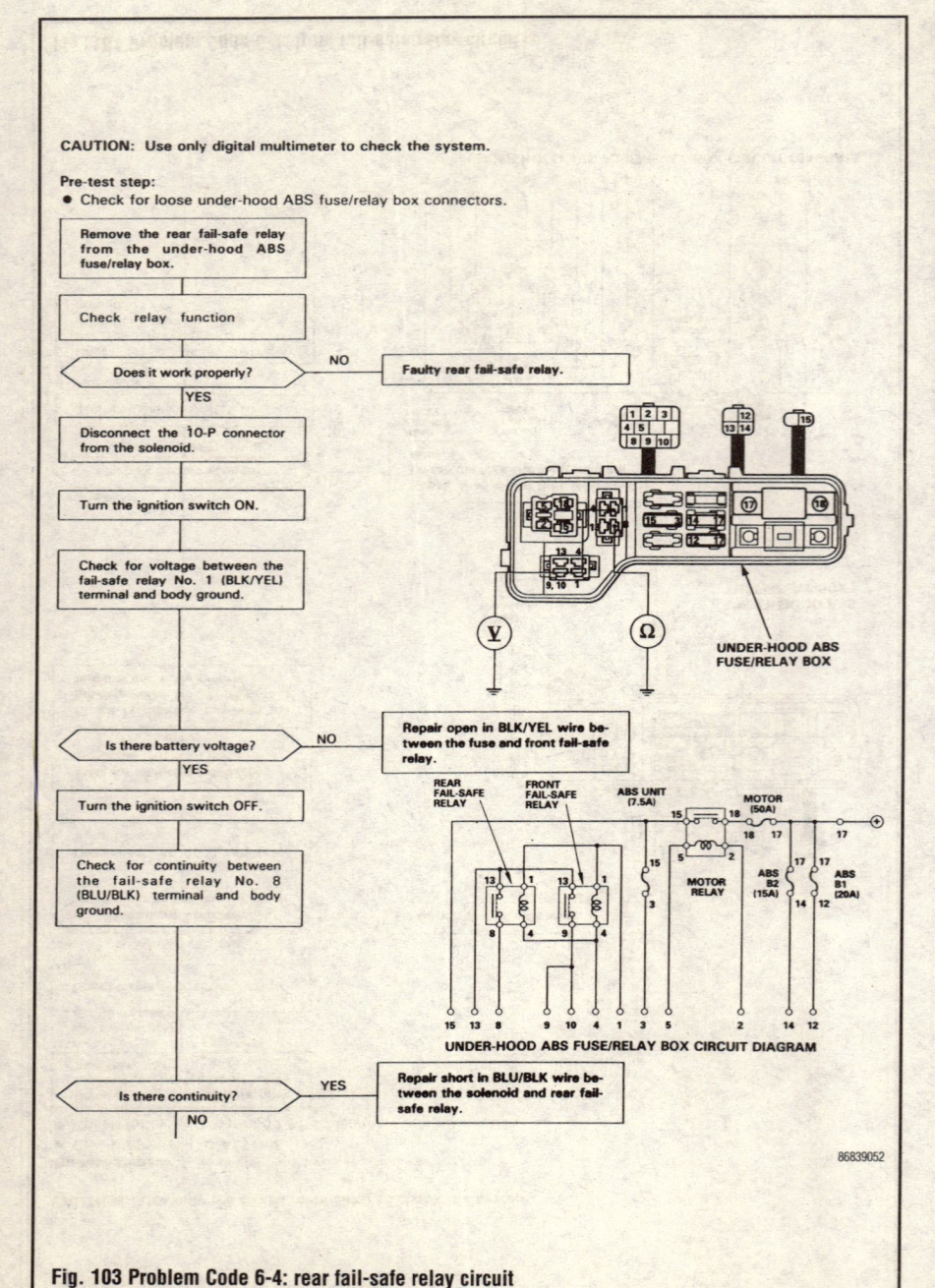

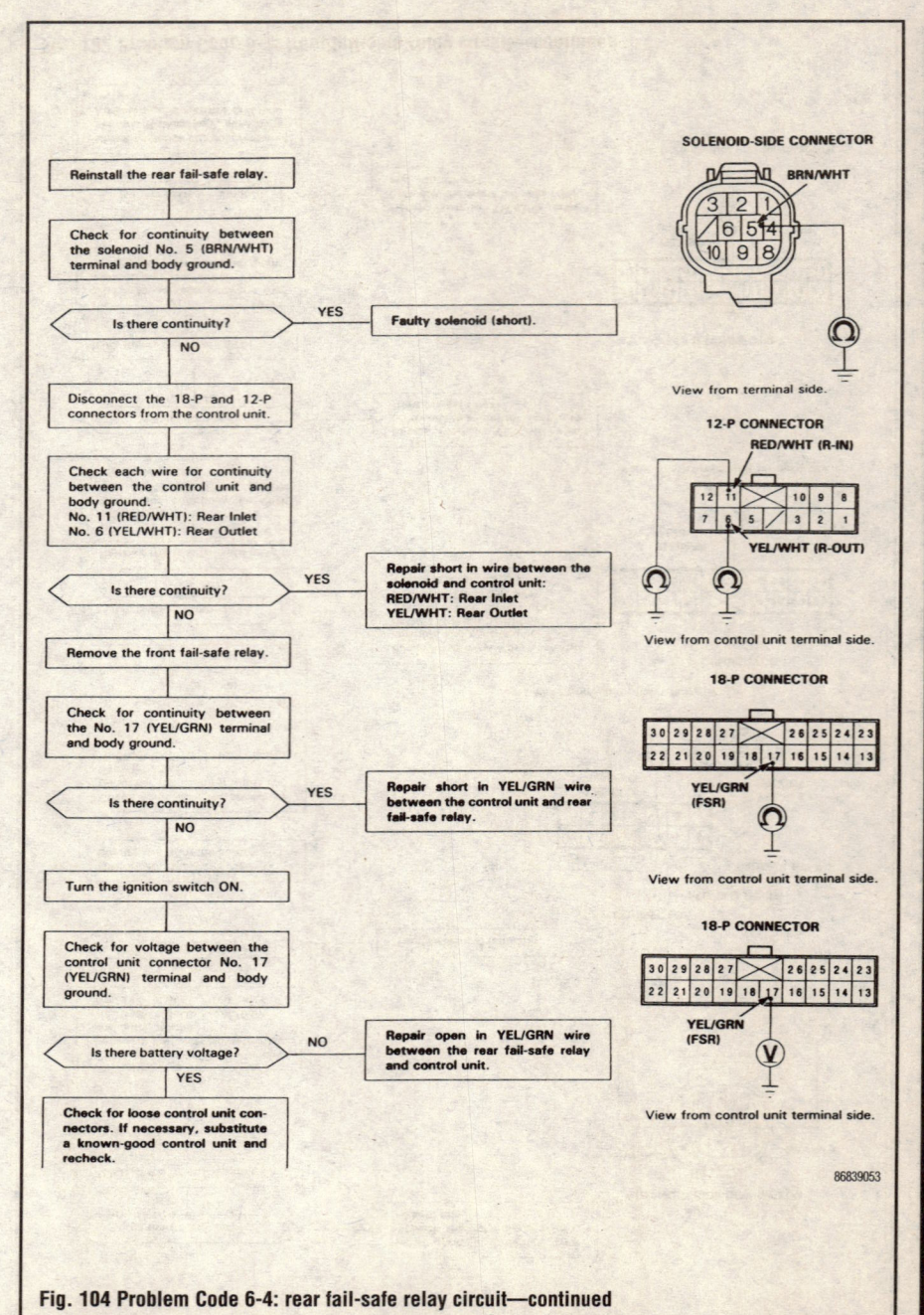

Fig. 104 Problem Code 6-4: rear fail-safe relay circuit—continued

Fig. 103 Problem Code 6-4: rear fail-safe relay circuit

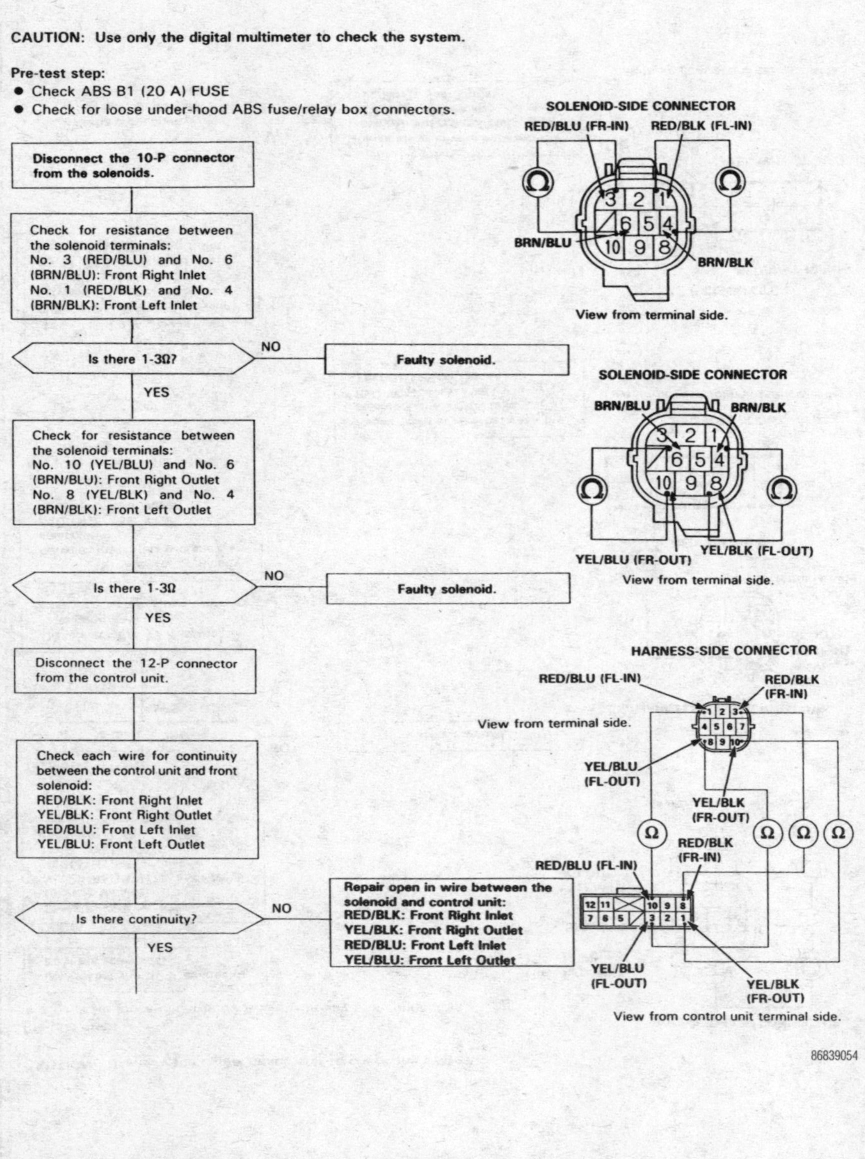

Brakes 9-31

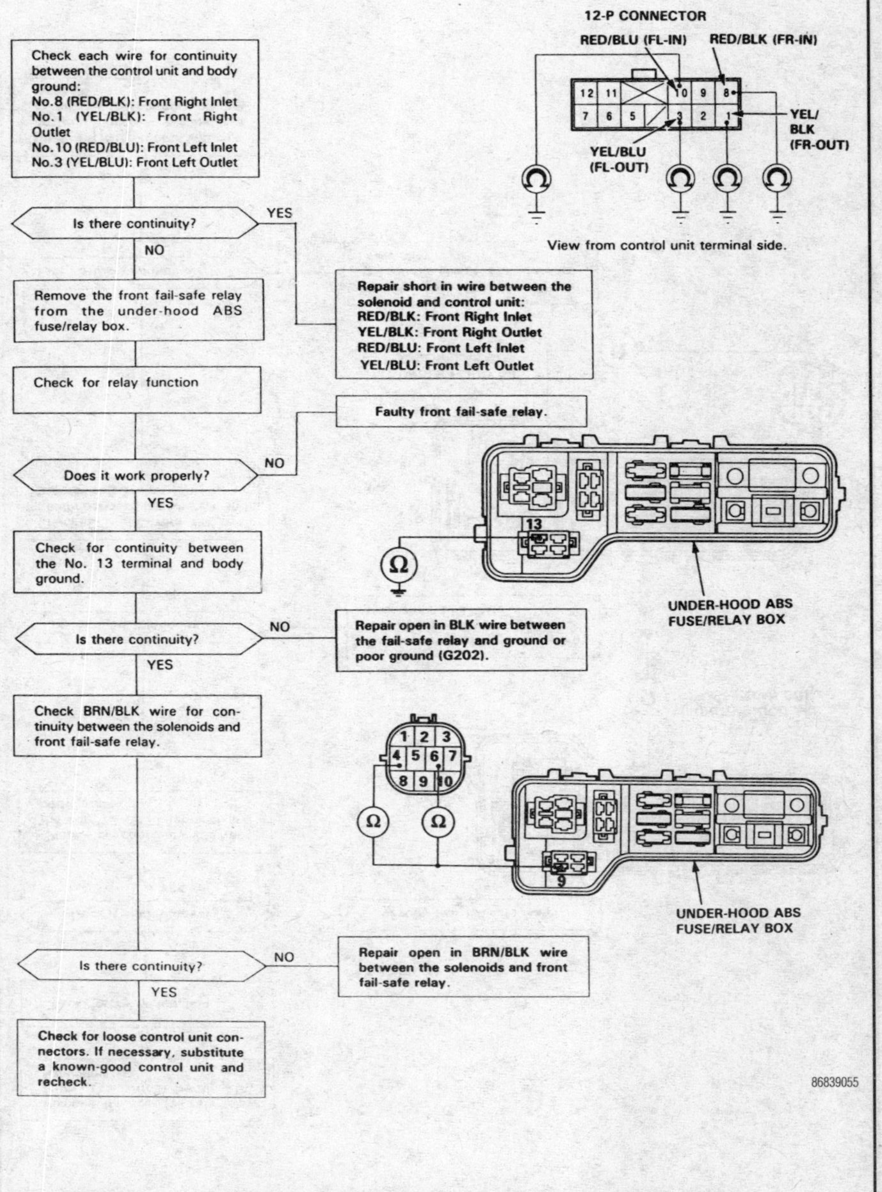

Fig. 106 Problem Code 7-1 and 7-2: front solenoid related problem—continued

Fig. 105 Problem Code 7-1 and 7-2: front solenoid related problem

9-32 BRAKES

Fig. 108 Problem Code 7-4: rear solenoid problem—continued

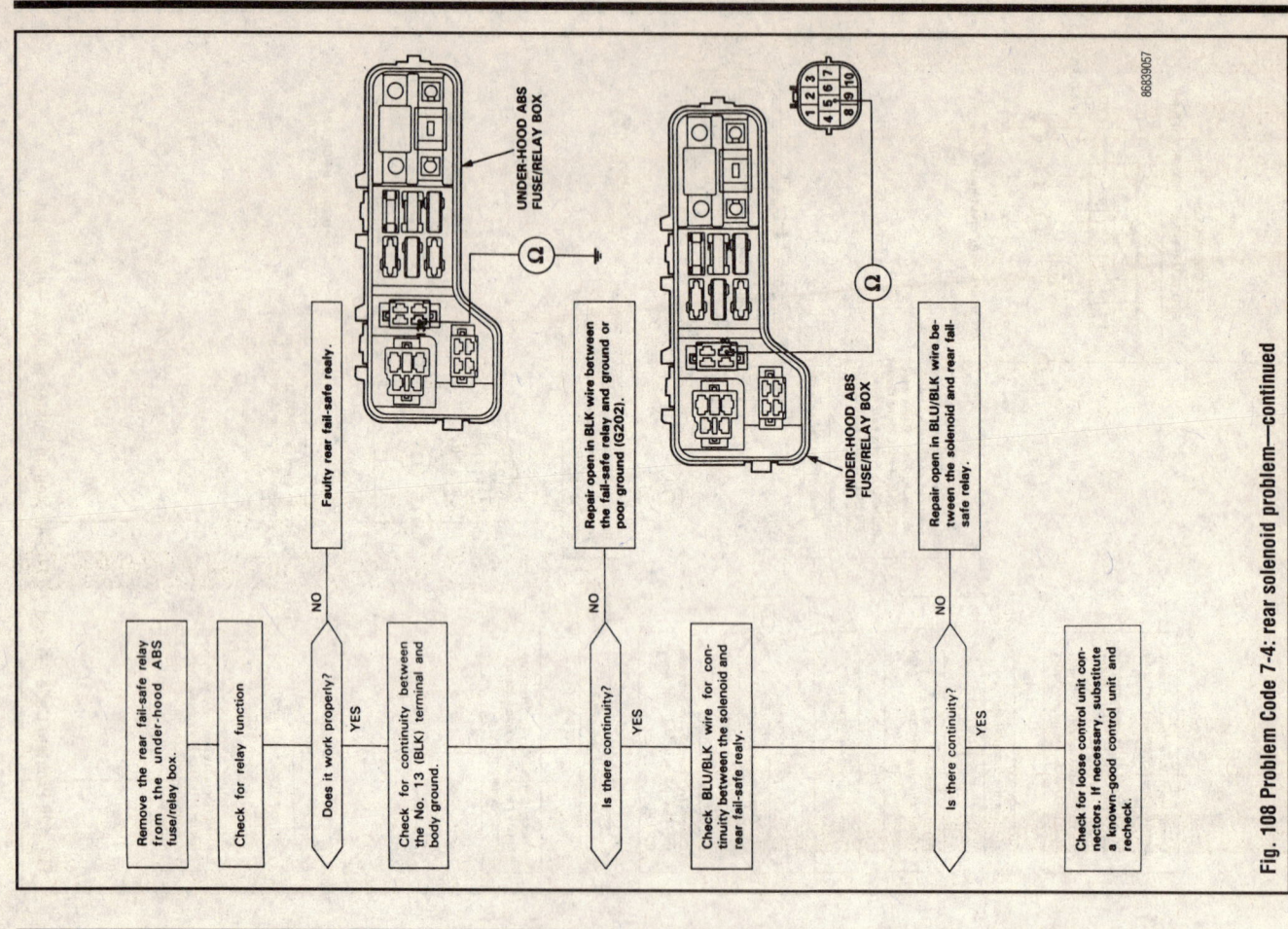

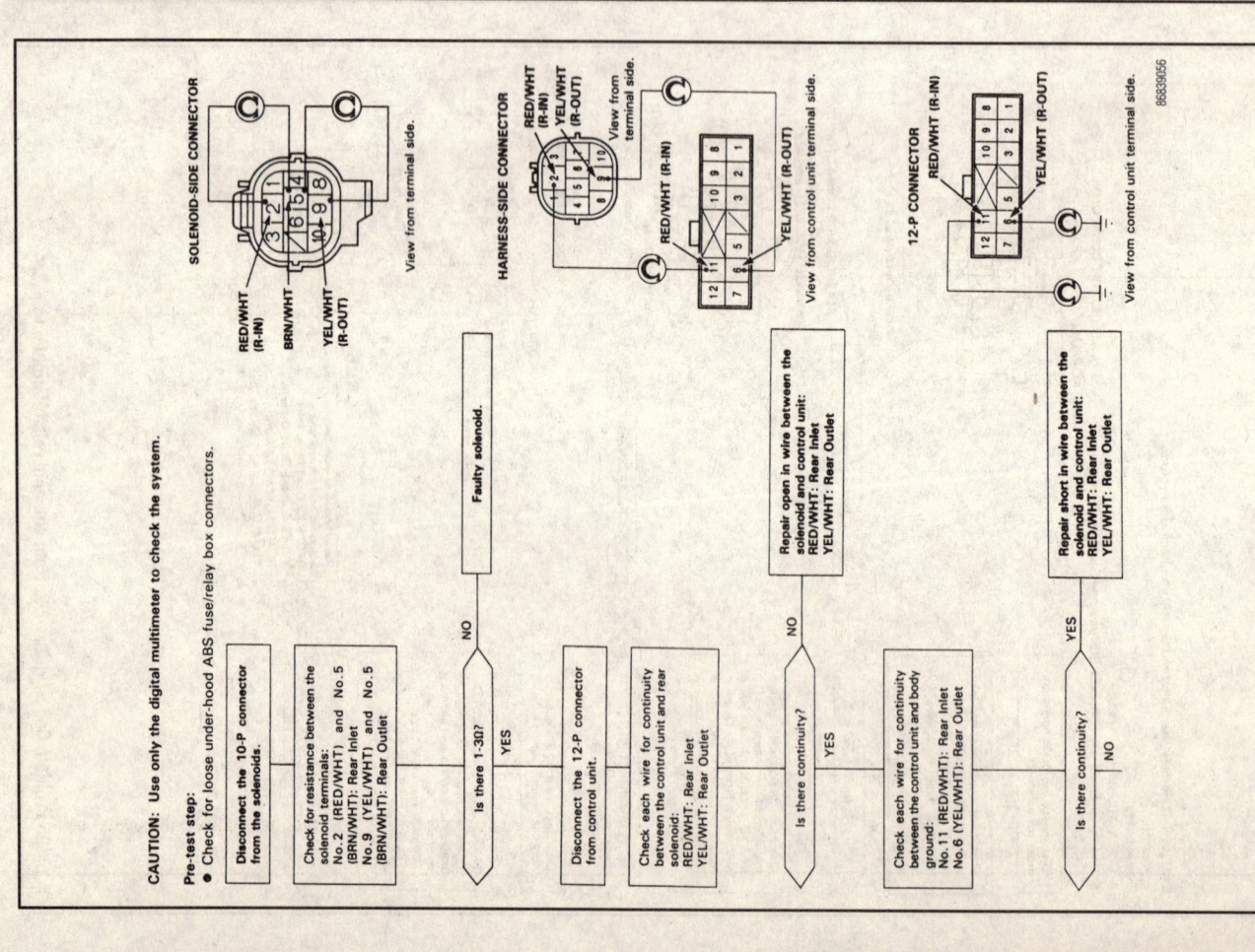

Fig. 107 Problem Code 7-4: rear solenoid problem

Brakes 9-33

Control Unit

REMOVAL & INSTALLATION

♦ See Figure 109

1. Disconnect the negative battery cable.
2. Make sure the ignition switch is **OFF**, then remove the right trunk side trim panel and unplug the control unit connector.
3. Installation by replacing the control unit and its mounting screws and connector.

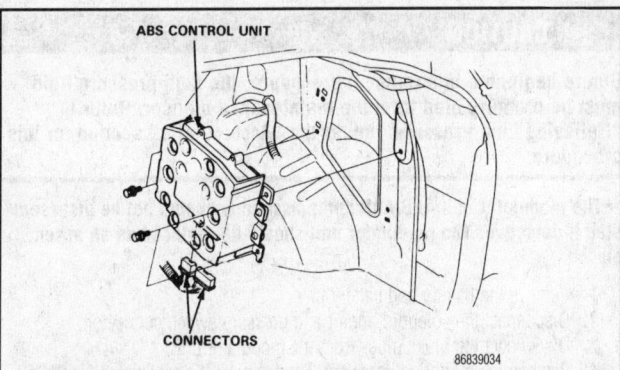

Fig. 109 ALB/ABS control unit, connectors and mounting location in the trunk

Wheel Sensor

The Wheel Sensors operate by means of a contactless magnetic pickup that reads the wheel speed generated by the Gear Pulsers.

REMOVAL & INSTALLATION

♦ See Figures 110 and 111

1. Disconnect the negative battery cable.
2. Raise and safely support the vehicle.
3. Remove the wheel and unbolt the sensor from the steering knuckle rear suspension member.

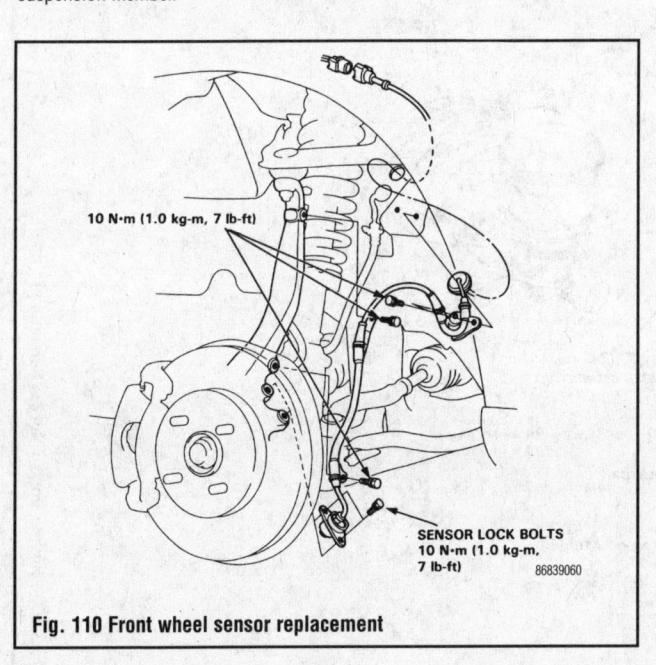

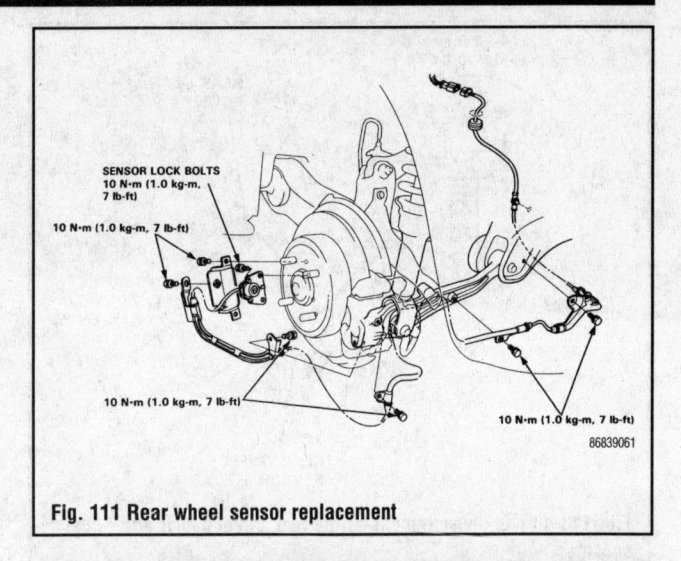

Fig. 110 Front wheel sensor replacement

Fig. 111 Rear wheel sensor replacement

Gear Pulsers

♦ See Figures 112 and 113

The Gear Pulsers are toothed rings that rotate with the road wheels to generate the signal picked up by the wheel sensors and sent as vital data to the ALB/ABS control unit.

INSPECTION

Front:
1. Check the pulser for chipped or damaged teeth and replace if necessary.
2. Measure the air gap all the way around between the gear pulser and the wheel sensor while manually rotating the halfshaft. The air gap should be between 0.02–0.04 in. (0.4–1.0mm) for proper operation.
3. If the air gap is greater than the specification, the knuckle is distorted and should be replaced.

Rear:
4. Remove the rear caliper assembly and the rear brake disc.
5. Check the gear pulser for chipped or damaged teeth and replace if necessary.
6. Measure the air gap all the way around between the gear pulser and wheel sensor by manually rotating the hub bearing unit. If the air gap exceeds 0.02–0.4 in. (0.4–1.0mm), the knuckle is probably distorted and should be replaced.

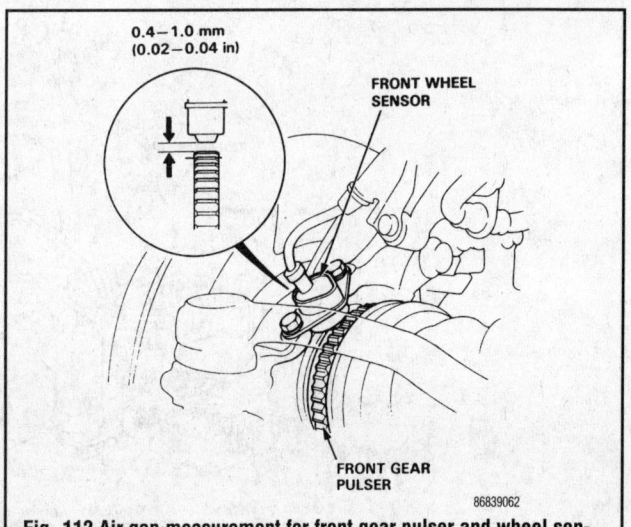

Fig. 112 Air gap measurement for front gear pulser and wheel sensor

9-34 BRAKES

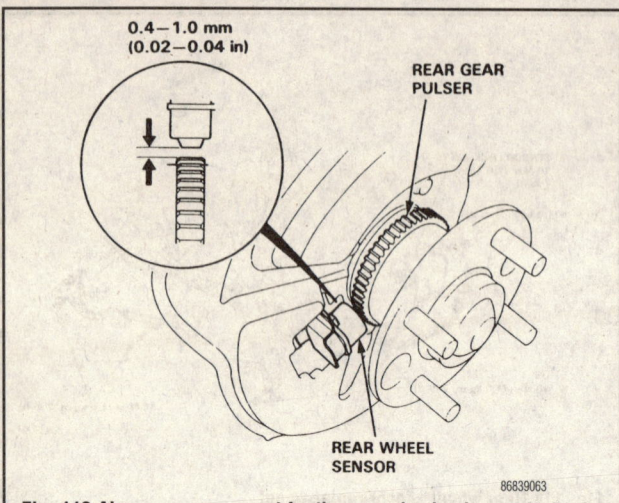

Fig. 113 Air gap measurement for the rear wheel pulser and wheel sensor

Modulator Unit/ALB/ABS Pump Assembly

RELIEVING LINE PRESSURE

◆ See Figures 114 and 115

✳✳ CAUTION

The accumulator contains high-pressure nitrogen gas. Do not puncture, expose to flame or attempt to disassemble the accumulator or it may explode, resulting in severe personal injury.

This procedure is necessary before performing certain repairs on the ALB/ABS.
1. Open the hood and remove the red cap from the modulator body.
2. Install the special tool, Bleeder T-Wrench, Honda part No. 07HAA–SG00101 on the maintenance bleeder and turn it slowly 90 degrees to collect the high-pressure fluid into the reservoir.
3. Turn the T-Wrench out one complete turn to thoroughly drain the fluid.
4. Retighten the maintenance bleeder, replace the red cap and safely discard the fluid.

REMOVAL & INSTALLATION

◆ See Figures 116 and 117

✳✳ CAUTION

Before beginning this removal procedure, the high-pressure fluid must be properly bled from the maintenance bleeder. Refer to "Relieving Line Pressure" under the "Accumulator" section for this procedure.

➥The modulator unit/ALB/ABS pump assembly should not be disassembled if defective. The modulator unit should be replaced as an assembly.

1. Remove the battery and battery tray.
2. Disconnect the solenoid, motor and pressure switch connectors.
3. Disconnect the brake pipes from the modulator unit.
4. Remove the mounting hardware. Then remove the modulator/ALB/ABS pump assembly from the frame.

To Install:

5. Replace the modulator/ALB/ABS pump assembly and connect with the mounting hardware.
6. Connect the brake pipes to the modulator unit.
7. Connect the solenoid, motor and pressure switch connectors.
8. Replace the battery tray and battery and bleed the system.

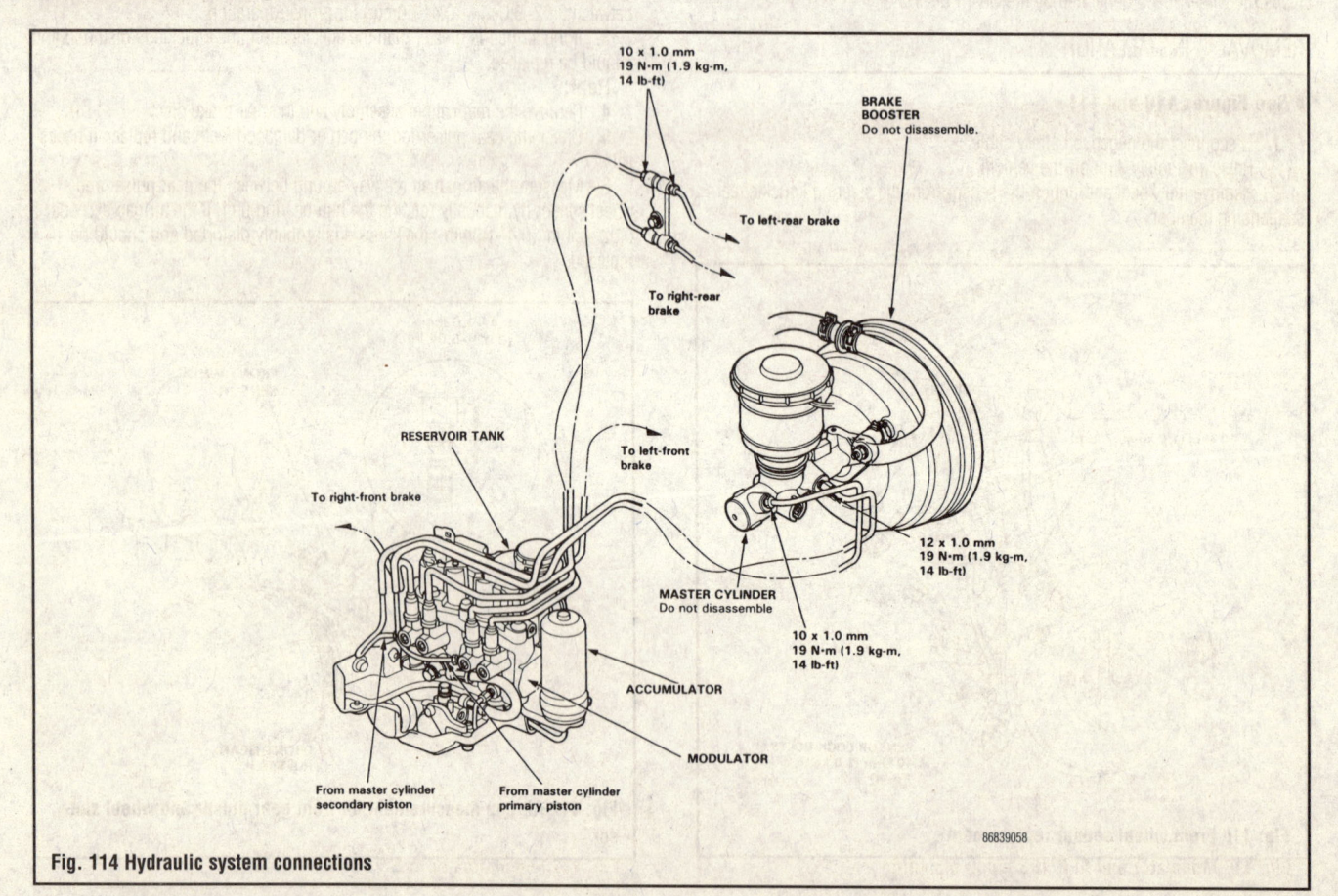

Fig. 114 Hydraulic system connections

Brakes 9-35

⚠ WARNING
- High-pressure fluid will squirt out if the shaded hose and pipe are removed.
- To drain high-pressure brake fluid, follow the procedure on this page.

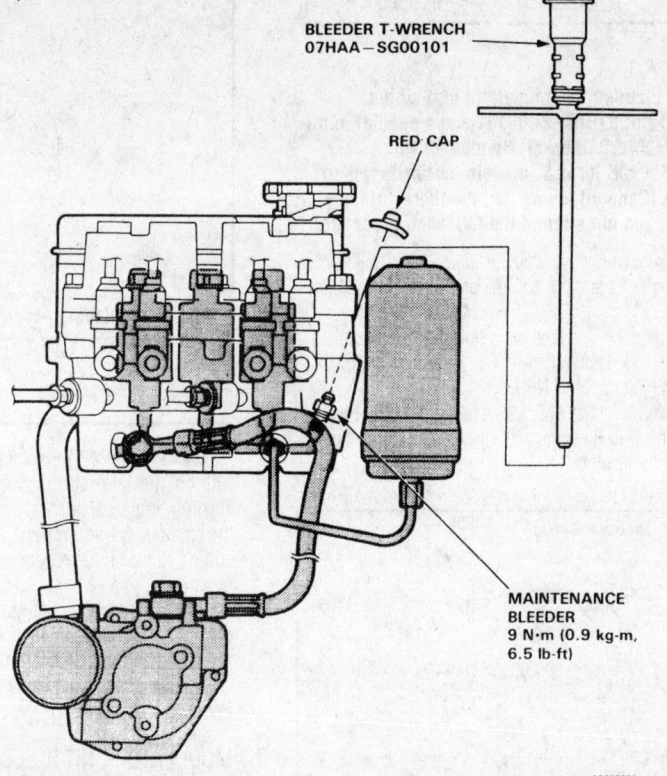

Fig. 115 Caution: If the system is not properly bled beforehand and any of the shaded components are disconnected, high pressure fluid will spray. Shown is the proper location and tool for relieving line pressure from the modulator unit and accumulator before maintenance procedures are performed. Exercise great care when using the bleeder tool

※ WARNING

Do not bend or damage the brake pipes. Cover the hose joints with shop towels or rags to prevent spills. Before reassembling, ensure all parts are clean and free of dust and contaminants. Do not mix brands or grades of brake fluid and use only clean DOT 3 or 4 brake fluid. When reconnecting, ensure there is no interference between the brake pipes and other parts.

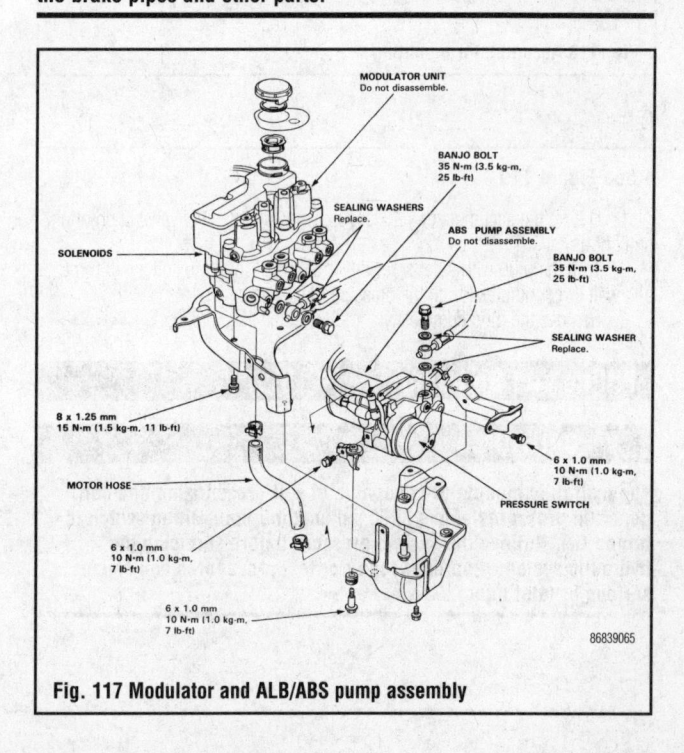

Fig. 116 Modulator and ALB/ABS pump mounting

Fig. 117 Modulator and ALB/ABS pump assembly

9-36 BRAKES

Accumulator

REMOVAL & INSTALLATION

▶ See Figure 118

➡ This procedure will require bleeding the modulator unit with a Bleeder T-Wrench, Honda tool No. 07HAA–SG00101 and a special multi-function diagnostic tool, the ALB/ABS Checker, Honda tool No. 07HAJ–AS0010A or 07HAJ–AS0010B for U.S. models, or for Canadian models, the ALB/ABS Checker (Canada) (same part numbers). Unless you have access to these tools, do not attempt the following procedure.

1. Loosen the flare nuts and remove the accumulator pipe.
2. Remove the three mounting bolts and the accumulator from the modulator unit.
3. Install a new accumulator, replace the three mounting bolts and connect the accumulator pipe. Tighten the 10 x 1.0mm flare nut to 14 ft. lbs. (19Nm). Tighten the 6 x 1.0mm flare nut to 7 ft. lbs. (10 Nm).
4. Bleed the high pressure line using the ALB/ABS Checker and the Bleeder T-Wrench as described "Modulator Unit Bleeding with the ALB/ABS Checker," in the "Filling and Bleeding" section, following.

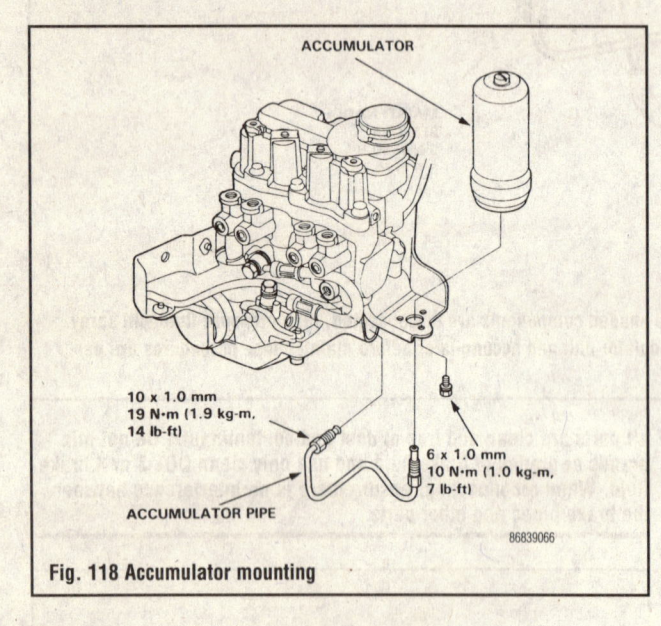

Fig. 118 Accumulator mounting

DISPOSAL

▶ See Figure 119

1. Secure the accumulator vertically in a vise so the relief valve is pointing straight up.
2. Turn the plug with a flat screwdriver SLOWLY for three and a half turns and wait three minutes for all the pressure to escape.
3. Remove the plug completely and discard the accumulator.

Filling and Bleeding

✳✳ CAUTION

The hydraulic modulator is capable of self-pressurizing and can generate pressures above 3000 psi anytime the ignition switch is turned ON. Relieve the system pressure before servicing the hydraulic system. Improper repair or test procedures can cause serious or fatal injury.

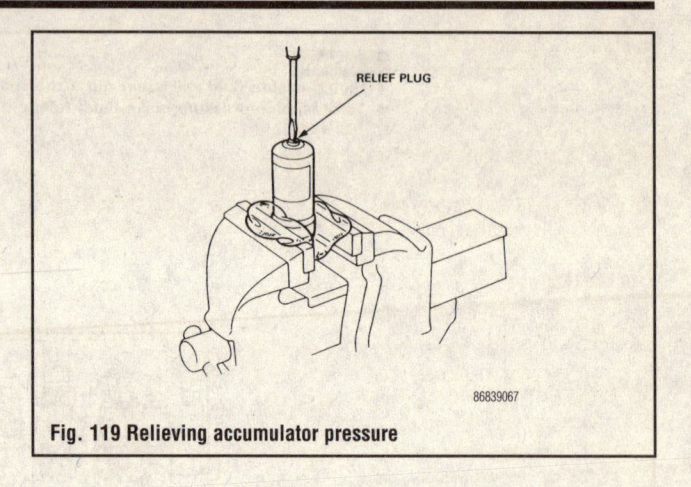

Fig. 119 Relieving accumulator pressure

FILLING THE SYSTEM

The reservoir on the hydraulic modulator is filled in the usual manner with no special procedures being necessary. Always wipe the cap and surrounding area clean of dirt and debris before opening the reservoir; the smallest bit of dirt may impair the operation of the system. When adding fluid, fill the reservoir only to the MAX line on the reservoir; do not overfill.

Only DOT 4 brake fluid must be used; silicone or DOT 5 fluid is specifically prohibited. Do not use any fluid which contains a petroleum base; these fluids will cause swelling and distortion of the rubber parts within the system. Do not use old or contaminated brake fluid. Do not reuse fluid which has been bled from the system.

MODULATOR UNIT BLEEDING WITH THE ALB/ABS CHECKER

1. Place the vehicle on a level surface and block the wheels. Put the transmission in neutral for manual transmission models or **P** for automatic models and release the parking brake.
2. Disconnect the 6-pin ALB/ABS inspection connector from the cross member under the passenger's seat and connect the 6-pin ALB/ABS inspection connector to the ALB/ABS Checker.
3. Fill the modulator reservoir to the MAX level and replace the cap.
4. Start the engine, allowing it to idle a few minutes, then turn it off. Check the fluid level in the modulator reservoir and refill to MAX, if necessary.
5. Bleed high-pressure fluid from the maintenance bleeder with the special tool.
6. Refill the modulator reservoir to the MAX level line and install the reservoir cap.
7. Turn the Mode Selector switch of the checker to 2.
8. While depressing the brake pedal firmly, push the Start Test switch to the modulator. There should be a kickback felt at the brake pedal. If not, repeat Steps 5-8.

➡ **Continue to depress the brake pedal firmly when operating the checker.**

9. Turn the Mode Selector to 3,4 and 5. Perform Step 8 for each of the test mode positions.
10. Refill the modulator reservoir to the MAX level line and install the reservoir cap.

✳✳ CAUTION

Before driving the car, disconnect the ALB/ABS Checker. Complete loss of braking ability could result in severe personal injury or death.

BRAKES 9-37

BLEEDING THE SYSTEM

Bleeding may be performed using either a pressure bleeder or the manual method. In either case an assistant will be required to depress the brake pedal. Extreme cleanliness must be observed at all times. If using the manual method, the fluid reservoir must be filled to the upper edge before bleeding begins. Do not allow the fluid level to drop below the MIN mark at any time. Do not reuse fluid released during bleeding.

Front Brakes

1. Turn the ignition switch **OFF**.
2. Relieve the brake system pressure, as described earlier in this section.
3. If using pressure bleeder equipment, connect it to the brake fluid reservoir and switch it on.
4. Connect a tight-fitting vinyl hose to the bleeder port of the caliper. If using pressure bleeding equipment, begin at the left front caliper. If using the manual method, begin on either side. Immerse the other end of the hose in a container of clean brake fluid.

➡**Use of a cap or cover on the container is recommended. The brake fluid may bleed with enough force to splash out of an open container.**

5. Open the bleeder screw. Have an assistant depress the brake pedal slowly until the fluid flows without bubbles.
6. Close the bleeder screw before the pedal is released.
7. Remove the vinyl tube from the caliper. Inspect and top off the fluid supply in the reservoir if necessary.
8. Repeat the procedure at the opposite wheel.

Rear Brakes

1. Turn the ignition switch **OFF**.
2. Relieve the brake system pressure, as described earlier in this section.
3. If using pressure bleeder equipment, connect it to the brake fluid reservoir and switch it on.
4. Connect a tight-fitting vinyl hose to the bleeder port of either caliper. Immerse the other end of the hose in a container of clean brake fluid.

➡**Use of a cap or cover on the container is recommended. The brake fluid may bleed with enough force to splash out of an open container.**

5. Open the bleeder screw. Have an assistant turn the ignition switch **ON**.
6. Press the lever of the proportioning valve towards the axle until brake fluid flows out without bubbles. Release the lever and close the bleeder screw.

➡**Running time of the ALB/ABS pump must not exceed 120 seconds at any one time. If this time is approached or exceeded, a minimum of 10 minutes cooling time is required before proceeding. Do not allow the fluid level to fall below the MIN line at any time.**

7. Switch the ignition **OFF** while transferring equipment. Remove the vinyl tube from the caliper. Inspect and top off the fluid supply in the reservoir if necessary.
8. Repeat the procedure at the opposite wheel.
9. Once both rear calipers are bled and the service equipment removed, switch the ignition **ON** until the pump shuts off.
10. Fill the brake fluid reservoir to the MAX line.

9-38 BRAKES

BRAKE SPECIFICATIONS
All measurements in inches unless noted

Year	Model		Master Cylinder Bore	Brake Disc Original Thickness	Brake Disc Minimum Thickness	Maximum Run-out	Brake Drum Original Inside Diameter	Max. Wear Limit	Maximum Machine Diameter	Min. Lining Front	Min. Lining Rear
1984	Civic/CRX		NA	-	[1]	0.004	7.07 [2]	-	-	0.120	0.080
1985	Civic/CRX		NA	-	[1]	0.004	7.07 [2]	-	-	0.120	0.080
1986	Civic/CRX		NA	-	[1]	0.004	7.07 [2]	-	-	0.120	0.080
1987	Civic/CRX		NA	-	[1]	0.004	7.07 [2]	-	-	0.120	0.080
1988	Civic/CRX		NA	-	- [3]	0.004	7.07 [2]	-	-	0.120	0.080
1989	Civic		NA	-	0.670	[4]	7.09	-	-	0.120	0.080
	Civic SW		NA	-	0.670	[4]	7.87	-	-	[5]	0.080
	CRX		NA	-	[1]	[4]	7.09	-	-	0.120	0.080
1990	Civic		NA	0.83	0.75	0.004	7.09	7.13	7.13	0.120	0.079
	Civic SW		NA	0.75	0.67	0.004	7.90	7.95	7.95	0.120	0.080
	CRX		NA	[6,7]	0.67 [4]	0.004	7.09	7.13	7.13	0.120	0.080
1991	Civic		NA	0.75 [8]	0.670 [9]	0.004	7.09	7.1	7.13	0.120	0.079
	Civic Wagon		NA	0.83 [10]	0.750 [11]	0.004	7.90	7.95	7.95	0.120	0.080
	CRX		NA	[12]	0.670 [13]	0.004	7.09	7.13	7.13	0.120	0.080
1992	Civic	F	NA	0.827	0.748	0.004	[14]	[15]	[16]	0.080	
		R		0.350	0.320	-	7.09	7.15	7.13		0.080
1993	Civic	F	NA	0.830	0.750	0.004	-	-	-	0.060	- [19]
		R		0.350	0.310	-	[17]	[18]	[18]	-	0.080
	del Sol	F	NA	0.830	0.750	0.004	-	-	-	0.060	- [19]
		R		0.350	0.310	-	7.09	7.13	7.13	-	0.080
1994	Civic	F	NA	0.830	0.750	0.004	-	-	-	0.060	- [19]
		R		0.350	0.310	-	[17]	[18]	[18]	-	0.080
	del Sol	F	NA	0.830	0.750	0.004	-	-	-	0.060	- [19]
		R		0.350	0.310	-	7.09	7.13	7.13	-	0.080
1995	Civic	F	NA	0.830	0.750	0.004	-	-	-	0.060	- [19]
		R	-	0.350	0.310	-	[17]	[18]	[18]	-	0.080
	del Sol	F	NA	0.830	0.750	0.004	-	-	-	0.060	- [19]
		R	-	0.350	0.310	-	7.09	7.13	7.13	-	0.080

NA: Not Available
F: Front
R: Rear
1. Civic 1300cc: 0.39
   Civic 1500cc: 0.59
   CRX Std. 0.67
   1300cc and HF: 0.35
2. Civic Wagon: 7.85
3. Civic/CRX HF: 0.590
   All others: 0.670
4. CRX DX: 0.750
   CRX HF: 0.590
5. 4WD: 0.120
   Except 4WD: 0.060
6. CRX DX.: 0.83
   CRX HF: 0.67
   CRX Si: 0.75
7. Rear: 0.39
8. Civic EX: 0.83
9. Civic EX: 0.750
10. 4WD: 0.75
11. CRX Std: 0.750
    CRX HF: 0.590
12. CRX Std.: 0.83
    CRX HF: 0.67
    CRX Si: 0.75
13. 4WD: 0.67
14. Except 1.5L 4 door models with automatic transaxle, and all 1.6L: 7.87
15. Except 1.5L 4 door models with automatic transaxle, and all 1.6L: 7.95
16. Except 1.5L 4 door models with automatic transaxle, and all 1.6L: 7.91
17. All manual transaxle models, except 1.6L Coupe 1.6L: 7.09
    All automatic transaxle models and 1.6L Coupe with manual transaxle: 7.87
18. All manual transaxle models, except 1.6L Coupe with manual trans.: 7.13
    All models with automatic trans. and 1.6L Coupe with manual trans.: 7.91
19. Rear disc brakes: 0.060

86839C01

BODY AND TRIM 10-3

ADJUSTMENT

♦ See Figures 5 and 6

Proper door alignment can be obtained by adjusting the door hinge and door lock striker. The door hinge and striker can be moved up and down, fore and aft in enlarged holes by loosening the attaching bolts.

➙The door should be adjusted for an even and parallel fit for the door opening and surrounding body panels.

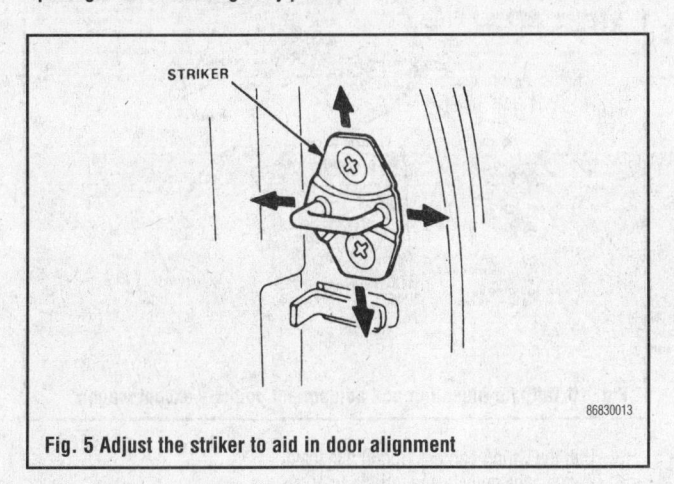

Fig. 5 Adjust the striker to aid in door alignment

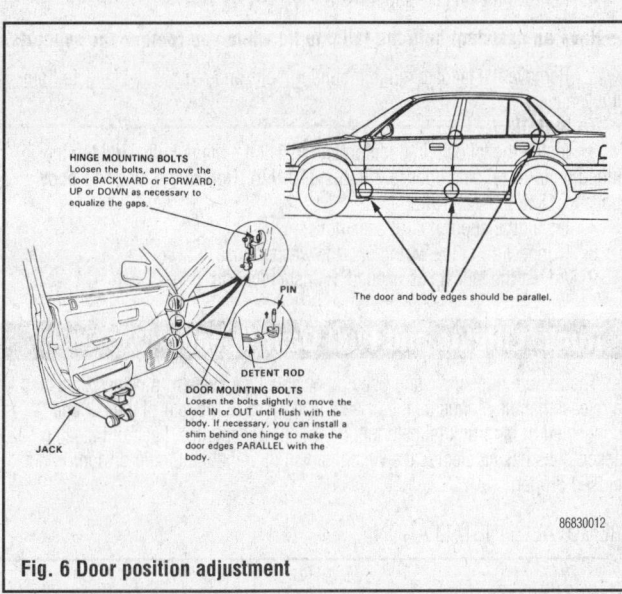

Fig. 6 Door position adjustment

Hood

REMOVAL & INSTALLATION

1. Open the hood and protect the body with covers to protect the painted surfaces.
2. Mark with a permanent felt tip pen the hood hinge locations on the hood for proper installation.
3. Holding both sides of the hood, unscrew the bolts securing the hinge to the hood. This operation requires a helper.
4. Install the hood in position, align the marks made during removal and install the retaining bolts.
5. Tighten the hood retaining bolts to 7 ft. lbs. (10 Nm). Adjust the hood as needed.

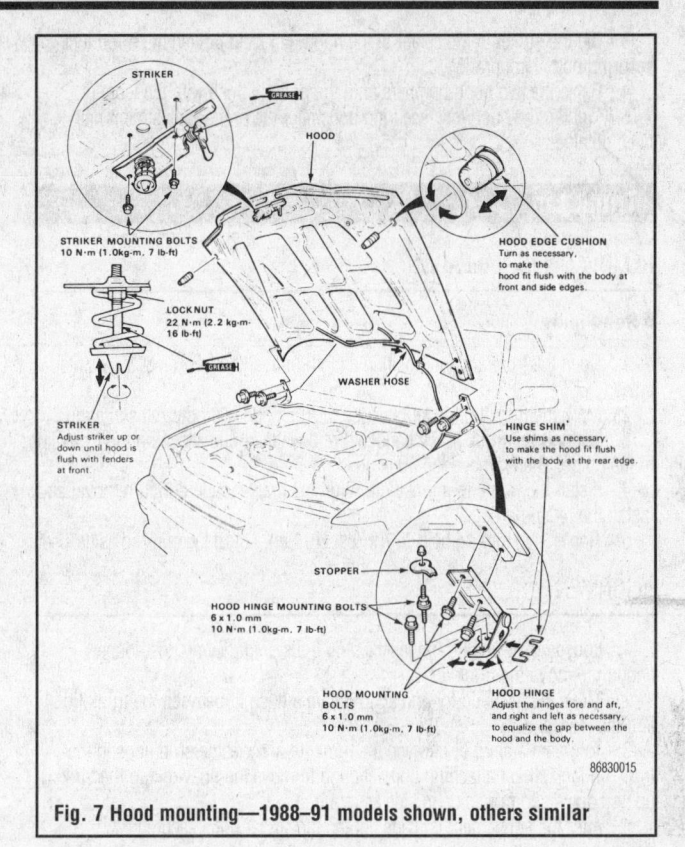

Fig. 7 Hood mounting—1988–91 models shown, others similar

ALIGNMENT

♦ See Figure 8

The hood can be adjusted with bolts attaching the hood to the hood hinges, hood lock mechanism and hood bumpers. Adjust the hood for an even fit between the front fenders.

1. Adjust the hood fore and aft by loosening the bolts attaching the hood to the hinge and repositioning hood.
2. Loosen the hood bumper lock nuts and lower bumpers until they do not contact the front of the hood when the hood is closed.

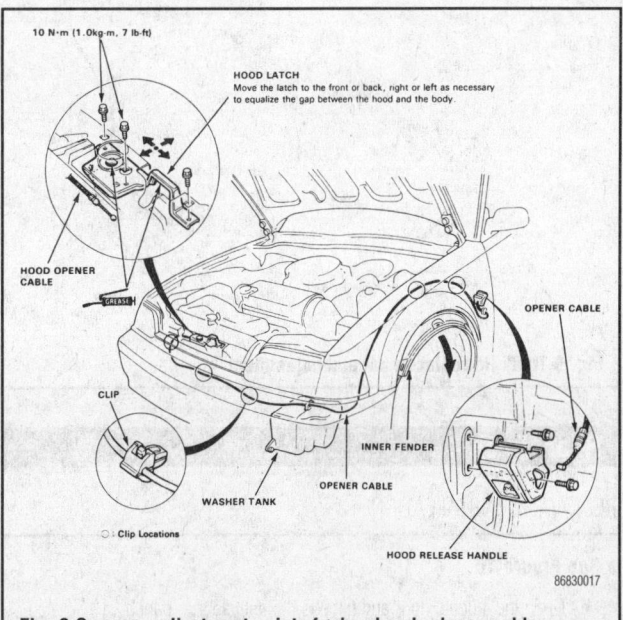

Fig. 8 Common adjustment points for hood and release cable

10-4 BODY AND TRIM

3. Set the striker at the center of the hood lock, and tighten the hood lock securing bolts temporarily.
4. Raise the two hood bumpers until the hood is flush with the fenders.
5. Tighten the hood lock securing bolts after the proper adjustment has been obtained.

Trunk Lid

REMOVAL & INSTALLATION

▶ See Figure 9

1. Open the trunk lid and position a cloth or cushion to protect the painted areas.
2. Mark the trunk lid hinge locations or trunk lid for proper reinstallation.
3. Support the trunk lid by hand and remove the bolts attaching the trunk lid to the hinge. Then remove the trunk lid.
4. Install the trunk lid in position. Align the marks made during removal and install the retaining bolts.
5. Tighten the trunk lid bolts to 7 ft. lbs. (10 Nm). Adjust the trunk lid as needed.

ALIGNMENT

1. Loosen the trunk lid hinge attaching bolts until they are just loose enough to move the trunk lid.
2. Move the trunk lid for and aft to obtain a flush fit between the trunk lid and the rear fender.
3. To obtain a snug fit between the trunk lid and weatherstrip, loosen the trunk lid lock striker attaching bolts enough to move the lid, working the striker up and down and from side to side as required.
4. After the adjustment is made tighten the striker bolts securely.

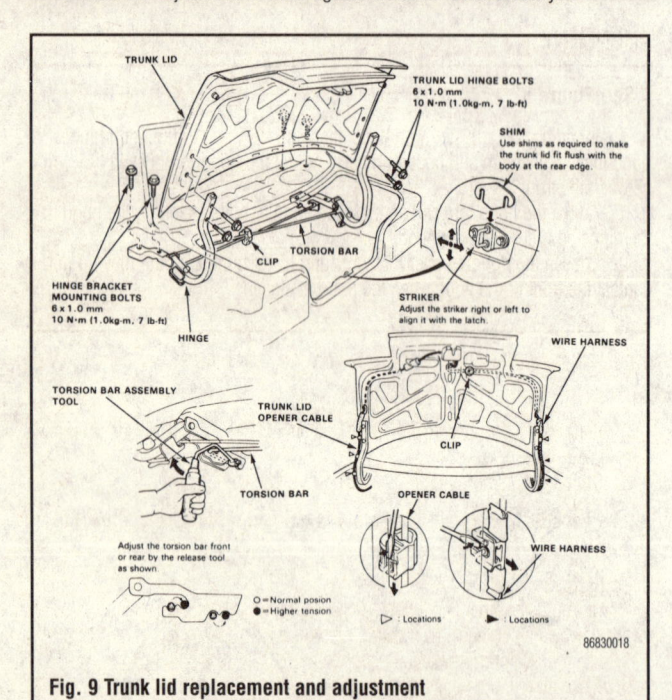

Fig. 9 Trunk lid replacement and adjustment

Rear Hatch

REMOVAL & INSTALLATION

▶ See Figure 10

1. Open the tailgate door and remove the tailgate trim panel.
2. Disconnect the electrical connectors from the rear windshield wiper motor, defroster and brake light. Disconnect the washer hose.

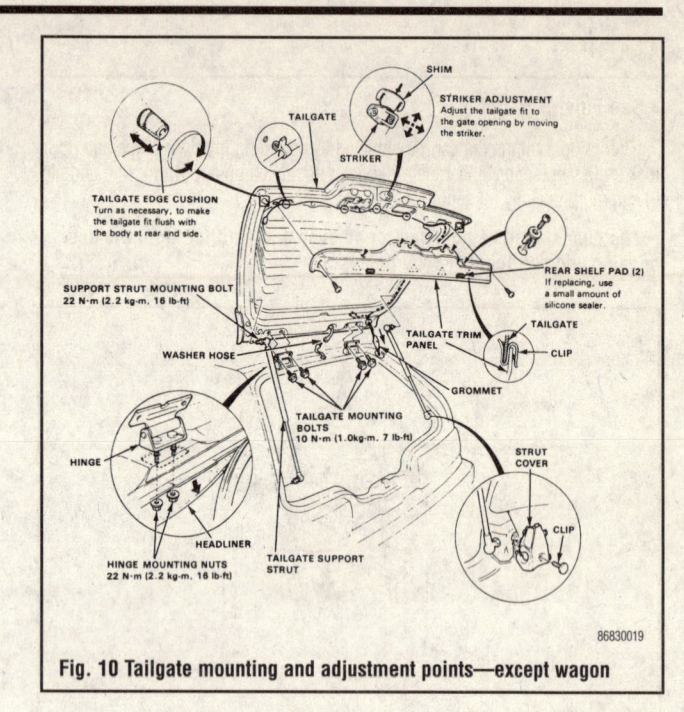

Fig. 10 Tailgate mounting and adjustment points—except wagon

3. Pull the wiring harness from the tailgate.
4. Remove the tailgate support struts.

➡ Have an assistant hold the tailgate lid while you remove the supports.

5. Remove the tailgate hinge mounting bolts and remove the tailgate from the vehicle.

To install:

6. Install the tailgate in position and install the hinge bolts. Tighten the hinge bolts-to-tailgate door to 7 ft. lbs. (10 Nm). Tighten the hinge-to-body bolts to 16 ft. lbs. (22 Nm).
7. Install the support strut assemblies.
8. Connect all of the wiring and the washer hose.
9. Adjust the tailgate as needed. Reinstall the trim panel.

Trunk, Liftgate and Fuel Door Opener

The remote opener system used, is a cable type system. A cable connected to a release handle alongside the drivers seat, goes to the rear of the car and is connected to the trunk/tailgate/fuel door latch assemblies. Each of the latch assemblies has an electrical switch that will illuminate if the lid or door is not closed properly.

REMOVAL & INSTALLATION

Opener Cables

▶ See Figure 11

1. Remove from the interior of the car, the left side door sill molding and roll the carpet back to expose the cables.
2. Remove the left quarter trim panel and the left rear trim or trunk panel.
3. Remove the opener cover screws and the opener. Remove the release handles.
4. Remove the cables from their retaining clip.
5. Disconnect the cable at either the trunk/hatch release or the fuel door release.
6. Disconnect the cable at the release lever. Remove the cable from the vehicle.

To install:

7. Install the new cable in position and connect it to the trunk/hatch or fuel door release. Connect the cable to the release handle.
8. Make sure the cable is seated in the retaining clips properly. Install the trim panels that were removed.
9. Connect the release lever handles and install the cover.

BODY AND TRIM 10-5

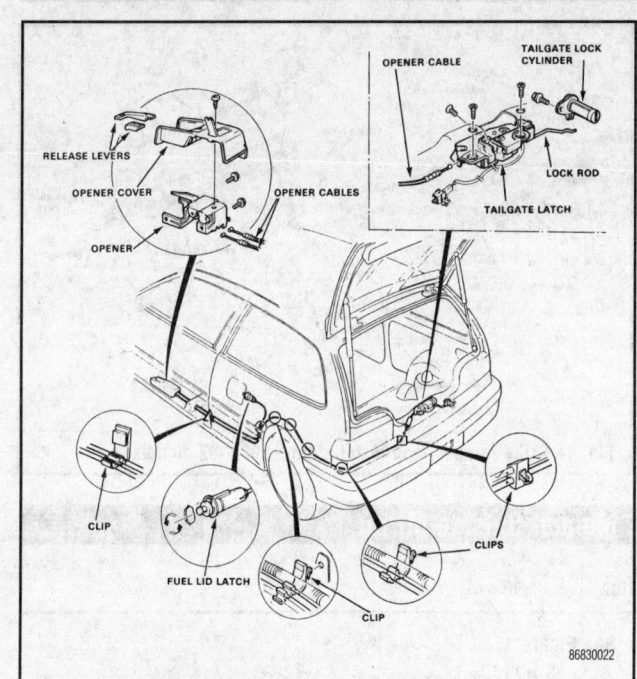

Fig. 11 The remote trunk and fuel door opener cables run along the driver's side of the vehicle

Bumpers

REMOVAL & INSTALLATION

Front

▶ See Figure 12

1984–87 MODELS

1. Remove the bumper lower mount bolts.
2. Remove the bumper mount bolts. Disconnect the electrical leads from the lights.

3. Slide the bumper assembly forward off of the front of the car.

➡ On each side of the front fender is a locating pin, be sure to slide the bumper off of this pin.

To Install:

4. Install the bumper by sliding back onto the car. Make sure that the inner part of the bumper catches the locating pins on the fender.
5. Install the mounting bolts, tighten them to 16 ft. lbs. (22 Nm).
6. Make sure that all bolts are tightened.

1988–91 MODELS

1. Remove the right and left turn signal assemblies.
2. Remove the two mounting screws on each side of the car that hold the bumper to the inner fender well.
3. Remove the two lower mounting bolts and the four mounting bolts.
4. Lift the bumper assembly slightly and slide it forward off of the vehicle.

➡ On each side of the front fender is a locating slide assembly, be sure to slide the bumper off of this.

To Install:

5. Install the bumper by sliding back into position. Make sure that the slide clips catch. The bumper needs to be lifted slightly so that the beam hooks catch the body.
6. Install the mounting bolts and tighten them to 16 ft. lbs. (22 Nm).
7. Install the turn signal assemblies. Make sure that all pieces are aligned.

1992–95 MODELS

1. Remove the upper bumper stiffener.
2. Remove the front spoiler on Hatchback VX, Civic Coupe models.
3. Remove the mounting screws on each side of the car that hold the bumper to the inner fender well.
4. Remove the two lower mounting bolts and the four bumper beam mounting bolts.
5. Lift the bumper assembly slightly and slide it forward off of the vehicle.

➡ On each side of the front fender is a locating slide assembly, be sure to slide the bumper off of this.

To Install:

6. Install the bumper by sliding back into position. Make sure that the slide clips catch. The bumper needs to be lifted slightly so that the beam hooks catch the body.
7. Install the bumper beam mounting bolts and tighten them to 16 ft. lbs. (22 Nm).

Rear Bumper

1984–87 MODELS

1. Remove the bumper lower mount bolts.
2. Remove the bumper mount bolts. Disconnect the electrical leads from the license plate light.
3. Slide the bumper assembly forward off of the car.

➡ On each side of the rear fender is a locating pin, be sure to slide the bumper off of this pin.

To Install:

4. Install the bumper by sliding back onto the car. Make sure that the inner part of the bumper catches the locating pins on the fender.
5. Install the mounting bolts, tighten them to 16 ft. lbs. (22 Nm).
6. Make sure that all bolts are tightened.

1988–91 MODELS

▶ See Figure 13

1. Remove the rear trunk/hatch cover panels. Disconnect the license plate light connector from the trunk/hatch area.
2. Remove the bumper retaining bolts from the fender well area (2).
3. Remove the screws and then remove the under protector from the floor area.
4. Remove the two lower mounting bolts from the trunk/hatch floor area.
5. Remove the two upper retaining bolts from the rear of the trunk/hatch area.

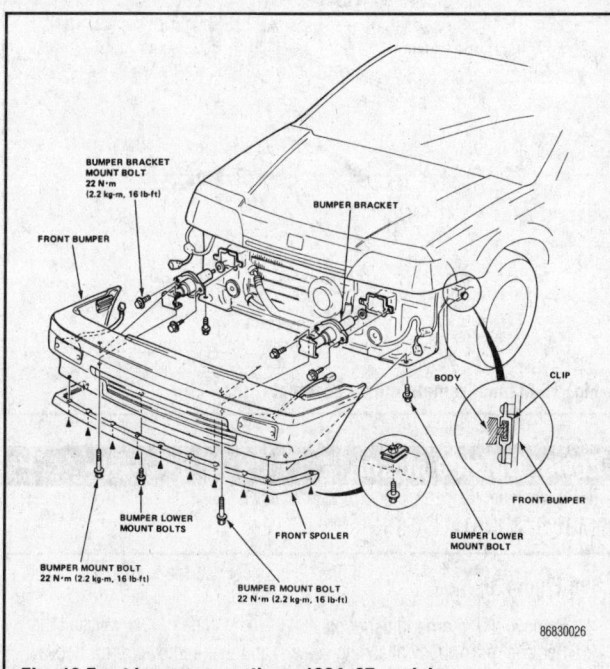

Fig. 12 Front bumper mounting—1984–87 models

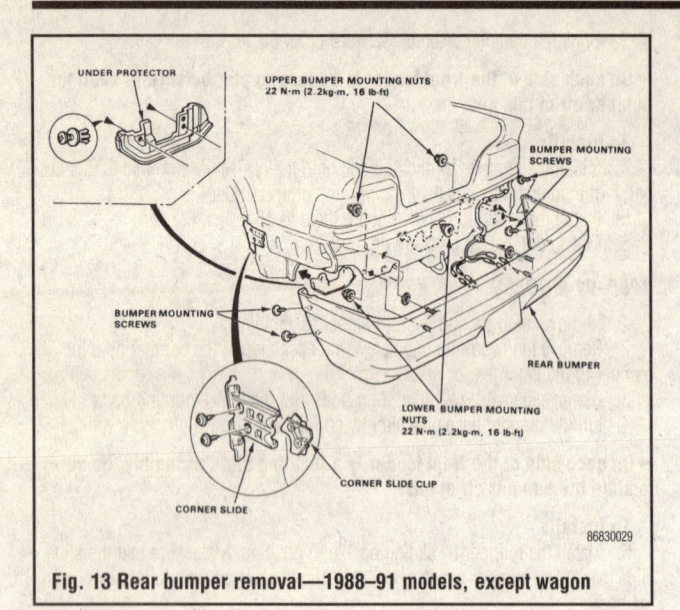

10-6 BODY AND TRIM

Fig. 13 Rear bumper removal—1988–91 models, except wagon

6. Remove the bumper by sliding it to the rear. Make sure that the bumper slides off of the bumper slide clips.

➡ On each side of the rear fender is a locating slide assembly, be sure to slide the bumper off of this.

To Install:

7. Install the bumper into position by sliding it into position.
8. Install the retaining bolts and tighten to 16 ft. lbs. (22 Nm).
9. Connect the license plate light assembly.

1992–95 MODELS

1. Remove the rear trunk/hatch cover panels. Disconnect the license plate light connector from the trunk/hatch area.
2. Remove the bumper retaining bolts from the fender well area.
3. Remove the screws and then remove the under protector from the floor area.
4. Remove the two lower mounting bolts from the trunk/hatch floor area.
5. Remove the two upper retaining bolts from the rear of the trunk/hatch area.
6. Remove the bumper by sliding it to the rear. Make sure that the bumper slides off of the bumper slide clips.

➡ On each side of the rear fender is a locating slide assembly, be sure to slide the bumper off of this.

To Install:

7. Install the bumper into position by sliding it into position.
8. Install the retaining bolts and tighten to 16 ft. lbs. (22 Nm).
9. Connect the license plate light assembly.

Front Grille Assembly

REMOVAL & INSTALLATION

1984–87 Models

▶ See Figure 14

1. Open the hood and remove the screws that run along the top of the grille.
2. Remove the screws from the front of the grille.
3. Remove the headlight garnish and remove the grille.
4. Install the assembly and install all of the screws.

1988–95 Models

1988–95 models do not have a removable grille assembly. When the body style was changed in 1988, the grille was removed in order to make the hood line lower and more aerodynamic.

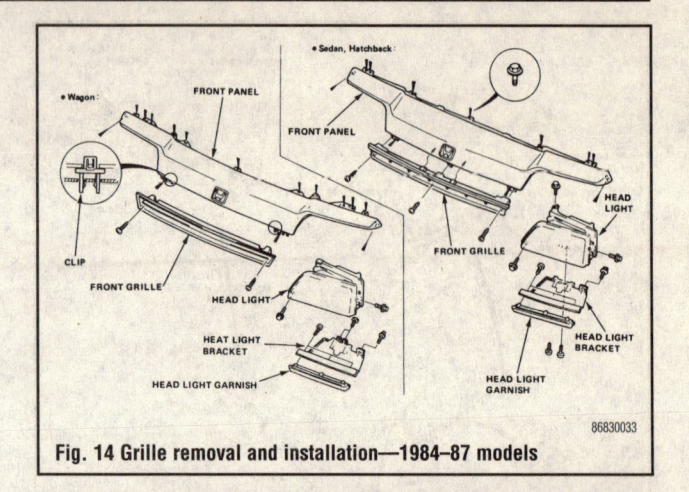

Fig. 14 Grille removal and installation—1984–87 models

Manual Outside Mirror

REMOVAL & INSTALLATION

▶ See Figure 15

1. Remove control knob handle.
2. Remove door corner finisher panel.
3. Remove mirror body attaching screws, and then remove mirror body
4. For 1994–91 models, install the mirror assembly and tighten the screws. For 1992–95 models, the nuts. Tightening nuts for 1992–95 models to 3.6 ft. lbs. (5 Nm).
5. Replace the cover panel.

➡ Apply sealer to the rear surface of door corner finisher panel during installation to prevent water leak. Check for water leaks with a source of low pressure water after installation is complete.

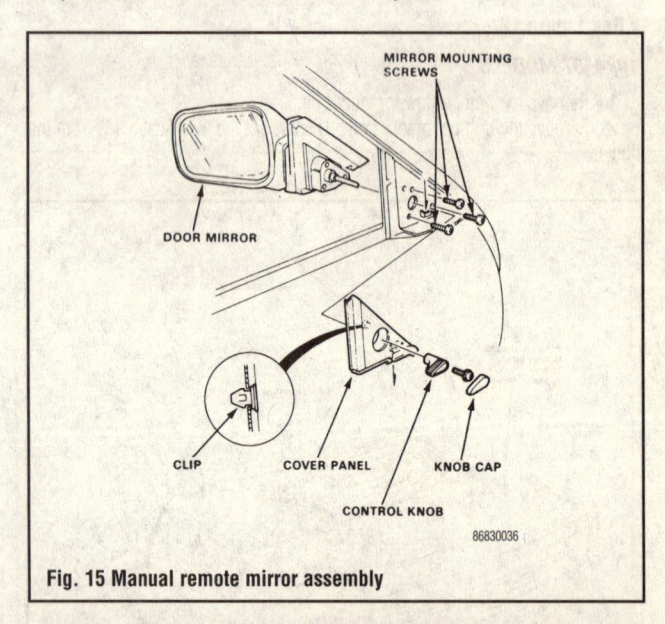

Fig. 15 Manual remote mirror assembly

Power Outside Mirror

REMOVAL & INSTALLATION

▶ See Figure 16

1. Remove door corner finisher panel.
2. Remove mirror body attaching screws, and then remove mirror body
3. Disconnect the electrical connection.

BODY AND TRIM 10-7

Fig. 16 Power remote mirror assembly

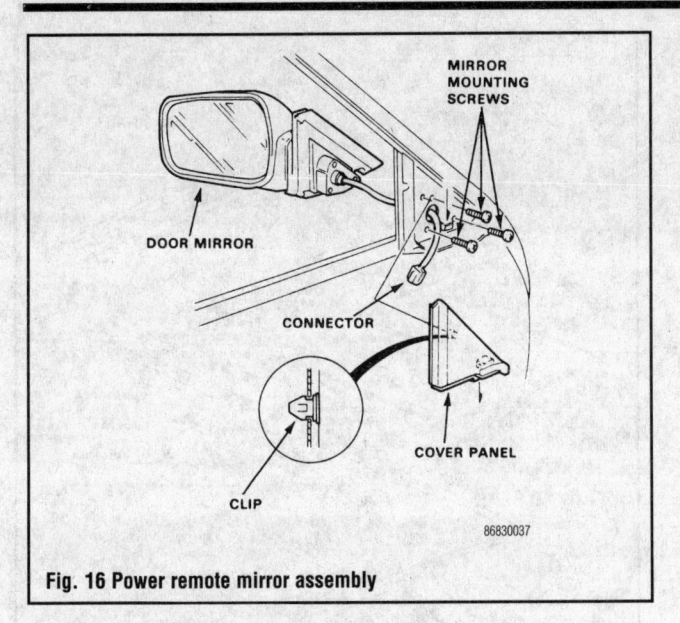

➡ It may be necessary to remove the door trim panel to gain access to the electrical connection.

To Install:
4. Position the mirror in place and replace the three retaining screws or nuts.
5. Connect the connector and replace the panel.

Power Mirror Switch

♦ See Figure 17

The power mirrors are controlled by a single switch assembly, located on the instrument panel. The motors that operate the mirrors are part of the mirror assembly and cannot be replaced separately.

The mirror switch consists of a left-right change over select knob and control knobs. The switch is ready to function only when the ignition switch is in the **ACC** or **ON** position. Movement of the mirror is accomplished by the motors, located in the mirror housing.

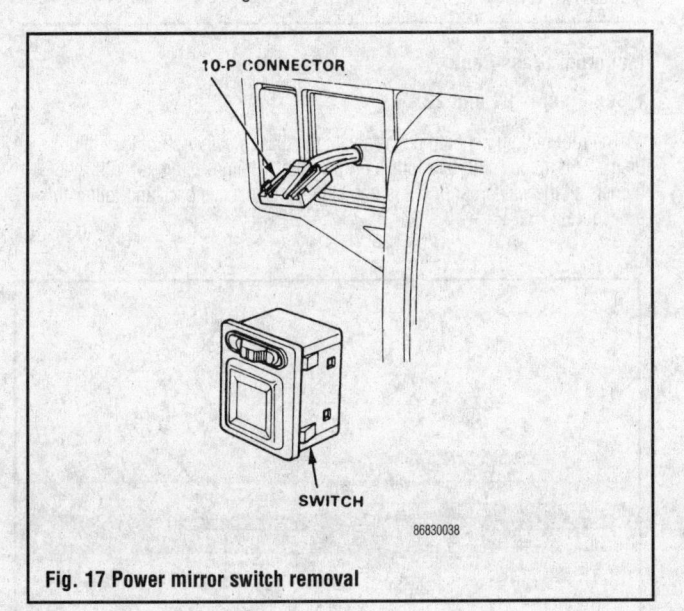

Fig. 17 Power mirror switch removal

REMOVAL & INSTALLATION

1. Disconnect the negative battery cable.
2. Remove the lower instrument panel trim cover.
3. Remove the screw retaining the switch and, using a small pry bar, remove the switch from the instrument panel.
4. Disconnect the electrical lead and remove the switch from the vehicle.

To Install:
5. Plug in the electrical lead to the switch and push it into the instrument panel.
6. Install the retaining screw and install the lower trim panel.
7. Connect the battery cable.

Antenna

REMOVAL & INSTALLATION

1. Remove antenna mounting nut or screws.
2. Disconnect the antenna lead at the radio.
3. Tie a cord to the end on the antenna lead.
4. Remove antenna from vehicle while fishing out the cord.
5. Install using the cord to pull the new antenna lead through the body.
6. Connect the lead at the radio.
7. Re-mount the antenna and attach with the mounting screws or nuts.

Fender

REMOVAL & INSTALLATION

1. Raise and support the front of the vehicle so that it is just off of the ground.
2. Remove the tire and wheel assembly.
3. Remove the inner fender cover assembly bolts and remove it.
4. Remove the fender mounting bolts.

➡ The fender is joined to the body with sealer at some points. When removing the fender, a small putty knife will help to break the seal in these areas.

5. Remove the fender from the vehicle. Be careful of the painted surfaces.

To install:
6. Apply body sealer (available from Honda) to the correct location. Install the fender in position, do not tighten the bolts. Check the fender-to-hood-to-door clearance. Once the clearance is correct and even, tighten all of the mounting bolts.
7. Install the inner fender assembly and its mounting bolts.
8. Install the wheel and tire assembly. Lower the vehicle.

Electric Sunroof/Moonroof

♦ See Figures 18, 19 and 20

➡ Although "sunroof" and "moonroof" are terms which are often interchanged, generally speaking, the sunroof is steel and must be opened to see through while the moonroof assembly is glass with a sunshade and need not be opened to see through. Both are equipped with water drain tubes. It is important to keep these drain tubes open. The preferred method is to blow compressed air through the drain tubes at regular intervals, in order to keep the drain tubes clear.

The sunroof or moonroof is operated by a switch, which is usually located on the left side of the instrument panel. The system consists of the switch, motor drive cables and relay(s), which are located in a fuse/relay junction block in the engine compartment.

10-8 BODY AND TRIM

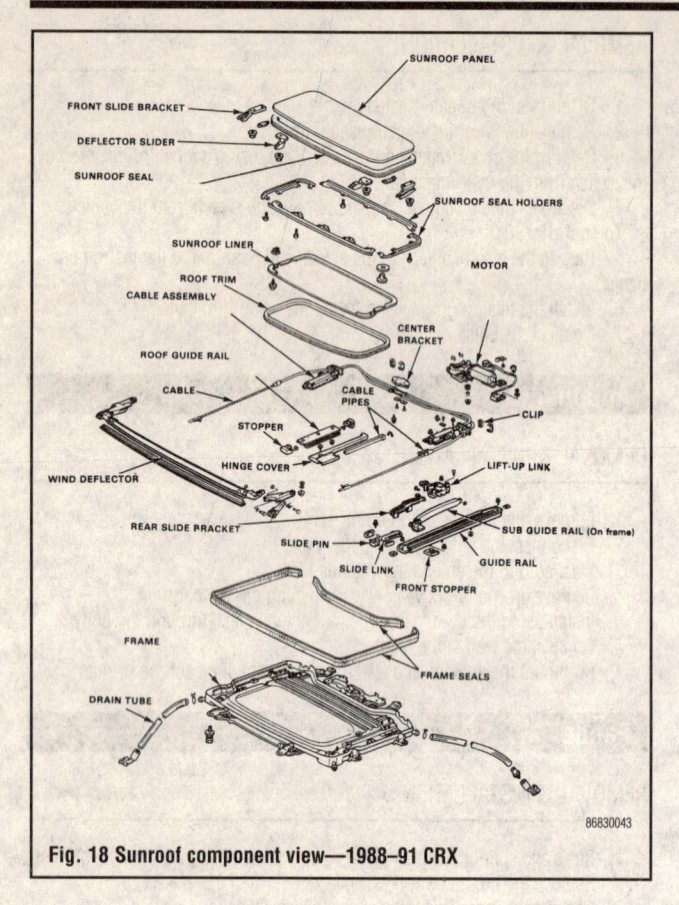

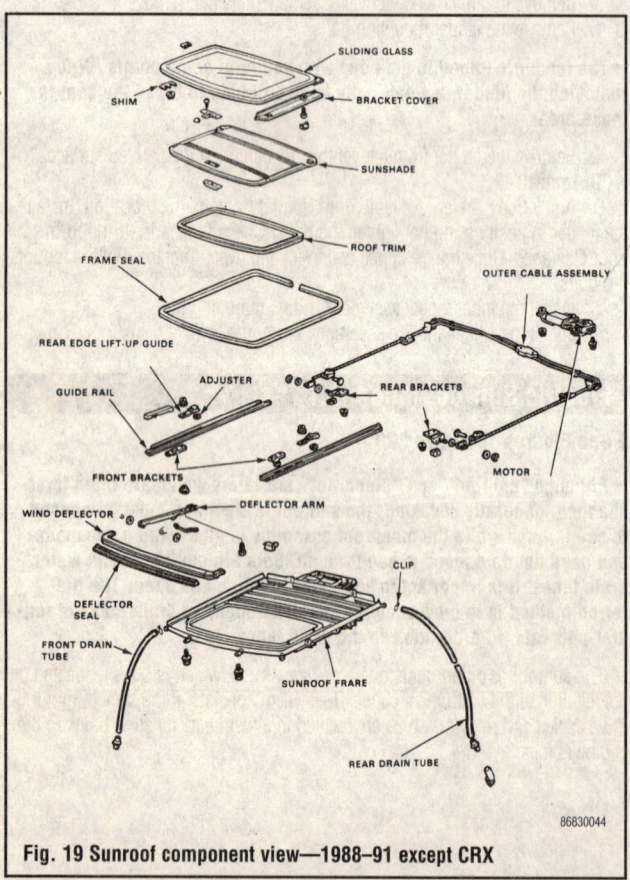

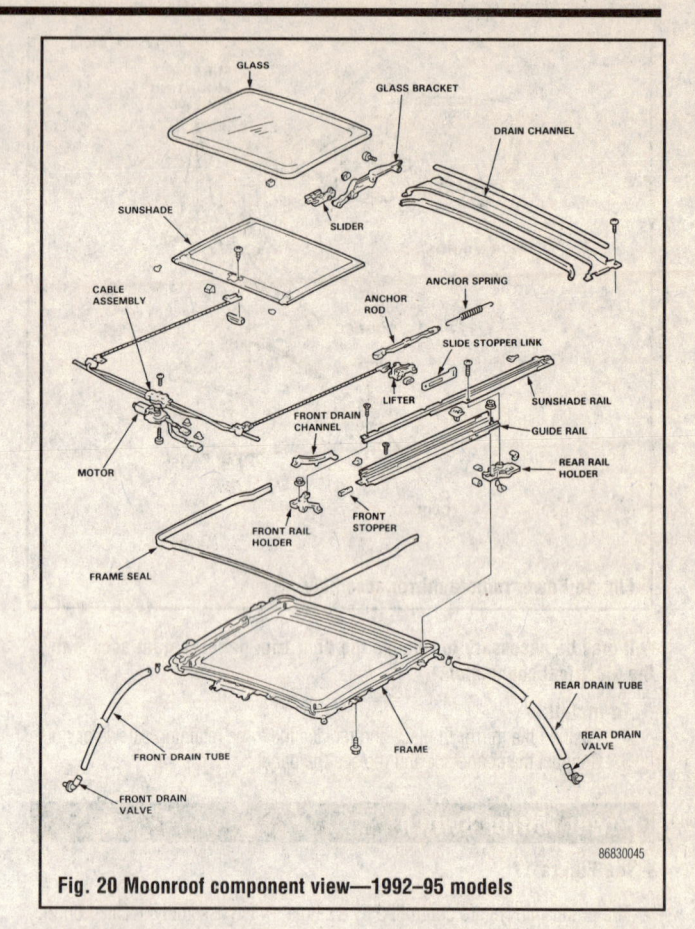

Fig. 18 Sunroof component view—1988-91 CRX

Fig. 19 Sunroof component view—1988-91 except CRX

Fig. 20 Moonroof component view—1992-95 models

Several fuses, located in a fuse/relay junction block in the engine compartment and dash fuse box, are used to power/protect the system. The powered roof can be closed manually, (should it be necessary) by removing the headliner plug, insert the handle and turn the gear to close the sunroof.

ADJUSTMENTS

Moonroof Glass Panel

▶ See Figures 21 and 22

The roof molding should be even with the glass weather strip, to within 0mm–2.5mm for 1988-91 models or to within 1.3mm–2.3mm for 1992-95 models all the way around. If it is not, slide the sunshade back and follow this procedure.

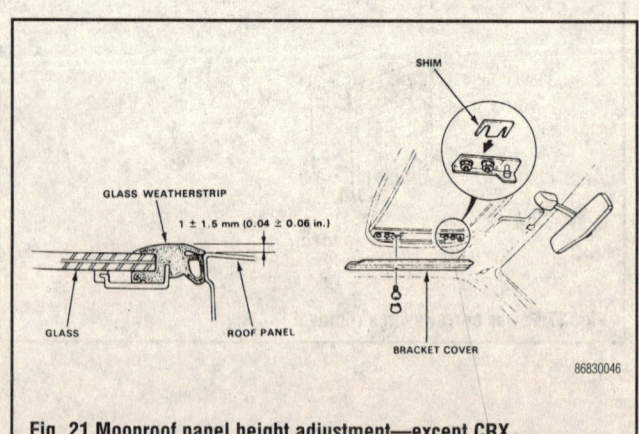

Fig. 21 Moonroof panel height adjustment—except CRX

BODY AND TRIM 10-9

Fig. 22 Glass height adjustment—1992–95 models

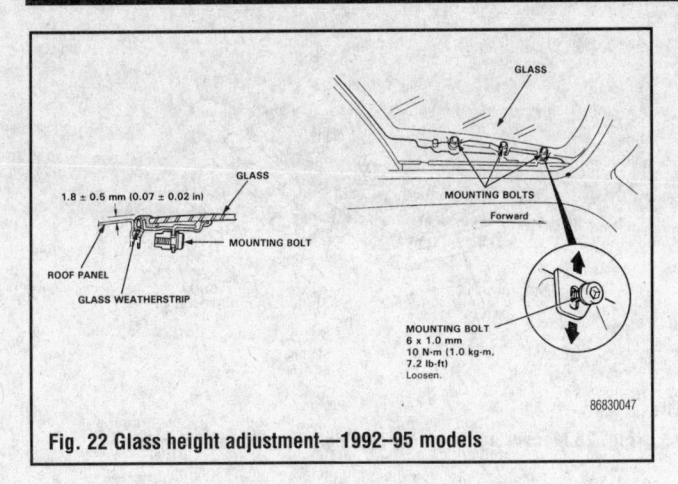

1. Pry the plug out of the glass mount bracket cover, remove the screw, then slide the cover off to the rear.
2. Loosen the mount bracket nuts and install shims between the glass frame and bracket. Repeat this on the other side if necessary.

Side Clearance

If the glass weather strip fits too tightly against the roof molding on either side when closed, slide the sunshade back and follow this procedure.
1. Pry the plug out of each mount bracket cover, remove the screw, then slide the cover off to the rear.
2. Loosen all eight mount bracket nuts. Move the glass right or left as necessary and tighten the mount bracket nuts.

Rear Edge Closing

Open the glass approximately 1 ft., then close it to check where the rear edge begins to rise. If it rises too soon and seats too tightly against the roof molding or too late and does not seat tightly enough, adjust as follows:
1. Open the glass fully. Remove the rail covers from both sides and loosen the lift-up guide screws.
2. Move the guide forward or backward, then tighten the screws and recheck the roof closing. The guides have notches 1.5mm each and can be adjusted two notches forward or backward.

REMOVAL & INSTALLATION

Sunroof Glass and Sunshade

1. Slide the sunshade all the way back. Pry the plug out of each mount bracket cover, remove the screw, then slide the cover off to the rear.
2. Close the glass fully. Remove the nuts from the front and rear mounts on both sides.
3. Remove the glass by lifting it up and pulling it towards the front of the vehicle. Once the glass is removed, pull the sunshade out. When removing the sunshade, it is correct to bend the sunshade slightly to aid in the removal.
4. To install replace the glass and shade and replace the mount nuts.
5. Replace the plugs to the mount bracket cover and check for fit and proper function.

Sunroof Motor, Drain Tube and Frame

▶ See Figure 23

1. Remove the headliner from inside of the vehicle.
2. Remove the sunroof motor by removing two bolts and three nuts from the bottom of the motor mount plate. Disconnect the motor wire harness at the connector and remove the motor.
3. Slide back the drain tube clamps and remove the drain tubes. Remove the 11 mounting bolts from the sunroof frame and remove the frame from the vehicle.
4. To install, insert the frame's rear pins into the body holes, then install the rest of the assembly in the reverse order of the removal procedure.

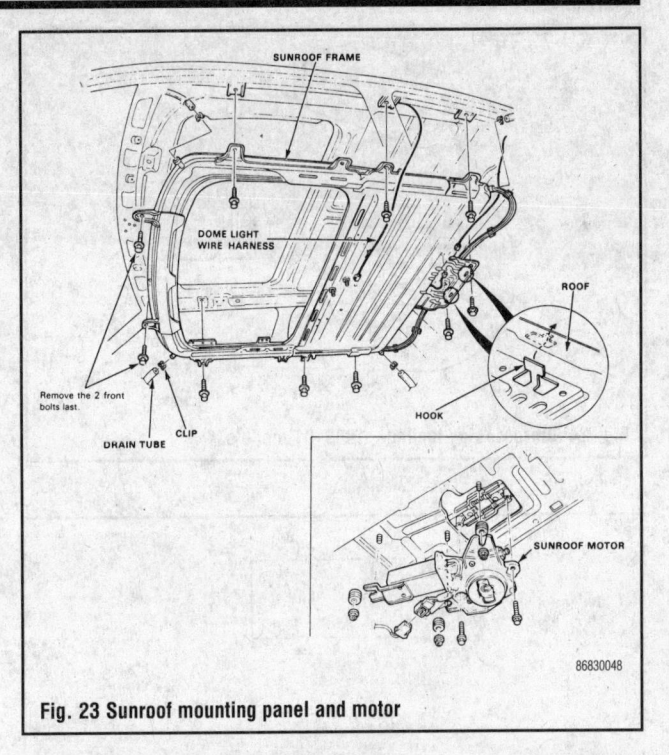

Fig. 23 Sunroof mounting panel and motor

➡ Before installing the sunroof motor, measure the effort required to close the sliding panel using a suitable spring scale. If the load is over 22 lbs. (10 kg), check the side clearance and the glass height adjustment. Be sure when using the spring scale to protect the leading edge of the sunroof with a shop rag.

Sunroof Cable Replacement

With the sun roof out of the vehicle, remove the guide rail mounting nuts, lift off the guide rails and remove the cables with the rear mounts attached. Be sure to fill the groove in each grommet with a suitable sealant and apply a suitable grease to the inner cable.

Wind Deflector

A gap between the deflector seal and roof molding will cause wind noise when driving at high speed with the roof opening.
1. Open the sunroof and pry the rail covers off of both sides. Loosen the deflector mounting nuts. The wind deflector can be adjusted 0.008 in. (2.0mm) forward or backward.
2. Adjust the deflector forward or backward so that the edge of its seal touches the roof molding evenly.
3. The height of the deflector when opened can not be adjusted. If it is damaged or deformed, replace it.

Sunroof Rear Mount Bracket Disassembly

1. Remove the side guides from the rear mount brackets. It is advisable to replace the guides with new ones whenever they are disassembled.
2. Pry the E-clip off of the pin and remove the rear mount bracket from the cable.
3. Assembly is the reverse order of the disassembly procedure.

COMPONENT TESTING

Sunroof Relay

▶ See Figures 24 and 25

1988–91 MODELS

1. Remove relay from dash relay holder.
2. There should be no continuity between terminals **A** and **B**.
3. Apply battery voltage to terminal **C** and ground terminal **D**.

10-10 BODY AND TRIM

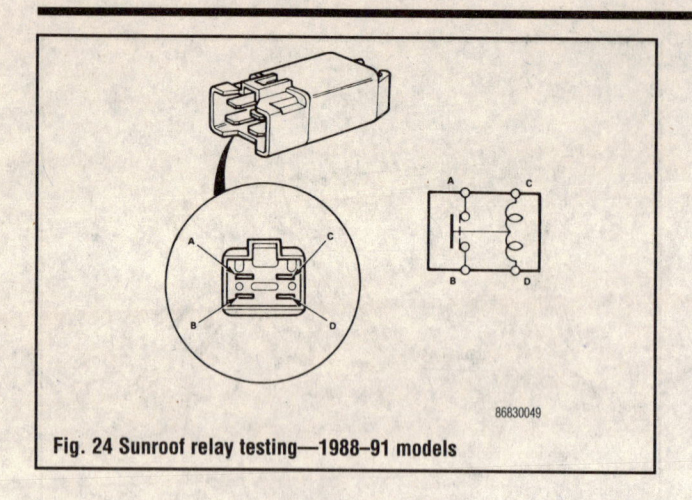

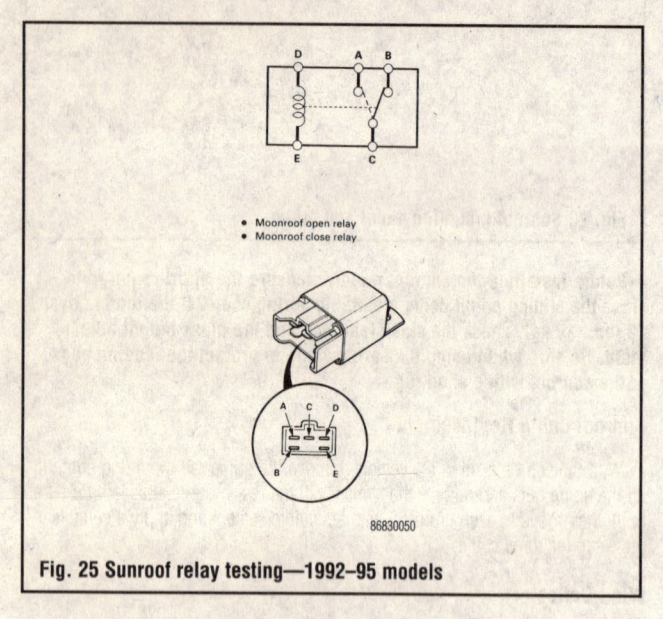

Fig. 24 Sunroof relay testing—1988–91 models

Fig. 25 Sunroof relay testing—1992–95 models

4. There should be continuity between terminals **A** and **B**.
5. If either test fails, replace the relay.

1992–95 MODELS

➡There is a relay for opening and one for closing the moonroof.

1. Remove the relay from its socket.
2. There should be continuity between terminals **A** and **C** when power and ground are applied to the **D** terminal and **E** terminal.
3. There should be continuity between terminals **B** and **C** terminals when the power is disconnected.
4. If either test fails, replace relay.

Moonroof Closing Force Check (With the Motor Installed)

♦ See Figure 26

1. After installing all removed parts, have a helper hold the switch to close the sunroof while measuring the force required to stop the sunroof with a suitable spring scale.
2. Read the force on the scale as soon as the glass stops moving, then immediately release the switch and spring scale. The closing force should be 44–56 lbs. (20–25.5 kg) for 1988–91 models and for 1992–95 models, the closing force should be 44–66 lbs. (20–30 kg).
3. If the force required to stop the sunroof is not within specifications, adjust it, by turning the sunroof motor clutch adjusting nut. Turn clockwise to increase the force and counterclockwise to decrease the force.
4. After the proper adjustment has been made, install a new lockwasher and bend it against the flat on the adjusting nut.

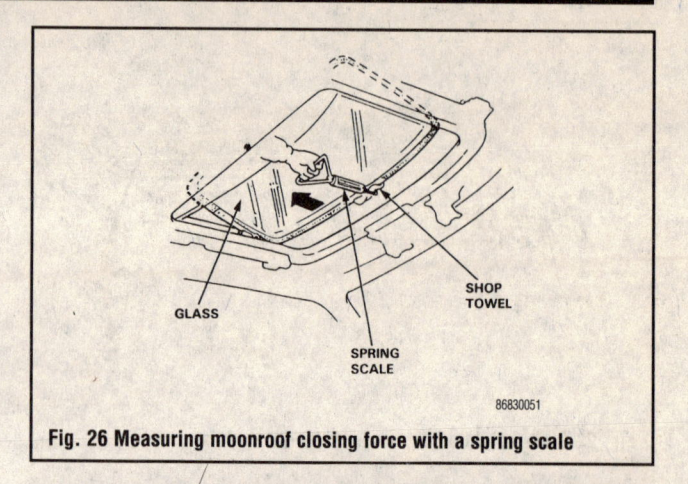

Fig. 26 Measuring moonroof closing force with a spring scale

del Sol Targa Roof

♦ See Figure 27

REMOVAL & INSTALLATION

Rood Side Weatherstrip/Retainer

♦ See Figure 28

1. Remove the roof and store in a safe place.
2. Using a small prytool, unclip the interior roof side trim.
3. Remove the weatherstrip mounting screws and peel off the weatherstrip.
4. Remove the retainer mounting screws and remove the retainer.

To Install:

5. Install the new weatherstrip with clear sealant (cemedine #8500) to the areas shown in the drawing. Glue both ends of the weatherstrip making sure there are no twists before gluing.
6. Install the retainer and tighten the screws progressively in the numbered order shown in the drawing.

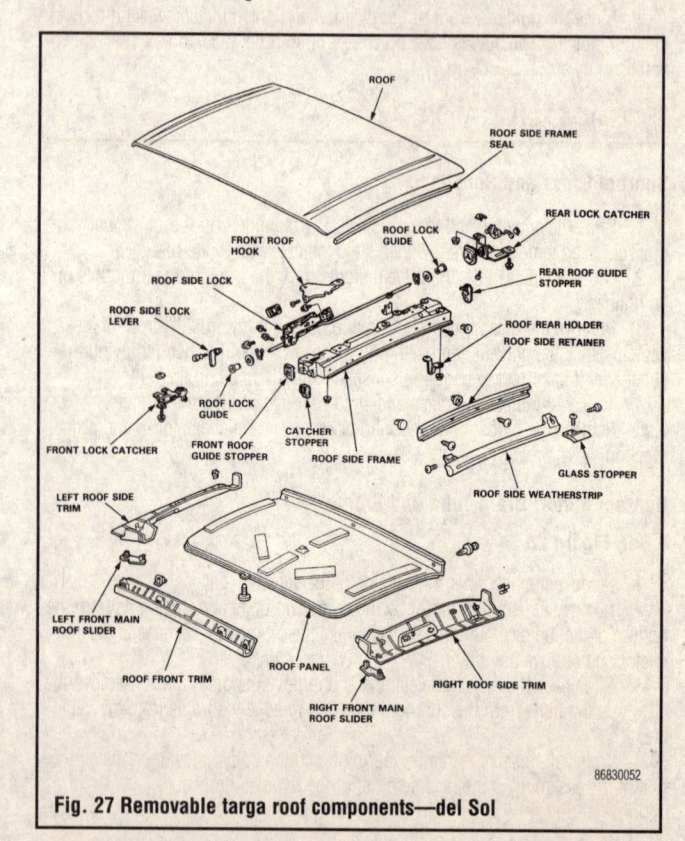

Fig. 27 Removable targa roof components—del Sol

BODY AND TRIM 10-11

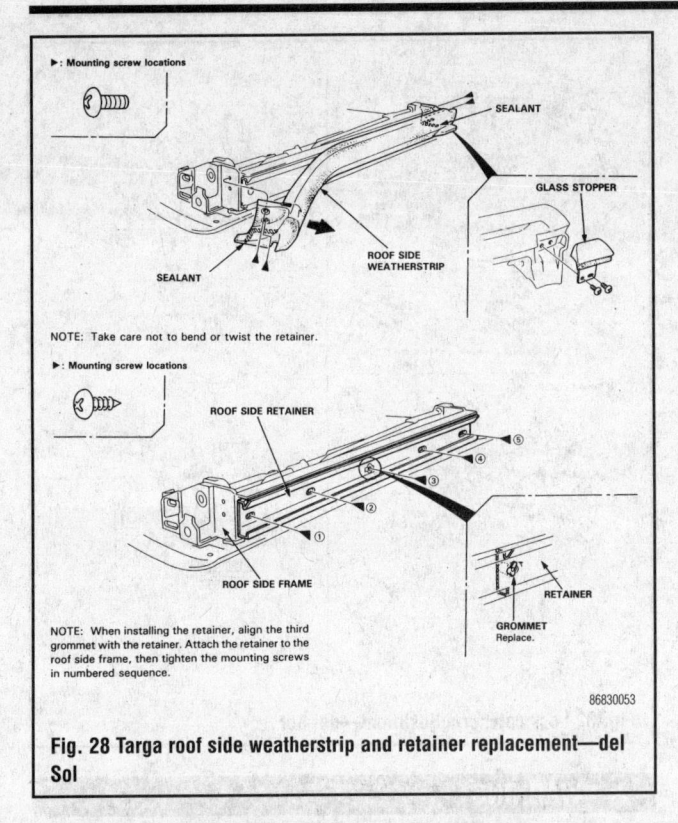

Fig. 28 Targa roof side weatherstrip and retainer replacement—del Sol

7. Replace the interior trim panel using new clips if necessary.
8. Install the roof and check for leaks by spraying low-pressure water over the roof.

Roof Side Frame

♦ See Figure 29

1. Remove the roof and store it in a safe location.
2. Remove the weatherstrip retainer as described in this Section.
3. Remove the side lock assembly.
4. Remove the mounting nuts to the roof side frame and remove the frame. If necessary, remove the component parts from the roof side frame.

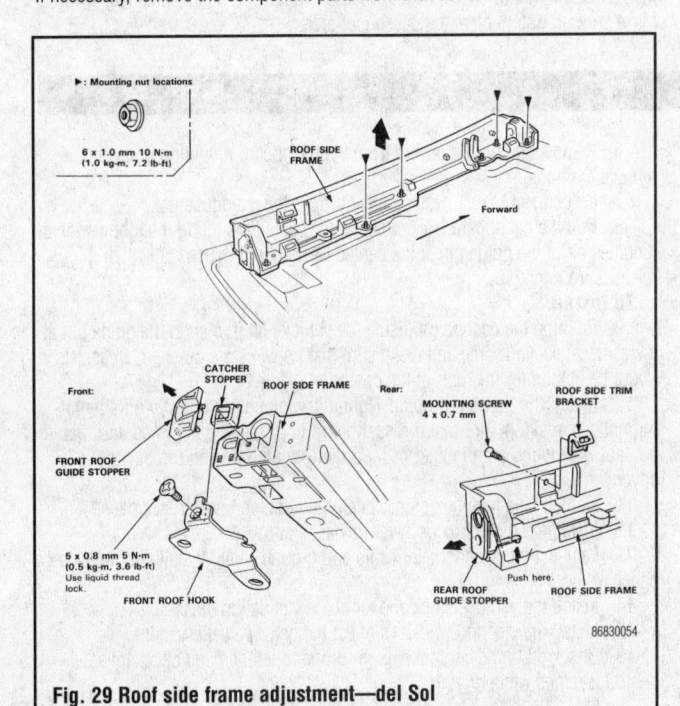

To Install:
5. Replace the roof side frame and tighten the nuts to 7.2 ft. lbs. (10 Nm).
6. Adjust the roof side frame as found in the procedure in this Section and test for water leakage.
7. Replace the roof side trim using new clips if necessary.

Front Lock Catcher

♦ See Figure 30

1. Remove the roof with the help of an assistant.
2. Remove the front pillar trim clips with a small prytool and remove the trim.
3. Remove the three bolts securing the front lock catcher and remove.
4. To install, place the front lock catcher in position and replace the bolts. Tighten them to 16 ft. lbs. (22 Nm).
5. Replace the front pillar trim by inserting the bottom edge into the gap at the dashboard. Use new clips if necessary.
6. Replace the roof and test for proper operation.

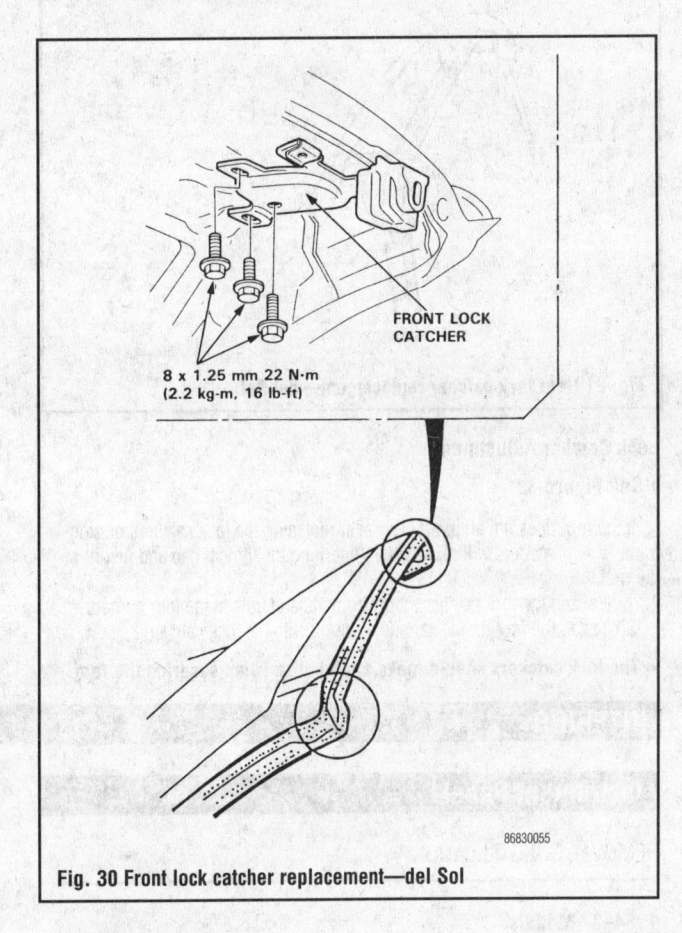

Fig. 29 Roof side frame adjustment—del Sol

Fig. 30 Front lock catcher replacement—del Sol

Rear Lock Catcher

♦ See Figure 31

1. Remove the rear ceiling light lens.
2. Remove the rear ceiling trim panel clips with a small prytool and remove the panel.
3. Unclip and remove the quarter trim panels using a small prytool.
4. Disconnect the connector from the rear lock catcher.
5. Remove the one nut and one bolt that fasten the rear lock catcher and remove the lock catcher.

To Install:
6. To install, replace the lock catcher and secure the nut and the bolt, tightening them to 16 ft. lbs. (22 Nm).
7. Connect the electrical connector.
8. Replace the interior trim panels using new clips if necessary.

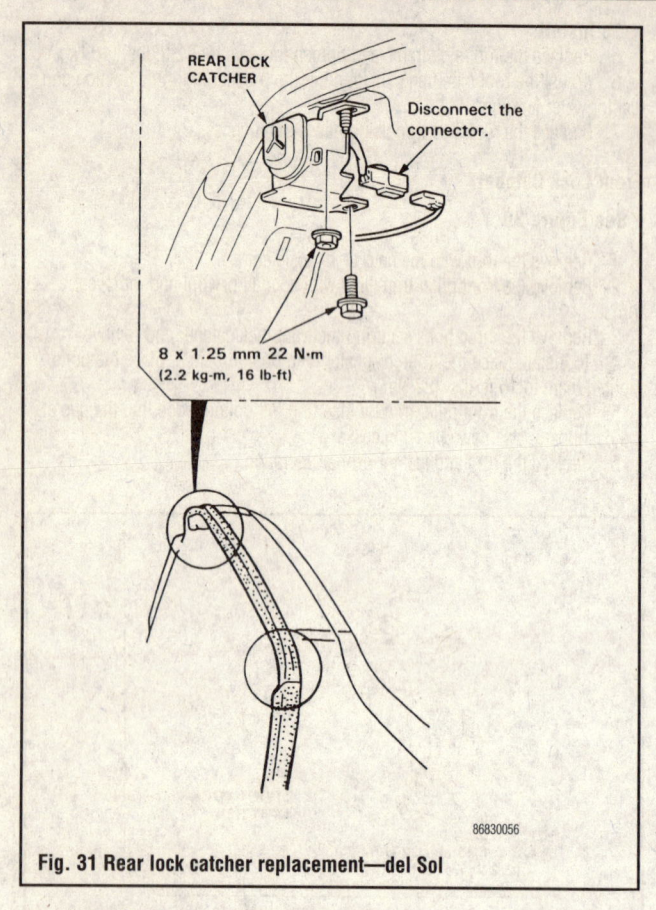

10-12 BODY AND TRIM

Fig. 31 Rear lock catcher replacement—del Sol

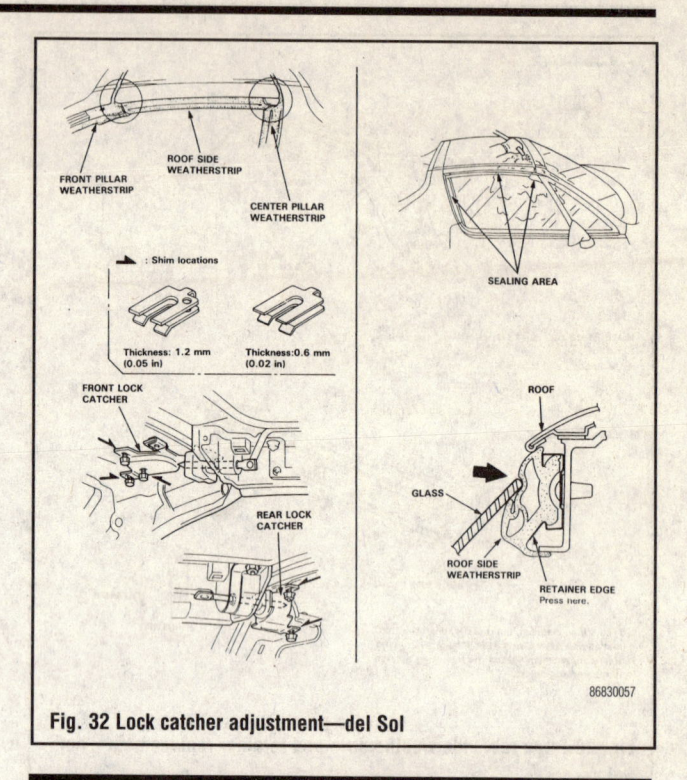

Fig. 32 Lock catcher adjustment—del Sol

Lock Catcher Adjustment

♦ See Figure 32

If the roof does not align properly after replacing the lock catchers or side frame, this procedure will allow you to fine-tune the fitment gap and height of the roof.

1. Loosen the lock catcher mounting bolts and nuts at the four corners.
2. Install the roof on the car and secure it with the lock catchers.

➙ The lock catchers should make a faint click when securing the roof.

※※ WARNING
Do not press the retainer too hard in one spot.

3. Tighten the lock catcher bolts and nut at each of the four corners.
4. Check that the height of the installed roof is correct in relation to the body and that the weatherstrips fit flush. If so, this procedure is complete after checking for water leaks.
5. If the fit is not satisfactory, loosen the lock catchers and adjust right and left by moving the fasteners in the elongated holes. If necessary, use shims (available from Honda in either 0.6mm or 1.2mm thickness) to adjust the height.
6. When satisfied with the adjusted fit, tighten the lock catchers and spray water over the roof to check for water leaks. Do not use a high-pressure sprayer. Make sure no water enters the passenger area.

INTERIOR

Instrument Panel

REMOVAL & INSTALLATION

1984–87 Models

♦ See Figures 33 and 34

1. Disconnect the negative battery cable.
2. Remove the following components:
   a. Steering wheel
   b. Heater/defroster ducts
   c. Right and left side end caps
   d. On Civic Wagon, the right and left defroster garnish
   e. Clock
   f. Ashtray assembly
3. Remove the lower dashboard panel and disconnect the speedometer cable and the ground wire.
4. Remove the glove box assembly. On wagon models, remove the tray under the dashboard.
5. Remove the temperature control panel face plate. Remove the radio, if equipped.
6. Remove the plastic covers from the dashboard mounting bolts and remove the bolts.
7. Disconnect the harness connector from the main fuse box.
8. Pull the dashboard back, while at the same time lifting it slightly so that it comes off of the guide pin at the center. Be careful it doesn't fall when it releases from the pin.

To install:

9. Position the dashboard inside the vehicle, lift it to align the guide pin at the center. Take your time and make sure that no wires are pinched, an assistant might be helpful for this part of the operation.
10. Install the mounting bolts. Tighten the mounting bolts in an alternating pattern, making sure that the dashboard is properly positioned and that no cables or wires are pinched or incorrectly routed. Install the bolt covers.
11. Connect the fuse box plug. Install the radio and heater face plate.
12. Install the glove box or lower tray assembly.
13. Connect the speedometer cable and ground cable. Install the lower dash panel.
14. Install the ashtray, clock and all of the ventilation ducts.
15. Install the steering wheel. Connect the negative battery cable.
16. Check all of the components for proper operation and check the dashboard for rattles while driving.

BODY AND TRIM   10-13

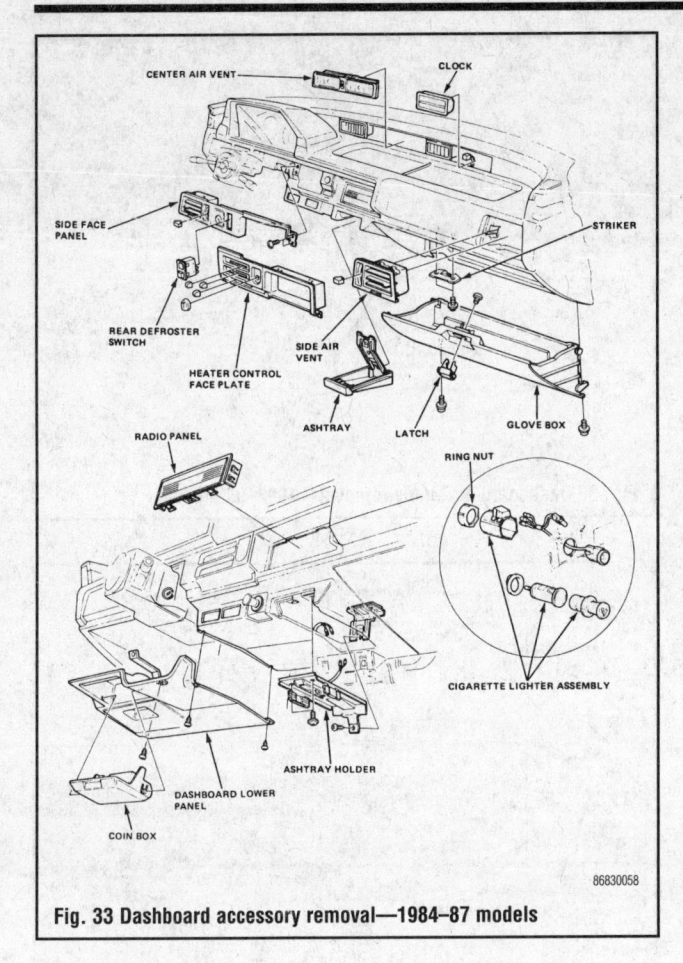

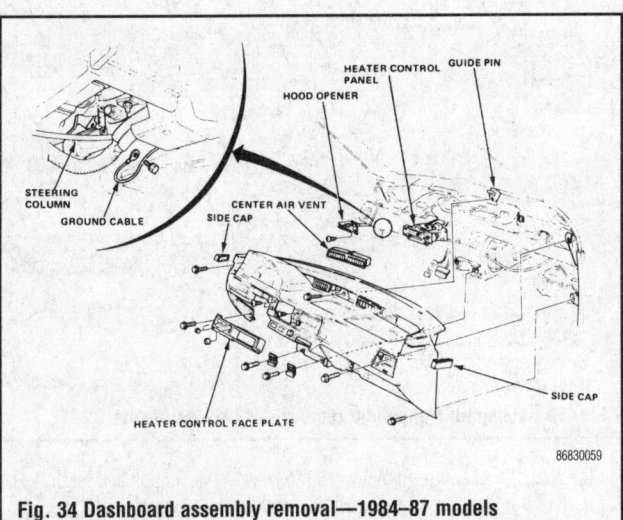

Fig. 33 Dashboard accessory removal—1984–87 models

Fig. 34 Dashboard assembly removal—1984–87 models

Fig. 35 Dashboard assembly removal—1988–91 sedan

Fig. 36 Dashboard assembly removal—1988–91 hatchback and CRX

1988–91 Models

♦ See Figures 35, 36 and 37

1. Disconnect the negative battery cable.
2. Slide the seats all the way back and remove the center console assembly.
3. Remove the fuse box lid and the knee bolster.
4. Disconnect the connectors from the fuse box, it may be necessary to lower the fuse box out of the way first. The fuse box can be lowered by removing the two mounting nuts.
5. Disconnect the ground cable at the right of the steering column and disconnect the power mirror switch, if equipped.

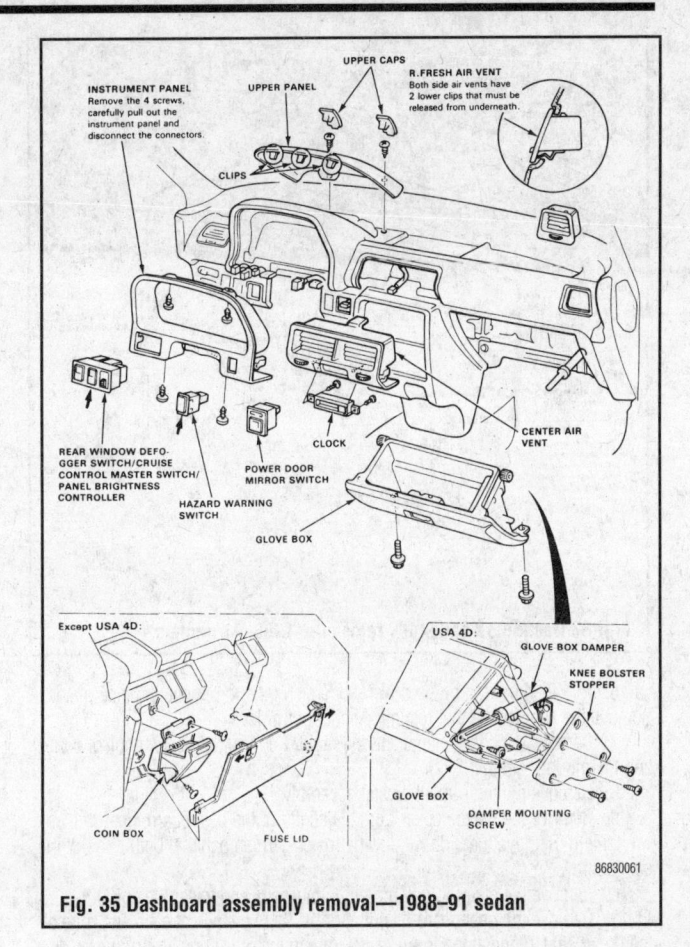

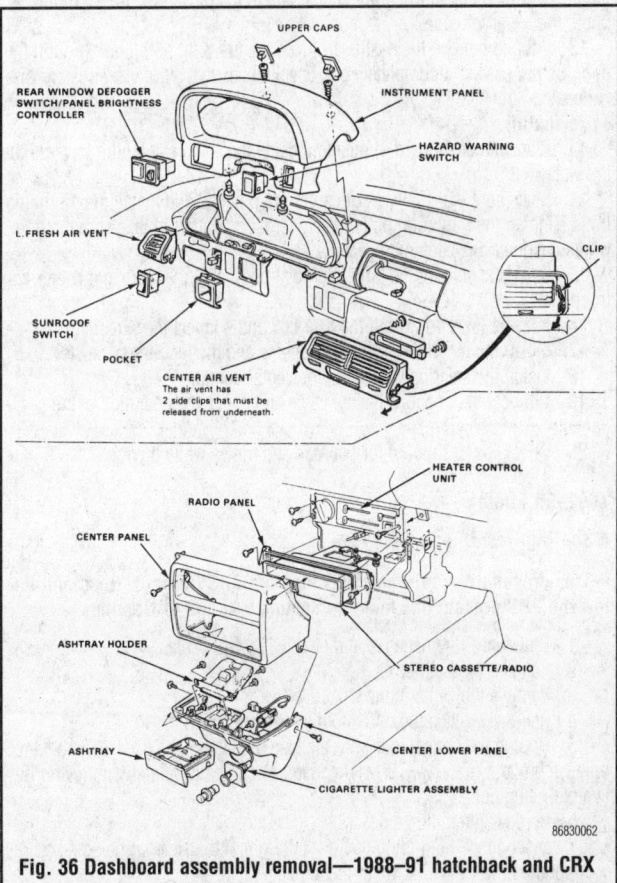

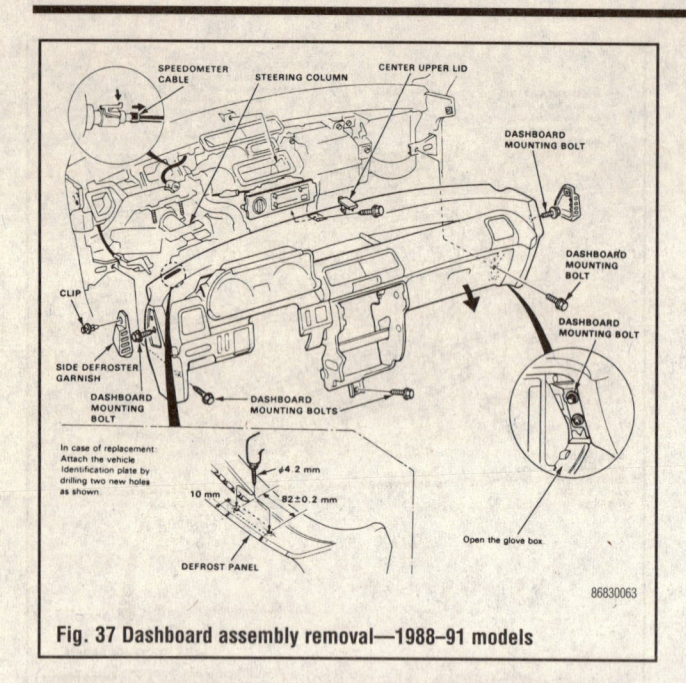

10-14 BODY AND TRIM

Fig. 37 Dashboard assembly removal—1988–91 models

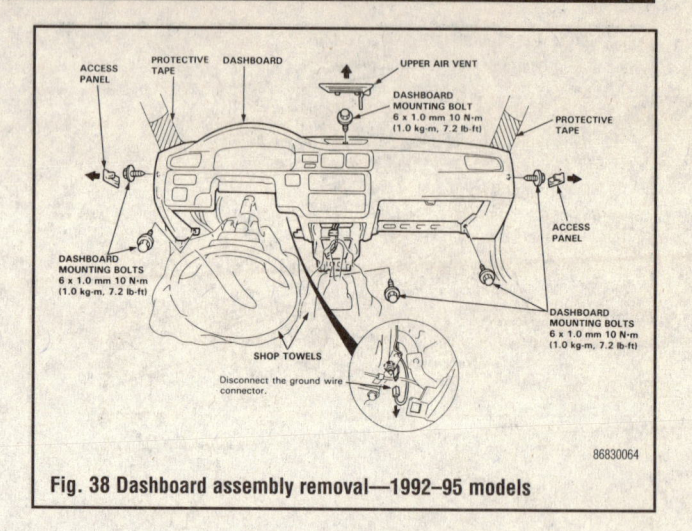

Fig. 38 Dashboard assembly removal—1992–95 models

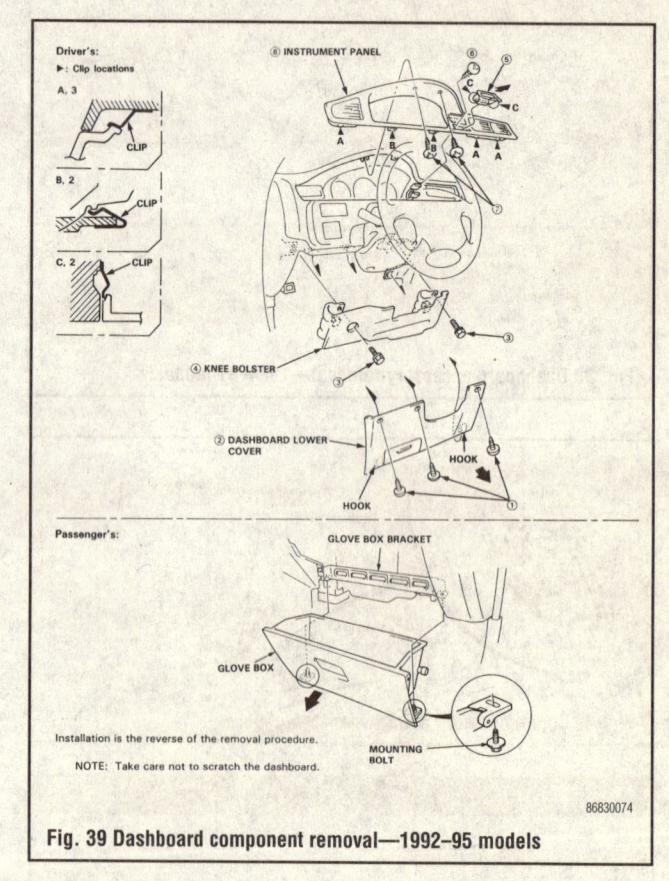

Fig. 39 Dashboard component removal—1992–95 models

6. Pull the knob off of the driver's side air vent and remove the face plate. Remove the two screws that retain the vent control lever.
7. Remove the center panel and the radio. Remove the heater control panel and the mounting panel.
8. Remove the instrument cluster assembly.
9. Remove the plastic cover from the center of the dashboard, near the windshield. Remove the side window defroster garnish panel at both ends of the dashboard.
10. Remove the steering column mounting bolts and lower the column.
11. Remove the dashboard retaining bolts; there are two at each end, one at the upper center, one at the lower center, one in the glove box area and one under the lower left side.
12. Lift and remove the dashboard from the fire wall, pull it out through the door. Before pulling it completely out, check to see that all of the wires are disconnected.

To install:
13. Place the dashboard in position inside the vehicle and lift it towards the fire wall.
14. Make sure that no wires or cables are pinched and install the mounting bolts. Tighten the mounting bolts in an alternating pattern. Make sure that the whole dashboard is centered properly.
15. Raise the steering column into position. Install the radio and heater control panel.
16. If it was lowered, install the fuse box and connect the wires.
17. Install the remaining trim components and the instrument cluster.
18. Install the steering wheel and the center console.
19. Connect the negative battery cable and check the operation of the components.
20. Check the dashboard for squeaks and rattles while driving.

1992–95 Models

▶ See Figures 38 and 39

➥This procedure requires deactivating the SRS (airbag). Read and follow the SRS precautions found in Section 6 before proceeding.

1. Disconnect the negative and positive battery cables.
2. Remove the front seats.
3. Remove the center console assembly.
4. Remove the fuse box lid and the knee bolster.
5. Unplug the connectors from the fuse box, it may be necessary to lower the fuse box out of the way first. The fuse box can be lowered by removing the two mounting nuts.
6. Remove the radio as described in Section 6.
7. Disconnect the ground cable at the right of the steering column and disconnect the power mirror switch, if equipped.
8. Wrap the steering column with a protective towel, unbolt and lower it.
9. Remove the access panel, the nuts, and the passenger's airbag, if equipped.
10. Remove the access panels the side and upper air vent.
11. Remove the power mirror control switch and unplug the connector.
12. Disconnect the under-dash fuse/relay box.
13. Disconnect the air mix control cable.
14. Remove the dashboard mounting bolts. Lift and remove the dashboard. Take care to not scratch it during removal.

➥Wrap the A/T gear position indicator panel and the lever, if applicable.

To install:
15. Place the dashboard in position inside the vehicle and lift it towards the fire wall.
16. Make sure that no wires or cables are pinched and install the mounting

BODY AND TRIM 10-15

bolts. Tighten the mounting bolts in an alternating pattern. Make sure that the whole dashboard is centered properly.

17. Raise the steering column into position. Install the radio and heater control panel.
18. Install the fuse box and connect the wires.
19. Install the remaining trim components and the instrument cluster.
20. Install the steering wheel and the center console.
21. Connect the negative and positive battery cable and check the operation of the components.
22. Check the dashboard for squeaks and rattles while driving.

Center Console

REMOVAL & INSTALLATION

♦ See Figure 40

1. Remove the shifter knob on manual transaxle equipped vehicles.
2. Remove the screws that retain the console assembly.

➥On 4-door models, the passengers seat may have to be removed.

3. Remove the console from the vehicle by pulling it up and out of the vehicle. On CRX models, the console will have to be tilted up from the back and pulled back.
4. When installing the console, make sure that it is properly positioned before tightening the screws.

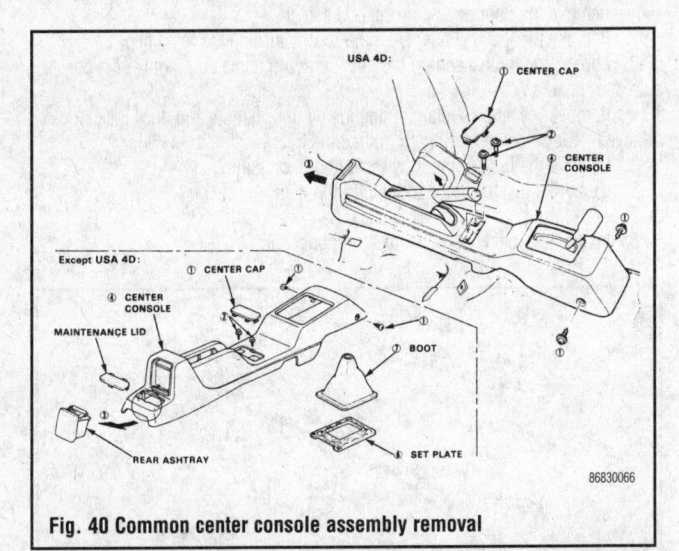

Fig. 40 Common center console assembly removal

Door Panel

REMOVAL & INSTALLATION

♦ See Figures 41 thru 48

1. Remove the window regulator handle with a pry tool or by removing the clip that retains it. This clip is tight and may be difficult to remove. To ease removal, make a tool out of an old wire hanger, take a length of the hanger and put a small hook shaped bend on the end. Use the tool to pull the clip off.
2. Remove the door handle trim plate.
3. On models equipped, remove the seat belt lap guide from the door panel.
4. Remove the screw from the arm rest pocket and carefully pry the pocket out. Remove the stereo speaker cover from the panel by prying out the beauty caps and unscrewing the phillips screws.
5. Pull the panel away from the door. The panel is retained by plastic clips, to ease removal use a trim removal tool (available at most auto parts stores). Try not to bend the panel to much when removing it or it will be damaged. On models equipped with power windows, disconnect the wiring from the switch before removing the panel completely.

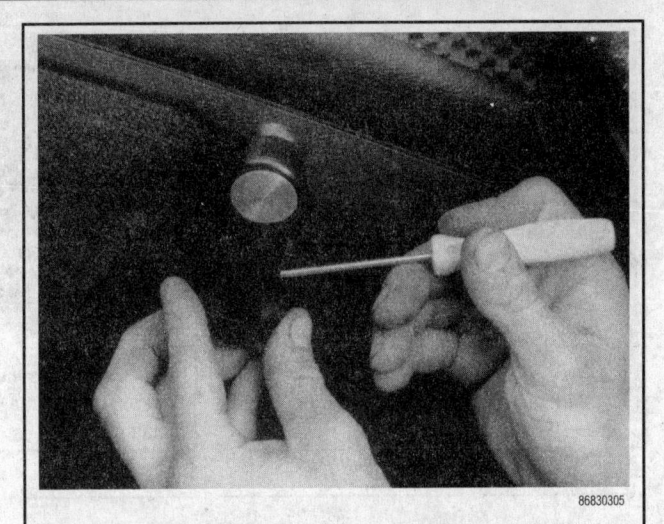

Fig. 41 Here the door handle clip is being removed with a small pry-tool—1984–87 models

Fig. 42 The clip came free without undue struggle—1984–87 models

Fig. 43 The handle uses this thin shim between it and the door—1984–87 models

10-16 BODY AND TRIM

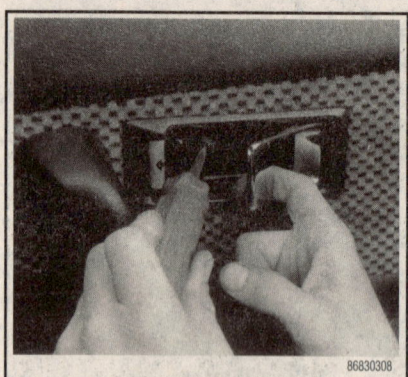

Fig. 44 Removing the door handle trim plate—1984–87 models

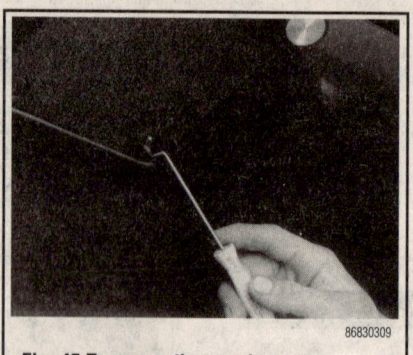

Fig. 45 To access the speaker cover screws, remove the beauty caps with a small prytool

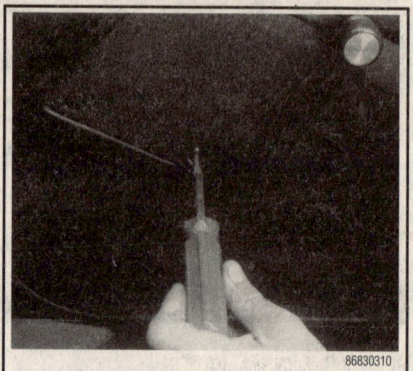

Fig. 46 Unscrewing the speaker cover from the door panel

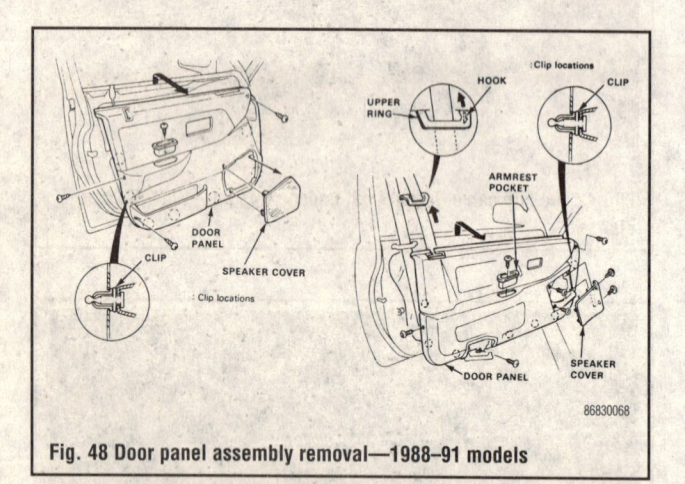

Fig. 47 Door panel assembly removal—1984–87 models

Fig. 48 Door panel assembly removal—1988–91 models

6. Install the panel in position and reinstall all of the components removed. When positioning the panel, it will have to be pushed into place until the retaining clips catch.

Headliner

REMOVAL & INSTALLATION

The headliner assembly is retained by clips at all of the corners. To remove the headliner, remove the sun visors, the inside mirror the A pillar trim and the dome light (on models with a sunroof also remove the sunroof trim). Release the headliner by carefully bring it downward. Use very light pressure when doing this or the panel will break.

When installing the headliner, make sure that it is properly aligned along the windshield and push it into place. Install the visors, mirror and dome light.

Manual Door Locks

REMOVAL & INSTALLATION

♦ See Figure 49

1. Remove the door panel and sealing screen.
2. Remove the lock cylinder from the rod by turning the resin clip.
3. Loosen the nuts attaching the outside door handle and remove the outside door handle.
4. Remove the screws retaining the inside door handle and door lock, and remove the door lock assembly from the hole in the inside of the door.
5. Remove the lock cylinder by removing the retaining clip.
6. Install the lock cylinder and clip to the door.
7. Install the door lock assembly and handles.
8. Install door panel and all attaching parts.

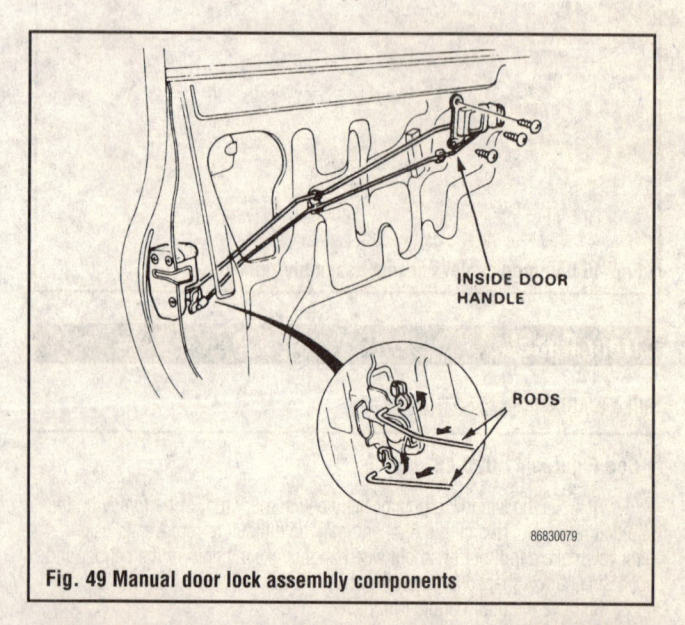

Fig. 49 Manual door lock assembly components

Power Door Locks

The power door locking system consists of switches, actuators and relays. Control switch(es) are used to operate the system. Actuators are used to raise and lower the door lock buttons. These actuators are mounted inside the door assembly and are electrically operated once the switch is depressed. A control unit or functional relay is used to allow the system to regulate current, to function and to align all the actuators and switches with one another.

BODY AND TRIM 10-19

> ※※ **CAUTION**
>
> The regulator gear will move suddenly when the motor is removed, because the regulator spring is tensioned against the gear.

To Install:

10. Install the window regulator with the sector gear in the proper position as marked earlier. Lubricate the sector and gear and rollers prior to installation.
11. Connect the power window harness, and screw and clip the door panel back in place.
12. Attach the armrest and speaker cover.

Windshield and Fixed Glass

REMOVAL & INSTALLATION

If your windshield, or other fixed window, is cracked or chipped, you may decide to replace it with a new one yourself. However, there are two main reasons why replacement windshields and other window glass should be installed only by a professional automotive glass technician: safety and cost.

The most important reason a professional should install automotive glass is for safety. The glass in the vehicle, especially the windshield, is designed with safety in mind in case of a collision. The windshield is specially manufactured from two panes of specially-tempered glass with a thin layer of transparent plastic between them. This construction allows the glass to "give" in the event that a part of your body hits the windshield during the collision, and prevents the glass from shattering, which could cause lacerations, blinding and other harm to passengers of the vehicle. The other fixed windows are designed to be tempered so that if they break during a collision, they shatter in such a way that there are no large pointed glass pieces. The professional automotive glass technician knows how to install the glass in a vehicle so that it will function optimally during a collision. Without the proper experience, knowledge and tools, installing a piece of automotive glass yourself could lead to additional harm if an accident should ever occur.

Cost is also a factor when deciding to install automotive glass yourself. Performing this could cost you much more than a professional may charge for the same job. Since the windshield is designed to break under stress, an often life saving characteristic, windshields tend to break VERY easily when an inexperienced person attempts to install one. Do-it-yourselfers buying two, three or even four windshields from a salvage yard because they have broken them during installation are common stories. Also, since the automotive glass is designed to prevent the outside elements from entering your vehicle, improper installation can lead to water and air leaks. Annoying whining noises at highway speeds from air leaks or inside body panel rusting from water leaks can add to your stress level and subtract from your wallet. After buying two or three windshields, installing them and ending up with a leak that produces a noise while driving and water damage during rainstorms, the cost of having a professional do it correctly the first time may be much more alluring. We here at Chilton, therefore, advise that you have a professional automotive glass technician service any broken glass on your vehicle.

WINDSHIELD CHIP REPAIR

▶ See Figures 57 and 58

➥Check with your state and local authorities on the laws for state safety inspection. Some states or municipalities may not allow chip repair as a viable option for correcting stone damage to your windshield.

Although severely cracked or damaged windshields must be replaced, there is something that you can do to prolong or even prevent the need for replacement of a chipped windshield. There are many companies which offer windshield chip repair products, such as Loctite's® Bullseye™ windshield repair kit. These kits usually consist of a syringe, pedestal and a sealing adhesive. The syringe is mounted on the pedestal and is used to create a vacuum which pulls the plastic layer against the glass. This helps make the chip transparent. The adhesive is then injected which seals the chip and helps to prevent further stress cracks from developing

➥Always follow the specific manufacturer's instructions.

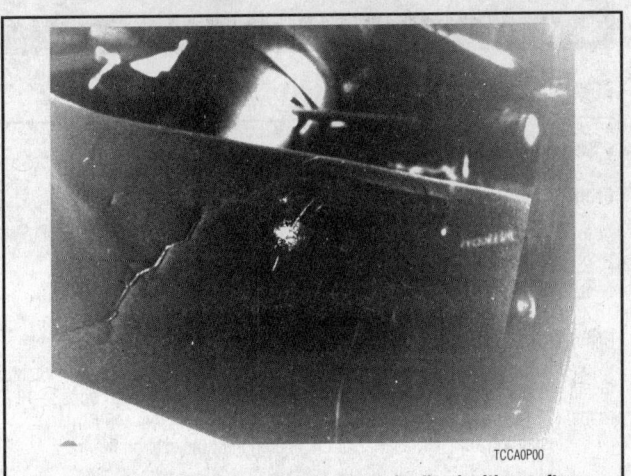

Fig. 57 Small chips on your windshield can be fixed with an aftermarket repair kit, such as the one from Loctite®

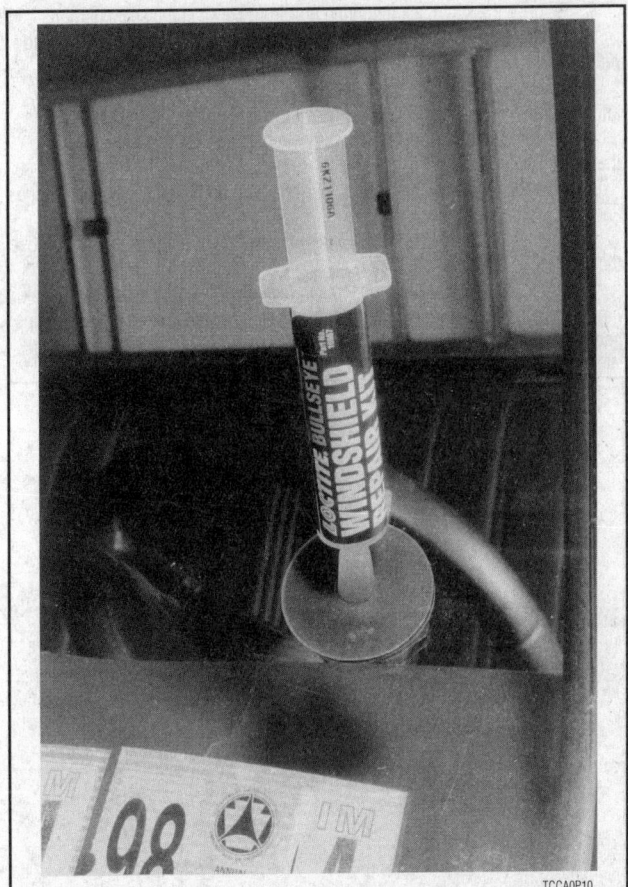

Fig. 58 Most kits use a self-stick applicator and syringe to inject the adhesive into the chip or crack

Inside Rear View Mirror

REMOVAL & INSTALLATION

Remove the screw cover from the mirror and remove the mounting screws. Slide the mirror from the bracket. When installing the new mirror, make sure that it is positioned properly on the mounting plate.

10-20 BODY AND TRIM

Seats

REMOVAL & INSTALLATION

♦ See Figures 59 and 60

Front

1. Remove the seat adjuster cover screws and remove the cover.
2. Remove the seat track mounting bolts.
3. Fold the seat back forward and tilt the seat assembly back.
4. Disconnect the electrical leads and remove the seat from the vehicle. Remove components from the seat as needed.
5. Install the seat assembly into the vehicle. Connect the electrical leads and install the seat track mounting bolts. Tighten the three 8 x 1.25mm bolts to 16 ft. lbs. (22 Nm) and the single 10 x 1.25mm bolts to 29 ft. lbs. (40 Nm).
6. Install the adjuster cover. Test the operation of the seat on the track.

Rear

ALL EXCEPT 1984–87 CIVIC SEDAN

1. Remove the seat back mounting bolt from between the seat cushions.
2. Lift the lower part of the seat at the front to release it from the clips and pull it from the vehicle.
3. Remove the two side seat back mounting bolts, 1988–91 Civic Sedan, and pull the side seat backs up then out of the vehicle.
4. Tilt the seat back forward and remove the clips that retain the carpet to the rear of the seat back.
5. On sedan models, remove the four retaining bolts at the seat back hinges and remove the seat back.
6. On hatchback models, remove the rear seat pivot hinge clips and remove the seat backs one at a time.
7. Install the seat back components in reverse order. Tighten the seat back hinge bolts on sedan models to 7 ft. lbs. (10 Nm). Install all of the plastic clips and make sure that they are completely seated.

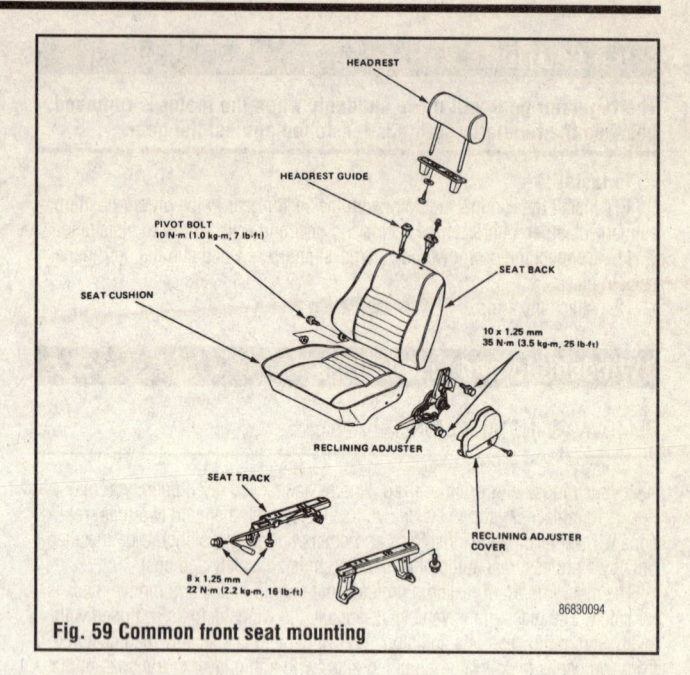

Fig. 59 Common front seat mounting

1984–87 CIVIC SEDAN

1. Remove the seat back mounting bolt from between the seat cushions.
2. Lift the lower part of the seat at the front to release it from the clips and pull it from the vehicle.
3. Remove the two seat back mounting bolts and pull the seat back up then out of the vehicle.
4. Install the seat back in position and install the mounting bolts, tighten them to 7 ft. lbs. (10 Nm).
5. Install the lower seat cushion, making sure it clicks into position and tighten the retaining bolt to 7 ft. lbs. (10 Nm).

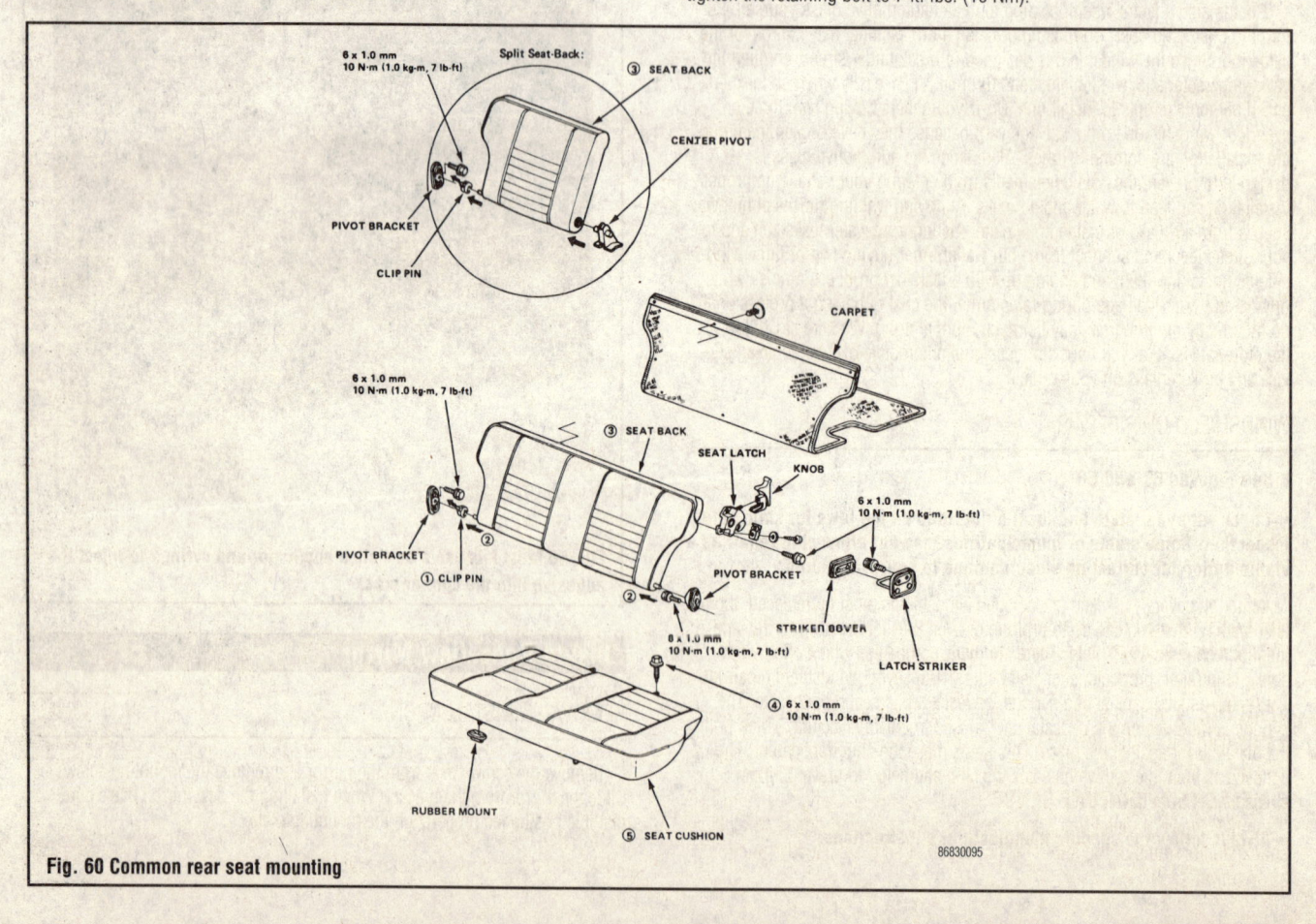

Fig. 60 Common rear seat mounting

GLOSSARY

AIR/FUEL RATIO: The ratio of air-to-gasoline by weight in the fuel mixture drawn into the engine.

AIR INJECTION: One method of reducing harmful exhaust emissions by injecting air into each of the exhaust ports of an engine. The fresh air entering the hot exhaust manifold causes any remaining fuel to be burned before it can exit the tailpipe.

ALTERNATOR: A device used for converting mechanical energy into electrical energy.

AMMETER: An instrument, calibrated in amperes, used to measure the flow of an electrical current in a circuit. Ammeters are always connected in series with the circuit being tested.

AMPERE: The rate of flow of electrical current present when one volt of electrical pressure is applied against one ohm of electrical resistance.

ANALOG COMPUTER: Any microprocessor that uses similar (analogous) electrical signals to make its calculations.

ARMATURE: A laminated, soft iron core wrapped by a wire that converts electrical energy to mechanical energy as in a motor or relay. When rotated in a magnetic field, it changes mechanical energy into electrical energy as in a generator.

ATMOSPHERIC PRESSURE: The pressure on the Earth's surface caused by the weight of the air in the atmosphere. At sea level, this pressure is 14.7 psi at 32°F (101 kPa at 0°C).

ATOMIZATION: The breaking down of a liquid into a fine mist that can be suspended in air.

AXIAL PLAY: Movement parallel to a shaft or bearing bore.

BACKFIRE: The sudden combustion of gases in the intake or exhaust system that results in a loud explosion.

BACKLASH: The clearance or play between two parts, such as meshed gears.

BACKPRESSURE: Restrictions in the exhaust system that slow the exit of exhaust gases from the combustion chamber.

BAKELITE: A heat resistant, plastic insulator material commonly used in printed circuit boards and transistorized components.

BALL BEARING: A bearing made up of hardened inner and outer races between which hardened steel balls roll.

BALLAST RESISTOR: A resistor in the primary ignition circuit that lowers voltage after the engine is started to reduce wear on ignition components.

BEARING: A friction reducing, supportive device usually located between a stationary part and a moving part.

BIMETAL TEMPERATURE SENSOR: Any sensor or switch made of two dissimilar types of metal that bend when heated or cooled due to the different expansion rates of the alloys. These types of sensors usually function as an on/off switch.

BLOWBY: Combustion gases, composed of water vapor and unburned fuel, that leak past the piston rings into the crankcase during normal engine operation. These gases are removed by the PCV system to prevent the buildup of harmful acids in the crankcase.

BRAKE PAD: A brake shoe and lining assembly used with disc brakes.

BRAKE SHOE: The backing for the brake lining. The term is, however, usually applied to the assembly of the brake backing and lining.

BUSHING: A liner, usually removable, for a bearing; an anti-friction liner used in place of a bearing.

CALIPER: A hydraulically activated device in a disc brake system, which is mounted straddling the brake rotor (disc). The caliper contains at least one piston and two brake pads. Hydraulic pressure on the piston(s) forces the pads against the rotor.

CAMSHAFT: A shaft in the engine on which are the lobes (cams) which operate the valves. The camshaft is driven by the crankshaft, via a belt, chain or gears, at one half the crankshaft speed.

CAPACITOR: A device which stores an electrical charge.

CARBON MONOXIDE (CO): A colorless, odorless gas given off as a normal byproduct of combustion. It is poisonous and extremely dangerous in confined areas, building up slowly to toxic levels without warning if adequate ventilation is not available.

CARBURETOR: A device, usually mounted on the intake manifold of an engine, which mixes the air and fuel in the proper proportion to allow even combustion.

CATALYTIC CONVERTER: A device installed in the exhaust system, like a muffler, that converts harmful byproducts of combustion into carbon dioxide and water vapor by means of a heat-producing chemical reaction.

CENTRIFUGAL ADVANCE: A mechanical method of advancing the spark timing by using flyweights in the distributor that react to centrifugal force generated by the distributor shaft rotation.

CHECK VALVE: Any one-way valve installed to permit the flow of air, fuel or vacuum in one direction only.

CHOKE: A device, usually a moveable valve, placed in the intake path of a carburetor to restrict the flow of air.

CIRCUIT: Any unbroken path through which an electrical current can flow. Also used to describe fuel flow in some instances.

CIRCUIT BREAKER: A switch which protects an electrical circuit from overload by opening the circuit when the current flow exceeds a predetermined level. Some circuit breakers must be reset manually, while most reset automatically.

COIL (IGNITION): A transformer in the ignition circuit which steps up the voltage provided to the spark plugs.

COMBINATION MANIFOLD: An assembly which includes both the intake and exhaust manifolds in one casting.

GLOSSARY

COMBINATION VALVE: A device used in some fuel systems that routes fuel vapors to a charcoal storage canister instead of venting them into the atmosphere. The valve relieves fuel tank pressure and allows fresh air into the tank as the fuel level drops to prevent a vapor lock situation.

COMPRESSION RATIO: The comparison of the total volume of the cylinder and combustion chamber with the piston at BDC and the piston at TDC.

CONDENSER: 1. An electrical device which acts to store an electrical charge, preventing voltage surges. 2. A radiator-like device in the air conditioning system in which refrigerant gas condenses into a liquid, giving off heat.

CONDUCTOR: Any material through which an electrical current can be transmitted easily.

CONTINUITY: Continuous or complete circuit. Can be checked with an ohmmeter.

COUNTERSHAFT: An intermediate shaft which is rotated by a mainshaft and transmits, in turn, that rotation to a working part.

CRANKCASE: The lower part of an engine in which the crankshaft and related parts operate.

CRANKSHAFT: The main driving shaft of an engine which receives reciprocating motion from the pistons and converts it to rotary motion.

CYLINDER: In an engine, the round hole in the engine block in which the piston(s) ride.

CYLINDER BLOCK: The main structural member of an engine in which is found the cylinders, crankshaft and other principal parts.

CYLINDER HEAD: The detachable portion of the engine, usually fastened to the top of the cylinder block and containing all or most of the combustion chambers. On overhead valve engines, it contains the valves and their operating parts. On overhead cam engines, it contains the camshaft as well.

DEAD CENTER: The extreme top or bottom of the piston stroke.

DETONATION: An unwanted explosion of the air/fuel mixture in the combustion chamber caused by excess heat and compression, advanced timing, or an overly lean mixture. Also referred to as "ping".

DIAPHRAGM: A thin, flexible wall separating two cavities, such as in a vacuum advance unit.

DIESELING: A condition in which hot spots in the combustion chamber cause the engine to run on after the key is turned off.

DIFFERENTIAL: A geared assembly which allows the transmission of motion between drive axles, giving one axle the ability to turn faster than the other.

DIODE: An electrical device that will allow current to flow in one direction only.

DISC BRAKE: A hydraulic braking assembly consisting of a brake disc, or rotor, mounted on an axle, and a caliper assembly containing, usually two brake pads which are activated by hydraulic pressure. The pads are forced against the sides of the disc, creating friction which slows the vehicle.

DISTRIBUTOR: A mechanically driven device on an engine which is responsible for electrically firing the spark plug at a predetermined point of the piston stroke.

DOWEL PIN: A pin, inserted in mating holes in two different parts allowing those parts to maintain a fixed relationship.

DRUM BRAKE: A braking system which consists of two brake shoes and one or two wheel cylinders, mounted on a fixed backing plate, and a brake drum, mounted on an axle, which revolves around the assembly.

DWELL: The rate, measured in degrees of shaft rotation, at which an electrical circuit cycles on and off.

ELECTRONIC CONTROL UNIT (ECU): Ignition module, module, amplifier or igniter. See Module for definition.

ELECTRONIC IGNITION: A system in which the timing and firing of the spark plugs is controlled by an electronic control unit, usually called a module. These systems have no points or condenser.

END-PLAY: The measured amount of axial movement in a shaft.

ENGINE: A device that converts heat into mechanical energy.

EXHAUST MANIFOLD: A set of cast passages or pipes which conduct exhaust gases from the engine.

FEELER GAUGE: A blade, usually metal, or precisely predetermined thickness, used to measure the clearance between two parts.

FIRING ORDER: The order in which combustion occurs in the cylinders of an engine. Also the order in which spark is distributed to the plugs by the distributor.

FLOODING: The presence of too much fuel in the intake manifold and combustion chamber which prevents the air/fuel mixture from firing, thereby causing a no-start situation.

FLYWHEEL: A disc shaped part bolted to the rear end of the crankshaft. Around the outer perimeter is affixed the ring gear. The starter drive engages the ring gear, turning the flywheel, which rotates the crankshaft, imparting the initial starting motion to the engine.

FOOT POUND (ft. lbs. or sometimes, ft.lb.): The amount of energy or work needed to raise an item weighing one pound, a distance of one foot.

FUSE: A protective device in a circuit which prevents circuit overload by breaking the circuit when a specific amperage is present. The device is constructed around a strip or wire of a lower amperage rating than the circuit it is designed to protect. When an amperage higher than that stamped on the fuse is present in the circuit, the strip or wire melts, opening the circuit.

GEAR RATIO: The ratio between the number of teeth on meshing gears.

GENERATOR: A device which converts mechanical energy into electrical energy.

HEAT RANGE: The measure of a spark plug's ability to dissipate heat from its firing end. The higher the heat range, the hotter the plug fires.

GLOSSARY 10-23

HUB: The center part of a wheel or gear.

HYDROCARBON (HC): Any chemical compound made up of hydrogen and carbon. A major pollutant formed by the engine as a byproduct of combustion.

HYDROMETER: An instrument used to measure the specific gravity of a solution.

INCH POUND (inch lbs.; sometimes in.lb. or in. lbs.): One twelfth of a foot pound.

INDUCTION: A means of transferring electrical energy in the form of a magnetic field. Principle used in the ignition coil to increase voltage.

INJECTOR: A device which receives metered fuel under relatively low pressure and is activated to inject the fuel into the engine under relatively high pressure at a predetermined time.

INPUT SHAFT: The shaft to which torque is applied, usually carrying the driving gear or gears.

INTAKE MANIFOLD: A casting of passages or pipes used to conduct air or a fuel/air mixture to the cylinders.

JOURNAL: The bearing surface within which a shaft operates.

KEY: A small block usually fitted in a notch between a shaft and a hub to prevent slippage of the two parts.

MANIFOLD: A casting of passages or set of pipes which connect the cylinders to an inlet or outlet source.

MANIFOLD VACUUM: Low pressure in an engine intake manifold formed just below the throttle plates. Manifold vacuum is highest at idle and drops under acceleration.

MASTER CYLINDER: The primary fluid pressurizing device in a hydraulic system. In automotive use, it is found in brake and hydraulic clutch systems and is pedal activated, either directly or, in a power brake system, through the power booster.

MODULE: Electronic control unit, amplifier or igniter of solid state or integrated design which controls the current flow in the ignition primary circuit based on input from the pick-up coil. When the module opens the primary circuit, high secondary voltage is induced in the coil.

NEEDLE BEARING: A bearing which consists of a number (usually a large number) of long, thin rollers.

OHM: (Ω) The unit used to measure the resistance of conductor-to-electrical flow. One ohm is the amount of resistance that limits current flow to one ampere in a circuit with one volt of pressure.

OHMMETER: An instrument used for measuring the resistance, in ohms, in an electrical circuit.

OUTPUT SHAFT: The shaft which transmits torque from a device, such as a transmission.

OVERDRIVE: A gear assembly which produces more shaft revolutions than that transmitted to it.

OVERHEAD CAMSHAFT (OHC): An engine configuration in which the camshaft is mounted on top of the cylinder head and operates the valve either directly or by means of rocker arms.

OVERHEAD VALVE (OHV): An engine configuration in which all of the valves are located in the cylinder head and the camshaft is located in the cylinder block. The camshaft operates the valves via lifters and pushrods.

OXIDES OF NITROGEN (NOx): Chemical compounds of nitrogen produced as a byproduct of combustion. They combine with hydrocarbons to produce smog.

OXYGEN SENSOR: Use with the feedback system to sense the presence of oxygen in the exhaust gas and signal the computer which can reference the voltage signal to an air/fuel ratio.

PINION: The smaller of two meshing gears.

PISTON RING: An open-ended ring with fits into a groove on the outer diameter of the piston. Its chief function is to form a seal between the piston and cylinder wall. Most automotive pistons have three rings: two for compression sealing; one for oil sealing.

PRELOAD: A predetermined load placed on a bearing during assembly or by adjustment.

PRIMARY CIRCUIT: the low voltage side of the ignition system which consists of the ignition switch, ballast resistor or resistance wire, bypass, coil, electronic control unit and pick-up coil as well as the connecting wires and harnesses.

PRESS FIT: The mating of two parts under pressure, due to the inner diameter of one being smaller than the outer diameter of the other, or vice versa; an interference fit.

RACE: The surface on the inner or outer ring of a bearing on which the balls, needles or rollers move.

REGULATOR: A device which maintains the amperage and/or voltage levels of a circuit at predetermined values.

RELAY: A switch which automatically opens and/or closes a circuit.

RESISTANCE: The opposition to the flow of current through a circuit or electrical device, and is measured in ohms. Resistance is equal to the voltage divided by the amperage.

RESISTOR: A device, usually made of wire, which offers a preset amount of resistance in an electrical circuit.

RING GEAR: The name given to a ring-shaped gear attached to a differential case, or affixed to a flywheel or as part of a planetary gear set.

ROLLER BEARING: A bearing made up of hardened inner and outer races between which hardened steel rollers move.

ROTOR: 1. The disc-shaped part of a disc brake assembly, upon which the brake pads bear; also called, brake disc. 2. The device mounted atop the distributor shaft, which passes current to the distributor cap tower contacts.

10-24 GLOSSARY

SECONDARY CIRCUIT: The high voltage side of the ignition system, usually above 20,000 volts. The secondary includes the ignition coil, coil wire, distributor cap and rotor, spark plug wires and spark plugs.

SENDING UNIT: A mechanical, electrical, hydraulic or electro-magnetic device which transmits information to a gauge.

SENSOR: Any device designed to measure engine operating conditions or ambient pressures and temperatures. Usually electronic in nature and designed to send a voltage signal to an on-board computer, some sensors may operate as a simple on/off switch or they may provide a variable voltage signal (like a potentiometer) as conditions or measured parameters change.

SHIM: Spacers of precise, predetermined thickness used between parts to establish a proper working relationship.

SLAVE CYLINDER: In automotive use, a device in the hydraulic clutch system which is activated by hydraulic force, disengaging the clutch.

SOLENOID: A coil used to produce a magnetic field, the effect of which is to produce work.

SPARK PLUG: A device screwed into the combustion chamber of a spark ignition engine. The basic construction is a conductive core inside of a ceramic insulator, mounted in an outer conductive base. An electrical charge from the spark plug wire travels along the conductive core and jumps a preset air gap to a grounding point or points at the end of the conductive base. The resultant spark ignites the fuel/air mixture in the combustion chamber.

SPLINES: Ridges machined or cast onto the outer diameter of a shaft or inner diameter of a bore to enable parts to mate without rotation.

TACHOMETER: A device used to measure the rotary speed of an engine, shaft, gear, etc., usually in rotations per minute.

THERMOSTAT: A valve, located in the cooling system of an engine, which is closed when cold and opens gradually in response to engine heating, controlling the temperature of the coolant and rate of coolant flow.

TOP DEAD CENTER (TDC): The point at which the piston reaches the top of its travel on the compression stroke.

TORQUE: The twisting force applied to an object.

TORQUE CONVERTER: A turbine used to transmit power from a driving member to a driven member via hydraulic action, providing changes in drive ratio and torque. In automotive use, it links the driveplate at the rear of the engine to the automatic transmission.

TRANSDUCER: A device used to change a force into an electrical signal.

TRANSISTOR: A semi-conductor component which can be actuated by a small voltage to perform an electrical switching function.

TUNE-UP: A regular maintenance function, usually associated with the replacement and adjustment of parts and components in the electrical and fuel systems of a vehicle for the purpose of attaining optimum performance.

TURBOCHARGER: An exhaust driven pump which compresses intake air and forces it into the combustion chambers at higher than atmospheric pressures. The increased air pressure allows more fuel to be burned and results in increased horsepower being produced.

VACUUM ADVANCE: A device which advances the ignition timing in response to increased engine vacuum.

VACUUM GAUGE: An instrument used to measure the presence of vacuum in a chamber.

VALVE: A device which control the pressure, direction of flow or rate of flow of a liquid or gas.

VALVE CLEARANCE: The measured gap between the end of the valve stem and the rocker arm, cam lobe or follower that activates the valve.

VISCOSITY: The rating of a liquid's internal resistance to flow.

VOLTMETER: An instrument used for measuring electrical force in units called volts. Voltmeters are always connected parallel with the circuit being tested.

WHEEL CYLINDER: Found in the automotive drum brake assembly, it is a device, actuated by hydraulic pressure, which, through internal pistons, pushes the brake shoes outward against the drums.

MASTER INDEX

ACCUMULATOR 9-36
   DISPOSAL 9-36
      REMOVAL AND INSTALLATION 9-36
ADDITIONAL EXHAUST EMISSION CONTROLS 4-6
   CARBURETED VEHICLES 4-6
   COMPONENT TESTING 4-7
ADJUSTMENT (ELECTRONIC IGNITION (1984-87)) 2-6
   RELUCTOR GAP 2-6
ADJUSTMENT (VALVE LASH) 2-18
ADJUSTMENTS (AUTOMATIC TRANSAXLE) 7-18
   SHIFT CABLE 7-18
   THROTTLE LINKAGE 7-19
ADJUSTMENTS (BRAKE OPERATING SYSTEM) 9-3
   BRAKE PEDAL FREE-PLAY 9-3
   REAR DRUM BRAKE ADJUSTMENT 9-3
ADJUSTMENTS (CLUTCH) 7-13
   PEDAL HEIGHT/FREE-PLAY ADJUSTMENT 7-13
ADJUSTMENTS (MANUAL TRANSAXLE) 7-2
   SHIFT LINKAGE 7-2
AIR CLEANER 1-13
   REMOVAL & INSTALLATION 1-13
AIR CONDITIONING COMPONENTS 6-11
   REMOVAL & INSTALLATION 6-11
AIR CONDITIONING SYSTEM 1-23
   PREVENTIVE MAINTENANCE 1-23
   SYSTEM INSPECTION 1-24
   SYSTEM SERVICE & REPAIR 1-23
ALTERNATOR 3-3
   PRECAUTIONS 3-3
   REMOVAL & INSTALLATION 3-4
ANTENNA 10-7
   REMOVAL & INSTALLATION 10-7
ANTI-LOCK BRAKES (ALB/ABS) 9-21
APPLICATION 7-25
AUTOMATIC TRANSAXLE 7-16
AUTOMATIC TRANSAXLES (FLUIDS AND LUBRICANTS) 1-30
   DRAIN AND REFILL 1-30
   FLUID RECOMMENDATIONS 1-30
   LEVEL CHECK 1-30
AVOIDING THE MOST COMMON MISTAKES 1-2
AVOIDING TROUBLE 1-2
BACK-UP LIGHT SWITCH 7-2
   REMOVAL & INSTALLATION 7-4
BASIC ELECTRICAL THEORY 6-2
   HOW DOES ELECTRICITY WORK: THE WATER
     ANALOGY 6-2
   OHM'S LAW 6-2
BASIC FUEL SYSTEM DIAGNOSIS 5-2
BATTERY (ENGINE ELECTRICAL) 3-4
   REMOVAL & INSTALLATION 3-4
BATTERY (ROUTINE MAINTENANCE) 1-16
   BATTERY FLUID 1-17
   CABLES 1-17
   CHARGING 1-18
   GENERAL MAINTENANCE 1-16
   PRECAUTIONS 1-16
   REPLACEMENT 1-18
BELTS 1-18
   ADJUSTING TENSION 1-19
   INSPECTION 1-18
   REMOVAL & INSTALLATION 1-19
BLEEDING THE BRAKE SYSTEM 9-8
BLOWER MOTOR 6-12
   REMOVAL & INSTALLATION 6-12
BLOWER MOTOR RESISTOR 6-12
   REMOVAL & INSTALLATION 6-12
BODY LUBRICATION 1-34
BOLTS, NUTS AND OTHER THREADED RETAINERS 1-6
BRAKE CALIPER (FRONT DISC BRAKES) 9-11
   OVERHAUL 9-12
   REMOVAL & INSTALLATION 9-11

10-26 MASTER INDEX

BRAKE CALIPER (REAR DISC BRAKES) 9-17
   REMOVAL & INSTALLATION 9-17
BRAKE DISC (REAR DISC BRAKES) 9-17
   INSPECTION 9-18
   REMOVAL & INSTALLATION 9-17
BRAKE DISC (ROTOR) (FRONT DISC BRAKES) 9-13
   INSPECTION 9-14
   REMOVAL & INSTALLATION 9-13
BRAKE DRUMS 9-15
   INSPECTION 9-15
   REMOVAL & INSTALLATION 9-15
BRAKE HOSES AND PIPES 9-8
   REMOVAL & INSTALLATION 9-8
BRAKE LIGHT SWITCH 9-4
   REMOVAL & INSTALLATION 9-4
   TESTING 9-4
BRAKE MASTER CYLINDER 1-32
   FLUID RECOMMENDATIONS 1-32
   LEVEL CHECK 1-32
BRAKE OPERATING SYSTEM 9-2
BRAKE PADS (FRONT DISC BRAKES) 9-9
   REMOVAL & INSTALLATION 9-9
BRAKE PADS (REAR DISC BRAKES) 9-17
   REMOVAL & INSTALLATION 9-17
BRAKE SHOES 9-15
   REMOVAL & INSTALLATION 9-15
BRAKE SPECIFICATIONS 9-38
BUMPERS 10-5
   REMOVAL & INSTALLATION 10-5
CABLE 9-19
   ADJUSTMENT 9-20
   REMOVAL & INSTALLATION 9-19
CAMSHAFT 3-45
   INSPECTION 3-46
   REMOVAL & INSTALLATION 3-45
CAMSHAFT SPROCKETS 3-45
   REMOVAL & INSTALLATION 3-45
CAPACITIES 1-40
CARBURETED FUEL SYSTEM 5-2
CARBURETOR ADJUSTMENTS 5-5
   ADJUSTMENTS 5-5
   TROUBLESHOOTING 5-5
CARBURETOR ASSEMBLY 5-5
   OVERHAUL 5-9
   REMOVAL & INSTALLATION 5-9
CATALYTIC CONVERTER 4-5
   DESCRIPTION 4-5
CATALYTIC CONVERTER AND/OR PIPE ASSEMBLY 3-52
   REMOVAL & INSTALLATION 3-54
CENTER CONSOLE 10-15
   REMOVAL & INSTALLATION 10-15
CHASSIS GREASING 1-34
CHECKING AND ADJUSTING TIMING 2-15
   1984-87 MODELS 2-16
   1988-91 MODELS 2-17
   1992-95 MODELS 2-17
CHECKING ENGINE COMPRESSION 3-6
CIRCUIT PROTECTION 6-30
CLOCK 6-26
   REMOVAL & INSTALLATION 6-26
CLUTCH 7-12
CLUTCH CABLE 7-13
   REMOVAL & INSTALLATION 7-13
CLUTCH MASTER CYLINDER (CLUTCH) 7-15
   OVERHAUL 7-16
   REMOVAL & INSTALLATION 7-15
CLUTCH MASTER CYLINDER (FLUIDS AND LUBRICANTS) 1-32

FLUID RECOMMENDATIONS 1-33
   LEVEL CHECK 1-33
COIL SPRINGS AND SHOCK ABSORBER 8-16
   REMOVAL & INSTALLATION 8-16
COMBINATION SWITCH (INSTRUMENTS AND SWITCHES) 6-26
   REMOVAL & INSTALLATION 6-26
COMBINATION SWITCH (STEERING) 8-21
   REMOVAL & INSTALLATION 8-21
COMPONENT REPLACEMENT (ELECTRONIC IGNITION (1984-87)) 2-6
   DISTRIBUTOR CAP 2-6
   DISTRIBUTOR SHAFT 2-10
   IGNITER UNIT 2-9
   PICK-UP COIL 2-8
   RELUCTOR 2-8
   ROTOR 2-6
   VACUUM ADVANCER UNIT 2-9
COMPONENT REPLACEMENT (PGM-IG IGNITION SYSTEM (1988-95)) 2-13
   IGNITER UNIT 2-13
   IGNITION COIL 2-13
   ROTOR 2-14
CONDENSER FAN 3-27
   REMOVAL & INSTALLATION 3-28
   TESTING 3-27
CONTROL ARMS 8-17
   REMOVAL & INSTALLATION 8-17
CONTROL CABLES 6-14
   ADJUSTMENT 6-14
   REMOVAL & INSTALLATION 6-15
CONTROL UNIT 9-33
   REMOVAL & INSTALLATION 9-33
COOLING FAN 3-28
   REMOVAL & INSTALLATION 3-28
   TESTING 3-28
COOLING SYSTEM 1-31
   DRAIN AND REFILL 1-31
   FLUID RECOMMENDATIONS 1-31
   FLUSHING AND CLEANING THE SYSTEM 1-32
   LEVEL CHECK 1-31
COVERAGE 1-9
CRANKSHAFT AND CONNECTING ROD SPECIFICATIONS 3-8
CRANKSHAFT AND MAIN BEARINGS 3-49
   CLEANING AND INSPECTION BEARING OIL CLEARANCE 3-50
   CRANKSHAFT END-PLAY/CONNECTING ROD SIDE PLAY 3-50
   CRANKSHAFT REPAIRS 3-51
   REMOVAL & INSTALLATION 3-49
CRUISE CONTROL (CHASSIS ELECTRICAL) 6-18
CRUISE CONTROL 6-17
CV-BOOTS 1-23
   INSPECTION 1-23
CV-JOINT AND BOOT 7-23
   OVERHAUL 7-23
CYLINDER HEAD 3-28
   CLEANING AND INSPECTION 3-35
   PRECAUTIONS 3-28
   REMOVAL & INSTALLATION 3-29
   RESURFACING 3-35
CYLINDER HEAD COVER 3-19
   REMOVAL & INSTALLATION 3-19
DAMPER STRUTS 8-16
   OVERHAUL 8-17
   REMOVAL & INSTALLATION 8-16
DEL SOL TARGA ROOF 10-10
   REMOVAL & INSTALLATION 10-10
DESCRIPTION AND OPERATION (ANTI-LOCK BRAKES) 9-21
DESCRIPTION AND OPERATION (ELECTRONIC IGNITION (1984-87)) 2-5
   PRECAUTIONS 2-6

MASTER INDEX   10-27

DESCRIPTION AND OPERATION (PGM-IG IGNITION SYSTEM (1988-95)) 2-12
    PRECAUTIONS 2-13
DETERMINING AXLE RATIO 7-25
DIAGNOSIS AND TESTING 2-14
    IGNITER UNIT 2-14
    IGNITION COIL 2-14
DIAGNOSTIC CHARTS 4-15
DIFFERENTIAL CARRIER 7-27
    REMOVAL & INSTALLATION 7-27
DISTRIBUTOR 3-2
    REMOVAL & INSTALLATION 3-2
DO'S 1-4
DON'TS 1-5
DOOR GLASS, REGULATOR AND POWER WINDOW MOTOR 10-18
    REMOVAL & INSTALLATION 10-18
DOOR PANEL 10-15
    REMOVAL & INSTALLATION 10-15
DOORS 10-2
    ADJUSTMENT 10-3
    REMOVAL & INSTALLATION 10-2
DRIVE AXLE—4WD WAGON 1-30
    DRAIN AND REFILL 1-31
    FLUID RECOMMENDATIONS 1-30
    LEVEL CHECK 1-31
DRIVELINE 7-22
DRIVEN DISC AND PRESSURE PLATE 7-13
    REMOVAL & INSTALLATION 7-13
DRIVESHAFT AND U-JOINTS 7-24
    REMOVAL & INSTALLATION 7-24
DUAL PROPORTIONING VALVE 9-7
    REMOVAL & INSTALLATION 9-7
ELECTRIC FUEL PUMP 5-17
    PRESSURE TESTING 5-17
    REMOVAL & INSTALLATION 5-18
ELECTRIC SUNROOF/MOONROOF 10-7
    ADJUSTMENTS 10-8
    COMPONENT TESTING 10-9
    REMOVAL & INSTALLATION 10-9
ELECTRICAL COMPONENTS 6-2
    CONNECTORS 6-4
    GROUND 6-3
    LOAD 6-4
    POWER SOURCE 6-2
    PROTECTIVE DEVICES 6-3
    SWITCHES & RELAYS 6-3
    WIRING & HARNESSES 6-4
ELECTRONIC ENGINE CONTROLS 4-9
ELECTRONIC IGNITION (1984-87) 2-5
EMISSION CONTROLS 4-2
ENGINE (ENGINE MECHANICAL) 3-11
    REMOVAL & INSTALLATION 3-11
ENGINE (FLUIDS AND LUBRICANTS) 1-28
    OIL AND FILTER CHANGE 1-29
    OIL LEVEL CHECK 1-28
ENGINE DESIGN 3-5
ENGINE ELECTRICAL 3-2
ENGINE IDENTIFICATION 1-12
ENGINE MECHANICAL 3-5
ENGINE OVERHAUL TIPS 3-5
    INSPECTION TECHNIQUES 3-5
    OVERHAUL TIPS 3-5
    REPAIRING DAMAGED THREADS 3-5
    TOOLS 3-5
ENGINE SERIAL NUMBER 1-9
ENGINE TUNE-UP SPECIFICATIONS 2-23
ENTERTAINMENT SYSTEM 6-18

EVAPORATIVE CHARCOAL CANISTER 1-16
    SERVICING 1-16
EXHAUST GAS RECIRCULATION (EGR) SYSTEM 4-4
    CARBURETED ENGINES 4-4
    FUEL INJECTED ENGINES 4-4
EXHAUST MANIFOLD 3-24
    REMOVAL & INSTALLATION 3-24
EXHAUST SYSTEM 3-52
EXTERIOR 10-2
FAN SWITCH 6-15
    REMOVAL & INSTALLATION 6-16
    TESTING 6-15
FASTENERS, MEASUREMENTS AND CONVERSIONS 1-6
FENDER 10-7
    REMOVAL & INSTALLATION 10-7
FILLING AND BLEEDING 9-36
    BLEEDING THE SYSTEM 9-37
    FILLING THE SYSTEM 9-36
    MODULATOR UNIT BLEEDING WITH THE ALB/ABS CHECKER 9-36
FIRING ORDERS 2-5
FLUID DISPOSAL 1-27
FLUIDS AND LUBRICANTS 1-27
FLYWHEEL/FLEX PLATE 3-51
    REMOVAL & INSTALLATION 3-51
FRONT DAMPER 8-5
    DISASSEMBLY AND INSPECTION 8-7
    INSPECTION 8-5
    REMOVAL & INSTALLATION 8-6
FRONT DISC BRAKES 9-9
FRONT GRILLE ASSEMBLY 10-6
    REMOVAL & INSTALLATION 10-6
FRONT SUSPENSION 8-2
FRONT TURN SIGNAL, SIDE MARKER AND LICENSE PLATE LIGHTS 6-29
    REMOVAL & INSTALLATION 6-29
FRONT WHEEL HUB, KNUCKLE AND SPINDLE 8-10
    REMOVAL & INSTALLATION 8-10
FUEL AND ENGINE OIL RECOMMENDATIONS 1-27
    FUEL 1-27
    OIL 1-27
FUEL EVAPORATIVE EMISSION CONTROL SYSTEM 4-2
    OPERATION 4-2
    SERVICE 4-2
FUEL FILTER 1-14
    REMOVAL & INSTALLATION 1-15
FUEL INJECTOR 5-18
    REMOVAL & INSTALLATION 5-18
FUEL PRESSURE REGULATOR 5-20
    REMOVAL & INSTALLATION 5-20
FUEL SYSTEM SERVICE PRECAUTIONS (CARBURETED FUEL SYSTEM) 5-3
FUEL SYSTEM SERVICE PRECAUTIONS (PROGRAMMED FUEL INJECTION SYSTEM) 5-12
FUEL TANK (CARBURETED FUEL SYSTEM) 5-4
    REMOVAL & INSTALLATION 5-4
FUEL TANK (PROGRAMMED FUEL INJECTION SYSTEM) 5-22
    REMOVAL & INSTALLATION 5-22
FUSES 6-30
    REPLACMENT 6-30
FUSIBLE LINKS 6-31
    REPLACEMENT 6-31
GAUGES 6-25
    REMOVAL & INSTALLATION 6-25
GEAR PULSERS 9-33
    INSPECTION 9-33
GEARSHIFT ASSEMBLY 7-2
    REMOVAL & INSTALLATION 7-2
GENERAL DESCRIPTION (BRAKE OPERATING SYSTEM) 9-2
    DISC BRAKES 9-3

10-28 MASTER INDEX

DRUM BRAKES 9-2
    HYDRAULIC SYSTEM 9-3
    HYDRAULIC SYSTEM 9-2
    POWER BRAKE BOOSTERS 9-3
GENERAL DESCRIPTION (CRUISE CONTROL) 6-17
    SERVICE PRECAUTIONS 6-18
GENERAL DESCRIPTION (PROGRAMMED FUEL INJECTION SYSTEM) 5-10
    AIR INTAKE SYSTEM 5-10
    ELECTRONIC CONTROL SYSTEM 5-11
    FUEL SYSTEM 5-12
GENERAL ENGINE SPECIFICATIONS 3-7
GENERAL INFORMATION (CARBURETED FUEL SYSTEM) 5-2
    ACCELERATOR PUMP 5-2
    AUTOMATIC CHOKE 5-2
    CARBURETOR HEAT RISER 5-3
    CHOKE OPENER SYSTEM 5-2
    POWER VALVE 5-3
    VACUUM CONTROLLED SECONDARY 5-3
GENERAL INFORMATION (HEATING AND AIR CONDITIONING) 6-11
GENERAL INFORMATION (IGNITION TIMING) 2-15
GENERAL INFORMATION (REAR SUSPENSION) 8-13
    1984-87 MODELS 8-13
    1988-95 MODELS 8-13
GENERAL INFORMATION (SUPPLEMENTAL RESTRAINT SYSTEM) 6-9
    1992-95 SUPPLEMENTAL RESTRAINT SYSTEM (SRS) PRECAUTIONS 6-9
    DISARMING THE SYSTEM 6-11
    ENABLING THE SYSTEM 6-11
    SYSTEM OPERATION 6-9
HALFSHAFTS 7-22
    REMOVAL & INSTALLATION 7-22
HEADLIGHT SWITCH 6-26
    REMOVAL & INSTALLATION 6-26
HEADLIGHTS 6-28
    REMOVAL & INSTALLATION 6-28
HEADLINER 10-16
    REMOVAL & INSTALLATION 10-16
HEATER CORE 6-12
    REMOVAL & INSTALLATION 6-12
HEATING AND AIR CONDITIONING 6-11
HISTORY 1-9
HONDA TRANSAXLE DESCRIPTION 7-2
HOOD 10-3
    ALIGNMENT 10-3
    REMOVAL & INSTALLATION 10-3
HOSES 1-20
    INSPECTION 1-20
    REMOVAL & INSTALLATION 1-20
HOW TO USE THIS BOOK 1-2
IDENTIFICATION (AUTOMATIC TRANSAXLE) 7-17
IDENTIFICATION (MANUAL TRANSAXLE) 7-2
IDLE MIXTURE 2-22
    ADJUSTMENT 2-22
IDLE SPEED 2-20
    ADJUSTMENT 2-20
IDLE SPEED AND MIXTURE 2-20
IGNITION COIL 3-2
IGNITION LOCK 8-22
    REMOVAL & INSTALLATION 8-22
IGNITION SWITCH 8-21
    REMOVAL & INSTALLATION 8-21
    REMOVAL & INSTALLATION 8-21
IGNITION TIMING 2-15
INSIDE REAR VIEW MIRROR 10-19
    REMOVAL & INSTALLATION 10-19

INSTRUMENT CLUSTER 6-23
    REMOVAL & INSTALLATION 6-23
INSTRUMENT PANEL 10-12
    REMOVAL & INSTALLATION 10-12
INSTRUMENTS AND SWITCHES 6-23
INTAKE MANIFOLD 3-22
    REMOVAL & INSTALLATION 3-22
INTERIOR 10-12
JACKING 1-35
LIGHTING 6-28
LOWER BALL JOINT 8-7
    INSPECTION 8-7
    LUBRICATION 8-7
    REMOVAL & INSTALLATION 8-7
LOWER CONTROL ARM 8-8
    REMOVAL & INSTALLATION 8-8
MAINTENANCE INTERVAL CHARTS 1-36
MAINTENANCE OR REPAIR? 1-2
MANUAL CONTROL HEAD 6-13
    REMOVAL & INSTALLATION 6-13
MANUAL DOOR LOCKS 10-16
    REMOVAL & INSTALLATION 10-16
MANUAL OUTSIDE MIRROR 10-6
    REMOVAL & INSTALLATION 10-6
MANUAL RACK AND PINION 8-24
    ADJUSTMENT 8-24
    OVERHAUL 8-25
    REMOVAL & INSTALLATION 8-25
MANUAL STEERING GEAR 1-33
    FLUID RECOMMENDATIONS 1-33
    INSPECTION 1-33
MANUAL TRANSAXLE 7-2
MANUAL TRANSAXLE (FLUIDS AND LUBRICANTS) 1-30
    DRAIN AND REFILL 1-30
    FLUID RECOMMENDATIONS 1-30
    LEVEL CHECK 1-30
MASTER CYLINDER 9-4
    OVERHAUL 9-5
    REMOVAL & INSTALLATION 9-4
MECHANICAL FUEL PUMP 5-4
    REMOVAL & INSTALLATION 5-4
    TESTING 5-5
MODULATOR UNIT/ALB/ABS PUMP ASSEMBLY 9-34
    RELIEVING LINE PRESSURE 9-34
    REMOVAL & INSTALLATION 9-34
MUFFLER ASSEMBLY 3-52
    REMOVAL & INSTALLATION 3-52
OIL PAN 3-39
    REMOVAL & INSTALLATION 3-39
OIL PUMP 3-41
    OIL PUMP OVERHAUL 3-42
    REMOVAL & INSTALLATION 3-41
PARKING BRAKE 9-19
PGM-IG IGNITION SYSTEM (1988-95) 2-12
PINION SEAL 7-27
    REMOVAL & INSTALLATION 7-27
PISTON AND RING SPECIFICATIONS 3-9
PISTONS AND CONNECTING RODS 3-46
    CLEANING AND INSPECTION 3-48
    CYLINDER BORE INSPECTION 3-48
    FITTING AND POSITIONING PISTON RINGS 3-48
    REMOVAL & INSTALLATION 3-46
POSITIVE CRANKCASE VENTILATION (PCV) (ROUTINE MAINTENANCE) 1-16
    INSPECTION 1-16
    REMOVAL & INSTALLATION 1-16
POSITIVE CRANKCASE VENTILATION (PCV) SYSTEM (EMISSION CONTROLS) 4-2

MASTER INDEX

OPERATION 4-2
   SERVICE 4-2
POWER BRAKE BOOSTER 9-7
   INSPECTION 9-7
   REMOVAL & INSTALLATION 9-7
POWER DOOR LOCKS 10-16
   COMPONENT TESTING 10-17
   REMOVAL & INSTALLATION 10-17
POWER MIRROR SWITCH 10-7
   REMOVAL & INSTALLATION 10-7
POWER OUTSIDE MIRROR 10-6
   REMOVAL & INSTALLATION 10-6
POWER STEERING GEAR 8-26
   ADJUSTMENT 8-26
   OVERHAUL 8-28
   REMOVAL & INSTALLATION 8-26
POWER STEERING PUMP (FLUIDS AND LUBRICANTS) 1-33
   FLUID RECOMMENDATIONS 1-33
   RESERVOIR LEVEL CHECK 1-33
POWER STEERING PUMP (STEERING) 8-30
   BELT ADJUSTMENT 8-30
   REMOVAL & INSTALLATION 8-30
   SYSTEM BLEEDING 8-30
PROGRAMMED FUEL INJECTION (PGM-FI) FAULT MEMORY 4-9
   CLEARING TROUBLE CODES 4-11
   DESCRIPTION 4-9
   ENTERING SELF-DIAGNOSTICS AND DIAGNOSTIC CODE
      DISPLAY 4-9
PROGRAMMED FUEL INJECTION (PGM-FI) SYSTEM 5-10
PROGRAMMED FUEL INJECTION (PGM-FI) SYSTEM (ELECTRONIC ENGINE CONTROLS) 4-13
   GENERAL DESCRIPTION 4-13
   SERVICE PRECAUTIONS 4-14
RADIATOR 3-26
   REMOVAL & INSTALLATION 3-26
RADIO 6-18
   REMOVAL & INSTALLATION 6-18
RADIUS ARM 8-8
   REMOVAL & INSTALLATION 8-8
REAR AXLE 7-25
REAR AXLE SHAFTS AND BEARINGS 7-26
   REMOVAL & INSTALLATION 7-26
REAR DISC BRAKES 9-17
REAR DRUM BRAKES 9-14
REAR HALFSHAFTS 7-26
REAR HATCH 10-4
   REMOVAL & INSTALLATION 10-4
REAR MAIN SEAL 3-49
   REMOVAL & INSTALLATION 3-49
REAR SUSPENSION 8-13
REAR WHEEL BEARINGS 8-19
   REPLACEMENT 8-19
REAR WINDOW DEFOGGER SYSTEM 6-27
RECIRCULATION CONTROL MOTOR 6-16
   REMOVAL & INSTALLATION 6-17
   TESTING 6-16
REGULATOR 3-4
RELAY, SENSORS AND COMPUTER LOCATIONS 6-8
RELAYS 6-31
   REPLACEMENT 6-31
RELIEVING FUEL SYSTEM PRESSURE 5-13
   COMPONENT TEST PROCEDURES 5-13
REMOVAL & INSTALLATION 6-27
   COMPONENT TESTING 6-27
ROCKER ARMS/SHAFTS 3-19
   REMOVAL & INSTALLATION 3-19
ROUTINE MAINTENANCE 1-13

SEATS 10-20
   REMOVAL & INSTALLATION 10-20
SERIAL NUMBER IDENTIFICATION 1-9
SERVICING YOUR VEHICLE SAFELY 1-4
SHOCK ABSORBERS 8-3
   INSPECTION 8-5
   REMOVAL & INSTALLATION 8-3
SLAVE CYLINDER 7-16
   OVERHAUL 7-16
   REMOVAL & INSTALLATION 7-16
   SYSTEM BLEEDING 7-16
SPARK PLUG WIRES 2-5
   TESTING & REPLACEMENT 2-5
SPARK PLUGS 2-2
   INSPECTION & CLEANING 2-4
   REMOVAL & INSTALLATION 2-3
SPECIAL TOOLS 1-4
SPECIFICATION CHARTS
   BRAKE SPECIFICATIONS 9-38
   CAPACITIES 1-40
   CRANKSHAFT AND CONNECTING ROD
      SPECIFICATIONS 3-8
   ENGINE IDENTIFICATION 1-12
   ENGINE TUNE-UP SPECIFICATIONS 2-23
   GENERAL ENGINE SPECIFICATIONS 3-7
   PISTON AND RING SPECIFICATIONS 3-9
   TORQUE SPECIFICATIONS 3-10
   VALVE SPECIFICATIONS 3-7
   WHEEL ALIGNMENT 8-13
SPEEDOMETER DRIVEN GEAR 7-12
   REMOVAL & INSTALLATION 7-12
STANDARD AND METRIC MEASUREMENTS 1-7
STARTER 3-4
   REMOVAL & INSTALLATION 3-4
STEERING 8-20
STEERING COLUMN 8-22
   REMOVAL & INSTALLATION 8-22
STEERING LINKAGE 8-23
   REMOVAL & INSTALLATION 8-23
STEERING WHEEL 8-20
   REMOVAL & INSTALLATION 8-20
STEREO SPEAKERS 6-19
   REMOVAL & INSTALLATION 6-20
SUPPLEMENTAL RESTRAINT SYSTEM (SRS) 6-9
TAIL LIGHTS, REVERSE LIGHTS, REAR MARKER AND SIGNAL
  LIGHTS 6-29
   REMOVAL & INSTALLATION 6-29
TEST EQUIPMENT 6-5
   JUMPER WIRES 6-5
   MULTIMETERS 6-5
   TEST LIGHTS 6-5
TESTING 6-6
   OPEN CIRCUITS 6-6
   RESISTANCE 6-7
   SHORT CIRCUITS 6-6
   VOLTAGE 6-6
   VOLTAGE DROP 6-7
THERMOSTAT 3-21
   REMOVAL & INSTALLATION 3-21
THERMOSTAT HOUSING 3-22
   REMOVAL & INSTALLATION 3-22
THROTTLE BODY 5-21
   REMOVAL & INSTALLATION 5-21
TIMING BELT AND TENSIONER 3-43
   REMOVAL & INSTALLATION 3-43
TIMING BELT COVER AND SEAL 3-42
   OIL SEAL REPLACEMENT 3-42
   REMOVAL & INSTALLATION 3-42

10-30 MASTER INDEX

TIMING BELTS 1-20
   INSPECTION 1-20
TIRES AND WHEELS 1-24
   INFLATION & INSPECTION 1-26
   TIRE DESIGN 1-25
   TIRE ROTATION 1-25
   TIRE STORAGE 1-26
TOOLS AND EQUIPMENT 1-2
TORQUE 1-6
   TORQUE ANGLE METERS 1-7
   TORQUE WRENCHES 1-7
TORQUE SPECIFICATIONS 3-10
TORSION BARS 8-2
   ADJUSTMENT 8-3
   REMOVAL & INSTALLATION 8-2
TOWING THE VEHICLE 1-34
TRAILER TOWING 1-34
TRANSAXLE (AUTOMATIC TRANSAXLE) 7-20
   REMOVAL & INSTALLATION 7-20
TRANSAXLE (MANUAL TRANSAXLE) 7-4
   REMOVAL & INSTALLATION 7-4
TRANSAXLE SERIAL NUMBER 1-9
TRANSFER CASE 1-31
TROUBLESHOOTING (ANTI-LOCK BRAKES) 9-21
   DIAGNOSTIC FLOW CHARTS 9-23
   GENERAL INFORMATION 9-21
   SYSTEM TESTING 9-22
TROUBLESHOOTING (ELECTRONIC IGNITION (1984-87)) 2-10
   IGNITER UNIT TEST 2-11
   IGNITION COIL TEST 2-10
   PICK-UP COIL TEST 2-11
   SYSTEM TEST 2-10
TROUBLESHOOTING CHARTS
   CRUISE CONTROL 6-18
TROUBLESHOOTING ELECTRICAL SYSTEMS 6-6
TRUNK LID 10-4
   ALIGNMENT 10-4
   REMOVAL & INSTALLATION 1-4
TRUNK, LIFTGATE AND FUEL DOOR OPENER 10-4
   REMOVAL & INSTALLATION 10-4
TUNE-UP PROCEDURES 2-2
UNDERSTANDING AND TROUBLESHOOTING ELECTRICAL SYSTEMS 6-2
UNDERSTANDING AUTOMATIC TRANSAXLES 7-16
   THE HYDRAULIC CONTROL SYSTEM 7-17
   THE PLANETARY GEARBOX 7-17
   THE SERVOS AND ACCUMULATORS 7-17
   THE TORQUE CONVERTER 7-16
UNDERSTANDING DRIVE AXLES 7-25
UNDERSTANDING THE CLUTCH 7-12
UNDERSTANDING THE MANUAL TRANSAXLE 7-2
UPPER BALL JOINTS 8-8

   INSPECTION 8-8
   REMOVAL & INSTALLATION 8-8
UPPER CONTROL ARM AND BUSHING 8-8
   REMOVAL & INSTALLATION 8-8
VACUUM DIAGRAMS 4-21
VALVE GUIDES 3-37
   KNURLING 3-39
   REMOVAL & INSTALLATION 3-38
VALVE LASH 2-18
VALVE SEATS 3-37
VALVE SPECIFICATIONS 3-7
VALVE SPRINGS 3-36
   INSPECTION 3-37
   REMOVAL & INSTALLATION 3-38
VALVES 3-35
   INSPECTION 3-36
   REFACING 3-36
   REMOVAL & INSTALLATION 3-35
VEHICLE IDENTIFICATION (CHASSIS) NUMBER 1-9
WASHER FLUID RESERVOIR AND WASHER PUMPS 6-23
   REMOVAL & INSTALLATION 6-23
WATER PUMP 3-28
   REMOVAL & INSTALLATION 3-28
WHEEL ALIGNMENT (SUSPENSION AND STEERING) 8-13
WHEEL ALIGNMENT (FRONT SUSPENSION) 8-11
   CAMBER 8-12
   CASTER 8-12
   TOE 8-12
WHEEL BEARINGS 1-34
WHEEL CYLINDERS 9-16
   OVERHAUL 9-16
   REMOVAL & INSTALLATION 9-16
WHEEL SENSOR 9-33
   REMOVAL & INSTALLATION 9-33
WHEELS 8-2
WHEELS AND TIRES 8-2
   REMOVAL & INSTALLATION 8-2
WHERE TO BEGIN 1-2
WINDSHIELD AND FIXED GLASS 10-19
   REMOVAL & INSTALLATION 10-19
   WINDSHIELD CHIP REPAIR 10-19
WINDSHIELD WIPER ARMS 6-21
   REMOVAL & INSTALLATION 6-21
WINDSHIELD WIPER FLUID 1-33
WINDSHIELD WIPER MOTOR 6-21
   REMOVAL & INSTALLATION 6-21
WINDSHIELD WIPER/WASHER SWITCH 6-26
   REMOVAL & INSTALLATION 6-26
WINDSHIELD WIPERS 1-24
   ELEMENT (REFILL) CARE & REPLACEMENT 1-24
WINDSHIELD WIPERS AND WASHERS 6-21
WIRE AND CONNECTOR REPAIR 6-8
WIRING DIAGRAMS 6-32